College Trigonometry

Second Edition

College Trigonometry

Second Edition

Stanley I. Grossman

University of Montana and University College London

SAUNDERS COLLEGE PUBLISHING

A Harcourt Brace Jovanovich College Publisher

Fort Worth Philadelphia San Diego New York Orlando Austin
San Antonio Toronto Montreal London Sydney Tokyo

THIS BOOK IS PRINTED ON **ACID-FREE, RECYCLED** PAPER 

Requests for permission to make copies of any part of the work should be mailed to:
Permissions, Harcourt Brace Jovanovich, Publishers, 8th Floor, Orlando, Florida 32887.

Text typeface: Times Roman
Compositor: York Graphic Services
Acquisitions Editor: Robert B. Stern
Developmental Editor: Sandra Kiselica
Copy Editor: York Production Services
Art Director: Christine Schueler
Art Assistant: Caroline McGowan
Text Designer: Rebecca Lemna
Cover Designer: Lawrence R. Didona
Text Artwork: York Production Services
Layout Artist: York Production Services
Production Manager: Bob Butler

Cover Credit: © 1982 Barbara Kasten/Courtesy John Weber Gallery

Printed in the United States of America

ISBN 0-03-052169-6
Library of Congress Catalog Card Number:
91-26866

To Aaron, Kerstin, and Erik

One learns by doing the thing; for though you think you know it, you have no certainty until you try.

SOPHOCLES

Preface

During the past few years the study of trigonometry has become increasingly important for students in almost every academic discipline for at least two reasons. First, periodic phenomena in biology, physics, and economics are described in terms of the trigonometric functions. Second, trigonometry is an indispensable prerequisite for calculus.

Many different kinds of students study trigonometry in college, but they fall mainly into two groups: recent high school graduates and returning students who may have been out of school for many years. Many students in both groups look upon trigonometry with less than unbridled enthusiasm. The comments "I never liked math" and "I haven't seen any math for years" are heard often by every trigonometry instructor.

College Trigonometry, Second Edition, contains all the trigonometric material that is needed to meet the requirements for other college disciplines and to satisfy prerequisites for further mathematical study. I have taught this topic many times at the University of Montana—a university with large numbers of both recent high school students and nontraditional students (the average age of our students is 27). Consequently, I wrote this book with the needs and concerns of these students in mind.

First, I have made the material more accessible by providing large numbers of examples with every step included. There should be no mystery in algebraic computations and this book has none. Second, I have provided many realistic applications so that students will understand that they are studying trigonometry because it is useful to them, not because it is a requirement for a degree. Finally, I have attempted to make the material interesting. In a number of cases I have included historical notes that should add spice to the subject.

Above all, it is my goal that students will both learn to appreciate the value of trigonometry and, most importantly, gain the confidence to succeed in their mathematical studies. I really believe that every university student can learn mathematics if he or she works hard doing problems and is given sufficient help and encouragement.

Changes in the Second Edition

I have made a number of changes which, I believe, will make this book more effective as a teaching tool. Here are the major ones:

1. Captions: Captions have been added to every example and figure in the text. This will make it easier for students to determine exactly what is being done in each example and will be very helpful when tackling the problem sets. The captions on figures clearly describe exactly what each figure illustrates.

2. Readiness Checks: Each problem set begins with up to five ''Readiness Check'' problems. These *true-false* and *multiple-choice* problems are intended to test students' understanding of the material they have just read and do not involve difficult computations. The student who can solve these is ready to tackle the exercise set that follows. Answers to the Readiness Check problems are given at the bottom of the last page on which they occur in each section—not at the back of the book.

3. New Exercises: The second edition has about 3,200 exercises—about 1,000 more than in the first edition. In addition, about 400 drill problems from the first edition have been replaced here.

4. New Graphs: This edition has almost 800 figures, about double the number of the previous one. These are used in new ways as well. See the discussion under ''Figures'' on page x.

5. New Introduction to the Trigonometric Functions: I changed the way I introduce the trigonometric functions. Now all six functions are defined with reference to a right triangle in Section 2.2. This is the way they were probably introduced in high school and should be more familiar to the student. In Section 2.3 I then define the sine and cosine functions using a circle of radius 1 and then extend this definition to circles of radius r (on page 80). Finally, I give the circular definitions of the other four trigonometric functions in Section 2.5. I believe that this change will make the study of trigonometry easier and more natural to students.

6. Analytic Geometry: A new chapter (Chapter 5) entitled Conic Sections and Parametric Equations has been added. Topics include the parabola, the ellipse, the hyperbola, translation and rotation of axes, and parametric equations. There are many applications of conics in the real world.

7. Two New Sections: In addition to the new chapter, there are two new sections:

a. The Dot Product of Two Vectors, Section 4.6.

b. Equations Involving Exponential and Logarithmic Functions, Section 7.5.

8. Use of the Graphing Calculator: Many students now have access to calculators that can, if used properly, provide accurate graphs of a great number of functions. In Appendix A I have shown students how to use their calculators effectively to draw graphs of functions, sketch conic sections and polar graphs, find zeros of polynomials, and solve other types of equations. Example 14 on page A.23 shows the limitations of such calculators by discussing a polynomial whose graph *cannot* accurately be sketched on a calculator. The appendix is written to be used with any calculator now available.

In nine sections within Chapters 3, 5 and 7, I have added problems that are intended for solution on a graphing calculator. I made the deliberate choice to limit the use of the graphing calculator to those sections where it is appropriate, rather than to integrate it throughout the text. This gives the instructor an option. I stress that ownership of a graphing calculator is *not required* for use with this book.

Organization

Chapter 1 is introductory and is intended as a review of the most important topics in college algebra. To understand trigonometry, it is necessary to know something about functions and their graphs. Therefore, a review of the material in Chapter 1 will help the student who has not been exposed to it recently.

The trigonometric functions are introduced in Chapter 2. As described above, the six trigonometric functions are defined using a right triangle in Section 2.2. In Section 2.3 $\sin \theta$ and $\cos \theta$ are first defined in terms of the unit circle and then redefined using a circle of radius r. This section also shows that the triangular and circular definitions give the same values if θ is an acute angle. Section 2.8 contains a number of important applications of right triangle trigonometry. These applications follow a discussion of inverse trigonometric functions (Section 2.7) because many applied problems cannot be solved if one does not know how to compute inverse trigonometric values.

Chapter 3 discusses trigonometric identities, graphs, and equations. There are now *two* sections on graphs involving sine and cosine functions (Sections 3.6 and 3.7). In the graphing sections as well as in Section 3.5, students are asked to solve problems on a graphing calculator.

Chapter 4 is devoted to applications of trigonometry. The central chapters of the book (Chapters 2, 3, and 4) contain five sections that are filled with unique and interesting applications (Sections 2.8, 3.3, 4.3, 4.5 and 4.6).

Chapter 5 is a new chapter on analytic geometry. Topic coverage was described under item #6 on the previous page.

Chapter 6 contains an introduction to complex numbers. Sections 6.2 and 6.3, on polar form and complex roots, present an important application of the two basic trigonometric functions.

Exponential and logarithmic functions are described in Chapter 7. Section 7.6 contains a large and diverse number of realistic applications of exponential growth and decay.

Features

Examples As a student, I learned algebra from seeing examples and doing exercises. There are 360 examples in this book. The examples include all the necessary steps so that students can see clearly how to get from "A" to "B." In many instances explanations are highlighted by colored notes to make steps easier to follow.

Exercises The text includes about 3,200 exercises—both drill and applied. More difficult problems are marked with an asterisk (*) and a few especially difficult ones are marked with two (**). In my opinion, exercises provide the most important tool in any undergraduate mathematics textbook. *If you don't work problems, you won't learn the mathematics*. Or, to quote a button popular at mathematics meetings,

MATH IS NOT A SPECTATOR SPORT.

Readiness Check Problems Each problem set contains multiple-choice and true-false questions that require relatively little computation. Answers to these problems appear on the bottom of the page on which the last such problem in the set appears. They are there to test whether the student understands the basic ideas in the section, and they should be done before tackling the more standard problems that follow.

Chapter Review Exercises At the end of each chapter I have provided a collection of review exercises. Any student able to do these exercises can feel confident that he or she understands the material in the chapter.

Chapter Summary Outlines A summary of the most important facts discussed in each chapter appears at the end of that chapter. Students should find these summaries useful, especially when studying for a test.

Applications We study trigonometry because of its great utility. This book has a large number and variety of applications. A list of these applications appears on pages xix–xx.

Figures There are approximately 800 figures in this book. Many of these appear in a standard way—as graphs of functions. Others are used in the problem sets to help the student understand how graphs change as functions change. There are also a number of new figures attached to applied problems to help the student visualize what he or she is expected to solve.

Use of the Graphing Calculator Students with access to a calculator that can draw graphs can learn how to use it more effectively by consulting Appendix A at the back of the book. For more details on this feature, see item #8 on page viii.

Warnings An important part of the teaching process is helping students to avoid making mistakes, especially those that are commonly made. In eighteen places in the book I provide warnings that illustrate common errors. Each warning illustrates the mistake and shows how it can be avoided.

Use of the Calculator Virtually all college and university students own or have access to a hand calculator. Problems that were computational monstrosities 20 years ago have become fairly easy with the aid of a calculator. I have used the calculator in many examples and have suggested its use in a number of exercises. Examples and problems that require the use of a calculator are marked with a [calculator icon].

Precalculus Many students taking trigonometry will go on to study calculus. In several places in the book I have provided examples that do arise in calculus and have labelled them as such.

Focuses Studying trigonometry can be fun. In this book students can read about real people from the history of mathematics in the sketches that are headed "Focus on . . ." In these thirteen focuses students can learn, for example, about the first known instance of irrational behavior (p. 2), the history of π (p. 6), the origin of the trigonometric functions in Greece and India (p. 86), and the astonishing use of trigonometry by the Babylonians (p. 74). One focus beginning on page 411 contains no history but, rather, asks students to think about the assumptions inherent in a mathematical model.

Accuracy The success of a mathematics textbook largely depends on its accuracy. Galleys and page proofs were carefully checked for accuracy by me and four other mathematicians: Lynne Kotrous and Ray Plankinton at Central Community College, Platte Campus, in Nebraska, Paul Allen at the University of Alabama and Bruce Sisko at Belleville Area College. George Bradley at Duquesne University wrote many of the replacement drill problems and provided solutions to them. Lynne Kotrous, Ray Plankinton, and I solved all the odd-numbered problems in the book. Finally, all three of us proofread the typeset answers to make sure they were accurate.

The result is a book and answer section that is as clean as is within human ability to compile. However, if you do find an error in an answer or in the text, please send it to the publisher or to me; it will be corrected in the next printing.

Chapter Interdependence The following chart indicates chapter interdependence—that is, which later chapters depend on the student's having mastered earlier material.

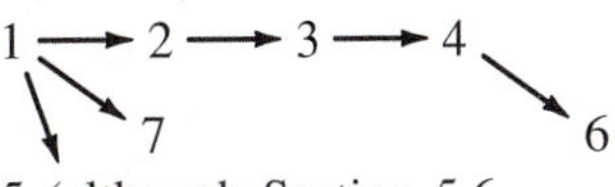

5 (although Section 5.6
uses material from Chapter 4)

Supplements

The answers to most odd-numbered problems appear at the back of the book. In addition, the following instructional aids are available from the publisher:

A **Student Solutions Manual** prepared by George Bradley and Daniel Barbush at Duquesne University contains chapter summaries and detailed solutions for all odd-numbered problems.

An **Instructor's Manual With Transparency Masters** also prepared by George Bradley and Daniel Barbush provides solutions for all the even-numbered problems and transparency masters of key figures from the text.

A **Computerized Test Bank,** of over 1200 multiple choice and open ended questions, prepared by Jan Wynn at Brigham Young University is available for IBM, Macintosh and Apple II computers. A **Printed Test Bank** of these questions is also available.

A & T Software is a computer software package referenced to *College Trigonometry*. Both interactive and tutorial, this software is available for use with IBM and compatible computers.

Videotaped Lectures prepared by Pat Stanley and Becki Bergs at Ball State University covers all the chapters of *College Trigonometry*. These lectures are referenced directly to the sections of this text.

A **Graphing Calculator Supplement** is available for purchase. Written by James Angelos of Central Michigan University, it explains how to use the Casio and TI graphing calculators and uses examples from this text.

An additional **Graphing Calculator Supplement,** written by Iris Fetta of Clemson University, also available for purchase, explores the use and value of the TI-81 graphing calculator.

Acknowledgements

Most of us really don't know what a book is like until it's been used in class and we get comments on how it works. I am grateful to the following individuals for their helpful comments on this second edition:

Mickie Ahlquist, Casper College
Paul Allen, University of Alabama
Daniel Anderson, University of Iowa
Ruth Berger, Memphis State University
Becki Bergs, Ball State University
John Bruha, University of Northern Iowa
Gary Crown, Wichita State University
Lucy Dechéne, Fitchburg State College
Kenneth Dodaro, Florida State University
Tod Feil, Denison University
Iris Fetta, Clemson University
Donald Goldsmith, Western Michigan University
Richard Haworth, Elon College
Lynne Kotrous, Central Community College, Platte Campus
H. T. Mathews, University of Tennessee, Knoxville
Pamela Matthews, Mount Hood Community College
Philip Montgomery, University of Kansas
Ray Plankinton, Central Community College, Platte Campus
Janice Rech, University of Nebraska, Omaha
Michael Schneider, Belleville Area College
Janina Udrys, Schoolcraft College
Ron Virden, Lehigh County Community College
Peter L. Waterman, Northern Illinois University
Carroll Wells, Western Kentucky University
Jan E. Wynn, Brigham Young University

The following reviewers made important contributions to the reliability and teachability of the first edition.

James Arnold, University of Wisconsin at Milwaukee
Daniel Barbush, Duquesne University
George Bradley, Duquesne University
Edgar Chandler, Paradise Valley College
Charles Cook, Tri-State University
Susan Danielson, University of New Orleans
Joe Diestel, Kent State University, Main Campus
Milton Eisner, Mount Vernon College
Susan Foreman, Bronx Community College
Jim Gussett, Longwood College
Barney Herron, Muskegon Community College
Lynne Kotrous, Central Community College, Platte Campus
David Logothetti, Santa Clara University
Reginald Luke, Middlesex Community College
Lyle Oleson, University of Wisconsin
Ray Plankinton, Central Community College, Platte Campus
Jean Rubin, Purdue University
Robert Sharpton, Miami Dade Community College, South Campus
Joseph Stokes, Western Kentucky University
George Szoke, University of Akron

I would like to thank and give credit to the following sources for the use of their original figures: on page 331 (cartoon), by permission of NEA, Inc.; on page 235 (cartoon), by permission of Bill Watterson and Universal Press Syndicate; and on page 28, reprinted by permission of the Wall Street Journal, © Dow Jones and Company, Inc., 1983. All Rights Reserved.

A few problems in this book first appeared in *Mathematics for the Biological Sciences* (New York: Macmillan, 1974) written by James E. Turner and me. I am grateful to Professor Turner for permission to use this material.

I would also like to thank Eric Hays, of Hellgate High School in Missoula, Montana, for providing some of the navigation and force problems that appear in Chapters 2 and 4.

I wrote a great deal of this book while I was a research associate at University College London. I am grateful to the Mathematics Department at UCL for providing office facilities, mathematical suggestions, and, especially, friendship, during my annual visits there.

The book was produced by York Production Services in York, Pennsylvania. I wish to acknowledge the spectacular job done by York, in general, and by my Production Editor Kirsten Kauffman, in particular.

Special thanks are due to the editorial and production staff at Saunders for the care and skill they brought to this process. Finally, I owe much to my Acquisitions Editor, Bob Stern, and my Developmental Editor, Sandi Kiselica, who provided much encouragement and help in determining the final form this book was to take.

Stanley I. Grossman

Contents

Chapter 3. Trigonometric Identities and Graphs 149

Chapter 4. Further Applications of Trigonometry 211

Chapter 5. Conic Sections and Parametric Equations 277

Index of Applications

Biological Sciences

Business and Economics

Physical Sciences

General Topics

Chapter 1

A Review of Algebra

The study of trigonometry involves a number of important ideas in algebra, including the related concepts of functions and graphs. The purpose of this introductory chapter is to review these ideas.

1.1 The Real Number System

The collection of real numbers, denoted by $\mathbb{R}$, consists of the sets of natural numbers, integers, rational numbers, and irrational numbers. Real numbers can be represented on a **number line** in such a way that each point corresponds to exactly one real number and each real number corresponds to one point on the line. The number 0 (zero) is placed. Then the positive real numbers are placed at regular intervals to the right of 0 and the negative real numbers are placed at regular intervals to the left of 0.

The **natural numbers** (also called **positive integers** or **counting numbers**) are the numbers of counting: 1, 2, 3, 4, . . . (the three dots indicate that the string of numbers goes on infinitely). The number 2 is placed one unit to the right of 1 on the number line, the number 3 one unit to the right of 2, and so on. The natural numbers are denoted by N.

The **integers** consist of the natural numbers, their negatives (called the **negative integers**), and the number 0. The collection of integers is denoted by Z.

−4 −3 −2 −1 0 1 2 3 4

Figure 1 The part of the number line that includes the integers from −4 to 4.

In Figure 1, we represent the integers 0, ± 1, ± 2, ± 3, and ± 4 on a **number line.**

A **rational number** is a real number that can be written as the quotient of two integers, where the integer in the denominator is not zero.

Definition of a Rational Number

$$r = \frac{m}{n} \text{ where } m \text{ and } n \text{ are integers and } n \neq 0$$

Every integer n is also a rational number because $n = n/1$. The collection of rational numbers is denoted by the symbol Q (for *quotient*).

EXAMPLE 1 ***Six Rational Numbers***

The following are rational numbers:

(a) $\frac{1}{2}$ (b) $-\frac{3}{4}$ (c) -5 (d) $0 = \frac{0}{1}$ (e) $-\frac{127}{105}$

(f) $0.721 = \frac{721}{1000}$

As in part (f), any terminating decimal is a rational number.

All rational numbers can be represented in an infinite number of ways. For example

$$\frac{1}{2} = \frac{2}{4} = \frac{4}{8} = \frac{3}{6} = \frac{125}{250} = \cdots$$

Usually, however, a rational number is written as m/n, where m and n have no common factors. That is, we will write the rational number in lowest terms.

Any real number that is not rational is called **irrational.** Examples of irrational numbers are $\pi = 3.141592653\ldots$ and $\sqrt{2} = 1.414213562\ldots$

FOCUS ON

The Discovery of Irrational Numbers

According to some historians of mathematics, the discoverer of irrational numbers (then called **incommensurable ratios**) was Hippasus of Metapontum, a follower of Pythagoras, who lived in the fifth century B.C. At the time of his discovery, Hippasus and fellow Pythagoreans were at sea. His colleagues were very upset that someone could produce a number that contradicted the Pythagorean belief that all phenomena on earth could be reduced to natural numbers or their ratios. As a result, Hippasus was thrown overboard. This is the first known instance of irrational behavior.

Sets of Real Numbers

We will often be interested in sets of numbers. A set of numbers is any well-defined† collection of numbers. For example, the integers and rational numbers are each sets of numbers. The collection of "large numbers" does

† A set of numbers is **well defined** if we can determine with certainty whether or not a number belongs to the set.

not constitute a set since it is not well defined. There is no universal agreement about whether a given number is or is not in this collection.

The numbers in a set are called **members** or **elements** of that set. If x is an element of the set A, we write $x \in A$ and read this notation as "x is an element of A" or "x belongs to A."

The elements of a set can be written with braces, { }. For example, the set A of integers between $\frac{1}{2}$ and $\frac{11}{2}$ can be written as

$$A = \left\{x \in \mathbb{R}: x \text{ is an integer and } \frac{1}{2} < x < \frac{11}{2}\right\}.$$

This notation is read as "A is the set of real numbers x such that x is an integer and x is between $\frac{1}{2}$ and $\frac{11}{2}$." Alternatively, we can write the same set as

$$A = \{1, 2, 3, 4, 5\}$$

NOTE The symbol $\notin$ is read "is not an element of." For example, if A is the set described above, then $0 \notin A$.

Intervals

There are certain sets of real numbers that are very important in applications.

(i) The **open interval** from a to b, denoted by (a, b), is the set of real numbers between a and b, *not including* the numbers a and b. We have

$$(a, b) = \{x: a < x < b\}$$

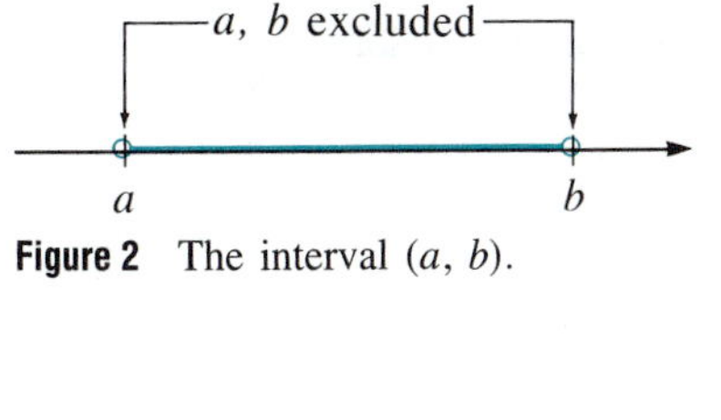

Figure 2 The interval (a, b).

Note that $a \notin (a, b)$ and $b \notin (a, b)$. The numbers a and b are called **endpoints** of the interval. This is depicted in Figure 2.

(ii) The **closed interval** from a to b, denoted by $[a, b]$, is the set of numbers between a and b, *including* the numbers a and b. We have

$$[a, b] = \{x: a \le x \le b\}$$

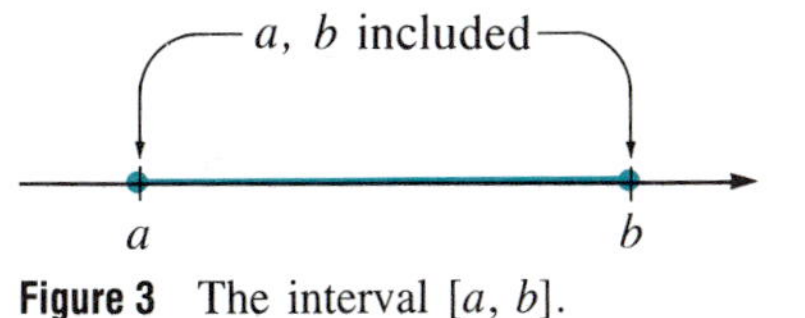

Figure 3 The interval $[a, b]$.

Note that $a \in [a, b]$ and $b \in [a, b]$.

As before, the numbers a and b are called **endpoints** of the interval. This situation is depicted in Figure 3.

NOTE As in Figures 2 and 3 we use a solid dot • to indicate that an endpoint is included and an open dot ∘ to indicate that the endpoint is not included.

EXAMPLE 2 *An Open Interval and a Closed Interval*

(a) $(-1, 5) = \{x: -1 < x < 5\}$
(b) $[0, 8] = \{x: 0 \le x \le 8\}$

Sometimes we will need to include one endpoint but not the other.

(iii) The **half-open interval** $[a, b)$ is given by

$$[a, b) = \{x: a \le x < b\}$$

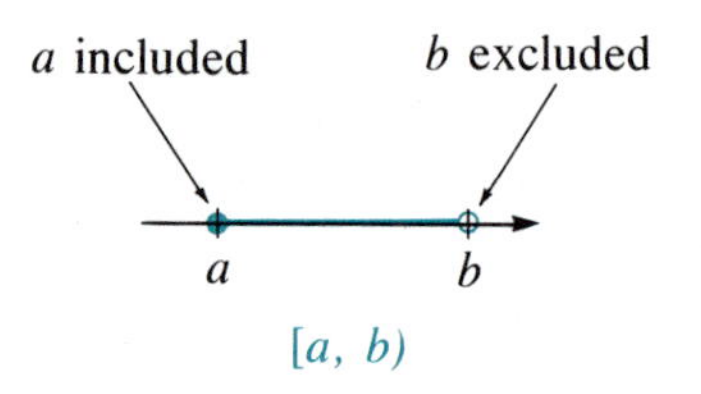

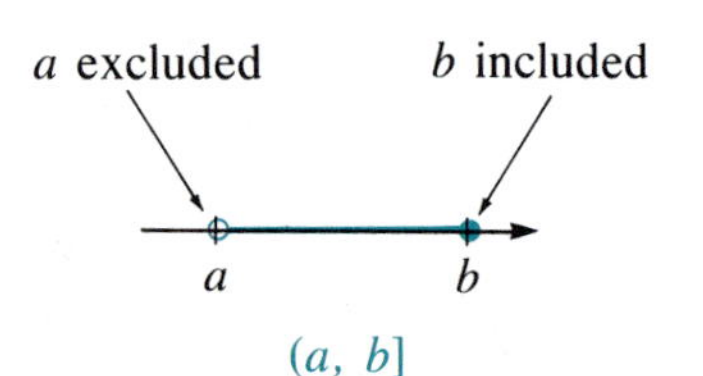

Figure 4 Two half-open intervals.

We include the endpoint $x = a$ but not the endpoint $x = b$. That is, $a \in [a, b)$ but $b \notin [a, b)$.

(iv) The **half-open interval** $(a, b]$ is given by

$$(a, b] = \{x\colon a < x \leq b\}$$

We have $a \notin (a, b]$, but $b \in (a, b]$. Here b is included, but a is not. See Figure 4.

EXAMPLE 3 *A Half-Open Interval*

$[-\frac{5}{2}, 8) = \{x\colon -\frac{5}{2} \leq x < 8\}$ is a half-open interval. This is shown in Figure 5.

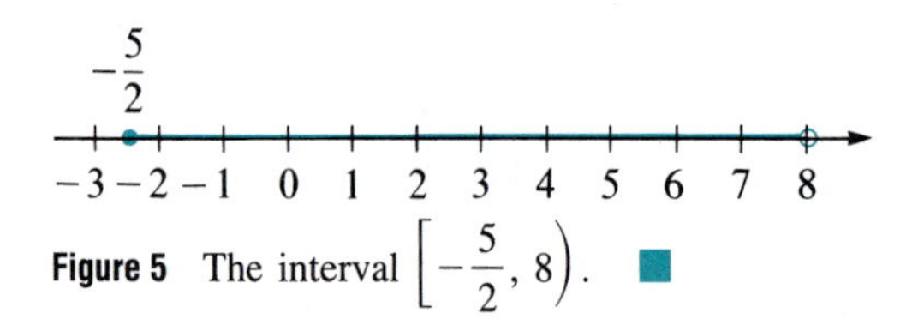

Figure 5 The interval $\left[-\frac{5}{2}, 8\right)$. ■

EXAMPLE 4 *A Half-Open Interval*

$(3, 10] = \{x\colon 3 < x \leq 10\}$ is a half-open interval. This interval is sketched in Figure 6.

0 1 2 3 4 5 6 7 8 9 10 11

Figure 6 The interval $(3, 10]$.

Infinite Intervals

Intervals may be infinite in length. We have

Six Types of Infinite Intervals

$$[a, \infty) = \{x\colon x \geq a\}$$
$$(a, \infty) = \{x\colon x > a\}$$
$$(-\infty, a] = \{x\colon x \leq a\}$$
$$(-\infty, a) = (x\colon x < a\}$$
$$(-\infty, \infty) = \mathbb{R} = \text{the set of real numbers}$$
$$[0, \infty) = \mathbb{R}^+ = \text{the set of nonnegative real numbers}$$

The symbols ∞ and $-\infty$, denoting infinity and minus infinity, respectively, are *not* real numbers and do not obey the usual laws of algebra, but they can be used for notational convenience. This idea is expressed in symbols as

$$[a, \infty) = \{x: a \le x < \infty\} \qquad \text{and} \qquad (-\infty, a) = \{x: -\infty < x < a\}$$

EXAMPLE 5 ***Four Infinite Intervals***

(a) $(-\infty, 2) = \{x: x < 2\}$
(b) $[4, \infty) = \{x: x \ge 4\}$
(c) $(-\infty, -\frac{7}{3}] = \{x: x \le -\frac{7}{3}\}$
(d) $(0, \infty) = \{x: x > 0\}$ = the positive real numbers

Absolute Value

The **absolute value** of a number a is the distance on the number line between that number and zero, and is written $|a|$. See Figure 7.

Thus 2 is 2 units from zero, so $|2| = 2$. The number -3 is 3 units from zero, so $|-3| = 3$.

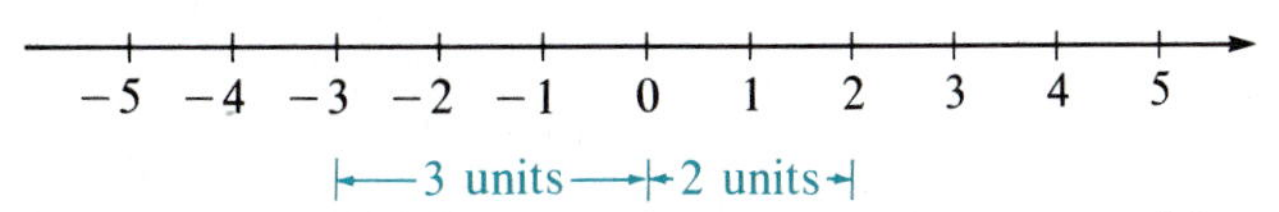

Figure 7 2 is 2 units from 0 and -3 is 3 units from 0, so $|2| = 2$ and $|-3| = 3$.

We may define

Definition of Absolute Value

If a is a real number, then the absolute value of a is given by

$$|a| = a \qquad \text{if } a \ge 0 \tag{1}$$

$$|a| = -a \qquad \text{if } a < 0 \tag{2}$$

Put another way: Let n be a *nonnegative* number. Then the absolute value of n is n, and the absolute value of $-n$ is also n.

The absolute value of a number is a nonnegative number. Note that, for example, $|5| = 5$ and $|-5| = -(-5) = 5$, so numbers that are negatives of one another have the same absolute value.

Properties of Absolute Value

$$|a| \ge 0 \text{ for every real number and } |a| = 0 \text{ only if } a = 0 \tag{3}$$

$$|-a| = |a| \tag{4}$$

$$|ab| = |a||b| \tag{5}$$

$$|a + b| \le |a| + |b| \qquad \text{Triangle inequality} \tag{6}$$

EXAMPLE 6 ***Illustration of the Properties of Absolute Value***

(a) $|-6| = |6| = 6$ Illustration of (4)
(b) $|(-2)(3)| = |-6| = 6 = 2 \cdot 3 = |-2||3|$ Illustration of (5)
(c) $|2 + 3| = 5 = 2 + 3 = |2| + |3|$ Illustration of triangle inequality (6) when $a > 0$ and $b > 0$
(d) $|-2 + (-3)| = |-5| = 5 = 2 + 3 = |-2| + |-3|$ Illustration of triangle inequality (6) when $a < 0$ and $b < 0$
(e) $|2 + (-3)| = |2 - 3| = |-1| = 1 < 2 + 3 = |2| + |-3|$

Illustration of triangle inequality when $a > 0$ and $b < 0$

(f) $|-2 + 3| = |1| = 1 < 2 + 3 = |-2| + |3|$

Illustration of triangle inequality when $a < 0$ and $b > 0$

Parts (iii)–(vi) illustrate a method of proof of the triangle inequality (see Problems 35–39).

The Number π

One of the most important numbers in trigonometry is the number π. Greek mathematicians, and Egyptian mathematicians before them, knew that the ratio of the circumference of a circle to the diameter of that circle is the same for any circle. This common ratio is denoted by π. That is,

$$\pi = \frac{\text{Circumference}}{\text{Diameter}} = \frac{C}{d}$$

or

$$C = \pi d = 2\pi r$$

where r denotes the radius of the circle.

We give a brief history of the fascinating number π.

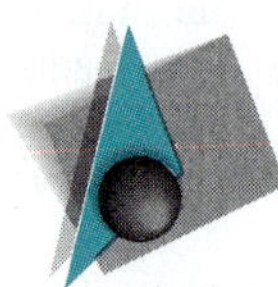

FOCUS ON

A Brief History of π

Interest in rational and irrational numbers goes back thousands of years. Euclid proved that $\sqrt{2}$ is irrational. However, a proof that π is irrational was not known until 1767 when it was published in a paper by the Swiss mathematician Johann Heinrich Lambert (1728–1777). In 1794, the great French mathematician Adrien-Marie Legendre (1752–1833) proved that π^2 is irrational.

However, the difficulties encountered in showing that π is irrational was just an interesting step in the search for properties of the number π, which began in antiquity. Estimates of π were given by perhaps the greatest mathematician of all time, Archimedes of Syracuse (287–212 B.C.). We describe the method below. Remember that π is defined as the ratio of the circumference of a circle to its diameter.

Archimedes' Method of Estimating π

Step 1 Inscribe a regular hexagon in a circle of radius $\frac{1}{2}$ (so the diameter is 1). See Figure 8. Because the six sides of the

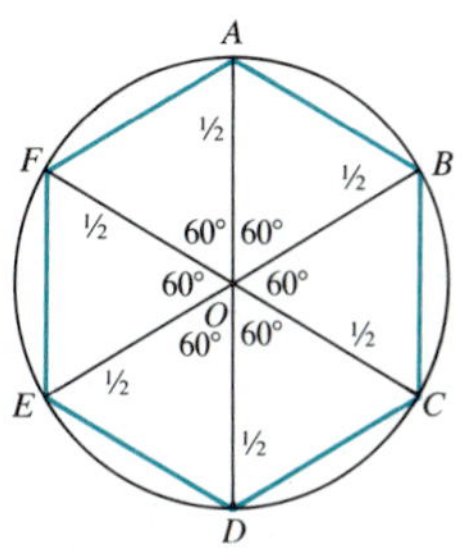

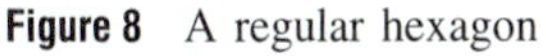
Figure 8 A regular hexagon

hexagon are all equal, the angles opposite them are equal. The six opposite angles add up to 360°, so each angle is 60°. Consider the triangle *AOB* redrawn in Figure 9. The other two angles are equal, so each must equal 60° (explain why). Thus *AOB* is an equilateral triangle, and the side $\overline{AB}$ has length $\frac{1}{2}$. We conclude that the circumference of the hexagon, which approximates the circumference of the circle, is $6(\frac{1}{2}) = 3$. Thus Archimedes' first estimate is given by

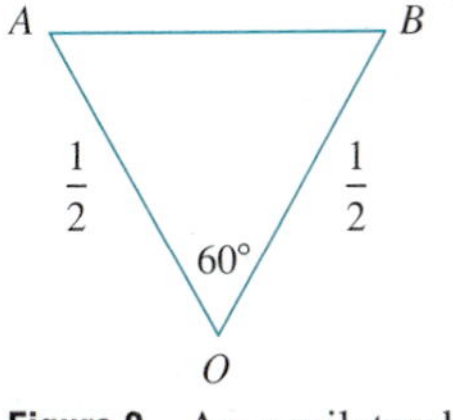

Figure 9 An equilateral triangle

$$\pi = \frac{\text{Circumference of circle}}{\text{Diameter of circle}}$$

$$\approx \frac{\text{Circumference of hexagon}}{\text{Diameter of circle}}$$

$$= \frac{3}{1} = 3$$

Step 2 Double the number of sides to form a regular 12-sided figure called a **dodecagon.** See Figure 10. Using the Pythagorean theorem, Archimedes was able to show that one side of the dodecagon has length $\frac{1}{2}\sqrt{2 - \sqrt{3}}$ (try to prove this). Then its circumference is $12\left[\frac{\sqrt{2 - \sqrt{3}}}{2}\right] = 6\sqrt{2 - \sqrt{3}} \approx 3.1058285$. This is our second estimate for π.

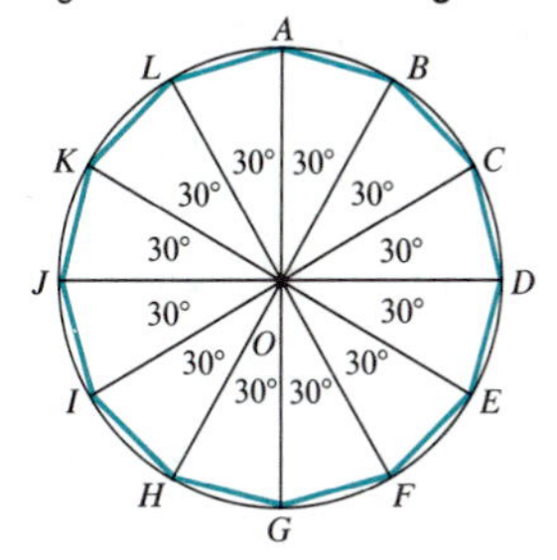

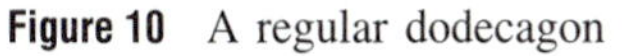
Figure 10 A regular dodecagon

It was not known at the time of Archimedes if π was rational or irrational, and his method was used (until more powerful tools were developed) in an attempt to find a repeating pattern in its decimal representation. One Dutch mathematician, Ludolph van Ceulen (1539–1610), reported (in 1596) using a polygon of 32 billion sides to yield a value of π to 20 decimal places. If you try some of the calculations, you will find that many square roots must be taken. The calculations by hand must have been extremely tedious. Nevertheless, no one came up with a substantially better way to estimate π until Euler in the 17th century.

Many people were fascinated by the value of π. The following sentence, written in the early part of the 20th century, gives a mnemonic device for memorizing π to 14 decimal places. The number of letters in each word equals the corresponding number in the expansion of π:

How I want a drink, alcoholic of course, after the heavy lectures involving quantum mechanics
($\pi \approx 3.14159265358979$)

In 1967, a computer found 500,000 decimal places of π in a little over a day. Computers have far surpassed this feat today. In any event, neither van Ceulen nor anyone else ever found a repeating pattern in the decimal expansion of π because such a pattern does not exist. Many mathematicians (and others) spent huge amounts of time in this fruitless search.

Mathematical proof does not, however, stop some individuals from making contrary discoveries. In 1897, a bill was presented to the Indiana legislature that would have established "a new and correct value for π." The value was certainly new, but quite incorrect.† The bill passed the Indiana House of Representatives on February 5, 1897, by a vote of 67 to 0. It was, fortunately, defeated in the Senate, but only on the second reading.

In 1931, a Cleveland businessman, Carl Theodore Heisel, published a book in which he "proved" that π is exactly equal to 256/81, the value used by the Egyptians about 4000 years earlier. His proof was, of course, quite wrong, but his attempt is a demonstration of the fascination the number π holds for us even today.

If you are interested in learning more about the history of π, you should read the fascinating book, *A History of π*, by Petr Beckman, The Golden Press, New York, 1971.

† The bill did not make it clear what this "new and correct" value was to be. However, the most logical interpretation of the bill gives this value as $\pi = 16/\sqrt{3} \approx 9.2376$ (see *A History of π*, p. 174).

Problems 1.1

Readiness Check

I. Which of the following represents a rational number?

a. 23 b. $\frac{17}{41}$ c. $\frac{-362}{471}$ d. 0.254

II. Which of the following does *not* represent the same set of real numbers?

a. $[-2, 1)$ b. $\{x: -3/2 \le x - 1/2 < 1/2\}$

c. 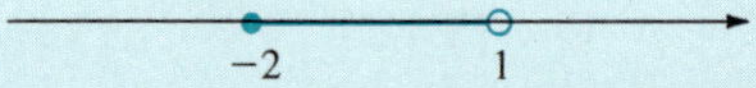

d. All the real numbers that are less than 1 but greater than or equal to -2.

III. Which of the following is equal to $|-3| - |-5|$?

a. -2 b. 5 c. 2 d. -8

In Problems 1–15, represent each set of inequalities, using interval notation.

1. $1 < x < 2$
2. $-3 < x < 5$
3. $0 < x < 8$
4. $-5 \le x \le 4$
5. $-2 \le x \le 0$
6. $3.7 \le x \le 5.2$
7. $0 < x \le 5$
8. $1 \le x < \sqrt{2}$
9. $-1.32 \le x < 4.16$
10. $x \ge 0$
11. $x < 0$
12. $x \le 0$
13. $x \le 3$
14. $x \ge -6$
15. $x < -5$

In Problems 16–26 a set of real numbers is depicted. Write an interval that describes the set.

16. 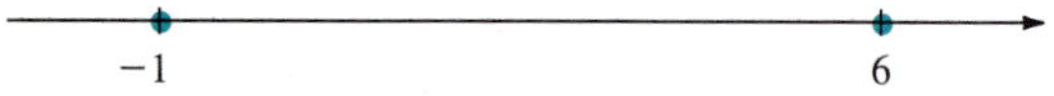

17. 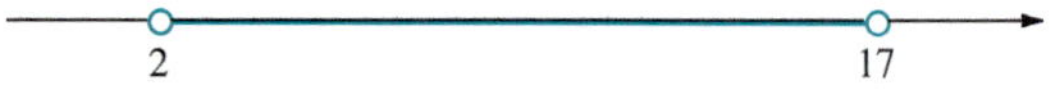

18. 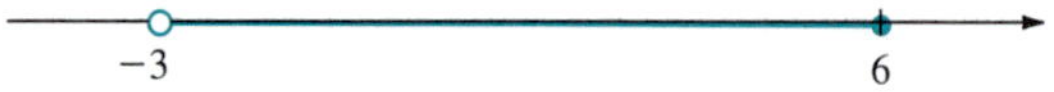

19. 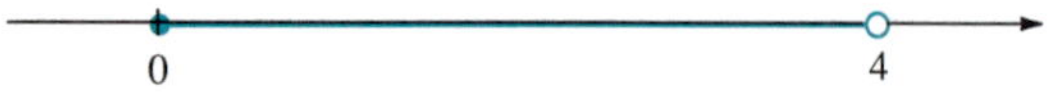

20. 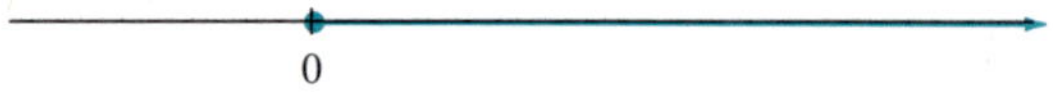

21. 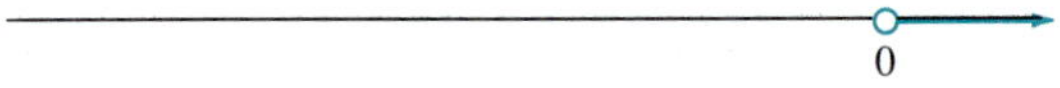

22. 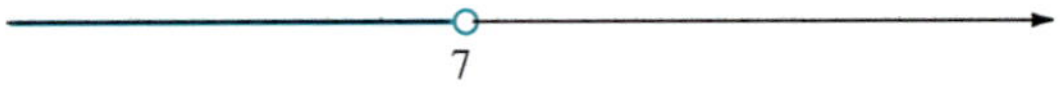

23. 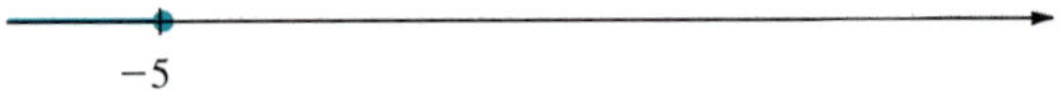

24. 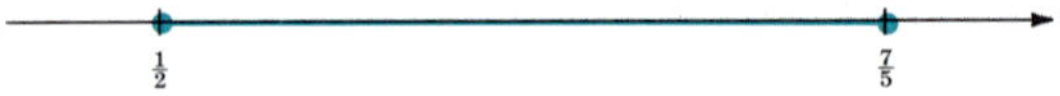

25. 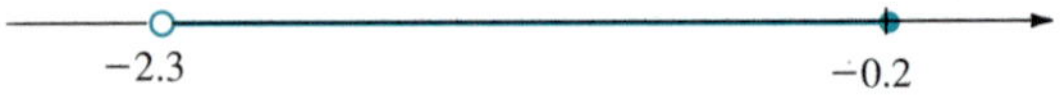

26. 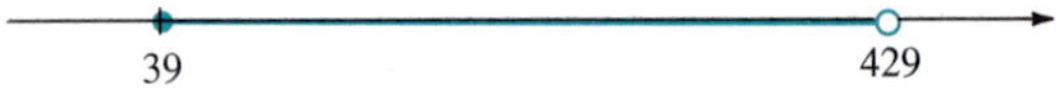

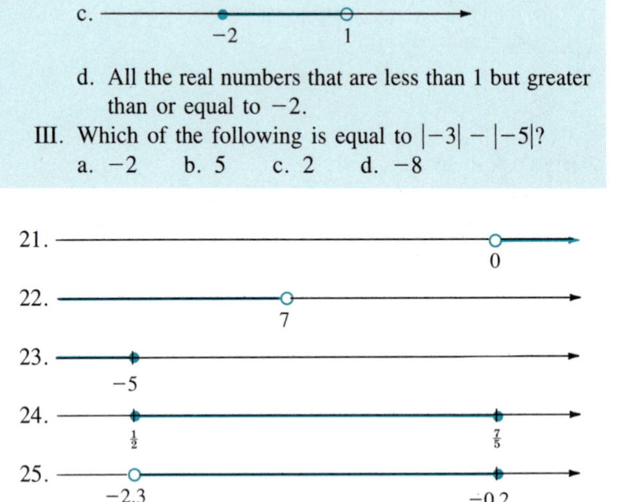

In Problems 27–32 compute each value.

27. $|7| - |3|$
28. $|3| - |-2|$
29. $|4| - |-9|$ 30. $|10| - |12|$
31. $|\pi - 2|$ 32. $|\pi - 7|$
33. Show that $|-a| = |a|$.
34. Show that $|ab| = |a||b|$. [Hint: Consider the separate cases $a > 0$ and $b > 0$, $a < 0$ and $b > 0$, $a < 0$ and $b < 0$, $a > 0$ and $b < 0$. What happens if $a = 0$ or $b = 0$?]
35. If $a > 0$ and $b > 0$, show that $|a + b| = |a| + |b|$.
36. If $a < 0$ and $b < 0$, show that $|a + b| = |a| + |b|$.
37. If $a < 0$ and $b > 0$, show that $|a + b| < |a| + |b|$.
38. If $a > 0$ and $b < 0$, show that $|a + b| < |a| + |b|$.
39. If $a = 0$ or $b = 0$, show that $|a + b| = |a| + |b|$.

Answers to Readiness Check

I. All of them II. b III. a

1.2 The Cartesian Coordinate System

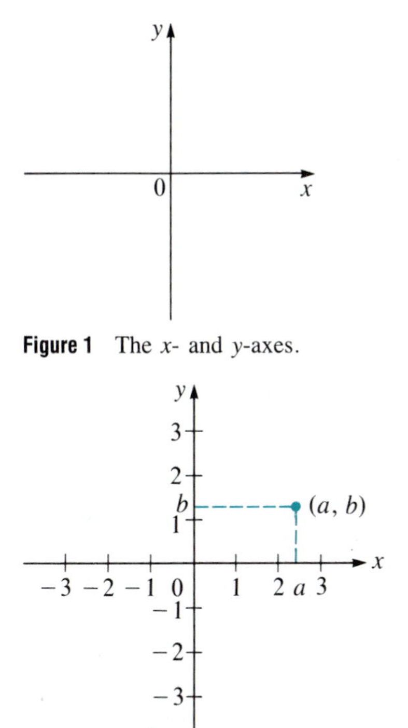

Figure 1 The x- and y-axes.

Figure 2 The point (a, b) in the xy-plane.

In this section we describe the most common way of representing points in a plane: the **Cartesian coordinate system.**† To form the Cartesian coordinate system, we draw two mutually perpendicular number lines as in Figure 1: one horizontal line and one vertical line. The horizontal line is called the ***x*-axis,** and the vertical line is called the ***y*-axis.** The point at which the lines meet is called the **origin** and is labeled 0.

To every point in the plane, we assign an **ordered pair** of numbers. The first element in the pair is called the ***x*-coordinate,** and the second element of the pair is called the ***y*-coordinate.**

The x-coordinate measures the number of units from the y-axis to the point. Points to the right of the y-axis have positive x-coordinates, and those to the left have negative x-coordinates.

The y-coordinate measures the number of units from the x-axis to the point. Points above the x-axis have a positive y-coordinate, and those below have a negative y-coordinate. Figure 2 shows a typical point (a, b), where $a > 0$ and $b > 0$.

Two ordered pairs, or points, are **equal** if their first elements are equal and their second elements are equal. Note that $(1, 0)$ and $(1, 1)$ are *different* because their second elements are different. Note too that $(1, 2)$ and $(2, 1)$ are different points.

In Figure 3, several different points are depicted.

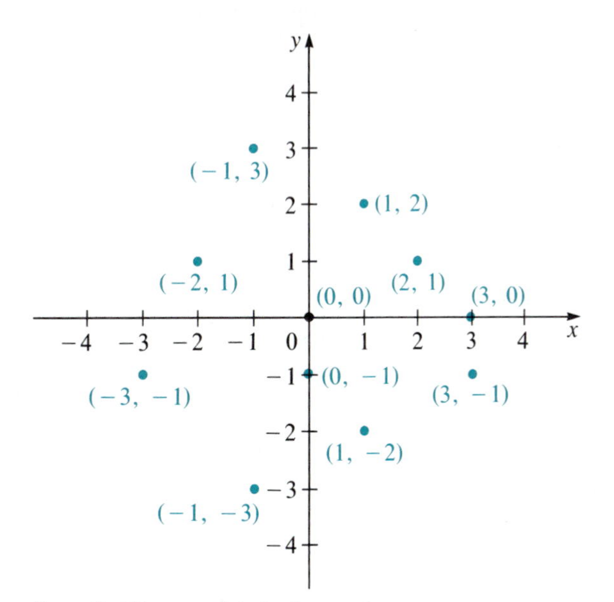

Figure 3 Eleven points in the xy-plane.

† See the biographical sketch on p. 13.

When points in the plane are represented by the Cartesian coordinate system, the plane is called the **Cartesian plane,** or the ***xy*-plane,** and is denoted by $\mathbb{R}^2$. We have

Definition of $\mathbb{R}^2$ (the *xy*-Plane)

$$\mathbb{R}^2 = \{(x, y): x \in \mathbb{R} \text{ and } y \in \mathbb{R}\}$$

Quadrants

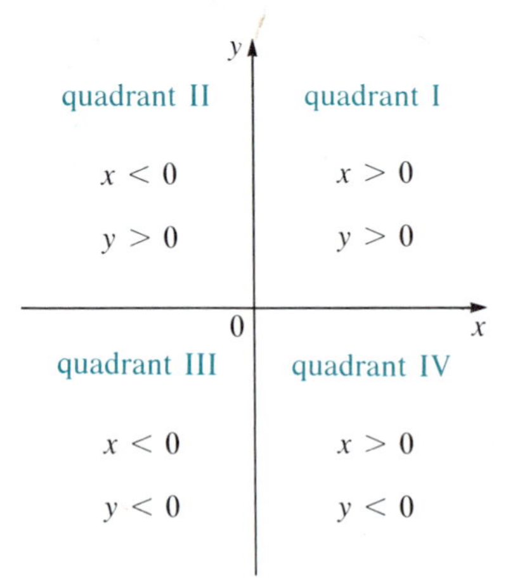

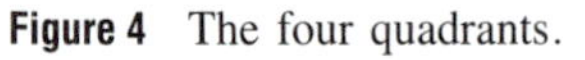

Figure 4 The four quadrants.

A glance at Figure 4 indicates that the x- and y-axes divide the xy-plane into four regions. These regions are called **quadrants** and are denoted as in the figure.

EXAMPLE 1 ***The Quadrants of Four Points***

(a) (1, 3) is in the first quadrant because $1 > 0$ and $3 > 0$.
(b) $(-4, -7)$ is in the third quadrant because $-4 < 0$ and $-7 < 0$.
(c) $(-2, 5)$ is in the second quadrant.
(d) $(7, -3)$ is in the fourth quadrant.

We now give a formula for finding the distance between two points in the plane.
Let (x_1, y_1) and (x_2, y_2) be two points in the xy-plane (see Figure 5).

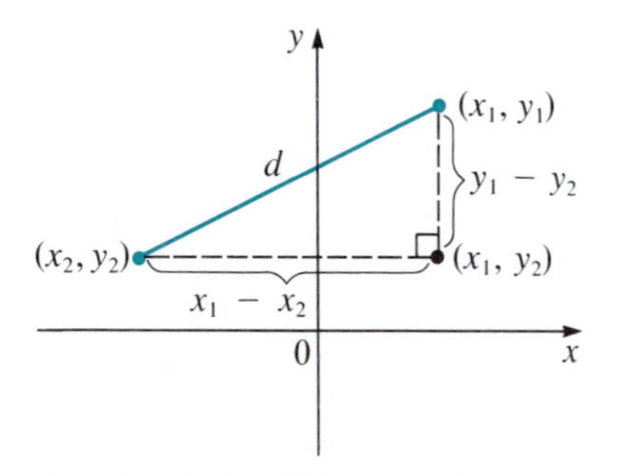

Figure 5 By the Pythagorean theorem, $(x_1 - x_2)^2 + (y_1 - y_2)^2 = d^2$.

Theorem: The Distance Formula

$$d = \sqrt{(x_1 - x_2)^2 + (y_1 - y_2)^2} \qquad (1)$$

EXAMPLE 2 ***Finding the Distance Between Two Points***

Find the distance between the points (2, 5) and $(-3, 7)$.

SOLUTION Let $(x_1, y_1) = (2, 5)$ and $(x_2, y_2) = (-3, 7)$. From (1),

$$d = \sqrt{(2 - (-3))^2 + (5 - 7)^2} = \sqrt{5^2 + (-2)^2} = \sqrt{29} \approx 5.385$$

Equations in Two Variables: Relations

Now that we have defined a coordinate system with two components, we will discuss **equations in two variables.** The set of ordered pairs that satisfy an equation in two variables is called a **relation.**

EXAMPLE 3 *Four Equations in Two Variables*

The following are equations in the two variables x and y:

(a) $y = 3x + 2$
(b) $x^2 + y^2 = 4$
(c) $y = \frac{1}{x}$
(d) $x^3y + \sqrt{x + y} = y$

Solution of an Equation in Two Variables

A **solution** to an equation in two variables is a point (x, y) whose coordinates satisfy the equation.

EXAMPLE 4 *Solutions to Two Equations*

(a) $(1, 5)$ is a solution to the equation $y = 3x + 2$ because

$$\underset{\substack{\downarrow \\ x}}{}\ 3 \cdot \overset{\substack{x \\ \downarrow}}{1} + 2 = \overset{\substack{y \\ \downarrow}}{5}$$

(b) $(\sqrt{3}, 1)$ and $(\sqrt{2}, -\sqrt{2})$ are solutions to the equation $x^2 + y^2 = 4$ because

$$(\underset{\substack{\uparrow \\ x}}{\sqrt{3}})^2 + \underset{\substack{\uparrow \\ y}}{1}^2 = 3 + 1 = 4$$

and

$$(\sqrt{2})^2 + (-\sqrt{2})^2 = 2 + 2 = 4$$

There's a cliché that states "a picture is worth a thousand words." In dealing with equations we can usually obtain answers more quickly and understand the answers better if we have a picture. Such a picture is called a graph.

Definition of a Graph

The **graph** of an equation in two variables is the set of all points in the xy-plane whose coordinates satisfy the equation.

In Sections 1.3 and 1.4 we will discuss general equations and graphs. Now we consider the equation of a circle.

Circle

A **circle** is defined as the set of all points in a plane at a given distance from a given point. The given point is called the **center** of the circle, and the common distance from the center is called the **radius.**

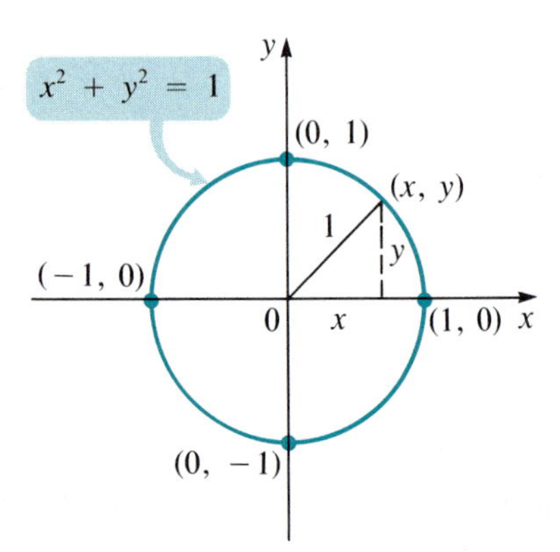

Figure 6 The unit circle.

EXAMPLE 5 *The Unit Circle*

Find an equation of the circle centered at the origin with radius 1.

SOLUTION If (x, y) is any point on the circle, then the distance from (x, y) to $(0, 0)$ is 1. From (1) we have

$$\sqrt{(x-0)^2 + (y-0)^2} = 1$$

Squaring both sides, we obtain

$$x^2 + y^2 = 1$$

This circle is sketched in Figure 6. It is called the **unit circle** and is of central importance in the study of the trigonometric functions.

We now discuss circles with other centers and other radii.

EXAMPLE 6 *Finding an Equation of a Circle Given Its Center and Radius*

Find an equation of the circle centered at $(3, -5)$ with radius 4.

SOLUTION If (x, y) is on the circle, then the distance from (x, y) to $(3, -5)$ is 4, so that

$$\sqrt{(x-3)^2 + (y+5)^2} = 4 \quad \text{or} \quad (x-3)^2 + (y+5)^2 = 16$$

This circle is sketched in Figure 7.

Figure 7 Circle centered at $(3, -5)$ with radius 4.

Figure 8 shows the graph of the circle centered at (h, k) with radius r. The point (x, y) is on this circle if and only if its distance from (h, k) is r. We have the following:

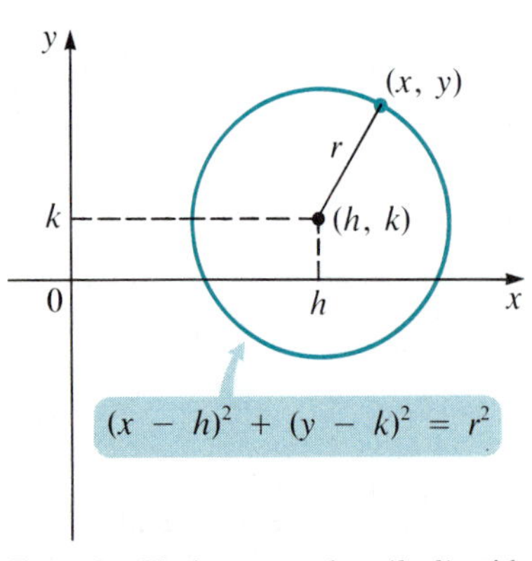

Figure 8 Circle centered at (h, k) with radius r.

Standard Equation of the Circle with Center at (h, k) and Radius r

$$(x-h)^2 + (y-k)^2 = r^2 \qquad (2)$$

EXAMPLE 7 *Showing That a Second Degree Equation in Two Variables Represents a Circle*

Show that the equation $x^2 - 6x + y^2 + 2y - 17 = 0$ represents a circle; find its center and radius.

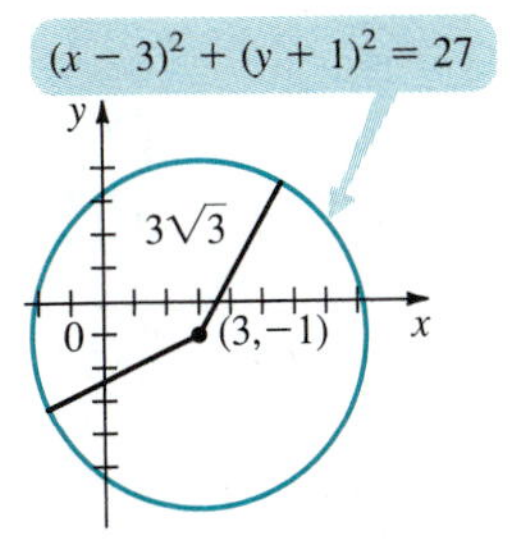

Figure 9 The circle centered at $(3, -1)$ with radius $3\sqrt{3}$.

SOLUTION We use the technique of completing the square.† We have

$$x^2 - 6x + y^2 + 2y = 17$$

$$(x^2 - 6x + 9) + (y^2 + 2y + 1) = 17 + 9 + 1 \qquad \text{Add } 9 + 1 \text{ to both sides and group terms}$$

$$(x-3)^2 + (y+1)^2 = 27 \qquad \text{Factor}$$

This is the equation of a circle with center at $(3, -1)$ and radius $\sqrt{27} = 3\sqrt{3} \approx 5.196$. It is sketched in Figure 9.

We close this section by noting that the graphs of quadratic equations in two variables — that is, equations having the form

$$ax^2 + bxy + cy^2 + dx + ey + f = 0 \qquad (3)$$

— are called **conic sections.** The equation of a circle is a special case of equation (3). Another special case is discussed in Example 1 in Section 1.4 (on p. 28). We will discuss conic sections in more generality in Chapter 5.

FOCUS ON

René Descartes (1596–1650)

The Cartesian plane is named after the great French mathematician and philosopher René Descartes. Born near the city of Tours in 1596, Descartes received his education first at the Jesuit school at La Flèche and later at Poitier, where he studied law. He had delicate health and, while still in school, developed the habit of spending the greater part of each morning working in bed. Later, he considered these morning hours the most productive period of the day.

At the age of 16, Descartes left school and moved to Paris, where he began his study of mathematics. In 1617, he joined the army of Maurice, Prince of Nassau. He also served with Duke Maximillian I of Bavaria and with the French army at the siege of La Rochelle.

Descartes was not a professional soldier, however, and his periods of military service were broken by periods of travel and study in various European cities. After leaving the army for good, he resettled in Paris to continue his mathematical studies and then moved to Holland, where he lived for 20 years.

Much stimulated by the scientists and philosophers he met in France, Holland, and elsewhere, Descartes later became known as the "father of modern philosophy." His statement "Cogito ergo sum" ("I think, therefore I am") played a central role in his philosophical writings.

Descartes's program for philosophical research was enunciated in his famous *Discours de la méthode pour bien conduire sa raison et chercher la vérité dans les sciences* (A Discourse on the Method of Rightly Conducting the Reason and Seeking Truth in the Sciences) published in 1637. This work was accompanied by three appendices: *La dioptrique* (in which the law of refraction — discovered by Snell — was first published), *Les météores* (which contained the first accurate explanation of the rainbow), and *La géométrie. La géométrie,* the third and most famous appendix, took up about a hundred pages of the *Discours.* One of the major achievements of *La géométrie* was that it connected figures of geometry with the equations of algebra. The work established Descartes as the founder of analytic geometry.

René Descartes
(David Smith Collection)

In 1649, Descartes was invited to Sweden by Queen Christina. He agreed, reluctantly, but was unable to survive the harsh, Scandinavian winter. He died in Stockholm in early 1650.

---

†Recall that this technique is sometimes used to solve the general quadratic equation $ax^2 + bx + c = 0$.

Problems 1.2

Readiness Check

I. Which of the following is true about the point $(0, b)$?
  a. It is located on the x-axis.
  b. It is located on the y-axis.
  c. It represents a vertical distance from the origin.
  d. If $b < 0$, its distance from the origin equals b^2.
  e. It is not possible to determine whether a, b, c, or d is the correct answer until the value of b is known.

II. Which of the following is the equation of a circle whose center is $(3, -4)$ and that passes through $(0, -2)$?
  a. $(x - 3)^2 + (y + 4)^2 = 45$
  b. $(x + 3)^2 + (y - 4)^2 = 45$
  c. $(x - 3)^2 + (y + 4)^2 = 13$
  d. $(x + 3)^2 + (y - 4)^2 = 40$

III. Which of the following is true about the Cartesian coordinate system?
  a. The origin is the point of intersection of the x- and y-axes.
  b. Every ordered pair is represented by a point on the x-axis.
  c. The ordered pair (a, b) is located in quadrant IV if $a < 0$ and $b > 0$.
  d. The vertical axis is called the x-axis.

In Problems 1–10 plot each point in the xy-plane. If the point is not on the x- or y-axis, determine the quadrant in which it lies.

1. $(3, -2)$
2. $(4, 3)$
3. $(2, 0)$
4. $(0, -5)$
5. $(-4, -1)$
6. $(-2, 3)$
7. $\left(\frac{1}{2}, \frac{1}{3}\right)$
8. $\left(\frac{1}{3}, -\frac{3}{2}\right)$
9. $\left(0, \frac{3}{4}\right)$
10. $\left(-\frac{2}{3}, \frac{7}{3}\right)$

In Problems 11–20 find the distance between the given points.

11. $(1, 3)$, $(4, 7)$
12. $(-7, 2)$, $(4, 3)$
13. $(-9, 4)$, $(0, -7)$
14. $\left(\frac{1}{2}, \frac{1}{3}\right)$, $\left(\frac{1}{3}, \frac{1}{2}\right)$
15. $(-3, -7)$, $(-1, -2)$
16. (a, b), (b, a)
17. $(0, 0)$, $(-c, d)$
18. $(1.3, 4.5)$, $(6.2, 3.4)$
19. $(14.13, -2.16)$, $(11.19, 5.71)$
20. $(0.0135, 0.0146)$, $(0.0723, 0.0095)$

In Problems 21–29, find an equation of the circle with the given center and radius, and sketch its graph.

21. $(0, 2)$, $r = 1$
22. $(2, 0)$, $r = 1$
23. $(1, 1)$, $r = \sqrt{2}$
24. $(1, -1)$, $r = 2$
25. $(-1, 4)$, $r = 5$
26. $\left(\frac{1}{2}, \frac{1}{3}\right)$, $r = \frac{1}{2}$
27. $(\pi, 2\pi)$, $r = \sqrt{\pi}$
28. $(4, -5)$, $r = 7$
29. $(3, -2)$, $r = 4$

In Problems 30–33 an equation is given. Show that it is an equation of a circle, and find the circle's center and radius.

30. $x^2 + 4x + y^2 - 2y + 1 = 0$
31. $x^2 + y^2 - 6y + 3 = 0$
32. $x^2 + 6x + y^2 + 4y + 9 = 0$
33. $2x^2 + 2x + 2y^2 - y - \frac{61}{8} = 0$

** 34. Show that the equation $x^2 + ax + y^2 + by + c = 0$ is the equation of a circle if and only if $a^2 + b^2 - 4c > 0$.

* 35. Find an equation for the unique circle that passes through the points $(0, -2)$, $(6, -12)$, and $(-2, -4)$.

Answers to Readiness Check

I. b II. c III. a

1.3 Functions

In Section 1.2 we discussed equations in two variables, and relations. Functions constitute an important class of relations. We now define a function and give some examples of functions.

Definition of Function, Domain and Range, and Image

A **function** is a rule that assigns to each member of one set (called the **domain** of the function) a unique member of another set. The set of all assigned members is called the **range** of the function. If x in the domain is assigned the element y in the range, then y is called the **image** of x.

For many functions the domain and range are both equal to $\mathbb{R}$, the set of real numbers.

You have seen many examples of functions.

Figure 1 The prices of four TV sets.

EXAMPLE 1 *The TV Price Function*

A certain appliance store sells TV sets. Each TV set is assigned a price as in Figure 1. This assignment defines a function because each TV set has exactly one price. The domain of the function is the set of TV sets in the store. The range is the set of prices of TV sets in the store. ■

EXAMPLE 2 *The Birthdate Function*

To each living person in the United States, assign his or her date of birth. This assignment defines a function because every person has exactly one date of birth. The domain of this function is the set of people living in the United States. The range of this function is the set of dates in history on which at least one person now living in the United States was born. ■

EXAMPLE 3 *The Squaring Function*

To each real number assign its square. For example, 4 is assigned to 2, 9 is assigned to -3, 1.69 is assigned to 1.3 and 2 is assigned to $\sqrt{2}$. This assignment defines a function because each real number has a unique square. The domain is the set of real numbers $\mathbb{R}$. The range is the set of nonnegative real numbers, denoted by $\mathbb{R}^+$.

NOTE A function is often denoted by a single letter, usually f, g, or h. If x is a member of the domain of f, say, then $f(x)$ denotes the image of x. The symbol $f(x)$ is read "f of x." We often write $y = f(x)$,† read "y equals f of x," to indicate that to each value x in the domain we assign the unique image y in the range. When we write $y = f(x)$, x is called the **independent variable,** and y is called the **dependent variable.**

†This notation was first used by the great Swiss mathematician Leonhard Euler (1707–1783) in the *Commentarii Academia Petropolitanae* (Petersburg Commentaries), published in 1734–1735.

EXAMPLE 4 *Using Function Notation*

(a) For the TV price function, if Brand X costs \$421, then

$$f(\text{Brand X}) = \$421$$

That is, the image of Brand X is \$421.

(b) For the squaring function, we have $f(2) = 2^2 = 4$, $f(-3) = (-3)^2 = 9$, $f(1.3) = 1.3^2 = 1.69$, and $f(\sqrt{2}) = (\sqrt{2})^2 = 2$. In general, $f(x) = x^2$ for every real number x.

To make this clearer, we write

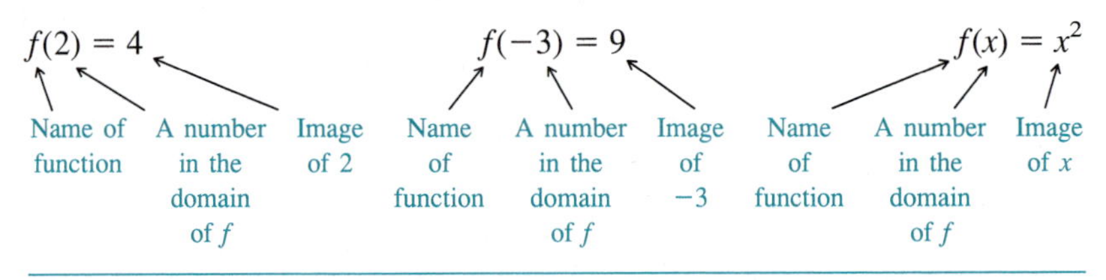

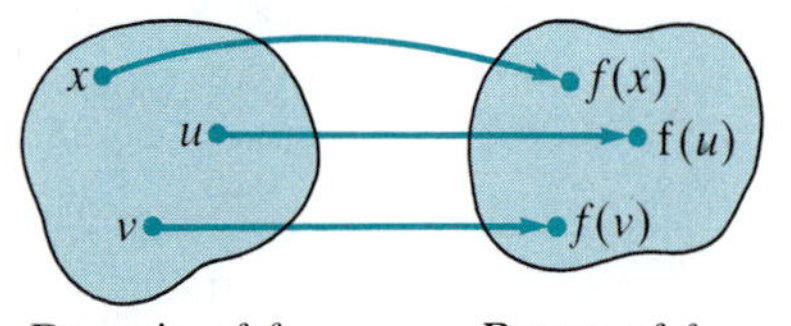

Figure 2 Illustration of domain and range.

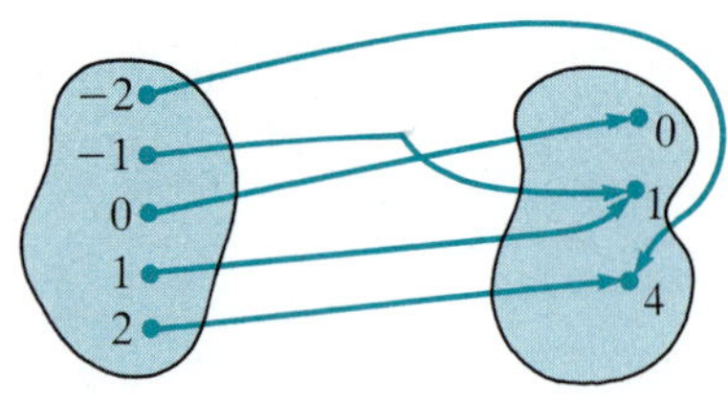

Figure 3 Some values of the squaring function $f(x) = x^2$.

Functions are often depicted pictorially as in Figure 2. The diagram illustrates, for example, that the function f takes a value x in the domain and assigns to it a unique value $f(x)$ in its range.

In Figure 3 we depict some of the values taken by the squaring function.

EXAMPLE 5 *Determining Whether a Rule Is a Function*

Let D and R be two sets. In Figure 4 we show four rules that assign members of D to members of R. Which of the four rules represent functions with domain D?

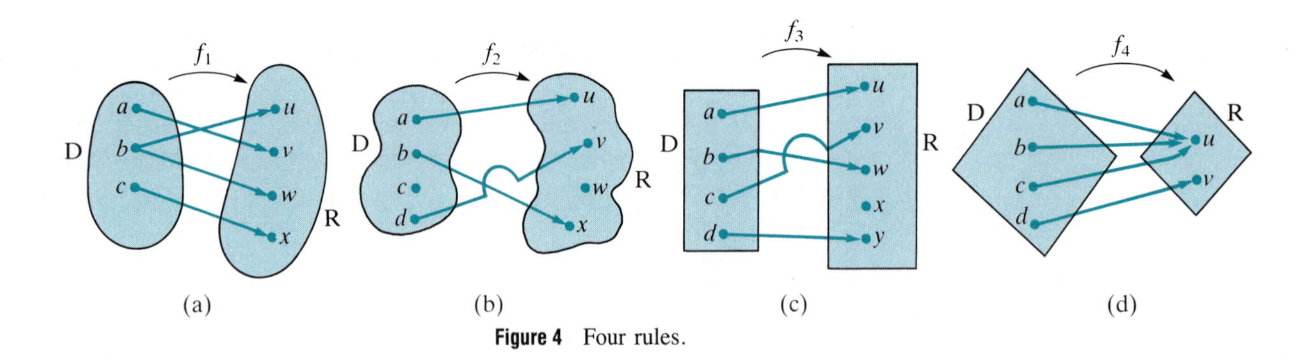

Figure 4 Four rules.

SOLUTION

(a) f_1 is not a function because it assigns two different members of R to one member (b) of D. That is, $f_1(b) = u$ and w. This contradicts the definition of a function. In order to be a function, f_1 must assign *exactly one* member of R to each member of D. If we delete the arrow from b to u or the arrow from b to w (but not both), then f will be a function with domain D.

(b) f_2 is not a function with domain D because f does not assign anything to c. In order to be a function with domain D, f must assign one member of R to *each* member of D.

(c) f_3 is a function because it assigns one member of R to each member of D. It is irrelevant that nothing in D is assigned to x. This may make x feel lonely, but it does not contradict the definition of a function. The range of f_3 is $\{u, v, w, y\}$.

(d) f_4 is a function because it assigns exactly one member of R to each member of D. The fact that u is assigned three times does not contradict the definition of a function.

For the remainder of this chapter we will look at functions for which the domains and ranges are sets of real numbers.

When the domain of a function is not given, we usually take the domain to be the set of all real values of x so that in the equation $y = f(x)$, $f(x)$ is a defined real number.

EXAMPLE 6 *A Linear Function*

Let $f(x) = 3x + 5$. Then since $3x + 5$ is defined for every real number x, the domain of f is $\mathbb{R}$, the set of all real numbers.

What is the range? To answer that question, we ask: For what values of y is $y = 3x + 5$ for some real number x? For example, for $y = 10$, we seek an x such that

$$\begin{aligned} 10 &= 3x + 5 \\ 10 - 5 &= 3x \\ 5 &= 3x \\ x &= \frac{5}{3} \end{aligned}$$

Thus $f(\frac{5}{3}) = 3 \cdot \frac{5}{3} + 5 = 5 + 5 = 10$ so 10 is the image of $\frac{5}{3}$ and 10 is in the range of f.

In fact, every real number y is in the range of f. If y is given, we seek an x such that

$$\begin{aligned} y &= 3x + 5 \\ y - 5 &= 3x \\ x &= \frac{y-5}{3} \end{aligned}$$

Then $f(x) = f\left(\frac{y-5}{3}\right) = 3\left(\frac{y-5}{3}\right) + 5 = y - 5 + 5 = y$ so y is the image of $\frac{y-5}{3}$, which means that y is in the range of f.

More generally, a **linear function** is a function that can be written in the form

$$y = f(x) = mx + b$$

The domain of this function is $\mathbb{R}$. If $m \neq 0$, we can solve for x in terms of y:

$$mx = y - b \quad \text{or} \quad x = \frac{y - b}{m}$$

That is, if $m \neq 0$, then for every real number y there is a unique x such that $y = mx + b$. Thus the range of f is $\mathbb{R}$ if $m \neq 0$.

EXAMPLE 7 *A Square Root Function*

Let $s = g(t) = \sqrt{3t + 1}$.

(a) Evaluate $g(0)$, $g(1)$, $g(8)$, $g(100)$, and $g(t^2 + 2)$.
(b) Find the domain of g.
(c) Find the range of g.

SOLUTION

(a) $g(0) = \sqrt{3 \cdot 0 + 1} = \sqrt{1} = 1$

$g(1) = \sqrt{3 \cdot 1 + 1} = \sqrt{4} = 2$

$g(8) = \sqrt{3 \cdot 8 + 1} = \sqrt{25} = 5$

$g(100) = \sqrt{3 \cdot 100 + 1} = \sqrt{301} \approx 17.35$

$g(t^2 + 2) = \sqrt{3(t^2 + 2) + 1} = \sqrt{3t^2 + 7}$

(b) $\sqrt{t}$ is defined as a real number as long as $t \geq 0$. Then, since the range must be a set of real numbers, $\sqrt{3t + 1}$ is defined if

$$\begin{aligned} 3t + 1 &\geq 0 \\ 3t &\geq -1 \\ t &\geq -\frac{1}{3} \end{aligned}$$

Thus domain of $g = \{t\colon t \geq -\frac{1}{3}\} = [-\frac{1}{3}, \infty)$.

(c) Suppose that s^* is in the range of g. Then there is a t such that $s^* = \sqrt{3t + 1} \geq 0$. So every number in the range of g is nonnegative. Moreover, if $s^* \geq 0$, then there is a t such that $s^* = \sqrt{3t + 1}$. To find that t, we work backward:

$$\begin{aligned} s^{*2} &= 3t + 1 \\ s^{*2} - 1 &= 3t \\ t &= \frac{s^{*2} - 1}{3} \end{aligned}$$

and

$$\sqrt{3t + 1} = \sqrt{3\left(\frac{s^{*2} - 1}{3}\right) + 1} = \sqrt{(s^{*2} - 1) + 1} = \sqrt{s^{*2}} = s^* \geq 0$$

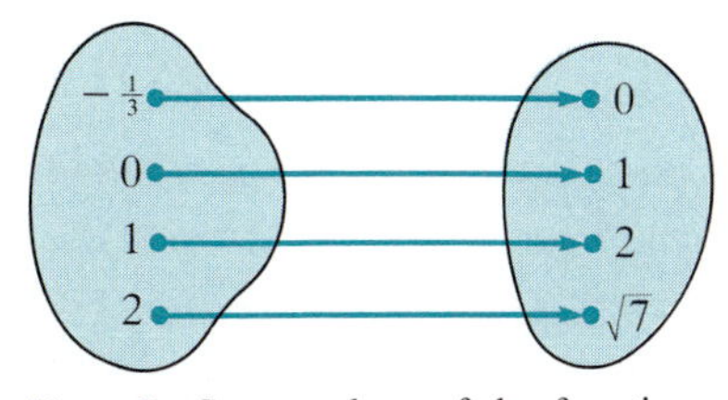

Figure 5 Some values of the function $g(t) = \sqrt{3t + 1}$.

This means that every nonnegative real number is the image of some real number t, so

$$\text{range of } g = \mathbb{R}^+ = \{s\colon s \geq 0\}\dagger$$

Here we used the letters t, s, and g instead of x, y, and f to denote the independent variable, the dependent variable, and the function. The symbol used to denote the function is unimportant. It's what the function *does* that is important.

In Figure 5 we depict some of the values taken by this square root function. ■

EXAMPLE 8 *A Quadratic Function*

Let $y = f(x) = x^2 + 4x + 5$.

(a) Evaluate $f(1)$, $f(-6)$, $f(4)$, and $f(x^3)$.
(b) Find the domain of f.
(c) Find the range of f.

SOLUTION

(a) $f(1) = 1^2 + 4 \cdot 1 + 5 = 10$
$f(-6) = (-6)^2 + 4(-6) + 5 = 17$
$f(4) = 4^2 + 4 \cdot 4 + 5 = 37$
$f(x^3) = (x^3)^2 + 4(x^3) + 5 = x^6 + 4x^3 + 5$

(b) The domain of $x^2 + 4x + 5$ is $\mathbb{R}$ because $x^2 + 4x + 5$ is defined for all real x.

(c) Suppose that y^* is in the range of f. Then, for some real number x,

$$\begin{aligned} y^* &= x^2 + 4x + 5 \\ &= (x^2 + 4x + 4) + 1 = (x + 2)^2 + 1 \geq 1 \quad \text{since } (x + 2)^2 \geq 0 \end{aligned}$$

Therefore, $y^* \geq 1$. Thus every number in the range of f is greater than or equal to 1. Moreover, if $y^* \geq 1$, we can find an x such that $y^* = x^2 + 4x + 5 = (x + 2)^2 + 1$. We work backward:

$$\begin{aligned} y^* &= (x + 2)^2 + 1 \\ (x + 2)^2 &= y^* - 1 \\ x + 2 &= \pm\sqrt{y^* - 1} \\ x &= -2 \pm \sqrt{y^* - 1} \end{aligned}$$

That is, if $y^* > 1$, there are two values of x for which $y^* = (x + 2)^2 + 1$. For example, if $y^* = 10$, then

$$x_1 = -2 + \sqrt{10 - 1} = -2 + 3 = 1 \quad \text{and} \quad x_2 = -2 - 3 = -5 \text{ satisfy}$$

$$(x_1 + 2)^2 + 1 = (x_2 + 2)^2 + 1 = 10 \qquad \text{(check this)}$$

We have shown that

$$\text{range of } x^2 + 4x + 5 \text{ is } \{y\colon y \geq 1\} = [1, \infty)$$ ■

† $\mathbb{R}^+$ is the set of nonnegative real numbers. See page 4.

EXAMPLE 9 *A Cubic Function*

Let $y = f(x) = x^3$.

(a) Compute $f(0)$, $f(2)$, $f(-2)$, $f(t^2)$, and $f\left(\frac{1}{t}\right)$.

(b) Find the domain of f.

(c) Find the range of f.

SOLUTION

(a) $f(0) = 0^3 = 0 \qquad f(2) = 2^3 = 8 \qquad f(-2) = (-2)^3 = -8$

$$f(t^2) = (t^2)^3 = t^6 \qquad f\left(\frac{1}{t}\right) = \left(\frac{1}{t}\right)^3 = \frac{1}{t^3}$$

(b) x^3 is defined for every real number x, so the domain is $\mathbb{R}$.

(c) If y^* is given, then $(\sqrt[3]{y^*})^3 = y^*$, so every real number is in the range. Note that the cube root of a negative number is negative, whereas the square root of a negative number is not defined.

WARNING Avoid the following common error in function evaluation. Let $f(x) = x^2 + 5x - 7$. Evaluate $f(2x)$.

Correct	Incorrect	
$f(2x) = (2x)^2 + 5(2x) - 7$	$f(2x) = 2x^2 + 5(2x) - 7$	$2x$ must be squared
$= 4x^2 + 10x - 7$	$= 2x^2 + 10x - 7$	$(2x)^2 \neq 2x^2$

this is the incorrect step

EXAMPLE 10 *Domain Restricted by Division by Zero*

Let $y = f(x) = \dfrac{1}{x-3}$. Find (a) the domain of f and (b) the range of f.

SOLUTION

(a) $\dfrac{1}{x-3}$ is not defined if $x = 3$ because we cannot divide by 0. Thus,

$$\text{domain of } \frac{1}{x-3} \text{ is } \{x: x \neq 3\}$$

(b) Let $y^* = \dfrac{1}{x-3}$. We solve for x:

$$xy^* - 3y^* = 1 \qquad \text{Multiply by } (x-3)$$

$$xy^* = 1 + 3y^*$$

$$x = \frac{1 + 3y^*}{y^*}$$

This is defined if $y^* \neq 0$. Thus

$$\text{range of } \frac{1}{x-3} \text{ is } \{y: y \neq 0\}$$

EXAMPLE 11 *A Constant Function*

Let $y = f(x) = 2$. Find

(a) $f(0)$, $f(-7)$, and $f(3.14)$.
(b) domain of f.
(c) range of f.

SOLUTION $f(x) = 2$ means that $f(x)$ is equal to 2 for every real number x. f is called a **constant function.** We have

(a) $f(0) = f(-7) = f(3.14) = 2$.
(b) domain of f is $\mathbb{R}$.
(c) range of f is 2 (this is the only image value). ■

EXAMPLE 12 *A Function Defined in Pieces*

A news vendor buys an out-of-town newspaper from a distributor. He must pay 20¢ (= \$0.20) per paper if he buys 100 or fewer papers, but only 18¢ (= \$0.18) per paper if he buys more than 100. Then his total price, $p(q)$, is

$$p(q) = \begin{cases} 0.20q, & 0 \le q \le 100 \\ 0.18q, & q > 100 \end{cases}$$

Thus,

$$\text{domain of } p \text{ is } \{\text{nonnegative integers}\}$$

and

$$\text{range of } p \text{ is } \{0, 0.2, 0.4, \cdots, 20, 0.18(101), 0.18(102), \cdots\}$$

In all the examples considered so far, we gave you a function and asked questions about it. However, not everything that looks like a function is actually a function. We illustrate this in the next example. Remember that to have a function, there must be a unique value of $f(x)$ for each x in the domain of f.

EXAMPLE 13 *A Relation That Does Not Define a Function*

Consider the equation of a circle of radius 2 (see p. 12):

$$x^2 + y^2 = 4$$

Can we write y as a function of x? Let us try to solve for y:

$$y^2 = 4 - x^2$$
$$y = \pm\sqrt{4 - x^2}$$

That is, if x is given (and $x^2 < 4$), then there are two values of y that satisfy $x^2 + y^2 = 4$. For example, if $x = 1$, then $y = \sqrt{3}$ and $y = -\sqrt{3}$ satisfy $x^2 + y^2 = 4$. But, in order to have a function, there must be a *unique* y for every x in the domain. Thus the relation $x^2 + y^2 = 4$ does *not* define a function. ■

EXAMPLE 14 *A Relation That Does Define a Function*

Let

$$x^3 + y^3 = 4$$

We can solve for y:

$$y^3 = 4 - x^3$$
$$y = \sqrt[3]{4 - x^3}$$

If $4 - x^3 \geq 0$, then $y \geq 0$; and if $4 - x^3 < 0$, then $y < 0$. Thus, for every x there is a unique y such that $x^3 + y^3 = 4$. This defines a function with domain and range equal to $\mathbb{R}$.

The Vertical Lines Test

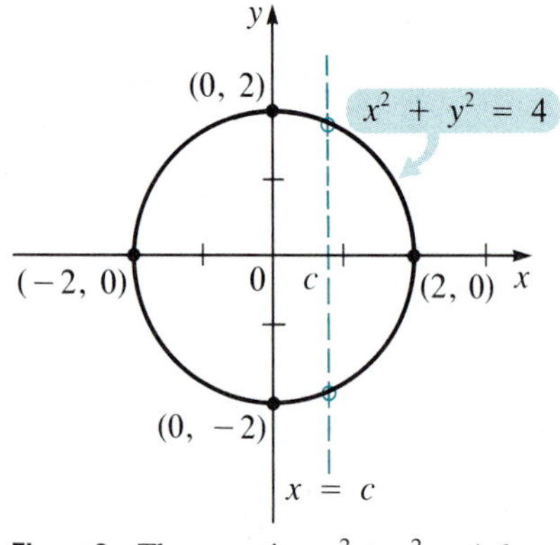

Figure 6 The equation $x^2 + y^2 = 4$ does not define a function because some vertical lines intersect its graph at two points.

In Examples 13 and 14 we saw that an equation in two variables may or may not define a function. There is a graphical test to determine this. Consider the graph of $x^2 + y^2 = 4$ in Figure 6.

We draw the vertical line $x = c$ with c in the interval $(-2, 2)$. This line intersects the circle at *two* points. This means that for every c in $(-2, 2)$ there are two y's such that $c^2 + y^2 = 4$. Thus we cannot write y as a function of x.

> **Vertical Lines Test**
>
> A set of points in the Cartesian plane is the graph of a function if every vertical line in the plane intersects the set of points in at most one point.

We summarize the reasons why the domain of a function may be restricted.

Reasons Why the Domain of a Function May Not Be $\mathbb{R}$

(a) Division by zero is not allowed (see Example 10).
(b) Even roots (square roots, fourth roots, and so on) of negative numbers do not exist as real numbers (see Example 7).
(c) The domain is restricted by the nature of the problem under consideration (see Example 12).

You will see a fourth kind of restriction when we discuss logarithmic functions in Section 7.3.

In Section 1.4 we will discuss the graphs of functions.

EXAMPLE 15 *An Example That Is Important in Calculus*

Let $f(x) = x^2$. Compute $\dfrac{f(x + \Delta x) - f(x)}{\Delta x}$ and simplify your answer. Assume that $\Delta x \neq 0$.

SOLUTION

$f(x + \Delta x) = (x + \Delta x)^2 = x^2 + 2x\Delta x + \Delta x^2$ and $f(x) = x^2$. Thus

Factor out Δx

$$\frac{f(x + \Delta x) - f(x)}{\Delta x} = \frac{(x^2 + 2x\Delta x + \Delta x^2) - x^2}{\Delta x} = \frac{2x\Delta x + \Delta x^2}{\Delta x} = \frac{\Delta x(2x + \Delta x)}{\Delta x}$$
$$= 2x + \Delta x \quad ■$$

EXAMPLE 16 ***Another Example Useful in Calculus***

Let $f(x) = \dfrac{1}{x}$. Compute $\dfrac{f(x + \Delta x) - f(x)}{\Delta x}$ and simplify your answer. Assume that $\Delta x \neq 0$.

SOLUTION $f(x + \Delta x) = \dfrac{1}{x + \Delta x}$ and $f(x) = \dfrac{1}{x}$ so

Multiply numerator and denominator by $x(x + \Delta x)$

$$\frac{f(x + \Delta x) - f(x)}{\Delta x} = \frac{\dfrac{1}{x + \Delta x} - \dfrac{1}{x}}{\Delta x} = \frac{\dfrac{x(x + \Delta x)}{x + \Delta x} - \dfrac{x(x + \Delta x)}{x}}{\Delta x(x)(x + \Delta x)}$$
$$= \frac{x - (x + \Delta x)}{\Delta x(x)(x + \Delta x)} = \frac{-\Delta x}{\Delta x(x)(x + \Delta x)}$$
$$= \frac{-1}{x(x + \Delta x)}$$

Definition of Even and Odd Functions

A function $f(x)$ is called **even** if

$$f(-x) = f(x)$$

and **odd** if

$$f(-x) = -f(x)$$

EXAMPLE 17 ***An Even Function and an Odd Function***

$f(x) = x^2$ is even because $f(-x) = (-x)^2 = x^2 = f(x)$.
$f(x) = x^3$ is odd because $f(-x) = (-x)^3 = (-1)^3x^3 = -x^3 = -f(x)$. ■

EXAMPLE 18 ***Determining Whether a Function Is Even or Odd***

Determine whether each function is even, odd, or neither.

(a) $\sqrt[3]{x}$ (b) $\sqrt{x}$ (c) $x^4 + x^5$ (d) $x^4 - x^6$

SOLUTION

(a) $f(-x) = \sqrt[3]{-x} = \sqrt[3]{(-1)x} = \sqrt[3]{-1}\sqrt[3]{x} = (-1)\sqrt[3]{x} = -\sqrt[3]{x} = -f(x)$, so $\sqrt[3]{x}$ is odd.

(b) $f(-x) = \sqrt{-x}$ is not defined if $x > 0$, so $\sqrt{x}$ is neither even nor odd.

(c) If $f(x) = x^4 + x^5$, then $f(-x) = (-x)^4 + (-x)^5 = x^4 - x^5$. But $-f(x) = -x^4 - x^5$. Thus $f(-x) \neq f(x)$ and $f(-x) \neq -f(x)$, so $f(x) = x^4 + x^5$ is neither even nor odd.

(d) If $f(x) = x^4 - x^6$, then $f(-x) = (-x)^4 - (-x)^6 = x^4 - x^6 = f(x)$, so $f(x) = x^4 - x^6$ is even.

One other fact about functions should be emphasized:

Two functions are different if their domains are different.

For example,

$$f(x) = x^2 \quad \text{and} \quad g(x) = x^2,\ x \geq 0$$

are *different* functions because the domain of f is $\mathbb{R}$, but the domain of g is $\{x\colon x \geq 0\} = \mathbb{R}^+$.

Problems 1.3

Readiness Check

I. Which of the following belongs to the range of $f(x) = -|x - 2|$?
a. 2 b. 1 c. -2
d. None of these because $f(x)$ is not a function.

II. Which of the following is the interval notation for the range of $f(x) = \sqrt{x + 2}$?
a. $[0, \infty)$ b. $[-2, \infty)$ c. $(-\infty, \infty)$ d. $(-\infty, 2]$

III. For which of the following is y a function of x?
a. $y = |x + 7|$
b. $x + y^2 = 4$ c. $y = f(x) = \begin{cases} x - 1, & \text{for } x \geq 2 \\ -4, & \text{for } x \leq 2 \end{cases}$
d. All the above are functions.

IV. Which of the following has domain $\{x\colon x \geq 0\} = \mathbb{R}^+$?
a. $f(x) = x^2$ b. $f(x) = \dfrac{1}{x}$
c. $f(x) = \dfrac{1}{x^2}$ d. $f(x) = \sqrt{x}$

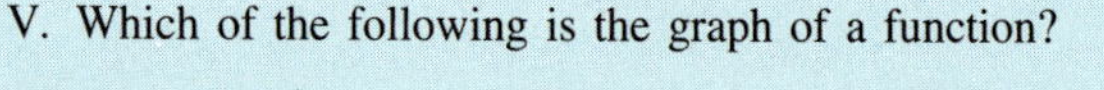

V. Which of the following is the graph of a function?

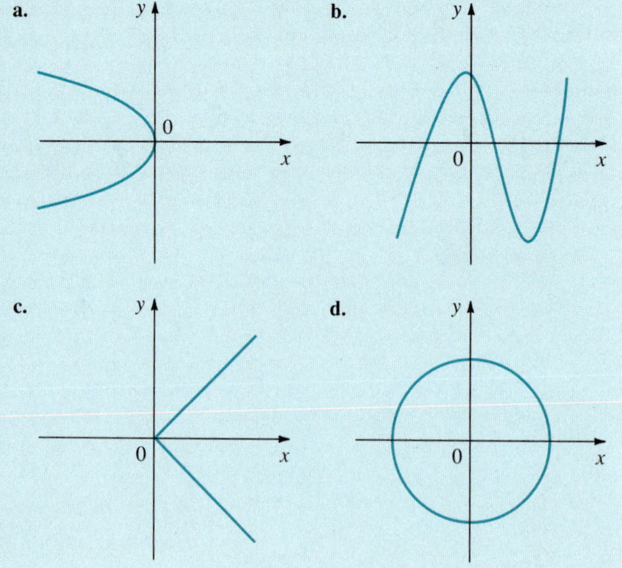

Answers to Readiness Check

I. c II. a III. a [The rule in c is not a function because 2 has two images.] IV. d V. b

In Problems 1–10 evaluate the given function at the given values.

1. $f(x) = 1/(1 + x)$; $f(0)$, $f(1)$, $f(-2)$, $f(-5)$, $f(x^2)$, and $f(\sqrt{x})$
2. $f(x) = 1 + \sqrt{x}$; $f(0)$, $f(1)$, $f(16)$, $f(25)$, $f(y^4)$, and $f\left(\frac{1}{y}\right)$
3. $f(x) = 2x^2 - 1$; $f(0)$, $f(2)$, $f(-3)$, $f\left(\frac{1}{2}\right)$, $f(\sqrt{w})$, and $f(w^5)$
4. $f(x) = \frac{1}{2x^3}$; $f(1)$, $f(\frac{1}{2})$, $f(-3)$, $f(1 + t)$, and $f(t^2 - 2)$
5. $f(x) = x^4$; $f(0)$, $f(2)$, $f(-2)$, $f(\sqrt{5})$, $f(s^{1/5})$, and $f(s - 1)$
6. $g(t) = t/(t - 2)$; $g(0)$, $g(1)$, $g(-1)$, $g(3)$, $g(v^2)$, and $g(v + 5)$
7. $g(t) = \sqrt{t + 1}$; $g(0)$, $g(-1)$, $g(3)$, $g(7)$, $g(n^3 - 1)$, and $g\left(\frac{1}{w}\right)$
8. $h(z) = \sqrt[3]{z}$; $h(0)$, $h(8)$, $h(-\frac{1}{27})$, $h(1000)$, $h(x^{300})$, and $h(-8p)$
9. $h(z) = z^2 - z + 1$; $h(0)$, $h(2)$, $h(10)$, $h(-5)$, $h(n^2)$, and $h\left(\frac{1}{n^3}\right)$
10. $h(z) = z^3 + 2z^2 - 3z + 5$; $h(0)$, $h(1)$, $h(-1)$, $h(2)$, $h(t^{1/4})$, and $h(t + 1)$

In Problems 11–20 which of the rules depicted in the graphs are functions with each domain equal to D? If the rule is not a function with domain D, explain why it is not.

11.–13.

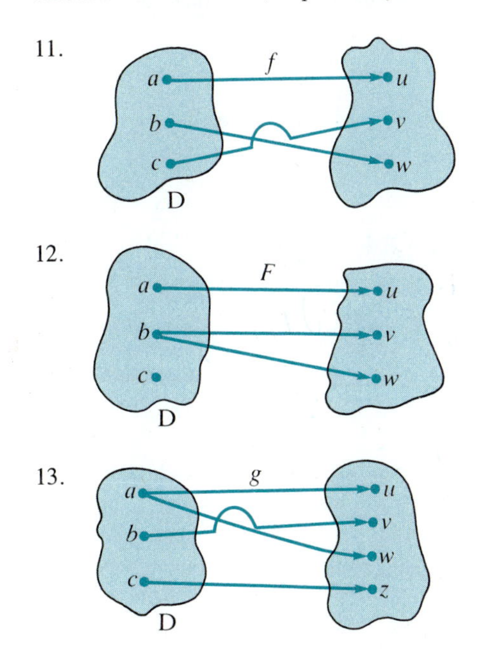

14.

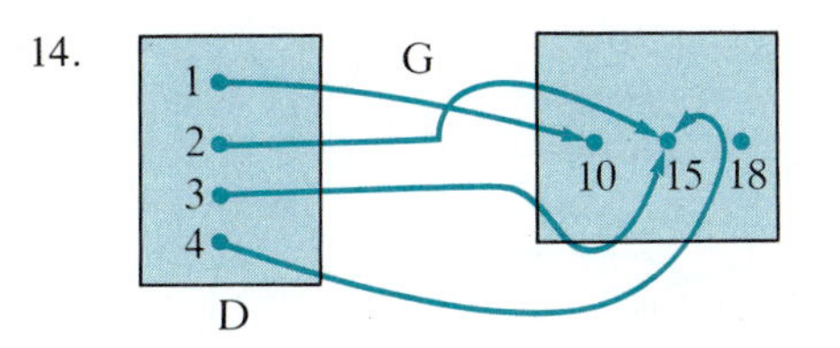

15.

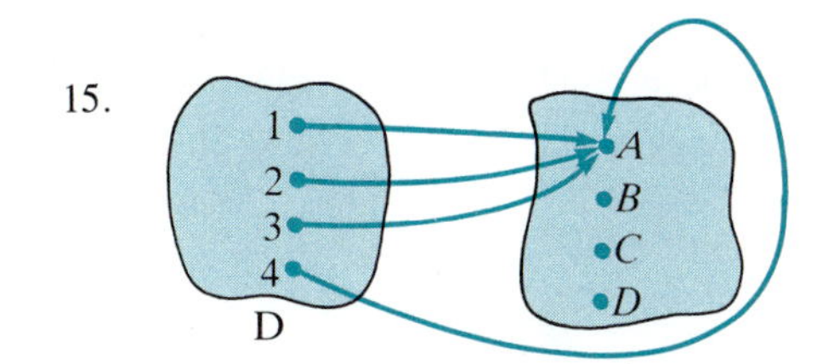

16.

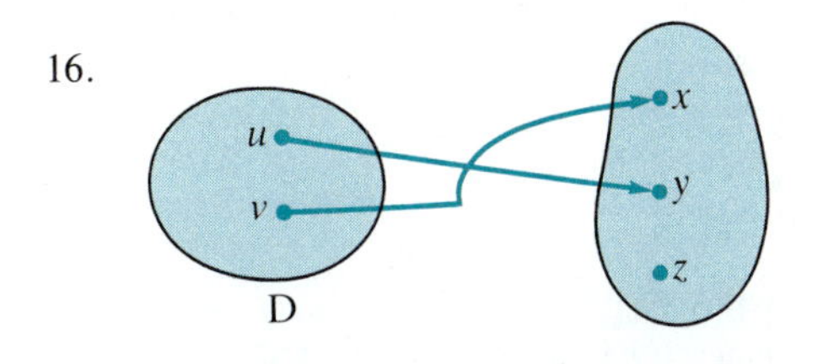

17.

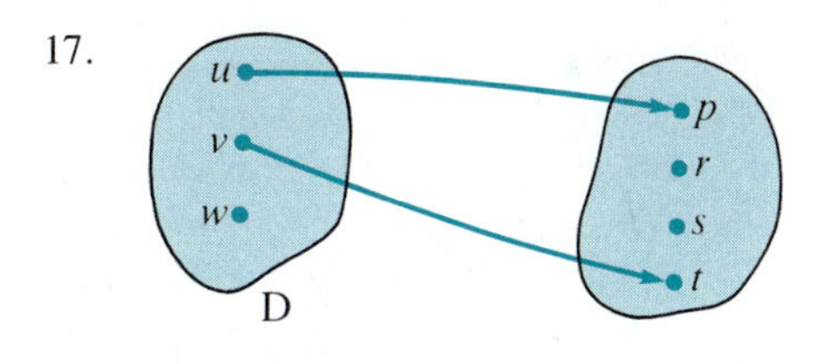

18.

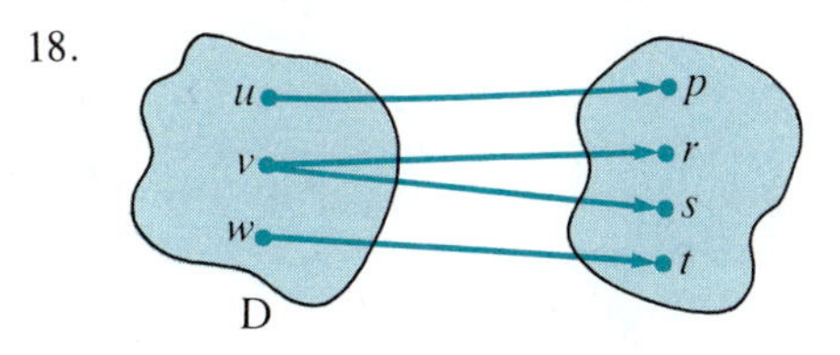

19.

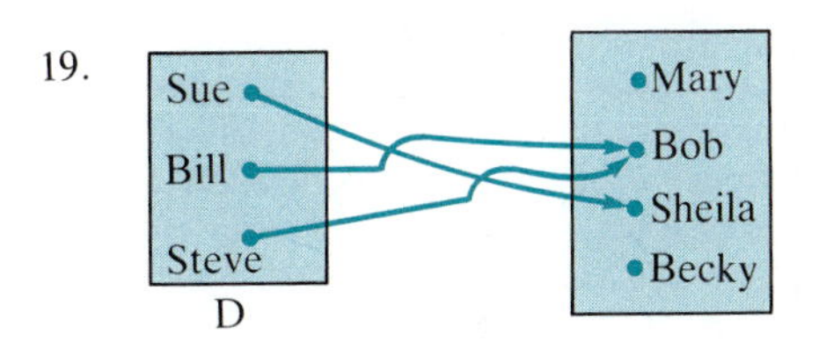

20.

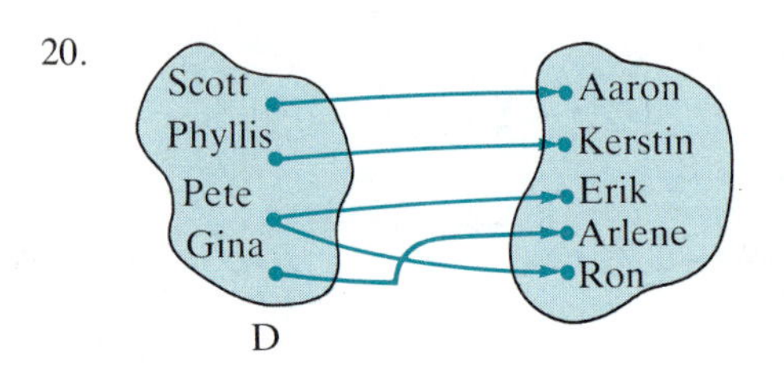

In Problems 21–32 an equation involving x and y is given. Determine whether y can be written as a function of x.

21. $2x + 3y = 6$
22. $\frac{x}{y} = 2$
23. $x^2 + 2y = 5$
24. $x - 3y^2 = 4$
25. $4x^2 + 2y^2 = 4$
26. $x^2 - y^2 = 1$
27. $\sqrt{x + y} = 1$
28. $y^2 + xy + 1 = 0$ [Hint: Use the quadratic formula.]
29. $y^3 - x = 0$
30. $y^4 - x = 0$
31. $y = |x|$
32. $y^2 = \frac{x}{x + 1}$
33. Explain why the equation $y^n - x = 0$ allows us to write y as a function of x if n is an odd integer but does not if n is an even integer. [Hint: First solve Problems 29 and 30.]

In Problems 34–43 the graph of an equation in two variables is given. Use the vertical lines test to determine whether the equation determines y as a function of x.

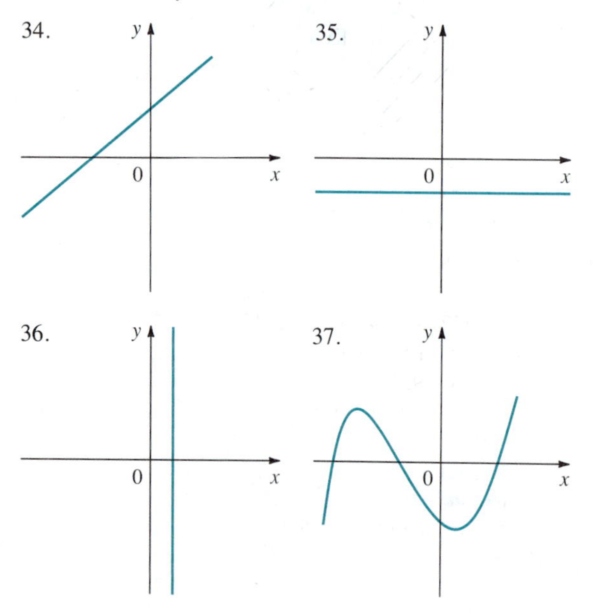

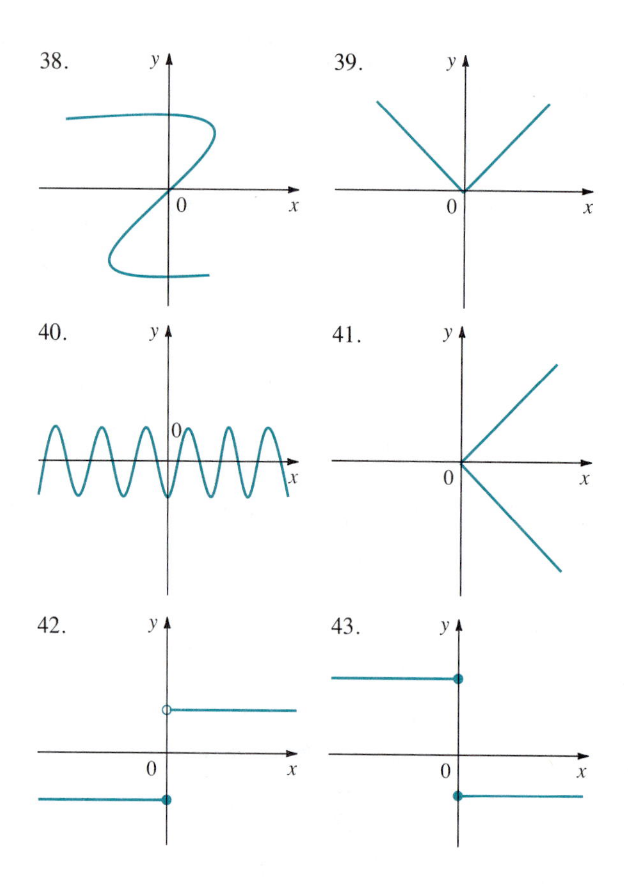

In Problems 44–57 find the domain and range of the given function.

44. $y = f(x) = 2x - 3$
45. $s = g(t) = 4t - 5$
46. $y = f(x) = 3x^2 - 1$
47. $v = h(u) = \frac{1}{u^2}$
48. $y = f(x) = x^3$
49. $y = f(x) = \frac{1}{x + 1}$
50. $s = g(t) = t^2 + 2t + 1$
51. $y = f(x) = \sqrt{x^3 - 1}$
52. $v = h(u) = |u - 2|$
53. $y = f(x) = \frac{1}{|x|}$
54. $h(x) = \sqrt{4 - x^2}$
55. $y = \begin{cases} 2x, & x \geq 0 \\ -x, & x < 0 \end{cases}$
56. $y = \begin{cases} x, & x \geq 1 \\ 1, & x < 1 \end{cases}$
57. $y = \begin{cases} x^3, & x > 0 \\ x^2, & x \leq 0 \end{cases}$
58. The **greatest integer function** $g(x) = [x]$ is defined by $[x]$ = the largest integer less than or equal to x. Compute
(a) $[3.1]$ (b) $[\pi]$ (c) $[-2.7]$ (d) $[16]$
(e) $\left[\frac{17}{9}\right]$ (f) $\left[-\frac{10}{3}\right]$
59. Find the domain and range of $g(x) = [x]$.

In Problems 60–63 use a calculator to evaluate the given function at the given values.

60. $f(x) = 1.25x^2 - 3.74x + 14.38$; $f(2.34)$, $f(-1.89)$, $f(10.6)$
61. $g(t) = t^3 - 0.74t^2 + 0.756t + 1.302$; $g(0.18)$, $g(3.95)$, $g(-11.62)$
62. $h(z) = \dfrac{z+3}{z^2-4}$; $h(38.2)$, $h(57.9)$, $h(238.4)$
63. $f(x) = \dfrac{x-1.6}{x+3.4} + \dfrac{x^2+5.8}{6.2-x^2}$; $f(5.8)$, $f(-23.4)$
64. Let $f(x) = \dfrac{1}{x-1}$. Find $f(t^2)$ and $f(3t+2)$.

Two Exercises Useful in Calculus

65. Let $f(x) = x^3$. Find $f(x + \Delta x)$ and $\dfrac{f(x+\Delta x) - f(x)}{\Delta x}$, where x denotes an arbitrary real number. Simplify your answer. This computation is very important in calculus.

* 66. Let $f(x) = \sqrt{x}$. Show, assuming that $\Delta x \neq 0$, that

$$\frac{f(x+\Delta x) - f(x)}{\Delta x} = \frac{1}{\sqrt{x+\Delta x} + \sqrt{x}}$$

[Hint: Multiply and divide by $\sqrt{x+\Delta x} + \sqrt{x}$.]

67. Let $f(x) = |x|/x$. Show that

$$f(x) = \begin{cases} 1, & x > 0 \\ -1, & x < 0 \end{cases}$$

Find the domain and range of f.

68. Describe a computational rule for expressing Fahrenheit temperature as a function of Centigrade temperature. [Hint: Pure water at sea level boils at 100°C and 212°F; it freezes at 0°C and 32°F. The graph of this function is a straight line.]

* 69. Consider the set of all rectangles whose perimeters are 50 cm. Once the width W of any one rectangle is measured, it is possible to compute the area of the rectangle. Verify this by producing an explicit expression for area A as a function of width W. Find the domain and the range of your function.

* 70. A spotlight shines on a screen; between them is an obstruction that casts a shadow. Suppose the screen is vertical, 20 m wide by 15 m high, and 50 m from the spotlight. Also suppose the obstruction is a square, 1 m on a side, and is parallel to the screen. Express the area of the shadow as a function of the distance from the light to the obstruction.

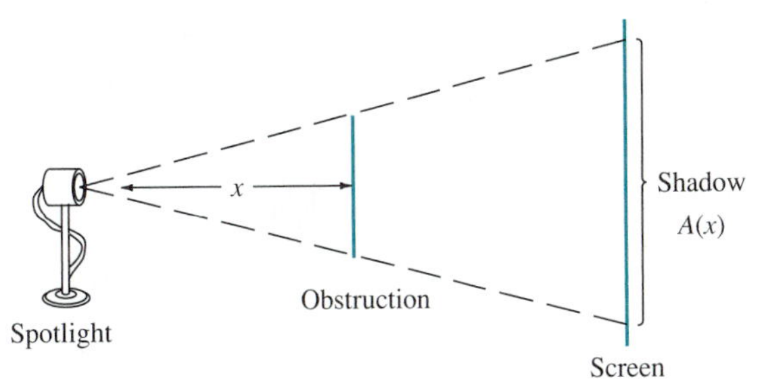

* 71. A baseball diamond is a square, 90 feet long on each side. Casey runs a constant 30 ft/sec whether he hits a ground ball or a home run. Today, in his first at-bat, he hit a home run. Write an expression for the function that measures his line-of-sight distance from second base as a function of the time t in seconds after he left home plate.

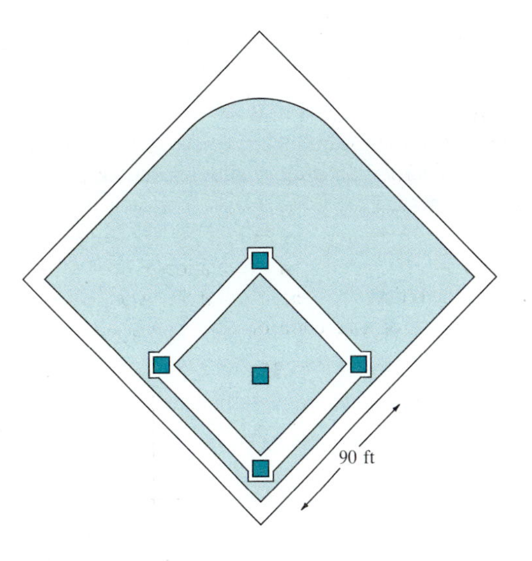

72. Let $f(x)$ be the fifth decimal place of the decimal expansion of x. For example, $f(\frac{1}{64}) = f(0.015624) = 2$, $f(98.786543210) = 4$, $f(-78.90123456) = 3$, and so on. Find the domain and range of f. [Note: Since, for example, $1.000\ldots = 0.9999\ldots$, assume, to avoid ambiguity, that every integer n is written as $1.000\ldots$, $2.000\ldots$, and so on.]

73. The Dow Jones closing averages for industrial stocks are given for the 3-month period from April 15 to July 15, 1983. Let April 18 (which was a Monday) be day 1 and July 15 be day 92. Let $A(t)$ be the closing average on day t. Find (a) $A(1)$, (b) $A(8)$, (c) $A(30)$, (d) $A(60)$, and

(e) $A(88)$. [Hint: Each vertical bar represents a business day.]

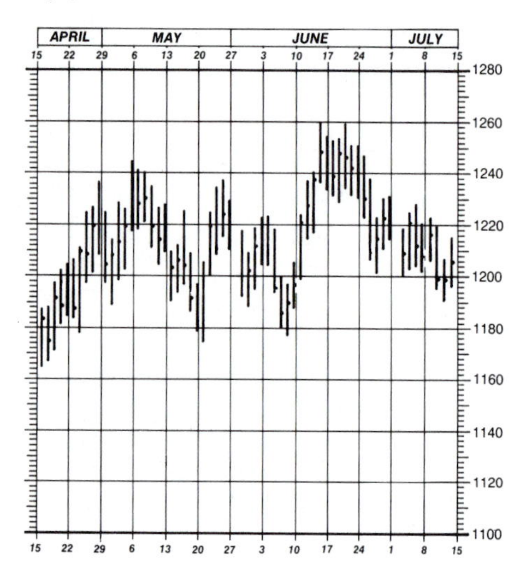

* 74. Alec, on vacation in Canada, found that he got a 25% premium on his U.S. money. When he returned, he discovered there was a 25% discount on converting his Canadian money back into U.S. currency. Describe each conversion function. Show that, after converting both ways, Alec lost money.

In Problems 75–92 determine whether the given function is even, odd, or neither.

75. $f(x) = x^2 - 1$
76. $f(x) = x^2 + x$
77. $f(x) = x^4 + x^2$
78. $f(x) = x^3 - x$
79. $f(x) = x^4 + x^2 + 2x$
80. $f(x) = \dfrac{1}{x}$
81. $f(x) = \dfrac{1}{x^2}$
82. $f(x) = \dfrac{1}{x+1}$
83. $f(x) = x^4 - x^2 + 2$
84. $f(x) = \dfrac{x}{x^3+1}$
85. $f(x) = \dfrac{x^2-2}{x^2+5}$
86. $f(x) = x^{1/3} - x^{1/5}$
87. $f(x) = x^{3/5}$
88. $f(x) = x^{17/4}$
89. $f(x) = \dfrac{x^3}{x^2+12}$
90. $f(x) = x^2 + x + 1$
91. $f(x) = x^3 + x + 1$
92. $f(x) = x^4 + 17x^2 - 5$

1.4 Graphs of Functions

Definition

The **graph** of the function f is the set of points

$$\{(x, f(x)) : x \in \text{domain of } f\}$$

EXAMPLE 1 *Graph of $f(x) = x^2$*

Sketch the graph of $y = f(x) = x^2$.

SOLUTION In Example 3 on p. 15 we saw that $f(x) = x^2$ is a function with domain $\mathbb{R}$ and range $\mathbb{R}^+$. The graph of this function is obtained by plotting all points of the form $(x, y) = (x, x^2)$.† First we note that when $f(x) = x^2$, $f(-x) = (-x)^2 = x^2 = f(x)$. That is, f is even. Thus, it is necessary to calculate $f(x)$ only for $x \geq 0$. For every $x > 0$, there is a value of $x < 0$ that gives

† Of course, we can't plot *all* points (there are an infinite number of them). Rather, we plot some sample points and connect them to obtain the sketch of the graph.

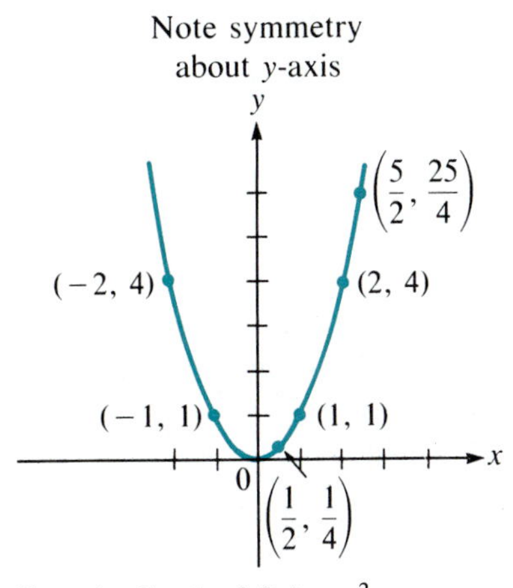

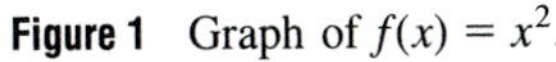
Figure 1 Graph of $f(x) = x^2$.

the same value of y. When a function is even, the graph of the function is **symmetric** about the y-axis. This means that the graph of $f(x)$ for x negative is the mirror image or reflection of the graph for x positive. Some values of $f(x)$ for $x \geq 0$ are shown in Table 1. The graph drawn in Figure 1 is a **parabola.**

Table 1

x	0	$\frac{1}{2}$	1	$\frac{3}{2}$	2	$\frac{5}{2}$	3	4	5
$f(x) = x^2$	0	$\frac{1}{4}$	1	$\frac{9}{4}$	4	$\frac{25}{4}$	9	16	25

EXAMPLE 2 *Graph of a Square Root Function*

Graph the function $f(x) = \sqrt{x}$.

SOLUTION First observe that f is defined only for $x \geq 0$. Note that $\sqrt{x} \geq 0$ for all $x \geq 0$. Table 2 gives values of $\sqrt{x}$. We plot these points and then join them to obtain the graph in Figure 2.

Table 2

x	0	0.5	1	2	3	4	5	10	15	20	25
$f(x) = \sqrt{x}$	0	0.707	1	1.414	1.732	2	2.236	3.162	3.873	4.472	5

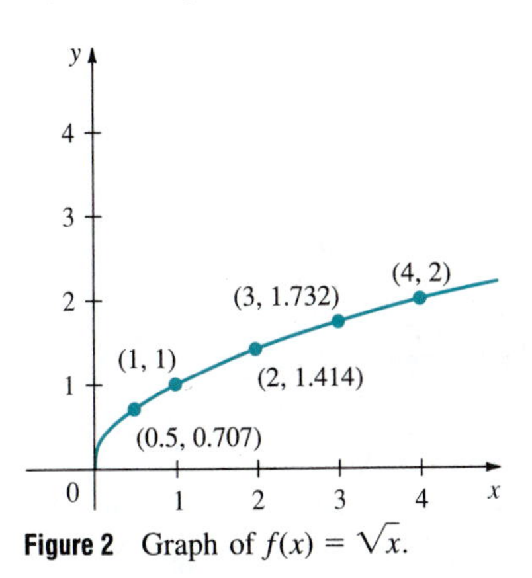

Figure 2 Graph of $f(x) = \sqrt{x}$.

EXAMPLE 3 *The Graph of a Linear Function*

Sketch the graph of the function defined by the equation $y = 2x + 1$.

SOLUTION Table 3 shows some points on the graph. We plot four of these points in Figure 3. It appears that the points lie on a straight line.

Table 3

x	$y = 2x + 1$	Corresponding Point
−5	−9	(−5, −9)
−4	−7	(−4, −7)
−3	−5	(−3, −5)
−2	−3	(−2, −3)
−1	−1	(−1, −1)
0	1	(0, 1)
1	3	(1, 3)
2	5	(2, 5)
3	7	(3, 7)
4	9	(4, 9)
5	11	(5, 11)

Figure 3 Graph of the linear function $y = 2x + 1$.

The following fact generalizes the last example:

The graph of a linear function is a straight line.

EXAMPLE 4 *Graph of $f(x) = x^3$*

Graph the function $f(x) = x^3$.

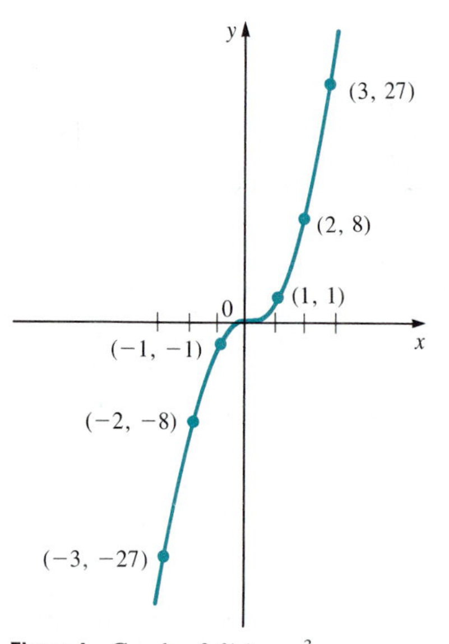

Figure 4 Graph of $f(x) = x^3$.

SOLUTION f is defined for all real numbers (domain of f is $\mathbb{R}$), and the cube of a negative number is negative, so we plot both positive and negative values for x. Some representative values are given in Table 4. We plot these points and join them to obtain the graph in Figure 4.

Table 4

x	0	$\frac{1}{2}$	$-\frac{1}{2}$	1	-1	2	-2	3	-3
$f(x) = x^3$	0	$\frac{1}{8}$	$-\frac{1}{8}$	1	-1	8	-8	27	-27

Since $f(-x) = (-x)^3 = -x^3 = -f(x)$, $f(x) = x^3$ is an odd function. In Figure 4, we say that the graph of x^3 for $x < 0$ is the graph of x^3 for $x > 0$ *reflected about the origin,* or that *the graph is symmetric about the origin.*

Before going further, we give a formal definition of symmetry.

Definition of Symmetry About the *x*-Axis, the *y*-Axis, and the Origin

(i) A graph is **symmetric about the *x*-axis** if whenever (x, y) is on the graph, $(x, -y)$ is also on the graph.

(ii) A graph is **symmetric about the *y*-axis** if whenever (x, y) is on the graph, $(-x, y)$ is also on the graph.

(iii) A graph is **symmetric about the origin** if whenever (x, y) is on the graph, $(-x, -y)$ is also on the graph.

The graph of $f(x) = x^2$ (Figure 1) is symmetric about the y-axis. The graph of $f(x) = x^3$ (Figure 4) is symmetric about the origin. The graphs in Figure 5 are symmetric about the x-axis.

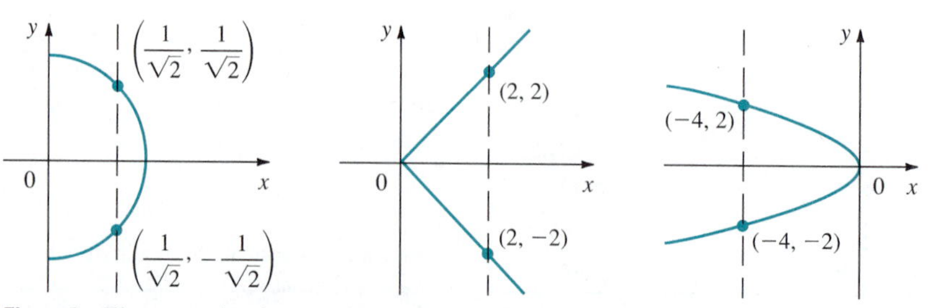

Figure 5 Three graphs that are symmetric about the x-axis.

Note that a graph that is symmetric about the x-axis is not the graph of a function because the vertical lines test is violated when $f(x) \neq 0$.

If $y = f(x)$ is a function, then it may be symmetric about the y-axis, symmetric about the origin, or neither.

Rules of Symmetry for a Function

(i) If $f(x)$ is an even function, then the graph of f is symmetric about the y-axis. To obtain the graph, first sketch it for $x \geq 0$, and then reflect the sketch about the y-axis.

(ii) If $f(x)$ is an odd function, then the graph of f is symmetric about the origin. To obtain the graph, first sketch it for $x \geq 0$, and then reflect the sketch about the origin (that is, reflect it first about the x-axis and then around the y-axis).

Suppose that f is an odd function. Then

$$f(0) = f(-0) \underset{\text{Because } f \text{ is odd}}{=} -f(0)$$

so

$$2f(0) = 0 \text{ or } f(0) = 0$$

That is,

If f is an odd function, then the graph of f passes through (0, 0).

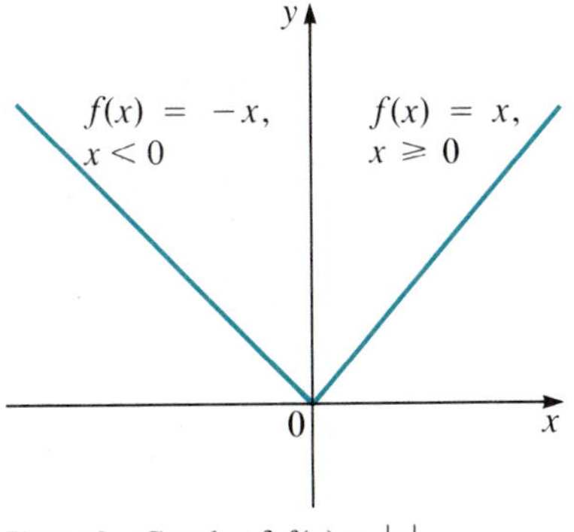

Figure 6 Graph of $f(x) = |x|$.

EXAMPLE 5 *Graph of the Absolute Value Function*

Sketch the graph of $y = |x|$.

SOLUTION From p. 5 we know that

$$f(x) = |x| = \begin{cases} x, & x \geq 0 \\ -x, & x < 0 \end{cases}$$

Also, since $f(-x) = |-x| = |x| = f(x)$, f is an even function and its graph is symmetric about the y-axis. Thus we need only sketch $f(x) = |x| = x$ for $x \geq 0$ and reflect it about the y-axis. This is done in Figure 6. ■

EXAMPLE 6 *An Equation That Does Not Determine a Function*

Consider the equation $y^2 = x$. The rule $f(x) = y$ where $y^2 = x$ does *not* determine y as a function of x, since for every $x > 0$ there are *two* values of y such that $y^2 = x$, namely, $y = \sqrt{x}$ and $y = -\sqrt{x}$. For example, if $x = 4$, then $y = 2$ and $y = -2$; both satisfy $y^2 = 4$. However, if we specify one of these values, say $g(x) = \sqrt{x}$, then we have a function. Here, domain of $g = \mathbb{R}^+$ and range of $g = \mathbb{R}^+$. We could obtain a second function, h, by choosing the negative square root. That is, the rule defined by $h(x) = -\sqrt{x}$ is a function with domain $\mathbb{R}^+$ and range $\mathbb{R}^-$ (the nonpositive real numbers). Note that both $g(x)$ and $h(x)$ pass through the origin.

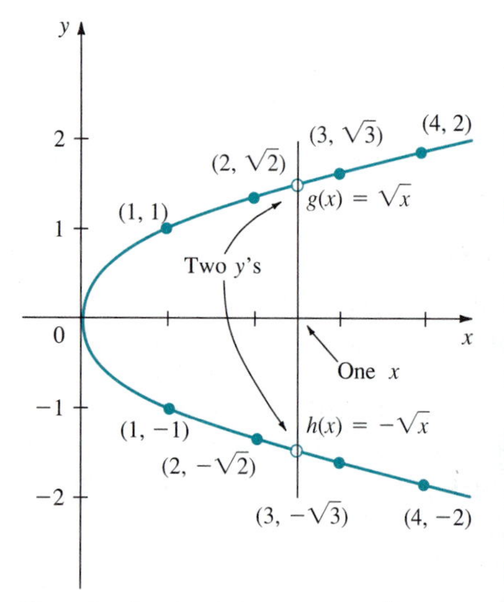

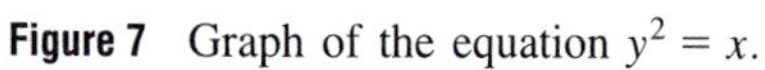

Figure 7 Graph of the equation $y^2 = x$.

We can look at these things in a different way. The graph of the equation $y^2 = x$ is given in Figure 7. The figure shows that for every positive x, there are two y's such that (x, y) is on the graph. This evidently violates the vertical lines test given on p. 22. This graph is symmetric about the x-axis. ■

Shifting the Graphs of Functions

Although more advanced methods (involving calculus) are needed to obtain the graphs of most functions (without plotting a large number of points), there are some techniques that make it a relatively simple matter to sketch certain functions based on known graphs.

EXAMPLE 7 *Shifting a Graph Vertically*

Sketch the graphs of $y = f_1(x) = x^2 + 1$ and $y = f_2(x) = x^2 - 2$.

SOLUTION In Figure 8(a), we have used the graph of $y = x^2$ obtained in Figure 1. To graph $y = x^2 + 1$ in Figure 8(b), we add 1 unit to every y value obtained in Figure 8(a); that is, we shift the graphs of $y = x^2$ up 1 unit. Analogously, for Figure 8(c) we shift the graph of $y = x^2$ down 2 units to obtain the graph of $y = x^2 - 2$.

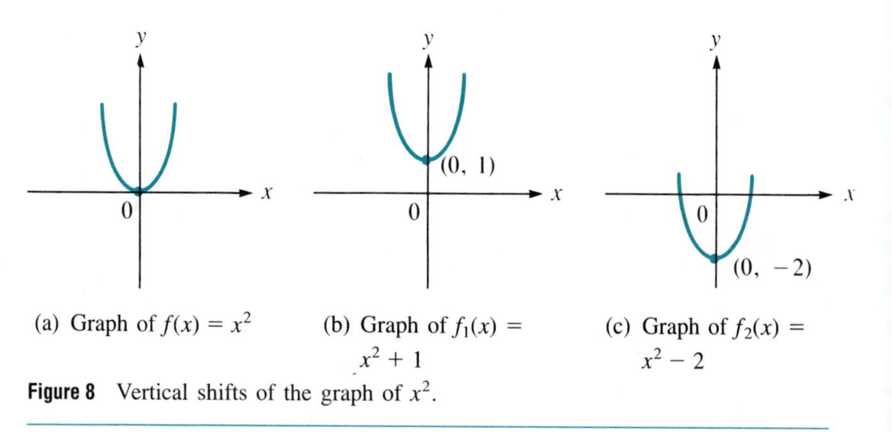

(a) Graph of $f(x) = x^2$ (b) Graph of $f_1(x) = x^2 + 1$ (c) Graph of $f_2(x) = x^2 - 2$

Figure 8 Vertical shifts of the graph of x^2.

Table 5
Values Taken by $(x - 1)^2$ Are the Values of x^2 Taken One Unit Later

x	x^2	$(x - 1)^2$
−5	25	36
−4	16	25
−3	9	16
−2	4	9
−1	1	4
0	0	1
1	1	0
2	4	1
3	9	4
4	16	9

The results of Example 7 can be generalized.

Vertical Shifts of Graphs

Let $y = f(x)$ and let $c > 0$.

(i) To obtain the graph of $y = f(x) + c$, shift the graph of $y = f(x)$ up c units.
(ii) To obtain the graph of $y = f(x) - c$, shift the graph of $y = f(x)$ down c units.

EXAMPLE 8 *Shifting a Graph Horizontally*

Sketch the graphs of $y = (x - 1)^2$ and $y = (x + 2)^2$.

Table 6
Values Taken by $(x + 2)^2$ Are the Values of x^2 Taken Two Units Earlier

x	x^2	$(x + 2)^2$
−5	25	9
−4	16	4
−3	9	1
−2	4	0
−1	1	1
0	0	4
1	1	9
2	4	16
3	9	25
4	16	36

SOLUTION Let us compare the functions x^2 and $(x - 1)^2$. For example, for the function $y = x^2$, $y = 0$ when $x = 0$, and for the function $y = (x - 1)^2$, $y = 0$ when $x = 1$. Similarly, $y = 4$ when $x = -2$ if $y = x^2$, and $y = 4$ when $x = -1$ if $y = (x - 1)^2$. By continuing in this manner, you can see that y values in the graph of $y = x^2$ are the same as y values in the graph of $y = (x - 1)^2$, except that they occur 1 unit to the right on the x-axis. Some representative values are given in Table 5. Thus, we find that the graph of $y = (x - 1)^2$ is the graph of $y = x^2$ *shifted 1 unit to the right*. This is indicated in Figure 9(b).

Similarly, in Figure 9(c) we find that the graph of $y = (x + 2)^2$ is the graph of $y = x^2$ *shifted 2 units to the left*. Some values are given in Table 6.

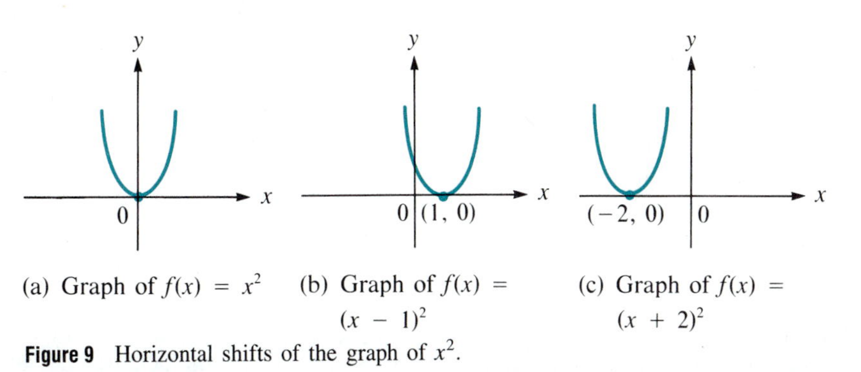

(a) Graph of $f(x) = x^2$ (b) Graph of $f(x) = (x - 1)^2$ (c) Graph of $f(x) = (x + 2)^2$

Figure 9 Horizontal shifts of the graph of x^2.

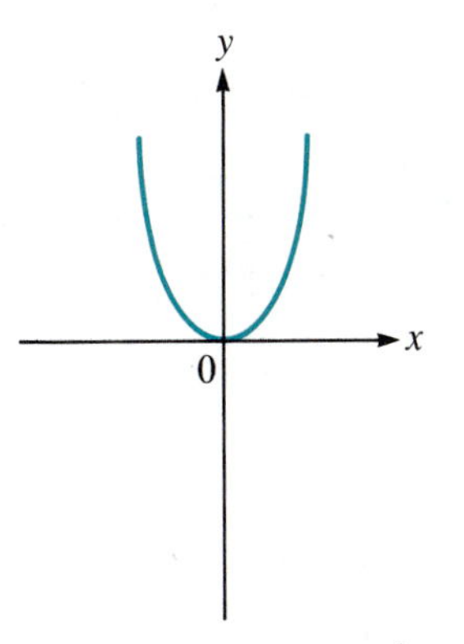

(a) Graph of $f(x) = x^2$

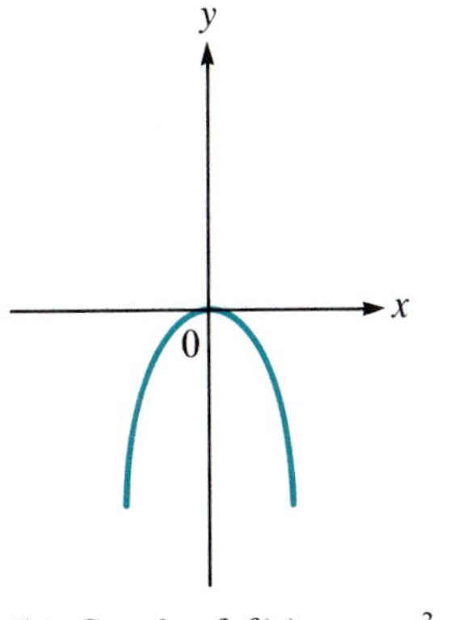

(b) Graph of $f(x) = -x^2$

Figure 10 Reflection of the graph of x^2 about the x-axis.

Horizontal Shifts of Graphs

Let $y = f(x)$ and let $c > 0$.

(iii) To obtain the graph of $y = f(x - c)$, shift the graph of $y = f(x)$ c units to the right.

(iv) To obtain the graph of $y = f(x + c)$, shift the graph of $y = f(x)$ c units to the left.

WARNING Be careful. $f(x + c) \neq f(x) + c$. For example, if $f(x) = x^2$, the graph of $f(x + 1) = (x + 1)^2 = x^2 + 2x + 1$ is the graph of x^2 shifted one unit to the left, but the graph of $f(x) + 1 = x^2 + 1$ is the graph of x^2 shifted 1 unit up. The graphs are not the same. ■

EXAMPLE 9 ***Reflecting a Graph About the x-Axis***

Sketch the graph of $y = -x^2$.

SOLUTION To obtain the graph of $y = -x^2$ from the graph of $y = x^2$, note that each y value is replaced by its negative so that the graph of $y = -x^2$ is the graph of $y = x^2$ *reflected about the x-axis* (that is, turned upside down). See Figure 10. ■

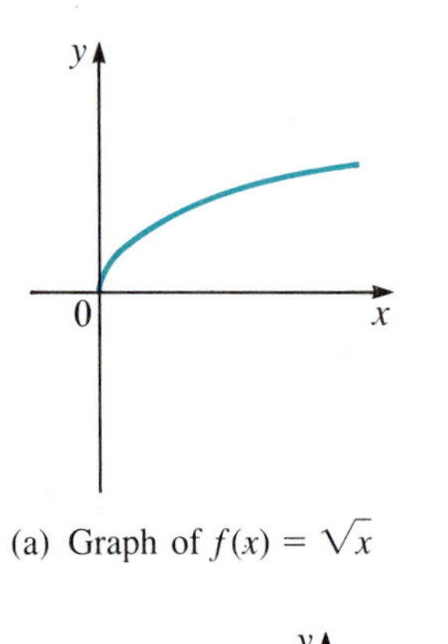

(a) Graph of $f(x) = \sqrt{x}$

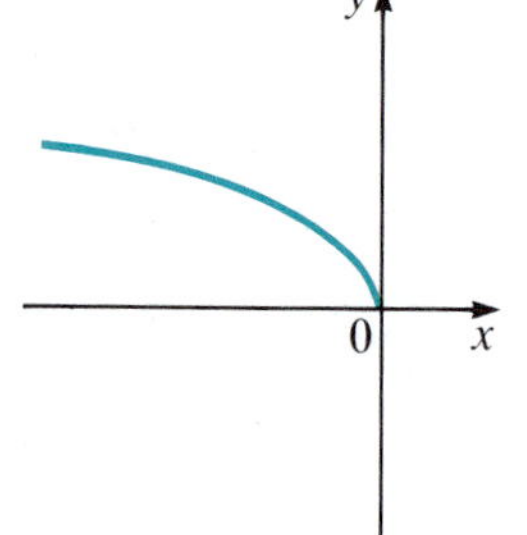

(b) Graph of $g(x) = \sqrt{-x}$

Figure 11 Reflection of the graph of $\sqrt{x}$ about the y-axis.

EXAMPLE 10 *Reflecting a Graph About the y-Axis*

Sketch the graph of $g(x) = \sqrt{-x}$.

SOLUTION $\sqrt{-x}$ is defined only when $x \leq 0$ (so $-x \geq 0$). In Table 7 we provide sample values of $\sqrt{-x}$ for $x \leq 0$.

Table 7

x	0	−0.5	−1	−2	−3	−4	−5	−10	−15	−20	−25
$g(x) = \sqrt{-x}$	0	0.707	1	1.414	1.732	2	2.236	3.162	3.873	4.472	5

These are the same values taken (in Table 2) by $f(x) = \sqrt{x}$ for $x \geq 0$. We see that $f(x) = \sqrt{x}$ and $g(x) = \sqrt{-x}$ take the same y values. $\sqrt{x}$ takes these values for $x \geq 0$ and $\sqrt{-x}$ takes them for $x \leq 0$. The two graphs are given in Figure 11. The graph of $g(x) = \sqrt{-x}$ is *the reflection about the y-axis* of the graph of $f(x) = \sqrt{x}$.

Note that $g(x) = \sqrt{-x} = f(-x)$.

Reflection About the x-Axis and the y-Axis

The graph of $y = -f(x)$ is obtained by reflecting the graph of $y = f(x)$ about the x-axis.

The graph of $y = f(-x)$ is obtained by reflecting the graph of $y = f(x)$ about the y-axis.

WARNING Don't confuse $f(-x)$ and $-f(x)$. They are usually *not* the same. For example, if $f(x) = x^2$, then $f(-2) = (-2)^2 = 4$, but $-f(2) = -2^2 = -4$. If $f(x) = 2x + 3$, then $f(-5) = 2(-5) + 3 = -10 + 3 = -7$, but $-f(5) = -(2 \cdot 5 + 3) = -13$. ■

We can use our shifting results to sketch any quadratic function with leading coefficient ± 1.

EXAMPLE 11 *Shifting and Reflecting a Graph*

Sketch the graph of $f(x) = -x^2 + 4x - 9$.

SOLUTION

$$\begin{aligned}
-x^2 + 4x - 9 &= -(x^2 - 4x + 9) && \text{Factor out } -1 \\
&= -(x^2 - 4x + 4 - 4 + 9) && \text{Complete the square by} \\
&= -[(x^2 - 4x + 4) + 5] && \text{adding and subtracting 4} \\
&= -[(x - 2)^2 + 5] && \text{Factor}
\end{aligned}$$

Starting with the graph of $y = x^2$, we obtain the graph of $y = -x^2 + 4x - 9$ in three steps, as in Figure 12.

(a) Graph of $y = x^2$ (original graph).

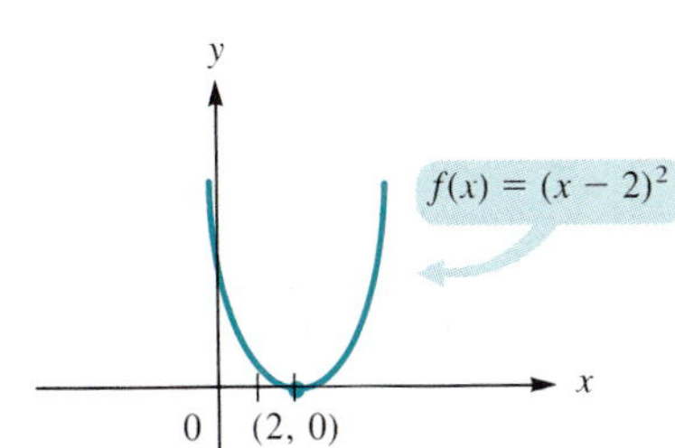

(b) Step 1: Shift graph of $f(x) = x^2$ right 2 units.

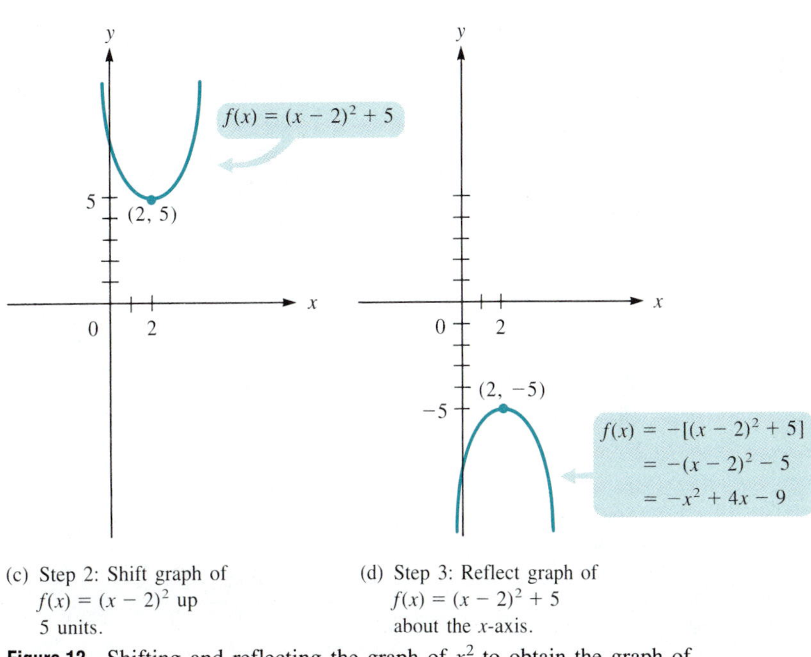

(c) Step 2: Shift graph of $f(x) = (x - 2)^2$ up 5 units.

(d) Step 3: Reflect graph of $f(x) = (x - 2)^2 + 5$ about the x-axis.

Figure 12 Shifting and reflecting the graph of x^2 to obtain the graph of $-x^2 + 4x - 9$.

EXAMPLE 12 *Reflecting a Graph About the y-Axis*

In Figure 13, the graph of an even function is given for $x \geq 0$. Complete the graph if the domain is $\mathbb{R}$.

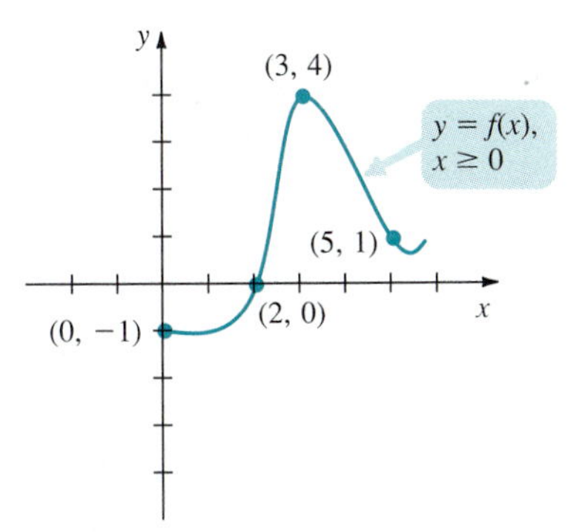

Figure 13 Graph of an even function for $x \geq 0$.

SOLUTION Since the function is even, we may obtain its graph, for $x \leq 0$, by reflecting the graph for $x \geq 0$ about the y-axis. This is done in Figure 14.

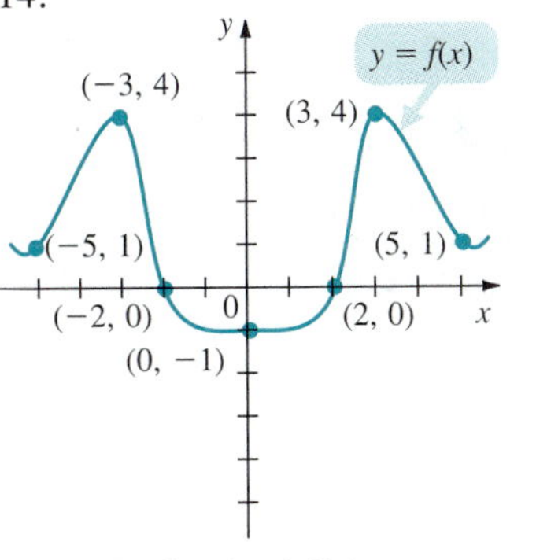

Figure 14 Graph of $f(x)$. ■

EXAMPLE 13 *Reflecting and Shifting a Graph*

The graph of a certain function $f(x)$ is given in Figure 15(a). Sketch the graph of $-f(3-x)$.

SOLUTION We do this in three steps:

(b) Reflect about the y-axis to obtain the graph of $f(-x)$ (Figure 15(b)).
(c) Shift to the right 3 units to obtain the graph of $f(-(x-3)) = f(3-x)$ (Figure 15(c)).
(d) Reflect about the x-axis to obtain the graph of $y = -f(3-x)$ (Figure 15(d)).

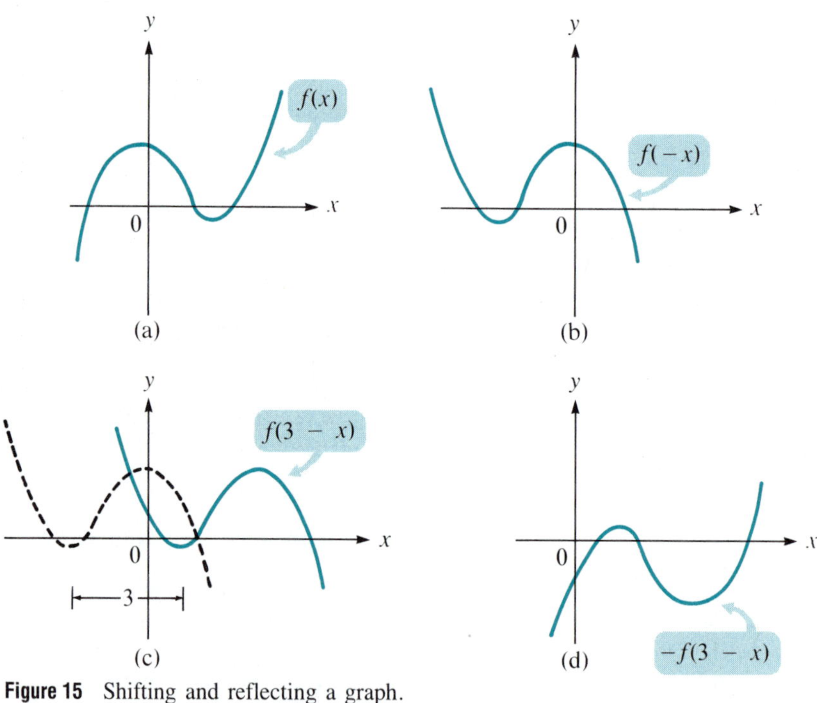

Figure 15 Shifting and reflecting a graph.

Problems 1.4

Readiness Check

I. Which of the following is true about an odd function f?
   a. Its graph is symmetric about the y-axis.
   b. Its graph is symmetric about the origin.
   c. $f(x) = f(-x)$
   d. If $(3, -2)$ is a point on the graph of f, then $(-3, -2)$ is also on the graph of $f(x)$.

II. Which of the following is graphed below?
   a. $f(x) = (x-2)^2 + 3$
   b. $g(x) = -(x+2)^2 - 3$
   c. $h(x) = -(x-2)^2 - 3$
   d. $k(x) = (x+2)^2 + 3$

y
0 x

III. The graph of which of the following equations would be obtained by shifting $y = f(x) = \sqrt{x}$ three units to the left?

a. $y = f(x) = \sqrt{x - 3}$ b. $y = f(x) = \sqrt{x} - 3$
c. $y = f(x) = \sqrt{x + 3}$ d. $y = f(x) = \sqrt{x} + 3$

IV. Which of the following would have a graph that is the reflection of the graph of $f(x) = x^2 - 2x + 9$ about the x-axis?

a. $f(x) = -x^2 + 2x - 9$ b. $f(x) = x^2 + 2x + 9$
c. $f(x) = -(x - 1)^2 + 10$
d. $f(x) = -[(x - 1)^2 - 8]$

V. Which of the following is the function that results from shifting the graph of $f(x) = x^2 - 4x + 3$ up 2 units and to the left 3 units?

a. $f(x) = x^2 + 8x + 6$
b. $f(x) = x^2 + 10x$
c. $f(x) = x^2 - 3x + 2$
d. $f(x) = x^2 + 2x + 2$

In Problems 1–7 sketch the graph of the given function by plotting some points (if necessary) and then connecting them. Check first for symmetry about the y-axis and the origin. Use a calculator when marked.

1. $f(x) = x^4$
2. $f(x) = x^2$
3. $f(x) = 5x^2$
4. $f(x) = -2x^2$
5. $f(x) = \sqrt[3]{x}$
6. $f(x) = x^3 + x$
7. $f(x) = \sqrt{x^2 + 1}$

In Problems 8–13 sketch the given function.

8. $y = x^2 + 3$
9. $y = x^2 - 5$
10. $y = (x + 3)^2$
11. $y = (x - 4)^2$
12. $y = 1 - x^2$
13. $y = (x - 1)^2 + 3$

In Problems 14–18 use the technique of Example 11 to sketch the given quadratic.

14. $y = x^2 - 4x + 7$
15. $y = x^2 + 8x + 2$
16. $y = x^2 + 3x + 4$
17. $y = -x^2 + 2x - 3$
18. $y = -x^2 - 5x + 8$

In Problems 19–24 use the graph in Figure 4 (p. 30) to obtain the graph of the given cubic.

19. $y = x^3 + 1$
20. $y = x^3 - 2$
21. $y = (x + 3)^3$
22. $y = -x^3$
23. $y = -(x + 2)^3$
24. $y = 3 - (x + 4)^3$
25. The graph of $f(x) = 1/x$ is given below. Sketch the graph of

(a) $\dfrac{1}{x + 3}$ (b) $\dfrac{1}{x - 4}$

(c) $3 + \dfrac{1}{x}$ (d) $2 - \dfrac{1}{x}$

* (e) $\dfrac{5x - 1}{x}$

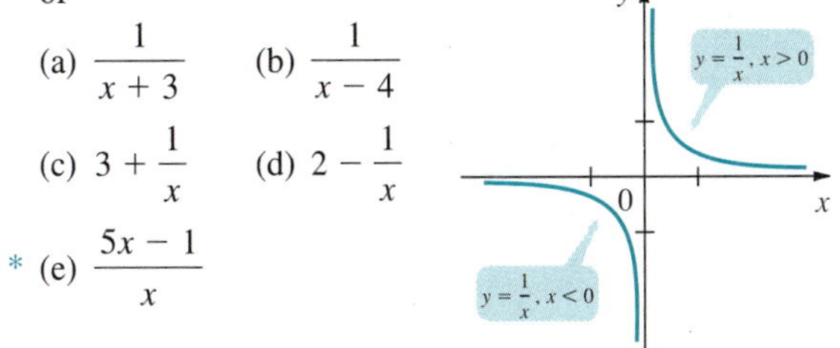

In each of Problems 26–34 the graph of a function is sketched. Obtain the graph of (a) $f(x - 2)$, (b) $f(x + 3)$, (c) $-f(x)$, (d) $f(-x)$, and (e) $f(2 - x) + 3$.

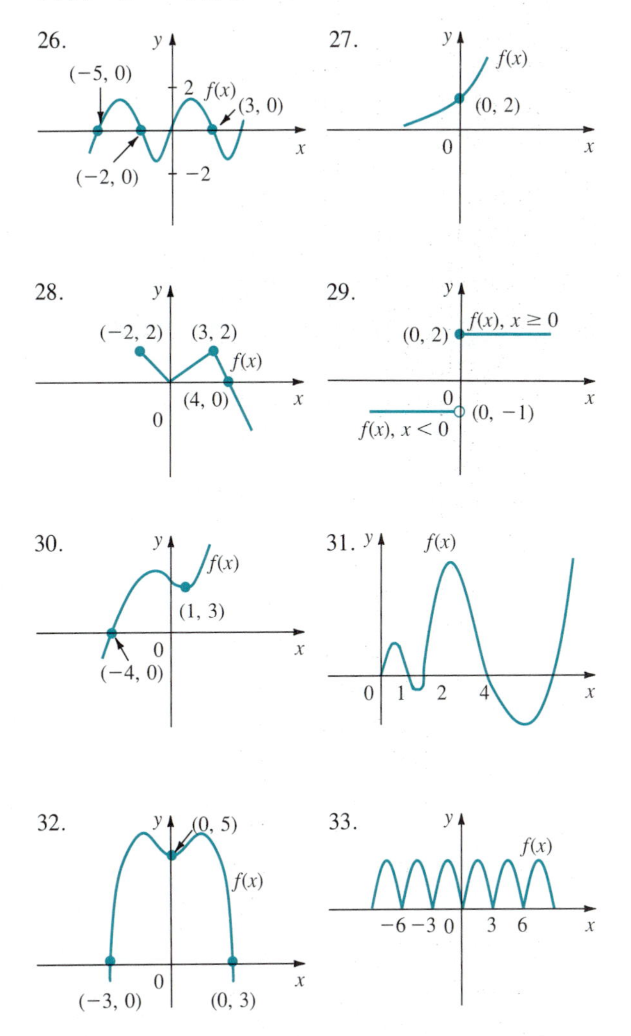

Answers to Readiness Check

I. b II. c III. c IV. a V. d

34.

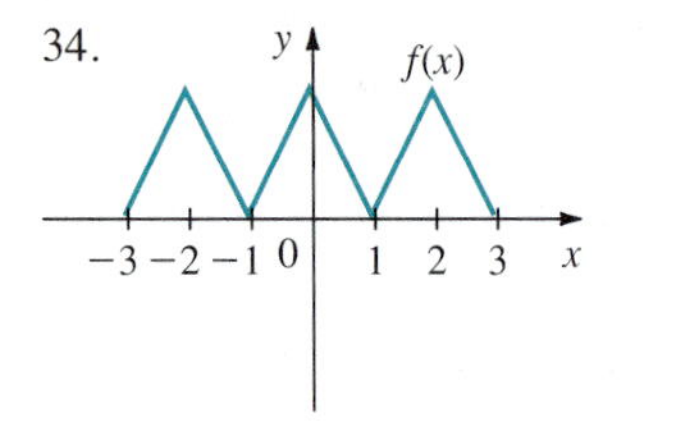

In Problems 35–42 the graph of an even or odd function is given for $x \geq 0$. Complete the graph of the function for $x < 0$.

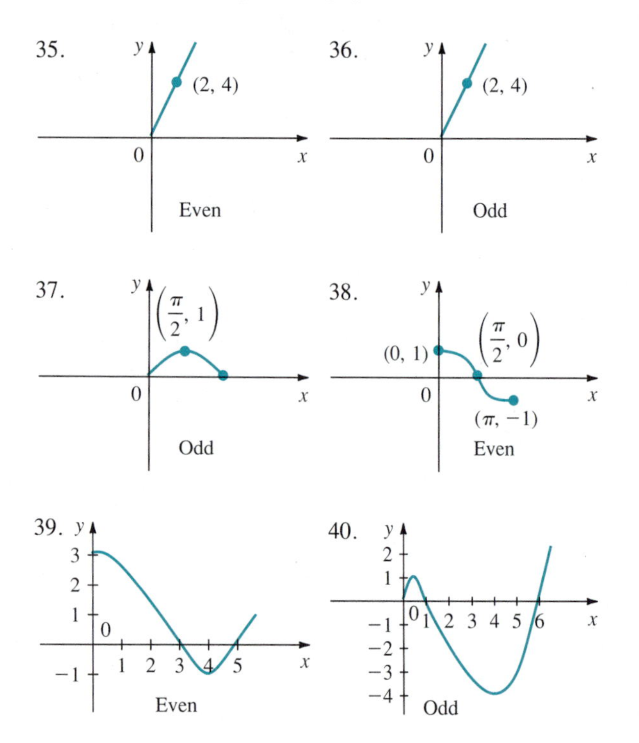

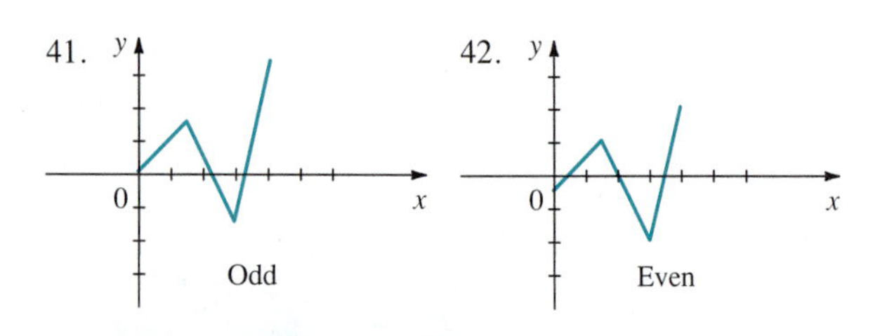

Graphing Calculator Problems

Before doing the following problems, read (or reread) Appendix A on using a graphing calculator.

43. Sketch the graph of $y = \sqrt[3]{x}$. Use the range values $-9 \leq x \leq 9$, $-2.5 \leq y \leq 2.5$, $x_{\text{scl}} = 2$, $y_{\text{scl}} = 1$.
44. Without clearing the graph of $\sqrt[3]{x}$, and without changing the range or scale values, sketch the graph of $y = \sqrt[3]{x-2}$.

In Problems 45–52 first clear previously sketched graphs, then sketch $y = \sqrt[3]{x}$ and, without clearing this graph, sketch the graph of the given function. It will be necessary to change the range values in some cases.

45. $y = \sqrt[3]{x+2}$
46. $y = \sqrt[3]{x} - 2$
47. $y = \sqrt[3]{x-2}$
48. $y = -\sqrt[3]{x}$
49. $y = \sqrt[3]{-x}$
50. $y = 3 - \sqrt[3]{x}$
51. $y = 2 + \sqrt[3]{x-1}$
52. $y = -1 + \sqrt[3]{2-x}$

In Problems 53–61 let $f(x) = -x^3 + 2x^2 + 5x - 6$. A graph of this function is given in Figure 3 in Appendix A for $-4 \leq x \leq 4$ and $-18 \leq y \leq 70$. Sketch each function after first changing the x- and y-range values in an appropriate way. (For example, if you shift 2 units to the right, then $-4 \leq x \leq 4$ is shifted to $-2 \leq x \leq 6$.)

53. $f(x-2)$
54. $f(x+3)$
55. $f(x)+1$
56. $f(x)-4$
57. $-f(x)$
58. $f(-x)$
59. $f(2-x)$
60. $-f(x)+3$
61. $-4 + f(1+x)$

1.5 Operations with Functions

We begin this section by showing how functions can be added, subtracted, multiplied, and divided.

Definition of the Sum, Difference, Product, and Quotient of Two Functions

Let f and g be two functions. Then

(a) The sum $f + g$ is defined by

$$(f+g)(x) = f(x) + g(x)$$

(b) The difference $f - g$ is defined by

$$(f - g)(x) = f(x) - g(x)$$

(c) The product $f \cdot g$ is defined by

$$(f \cdot g)(x) = f(x)g(x)$$

(d) The quotient f/g is defined by

$$\left(\frac{f}{g}\right)(x) = \frac{f(x)}{g(x)}, \text{ whenever } g(x) \neq 0$$

The functions $f + g, f - g$, and $f \cdot g$ are defined for each x for which both f and g are defined. We have,

$$\begin{aligned}(\text{domain of } f + g) &= (\text{domain of } f - g) = (\text{domain of } f \cdot g) \\ &= \{x\text{: } x \in \text{domain of } f \text{ and } x \in \text{domain of } g\}.\end{aligned}$$

That is, in order for $(f + g)(x)$ to be defined, we must have both $f(x)$ *and* $g(x)$ defined.

Finally, f/g is defined whenever both f and g are defined and $g(x) \neq 0$ (so that we do not divide by zero). This last fact is very important.

$$(\text{domain of } f/g) = \{x\text{: } x \in \text{domain of } f,\ x \in \text{domain of } g \text{ and } g(x) \neq 0\}.$$

When taking the quotient of two functions, make sure you are not dividing by zero.

EXAMPLE 1 *The Sum, Difference, Product, and Quotient of Two Functions*

Let $f(x) = \sqrt{1 + x}$ and $g(x) = 4 - x^2$. Find $(f + g)(x)$, $(f - g)(x)$, $(f \cdot g)(x)$, and $(f/g)(x)$ and determine the domain of each.

SOLUTION The domain of f is $\{x\text{: } x \geq -1\}$. Since the domain of g is $\mathbb{R}$, $f + g, f - g$, and $f \cdot g$ are defined for $\{x\text{: } x \geq -1\} = [-1, \infty)$. We have

(a) $(f + g)(x) = f(x) + g(x) = \sqrt{1 + x} + 4 - x^2$
(b) $(f - g)(x) = f(x) - g(x) = \sqrt{1 + x} - (4 - x^2) = \sqrt{1 + x} - 4 + x^2$
(c) $(f \cdot g)(x) = f(x)g(x) = \sqrt{1 + x}(4 - x^2)$
(d) $\left(\frac{f}{g}\right)(x) = \frac{f(x)}{g(x)} = \frac{\sqrt{1 + x}}{4 - x^2}$

Here the domain is different. The denominator $4 - x^2 = 0$ when $x = \pm 2$. The number -2 is not in the domain of f, so we don't worry about it. However, we must throw out 2. Thus

$$\text{domain of } \frac{f}{g} \text{ is } \{x\text{: } x \geq -1 \text{ and } x \neq 2\}$$

Composite Function

You will often need to deal with functions of functions. If f and g are functions, then their **composite function,** $f \circ g$, is defined as follows:

Definition of the Composite Function

$$(f \circ g)(x) = f(g(x))$$

and domain of $f \circ g$ is $\{x: x \in \text{domain of } g \text{ and } g(x) \in \text{domain of } f\}$.

That is, $(f \circ g)(x)$ is defined for every x such that $g(x)$ and $f(g(x))$ are defined. An illustration of the composition of two functions is given in the following figure.

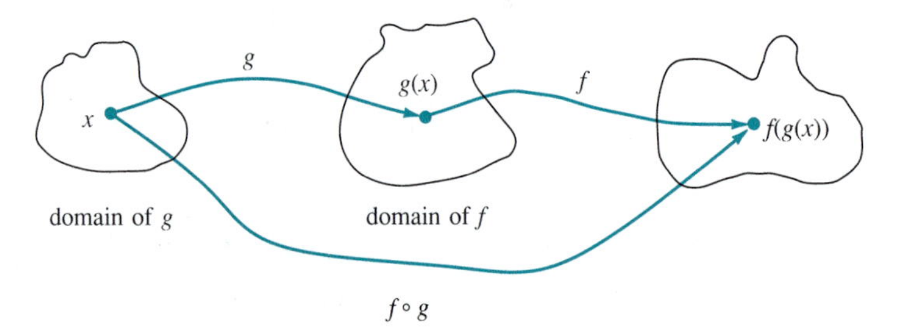

EXAMPLE 2 ***Finding the Composition of Two Functions***

Let $f(x) = \sqrt{x}$ and $g(x) = x^2 + 1$. Find $(f \circ g)(x)$ and $(g \circ f)(x)$ and determine their respective domains.

SOLUTION

$$(f \circ g)(x) = f(g(x)) = f(x^2 + 1) = \sqrt{x^2 + 1}$$

and

$$(g \circ f)(x) = g(f(x)) = g(\sqrt{x}) = (\sqrt{x})^2 + 1 = x + 1$$

Note that we must have $x \geq 0$ because $\sqrt{x}$ is defined (as a real number) only when $x \geq 0$. Now, the domain of f is $\mathbb{R}^+$, the domain of g is $\mathbb{R}$, and

$$\text{domain of } f \circ g \text{ is } \{x: g(x) = x^2 + 1 \in \text{domain of } f\}$$

But since $x^2 + 1 > 0$, $x^2 + 1 \in$ domain of f for every real x, so domain of $f \circ g$ is $\mathbb{R}$. On the other hand, domain of $g \circ f$ is $\mathbb{R}^+$ because f is defined only for $x \geq 0$.

WARNING It is *not* true, in general, that $(f \circ g)(x) = (g \circ f)(x)$. This is illustrated in Examples 2 and 3. ■

EXAMPLE 3 *Finding the Composition of Two Functions*

Let $f(x) = 3x - 4$ and $g(x) = x^3$. Find $(f \circ g)(x)$ and $(g \circ f)(x)$ and determine their respective domains.

SOLUTION

$$(f \circ g)(x) = f(g(x)) = f(x^3) = 3x^3 - 4$$

and

$$(g \circ f)(x) = g(f(x)) = g(3x - 4) = (3x - 4)^3$$

Here domain of $f \circ g$ = domain of $g \circ f = \mathbb{R}$. Note that the functions $f \circ g$ and $g \circ f$ are quite different. ■

EXAMPLE 4 *Finding a Function That Can Be Used to Obtain a Desired Composition*

Let $f(x) = 2x + 1$. Find a function $g(x)$ such that $(f \circ g)(x) = x^3$.

SOLUTION We must have $(f \circ g)(x) = f(g(x)) = 2g(x) + 1 = x^3$. Then $2g(x) = x^3 - 1$ and $g(x) = \dfrac{x^3 - 1}{2}$. ■

EXAMPLE 5 *A Technique Useful in Calculus*

Let $u(x) = \sqrt{x^3 + 5}$. Find three functions f, g, and h such that $u(x) = (f \circ g \circ h)(x)$.

SOLUTION We first observed that u is the square root of something. Since $(f \circ g \circ h)(x)$ is evaluated with f applied last, we set $f(x) = \sqrt{x}$. Next, to evaluate $x^3 + 5$, we first cube x and then add 5. Thus, if $h(x) = x^3$ and $g(x) = x + 5$, then

$$(g \circ h)(x) = g(h(x)) = g(x^3) = x^3 + 5$$

Finally,

$$(f \circ g \circ h)(x) = f((g \circ h)(x)) = f(x^3 + 5) = \sqrt{x^3 + 5}$$

So one solution is

$$f(x) = \sqrt{x}$$
$$g(x) = x + 5$$
$$h(x) = x^3$$

Problems 1.5

Readiness Check

I. Which of the following is true?
  a. The domain of $f \circ g$ is $\{x: x$ belongs to the domains of f and $g\}$.
  b. The domain of $f + g$ is the same as the domain of f/g for all f and g.
  c. The domain of a composite function cannot be determined until the composite function is written explicitly.
  d. The domain of $f + g$ is the same as the domain of $f - g$ for all f and g.

II. Which of the following has domain $= \{x: x \in \mathbb{R}, x > 0\}$ where $f(x) = \dfrac{-1}{x+1}$ and $g(x) = \sqrt{x}$?
  a. $f + g$ b. f/g c. $f \circ g$ d. $g \circ f$

III. Which of the following is true about $y = -8 + 6x - 2x^2$ if $f(x) = 2x^2 - x + 5$ and $g(x) = 5x - 3$?
  a. $y = (f + g)(x)$
  b. $y = (f - g)(x)$
  c. $y = (g - f)(x)$
  d. $y = (f + g)(-x)$

IV. Which of the following is the domain of $\dfrac{f}{g}$ if $f(x) = x^2 - 5x + 6$ and $g(x) = x - 3$?
  a. $\{x: x \in \mathbb{R}\}$
  b. $\{x: x \in \mathbb{R}$ and $x \neq 2, 3\}$
  c. $\{x: 2 \le x \le 3\}$
  d. $\{x: x \in \mathbb{R}$ and $x \neq 3\}$

In Problems 1–12 two functions, f and g, are given. Determine the functions $f + g$, $f - g$, $f \cdot g$, and f/g and find their respective domains.

1. $f(x) = 2x + 3, \quad g(x) = -3x$
2. $f(x) = 2x - 5, \quad g(x) = -3x + 4$
3. $f(x) = 4, \quad g(x) = 10$
4. $f(x) = -3, \quad g(x) = 0$
5. $f(x) = x, \quad g(x) = \dfrac{1}{x}$
6. $f(x) = x^2, \quad g(x) = x + 1$
7. $f(x) = \sqrt{x + 1}, \quad g(x) = \sqrt{1 - x}$
8. $f(x) = x^3 + x, \quad g(x) = \dfrac{1}{\sqrt{x + 1}}$
9. $f(x) = 1 + x^5, \quad g(x) = 1 - |x|$
10. $f(x) = \sqrt{1 + x}, \quad g(x) = \dfrac{1}{x^5}$
11. $f(x) = \sqrt[5]{x + 2}, \quad g(x) = \sqrt[4]{x - 3}$
12. $f(x) = \dfrac{x + 1}{x}, \quad g(x) = \dfrac{x}{x - 3}$

In Problems 13–24 find $f \circ g$ and $g \circ f$ and determine the domain of each.

13. $f(x) = 3x, \quad g(x) = x - 2$
14. $f(x) = x - 3, \quad g(x) = 2x + 1$
15. $f(x) = 5, \quad g(x) = 8$
16. $f(x) = 0, \quad g(x) = -2$
17. $f(x) = x, \quad g(x) = \dfrac{1}{2x}$
18. $f(x) = 2x^2, \quad g(x) = x + 5$
19. $f(x) = 2x - 4, \quad g(x) = x + 3$
20. $f(x) = \sqrt{x + 1}, \quad g(x) = x^3$
21. $f(x) = \dfrac{x}{2 - x}, \quad g(x) = \dfrac{x + 1}{x}$
22. $f(x) = |x|, \quad g(x) = -x$
23. $f(x) = \sqrt{1 - x}, \quad g(x) = \sqrt{x - 1}$
24. $f(x) = \begin{cases} x, & x \ge 0, \\ 2x, & x < 0, \end{cases}$ $\quad g(x) = \begin{cases} -3x, & x \ge 0 \\ 5x, & x < 0 \end{cases}$
25. Let $f(x) = 2x + 4$ and $g(x) = \frac{1}{2}x - 2$. Show that $(f \circ g)(x) = (g \circ f)(x) = x$. (When this occurs, we say that f and g are **inverse functions.**)
26. If $f(x) = -3x + 2$, find a function g such that $(f \circ g)(x) = (g \circ f)(x) = x$.
27. If $f(x) = x^2$, find two functions g_1 and g_2 such that $(f \circ g_1)(x) = (f \circ g_2)(x) = x^2 - 10x + 25$.
28. Let $h(x) = 1/\sqrt{x^2 + 1}$. Determine two functions f and g such that $f \circ g = h$.
29. Let $k(x) = (1 + \sqrt{x})^{5/7}$. Find the domain of k. Determine three functions f, g, and h such that $f \circ g \circ h = k$.
30. Let $h(x) = x^2 + x$, and let $f_1(x) = x^2 - x$, $g_1(x) = x + 1$, $f_2(x) = x^2 + 3x + 2$, and $g_2(x) = x - 1$. Show that $f_1 \circ g_1 = f_2 \circ g_2 = h$. This illustrates the fact that there is

Answers to Readiness Check

I. d II. b III. c IV. d

often more than one way to write a given function as the composition of two other functions.

31. Let f and g be the linear functions

$$f(x) = ax + b, \qquad g(x) = cx + d$$

Find conditions on a and b in order that $f \circ g = g \circ f$.

** 32. Each of the following functions satisfies an equation of the form $(f \circ f)(x) = x$ or $(f \circ f \circ f)(x) = x$ or $(f \circ f \circ f \circ f)(x) = x$, and so on. For each function, discover what type of equation is appropriate.
    (a) $A(x) = \sqrt[3]{1 - x^3}$
    (b) $B(x) = \sqrt[7]{23 - x^7}$
    (c) $C(x) = 1 - 1/x$, domain is $\mathbb{R} - \{0, 1\}$ (This is the set of all real numbers except 0 and 1.)
    (d) $D(x) = 1/(1 - x)$, domain is $\mathbb{R} - \{0, 1\}$
    (e) $E(x) = (x + 1)/(x - 1)$, domain is $\mathbb{R} - \{1\}$
    (f) $F(x) = (x - 1)/(x + 1)$, domain is $\mathbb{R} - \{-1, 0, 1\}$
    (g) $G(x) = (4x - 1)/(4x + 2)$,

    domain is $\mathbb{R} - \{-\frac{1}{2}, 0, \frac{1}{4}, \frac{1}{2}, 1\}$

* 33. A manufacturer of designer shirts determines that the demand function for her shirts is $x = D(p) = 400(50 - p)$, where p is the wholesale price she charges per shirt and x is the number of shirts she can sell at that price. Note that, as is common, the higher the price, the fewer shirts she can sell. Assume that the manufacturer's fixed cost is \$8000 and her material and labor costs amount to \$8 per shirt.
    (a) Determine the total cost function C as a function of p.
    (b) Determine the total revenue function R as a function of p.
    (c) Determine the profit function P as a function of p.
    (d) By completing the square, determine the price that yields the greatest profit. What is this maximum profit (or minimum loss)?

34. Let $f(x)$ be a function. Show that

$$g(x) = \tfrac{1}{2}[f(x) + f(-x)]$$

is an even function.

35. Let $f(x)$ be a function. Show that

$$h(x) = \tfrac{1}{2}[f(x) - f(-x)]$$

is an odd function.

36. Show that any function $f(x)$ whose domain is $\mathbb{R}$ can be written as the sum of an even function and an odd function. [Hint: Use the results of Problems 34 and 35.]

1.6 Inverse Functions

In Section 1.3 we defined a function as a rule: You give me an x in the domain and I'll give you $y = f(x)$. Sometimes it is necessary to reverse this procedure: You give me a y and I'll find the x for which $y = f(x)$.

Before giving a definition, we start with an example. Let $y = f(x) = 2x + 3$. What value of x leads to a given y? To answer this question, we solve the equation $y = 2x + 3$ for x.

$$\begin{aligned} y &= 2x + 3 \\ 2x &= y - 3 \\ x &= \frac{y - 3}{2} \end{aligned}$$

Because it is customary (but not necessary) to write y as a function of x, interchange x and y in the last step to obtain the function $y = (x - 3)/2$.

Consider the two functions $f(x) = 2x + 3$ and $g(x) = (x - 3)/2$. We compute

$$(f \circ g)(x) = f(g(x)) = f\left(\frac{x - 3}{2}\right) = 2\left(\frac{x - 3}{2}\right) + 3 = x - 3 + 3 = x$$

and

$$(g \circ f)(x) = g(f(x)) = g(2x + 3) = \frac{(2x + 3) - 3}{2} = \frac{2x}{2} = x$$

The functions $f(x) = 2x + 3$ and $g(x) = (x - 3)/2$ are said to be *inverses* of one another.

Definition of Inverse Functions

The functions f and g are **inverse functions** if the following conditions hold:

(i) For every x in the domain of g, $g(x)$ is in the domain of f and $f(g(x)) = x$.
(ii) For every x in the domain of f, $f(x)$ is in the domain of g and $g(f(x)) = x$.

In this case we write $f(x) = g^{-1}(x)$ or $g(x) = f^{-1}(x)$, and we say f is the inverse of g and g is the inverse of f.

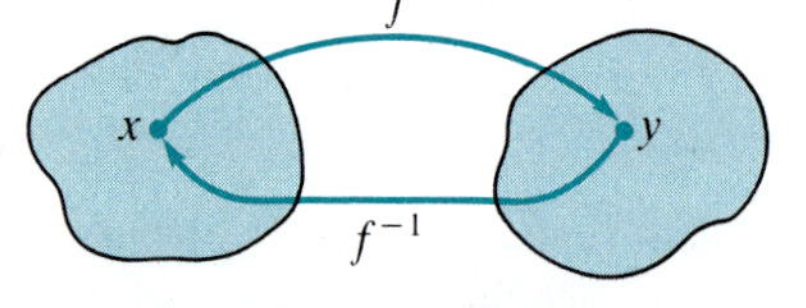

Figure 1 Pictorial representation of two inverse functions.

NOTE The domain of f^{-1} is the range of f, and the domain of f is the range of f^{-1}.

In the notation of Section 1.5, we may write

$$(f \circ f^{-1})(x) = x \quad \text{and} \quad (f^{-1} \circ f)(x) = x$$

In Figure 1 we give a pictorial representation of a function f and its inverse f^{-1}.

EXAMPLE 1 *Two Inverse Functions*

$f(x) = x^3$ and $g(x) = \sqrt[3]{x} = x^{1/3}$ are inverses because

$$f(g(x)) = f(x^{1/3}) = (x^{1/3})^3 = x$$

and

$$g(f(x)) = g(x^3) = (x^3)^{1/3} = x$$

We do not have to worry about domains here because both x^3 and $\sqrt[3]{x}$ are defined for every real number x.

Computing an Inverse Function

To compute an inverse function, we use the following procedure:

(i) Replace $f(x)$ with y.
(ii) Interchange x and y.
(iii) Solve for y in terms of x, if possible.
(iv) The resulting y is equal to $f^{-1}(x)$.
(v) Set the domain of f^{-1} equal to the range of f.

We use the phrase *if possible* in step (iii) because it is not always possible to solve explicitly for y as a *function* of x.

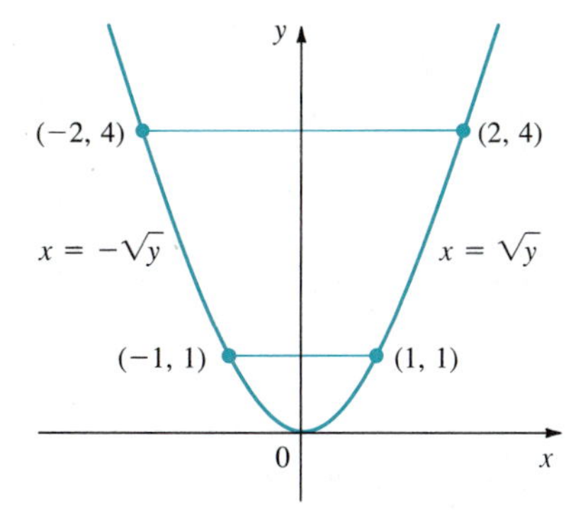

Figure 2 The graph of $y = x^2$. For each $y > 0$, there are two x's such that $y = x^2$.

EXAMPLE 2 *A Function That Does Not Have an Inverse*

Let $y = x^2$. Then $x = \pm\sqrt{y}$. This means that each nonzero value of y comes from *two* different values of x. See Figure 2. We recall from Section 1.3 that a function $y = f(x)$ can be thought of as a rule that assigns to each x in its domain a *unique* value of y. Suppose we try to define the inverse of x^2 by taking the positive square root. If $f(x) = x^2$ and $g(x) = \sqrt{x}$ are inverses, then $g(f(x)) = x$ for every x in the domain of f. But $-2 \in$ domain of $f\,(= \mathbb{R})$, and $g(f(-2)) = g((-2)^2) = g(4) = \sqrt{4} = 2 \neq -2$. Thus $g(f(x)) \neq x$ for every x in the domain of f.

To avoid problems like the one just encountered, we make the following definition.

Definition of a One-to-One Function

The function $y = f(x)$ is **one-to-one on the interval [*a*, *b*]** if whenever $x_1, x_2 \in [a, b]$ and $f(x_1) = f(x_2)$, then $x_1 = x_2$. That is, each value of $f(x)$ comes from only one value of x. If $x_1 \neq x_2$ implies that $f(x_1) \neq f(x_2)$, for every x_1 and x_2 in the domain of f, we say that f is, simply, one-to-one.

The definition of a one-to-one function is illustrated in Figure 3.

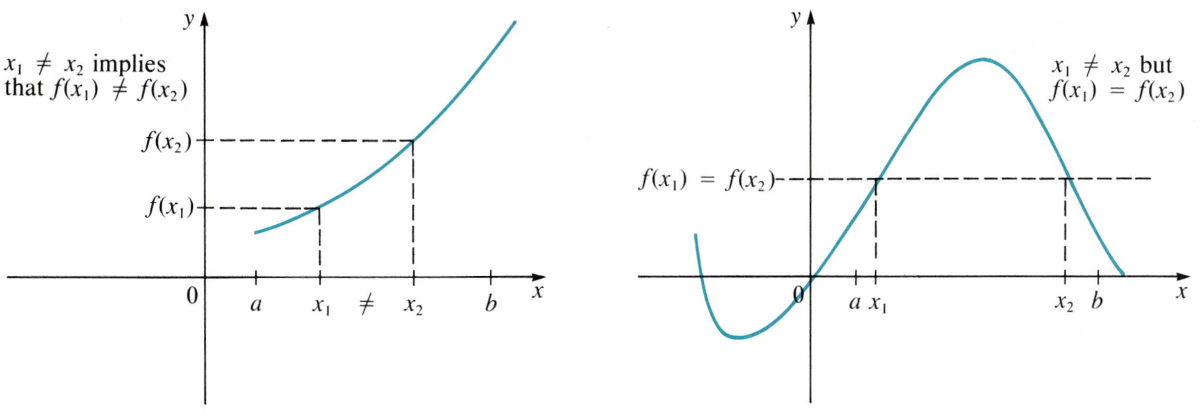

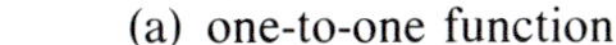

Figure 3 (a) one-to-one function (b) not a one-to-one function

It follows from the definition that

> If there are two distinct numbers x_1 and x_2 such that $f(x_1) = f(x_2)$, then f is *not* one-to-one.

In Section 1.3 we saw that an equation in x and y defined a function $y = f(x)$ if its graph satisfied the vertical lines test. We have a similar test for one-to-one functions.

Horizontal Lines Test

Let f be a function. If every horizontal line in the plane intersects the graph of f in at most one point, then f is one-to-one.

NOTE If f is a function, then the vertical lines test on p. 22 must also hold.

EXAMPLE 3 ***Illustration of the Horizontal Lines Test***

We can see, in Figure 4, that $f(x) = x^2$ is not one-to-one, but $f(x) = x^3$ and $f(x) = \dfrac{1}{x}$ are one-to-one.

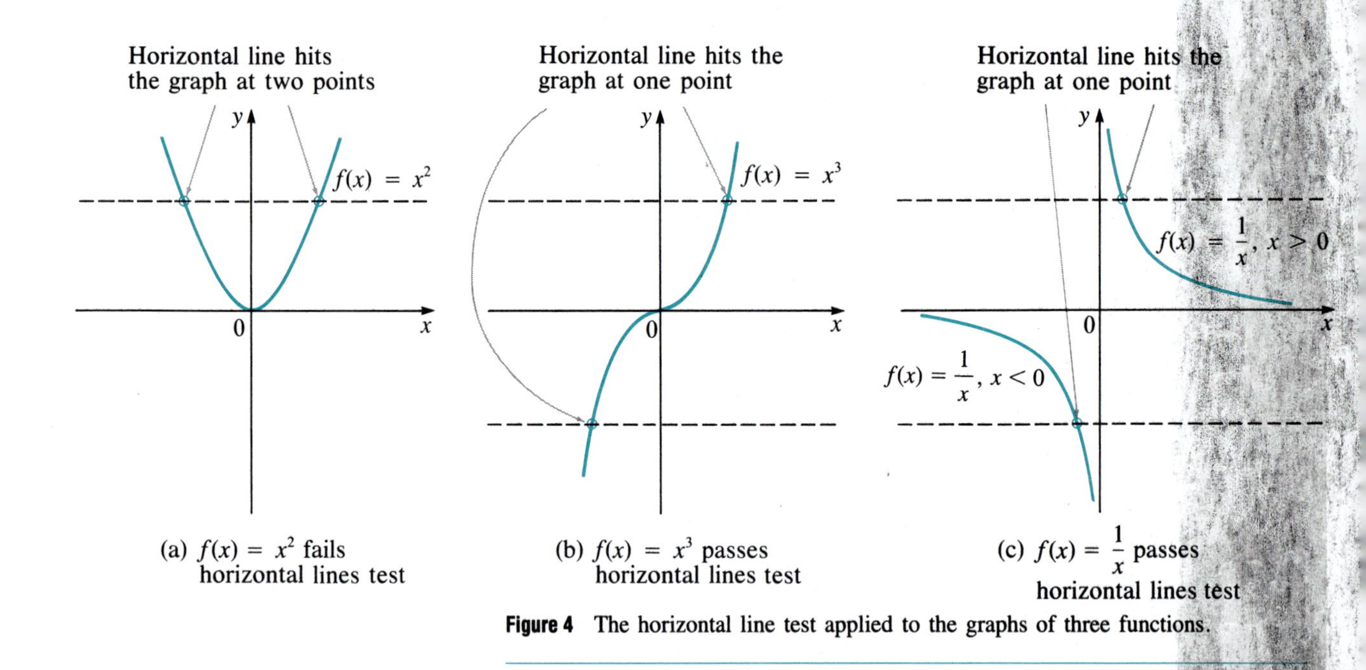

Figure 4 The horizontal line test applied to the graphs of three functions.

Why are we interested in one-to-one functions? Because if $y = f(x)$ is one-to-one, then every value of y comes from a unique value of x, so we can write x as a function of y and this new function is the inverse of y. We have shown the following:

Theorem

Every one-to-one function has an inverse.

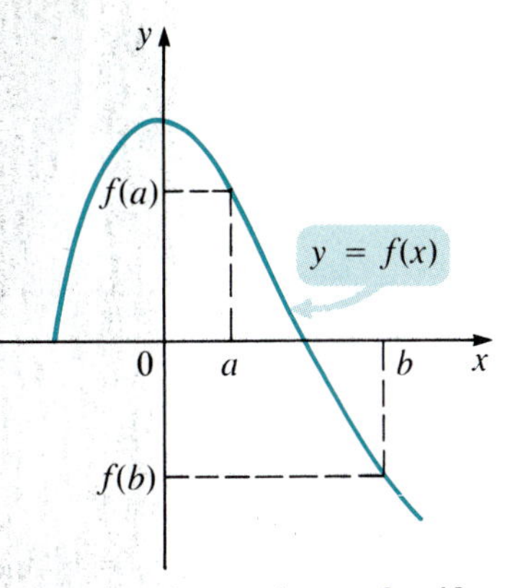

(a) $f(x)$ decreasing on $[a, b]$

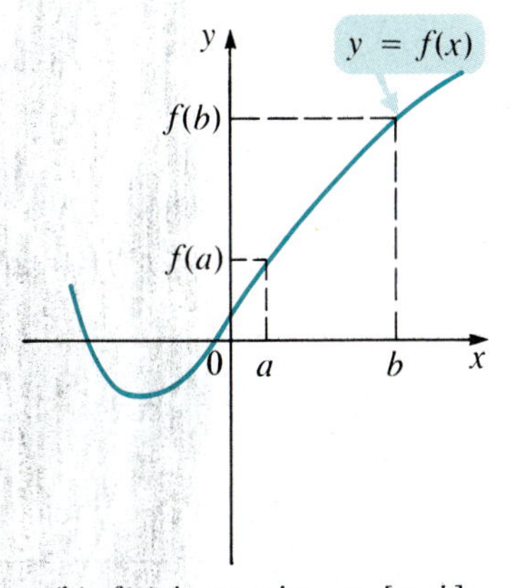

(b) $f(x)$ increasing on $[a, b]$

Figure 5

Increasing and Decreasing Functions

f is **increasing** on $[a, b]$ if for every pair of numbers x_1, x_2 in $[a, b]$ with $x_2 > x_1$, $f(x_2) > f(x_1)$.

f is **decreasing** on $[a, b]$ if, for every pair of numbers x_1, x_2 in $[a, b]$ with $x_2 > x_1$, $f(x_2) < f(x_1)$.

These definitions are illustrated in Figure 5.

Theorem

If f is increasing or decreasing on $[a, b]$, then f is one-to-one on $[a, b]$.

Proof

Suppose f is increasing, x_1 and x_2 are in $[a, b]$ and $x_1 \neq x_2$. If $x_2 > x_1$, then $f(x_2) > f(x_1)$. If $x_1 > x_2$, then $f(x_1) > f(x_2)$. In either case, $f(x_1) \neq f(x_2)$, so f is one-to-one. The proof in the case f is decreasing is similar. ■

EXAMPLE 4 *The Function $f(x) = x^2$ Is One-to-One When Its Domain Is Suitably Restricted*

From Figure 4a, we see that $f(x) = x^2$ is decreasing on $(-\infty, 0]$ and increasing on $[0, \infty)$. As we have seen, $f(x) = x^2$ is not one-to-one and does not have an inverse. However, the function $f_1(x) = x^2$ with domain $[0, \infty)$ does have an inverse as does the function $f_2(x) = x^2$ on $(-\infty, 0]$. ■

EXAMPLE 5 *Finding an Inverse Function*

Show that $f(x) = \sqrt{x}$ has an inverse given by $g(x) = x^2$, $x \geq 0$.

SOLUTION $y = \sqrt{x}$ is an increasing function for $x \geq 0$, so it has an inverse on $[0, \infty)$.

$$y = \sqrt{x}$$

$$x = \sqrt{y} \qquad \text{Interchange } x \text{ and } y$$

$$x^2 = y \qquad \text{Square both sides}$$

Thus $g(x) = f^{-1}(x) = x^2$, $x \geq 0$.

We need to have $x \geq 0$ to ensure that the domain of f^{-1} is equal to the range of f (the range of $f(x) = \sqrt{x}$ is $\mathbb{R}^+ = \{y\colon y \geq 0\}$).

Check $f(g(x)) = f(x^2) = \sqrt{x^2} = x \quad \text{if} \quad x \geq 0$

$$g(f(x)) = g(\sqrt{x}) = (\sqrt{x})^2 = x$$

NOTE The function $f(x) = x^2$ is not one-to-one and does not have an inverse, but the function $g(x) = x^2$, $x \geq 0$, which is a *different* function, does have the inverse $\sqrt{x}$.

A function changes if you change its domain.

EXAMPLE 6 *Finding an Inverse Function*

Show that $f(x) = \dfrac{1}{x}$ is one-to-one over its domain and compute its inverse.

SOLUTION $\dfrac{1}{x}$ is defined if $x \neq 0$. Suppose

Then
$$\frac{1}{x_1} = \frac{1}{x_2} \qquad \text{Neither } x_1 \text{ nor } x_2 = 0$$
$$x_2 = x_1 \qquad \text{Cross multiply}$$

Thus f is one-to-one. To find f^{-1}, we follow the steps on page 44:

$$y = \frac{1}{x} \qquad \text{Replace } f(x) \text{ with } y$$
$$x = \frac{1}{y} \qquad \text{Interchange } x \text{ and } y$$
$$y = \frac{1}{x} \qquad \text{Solve for } y \text{ in terms of } x$$

Thus $g(x) = f^{-1}(x) = \dfrac{1}{x}$. That is, the function $f(x) = \dfrac{1}{x}$, $x \neq 0$, is its own inverse.

We close this section by showing the relationship between the graph of a function f and the graph of its inverse function f^{-1}. The graphs of three functions and their inverses are sketched in Figure 6. It appears that the graphs of f and f^{-1} are symmetric about the line $y = x$. Let us see why this observation is true.

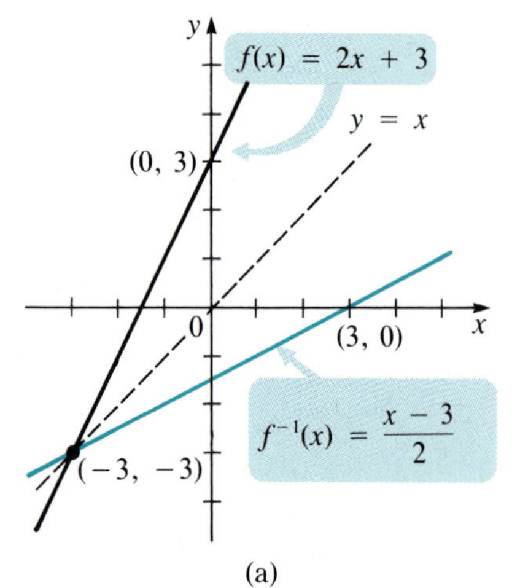

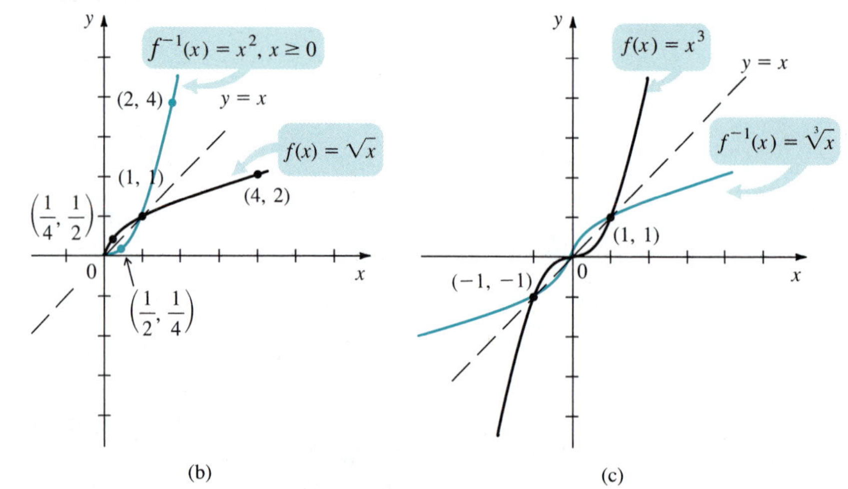

Figure 6 The graphs of three functions and their inverses.

Reflection Property

Suppose $y = f(x)$. Then, if f^{-1} exists, $x = f^{-1}(y)$. That is, (x, y) is in the graph of f if and only if (y, x) is in the graph of f^{-1}. In Figure 6(b), for example, $(\frac{1}{4}, \frac{1}{2})$, $(1, 1)$, and $(4, 2)$ are in the graph of $f(x) = \sqrt{x}$. The points $(\frac{1}{2}, \frac{1}{4})$, $(1, 1)$, and $(2, 4)$ are in the graph of $f^{-1}(x) = x^2, x \geq 0$. In Figure 6(b) if we fold the page along the line $y = x$, we find that the graphs of f and f^{-1} coincide. We see that

Reflection Property

The graphs of f and f^{-1} are **reflections** of one another about the line $y = x$.

This **reflection property,** as it is called, of the graphs of inverse functions enables us immediately to obtain the graph of f^{-1} once the graph of f is known.

Problems 1.6

Readiness Check

I. Which of the following is decreasing on $(-\infty, \infty)$?
a. $f(x) = -(x + 2)^2$ b. $f(x) = -|x - 1|$
c. $f(x) = -x^3$ d. $f(x) = \begin{cases} -2, & x \geq 0 \\ -|x|, & x < 0 \end{cases}$

II. Which of the following is a point on the graph of the inverse of the function graphed below?
a. (2, 4)
b. (4, −2)
c. (−4, 2)
d. (−2, −4)

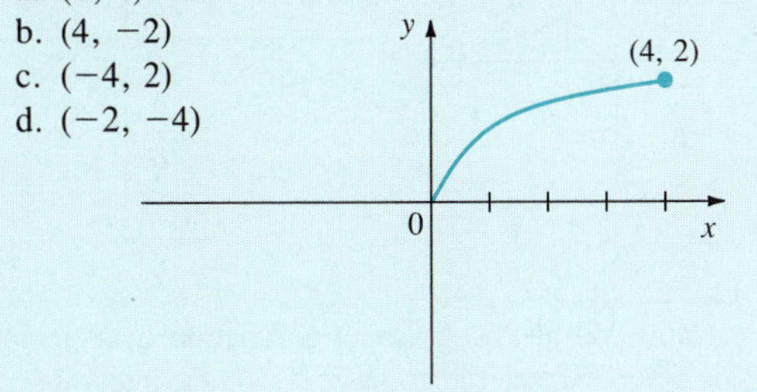

III. Which of the following is the inverse function for $y = f(x) = x^2 - 3$?
a. $f^{-1}(x) = x^2 + 3$
b. $f^{-1}(x) = \sqrt{x + 3}$
c. $f^{-1}(x) = \sqrt{x - 3}$
d. The function does not have an inverse on $(-\infty, \infty)$.

IV. Which of the following is *not* a one-to-one function?
a. $f(x) = |x + 1|$ b. $f(x) = x^3 + x$
c. $f(x) = 2x - 7$ d. $f(x) = \dfrac{1}{2x - 10}$

V. Which of the following is the inverse function of $g(x) = x^2 - 4,\ x \leq 0$?
a. $f(x) = -\sqrt{x + 4}$
b. $f(x) = \sqrt{x} - 4$
c. $f(x) = -\sqrt{x} - 4$
d. $g(x)$ does not have an inverse.

In Problems 1–24 show that the given function is one-to-one over its entire domain and find its inverse.

1. $f(x) = 2x - 1$
2. $f(x) = -7x + 4$
3. $f(x) = \dfrac{2}{3}x - \dfrac{1}{4}$
4. $f(x) = \dfrac{x + 5}{7}$
5. $f(x) = \dfrac{3 - 2x}{11}$
6. $f(x) = 3.862x - 1.803$
7. $f(x) = \dfrac{3}{x}$
8. $f(x) = \dfrac{-1}{2x}$

Answers to Readiness Check

I. c II. a III. d IV. a V. a

9. $f(x) = \dfrac{1}{x+1}$ 10. $f(x) = \dfrac{2}{x+5}$

11. $f(x) = \dfrac{3}{4-x}$ 12. $f(x) = -x^3$

13. $f(x) = \dfrac{4}{2+x^3}$ 14. $f(x) = \dfrac{3-x^3}{7}$

* 15. $f(x) = \sqrt{x+2}$ * 16. $f(x) = \sqrt{4x-5}$

* 17. $f(x) = \sqrt{1-2x}$ 18. $f(x) = \sqrt[3]{x-1}$

19. $f(x) = \dfrac{1}{\sqrt[3]{x-7}}$ 20. $f(x) = x^5$

* 21. $f(x) = \dfrac{x}{x+1}$ * 22. $f(x) = \dfrac{2x}{5-x}$

23. $f(x) = (x+3)^3$ 24. $f(x) = 1-(x-2)^3$

In Problems 25–38 determine intervals over which each function is one-to-one and find the inverse function over each such interval.

25. $f(x) = 1 + x^2$ 26. $f(x) = (1+x)^2$

27. $f(x) = (4-x)^2$ 28. $f(x) = 4-(7+x)^2$

29. $f(x) = x^4$ 30. $f(x) = \dfrac{1}{x^2}$

31. $f(x) = x^4 - 5$ * 32. $f(x) = \dfrac{x^2}{1+x^2}$

* 33. $f(x) = x^2 - 7x + 6$ [Hint: Complete the square.]

34. $f(x) = x^2 + 2x + 2$

* 35. $f(x) = x^4 + 4x^2 + 4$

36. $f(x) = x^{10}$

37. $f(x) = |x|$ 38. $f(x) = |x-3|$

In Problems 39–41 show that each function has an inverse function. Do not try to compute it.

* 39. $f(x) = x + x^3$

* 40. $f(x) = 1 + x + x^3 + x^5 + x^7$

* 41. $f(x) = \dfrac{1}{x^3+x+1}$

In Problems 42–51 the graph of a function is given. Determine whether or not the function is one-to-one over the given interval.

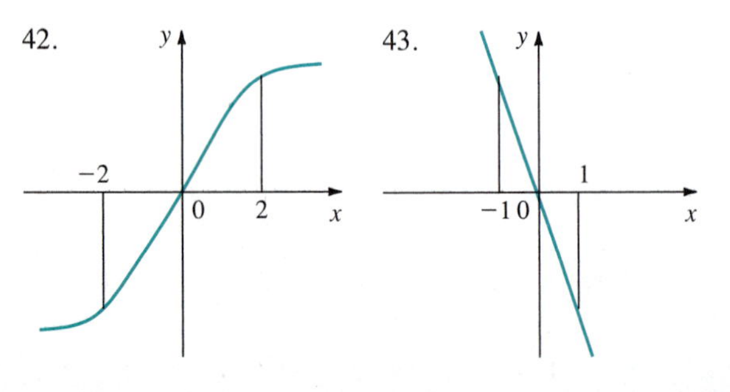

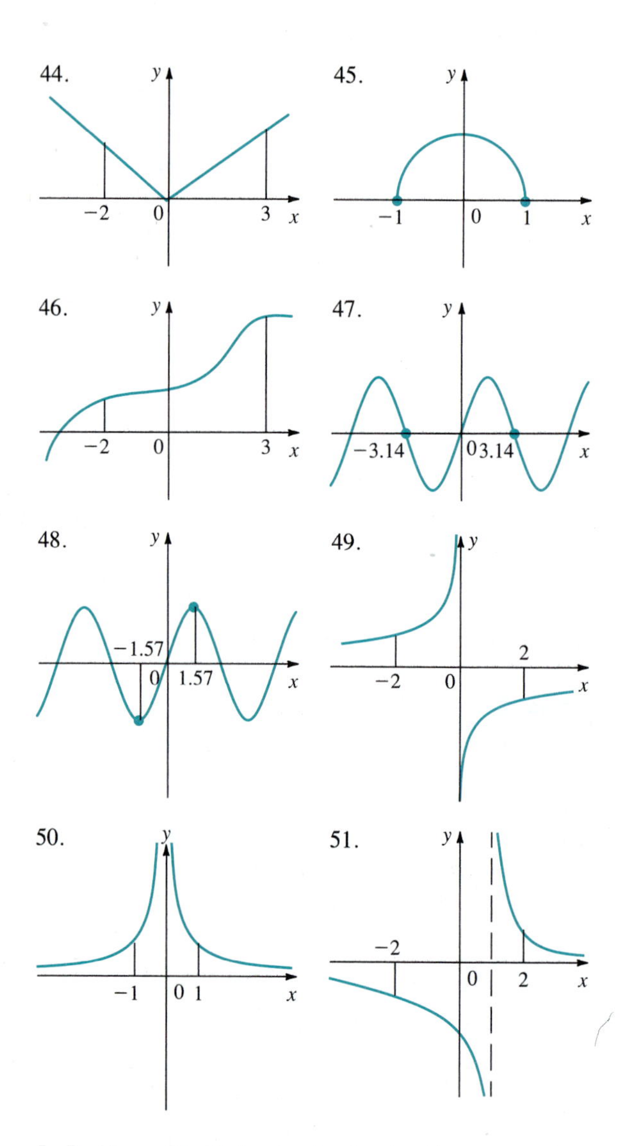

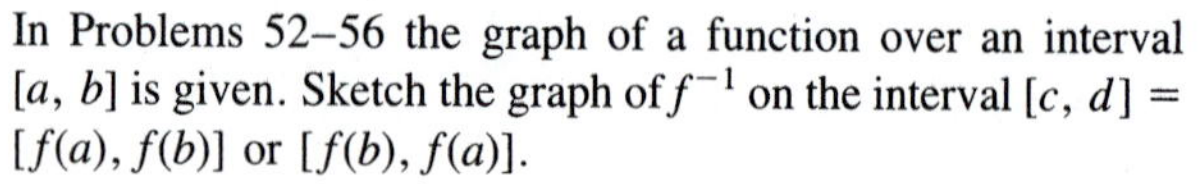

In Problems 52–56 the graph of a function over an interval $[a, b]$ is given. Sketch the graph of f^{-1} on the interval $[c, d] = [f(a), f(b)]$ or $[f(b), f(a)]$.

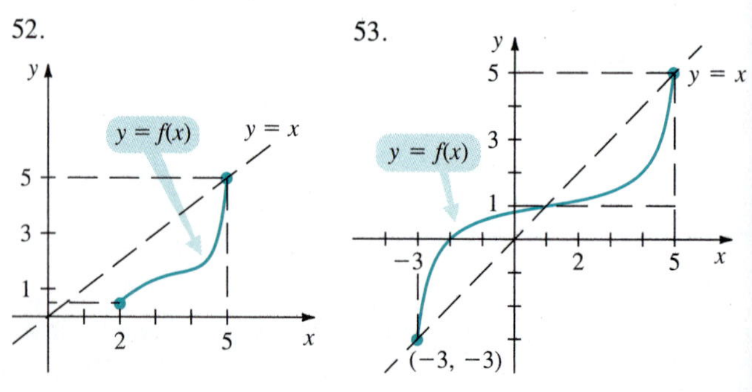

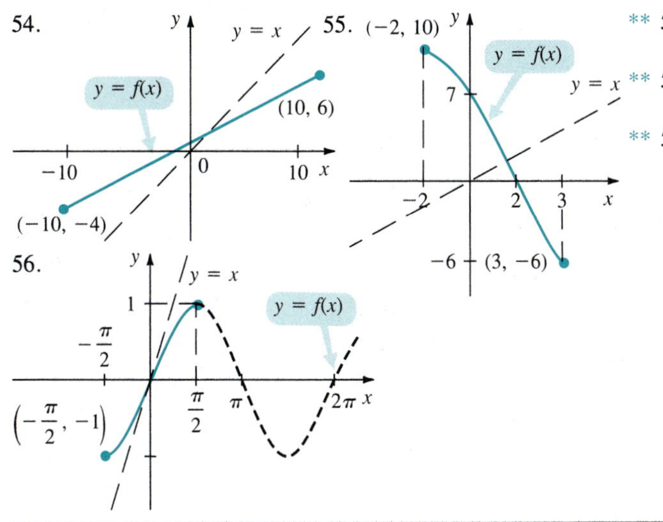

** 57. Prove that if a function is one-to-one, then its inverse is unique.

** 58. If f is one-to-one, show that the domain of f^{-1} is the range of f.

** 59. A function is called **continuous** on $[a, b]$ if its graph is unbroken—that is, if f is defined for all x in $[a, b]$ and if its graph has no gaps over $[a, b]$. The **intermediate value theorem** states that if f is continuous on $[a, b]$ and if c is any number with $f(a) < c < f(b)$ or $f(b) < c < f(a)$, then there is an $x \in (a, b)$ such that $f(x) = c$. That is, if f is continuous and assumes the values $f(a)$ and $f(b)$, then it assumes every value in between. Suppose that f is continuous and increasing with domain $[a, b]$. Prove the following:

(a) Range of $f = [f(a), f(b)]$.
(b) f^{-1} exists with domain $[f(a), f(b)]$.
(c) f^{-1} is an increasing function.

Summary Outline of Chapter 1

- **Real Numbers**

The **natural numbers,** denoted by N, are $1, 2, 3, \ldots$. p. 1
The **integers,** denoted by Z, are $0, \pm 1, \pm 2, \ldots$. p. 1
The **rational numbers,** denoted by Q, are all numbers that can be written p. 1

$$r = \frac{m}{n}$$

where m and n are integers and $n \neq 0$.
The **irrational numbers** consist of all real numbers that are not rational. p. 2

- **Properties of Absolute Value**

$|a| = a$ if $a \geq 0$ p. 5
$|a| = -a$ if $a < 0$ p. 5
$|-a| = |a|$ p. 5
$|ab| = |a||b|$ p. 5
$|a + b| \leq |a| + |b|$ p. 5

- **Intervals**

open interval $(a, b) = \{x: a < x < b\}$ p. 3
closed interval $[a, b] = \{x: a \leq x \leq b\}$ p. 3
half-open interval $[a, b) = \{x: a \leq x < b\}$ p. 3
half-open interval $(a, b] = \{x: a < x \leq b\}$ p. 4
$[a, \infty) = \{x: x \geq a\}$ p. 4
$(a, \infty) = \{x: x > a\}$ p. 4
$(-\infty, a] = \{x: x \leq a\}$ p. 4
$(-\infty, a) = \{x: x < a\}$ p. 4
$(-\infty, \infty) = \mathbb{R}$ = the set of real numbers p. 4
$[0, \infty) = \mathbb{R}^+$, the set of nonnegative numbers p. 4

- **Distance Formula:** The distance d between the points (x_1, y_1) and (x_2, y_2) is given by

$$d = \sqrt{(x_1 - x_2)^2 + (y_1 - y_2)^2}$$ p. 10

- **Graph:** The graph of an equation in two variables is the set of all points in the xy-plane whose coordinates satisfy the equation. p. 11

- **Unit Circle:** The circle centered at (0, 0) with radius 1. Its equation is $x^2 + y^2 = 1$. p. 12
- The equation $(x - h)^2 + (y - k)^2 = r^2$ is the equation of a circle centered at (h, k) with radius r. p. 12

- A **function** f is a rule that assigns to each member of one set (called the **domain** of the function) a unique member of another set (called the **range** of the function). We often write $y = f(x)$. Here y is called the **image** of x. x is called the **independent variable,** and y is the **dependent variable.** p. 15

- **Vertical Lines Test:** An equation in two variables defines a function if every vertical line in the plane intersects the graph of the equation in at most one point. p. 22
- The function $f(x)$ is **even** if $f(-x) = f(x)$ or **odd** if $f(-x) = -f(x)$. p. 23
- The **graph** of a function f is the set of points $\{(x, f(x)): x \in \text{domain of } f\}$. p. 28
- If $f(x)$ is even, then its graph is symmetric about the y-axis. p. 31
- If $f(x)$ is odd, then its graph is symmetric about the origin. p. 31

- **Shifting Graphs**

  The graph of $f(x) + c$ is the graph of $f(x)$ shifted up c units if $c > 0$ or down $|c|$ units if $c < 0$. p. 32
  The graph of $f(x - c)$ is the graph of $f(x)$ shifted c units to the right if $c > 0$ or $|c|$ units to the left if $c < 0$. p. 33
  The graph of $-f(x)$ is the graph of $f(x)$ reflected about the x-axis. p. 34
  The graph of $f(-x)$ is the graph of $f(x)$ reflected about the y-axis. p. 34

- **Function Operations**

  Sum $(f + g)(x) = f(x) + g(x)$ p. 38
  Difference $(f - g)(x) = f(x) - g(x)$ p. 39
  Product $(f \cdot g)(x) = f(x)g(x)$ p. 39
  Quotient $\left(\frac{f}{g}\right)(x) = \frac{f(x)}{g(x)}$ p. 39
  Composition $(f \circ g)(x) = f(g(x))$ p. 40

- **Inverse Function:** f and g are inverse functions if p. 44

  (i) for every $x \in$ domain of g, $g(x) \in$ domain of f and $f(g(x)) = x$
  (ii) for every $x \in$ domain of f, $f(x) \in$ domain of g and $g(f(x)) = x$
- $f(x)$ is **one-to-one** on $[a, b]$ if $f(x_1) = f(x_2)$ implies that $x_1 = x_2$ whenever x_1 and x_2 are in $[a, b]$. A one-to-one function f has an inverse denoted by f^{-1}. pp. 45, 46

- **Horizontal lines test:** f is one-to-one on its domain if every horizontal line in the plane intersects the graph of f in at most one point. p. 46
- f is **increasing** on $[a, b]$ if $f(x_2) > f(x_1)$ when $x_2 > x_1$. p. 47
- f is **decreasing** on $[a, b]$ if $f(x_2) < f(x_1)$ when $x_2 > x_1$. p. 47
- If f is increasing or decreasing on $[a, b]$, then f is one-to-one on $[a, b]$. p. 47

- **Reflection property:** The graphs of f and f^{-1} are reflections of one another about the line $y = x$. p. 49

Review Exercises for Chapter 1

In Exercises 1–5, find the distance between the given points.

1. $(2, -1), (3, 2)$
2. $(4, 0), (0, 4)$
3. $(-1, 1), (1, -1)$
4. $(2, 3), (-4, -5)$
5. $(0, 4), (0, -7)$
6. Find the equation of the circle centered at $(2, -3)$ with radius 4.
7. Show that $x^2 - 6x + y^2 + 10y + 32 = 0$ is the equation of a circle. Find its center and radius.

In Exercises 8–15, determine whether the given equation defines y as a function of x and, if so, find its domain and range.

8. $4x - 2y = 5$
9. $\dfrac{x^2 - y}{2} = 4$
10. $(x - 1)^2 + (y - 3)^2 = 4$
11. $y = \sqrt{x + 2}$
12. $3 = \dfrac{1 + x^2 + x^4}{2y}$
13. $y = \dfrac{x}{x^2 + 1}$
14. $y = \dfrac{x}{x^2 - 1}$
15. $y = \sqrt{x^2 - 6}$
16. For $y = f(x) = \sqrt{x^2 - 4}$, calculate $f(2)$, $f(-\sqrt{5})$, $f(x + 4)$, $f(x^3 - 2)$, and $f(-1/x)$.
17. If $y = f(x) = 1/x$, show that for $\Delta x \neq 0$,

$$\frac{f(x + \Delta x) - f(x)}{\Delta x} = -\frac{1}{x(x + \Delta x)}$$

In Problems 18–21, determine whether the given function is even, odd, or neither.

18. $f(x) = x^2 - 10$
19. $f(x) = \dfrac{1}{x^3}$
20. $f(x) = x^3 + 1$
21. $f(x) = \dfrac{x^2}{1 + x^4}$

In Problems 22–30, sketch the graph of the given function.

22. $f(x) = x^2 - 2$
23. $f(x) = x^3 + 1$
24. $f(x) = |x| - 1$
25. $f(x) = (x - 1)^2$
26. $f(x) = (x + 2)^3$
27. $f(x) = 5 - (x - 1)^3$
28. $f(x) = \dfrac{1}{x + 2}$
29. $f(x) = \begin{cases} 2, & x \geq 0 \\ 3, & x < 0 \end{cases}$
30. $f(x) = \begin{cases} 1, & x \leq 0 \\ x, & 0 < x \leq 2 \\ x^2, & x > 2 \end{cases}$
31. The graph of the function $y = f(x)$ is given below. Sketch the graph of $f(x - 3)$, $f(x) - 5$, $f(-x)$, $-f(x)$, and $4 - f(1 - x)$.

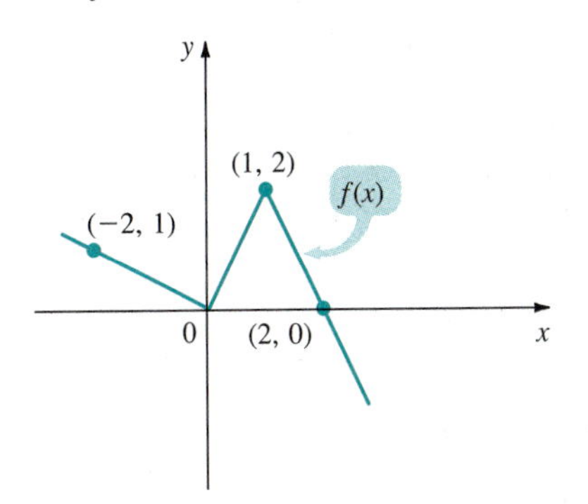

32. Repeat Problem 31 for the function graphed below.

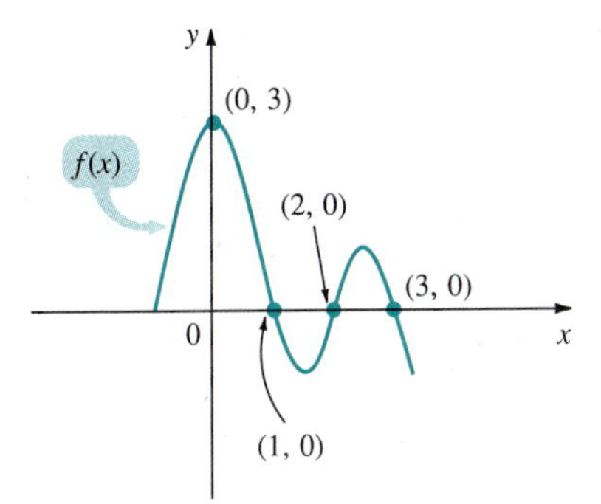

33. Let $f(x) = \sqrt{x + 1}$ and $g(x) = x^3$. Find $f + g$, $f - g$, $f \cdot g$, g/f, $f \circ g$, and $g \circ f$, and determine their respective domains.
34. Repeat Problem 33 for $f(x) = 1/x$ and $g(x) = x^2 - 4x + 3$.

In Exercises 35–40, explain why the given function is one-to-one on its domain and find its inverse.

35. $f(x) = 4x - 1$
36. $f(x) = -2x + 5$
37. $f(x) = \dfrac{2}{x}$
38. $f(x) = 3x^3 - 1$
39. $f(x) = \sqrt{x - 2}$
40. $f(x) = \sqrt[3]{x + 3}$

In Exercises 41 and 42 determine intervals over which each function is one-to-one and determine the inverse function over each such interval.

41. $f(x) = 3 + 2x^2$
42. $f(x) = x^2 + 4x + 1$
43. The graph of a function over an interval $[a, b]$ is given. Sketch the graph of f^{-1} on the interval $[f(a), f(b)]$.

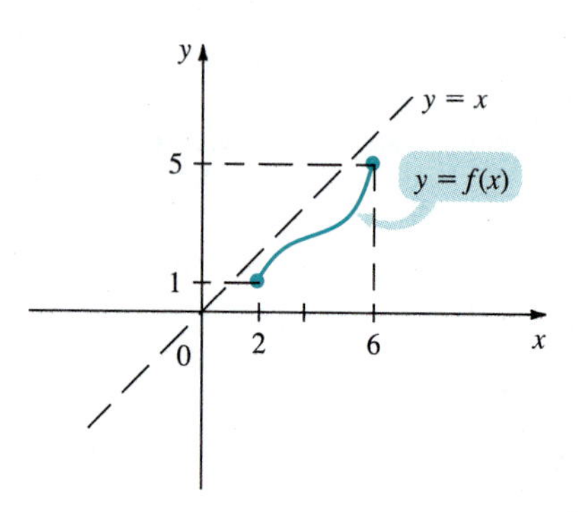

Chapter 2

The Trigonometric Functions

The word *trigonometry* comes from the Greek words *trigonon,* meaning ''triangle,'' and *metria,* meaning ''measurement.'' Originally trigonometry dealt with the measurement of the sides and angles of triangles and was developed to solve problems in astronomy. In fact, while results in astronomy date back over 3500 years, it wasn't until the thirteenth century that trigonometry and astronomy were considered separate subjects.

One of the earliest known uses of trigonometry is an Egyptian table that shows the relationship between the time of day and the length of the shadow cast by a vertical stick. The Egyptians knew that this shadow was longer in the morning, decreased to a minimum at noon, and increased thereafter until sundown. The rule that gives time of day as a function of shadow length is a forerunner of the tangent and cotangent functions (trigonometric functions) we study today.

In this chapter and the next two, we will describe the six trigonometric functions (sine, cosine, tangent, cotangent, secant, and cosecant) and show how they can be used in a wide variety of applications. We will show, too, how trigonometry involves a great deal more than the study of triangles.

Since trigonometry involves, first of all, the study of angles, we begin our discussion with a description of how angles are measured.

2.1 Degree and Radian Measure of an Angle

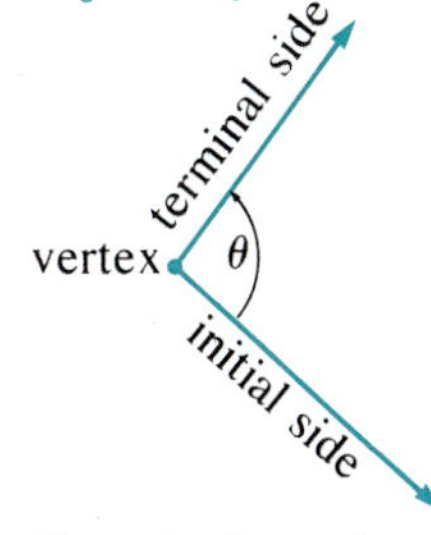

Figure 1 An angle.

An **angle** consists of two half-lines (rays) with a common initial point. The initial point is called the **vertex** of the angle; one half-line is called the **initial side,** and the other half-line is called the **terminal side** (see Figure 1).

We will usually denote angles by lowercase Greek letters. The most frequently used letters are α (alpha), β (beta), γ (gamma), δ (delta), and θ (theta). Of these, the most commonly used symbol for an angle is θ. Angles can be placed anywhere in the plane. However, for uniformity of measurement, we will usually place an angle in the **standard position** with its vertex at the origin and its initial side along the positive x-axis.

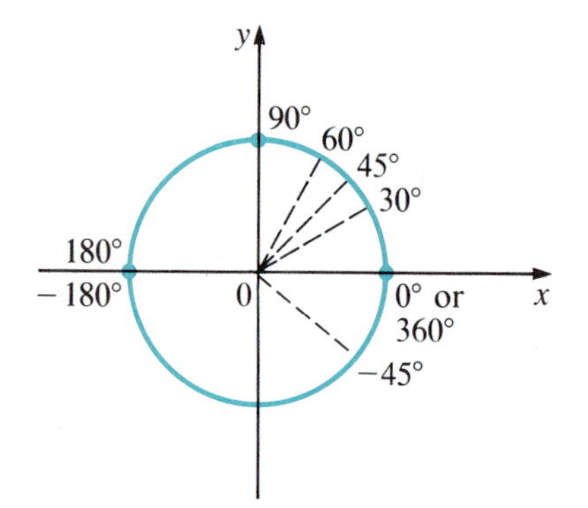

Figure 2 Some angles (measured in degrees).

Degree Measure of an Angle

Let an angle be in standard position. It is said to have the measure one **degree,** written 1°, if the angle is obtained by rotating its terminal side $\frac{1}{360}$ of a complete revolution in the positive (counterclockwise) direction. Thus, an angle obtained from one complete counterclockwise revolution has a measure of 360°; an angle obtained from half a complete counterclockwise revolution has a measure of 180°; an angle obtained from one quarter of a complete counterclockwise revolution has a measure of 90°, and so on. An angle obtained from half a complete revolution in the clockwise (negative) direction has a measure of −180°. If the terminal side is not rotated so that the initial and terminal sides coincide, then the angle has measure zero degrees, written 0°. Some angles are depicted in Figure 2.

An angle of **positive measure** is obtained when the terminal side is rotated counterclockwise from the initial side. An angle of **negative measure** is obtained when the terminal side is rotated clockwise from the initial side (see Figure 3).

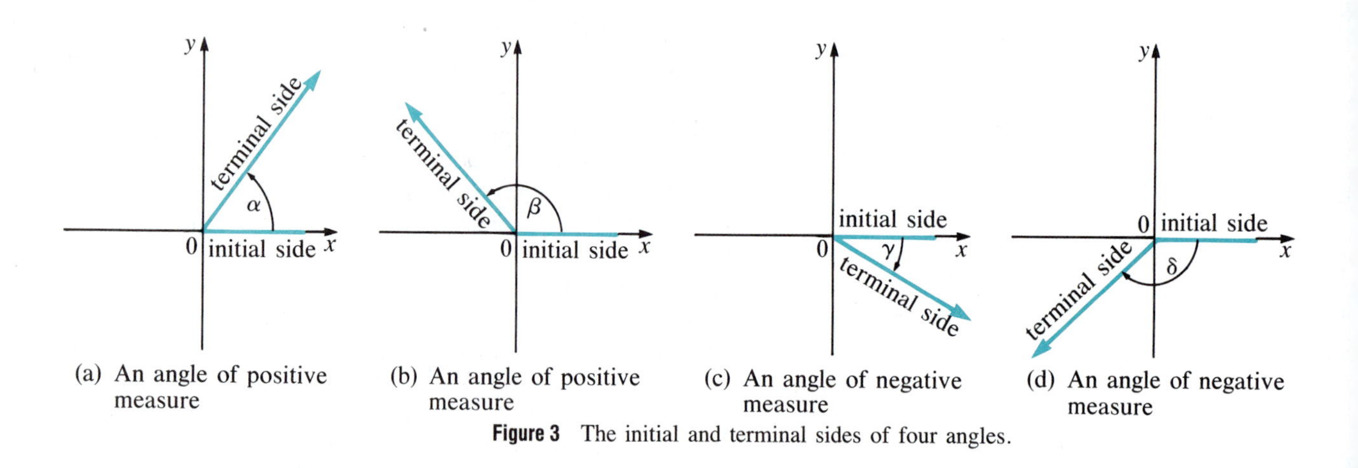

(a) An angle of positive measure

(b) An angle of positive measure

(c) An angle of negative measure

(d) An angle of negative measure

Figure 3 The initial and terminal sides of four angles.

An Important Distinction

There is a difference between an angle, which is a geometrical object, and its measure, which is a number of degrees (or radians). However, being perfectly precise sometimes leads to clumsy language. It is correct, but tedious, to say each time that "an angle has a measure of x degrees." Therefore, from now on, we will refer to an angle of 30° or an angle of −45°, for example, instead of saying "an angle of measure 30° or −45°." We will also refer to "positive angles" and "negative angles" rather than angles of positive or negative measure. However, you should always keep in mind that an angle and its measure are different things.

In Figure 4, we illustrate eight angles. Since each angle can be obtained as a rotation, we draw a circle representing 360° as a reference.

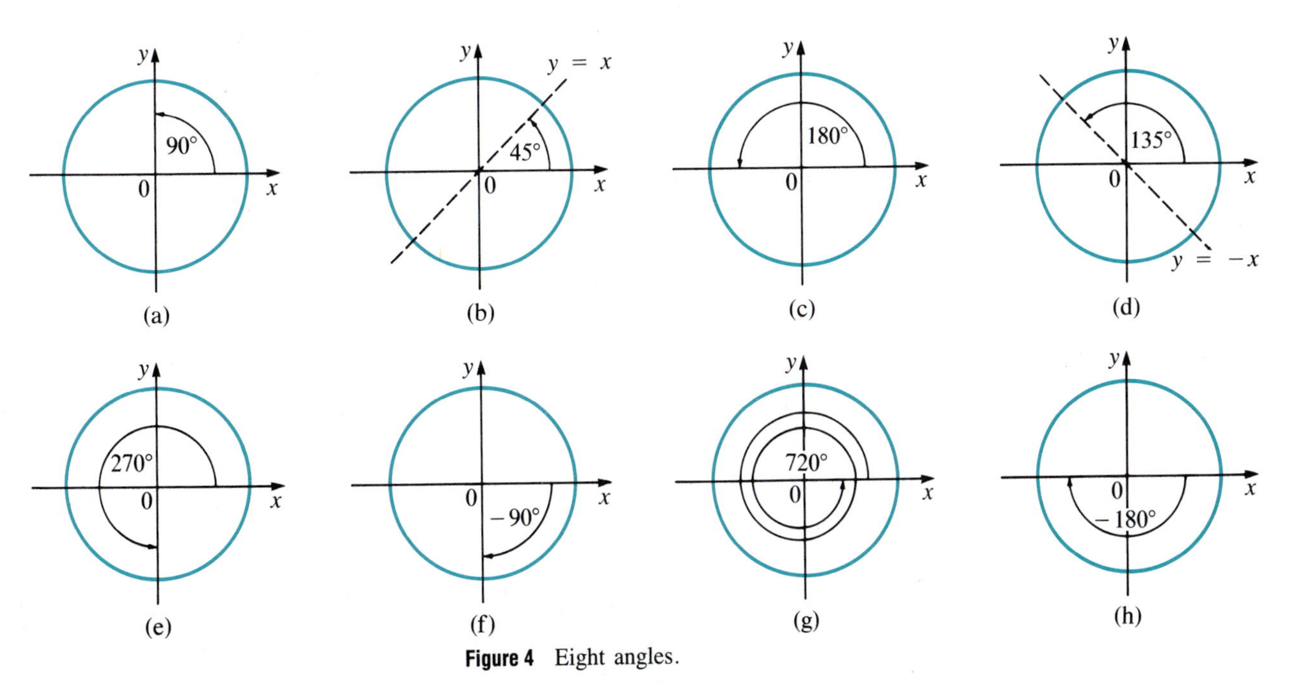

Figure 4 Eight angles.

Some Special Angles

In Figure 5, we illustrate six types of special angles.

a. A **right angle** is an angle of 90°.
b. A **straight angle** is an angle of 180°.
c. θ is an **acute angle** if $0 < \theta < 90°$.
d. θ is an **obtuse angle** if $90° < \theta < 180°$.
e. Two acute angles, θ_1 and θ_2, are **complementary** if $\theta_1 + \theta_2 = 90°$. We say that θ_2 is the **complement** of θ_1, and vice versa.
f. Two positive angles, θ_1 and θ_2, are **supplementary** if $\theta_1 + \theta_2 = 180°$. We say that θ_2 is the **supplement** of θ_1, and vice versa.

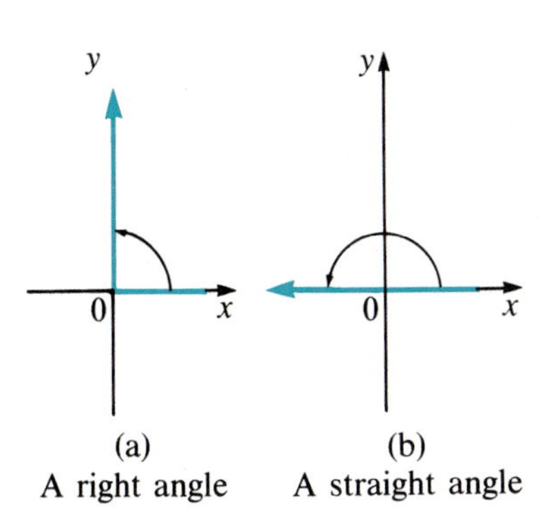

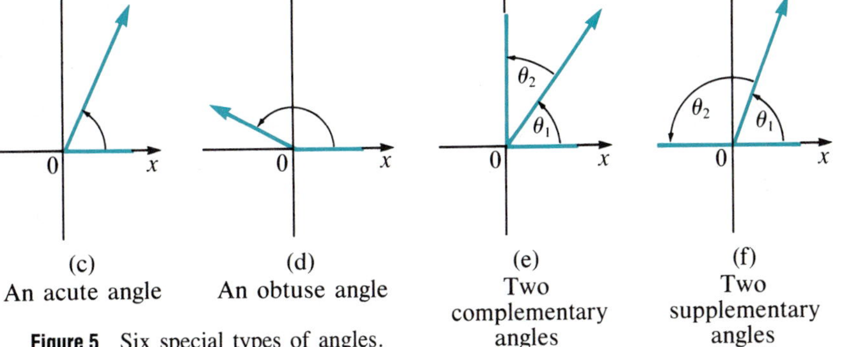

Figure 5 Six special types of angles.

EXAMPLE 1 *Finding the Complement and Supplement of an Angle*

Find (a) the complement of 32° and (b) the supplement of 32°.

SOLUTION Let $c°$ denote the measure of the complement and $s°$ the measure of its supplement.

(a) $32° + c° = 90°$, so

$$c° = \text{complement of } 32° = 90° - 32° = 58°$$

(b) $32° + s° = 180°$, so

$$s° = \text{supplement of } 32° = 180° - 32° = 148°$$

Recall that the x- and y-axes divide the plane into four quadrants. From Figure 6, we have Table 1.

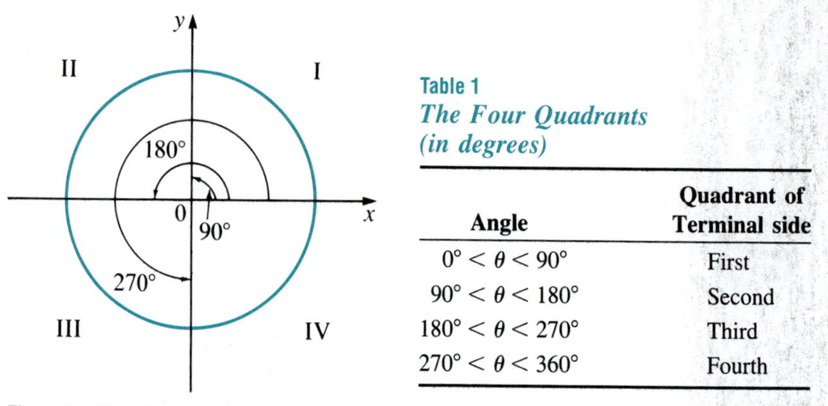

Table 1
The Four Quadrants (in degrees)

Angle	Quadrant of Terminal side
$0° < \theta < 90°$	First
$90° < \theta < 180°$	Second
$180° < \theta < 270°$	Third
$270° < \theta < 360°$	Fourth

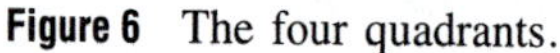

Figure 6 The four quadrants.

Two angles θ_1 and θ_2 are said to be **coterminal** if they have the same vertex and the same initial and terminal sides. We see from Figure 7 that 30°,

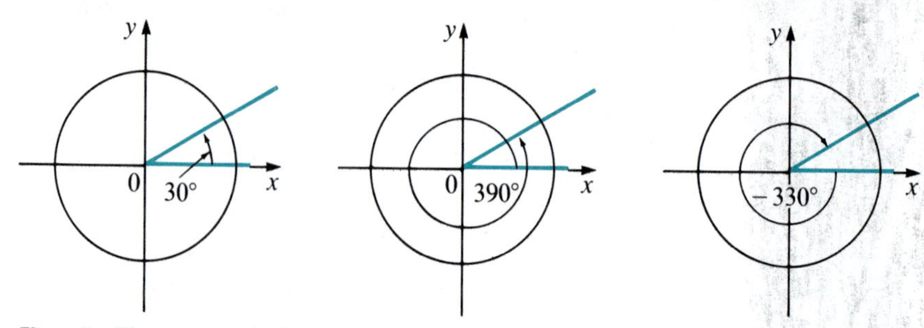

Figure 7 Three coterminal angles.

390°, and −330° are coterminal. In fact, since rotating a ray 360° (one complete revolution) brings us back to the same ray, we see that

$$\theta,\ \theta + 360°,\ \text{and}\ \theta - 360°\ \text{are coterminal for every angle}\ \theta.$$

EXAMPLE 2 *Finding Coterminal Angles*

Find two positive angles and two negative angles that are coterminal with $\theta = 45°$.

SOLUTION $45° + 360° = 405°$ and $45° - 360° = -315°$ are coterminal with 45°. In addition, $405° + 360° = 765°$ and $-315° - 360° = -675°$ are coterminal with 45°. We can obtain more coterminal angles by continuing to add or subtract additional multiples of 360°.

There are two ways to write parts of a degree. The first is to use a fraction or a decimal.† For example, 37.4° = 37° + four tenths of a degree. However, the Greeks (and the Babylonians before them) found it more convenient to divide each degree into 60 parts, called **minutes,** and each minute into 60 parts, called **seconds.** Minutes and seconds are used today in navigation. Minutes are written with a prime, seconds with two primes. For example, 23 minutes = 23′, and 47 seconds = 47″.

$$1\ \text{minute} = 1' = \frac{1}{60}^{\circ} \quad \text{and} \quad 1\ \text{second} = 1'' = \frac{1}{60}'$$

$$1° = 60' \qquad\qquad 1' = 60''$$

EXAMPLE 3 *Converting the Decimal Part of a Degree to Minutes and Seconds*

Convert 42.43° into degrees, minutes, and seconds.

SOLUTION $42.43° = 42° + 0.43°$. We ask, 0.43° is what fraction of $1° = 60'$? That is, $0.43° = \dfrac{x}{60}$ minutes, so

$$x = 60 \times 0.43\ \text{minutes} = 25.8'$$

Similarly,

$$0.8' = (60 \times 0.8)\ \text{seconds} = 48''$$

Thus

$$42.43° = 42° + 25' + 48'' = 42°25'48''$$

†Most hand-held calculators represent the fractional part of a degree by a decimal.

EXAMPLE 4 *Converting Minutes and Seconds to a Decimal*

Write $121°37'08''$ as a decimal. Give your answer correct to 4 decimal places.

SOLUTION $121°37'08'' = 121° + 37' + 8''$. But

$$1' = \frac{1}{60}^{\circ} \quad \text{so} \quad 37' = \frac{37}{60}^{\circ} \approx 0.616667°$$

Similarly

$$1'' = \frac{1}{60}' = \frac{1}{60}\left(\frac{1}{60}\right)^{\circ} = \frac{1}{3600}^{\circ} \quad \text{so} \quad 8'' = \frac{8}{3600}^{\circ} \approx 0.002222°$$

We add:

$$\begin{aligned} 121°37'08'' = \; & 121.0000 \\ & +0.616667 \\ & \underline{+0.002222} \\ & 121.618889 \end{aligned}$$

The answer, correct to 4 decimal places, is 121.6189°.

In problems when, as here, you are asked to round to a certain number of decimal places, it is better to carry all calculations to at least two more decimal places and round at the end. This reduces the roundoff error at each intermediate step.

We summarize the results of Examples 3 and 4.

To Convert an Angle x Between 0° and 1° to Minutes and Seconds

(1) Multiply by 60. This gives the number of minutes.
(2) Subtract the whole number of minutes and multiply the result by 60. This gives the number of seconds.

To Convert Minutes and Seconds to Degrees

(1) Divide seconds by 60 to obtain a fraction of one minute.
(2) Add this to the number of minutes and divide the result by 60 to obtain a fraction of one degree:

$$\frac{1}{60}\left[\frac{\text{seconds}}{60} + \text{minutes}\right] = \text{fraction of one degree}$$

In most places in this book, we will represent degrees by the decimal notation. When we round, decimals will usually be rounded to 4 places, and minutes and seconds will be rounded to the nearest minute or second.

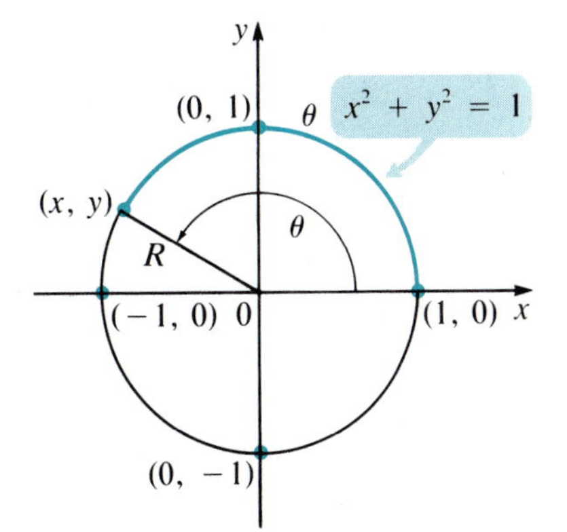

Figure 8 The unit circle. The radian measure of θ is the length of the arc extending from (1, 0) to (x, y).

Radian Measure

There is another way to measure angles which, in many applications, is more useful than measurement in degrees. Recall from page 12 that the unit circle is the circle centered at the origin with radius 1. The coordinates of every point on the unit circle satisfy the equation $x^2 + y^2 = 1$.

Definition of Radian Measure

In Figure 8, we have drawn the unit circle. Let R denote the radial line segment that makes an angle of θ with the positive x-axis. That is, θ is the angle whose initial side is the positive x-axis and whose terminal side contains R. Let (x, y) denote the point at which this radial line intersects the unit circle. Then the **radian measure** of the angle is equal to the numerical value of the length of the arc of the unit circle from the point (1, 0) to the point (x, y).

Since the circumference of a circle is $2\pi r$, where r is its radius, the circumference of the unit circle $(r = 1)$ is 2π. Thus

$$360^\circ = 2\pi \text{ radians} \tag{1}$$

The equality here means that an angle of degree measure 360° has radian measure 2π. Then, since $180^\circ = \frac{1}{2}(360^\circ)$, $180^\circ = \frac{1}{2}(2\pi) = \pi$ radians. In general, θ (in degrees) is to 360° as θ (in radians) is to 2π radians. Thus

$$\frac{\theta \text{ in degrees}}{360^\circ} = \frac{\theta \text{ in radians}}{2\pi}, \text{ or}$$

Converting Between Degrees and Radians

$$\theta \text{ (degrees)} = \frac{180}{\pi}\,\theta \text{ (radians)} \tag{2}$$

and

$$\theta \text{ (radians)} = \frac{\pi}{180}\,\theta \text{ (degrees)} \tag{3}$$

EXAMPLE 5 *Converting from Degrees to Radians*

Convert 45° and 270° from degrees to radians.

SOLUTION From (3),

$$\text{radian measure of } 45^\circ = \frac{\pi}{180} \times 45 = \pi \times \frac{45}{180} = \pi \times \frac{1}{4} = \frac{\pi}{4} \text{ radians}$$

$$\text{radian measure of } 270^\circ = \frac{\pi}{180} \times 270 = \pi \times \frac{270}{180} = \pi \times \frac{3}{2} = \frac{3\pi}{2} \text{ radians}$$

We can calculate that 1 radian $= 180/\pi \approx 57.3°$ and $1° = \pi/180 \approx 0.0175$ radians. The degree and radian measures of some standard angles are given in Table 2.

Table 2

θ (degrees)	0	90	180	270	360	45	30	60	-90	135	120	720
θ (radians)	0	$\frac{\pi}{2}$	π	$\frac{3\pi}{2}$	2π	$\frac{\pi}{4}$	$\frac{\pi}{6}$	$\frac{\pi}{3}$	$\frac{-\pi}{2}$	$\frac{3\pi}{4}$	$\frac{2\pi}{3}$	4π

Radian measure is more convenient when discussing trigonometric functions that arise in applications having nothing at all to do with angles (see, for example, Section 3.7). In addition, calculus *requires* radian measure for all angles.

In Table 1, we saw that different angles may have their terminal sides in different quadrants. Table 3 repeats Table 1 with radian measure instead of degree measure.

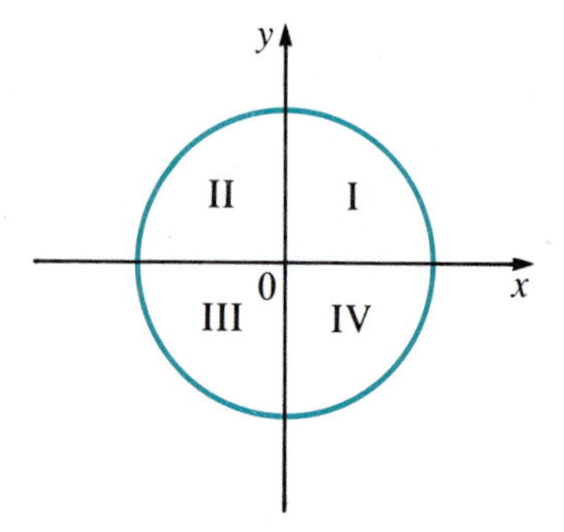

Figure 9 The four quadrants.

Table 3
Angles in the Four Quadrants (in radians)

Angle	Quadrant of Terminal Side	Decimal Approximation
$0 < \theta < \frac{\pi}{2}$	First	$0 < \theta < 1.5708$
$\frac{\pi}{2} < \theta < \pi$	Second	$1.5708 < \theta < 3.1416$
$\pi < \theta < \frac{3\pi}{2}$	Third	$3.1416 < \theta < 4.7124$
$\frac{3\pi}{2} < \theta < 2\pi$	Fourth	$4.7124 < \theta < 6.2832$

EXAMPLE 6 *Converting from Degrees to Radians*

Convert from degrees to radians. Round to 4 decimal places.

(a) 67° (b) 275° (c) −100°

SOLUTION In (3), we use the value for π given in a calculator: $\pi = 3.141592654$.

$$\text{(a) } \theta \text{ (radians)} = \frac{\pi}{180} \times 67 = \frac{67}{180}\pi \text{ radians} \approx 0.372222222\pi$$
$$= 1.169370599 \approx 1.1694 \text{ radians}$$

(b) θ (radians) $= \dfrac{\pi}{180} \times 275 = \dfrac{275}{180}\pi$ radians $\approx 1.527777778\pi$
$= 4.799655443 \approx 4.7997$ radians

(c) θ (radians) $= \dfrac{\pi}{180} \times (-100) = -\dfrac{100}{180}\pi$ radians $= -\dfrac{5}{9}\pi \approx$
$-0.55555556\pi \approx -1.745329252 \approx -1.7453$ radians ■

EXAMPLE 7 *Converting from Radians to Degrees*

Convert from radians to degrees. Round to 4 decimal places of accuracy.

(a) $\frac{1}{2}$ radian (b) -1.2 radians (c) 16.5 radians

SOLUTION We use equation (2) and the value $\pi \approx 3.141592654$.

(a) θ (degrees) $= \dfrac{180}{\pi} \times \dfrac{1}{2} = \dfrac{90^\circ}{\pi} \approx 28.6479^\circ$

(b) θ (degrees) $= \dfrac{180}{\pi} \times (-1.2) = -\dfrac{216^\circ}{\pi} \approx -68.7549^\circ$

(c) θ (degrees) $= \dfrac{180}{\pi} \times 16.5 = \dfrac{2970^\circ}{\pi} \approx 945.3804^\circ$

NOTATION FOR DEGREES AND RADIANS The symbol ° is used to denote degrees. However, no symbol is commonly used to denote radians. The custom is to denote 45 degrees by 45° but $\pi/4$ radians simply by $\pi/4$. Thus, if an angle is written without a degree symbol, then you can assume that the angle is given in radians.

Let C_r denote the circle of radius r centered at the origin (see Figure 10). If OP denotes a radial line segment as pictured in the figure, then OP cuts an arc from C_r of length L. Let θ be the positive angle from the positive x-axis to OP. If $\theta = 360^\circ$, then $L = 2\pi r$. If $\theta = 180^\circ$, then $L = \pi r$. In fact, it follows from the figure that θ is the same fraction of 360° that L is of the whole circumference; that is,

$$\frac{\theta^\circ}{360^\circ} = \frac{L}{2\pi r} \tag{4}$$

or

$$\theta^\circ = 360^\circ \frac{L}{2\pi r} \tag{5}$$

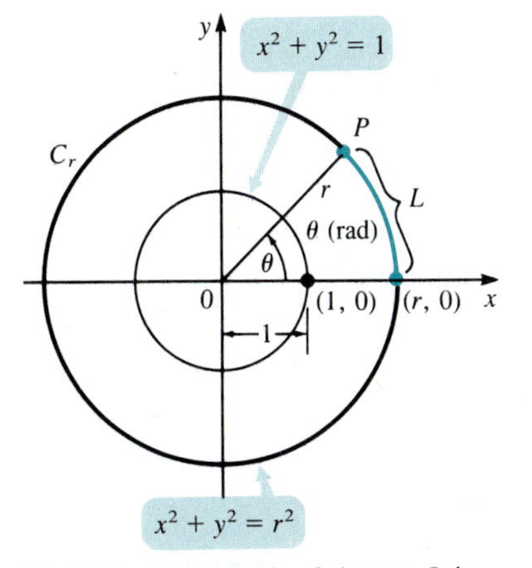

Figure 10 The length of the arc L is equal to $r\theta$, where θ is measured in radians.

If we measure θ in radians, then 2π radians replaces 360° and (4) becomes

$$\frac{\theta}{2\pi} = \frac{L}{2\pi r} \tag{6}$$

or

$$\theta = \frac{2\pi L}{2\pi r} = \frac{L}{r} \tag{7}$$

Finally, rewriting (7), we obtain

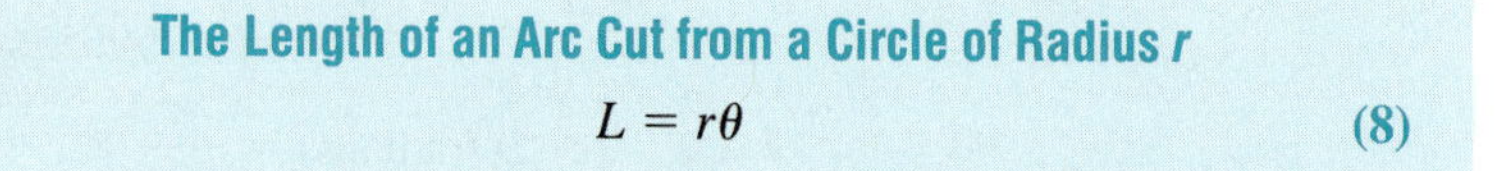

The Length of an Arc Cut from a Circle of Radius r

$$L = r\theta \tag{8}$$

That is, if θ is measured in radians, then the angle θ "cuts" from the circle of radius r centered at the origin an arc of length $r\theta$. Note that if $r = 1$, then (8) reduces to $L = \theta$, which is the definition of the radian measure of an angle.

EXAMPLE 8 *Computing the Length of an Arc*

What is the length of an arc cut from the circle of radius 4 cm centered at the origin by an angle of (a) 45°, (b) 60°, (c) 270°?

SOLUTION From (8), we find that $L = 4\theta$ cm, where θ is the radian measure of the angle. We therefore convert each degree measure to radian measure and then multiply the result by $r = 4$.

(a) $L = 4 \cdot \dfrac{\pi}{4} = \pi$ cm

(b) $L = 4 \cdot \dfrac{\pi}{3} = \dfrac{4\pi}{3}$ cm

(c) $L = 4 \cdot \dfrac{3\pi}{2} = 6\pi$ cm

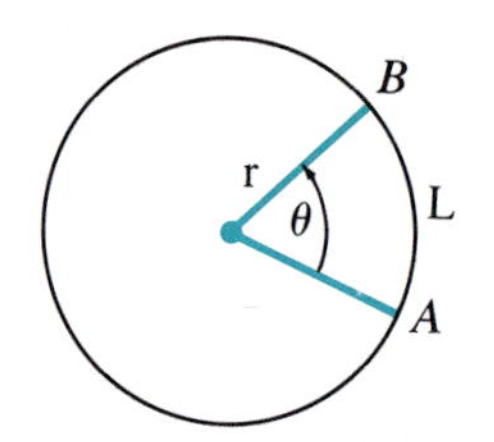

Figure 11 A central angle.

A **central angle** θ of a circle is an angle whose vertex is at the center of the circle (see Figure 11). We say that θ **subtends** the arc AB as in Figure 11. The computations that led to equation (8) are unchanged if the circle is not drawn on the x- and y-axes. Therefore, using (8), we conclude that

Radian Measure of a Central Angle

If a central angle θ subtends (cuts off) an arc of length L in a circle of radius r, then the radian measure of θ is given by

$$\theta = \frac{L}{r} \tag{9}$$

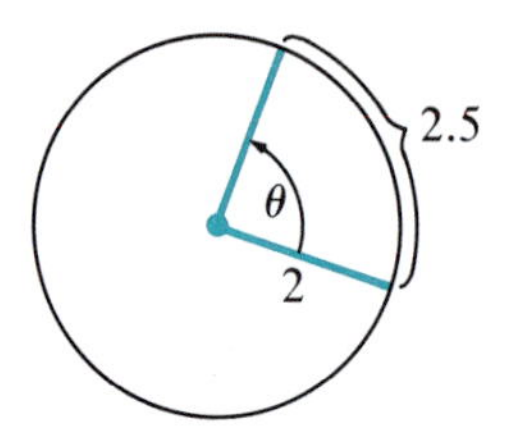

Figure 12 A central angle.

EXAMPLE 9 *Finding a Central Angle*

Find the radian and degree measures of a central angle that subtends an arc of length 2.5 in from a circle of radius 2 in. (See Figure 12.)

SOLUTION From (9),

$$\theta \text{ (radians)} = \frac{L}{r} = \frac{2.5 \text{ in}}{2 \text{ in}} = 1.25$$

Then

$$\theta \text{ (degrees)} = \frac{180^\circ}{\pi}\theta \text{ (radians)} = \frac{180^\circ}{\pi} \times 1.25$$
$$\approx 71.6197^\circ$$

As before, we carried all computations to 4 decimal places.

Circular Motion

There are many situations in which an object moves on a circle. Two examples are a point on the rim of a flywheel and a point on the edge of a phonograph record. The most common unit of circular motion is **revolutions per minute,** abbreviated **rpm.** In one revolution, a radial line sweeps out an angle of 2π radians. We define the **angular speed,** ω, to be the angle swept out per unit time. (ω is the lowercase Greek omega.)

$$\text{in 1 rpm, } \omega = 2\pi \text{ radians/minute}$$
$$\text{in 2 rpm, } \omega = 2(2\pi) = 4\pi \text{ radians/minute}$$

Angular Speed

If an object is moving on a circle at n rpm, then

$$\text{angular speed} = \omega = 2\pi n \text{ radians/minute} \tag{10}$$

The **linear speed,** v, of an object is the distance traveled by the object in one unit of time. If the radius of the circle is r, then the object travels a distance of $2\pi r$ in one revolution. Thus

Linear Speed

If an object is moving on a circle at n rpm, then

$$\text{linear speed} = v = 2\pi nr \text{ units/minute} \tag{11}$$

If r is measured in feet, then $v = 2\pi nr$ ft/min. If r is measured in meters, then $v = 2\pi nr$ m/min, and so on.

EXAMPLE 10 *Computing Angular and Linear Speed*

A bicycle wheel of diameter 26 in rotates 400 times each minute. Find the angular and linear speed of the wheel.

SOLUTION

(a) Here $n = 400$ rpm, so, from (10),

$$\omega = 2\pi(400) = 800\pi \approx 2513.27 \text{ radians/minute}$$

(b) The radius is half the diameter, so $r = 13$ in and, from (11),

$$v = 2\pi n(13) = (800\pi)(13) \approx 32{,}672.56 \text{ in/min}$$

This is the answer. However, inches per minute is not a very common way to write a speed. We convert to miles per hour in three steps:

1. Divide by 12 to get ft/min.
2. Divide by 5280 to get miles/min.
3. Multiply by 60 to get miles/hr.

We compute $$\frac{32{,}672.56}{(12)(5280)}(60) = 30.94 \text{ miles/hr}$$

Problems 2.1

Readiness Check

I. Which of the following is true about an angle θ in standard position?
   a. Its terminal side must be in quadrant I.
   b. Its rotation must be counterclockwise.
   c. Its vertex must be on the positive x-axis.
   d. Its initial side must be on the positive x-axis.

II. Which of the following is true about an angle $\theta = -100°$ drawn in standard position?
   a. Its radian measure is $\frac{5\pi}{9}$.
   b. Its terminal side is in quadrant III.
   c. $-260°$ is coterminal to θ.
   d. It is a straight angle.

III. Which of the following pairs of angles are coterminal with an angle of $5\pi/4$ radians?
   a. $\frac{\pi}{4}, -\frac{7\pi}{4}$ b. $\frac{3\pi}{4}, -\frac{13\pi}{4}$
   c. $\frac{13\pi}{4}, -\frac{3\pi}{4}$ d. $\frac{7\pi}{4}, -\frac{\pi}{4}$

IV. Which of the following is true about the angle β shown in the figure?

   a. $-\pi < \beta < -\frac{\pi}{2}$
   b. $\pi < \beta < \frac{3\pi}{2}$
   c. $\frac{\pi}{2} < \beta < \pi$
   d. $-\frac{3\pi}{2} < \beta < -\pi$

V. Which of the following is the point of intersection of the terminal side of $\theta = -\pi/2$ and the unit circle?
   a. $(1, 0)$ b. $(0, 1)$ c. $(-1, 0)$ d. $(0, -1)$

Answers to Readiness Check

I. d II. b III. c IV. a V. d

In Problems 1–14 determine the quadrant of the given angle in standard position. Measurement is in radians unless a degree sign (°) is present.

1. 50°	2. 100°	3. 200°	4. 300°	5. 1
6. 2	7. 3	8. 4	9. 5	10. 6
11. 176°	12. 4.8	13. 2.18	14. 266°	

In Problems 15–26 find a positive angle and a negative angle that are coterminal with the given angle. If the given angle is written in degrees, write your answers in degrees; if it is written in radians, write your answers in radians.

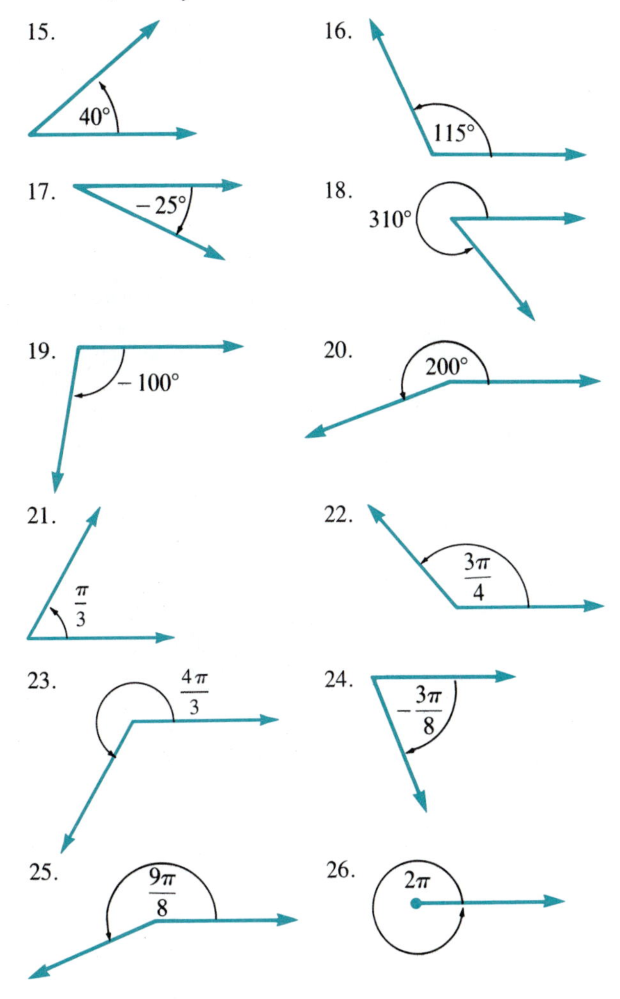

In Problems 27–32 find the complement of each angle.

27. 47°	28. 3°	29. 66.6°
30. 88.3°	31. 46°15′	32. 58°37′18″

In Problems 33–38 find the supplement of each angle.

33. 21°	34. 78.4°	35. 110°
36. 174°	37. 48°22′	38. 102°05′21″

In Problems 39–42 write each angle using degrees, minutes, and seconds. Round to the nearest second.

39. 42.35°	40. 61.318°
41. 121.425°	42. 213.8156°

In Problems 43–48 write each angle as a decimal. Round to 4 decimal places.

43. 24°30′15″	44. 100°45′30″	45. −20°10′50″
46. 231°47′19″	47. −48°17′39″	48. 311°09′02″

In Problems 49–54 write each angle as a multiple of π radians. $\left[\text{For example, } 30° = \frac{\pi}{6} \text{ radians.}\right]$

49. 10°	50. 75°	51. 150°
52. 305°	53. 540°	54. 900°

In Problems 55–58 convert from degrees to radians. Round to 4 decimal places.

55. 27° 56. 137° 57. −71° 58. 297°

In Problems 59–70 convert from radians to degrees. Round to 4 decimal places.

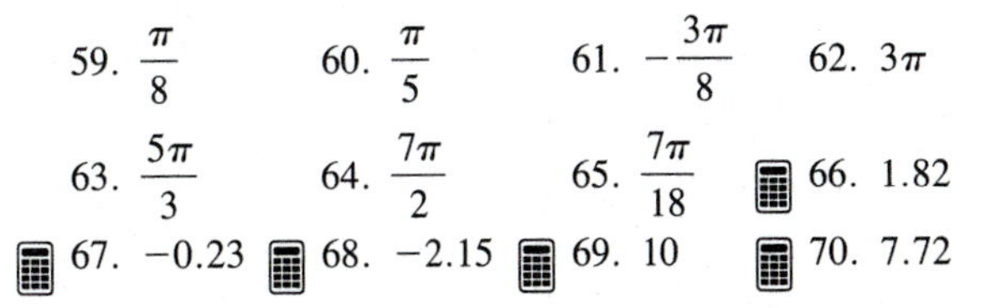

59. $\frac{\pi}{8}$ 60. $\frac{\pi}{5}$ 61. $-\frac{3\pi}{8}$ 62. 3π

63. $\frac{5\pi}{3}$ 64. $\frac{7\pi}{2}$ 65. $\frac{7\pi}{18}$ 66. 1.82

67. −0.23 68. −2.15 69. 10 70. 7.72

In Problems 71–80 an arc in a circle of radius r subtends an angle θ. Find the length of the arc.

71. $\theta = \frac{\pi}{3}$, $r = 1$ cm 72. $\theta = \frac{3\pi}{4}$, $r = 1$ cm

73. $\theta = 2.8$, $r = 1$ m 74. $\theta = \frac{\pi}{8}$, $r = 3$ ft

75. $\theta = \frac{7\pi}{3}$, $r = 4$ in 76. $\theta = 0.75$, $r = 12$ m

77. $\theta = 75°$, $r = 2$ ft [Hint: First convert to radians.]

78. $\theta = 122°$, $r = 3.5$ mm 79. $\theta = 200°$, $r = 7$ m

80. $\theta = 5°$, $r = 100$ in

In Problems 81–84 find the radian and degree measure of a central angle that subtends an arc of length L from a circle of radius r.

81. $L = 5$ cm, $r = 3$ cm
82. $L = \frac{1}{2}$ m, $r = 4$ m
83. $L = 15$ in, $r = 2.5$ in
84. $L = 2.7$ ft, $r = 3.8$ ft

Problems 85–102 require the use of a calculator.

85. A record 7 inches in diameter is rotating on a turntable at a rate of 45 rpm.
    (a) What is the angular speed of the record?
    (b) What angle is swept through in 3 seconds?
    (c) What is the linear speed of a point on the rim?
    (d) What is the distance traveled by the point in 3 seconds?
86. A rock, attached to the end of a rope 2 meters long, is swung in a circle at a rate of 40 rpm.
    (a) What is the angular speed of the rock?
    (b) What is its linear speed?
    (c) How far does the rock travel in 8 seconds?
87. Answer the questions in Problem 86 if the rope is $\frac{3}{4}$ m long.
88. A truck tire has a diameter of 54 in. If the tire is rotating 750 rpm, how fast is the truck moving in miles per hour?
*89. If the truck in Problem 88 is moving at 55 miles/hr, how many times is one of its tires rotating each minute?
90. What is the smaller angle (in radians) between the hour hand and the minute hand of a clock at (a) 5 P.M.? (b) 6:30 P.M.?
91. The minute hand of a clock is 4 in long. How far does the end of the minute hand travel in 2 hours?
*92. The pendulum of a grandfather clock is 2 ft long and swings in a circular arc. It takes 1 second to swing from one end of the arc to the other. In this time, the end of the pendulum moves 3 feet. What is the measure of the angle it sweeps out in 1 second?

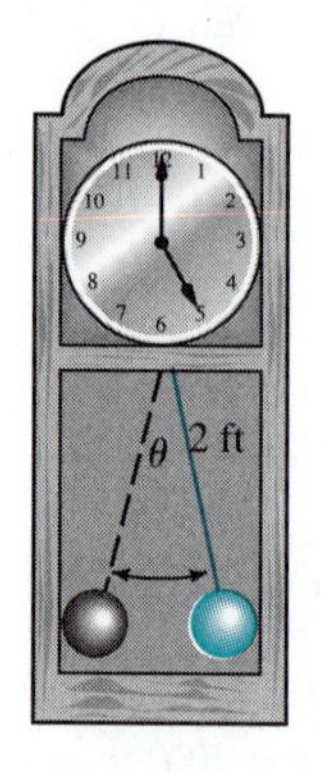

93. A paddle wheel with a radius of $1\frac{1}{2}$ meters is partially immersed in a flowing river and connected to a power generator. The wheel rotates 12 times each minute. Because of resistance, the wheel moves at half the speed of the current. What is the speed of the current in m/min? Convert your answer to km/hr.
94. The moon revolves around the earth in an elliptical orbit every 27.32 days. Its closest distance to the earth (the *perigee*) is 221,463 miles, and the maximum distance (the *apogee*) is 252,710 miles. The average distance is 238,857 miles. In this problem, we treat the moon's orbit as circular with a radius of 238,857 miles and center at the center of the earth.
    (a) What is the angle swept out by a line joining the earth and moon in 10 days?
    (b) in 60 days?
    (c) What is the linear speed of the moon in relation to the earth in miles/hr?
    (d) How far does the moon travel around the earth in 30 days?
    (e) in one year?
95. Ganymede is the largest moon of Jupiter. Its average distance from Jupiter is 1,169,820 miles and its period of revolution is 16.69 days.
    (a) What is the angle swept out by a line joining Jupiter and Ganymede in 5 days?
    (b) What is the linear speed of Ganymede around Jupiter?
    (c) How far does it travel around Jupiter in one of Jupiter's years? [Hint: Jupiter revolves about the sun once every 11.86 years.]
96. A belt rotates, without slipping, around a pulley with a diameter of 3 feet. How many radians has the pulley rotated when the belt has moved 12 feet around it?
97. For purposes of assigning latitude, the earth is thought of as a sphere with radius approximately equal to 3960 miles. Through a point on the earth, draw a circle (called a **great circle**) whose center is at the center of the earth. The angle between the plane of this circle and the plane of the equator is called the **latitude** of the point. The latitude of New York City is 40°45′N. How far is New York City from the North Pole?

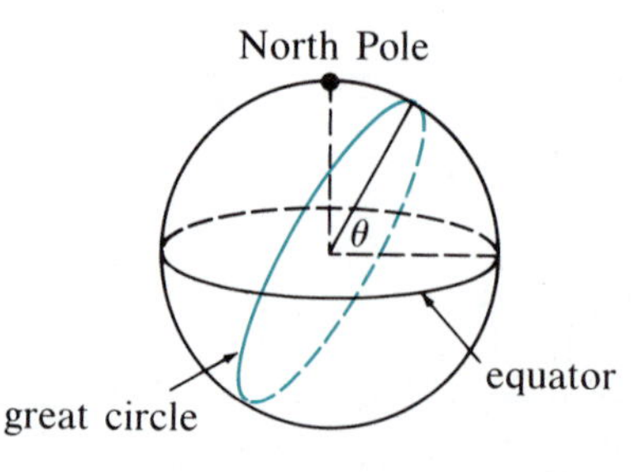

98. How far is Miami, Florida (latitude 25°47′), from the North Pole?

99. How far is Santa Fe, New Mexico (latitude 35°41′), from the equator?
100. How far is Seattle, Washington (latitude 47°37′), from the South Pole?
101. City A is 500 miles due north of city B. What is the difference in their latitudes?
102. A **nautical mile** is defined as the length of an arc on a great circle on the earth whose central angle is 1 minute. What is the length (in feet) of a nautical mile, assuming that the radius of the earth is 3960 miles?

2.2 Trigonometric Functions of Acute Angles

There are two ways to define the trigonometric functions: using triangles or circles. Historically, the six trigonometric functions were defined with reference to a right triangle. If you had trigonometry in high school, you probably began your course by studying trigonometric functions of acute angles in a right triangle.

In this section we define the six trigonometric functions in terms of a right triangle. Since the sum of the angles of a triangle is 180°, and the right angle is 90°, the other two angles must add up to 90°, and so each is acute (and the two are complements of each other). This means that by using right triangles we can define the trigonometric functions only for values of θ where $0° < \theta < 90°$. This is a severe limitation. In Sections 2.3 and 2.5, we will define these functions for all real values of θ by using circles.

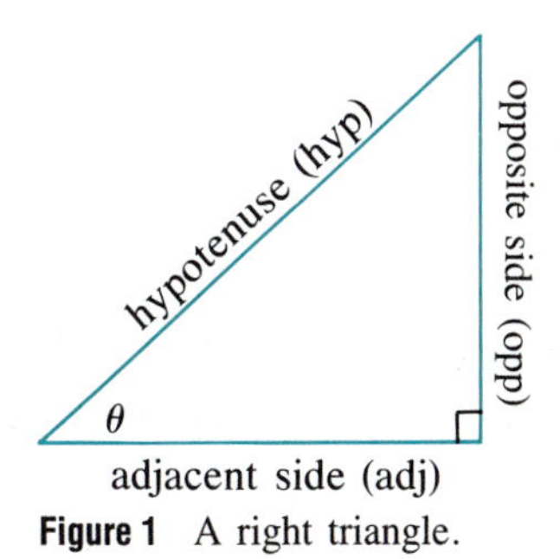

Figure 1 A right triangle.

Consider the angle θ in the right triangle in Figure 1. The side opposite θ is labeled "opp," the side adjacent to θ (which is not the hypotenuse) is labeled "adj," and the hypotenuse (the side opposite the right angle) is labeled "hyp." Then we define

Right Triangle Definitions of Trigonometric Functions

$$\sin\theta = \frac{\text{opposite}}{\text{hypotenuse}} = \frac{\text{opp}}{\text{hyp}} \qquad \csc\theta = \frac{\text{hypotenuse}}{\text{opposite}} = \frac{\text{hyp}}{\text{opp}}$$

$$\cos\theta = \frac{\text{adjacent}}{\text{hypotenuse}} = \frac{\text{adj}}{\text{hyp}} \qquad \sec\theta = \frac{\text{hypotenuse}}{\text{adjacent}} = \frac{\text{hyp}}{\text{adj}} \tag{1}$$

$$\tan\theta = \frac{\text{opposite}}{\text{adjacent}} = \frac{\text{opp}}{\text{adj}} \qquad \cot\theta = \frac{\text{adjacent}}{\text{opposite}} = \frac{\text{adj}}{\text{opp}}$$

The equation $f(x) = g(x)$ is called an **identity** if $f(x) = g(x)$ at every x for which both $f(x)$ and $g(x)$ are defined. In the next chapter we will discuss a large number of trigonometric identities. Using the definitions (1), we can establish some fundamental identities right away.

Four Basic Identities

$$\tan\theta = \frac{\sin\theta}{\cos\theta} \qquad \cot\theta = \frac{\cos\theta}{\sin\theta} = \frac{1}{\tan\theta}$$

$$\sec\theta = \frac{1}{\cos\theta} \qquad \csc\theta = \frac{1}{\sin\theta} \tag{2}$$

Proof

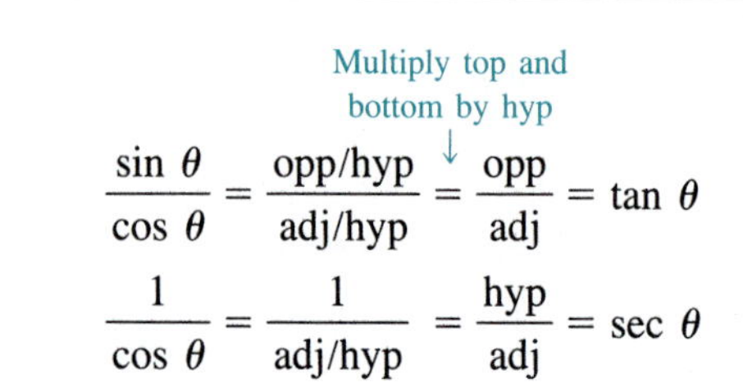

$$\frac{\sin\theta}{\cos\theta} = \frac{\text{opp/hyp}}{\text{adj/hyp}} = \frac{\text{opp}}{\text{adj}} = \tan\theta$$

$$\frac{1}{\cos\theta} = \frac{1}{\text{adj/hyp}} = \frac{\text{hyp}}{\text{adj}} = \sec\theta$$

The other two identities are proved in a similar way. ■

EXAMPLE 1 *Using the Basic Identities*

In Section 2.3 (Example 2 on page 81) we will show that $\sin 60° = \dfrac{\sqrt{3}}{2}$ and $\cos 60° = \frac{1}{2}$. Use these values to compute tan 60°, cot 60°, sec 60° and csc 60°.

SOLUTION Using (2), we have

$$\tan 60° = \frac{\sin 60°}{\cos 60°} = \frac{\sqrt{3}/2}{1/2} = \sqrt{3}$$

$$\cot 60° = \frac{1}{\tan 60°} = \frac{1}{\sqrt{3}}$$

$$\sec 60° = \frac{1}{\cos 60°} = \frac{1}{1/2} = 2$$

$$\csc 60° = \frac{1}{\sin 60°} = \frac{1}{\sqrt{3}/2} = \frac{2}{\sqrt{3}}$$

Cofunctions

We remind you (see p. 57) that two acute angles α and β are *complements* of each other if $\alpha + \beta = 90°$. In a right triangle, the two angles that are not right angles are complements of each other. If one angle is θ, the other is $90° - \theta$.

Theorem: Trigonometric Functions of Complementary Angles

Let θ be an acute angle. Then

$\sin(90° - \theta) = \cos\theta$ (3) $\quad$ $\cot(90° - \theta) = \tan\theta$ (6)

$\cos(90° - \theta) = \sin\theta$ (4) $\quad$ $\sec(90° - \theta) = \csc\theta$ (7)

$\tan(90° - \theta) = \cot\theta$ (5) $\quad$ $\csc(90° - \theta) = \sec\theta$ (8)

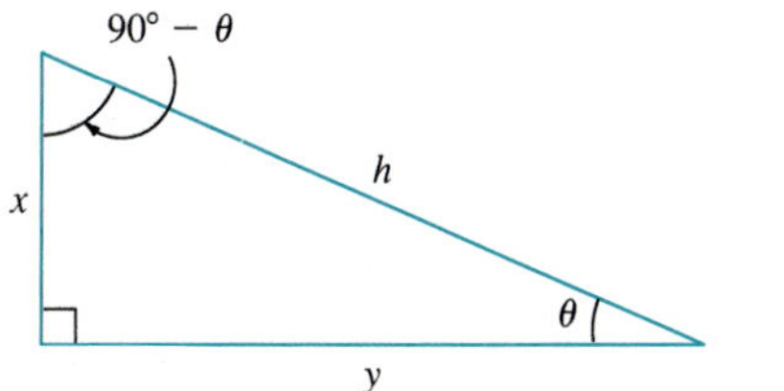

Figure 2 In this right triangle $\sin(90° - \theta) = \dfrac{\text{opp}}{\text{hyp}} = \dfrac{y}{h}$, and $\cos\theta = \dfrac{\text{adj}}{\text{hyp}} = \dfrac{y}{h}$.

Proof

From the triangle in Figure 2 we see that

$$\sin(90° - \theta) = \frac{\text{opp}}{\text{hyp}} = \frac{y}{h} \quad \text{and} \quad \cos\theta = \frac{\text{adj}}{\text{hyp}} = \frac{y}{h}$$

so

$$\sin(90° - \theta) = \cos\theta.$$

Similarly,

$$\tan(90° - \theta) = \frac{\text{opp}}{\text{adj}} = \frac{y}{x} \quad \text{and} \quad \cot\theta = \frac{\text{adj}}{\text{opp}} = \frac{y}{x}$$

so

$$\tan(90° - \theta) = \cot\theta$$

The other four identities are proved in a similar way. ■

The following pairs are called **cofunctions.**

Three Pairs of Cofunctions

$\sin\theta$	and	$\cos\theta$
$\tan\theta$	and	$\cot\theta$
$\sec\theta$	and	$\csc\theta$

The reason for the term *cofunction* is, for example, that sine of θ is equal to the *co*sine of the *co*mplement of θ. Originally, mathematicians wrote sineco θ and this, eventually, turned into the term cosine.

EXAMPLE 2 *Using Cofunctions to Compute Trigonometric Values*

In Example 1 we used the fact that $\sin 60° = \dfrac{\sqrt{3}}{2}$ and $\cos 60° = \dfrac{1}{2}$ to compute $\tan 60°$, $\cot 60°$, $\sec 60°$ and $\csc 60°$. Use these values to compute $\sin 30°$, $\cos 30°$, $\tan 30°$, $\cot 30°$, $\sec 30°$ and $\csc 30°$.

SOLUTION From (4), we have

$$\sin 30° = \cos(90° - 30°) = \cos 60° = \frac{1}{2}$$

Similarly, we compute

$$\cos 30° = \sin(90° - 30°) = \sin 60° = \frac{\sqrt{3}}{2}$$

$$\tan 30° = \cot(90° - 30°) = \cot 60° = \frac{1}{\sqrt{3}}$$

Alternatively, from (2)

$$\tan 30^\circ = \frac{\sin 30^\circ}{\cos 30^\circ} = \frac{1/2}{\sqrt{3}/2} = \frac{1}{\sqrt{3}}$$

$$\cot 30^\circ = \frac{1}{\tan 30^\circ} = \frac{1}{1/\sqrt{3}} = \sqrt{3}$$

$$\sec 30^\circ = \frac{1}{\cos 30^\circ} = \frac{1}{\sqrt{3}/2} = \frac{2}{\sqrt{3}}$$

$$\csc 30^\circ = \frac{1}{\sin 30^\circ} = \frac{1}{1/2} = 2$$ ■

EXAMPLE 3 ***Trigonometric Functions of Cofunctions***

From the theorem on p. 70, we see that

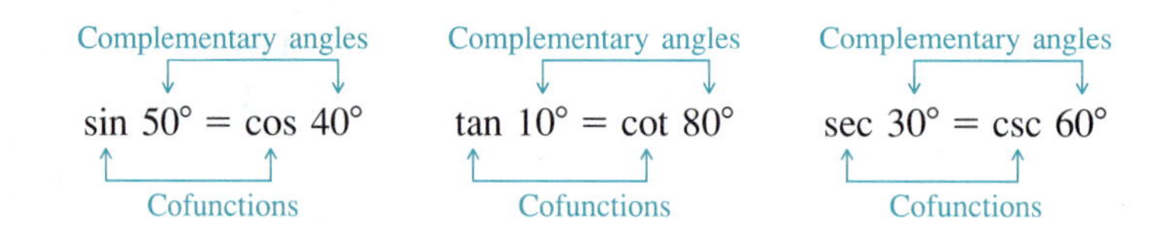

Finding the Values of Five Functions if One Is Given

If θ is an acute angle and the value of one of the six trigonometric functions at θ is given, then we can compute the values of the other five functions.

EXAMPLE 4 ***Finding Values for Five Trigonometric Functions if cos θ Is Given***

If $\cos\theta = \frac{4}{9}$ and θ is an acute angle, find $\sin\theta$, $\tan\theta$, $\cot\theta$, $\sec\theta$, and $\csc\theta$.

SOLUTION Since $\cos\theta = \frac{4}{9} = \frac{\text{adj}}{\text{hyp}}$, we draw, in Figure 3, a right triangle, label one acute angle θ, and set the adjacent side equal to 4 and the hypotenuse equal to 9. Then, from the Pythagorean theorem, the other leg (side) has length $\sqrt{9^2 - 4^2} = \sqrt{81 - 16} = \sqrt{65}$. Thus

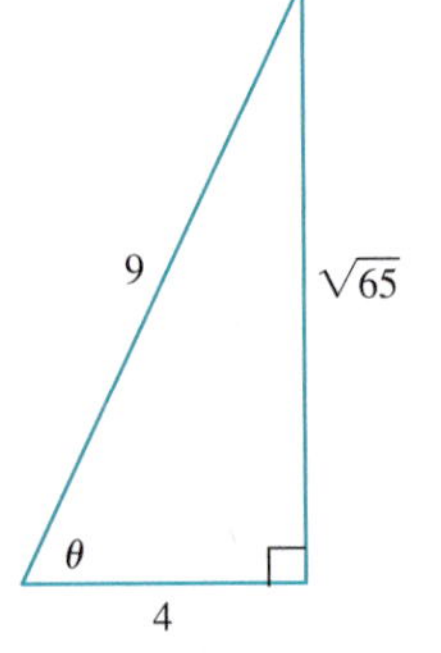

Figure 3 In this triangle $\cos\theta = \frac{4}{9}$.

$$\sin\theta = \frac{\text{opp}}{\text{hyp}} = \frac{\sqrt{65}}{9} \qquad \tan\theta = \frac{\text{opp}}{\text{adj}} = \frac{\sqrt{65}}{4}$$

$$\cot\theta = \frac{\text{adj}}{\text{opp}} = \frac{4}{\sqrt{65}} \qquad \sec\theta = \frac{\text{hyp}}{\text{adj}} = \frac{9}{4}$$

$$\csc\theta = \frac{\text{hyp}}{\text{opp}} = \frac{9}{\sqrt{65}}$$ ■

EXAMPLE 5 ***Finding Values for Five Trigonometric Functions if tan θ Is Given***

If $\tan\theta = 3$ and $0^\circ < \theta < 90^\circ$, find values for the other five functions.

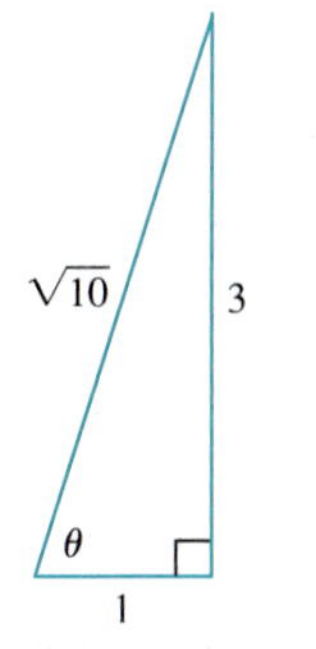

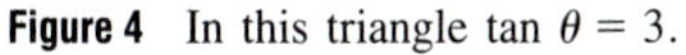
Figure 4 In this triangle $\tan\theta = 3$.

SOLUTION $\tan\theta = 3 = \dfrac{3}{1} = \dfrac{\text{opp}}{\text{adj}}$, so we draw, in Figure 4, a right triangle with opposite side 3 and adjacent side 1. Then $\text{hyp}^2 = 3^2 + 1^2 = 10$ and $\text{hyp} = \sqrt{10}$, so from the triangle we find that

$$\sin\theta = \frac{3}{\sqrt{10}} \qquad \cos\theta = \frac{1}{\sqrt{10}} \qquad \cot\theta = \frac{1}{3}$$

$$\sec\theta = \sqrt{10} \qquad \csc\theta = \frac{\sqrt{10}}{3}$$

A Trigonometric Technique That Is Useful in Calculus

The next two examples typify a type of problem that often arises in calculus.

EXAMPLE 6 *Finding sin θ and sec θ if x = b tan θ*

Suppose that $x = b\tan\theta$. Write $\sin\theta$ and $\sec\theta$ in terms of b and x. Assume that $x > 0$, $b > 0$ and that θ is an acute angle.

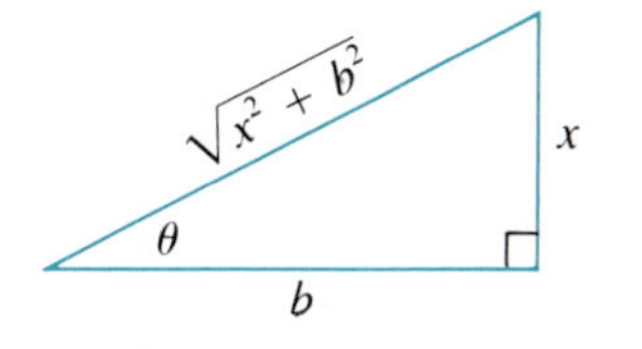

Figure 5 In this triangle $\tan\theta = \dfrac{x}{b}$.

SOLUTION We see that $\tan\theta = \dfrac{x}{b}$, so we draw a right triangle with the opposite side equal to x and the adjacent side equal to b, as in Figure 5. Then, from the Pythagorean theorem, $\text{hyp} = \sqrt{x^2 + b^2}$, so

$$\sin\theta = \frac{\text{opp}}{\text{hyp}} = \frac{x}{\sqrt{x^2 + b^2}} \quad \text{and} \quad \sec\theta = \frac{\text{hyp}}{\text{adj}} = \frac{\sqrt{x^2 + b^2}}{b}$$

■

EXAMPLE 7 *Finding tan θ and sec θ if y = √5 sin θ*

Let $y = \sqrt{5}\sin\theta$. Find $\tan\theta$ and $\sec\theta$ in terms of y. Assume that $y > 0$ and that θ is an acute angle.

SOLUTION $\sin\theta = \dfrac{y}{\sqrt{5}} = \dfrac{\text{opp}}{\text{hyp}}$ as in Figure 6. Then

$$\text{adj}^2 = \text{hyp}^2 - \text{opp}^2 = (\sqrt{5})^2 - y^2 = 5 - y^2$$

so

$$\text{adj} = \sqrt{5 - y^2}$$

Then

$$\tan\theta = \frac{\text{opp}}{\text{adj}} = \frac{y}{\sqrt{5 - y^2}}$$

and

$$\sec\theta = \frac{\text{hyp}}{\text{adj}} = \frac{\sqrt{5}}{\sqrt{5 - y^2}} = \sqrt{\frac{5}{5 - y^2}}$$

NOTE These expressions are defined when $5 - y^2 > 0$ or since $y > 0$, $0 < y < \sqrt{5}$. Explain why it must be the case that $y < \sqrt{5}$.

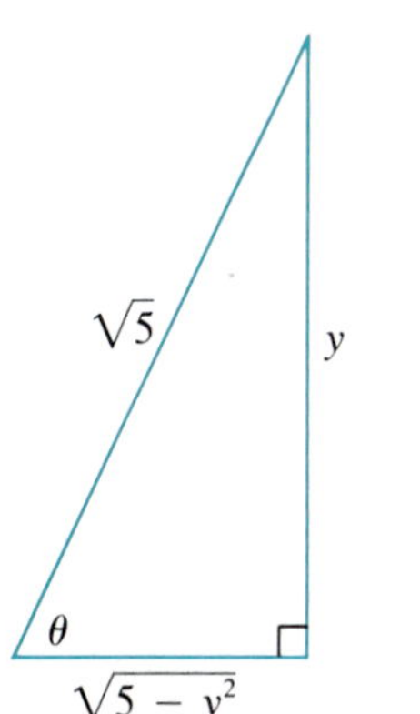

Figure 6 In this triangle $\sin\theta = \dfrac{y}{\sqrt{5}}$.

FOCUS ON

Babylonian Trigonometry

Historians credit the Greeks with the first formal development of the trigonometric functions. However, they were not the first to use these functions.

Archaeologists working in Mesopotamia, in a region now part of Iraq, have, since the early nineteenth century, dug up approximately 500,000 clay tablets inscribed in cuneiform, the writing of the ancient Babylonians. Scholars learned how to decipher these tablets around 1850. Since then, we have learned a great deal about Babylonian culture and civilization, a civilization that was apparently highly sophisticated.

Of the half-million tablets found, about 300 dealt with mathematical topics that we today categorize as computational, geometric, or trigonometric. One of the most remarkable of these tablets inscribed between 1900 B.C. and 1600 B.C. is called *Plimpton 322* (see the figure). The name refers to catalogue number 322 in the G. A. Plimpton collection at Columbia University.

The tablet contains four columns of numbers, with the one on the extreme left somewhat incomplete. When these numbers are written in base 10 and a few missing numbers are filled in, we arrive at the table on the right (except for the column on the extreme right, which we will explain shortly).

$(\text{hyp/adj})^2$	adj	hyp		θ (degrees)
2.0169	119	169	1	45.2
2.0536	3,367	4,825	2	45.7
2.0884	4,601	6,649	3	46.2
2.1284	12,709	18,541	4	46.7
2.2270	65	97	5	47.9
2.2736	319	481	6	48.5
2.3889	2,291	3,541	7	49.7
2.4436	799	1,249	8	50.2
2.5560	481	769	9	51.3
2.7061	4,961	8,161	10	52.6
2.7778	45	75	11	53.1
3.0432	1,679	2,929	12	55.0
3.2221	161	289	13	56.1
3.3243	1,771	3,229	14	56.7
3.5829	56	106	15	58.1

Plimpton 322 (*Columbia University*)

This table is really remarkable considering that it was made almost 4000 years ago. The fourth column from the left contains a simple numbering of the lines. The columns labeled adj and hyp give one side and the hypotenuse of a right triangle, as in Figure 7.

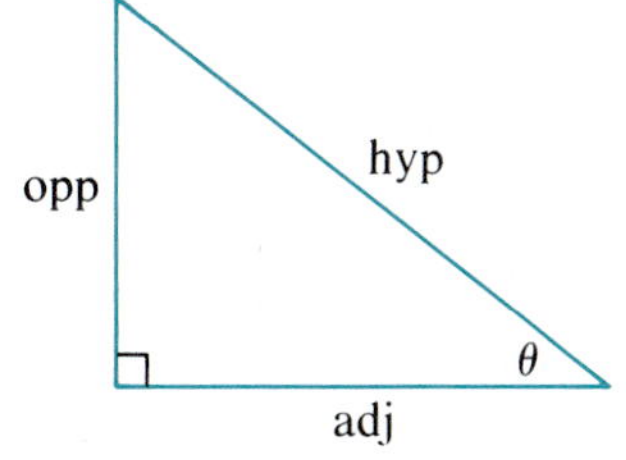

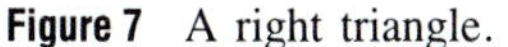

Figure 7 A right triangle.

Each right triangle has sides that are integers.

For example, if adj = 119 and hyp = 169 as in line 1, then

$$\begin{aligned} \text{adj}^2 + \text{opp}^2 &= \text{hyp}^2 \\ 119^2 + \text{opp}^2 &= 169^2 \\ \text{opp}^2 &= 169^2 - 119^2 \\ &= 28561 - 14161 = 14400 \\ \text{opp} &= \sqrt{14400} = 120 \end{aligned}$$

In line 2,

$$\begin{aligned} \text{opp}^2 &= 4825^2 - 3367^2 \\ &= 23{,}280{,}625 - 11{,}336{,}689 = 11{,}943{,}936 \\ \text{opp} &= \sqrt{11{,}943{,}936} = 3456 \end{aligned}$$

The numbers in the leftmost column are approximately equal to $\left(\frac{\text{hyp}}{\text{adj}}\right)^2$. But $\left(\frac{\text{hyp}}{\text{adj}}\right)^2$ is the square of the secant of the angle between adj and hyp. In the rightmost column we give the acute angles whose secants are $\frac{\text{hyp}}{\text{adj}}$.

Thus we see that almost 4000 years ago, the Babylonians had a table of secants for angles between 45° and 58°. It is believed that they had similar tables for other angles as well.

For further and highly readable discussions of Plimpton 322 and other topics in Babylonian mathematics see Howard Eves, *An Introduction to the History of Mathematics,* 6th ed., Saunders, Philadelphia, 1990, p. 44, and Jöran Friberg, "Methods and Traditions of Babylonian Mathematics," *Historia Mathematica* 8, No. 3 (Aug. 1981), 277–318.

The Origin of Terms

As we just saw, tables of the secant function have long been known. The tangent function was introduced into trigonometry by the Islamic mathematician Abû'l-Wefâ (940–998), who was born in the Persian mountain region of Khorâsân. Abû'l-Wefâ computed tables of sines and tangents for 15-minute intervals.

The words *tangent* and *secant* were not used in the trigonometric sense until the sixteenth century, but the names come from geometry. In Figure 8 we draw a unit circle. The line segment *AB*

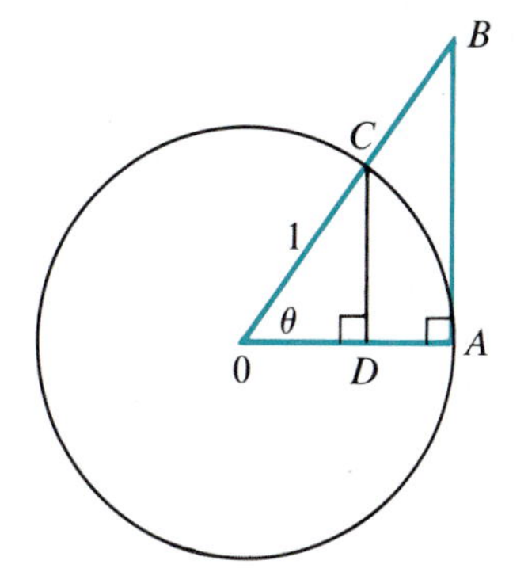

Figure 8 A circle of radius 1.

is tangent to the circle at *A*. It is a fact from Euclidean geometry that the tangent line to a circle is perpendicular to the radial line at the point of intersection. Thus angle *A* is a right angle. Then

$$\begin{aligned} \tan\theta &= \frac{\text{opp}}{\text{adj}} = \frac{\overline{AB}}{\overline{OA}} = \frac{\overline{AB}}{1} = \overline{AB} \\ &= \text{length of tangent line segment } \overline{AB} \end{aligned}$$

The line segment *OB* is called a *secant line* to the curve. We have

$$\begin{aligned} \sec\theta &= \frac{\text{hyp}}{\text{adj}} = \frac{\overline{OB}}{\overline{OA}} = \frac{\overline{OB}}{1} = \overline{OB} \\ &= \text{length of the secant line} \end{aligned}$$

This explains how the secant function got its name.

The first use of the words *tangent* and *secant* as trigonometric functions is attributed to the mathematician and physician Thomas Finck, a native of Flensburg, Germany, who used the terms in his book *Geometria Rotundi,* published in Basel, Switzerland, in 1583.

Finally, the names *cosine* and *cotangent,* which were abbreviations of the terms *complemental sine* and *complemental tangent* were first used by the British mathematician Edmund Gunter (1581–1626) in 1620. Gunter is best known for his tables of the common logarithms of sines and tangents for every minute to seven decimal places. (Life before calculators was indeed hard.) The origin of the term *sine* is especially interesting. It is given on p. 87.

Problems 2.2

Readiness Check

I. Which of the following is the right triangle definition of $\csc\theta$ if θ is an acute angle in a right triangle?

a. $\dfrac{\text{opp}}{\text{adj}}$ b. $\dfrac{\text{hyp}}{\text{opp}}$ c. $\dfrac{\text{hyp}}{\text{adj}}$ d. $\dfrac{\text{adj}}{\text{opp}}$

II. Which of the following is $\tan\theta$ for the right triangle shown below?

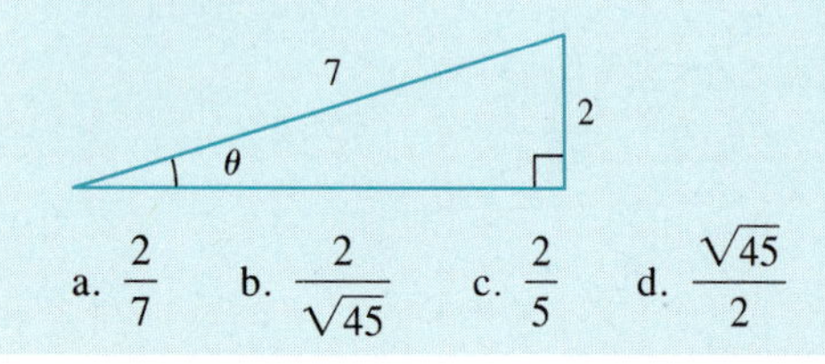

a. $\dfrac{2}{7}$ b. $\dfrac{2}{\sqrt{45}}$ c. $\dfrac{2}{5}$ d. $\dfrac{\sqrt{45}}{2}$

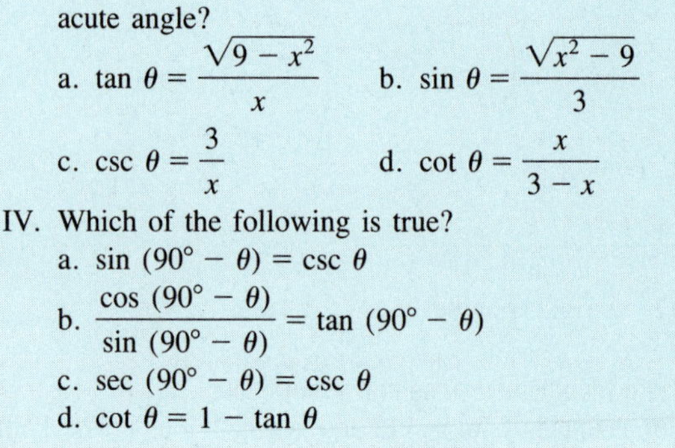

III. Which of the following is true if $\cos\theta = \dfrac{x}{3}$ and θ is an acute angle?

a. $\tan\theta = \dfrac{\sqrt{9-x^2}}{x}$ b. $\sin\theta = \dfrac{\sqrt{x^2-9}}{3}$

c. $\csc\theta = \dfrac{3}{x}$ d. $\cot\theta = \dfrac{x}{3-x}$

IV. Which of the following is true?

a. $\sin(90° - \theta) = \csc\theta$

b. $\dfrac{\cos(90° - \theta)}{\sin(90° - \theta)} = \tan(90° - \theta)$

c. $\sec(90° - \theta) = \csc\theta$

d. $\cot\theta = 1 - \tan\theta$

In Problems 1–10 a right triangle depicting an angle θ is given. Find $\sin\theta$, $\cos\theta$, $\tan\theta$, $\sec\theta$, $\csc\theta$, and $\cot\theta$.

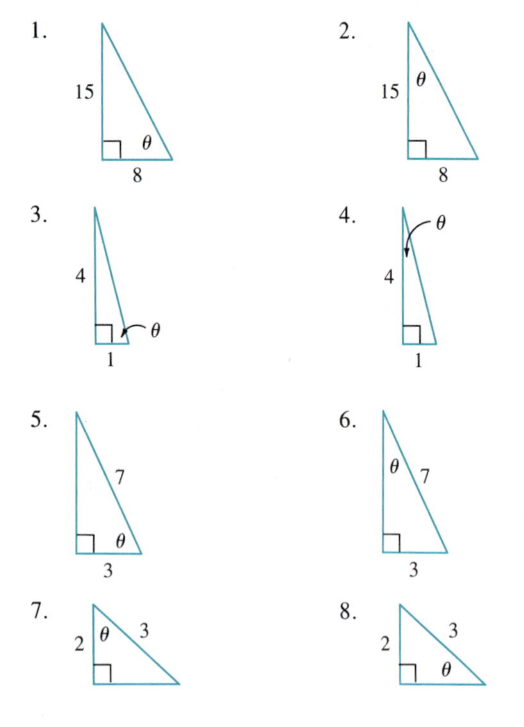

In Problems 11–28 a trigonometric function of the *acute* angle θ is given. (a) Draw a triangle that illustrates the situation. (b) Find the exact values of the other five trigonometric functions. Do not use a calculator.

11. $\sin\theta = \frac{3}{5}$
12. $\tan\theta = \frac{5}{12}$
13. $\cot\theta = \frac{12}{5}$
14. $\sec\theta = 5$
15. $\cos\theta = \frac{2}{3}$
16. $\csc\theta = \frac{7}{2}$
17. $\cot\theta = \frac{4}{7}$
18. $\tan\theta = 10$
19. $\csc\theta = 1.5$
20. $\sin\theta = 0.7$
21. $\tan\theta = 4.5$
22. $\csc\theta = 1\frac{1}{3}$
23. $\sin\theta = 0.3$
24. $\tan\theta = 100$
25. $\sec\theta = 20$
26. $\sin\theta = 0.02$
27. $\csc\theta = \frac{8}{5}$
28. $\csc\theta = \frac{9}{7}$

29. Find $\tan\theta$ if $\cos\theta = \dfrac{x}{2}$, $x > 0$.
30. Find $\sin\theta$ if $\cot\theta = \sqrt{2}r$, $r > 0$.
31. Find $\sec\theta$ if $4\cos\theta = y$, $y > 0$.
32. Find $\csc\theta$ if $x\tan\theta = y$, $x, y > 0$.
33. Find $\cos\theta$ if $\sqrt{3}\sin\theta = u$, $u > 0$.
34. Find $\tan\theta$ if $\sec\theta = \sqrt{2}/y$, $y > 0$.
35. Find $\sin\theta$ if $\sqrt{2}\cos\theta = 7$.
36. Use the right triangle in Figure 1 and the Pythagorean theorem to show that $\sin^2\theta + \cos^2\theta = 1$.
37. Show that $1 + \tan^2\theta = \sec^2\theta$.
38. Show that $1 + \cot^2\theta = \csc^2\theta$.

Answers to Readiness Check

I. b II. b III. a IV. c

2.3 The Sine and Cosine Functions of a Real Variable

As we saw in the last section, when the trigonometric functions are defined in terms of a right triangle, they are defined only when θ is an acute angle. In this section we use a circle to define $\sin\theta$ and $\cos\theta$ where θ can be any real number. In Section 2.5 we will define $\tan\theta$, $\sec\theta$, $\csc\theta$, and $\cot\theta$ for any real number.

We begin with the unit circle; this is the circle of radius 1 centered at the origin (see Figure 1). An angle θ in standard position uniquely determines a

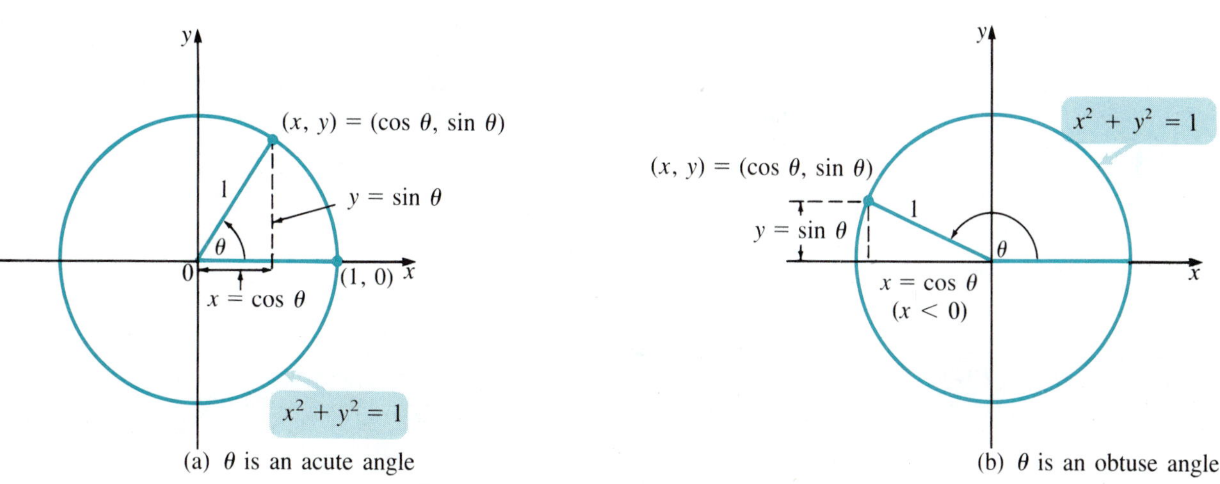

Figure 1 In each unit circle, $\sin\theta = y$ and $\cos\theta = x$.

point (x, y) where the radial line (terminal side of the angle) intersects the circle. We then define

Circular Definitions of the Cosine and Sine Functions

$$\text{cosine } \theta = x \qquad \text{and} \qquad \text{sine } \theta = y \tag{1}$$

These are the two basic trigonometric (or circular) functions, usually abbreviated as **cos θ** and **sin θ.** Now

$$(\sin\theta)^2 = \sin^2\theta = y^2$$

and

$$(\cos\theta)^2 = \cos^2\theta = x^2$$

Since an equation of the circle is $x^2 + y^2 = 1$, we have

$$\sin^2\theta + \cos^2\theta = y^2 + x^2 = 1.$$

That is,

A Basic Trigonometric Identity

$$\sin^2\theta + \cos^2\theta = 1 \tag{2}$$

Equation (2) comes up often. It's worth memorizing.

We stress that the notation $\sin^2\theta$ stands for $(\sin\theta)^2 = (\sin\theta)(\sin\theta)$.

The functions $\cos\theta$ and $\sin\theta$ are defined for every real *number* θ. Although they are defined in terms of angles, these functions are used in many applications that have nothing at all to do with angles. You will see some of the applications in Section 3.7 and elsewhere.

The definitions of $\sin\theta$ and $\cos\theta$ using the unit circle enable us to determine quickly values of these functions for several numbers. Look at Figure 2. In each graph, one value of $\cos\theta$ and one value of $\sin\theta$ are depicted. For

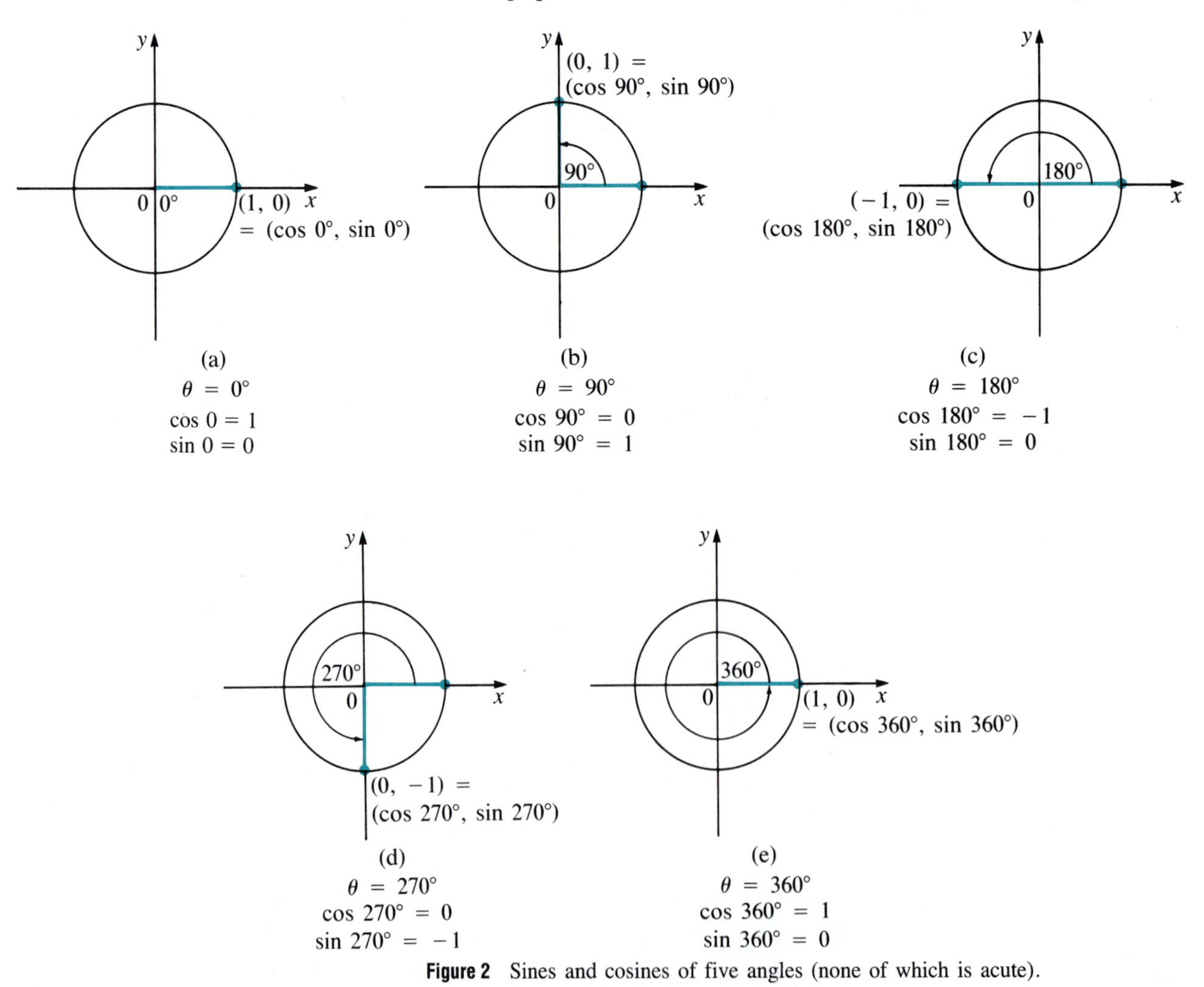

Figure 2 Sines and cosines of five angles (none of which is acute).

example, in (b), we see that the terminal side of 90° intersects the unit circle at (0, 1). Thus

$$\cos 90^\circ = x\text{-coordinate of } (0, 1) = 0$$

and

$$\sin 90^\circ = y\text{-coordinate of } (0, 1) = 1$$

In radians, we write these results as

$$\cos \frac{\pi}{2} = 0 \qquad \text{and} \qquad \sin \frac{\pi}{2} = 1$$

Table 1

θ (degrees)	0	90	180	270	360
θ (radians)	0	$\pi/2$	π	$3\pi/2$	2π
$\cos\theta$	1	0	-1	0	1
$\sin\theta$	0	1	0	-1	0

The values of $\cos\theta$ and $\sin\theta$ that can be derived from Figure 2 are given in Table 1.

Before continuing, we make an important observation. In Figure 1, we see that the x- and y-coordinates of a point on the unit circle vary between -1 and 1. We have the following:

Range of the Cosine and Sine Functions

For every real number θ,

$$-1 \le \cos\theta \le 1 \tag{3}$$

$$-1 \le \sin\theta \le 1 \tag{4}$$

That is,

The range of $\cos\theta$ and $\sin\theta$ is $[-1, 1]$.

Thus if a problem begins, suppose $\sin\theta = 2$ or $\cos\theta = -1.5$, for example, you can stop right there since it is *impossible* that $\sin\theta = 2$ or $\cos\theta = -1.5$. We stress this point with a warning.

WARNING If in a calculation you find that $|\sin\theta| > 1$ or $|\cos\theta| > 1$, then you may have made a mistake. Go back and check your calculation. ■

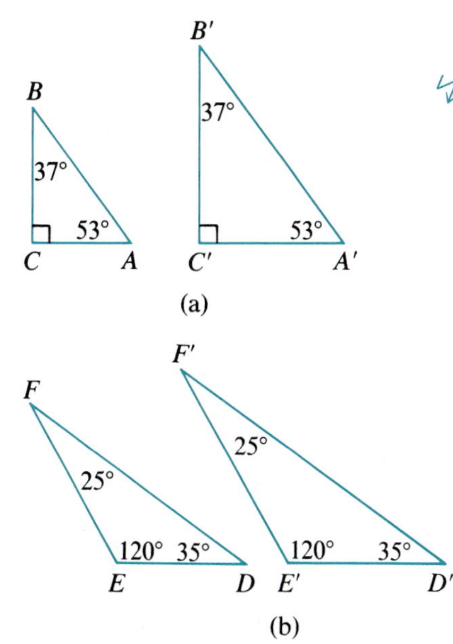

Figure 3 Two pairs of similar triangles.

We would like to show that the triangular definitions of $\sin\theta$ and $\cos\theta$ given in Section 2.2 (on page 69) and the circular definitions given in (1) give the same values when $0 < \theta < 90^\circ$ (that is, when θ is an acute angle). To do so, we define similar triangles.

Two triangles are said to be **similar** if they have equal angles (that is, angles with the same measure). The two pairs of triangles in Figure 3 are similar.

If two triangles are similar, then a pair of equal angles, one in each triangle, are called **corresponding angles.** Two sides opposite corresponding angles are called **corresponding sides.** An important fact from plane geometry is

Ratios of the lengths of corresponding sides of similar triangles are equal.

We use the notation $AB \sim CD$ to indicate that AB corresponds to CD.

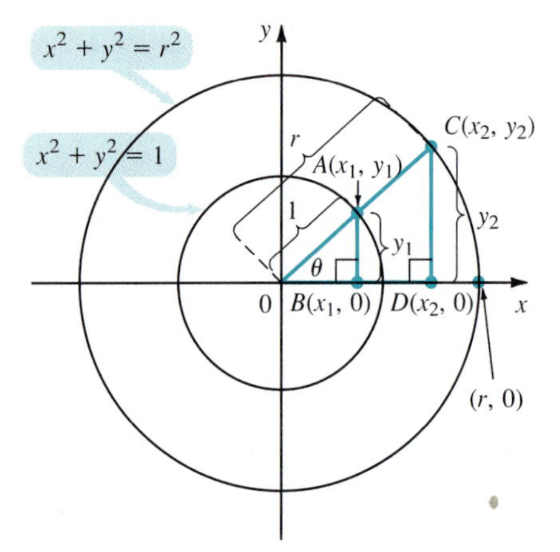

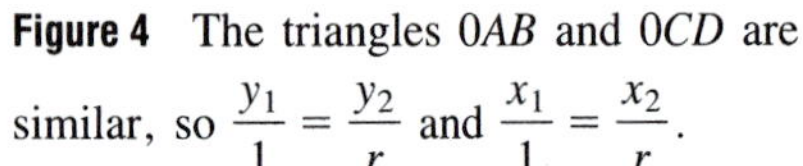
Figure 4 The triangles $0AB$ and $0CD$ are similar, so $\frac{y_1}{1} = \frac{y_2}{r}$ and $\frac{x_1}{1} = \frac{x_2}{r}$.

In Figure 3(a), $AB \sim A'B'$, $BC \sim B'C'$, and $AC \sim A'C'$. Then

$$\frac{\overline{AB}}{\overline{BC}} = \frac{\overline{A'B'}}{\overline{B'C'}}, \quad \frac{\overline{AB}}{\overline{AC}} = \frac{\overline{A'B'}}{\overline{A'C'}}, \quad \text{and} \quad \frac{\overline{BC}}{\overline{AC}} = \frac{\overline{B'C'}}{\overline{A'C'}}$$

In Figure 3(b), $EF \sim E'F'$, and so on. For example

$$\frac{\overline{EF}}{\overline{ED}} = \frac{\overline{E'F'}}{\overline{E'D'}}$$

Here $\overline{AB}$ denotes the length of AB, and so on.

In Figure 4, two circles centered at the origin are drawn. The first has radius 1 (the unit circle); the second has radius r. In the figure, the two triangles OAB and OCD are similar because they share the angle θ and each has a right angle. Thus the ratios of corresponding sides are equal. That is,

$$\frac{\overline{AB}}{\overline{OA}} = \frac{\overline{CD}}{\overline{OC}} \quad \text{and} \quad \frac{\overline{OB}}{\overline{OA}} = \frac{\overline{OD}}{\overline{OC}}$$

or

$$\frac{y_1}{1} = \frac{y_2}{r} \quad \text{and} \quad \frac{x_1}{1} = \frac{x_2}{r}$$

But, according to (1), $\quad y_1 = \sin\theta \quad$ and $\quad x_1 = \cos\theta$

so

$$\sin\theta = \frac{y_2}{r} \quad \text{and} \quad \cos\theta = \frac{x_2}{r}$$

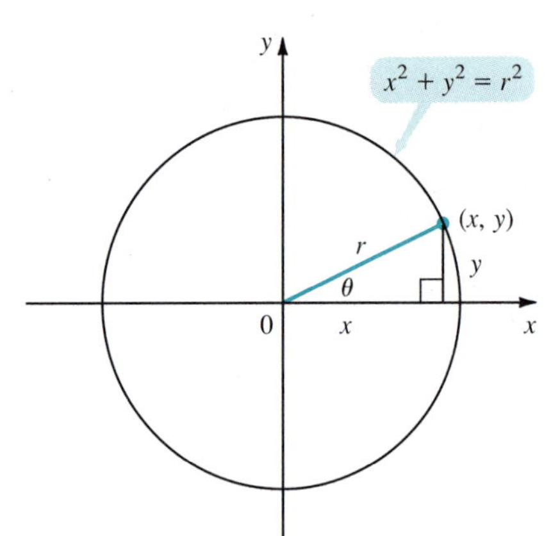

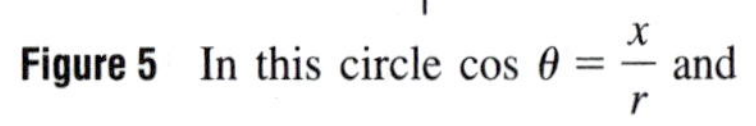
Figure 5 In this circle $\cos\theta = \frac{x}{r}$ and $\sin\theta = \frac{y}{r}$.

We have shown the following:

An angle θ in standard position uniquely determines a point (x, y) where the radial line (terminal side of the angle) intersects a circle of radius r centered at the origin. Then

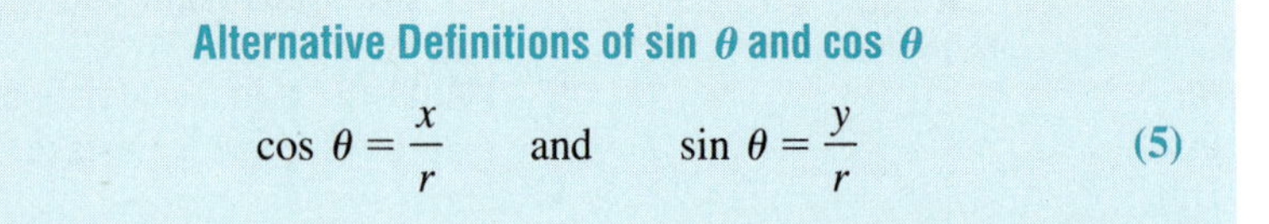
Alternative Definitions of sin θ and cos θ

$$\cos\theta = \frac{x}{r} \quad \text{and} \quad \sin\theta = \frac{y}{r} \qquad (5)$$

This alternative definition is sometimes useful in applications. It is illustrated in Figure 5.

The Triangular and Circular Definitions Give the Same Values for $0° < \theta < 90°$

We now show that these "circular" definitions of $\sin\theta$ and $\cos\theta$ have the same values as the "triangular" definitions given earlier. To demonstrate this, we place the triangle as in Figure 6 and draw the circle centered at the origin whose radius is equal to the length of the hypotenuse of the triangle.

From equation (5):

$$\cos\theta = \frac{x}{r} \quad \text{and} \quad \sin\theta = \frac{y}{r}$$

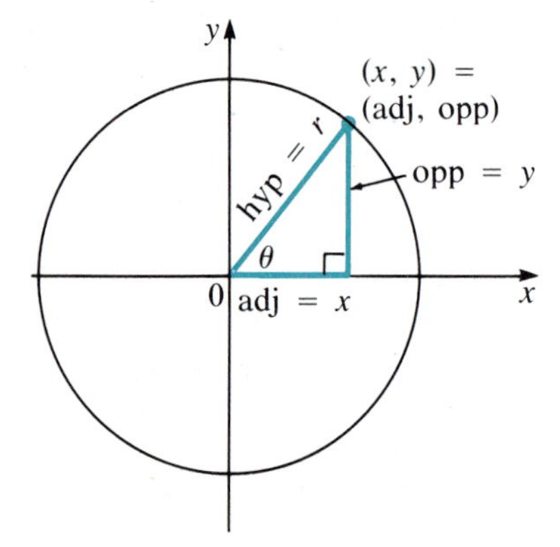

Figure 6 A circle of radius r with θ an acute angle.

From the triangular definition (p. 69):

$$\cos\,\theta = \frac{\text{adj}}{\text{hyp}} \qquad \text{and} \qquad \sin\,\theta = \frac{\text{opp}}{\text{hyp}}$$

Since adj $= x$, opp $= y$, and hyp $= r$, we see that these definitions yield the same values for $\sin\,\theta$ and $\cos\,\theta$ for $0° < \theta < 90°$.

We will explore a number of facts about $\sin\,\theta$ and $\cos\,\theta$ in this and the next section. First, we compute some additional values of these functions.

EXAMPLE 1 ***Computing cos 45° and sin 45°***

Compute cos 45° and sin 45°.

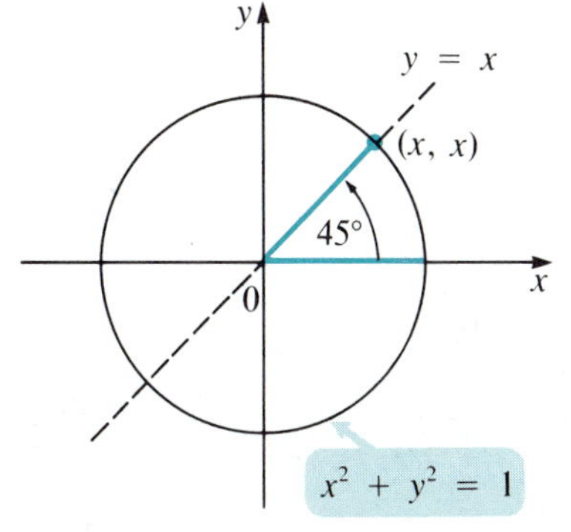

Figure 7 The terminal side of a 45° angle lies on the line $y = x$.

SOLUTION In Figure 7, we draw the unit circle and the angle 45°. Since 45° is one half of 90°, the terminal side of 45° bisects the first quadrant. This terminal side lies on the line $y = x$. In other words, the y-coordinate of any point on the terminal side will equal the x-coordinate. Hence, the point at which the terminal side intersects the unit circle has coordinates (x, x). But (x, x) is on the unit circle, so $x^2 + x^2 = 1$ or $2x^2 = 1$ and $x^2 = \frac{1}{2}$. Since (x, x) is in the first quadrant, $x > 0$ and $x = 1/\sqrt{2}$.

Thus, $(x, y) = \left(\dfrac{1}{\sqrt{2}}, \dfrac{1}{\sqrt{2}}\right)$, so

$$\cos 45° = x\text{-coordinate} = \frac{1}{\sqrt{2}} = \frac{\sqrt{2}}{2}$$

and

$$\sin 45° = y\text{-coordinate} = \frac{1}{\sqrt{2}} = \frac{\sqrt{2}}{2}$$ ■

EXAMPLE 2 ***Computing cos 60°, sin 60°, cos 30°, and sin 30°***

Compute cos 60°, sin 60°, cos 30°, and sin 30°.

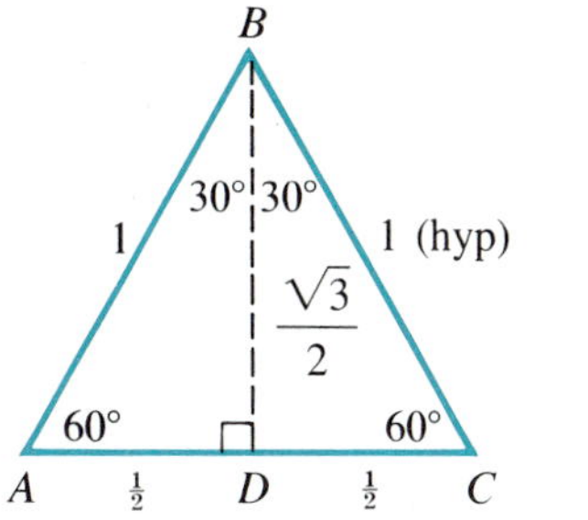

Figure 8 An equilateral triangle.

SOLUTION Look at the equilateral triangle in Figure 8. The line BD is drawn to bisect the 60° angle at B. From geometry, we know that BD also is the perpendicular bisector of the side AC. Since $\overline{AC} = 1$, we have $\overline{AD} = \overline{DC} = \frac{1}{2}$. From the Pythagorean theorem,

$$\overline{AB}^2 = \overline{AD}^2 + \overline{BD}^2$$

$$1^2 = \left(\frac{1}{2}\right)^2 + \overline{BD}^2$$

$$1 = \frac{1}{4} + \overline{BD}^2$$

$$\frac{3}{4} = \overline{BD}^2$$

$$\overline{BD} = \frac{\sqrt{3}}{2}$$

Then, from the triangular definition, we see in Figure 8 that

$$\cos 60^\circ = \frac{\text{adj}}{\text{opp}} = \frac{1/2}{1} = \frac{1}{2} \qquad \sin 60^\circ = \frac{\text{opp}}{\text{hyp}} = \frac{\sqrt{3}/2}{1} = \frac{\sqrt{3}}{2}$$

Then, using equations (3) and (4) on p. 70

$$\cos 30^\circ = \sin (90^\circ - 30^\circ) = \sin 60^\circ = \frac{\sqrt{3}}{2}$$

and

$$\sin 30^\circ = \cos (90^\circ - 30^\circ) = \cos 60^\circ = \frac{1}{2}$$

Let us summarize our results to this point. In Table 2, we give the values we know for $0 \le \theta \le 90^\circ$.

The values in Table 2 will come up many times in this book and in any course you have later that uses trigonometry. Memorize them!

Table 2

θ (degrees)	0	30	45	60	90
θ (radians)	0	$\frac{\pi}{6}$	$\frac{\pi}{4}$	$\frac{\pi}{3}$	$\frac{\pi}{2}$
$\cos\theta$	1	$\frac{\sqrt{3}}{2}$	$\frac{1}{\sqrt{2}} = \frac{\sqrt{2}}{2}$	$\frac{1}{2}$	0
$\sin\theta$	0	$\frac{1}{2}$	$\frac{1}{\sqrt{2}} = \frac{\sqrt{2}}{2}$	$\frac{\sqrt{3}}{2}$	1

There is an easy way to remember the values of $\sin\theta$ for $\theta = 0, \frac{\pi}{6}, \frac{\pi}{4}, \frac{\pi}{3}$, and $\frac{\pi}{2}$:

$$\sin 0 = \sin 0^\circ = \sqrt{\frac{0}{4}}$$

$$\sin \frac{\pi}{6} = \sin 30^\circ = \sqrt{\frac{1}{4}}$$

$$\sin \frac{\pi}{4} = \sin 45^\circ = \sqrt{\frac{2}{4}}$$

$$\sin \frac{\pi}{3} = \sin 60^\circ = \sqrt{\frac{3}{4}}$$

$$\sin \frac{\pi}{2} = \sin 90^\circ = \sqrt{\frac{4}{4}}$$

If two angles have their initial sides on the positive x-axis and are coterminal, then their terminal sides coincide and intersect the unit circle at the same point. Thus we have the following result.

Theorem

If θ_1 and θ_2 are coterminal, then

$$\cos \theta_1 = \cos \theta_2 \quad \text{and} \quad \sin \theta_1 = \sin \theta_2$$

EXAMPLE 3 *Sines and Cosines of Coterminal Angles*

Since $\theta_1 = 90°$ and $\theta_2 = 450° = 90° + 360°$ are coterminal, we see that

$$\cos 450° = \cos 90° = 0 \quad \text{and} \quad \sin 450° = \sin 90° = 1$$

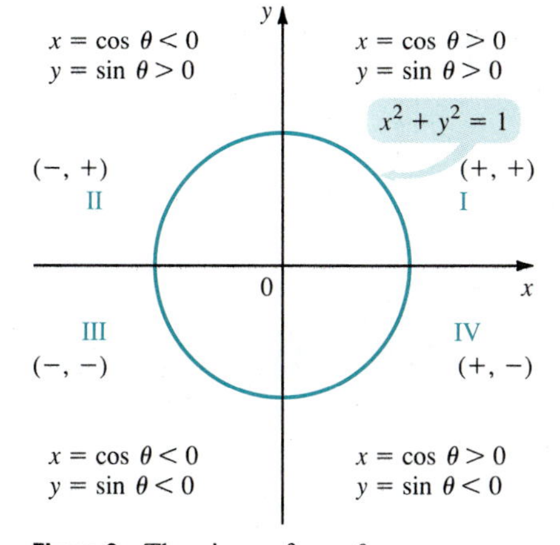

Figure 9 The signs of $\cos\theta$ and $\sin\theta$ in each quadrant.

We can use formula (2) to check the accuracy of the values in Table 2. We know that for every θ, $\sin^2\theta + \cos^2\theta = 1$. We verify that

$$\sin^2 45° + \cos^2 45° = \left(\frac{1}{\sqrt{2}}\right)^2 + \left(\frac{1}{\sqrt{2}}\right)^2 = \frac{1}{2} + \frac{1}{2} = 1$$

$$\sin^2 60° + \cos^2 60° = \left(\frac{\sqrt{3}}{2}\right)^2 + \left(\frac{1}{2}\right)^2 = \frac{3}{4} + \frac{1}{4} = 1$$

$$\sin^2 30° + \cos^2 30° = \left(\frac{1}{2}\right)^2 + \left(\frac{\sqrt{3}}{2}\right)^2 = \frac{1}{4} + \frac{3}{4} = 1$$

A glance at Figure 9 tells us the signs of $\cos\theta$ and $\sin\theta$ in each quadrant. For example, in the second quadrant, $x < 0$ and $y > 0$, so $\cos\theta = x < 0$ and $\sin\theta = y > 0$.

EXAMPLE 4 *Finding $\sin\theta$ if $\cos\theta$ Is Given*

Suppose that θ is in the first quadrant with $\cos\theta = \frac{1}{3}$. Find $\sin\theta$.

SOLUTION There are two ways to solve this problem.

Method 1 Use formula (2):

$$\sin^2\theta + \cos^2\theta = 1$$

$$\sin^2\theta + \left(\frac{1}{3}\right)^2 = 1$$

$$\sin^2\theta + \frac{1}{9} = 1$$

$$\sin^2\theta = \frac{8}{9}$$

$$\sin\theta = \pm\sqrt{\frac{8}{9}} = \pm\frac{2\sqrt{2}}{3}$$

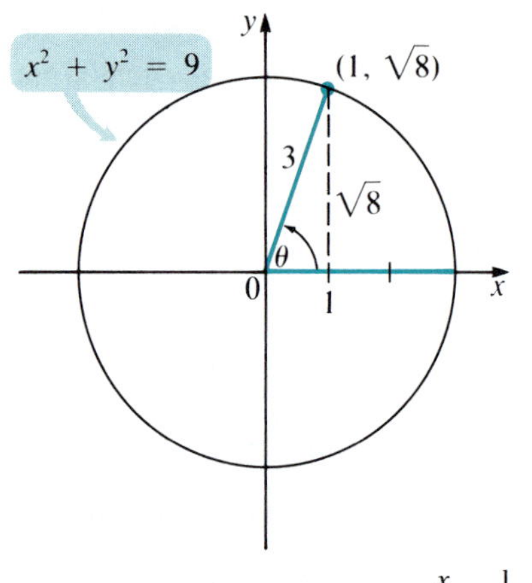

Figure 10 In this circle $\cos\theta = \frac{x}{r} = \frac{1}{3}$.

But $\sin\theta > 0$ because θ is in the first quadrant. Thus, $\sin\theta = \frac{2\sqrt{2}}{3}$.

Method 2 Draw a picture as in Figure 10, we have drawn a circle of radius 3. Since $\cos\theta = \frac{x}{r} = \frac{1}{3}$, θ is in the first quadrant, and $r = 3$, the terminal side of θ intersects the circle at a point whose x-coordinate is 1. Then, since $x^2 + y^2 = 3^2 = 9$, $y^2 = 9 - x^2$. When $x = 1$, $y^2 = 9 - 1 = 8$ and $y = \pm\sqrt{8}$. We conclude that $y = \sqrt{8}$ since θ is in the first quadrant. Thus

$$\sin\theta = \frac{y}{r} = \frac{\sqrt{8}}{3} = \frac{2\sqrt{2}}{3}$$

The fact that $y = \sqrt{8}$ in Figure 10 also follows directly from the Pythagorean theorem.

NOTE In this problem you can draw a triangle instead of a circle, as in Section 2.2. However, triangles work only when θ is in the first quadrant. Circles work for any real value of θ. ■

EXAMPLE 5 *Finding $\cos\theta$ if $\sin\theta$ Is Given*

Find $\cos\theta$ if $\sin\theta = 0.45$ and θ is in the second quadrant.

SOLUTION Again, there are two ways to do the problem.

Method 1 Use formula (2).

$$\begin{aligned}
\sin^2\theta + \cos^2\theta &= 1 \\
(0.45)^2 + \cos^2\theta &= 1 \\
0.2025 + \cos^2\theta &= 1 \\
\cos^2\theta &= 0.7975 \\
\cos\theta &= \pm\sqrt{0.7975} \approx \pm 0.8930
\end{aligned}$$

But $\cos\theta < 0$ in the second quadrant, so $\cos\theta = -0.8930$.

Method 2 Draw a picture. In Figure 11 we have drawn a circle of radius 1. We did this because $0.45 = \frac{0.45}{1}$. Since $\sin\theta = \frac{y}{r} = \frac{0.45}{1}$ and θ is in the second quadrant, θ is the angle whose terminal side intersects the unit circle at a point whose y-coordinate is 0.45 and whose x-coordinate is negative. Then, since $x^2 + y^2 = 1$,

$$x^2 = 1 - y^2 = 1 - 0.45^2 = 1 - 0.2025 = 0.7975$$

and

$$x = -\sqrt{0.7975} \approx -0.8930$$

Finally,

$$\cos\theta = \frac{x}{r} = \frac{-0.8930}{1} = -0.8930$$

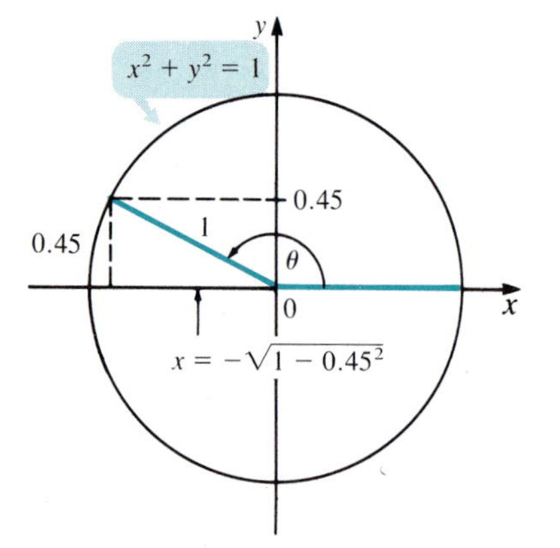

Figure 11 In this circle $\sin\theta = \frac{y}{r} = 0.45$.

NOTE In the last two examples, the problem was solved in two ways. It's up to you to decide which method to use. However, the picture method is preferable since it shows you what is happening. There is little insight gained by applying a formula.

Using a Calculator

Every scientific calculator has sine and cosine keys that provide values of $\sin\theta$ and $\cos\theta$ for every real number θ. Each such calculator also computes in either degree or radian mode. To select the mode, press the appropriate key (usually [DEG] or [DRG]) until the mode you want (deg or rad) is displayed.

WARNING When using a calculator to compute $\sin\theta$ and $\cos\theta$, make certain that:

If θ is given in degrees, the calculator is set to the degree mode.
If θ is given in radians, the calculator is set to the radian mode. ■

EXAMPLE 6 *Obtaining Values of sin θ and cos θ on a Calculator (θ in Degrees)*

Compute (a) $\sin 80°$, (b) $\cos 223°$, (c) $\sin 223°$, (d) $\cos -54°$.

SOLUTION First set the mode to degree (deg). Then press the appropriate keys.

(a) [8] [0] [sin] 0.984807753 is displayed†
(b) [2] [2] [3] [cos] −0.731353701 is displayed
(c) [2] [2] [3] [sin] −0.68199836 is displayed
(d) [5] [4] [+/−] [cos] 0.587785252 is displayed‡

The minus signs in $\cos 223°$ and $\sin 223°$ should not be surprising because 223° is in the third quadrant ($180° < \theta < 270°$ defines the third quadrant), and both $\sin\theta$ and $\cos\theta$ are negative in the third quadrant. As a further check, observe that

$$\begin{aligned}\cos^2 223° + \sin^2 223° &= (-0.731353701)^2 + (-0.68199836)^2\\ &= 0.534878236 + 0.465121763\\ &= 0.999999999\end{aligned}$$

This is okay. We have just a tiny bit of roundoff error.

---

†On some calculators, the [sin] key or the [cos] key must be pressed before the number (in degrees or radians) is entered.

‡Some calculators have the key [CHS] (for "change sign") instead of a [+/−] key.

EXAMPLE 7 *Obtaining Values of sin θ and cos θ on a Calculator (θ in Radians)*

Compute (a) sin 1, (b) cos 2.4, (c) sin 2.4.

SOLUTION First set the mode to radian (rad). Then

(a) [1] [sin] 0.841470984 is displayed
(b) [2] [.] [4] [cos] −0.737393715 is displayed
(c) [2] [.] [4] [sin] 0.67546318 is displayed

Note that 2.4 is in the second quadrant (see Table 3 on p. 62), so

$$\cos 2.4 < 0 \qquad \text{and} \qquad \sin 2.4 > 0$$

NOTE In Table 4 at the back of the book we provide tables of sin θ and cos θ for those of you who are working without calculators.

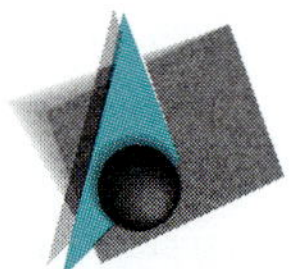

FOCUS ON
The History of Trigonometry

We mentioned in the introduction that one early use of trigonometric ideas was in the design of a simple clock. These ideas were known to the Egyptians and the Babylonians in the Middle East at least 3500 years ago. However, as we mentioned earlier the first mathematicians formally to develop ideas in trigonometry were Greek. The Greeks used trigonometry to solve a variety of important problems, including telling time, predicting the paths of celestial bodies (sun, moon, planets, stars), navigating on the earth, and designing calendars.

The earliest trigonometry was spherical. That is, it was carried out on the surface of a sphere (like, approximately, the earth). However, many of the ideas in spherical trigonometry extend to the plane trigonometry we study in this book.

Hipparchus of Rhodes (ca. 180–ca. 125 B.C.) is considered by many to be the founder of trigonometry. Hipparchus was an eminent astronomer who made a number of important discoveries at his observatory. Among them was his calculation of the average lunar month to within 1 second of the value accepted today.

In his computations, Hipparchus defined what he called the **chord function,** crd θ. To describe this function, consider the circle drawn in Figure 12. In the circle, crd θ denotes the length of the chord that joins the two ends of the arc of length θ. (Remember, the radian measure of the central angle θ in a circle of radius 1 is the numerical value of the length of the arc subtended by θ.)

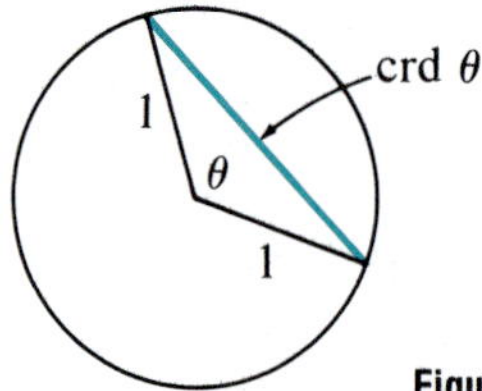

Figure 12 A circle of radius 1 with central angle θ.

It is not difficult to show that crd θ is related to sin θ by the formula

$$\text{crd } \theta = 2 \sin \frac{\theta}{2} \tag{6}$$

You are asked to prove this in Problem 63.

One of the most outstanding astronomers at the beginning of the Christian era was Claudius Ptolemy (ca. 85–ca A.D. 165). Ptolemy's most influential work was his *Almagest.* This was named by later Islamic astronomers and means, in Arabic, "the greatest." The *Almagest* is based primarily on the work of Hipparchus, but it exerted much more influence because of the clarity and elegance of its style. The *Almagest* was the standard reference work on astronomy until the publication of works by Copernicus and Kepler in the sixteenth century.

In the *Almagest,* Ptolemy compiled a very accurate table of chords which gives, essentially (according to (6)), the sines of angles between 0° and 45° in fifteen-minute intervals. He also derived a number of identities that led to the trigonometric identities that we study in Chapter 8.

Long after the publication of the *Almagest,* some unknown mathematician (or, more likely, astronomer) thought to use half the central angle in his computations, and the sine function was born. (Note that, in formula (6), $\theta/2$ is the variable in the sine function.) The first tables of values of sin θ originated in India in the fourth century. One of these occurs in the obscure *Surya Siddhanta* published in Sanskrit verse. This work was a compendium of astronomy that freely used Greek ideas. It is likely that Greek scientific works came to India as part of the extensive Roman trade with south India by way of the Red Sea and the Indian Ocean.

The study of trigonometry soon spread to many parts of the world, with the notable exception of Western Europe. From the ninth to the fourteenth centuries, many of the significant advances in trigonometry were carried out by Islamic astronomers and mathematicians. It was not until the fifteenth century that trigonometry was studied in Western Europe, and by then most of the results we use today were already known.

The origin of the term *sine* is interesting. The original word was the Sanskrit *ardhajya* meaning "half chord." It was first used in India by Aryabhata the Elder around A.D. 500. Islamic astronomers shortened the word and transliterated it as *jyb*. In Arabic, vowels are not spelled out. The word was pronounced as *jayb*, which in Arabic means "pocket" or "gulf." Fifteenth-century Europeans then translated the term into the Latin word for gulf, which is *sinus*. This, when translated into English, is the word we use today.

Problems 2.3

Readiness Check

I. Which of the following represents $\sin\theta$?
   a. x, on the unit circle
   b. y, on the unit circle
   c. $\frac{r}{y}$
   d. $\cos\theta - 1$

II. Which of the following is the quadrant containing the terminal side of θ in standard position if $\sin\theta < 0$ and $\cos\theta < 0$?
   a. I b. II c. III d. IV

III. Which of the following is negative?
   a. $\cos(-30°)$ b. $\cos\frac{3\pi}{2}$
   c. $\sin\frac{5\pi}{4}$ d. $\sin(-200°)$

IV. Which of the following pairs of angles have equal sine values?
   a. 50° and 410° b. 110° and −110°
   c. 90° and 270° d. 240° and 330°

In Problems 1–10 a circle is drawn. Determine $\cos\theta$ and $\sin\theta$.

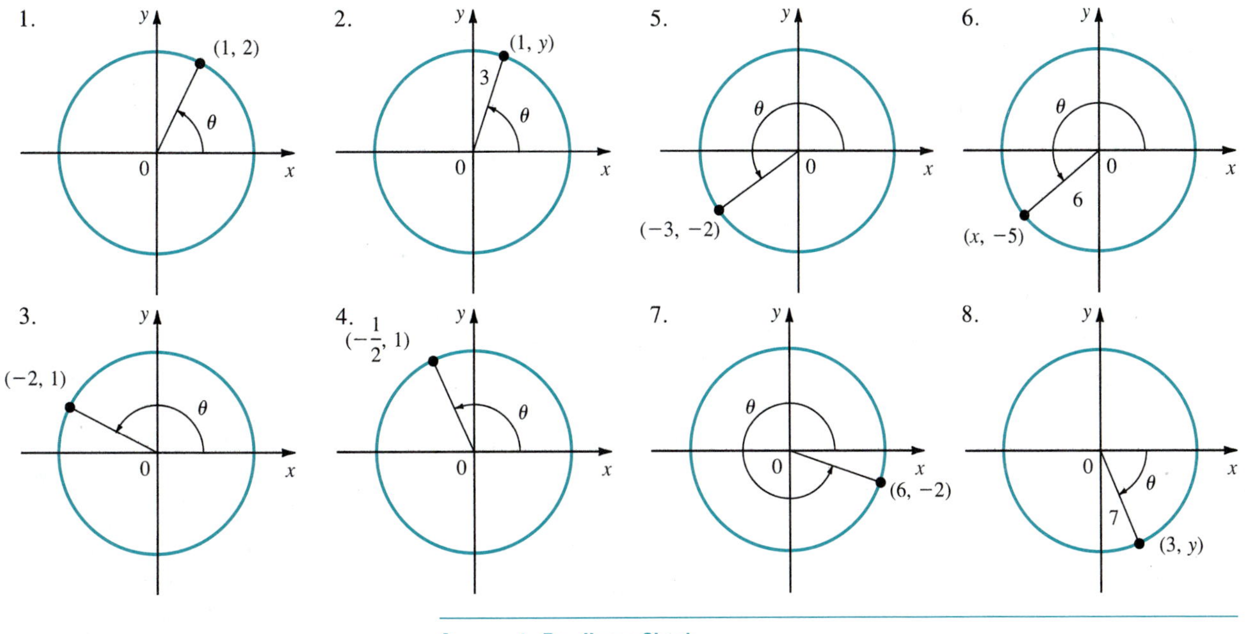

Answers to Readiness Check

I. b II. c III. c IV. a

9.

10.

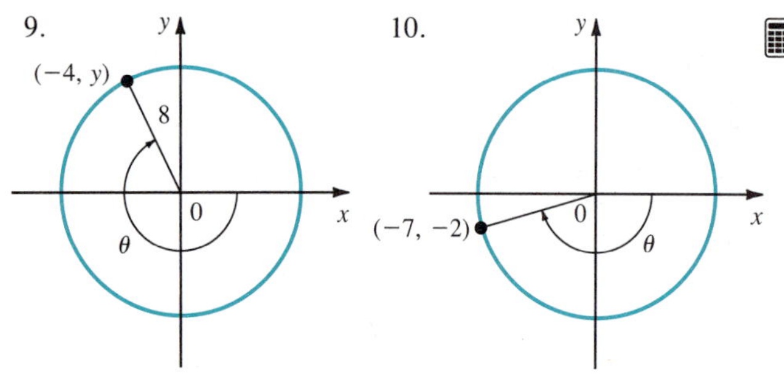

In Problems 11–22 (a) find the point at which the terminal side of the given angle θ intersects the unit circle and (b) use this point to find $\cos\theta$ and $\sin\theta$.

11. $-180°$ 12. 6π 13. 5π 14. $540°$
15. $\dfrac{5\pi}{2}$ 16. $\dfrac{15\pi}{2}$ 17. $-90°$ 18. $450°$
19. $\dfrac{9\pi}{4}$ 20. $-\dfrac{5\pi}{3}$ 21. $390°$ 22. $420°$

In Problems 23–40 determine the sign of $\cos\theta$ and $\sin\theta$ for the given value of θ. Do not use a calculator.

23. $55°$ 24. $312°$ 25. $247°$ 26. $115°$
27. $161°$ 28. $-37°$ 29. $-200°$ 30. $500°$
31. $1500°$ 32. 1.3 33. 2.1 34. 3.2
35. 8.9 36. 4.8 37. $\dfrac{15\pi}{16}$ 38. $\dfrac{-13\pi}{8}$
39. 10 40. 48

In Problems 41–46 the quadrant of θ is given. If $\sin\theta$ is given, find $\cos\theta$. If $\cos\theta$ is given, compute $\sin\theta$.

41. $\sin\theta = 0.6$, I 42. $\sin\theta = 0.8$, II
43. $\cos\theta = -0.75$, III 44. $\cos\theta = -0.75$, II
45. $\sin\theta = -0.38$, IV 46. $\sin\theta = -0.38$, II
47. Find (if possible) $\cos\theta$ if θ is in the first quadrant and $4\sin\theta = 5$.
48. Find (if possible) $\sin\theta$ if θ is in the third quadrant and $\dfrac{\cos\theta}{2} + 0.65 = 0$.

In Problems 49–52 determine the quadrant of θ.

49. $\sin\theta \approx -0.2347$, $\cos\theta \approx 0.9721$
50. $\sin\theta \approx -0.8273$, $\cos\theta \approx -0.5618$
51. $\sin\theta \approx 0.6831$, $\cos\theta \approx 0.7303$
52. $\sin\theta \approx 0.3296$, $\cos\theta \approx -0.9441$

In Problems 53–56 use a calculator to compute $\sin\theta$ and $\cos\theta$ and verify that $\sin^2\theta + \cos^2\theta = 1$.

53. $\theta = 37°$ 54. $\theta = 121°$
55. $\theta = 263°$ 56. $\theta = 27.5°$

In calculus, it is shown that if θ is measured in radians then

$$\cos\theta \approx 1 - \frac{\theta^2}{2} + \frac{\theta^4}{24} - \frac{\theta^6}{720}$$

and

$$\sin\theta \approx \theta - \frac{\theta^3}{6} + \frac{\theta^5}{120} - \frac{\theta^7}{5040}$$

These approximations become more accurate as θ gets closer to 0.

In Problems 57–62 use these formulas to approximate $\sin\theta$ and $\cos\theta$. [Hint: If θ is given in degrees, it is first necessary to convert to radians.] Then use your calculator to evaluate each expression to 6 decimal places and compute the error in your approximation.

57. $\theta = 1.5$ 58. $\theta = 0.7$ 59. $\theta = 50°$
60. $\theta = 35°$ 61. $\theta = -0.9$ 62. $\theta = -25°$
63. In order to derive values of the chord function, crd θ, place the unit circle so that its center is at the origin and half of θ is in the first quadrant, as in the figure below. Show that $\sin\dfrac{\theta}{2} = \dfrac{1}{2}\,\text{crd}\,\theta$ and that $\text{crd}\,\theta = 2\sin\dfrac{\theta}{2}$.

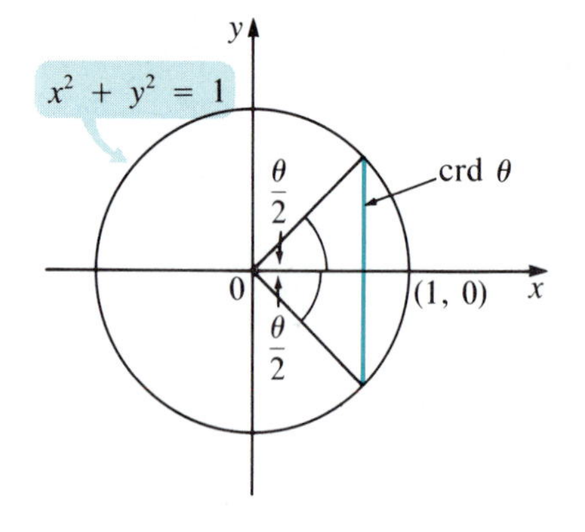

In Problems 64–72 find the length of the chord subtended by θ. That is, compute crd θ.

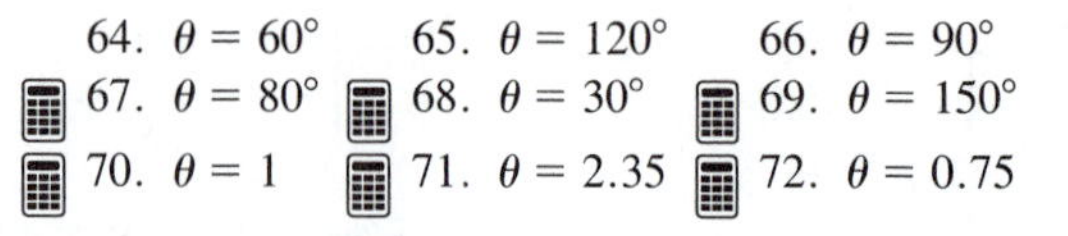

64. $\theta = 60°$ 65. $\theta = 120°$ 66. $\theta = 90°$
67. $\theta = 80°$ 68. $\theta = 30°$ 69. $\theta = 150°$
70. $\theta = 1$ 71. $\theta = 2.35$ 72. $\theta = 0.75$

73. The arrow in a spinner is 1 foot long with its center at the origin. Assume that, before spinning, the spinner points horizontally to the right and that it spins counterclockwise. Find, approximately, the coordinates of the point at which the arrow rests after it has
    (a) revolved exactly $2\frac{1}{2}$ times
    (b) revolved exactly $7\frac{3}{4}$ times
    * (c) traveled 2 ft
    * (d) traveled 20 ft

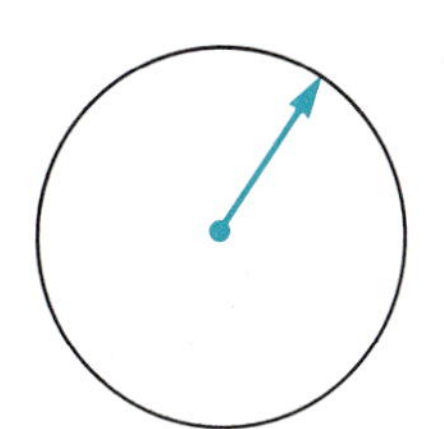

2.4 Variation, Periodicity, Basic Identities, and Graphs of sin θ and cos θ

A great deal of information about the functions sin θ and cos θ can be obtained by looking at the graph of the unit circle. In this section, we discuss a variety of facts; each can be proved by referring to the circle.

Periodicity of sin θ and cos θ

A function f is said to be **periodic of period T** if T is the *smallest* positive number such that

$$f(x + T) = f(x) \tag{1}$$

for every x in the domain of f.

Theorem 1: sin θ and cos θ Are Periodic of Period 2π (360°)

Let θ be a real number.

(a) If θ is measured in radians, then

$$\cos(\theta + 2\pi) = \cos\theta \tag{2}$$

$$\sin(\theta + 2\pi) = \sin\theta \tag{3}$$

(b) If θ is measured in degrees, then

$$\cos(\theta + 360°) = \cos\theta \tag{4}$$

$$\sin(\theta + 360°) = \sin\theta \tag{5}$$

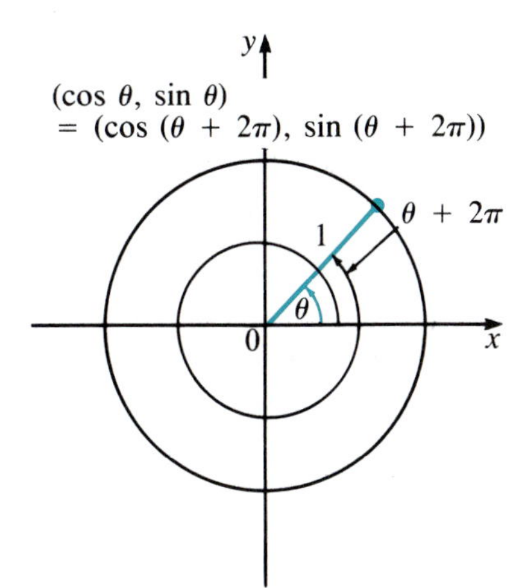

Figure 1 When we go once around the unit circle, we return to the same values of cos θ ($= x$) and sin θ ($= y$).

Sketch of Proof

Look at Figure 1. θ and $\theta + 2\pi$ have the same initial side (the positive x-axis). The terminal side of $\theta + 2\pi$ is obtained by revolving the terminal side of θ exactly once around the circle. Thus, θ and $\theta + 2\pi$ have the same terminal side and so are coterminal. By the theorem on p. 83, we see that $\sin\theta = \sin(\theta + 2\pi)$ and $\cos\theta = \cos(\theta + 2\pi)$. Note that if we revolve the terminal side of θ by less than 2π radians, then we do not get back to the same point. Since 2π is the *smallest* number for which $\sin(\theta + T) = \sin\theta$ and $\cos(\theta + T) = \cos\theta$ for all θ, the period of sin θ and cos θ is 2π. ■

EXAMPLE 1 ***Computing Values of sin θ and cos θ for θ > 2π (θ > 360°)***

Compute (a) $\cos \dfrac{9\pi}{4}$ and (b) $\sin 390°$.

SOLUTION

(a) $\cos \dfrac{9\pi}{4} = \cos\left(\dfrac{\pi}{4} + 2\pi\right) = \cos \dfrac{\pi}{4} = \dfrac{1}{\sqrt{2}}$

(b) $\sin 390° = \sin (30° + 360°) = \sin 30° = \dfrac{1}{2}$

Variation of sin θ and cos θ

Look at Figure 2. The arrows on the circle indicate the direction we move as θ increases from 0 to 2π. Let us follow the x-coordinate. When $\theta = 0$, $x = 1$.

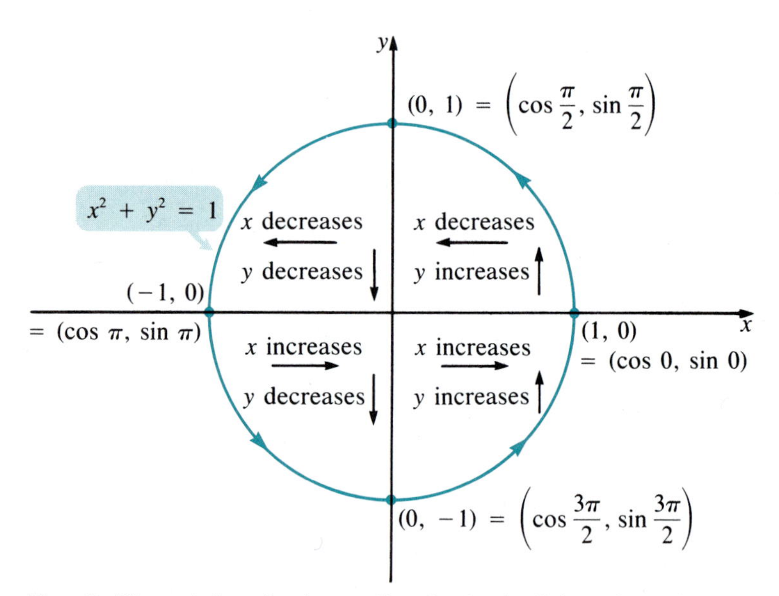

Figure 2 The variation of x (= cos θ) and y (= sin θ) in each quadrant.

As θ increases from 0 to $\pi/2$, x decreases from 1 to 0. As θ increases from $\pi/2$ to π, x decreases from 0 to -1, its minimum value. Continuing, we see that as θ increases from π to $3\pi/2$, x increases from -1 to 0. Finally, as θ increases from $3\pi/2$ to 2π, x increases from 0 to 1.

The x values are the values of cos θ. Since cos θ is periodic of period 2π, the cycle repeats as θ increases from 2π to 4π or from -2π to 0.

If we look at the y-coordinate, then we see how sin θ varies as θ increases from 0 to 2π. The results are given in Table 1.

Table 1

As θ Increases	$\cos\theta$	$\sin\theta$
from 0 to $\pi/2$	← decreases from 1 to 0	↑ increases from 0 to 1
from $\pi/2$ to π	← decreases from 0 to -1	↓ decreases from 1 to 0
from π to $3\pi/2$	→ increases from -1 to 0	↓ decreases from 0 to -1
from $3\pi/2$ to 2π	→ increases from 0 to 1	↑ increases from -1 to 0

Graphs of cos θ and sin θ

We can use the information in Table 1 to obtain graphs of $\cos\theta$ and $\sin\theta$. This is done in Figures 3 and 4. In Figure 3(a), we graph $u = \cos\theta$ for θ between 0 and 2π. In Figure 3(b), we extend the graph periodically to obtain values of θ outside the interval $[0, 2\pi]$. We construct analogous graphs in Figure 4 for $u = \sin\theta$.

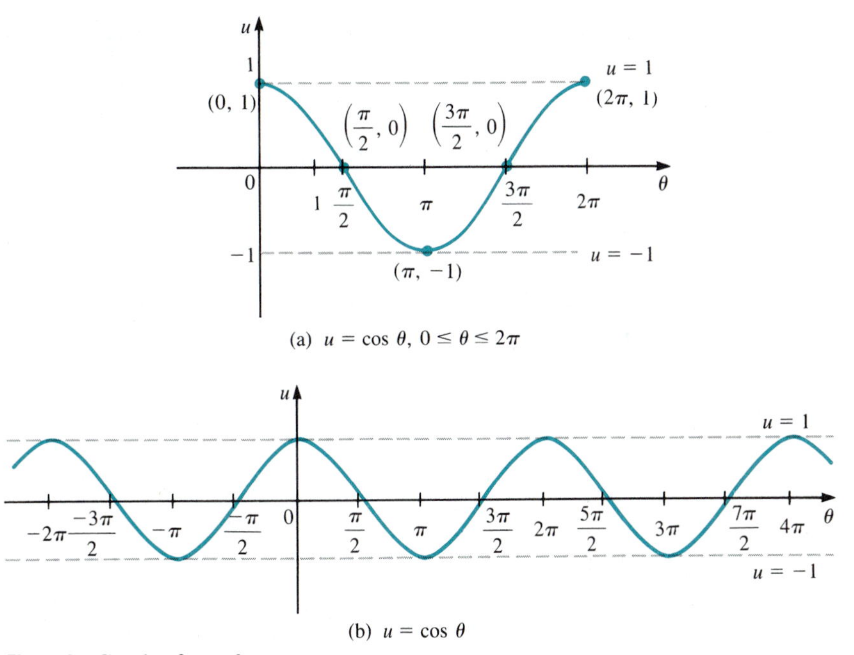

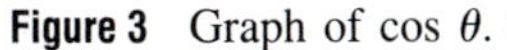
Figure 3 Graph of $\cos\theta$.

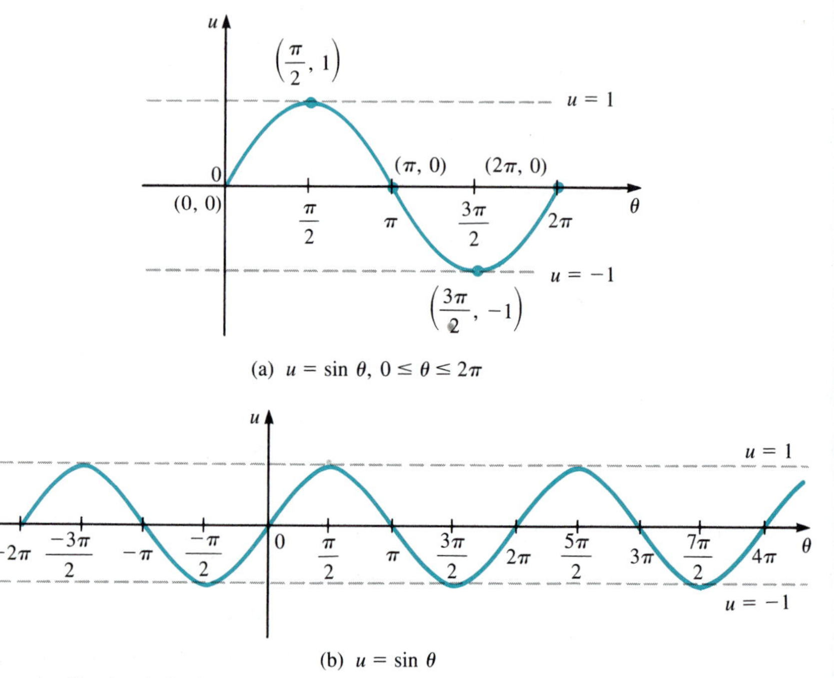

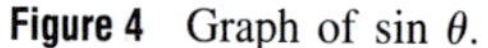

Figure 4 Graph of $\sin\theta$.

Recall from Section 1.3 (p. 23) that a function f is **even** if $f(-x) = f(x)$ or **odd** if $f(-x) = -f(x)$.

Theorem 2: cos*θ* Is an Even Function and sin*θ* Is an Odd Function

That is, for every real number θ

$$\cos(-\theta) = \cos\theta \tag{6}$$

$$\sin(-\theta) = -\sin\theta \tag{7}$$

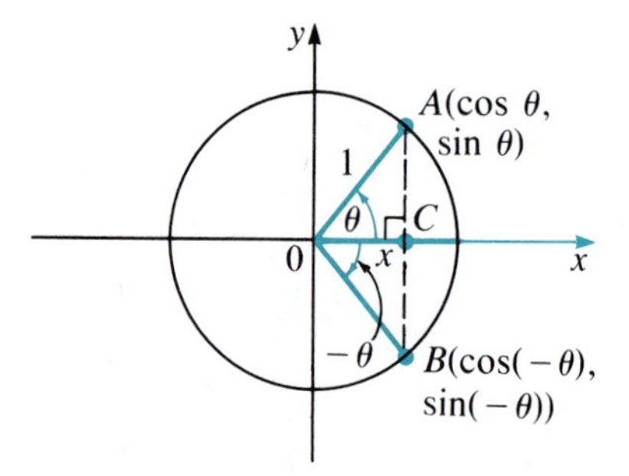

Figure 5 θ and $-\theta$ produce the same x-coordinates; the y-coordinates are negatives of each other.

Sketch of Proof

Look at Figure 5. The points A and B have the same x-coordinate, so $\cos\theta = \cos(-\theta)$. From the Pythagorean theorem

$$\overline{AC}^2 + x^2 = 1 \qquad \text{or} \qquad \overline{AC} = \sqrt{1 - x^2}$$

and

$$\overline{BC}^2 + x^2 = 1 \qquad \text{or} \qquad \overline{BC} = \sqrt{1 - x^2}$$

Thus, $\overline{AC} = \overline{BC}$, and, since the y-coordinate of B is negative, the y-coordinate of B is the negative of the y-coordinate of A. That is, $\sin(-\theta) = -\sin\theta$. ■

EXAMPLE 2 *Computing cos (−θ) and sin (−θ)*

Compute (a) $\cos(-30°)$ and (b) $\sin(-\pi/4)$.

SOLUTION

(a) $\cos(-30°) = \cos 30° = \dfrac{\sqrt{3}}{2}$

(b) $\sin\left(-\dfrac{\pi}{4}\right) = -\sin\dfrac{\pi}{4} = -\dfrac{1}{\sqrt{2}}$

Other facts follow readily from looking at an appropriate graph. We list some of them here with an indication of why they are true.

Adding or Subtracting π or 180°

If θ is measured in radians, then

$$\cos(\theta \pm \pi) = -\cos\theta \tag{8}$$

$$\sin(\theta \pm \pi) = -\sin\theta \tag{9}$$

If θ is measured in degrees, then

$$\cos(\theta \pm 180°) = -\cos\theta \tag{10}$$

$$\sin(\theta \pm 180°) = -\sin\theta \tag{11}$$

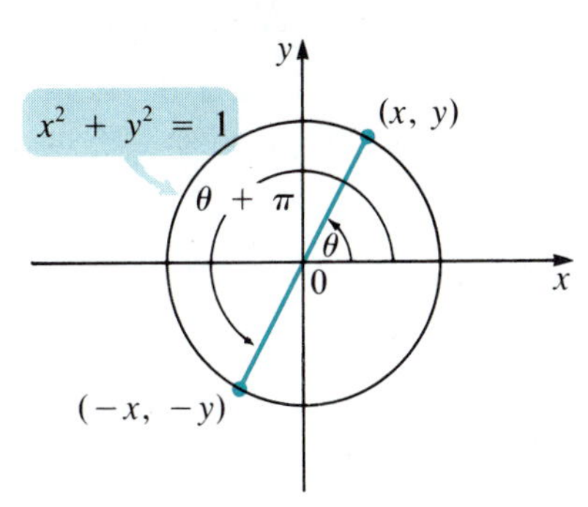

Figure 6 The x- and y-coordinates of $\theta + \pi$ are the negatives of the x- and y-coordinates of θ.

Sketch of Proof

Look at Figure 6. If the terminal side of θ intersects the unit circle at (x, y), then the terminal side of $\theta + \pi$ intersects the unit circle at $(-x, -y)$. So

$$\cos(\theta + \pi) = -x = -\cos\theta$$

and

$$\sin(\theta + \pi) = -y = -\sin\theta$$

Also, $\cos(\theta - \pi) \underset{\text{From (2)}}{=} \cos(\theta - \pi + 2\pi) = \cos(\theta + \pi) = -\cos\theta$. Similarly, $\sin(\theta - \pi) = \sin(\theta + \pi) = -\sin\theta$.

For more details, see Problem 55. ■

EXAMPLE 3 ***Using the Fact That $\sin(\theta + \pi) = -\sin\theta$ and $\cos(\theta + \pi) = -\cos\theta$***

Compute (a) $\sin 4\pi/3$ and (b) $\cos 225°$.

SOLUTION

(a) $\sin\dfrac{4\pi}{3} = \sin\left(\dfrac{\pi}{3} + \pi\right) = -\sin\dfrac{\pi}{3} = -\dfrac{\sqrt{3}}{2}$

(b) $\cos 225° = \cos(45° + 180°) = -\cos 45° = -1/\sqrt{2}$

Sine and Cosine of Supplementary Angles

Suppose that $0 < \theta < \pi$. If θ is measured in radians, then

$$\cos(\pi - \theta) = -\cos\theta \tag{12}$$

$$\sin(\pi - \theta) = \sin\theta \tag{13}$$

If θ is measured in degrees, then

$$\cos(180^\circ - \theta) = -\cos\theta \tag{14}$$

$$\sin(180^\circ - \theta) = \sin\theta \tag{15}$$

Proof

There are at least two ways to prove these relations. One way is to observe that $\pi - \theta = -(\theta - \pi)$.

Thus

$$\cos(\pi - \theta) = \cos(-(\theta - \pi)) \overset{\text{From (6)}}{=} \cos(\theta - \pi) \overset{\text{From (8)}}{=} -\cos\theta$$

and

$$\sin(\pi - \theta) = \sin(-(\theta - \pi)) \overset{\text{From (7)}}{=} -\sin(\theta - \pi) \overset{\text{From (9)}}{=} -(-\sin\theta) = \sin\theta$$

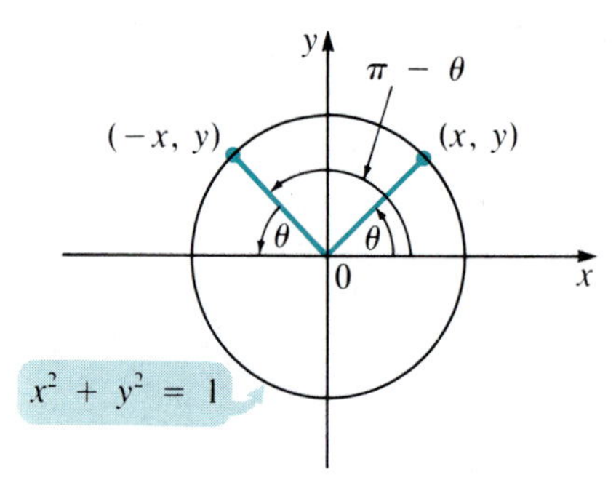

Figure 7 The x-coordinate of $\pi - \theta$ is the negative of the x-coordinate of θ. The y-coordinates are the same.

The other way is to look at Figure 7. Both θ and $\pi - \theta$ have the same y-coordinate (sine), but their x-coordinates (cosines) are negatives of one another. ■

EXAMPLE 4 ***Using the Fact That $\cos(\pi - \theta) = -\cos\theta$ and $\sin(\pi - \theta) = \sin\theta$***

Compute (a) $\cos 150^\circ$ and (b) $\sin \dfrac{3\pi}{4}$.

SOLUTION (a) $\cos 150^\circ = \cos(180^\circ - 30^\circ) = -\cos 30^\circ = -\dfrac{\sqrt{3}}{2}$

(b) $\sin \dfrac{3\pi}{4} = \sin\left(\pi - \dfrac{\pi}{4}\right) = \sin\dfrac{\pi}{4} = \dfrac{1}{\sqrt{2}} = \dfrac{\sqrt{2}}{2}$

Sine and Cosine of Complementary Angles

Let θ be an acute angle. If θ is measured in radians, then

$$\cos\left(\frac{\pi}{2} - \theta\right) = \sin\theta \tag{16}$$

$$\sin\left(\frac{\pi}{2} - \theta\right) = \cos\theta \tag{17}$$

If θ is measured in degrees, then

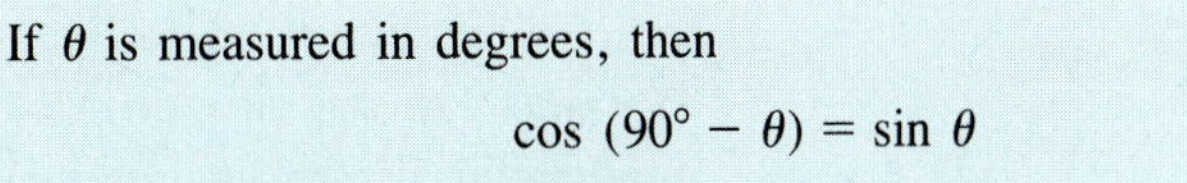

$$\cos(90° - \theta) = \sin\theta \tag{18}$$

$$\sin(90° - \theta) = \cos\theta \tag{19}$$

Proof

We proved this in Section 2.2 (p. 71). ■

EXAMPLE 5 *Illustrating the Fact That* $\cos\left(\frac{\pi}{2} - \theta\right) = \sin\theta$ *and* $\sin\left(\frac{\pi}{2} - \theta\right) = \cos\theta$

We observe that $\frac{\pi}{3} + \frac{\pi}{6} = \frac{\pi}{2}$, so as we already know, $\cos\frac{\pi}{3} = \sin\left(\frac{\pi}{2} - \frac{\pi}{3}\right) = \sin\frac{\pi}{6} = \frac{1}{2}$ and $\cos\frac{\pi}{6} = \sin\left(\frac{\pi}{2} - \frac{\pi}{6}\right) = \sin\frac{\pi}{3} = \frac{\sqrt{3}}{2}$. ■

EXAMPLE 6 *Sines and Cosines of Complementary Angles*

Given that sin 34° = 0.5592, find an acute angle θ such that $\cos\theta = 0.5592$.

SOLUTION $0.5592 = \sin 34° = \cos(90° - 34°) = \cos 56°$. The answer is 56°. ■

EXAMPLE 7 *Using the Fact That* $\cos\theta = \sin\left(\frac{\pi}{2} - \theta\right)$

Given that cos 1.2 = 0.3624, find two numbers α and β in $(0, 2\pi)$ such that $\sin\alpha = \sin\beta = 0.3624$. Use the value $\pi = 3.1416$.

SOLUTION $0.3624 = \cos 1.2 = \sin\left(\frac{\pi}{2} - 1.2\right) = \sin\left(\frac{3.1416}{2} - 1.2\right)$

$$= \sin(1.5708 - 1.2) = \sin 0.3708$$

Thus, one value is $\alpha = 0.3708$

To find the other value, recall that $\sin\theta > 0$ in the first and second quadrants. Since $\alpha = 0.3708$ is in the first quadrant, we seek a number β in the second quadrant such that $\sin\beta = \sin\alpha$. From (13), $\sin(\pi - \alpha) = \sin\alpha$. Thus

$$\beta = \pi - \alpha = 3.1416 - 0.3708 = 2.7708$$

The two numbers are 0.3708 and 2.7708. You should verify on a calculator that $\sin 0.3708 = \sin 2.7708 = 0.3624$ to 4 decimal places.

Reference Angles

The angles 0, $\pi/2 = 90°$, $\pi = 180°$, and $3\pi/2 = 270°$ are called **axial** (or **quadrantal**) because their terminal sides do not lie in any quadrant. (Each lies on the x- or y-axis). Angles coterminal to one of these are also called axial.

Let θ be an angle in standard position that is not axial. Then the **reference angle** of θ, denoted by θ_r, is the positive acute angle that the terminal side of θ makes with the x-axis.

EXAMPLE 8 *Finding Six Reference Angles*

Find the reference angles for (a) 125°, (b) 210°, (c) 320°, (d) $5\pi/6$, (e) $5\pi/4$, (f) 6.

SOLUTION In Figure 8, we draw the appropriate pictures.

We have (a) $\theta_r = 55°$, (b) $\theta_r = 30°$, (c) $\theta_r = 40°$, (d) $\theta_r = \pi/6$, (e) $\theta_r = \pi/4$, (f) $\theta_r = 2\pi - 6 \approx 0.2832$.

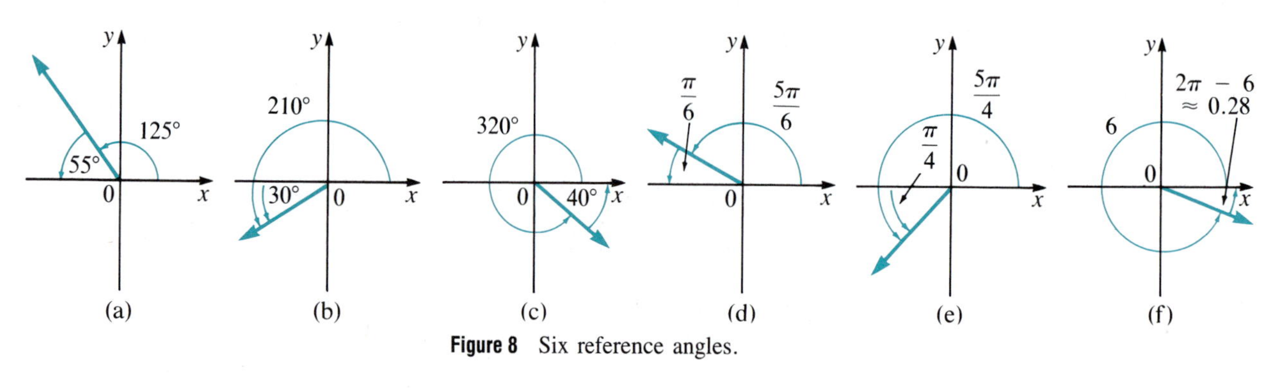

Figure 8 Six reference angles.

From Figure 9, we find that

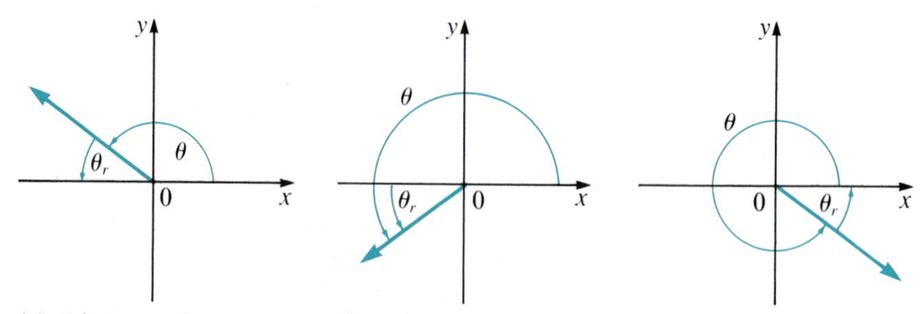

(a) θ in second quadrant (b) θ in third quadrant (c) θ in fourth quadrant

Figure 9 Depicting reference angles in each quadrant.

If θ is in the second quadrant, then $\theta_r = \pi - \theta = 180° - \theta =$ supplement of θ.

If θ is in the third quadrant, then $\theta_r = \theta - \pi = \theta - 180°$.

If θ is in the fourth quadrant, then $\theta_r = 2\pi - \theta = 360° - \theta$.

Thus,

If θ is in the second quadrant, then $\theta_r = \pi - \theta$ and $\theta = \pi - \theta_r$, so

$$\sin\theta = \underset{\text{From (13)}}{\sin(\pi - \theta_r)} = \sin\theta_r$$

and

$$\cos\theta = \underset{\text{From (12)}}{\cos(\pi - \theta_r)} = -\cos\theta_r$$

If θ is in the third quadrant, then $\theta_r = \theta - \pi$ and $\theta = \theta_r + \pi$, so

$$\sin\theta = \underset{\text{From (9)}}{\sin(\theta_r + \pi)} = -\sin\theta_r$$

and

$$\cos\theta = \underset{\text{From (8)}}{\cos(\theta_r + \pi)} = -\cos\theta_r$$

If θ is in the fourth quadrant, then $\theta_r = 2\pi - \theta$ and $\theta = 2\pi - \theta_r$, so

$$\sin\theta = \sin(2\pi - \theta_r) \overset{\text{From (3)}}{=} \sin(-\theta_r) \overset{\text{From (7)}}{=} -\sin\theta_r$$

and

$$\cos\theta = \cos(2\pi - \theta_r) \overset{\text{From (2)}}{=} \cos(-\theta_r) \overset{\text{From (6)}}{=} \cos\theta_r$$

In sum, we have

Reference Angle Theorem

If θ_r is the reference angle of θ, then

$$\sin\theta = \sin\theta_r \quad \text{or} \quad \sin\theta = -\sin\theta_r$$

and

$$\cos\theta = \cos\theta_r \quad \text{or} \quad \cos\theta = -\cos\theta_r$$

The sign is determined by the quadrant of θ.

EXAMPLE 9 *Using the Reference Angle Theorem*

Determine $\sin\theta$ and $\cos\theta$ for (a) $\theta = 210°$, (b) $\theta = \dfrac{2\pi}{3}$, (c) $\theta = \dfrac{7\pi}{4}$.

SOLUTION

(a) $\theta = 210°$ is in the third quadrant, so $\theta_r = 210° - 180° = 30°$. Thus,

$\sin 210°$ and $\cos 210°$ are negative, so

$$\sin 210° = -\sin 30° = -\frac{1}{2}$$

and

$$\cos 210° = -\cos 30° = -\frac{\sqrt{3}}{2}$$

(b) $\theta = \frac{2\pi}{3}$ is in the second quadrant so $\theta_r = \pi - \frac{2\pi}{3} = \frac{\pi}{3}$. Then $\sin\left(\frac{2\pi}{3}\right) > 0$, $\cos\left(\frac{2\pi}{3}\right) < 0$, so

$$\sin \frac{2\pi}{3} = \sin \frac{\pi}{3} = \frac{\sqrt{3}}{2}$$

and

$$\cos \frac{2\pi}{3} = -\cos \frac{\pi}{3} = -\frac{1}{2}$$

(c) $\theta = \frac{7\pi}{4}$ is in the fourth quadrant and $\theta_r = 2\pi - \frac{7\pi}{4} = \frac{\pi}{4}$. Then $\sin \theta < 0$ and $\cos \theta > 0$, so

$$\sin \frac{7\pi}{4} = -\sin \frac{\pi}{4} = -\frac{1}{\sqrt{2}}$$

and

$$\cos \frac{7\pi}{4} = \cos \frac{\pi}{4} = \frac{1}{\sqrt{2}}$$

Table 2 gives $\sin \theta$ and $\cos \theta$ for values of θ in $[0, 2\pi]$.

Table 2

θ (radians)	0	$\frac{\pi}{6}$	$\frac{\pi}{4}$	$\frac{\pi}{3}$	$\frac{\pi}{2}$	$\frac{2\pi}{3}$	$\frac{3\pi}{4}$	$\frac{5\pi}{6}$	π	$\frac{7\pi}{6}$	$\frac{5\pi}{4}$	$\frac{4\pi}{3}$	$\frac{3\pi}{2}$	$\frac{5\pi}{3}$	$\frac{7\pi}{4}$	$\frac{11\pi}{6}$	2π
θ (degrees)	0	30	45	60	90	120	135	150	180	210	225	240	270	300	315	330	360
$\sin \theta$	0	$\frac{1}{2}$	$\frac{1}{\sqrt{2}}$	$\frac{\sqrt{3}}{2}$	1	$\frac{\sqrt{3}}{2}$	$\frac{1}{\sqrt{2}}$	$\frac{1}{2}$	0	$-\frac{1}{2}$	$-\frac{1}{\sqrt{2}}$	$-\frac{\sqrt{3}}{2}$	-1	$-\frac{\sqrt{3}}{2}$	$-\frac{1}{\sqrt{2}}$	$-\frac{1}{2}$	0
$\cos \theta$	1	$\frac{\sqrt{3}}{2}$	$\frac{1}{\sqrt{2}}$	$\frac{1}{2}$	0	$-\frac{1}{2}$	$-\frac{1}{\sqrt{2}}$	$-\frac{\sqrt{3}}{2}$	-1	$-\frac{\sqrt{3}}{2}$	$-\frac{1}{\sqrt{2}}$	$-\frac{1}{2}$	0	$\frac{1}{2}$	$\frac{1}{\sqrt{2}}$	$\frac{\sqrt{3}}{2}$	1

We have seen that $\sin 0 = \sin \pi = 0$ and $\cos \frac{\pi}{2} = \cos \frac{3\pi}{2} = 0$. These are the only zeros of $\sin \theta$ and $\cos \theta$ in the interval $[0, 2\pi)$. However, both functions are periodic of period 2π. Thus $\sin \theta = 0$ when $\theta = 0, \pi, 0 + 2\pi,$

$0 - 2\pi,\quad \pi + 2\pi,\quad \pi - 2\pi, \ldots,$ and $\cos\theta = 0$ when $\theta = \frac{\pi}{2}, \frac{3\pi}{2},$

$\frac{\pi}{2} + 2\pi, \frac{\pi}{2} - 2\pi, \frac{3\pi}{2} + 2\pi, \frac{3\pi}{2} - 2\pi, \ldots.$

That is,

Zeros of sin θ and cos θ

The zeros of $\sin\theta$ are $0, \pm\pi, \pm 2\pi, \ldots$

$= n\pi$ for $n = 0, \pm 1, \pm 2, \ldots$

$=$ integer multiples of π

and

the zeros of $\cos\theta$ are $\pm\frac{\pi}{2}, \pm\frac{3\pi}{2}, \pm\frac{5\pi}{2}, \ldots$

$= \frac{\pi}{2} + n\pi, \qquad n = 0, \pm 1, \pm 2, \ldots,$

$= (2n + 1)\frac{\pi}{2}, \qquad n = 0, \pm 1, \pm 2, \ldots,$

$=$ odd multiples of $\frac{\pi}{2}$

Problems 2.4

Readiness Check

I. Which of the following is true about $y = f(\theta) = \cos\theta$?
a. It is an odd function.
b. It decreases as θ increases from $\frac{\pi}{2}$ to π.
c. It increases as θ increases from 0 to $\frac{\pi}{2}$.
d. It has a range of all real numbers.

II. Which of the following is true about the sine function?
a. Its domain is [−1, 1].
b. It has a period of π.
c. It is an even function.
d. It increases as θ increases from $\frac{3\pi}{2}$ to 2π.

III. Which of the following is equal to sin 80°?
a. sin 10°
b. sin (−10°)
c. cos (−10°)
d. cos 80°

IV. Which of the following are all the θ values in $[0, 2\pi]$ such that $\cos\theta = \frac{1}{2}$? [Do not use a calculator.]
a. $\frac{\pi}{6}, \frac{11\pi}{6}$ b. $\frac{\pi}{3}, \frac{5\pi}{3}$
c. $\frac{\pi}{6}, \frac{5\pi}{6}$ d. $\frac{\pi}{3}, \frac{2\pi}{3}$

Answers to Readiness Check

I. b II. d III. c IV. b

In Problems 1–10 find the reference angle for each given angle.

1. $\theta = \frac{2\pi}{3}$ 2. $\theta = \frac{11\pi}{6}$ 3. $\theta = \frac{5\pi}{3}$
4. $\theta = \frac{7\pi}{4}$ 5. $\theta = \frac{21\pi}{10}$ 6. $\theta = 100°$
7. $\theta = 200°$ 8. $\theta = 300°$ 9. $\theta = 197°$
10. $\theta = 238°$

In Problems 11–38 find $\sin\theta$ and $\cos\theta$ exactly. Do not use a calculator or a table.

11. $\theta = \frac{5\pi}{6}$ 12. $\theta = \frac{7\pi}{6}$ 13. $\theta = -\frac{3\pi}{2}$
14. $\theta = \frac{2\pi}{3}$ 15. $\theta = 210°$ 16. $\theta = -45°$
17. $\theta = -90°$ 18. $\theta = 240°$ 19. $\theta = \frac{11\pi}{4}$
20. $\theta = \frac{13\pi}{3}$ 21. $\theta = \frac{11\pi}{6}$ 22. $\theta = \frac{14\pi}{3}$
23. $\theta = 540°$ 24. $\theta = 900°$ 25. $\theta = 960°$
26. $\theta = 945°$ 27. $\theta = -\frac{9\pi}{4}$ 28. $\theta = 100\pi$
29. $\theta = 101\pi$ 30. $\theta = \frac{59\pi}{2}$ 31. $\theta = \frac{61\pi}{2}$
32. $\theta = \frac{63\pi}{2}$ 33. $\theta = \frac{167\pi}{2}$ 34. $\theta = \frac{1003\pi}{6}$
35. $\theta = 3660°$ 36. $\theta = 1845°$ 37. $\theta = 1770°$
38. $\theta = 3{,}600{,}150°$

In Problems 39–46 state whether $\cos\theta$ and $\sin\theta$ increase or decrease as θ increases from the first value to the second.

39. $\frac{\pi}{2} \to \frac{3\pi}{4}$ 40. $\frac{2\pi}{3} \to \pi$ 41. $\frac{5\pi}{2} \to 3\pi$
42. $-\frac{\pi}{2} \to 0$ 43. $\frac{5\pi}{4} \to \frac{3\pi}{2}$ 44. $7\pi \to \frac{15\pi}{2}$
45. $-\frac{9\pi}{2} \to -4\pi$ 46. $\frac{9\pi}{2} \to 5\pi$

In Problems 47–50 use the value $\pi = 3.1416$.

47. Given that $\sin 0.95 = 0.8134$, find two numbers α and β in $(0, 2\pi)$ such that $\cos\alpha = \cos\beta = 0.8134$.
48. Given that $\cos 0.2 = 0.9801$, find two numbers α and β in $(0, 2\pi)$ such that $\sin\alpha = \sin\beta = 0.9801$.
49. In Problem 47 find two numbers α and β in $(0, 2\pi)$ such that $\cos\alpha = \cos\beta = -0.8134$.
50. In Problem 48 find two numbers α and β in $(0, 2\pi)$ such that $\sin\alpha = \sin\beta = -0.9801$.
51. Find the four smallest positive numbers that satisfy $\cos\theta = \frac{1}{2}$.
52. Find the four smallest positive numbers such that $\cos\theta = 0$.
53. Find the five smallest positive numbers such that $\sin\theta = \frac{1}{\sqrt{2}}$.
54. Find the three smallest (in absolute value) negative numbers such that $\sin\theta = -1$.
55. In the figure below, show that the triangles COD and AOB are congruent. Explain why this proves that $\cos(\theta + \pi) = -\cos\theta$ and $\sin(\theta + \pi) = -\sin\theta$.

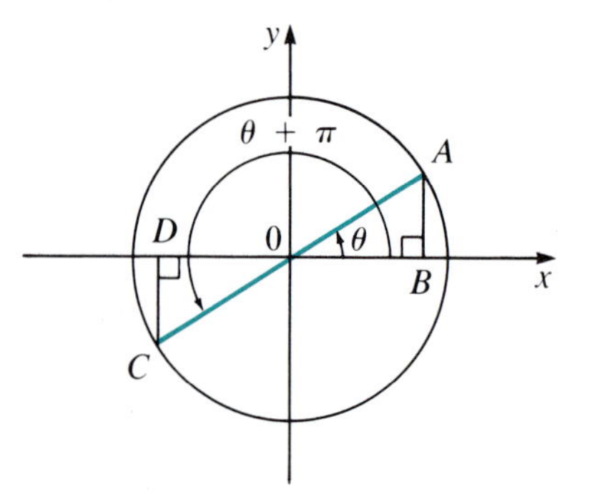

* 56. Prove that $\sin\left(\theta + \frac{\pi}{2}\right) = \cos\theta$

$\left[\text{Hint: } \theta + \frac{\pi}{2} = \pi - \left(\frac{\pi}{2} - \theta\right)\right]$

* 57. Prove that $\cos\left(\theta + \frac{\pi}{2}\right) = -\sin\theta$.

58. What values of θ in $[0, \pi]$ satisfy $\cos\theta \geq \frac{1}{2}$?
59. What values of θ in $[0, \pi]$ satisfy $\sin\theta < \frac{1}{2}$?
60. Show that $\cos(2\pi - \theta) = \cos\theta$ and $\sin(2\pi - \theta) = -\sin\theta$.

Reread the material in Section 1.4 on the vertical and horizontal shifts of graphs of functions. Then draw the graphs of the functions given in Problems 61–70.

61. $u = \sin\left(\theta - \frac{\pi}{4}\right)$ 62. $u = \cos(\theta + \pi)$
63. $u = \cos\left(\theta - \frac{3\pi}{4}\right)$ 64. $u = \sin\left(\theta + \frac{\pi}{2}\right)$

65. $u = \sin\left(\theta + \frac{3\pi}{2}\right)$

66. $u = \sin \theta + 2$

67. $u = \sin \theta - 1$

68. $u = \cos \theta + 1$

69. $u = \cos \theta - 4$

* 70. $u = \sin (\theta - 1)$

71. Find all numbers θ such that $\sin \theta = 1$.

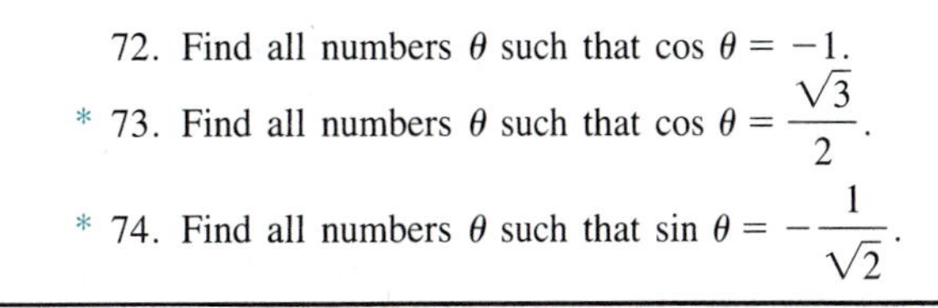

72. Find all numbers θ such that $\cos \theta = -1$.

* 73. Find all numbers θ such that $\cos \theta = \frac{\sqrt{3}}{2}$.

* 74. Find all numbers θ such that $\sin \theta = -\frac{1}{\sqrt{2}}$.

2.5 The Tangent, Cotangent, Secant, and Cosecant Functions of a Real Variable

(a) $0 < \theta < \frac{\pi}{2}$

(b) $\pi < \theta < \frac{3\pi}{2}$

Figure 1 The unit circle.

In Section 2.2 we used right triangles to define six trigonometric functions of an acute angle. In Section 2.3 we used the unit circle to define $\sin \theta$ and $\cos \theta$ where θ is a real number. We now use the unit circle to define $\tan \theta$, $\cot \theta$, $\sec \theta$, and $\csc \theta$.

As in Section 2.3, we begin with the circle of radius 1 centered at the origin (the unit circle) and let (x, y) denote the point where the terminal side of θ intersects this circle (see Figure 1). The functions **tangent** of θ, **cotangent** of θ, **secant** of θ, and **cosecant** of θ, which we abbreviate as **tan $\boldsymbol{\theta}$, cot $\boldsymbol{\theta}$, sec $\boldsymbol{\theta}$,** and **csc $\boldsymbol{\theta}$** are defined as follows:

Unit Circle Definitions of the Tangent, Cotangent, Secant, and Cosecant Functions

$$\tan \theta = \frac{y}{x},\ x \neq 0 \qquad \cot \theta = \frac{x}{y},\ y \neq 0$$

$$\sec \theta = \frac{1}{x},\ x \neq 0 \qquad \csc \theta = \frac{1}{y},\ y \neq 0 \tag{1}$$

Since $\cos \theta = x$ and $\sin \theta = y$, we immediately obtain the important relationships

$$\tan \theta = \frac{y}{x} = \frac{\sin \theta}{\cos \theta} \qquad \sec \theta = \frac{1}{x} = \frac{1}{\cos \theta}$$

$$\cot \theta = \frac{x}{y} = \frac{\cos \theta}{\sin \theta} \qquad \csc \theta = \frac{1}{y} = \frac{1}{\sin \theta}$$

That is,

Writing Four Functions in Terms of sin θ and cos θ

$$\tan \theta = \frac{\sin \theta}{\cos \theta} \qquad \cot \theta = \frac{\cos \theta}{\sin \theta} = \frac{1}{\tan \theta}$$

$$\sec \theta = \frac{1}{\cos \theta} \qquad \csc \theta = \frac{1}{\sin \theta} \tag{2}$$

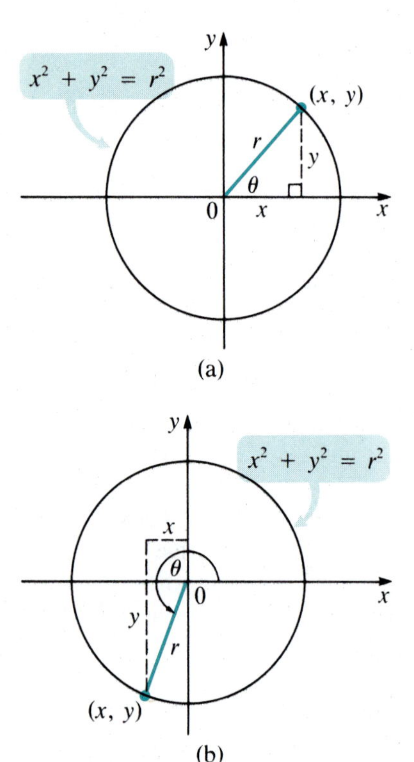

Figure 2 Circles of radius r centered at the origin.

The definitions (1) and relationships (2) should be memorized.

In Section 2.3 (page 80) we used similar triangles to give alternative definitions of $\sin\theta$ and $\cos\theta$ using a circle of radius r centered at the origin. These alternative definitions are sometimes useful in applications. We provide them here for the functions $\tan\theta$, $\cot\theta$, $\sec\theta$, and $\csc\theta$.

Using the circle in Figure 2, we have

Alternative Definitions of the Tangent, Cotangent, Secant, and Cosecant Functions

$$\tan\theta = \frac{y}{x},\ x \neq 0 \qquad \cot\theta = \frac{x}{y},\ y \neq 0$$
$$\sec\theta = \frac{r}{x},\ x \neq 0 \qquad \csc\theta = \frac{r}{y},\ y \neq 0 \tag{1'}$$

NOTE We saw on p. 80 that the triangular and circular definitions of $\sin\theta$ and $\cos\theta$ give the same values for $0 < \theta < 90°$. Since the other four trigonometric functions can be written in terms of $\sin\theta$ and $\cos\theta$ (by (2)), we see that the triangular and circular definitions of all six trigonometric functions give the same values for $0 < \theta < 90°$.

EXAMPLE 1 *Computing Trigonometric Function Values of $\frac{\pi}{6}$*

Let $\theta = \frac{\pi}{6}$. Compute $\tan\theta$, $\cot\theta$, $\sec\theta$, and $\csc\theta$.

SOLUTION
$$\tan\frac{\pi}{6} = \frac{\sin \pi/6}{\cos \pi/6} = \frac{1/2}{\sqrt{3}/2} = \frac{1}{\sqrt{3}}$$
$$\cot\frac{\pi}{6} = \frac{1}{\tan \pi/6} = \sqrt{3}$$
$$\sec\frac{\pi}{6} = \frac{1}{\cos \pi/6} = \frac{1}{\sqrt{3}/2} = \frac{2}{\sqrt{3}}$$
$$\csc\frac{\pi}{6} = \frac{1}{\sin \pi/6} = \frac{1}{1/2} = 2 \quad ■$$

EXAMPLE 2 *Computing Trigonometric Function Values of $\frac{\pi}{4}$*

Let $\theta = \frac{\pi}{4}$. Compute $\tan\theta$, $\cot\theta$, $\sec\theta$, and $\csc\theta$.

SOLUTION

$$\tan\frac{\pi}{4} = \frac{\sin \pi/4}{\cos \pi/4} = \frac{1/\sqrt{2}}{1/\sqrt{2}} = 1$$

$$\cot\frac{\pi}{4} = \frac{1}{\tan \pi/4} = \frac{1}{1} = 1$$

$$\sec\frac{\pi}{4} = \frac{1}{\cos \pi/4} = \frac{1}{1/\sqrt{2}} = \sqrt{2}$$

$$\csc\frac{\pi}{4} = \frac{1}{\sin \pi/4} = \frac{1}{1/\sqrt{2}} = \sqrt{2}$$

One of the first rules you learned in algebra was *you cannot divide by zero*. Therefore, since $\tan\theta = \dfrac{\sin\theta}{\cos\theta}$ and $\sec\theta = \dfrac{1}{\cos\theta}$,

Domain of the Tangent and Secant Functions

$\tan\theta$ and $\sec\theta$ are **not** defined when $\cos\theta = 0$; that is, when

$$\theta = \pm\frac{\pi}{2},\ \pm\frac{3\pi}{2},\ \pm\frac{5\pi}{2},\ \ldots$$ (see p. 99)

$$= \frac{\pi}{2} + n\pi, \qquad n = 0, \pm 1, \pm 2, \ldots.$$

Thus domain of $\tan\theta$ = domain of $\sec\theta$

$$= \left\{\theta\colon \theta \neq \frac{\pi}{2} + n\pi,\ n = 0, \pm 1, \pm 2, \ldots\right\}$$

Analogously, since $\cot\theta = \dfrac{\cos\theta}{\sin\theta}$ and $\csc\theta = \dfrac{1}{\sin\theta}$,

Domain of the Cotangent and Cosecant Functions

$\cot\theta$ and $\csc\theta$ are **not** defined when $\sin\theta = 0$; that is, when $\theta = 0$, $\pm\pi, \pm 2\pi, \ldots = n\pi, \qquad n = 0, \pm 1, \pm 2, \ldots.$

Thus domain of $\cot\theta$ = domain of $\csc\theta$

$$= \{\theta\colon \theta \neq n\pi,\ n = 0, \pm 1, \pm 2, \ldots\}$$

Since we are primarily concerned with the interval $[0, 2\pi)$, we list below the values that are not defined.

WARNING These values for θ in $[0, 2\pi)$ are not defined.

$$\tan\frac{\pi}{2} = \tan 90^\circ \qquad \tan\frac{3\pi}{2} = \tan 270^\circ$$

$$\cot 0 = \cot 0^\circ \qquad \cot \pi = \cot 180^\circ$$

$$\sec\frac{\pi}{2} = \sec 90^\circ \qquad \sec\frac{3\pi}{2} = \sec 270^\circ$$

$$\csc 0 = \csc 0^\circ \qquad \csc \pi = \csc 180^\circ$$

Signs of tan θ, cot θ, sec θ, and csc θ

In Figure 9 on p. 83 we illustrated the signs of the functions $\cos\theta$ and $\sin\theta$ (see also Figure 3). Since $\tan\theta = \dfrac{\sin\theta}{\cos\theta}$ and $\cot = \dfrac{\cos\theta}{\sin\theta}$, we see that

tan θ and cot θ are positive when cos θ and sin θ have the same sign, and are negative when cos θ and sin θ have opposite signs.

In addition, since $\sec\theta = \dfrac{1}{\cos\theta}$ and $\csc\theta = \dfrac{1}{\sin\theta}$, we know that

sec θ has the same sign as cos θ, and csc θ has the same sign as sin θ.

Using these facts, we obtain the information in Table 1.

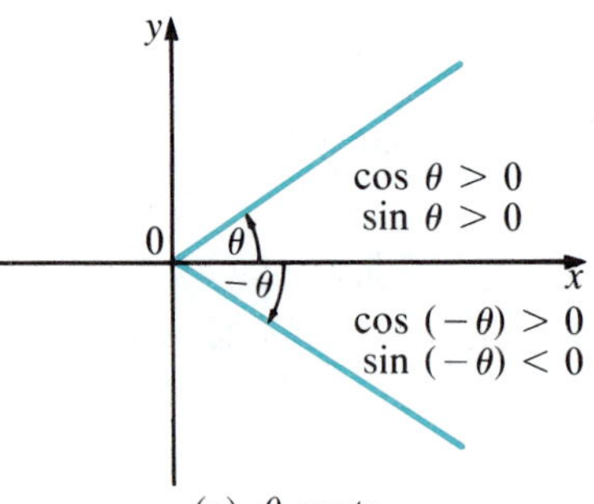

(a) θ acute

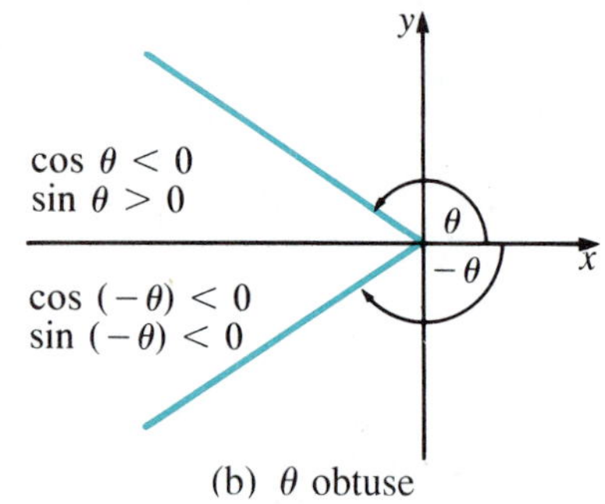

(b) θ obtuse

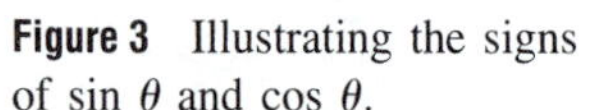

Figure 3 Illustrating the signs of sin θ and cos θ.

Table 1

Quadrant	cos θ	sin θ	tan θ	cot θ	sec θ	csc θ
I	positive	positive	positive	positive	positive	positive
II	negative	positive	negative	negative	negative	positive
III	negative	negative	positive	positive	negative	negative
IV	positive	negative	negative	negative	positive	negative

Even and Odd Functions

In Section 2.4 (Theorem 2), we saw that $\cos(-\theta) = \cos\theta$ and $\sin(-\theta) = -\sin\theta$. That is, $\cos\theta$ is an even function and $\sin\theta$ is an odd function.

Theorem: tan θ, cot θ, and csc θ Are Odd Functions, and sec θ Is an Even Function

Whenever each function is defined

$$\tan(-\theta) = -\tan\theta \tag{3}$$

$$\cot(-\theta) = -\cot\theta \tag{4}$$

$$\sec(-\theta) = \sec\theta \tag{5}$$

$$\csc(-\theta) = -\csc\theta \tag{6}$$

Proof

$$\tan(-\theta) = \frac{\sin(-\theta)}{\cos(-\theta)} = \frac{-\sin\theta}{\cos\theta} = -\frac{\sin\theta}{\cos\theta} = -\tan\theta$$

$$\cot(-\theta) = \frac{1}{\tan(-\theta)} = \frac{1}{-\tan\theta} = -\cot\theta$$

$$\sec(-\theta) = \frac{1}{\cos(-\theta)} = \frac{1}{\cos\theta} = \sec\theta$$

$$\csc(-\theta) = \frac{1}{\sin(-\theta)} = \frac{1}{-\sin\theta} = -\csc\theta \quad \blacksquare$$

EXAMPLE 3 ***Computing Trigonometric Function Values of*** $-\frac{\pi}{6}$

Let $\theta = -\frac{\pi}{6}$. Compute $\tan\theta$, $\cot\theta$, $\sec\theta$, and $\csc\theta$.

SOLUTION Using the values obtained in Example 1, we have

$$\tan\left(-\frac{\pi}{6}\right) = -\tan\frac{\pi}{6} = -\frac{1}{\sqrt{3}}$$

$$\cot\left(-\frac{\pi}{6}\right) = -\cot\frac{\pi}{6} = -\sqrt{3}$$

$$\sec\left(-\frac{\pi}{6}\right) = \sec\frac{\pi}{6} = \frac{2}{\sqrt{3}}$$

$$\csc\left(-\frac{\pi}{6}\right) = -\csc\frac{\pi}{6} = -2 \quad \blacksquare$$

EXAMPLE 4 ***Computing Trigonometric Function Values of*** $\frac{2\pi}{3}$

Compute $\tan\frac{2\pi}{3}$, $\cot\frac{2\pi}{3}$, $\sec\frac{2\pi}{3}$, and $\csc\frac{2\pi}{3}$.

SOLUTION In Example 9 on p. 98 we found that $\sin\frac{2\pi}{3} = \frac{\sqrt{3}}{2}$ and $\cos\frac{2\pi}{3} = -\frac{1}{2}$. Then

$$\tan\frac{2\pi}{3} = \frac{\sin\frac{2\pi}{3}}{\cos\frac{2\pi}{3}} = \frac{\sqrt{3}/2}{-1/2} = -\sqrt{3} \qquad \cot\frac{2\pi}{3} = \frac{1}{\tan\frac{2\pi}{3}} = -\frac{1}{\sqrt{3}}$$

$$\sec\frac{2\pi}{3} = \frac{1}{\cos\frac{2\pi}{3}} = -2 \qquad \csc\frac{2\pi}{3} = \frac{1}{\sin\frac{2\pi}{3}} = \frac{2}{\sqrt{3}}$$

In Examples 1, 2, and 4 we used the identities (2) to compute a number of values. In Table 2, we give values for all six trigonometric functions for various numbers in $[0, 2\pi]$.

Table 2

θ (radians)	0	$\frac{\pi}{6}$	$\frac{\pi}{4}$	$\frac{\pi}{3}$	$\frac{\pi}{2}$	$\frac{2\pi}{3}$	$\frac{3\pi}{4}$	$\frac{5\pi}{6}$	π	$\frac{7\pi}{6}$	$\frac{5\pi}{4}$	$\frac{4\pi}{3}$	$\frac{3\pi}{2}$	$\frac{5\pi}{3}$	$\frac{7\pi}{4}$	$\frac{11\pi}{6}$	2π
θ (degrees)	0	30	45	60	90	120	135	150	180	210	225	240	270	300	315	330	360
$\cos\theta$	1	$\frac{\sqrt{3}}{2}$	$\frac{1}{\sqrt{2}}$	$\frac{1}{2}$	0	$-\frac{1}{2}$	$-\frac{1}{\sqrt{2}}$	$-\frac{\sqrt{3}}{2}$	-1	$-\frac{\sqrt{3}}{2}$	$-\frac{1}{\sqrt{2}}$	$-\frac{1}{2}$	0	$\frac{1}{2}$	$\frac{1}{\sqrt{2}}$	$\frac{\sqrt{3}}{2}$	1
$\sin\theta$	0	$\frac{1}{2}$	$\frac{1}{\sqrt{2}}$	$\frac{\sqrt{3}}{2}$	1	$\frac{\sqrt{3}}{2}$	$\frac{1}{\sqrt{2}}$	$\frac{1}{2}$	0	$-\frac{1}{2}$	$-\frac{1}{\sqrt{2}}$	$-\frac{\sqrt{3}}{2}$	-1	$-\frac{\sqrt{3}}{2}$	$-\frac{1}{\sqrt{2}}$	$-\frac{1}{2}$	0
$\tan\theta = \frac{\sin\theta}{\cos\theta}$	0	$\frac{1}{\sqrt{3}}$	1	$\sqrt{3}$	*	$-\sqrt{3}$	-1	$-\frac{1}{\sqrt{3}}$	0	$\frac{1}{\sqrt{3}}$	1	$\sqrt{3}$	*	$-\sqrt{3}$	-1	$-\frac{1}{\sqrt{3}}$	0
$\cot\theta = \frac{\cos\theta}{\sin\theta}$	*	$\sqrt{3}$	1	$\frac{1}{\sqrt{3}}$	0	$-\frac{1}{\sqrt{3}}$	-1	$-\sqrt{3}$	*	$\sqrt{3}$	1	$\frac{1}{\sqrt{3}}$	0	$-\frac{1}{\sqrt{3}}$	-1	$-\sqrt{3}$	*
$\sec\theta = \frac{1}{\cos\theta}$	1	$\frac{2}{\sqrt{3}}$	$\sqrt{2}$	2	*	-2	$-\sqrt{2}$	$-\frac{2}{\sqrt{3}}$	-1	$-\frac{2}{\sqrt{3}}$	$-\sqrt{2}$	-2	*	2	$\sqrt{2}$	$\frac{2}{\sqrt{3}}$	1
$\csc\theta = \frac{1}{\sin\theta}$	*	2	$\sqrt{2}$	$\frac{2}{\sqrt{3}}$	1	$\frac{2}{\sqrt{3}}$	$\sqrt{2}$	2	*	-2	$-\sqrt{2}$	$-\frac{2}{\sqrt{3}}$	-1	$-\frac{2}{\sqrt{3}}$	$-\sqrt{2}$	-2	*

*Undefined.

Two More Identities

We start with the basic identity (equation (2) on p. 78)

$$\sin^2\theta + \cos^2\theta = 1 \tag{7}$$

For $\cos\theta \neq 0$, we divide both sides of (7) by $\cos^2\theta$ to obtain

$$\frac{\sin^2\theta}{\cos^2\theta} + \frac{\cos^2\theta}{\cos^2\theta} = \frac{1}{\cos^2\theta}$$

$$\left(\frac{\sin\theta}{\cos\theta}\right)^2 + 1 = \left(\frac{1}{\cos\theta}\right)^2$$

A Basic Trigonometric Identity Involving $\tan\theta$ and $\sec\theta$

$$\tan^2\theta + 1 = \sec^2\theta \tag{8}$$

If we divide both sides of (7) by $\sin^2\theta$, for $\sin\theta \neq 0$, we obtain

$$\frac{\sin^2\theta}{\sin^2\theta} + \frac{\cos^2\theta}{\sin^2\theta} = \frac{1}{\sin^2\theta}$$

A Basic Trigonometric Identity Involving cot θ and csc θ

$$1 + \cot^2 \theta = \csc^2 \theta \tag{9}$$

Using identities (7), (8), and (9), we can compute the values of $\sin \theta$, $\cos \theta$, $\tan \theta$, $\cot \theta$, $\sec \theta$, and $\csc \theta$ if the quadrant of θ and just one of the values are known.

EXAMPLE 5 *Finding Five Trigonometric Function Values When csc θ and the Quadrant of θ Are Given*

Find $\sin \theta$, $\cos \theta$, $\tan \theta$, $\cot \theta$, and $\sec \theta$ if $\csc \theta = -2.3700$ and θ is in the third quadrant. Round to 4 decimal places.

SOLUTION In the third quadrant, $\sin \theta < 0$, $\cos \theta < 0$, $\tan \theta > 0$, $\cot \theta > 0$, and $\sec \theta < 0$. Then

$$\sin \theta = \frac{1}{\csc \theta} = \frac{1}{-2.37} \approx -0.4219$$

$$\begin{aligned} \cos^2 \theta &= 1 - \sin^2 \theta = 1 - (0.4219)^2 \\ &= 1 - 0.17799961 \approx 0.8220 \end{aligned}$$

$$\cos \theta \approx \underset{\substack{\uparrow \\ \cos \theta < 0}}{-}\sqrt{0.8220} \approx -0.9066$$

$$\sec \theta = \frac{1}{\cos \theta} \approx \frac{1}{-0.9066} \approx -1.1030$$

$$\begin{aligned} \cot^2 \theta &\underset{\text{From (9)}}{=} \csc^2 \theta - 1 = (-2.37)^2 - 1 \\ &= 5.6169 - 1 = 4.6169 \end{aligned}$$

$$\cot \theta \underset{\cot \theta > 0}{=} \sqrt{4.6169} \approx 2.1487$$

$$\tan \theta = \frac{1}{\cot \theta} \approx \frac{1}{2.1487} \approx 0.4654$$

Note that we could also obtain $\tan \theta$ and $\cot \theta$ from

$$\tan \theta = \frac{\sin \theta}{\cos \theta} \approx \frac{-0.4219}{-0.9066} \approx 0.4654$$

and

$$\cot \theta = \frac{\cos \theta}{\sin \theta} \approx \frac{-0.9066}{-0.4219} \approx 2.1489$$

There is a slight discrepancy in this last value because of roundoff error.

EXAMPLE 6 ***Finding Four Trigonometric Function Values When cos θ and cot θ Are Given***

If $\cos\theta = -\dfrac{2}{3}$ and $\cot\theta = -\dfrac{2}{\sqrt{5}}$, find $\sin\theta$, $\tan\theta$, $\sec\theta$, and $\csc\theta$.

SOLUTION Since $\cos\theta < 0$ and $\cot\theta < 0$, θ is in the second quadrant. Then

$$\sin^2\theta = 1 - \cos^2\theta = 1 - \left(-\frac{2}{3}\right)^2 = 1 - \frac{4}{9} = \frac{5}{9}$$

In the second quadrant, $\sin\theta > 0$, so

$$\sin\theta = \sqrt{\frac{5}{9}} = \frac{\sqrt{5}}{3}$$

Continuing, we obtain

$$\tan\theta = \frac{1}{\cot\theta} = -\frac{\sqrt{5}}{2}$$

$$\sec\theta = \frac{1}{\cos\theta} = -\frac{3}{2}$$

$$\csc\theta = \frac{1}{\sin\theta} = \frac{3}{\sqrt{5}}$$

As in Examples 4 and 5 in Section 2.3, there is another way to solve this problem. We draw a circle of radius r. We have

$$\cos\theta = \frac{x}{r} = \frac{-2}{3} \qquad \text{and} \qquad \cot\theta = \frac{x}{y} = \frac{-2}{\sqrt{5}}$$

Then, since $r > 0$, we may choose $x = -2$, $y = \sqrt{5}$, and $r = 3$. Thus

$$\sin\theta = \frac{y}{r} = \frac{\sqrt{5}}{3}$$

$$\tan\theta = \frac{y}{x} = \frac{\sqrt{5}}{-2} = -\frac{\sqrt{5}}{2}$$

$$\sec\theta = \frac{r}{x} = \frac{3}{-2} = -\frac{3}{2}$$

$$\csc\theta = \frac{r}{y} = \frac{3}{\sqrt{5}}$$

Since the point $(x, y) = (-2, \sqrt{5})$ is in the second quadrant, θ is in the second quadrant. It is sketched in Figure 4.

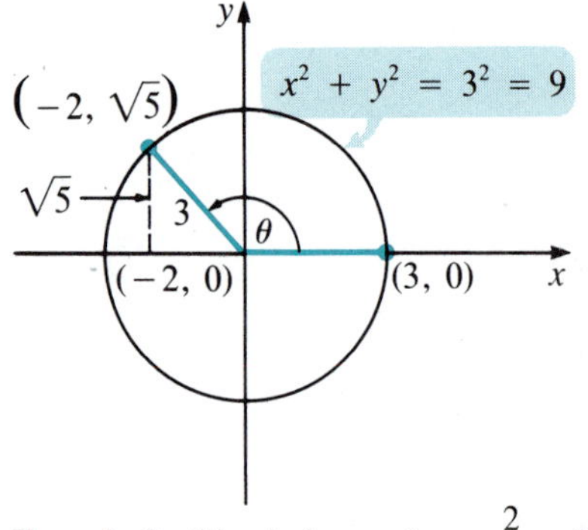

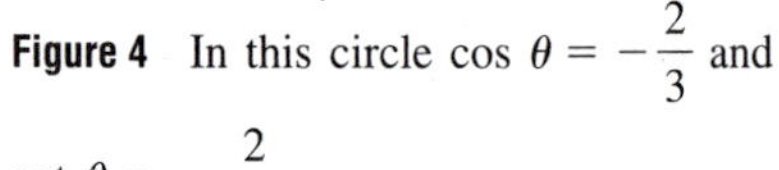
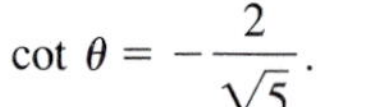

Figure 4 In this circle $\cos\theta = -\dfrac{2}{3}$ and $\cot\theta = -\dfrac{2}{\sqrt{5}}$.

Slopes and Tangents

If L is a straight line and (x_1, y_1) and (x_2, y_2) are two points on L, then

$$m = \text{slope of } L = \frac{y_2 - y_1}{x_2 - x_1} = \frac{\Delta y}{\Delta x}$$

Theorem

Let θ be the angle that the line L makes with the positive x-axis. Then the slope of L is given by

$$m = \tan\theta \tag{10}$$

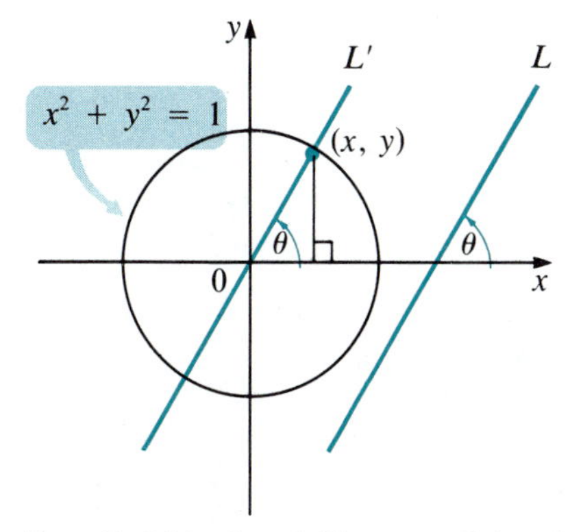

Figure 5 Lines L and L' are parallel and therefore have the same slope.

Proof

Look at the line L' parallel to the given line that passes through the origin, and draw the unit circle around that new line (see Figure 5).

Because L' and L are parallel, they have the same slope. Two points on L' are $(0, 0)$ and (x, y). Then

$$m = \text{slope of } L = \text{slope of } L' = \frac{\Delta y}{\Delta x} = \frac{y - 0}{x - 0} = \frac{y}{x} = \tan\theta \quad ■$$

Using a Calculator

Most calculators that have [sin] and [cos] keys also have a tangent key tan . The calculator also has a reciprocal key [1/x]. This key is used to find $\cot\theta$, $\sec\theta$, and $\csc\theta$.

To find $\cot\theta$:	enter θ press [tan] [1/x]†
To find $\sec\theta$:	enter θ press [cos] [1/x]
To find $\csc\theta$:	enter θ press [sin] [1/x]

Table 4 at the back of the book has tables of $\sin\theta$, $\cos\theta$, and $\tan\theta$ for those of you who are working without calculators.

TWO WARNINGS Remember to set the calculator to the appropriate mode: degree or radian.

You can't divide by zero. If you try to compute something like $\tan\frac{\pi}{2}$, you will get an error message since $\tan\frac{\pi}{2}$ is not defined. ■

† On some calculators the [sin], [cos], or [tan] key must be pressed before the number (in degrees or radians) is entered.

EXAMPLE 7 *Obtaining Four Trigonometric Function Values on a Calculator (Radian Mode)*

Compute (a) tan 3, (b) sec 0.78, (c) cot 4.3, and (d) csc 5.16.

SOLUTION First set the mode to radian (rad). Then press the appropriate keys.

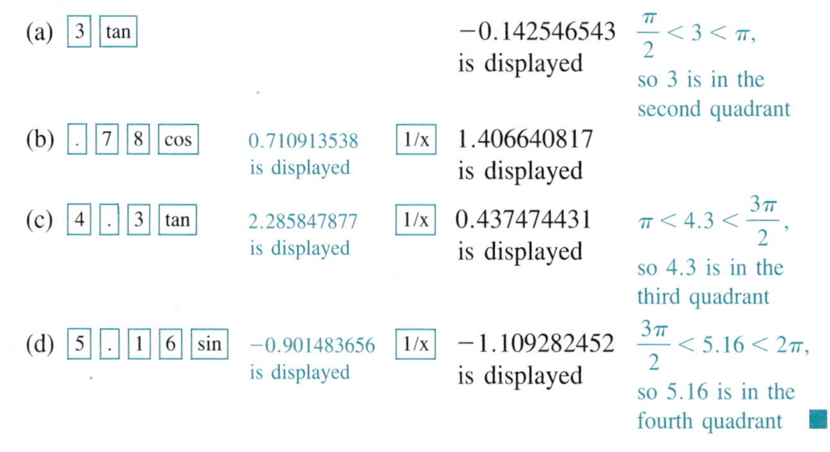

EXAMPLE 8 *Obtaining Three Trigonometric Function Values on a Calculator (Degree Mode)*

Compute (a) cot 62°, (b) sec 116°, and (c) csc 203°.

SOLUTION First set the mode to degree (deg). Then press the appropriate keys.

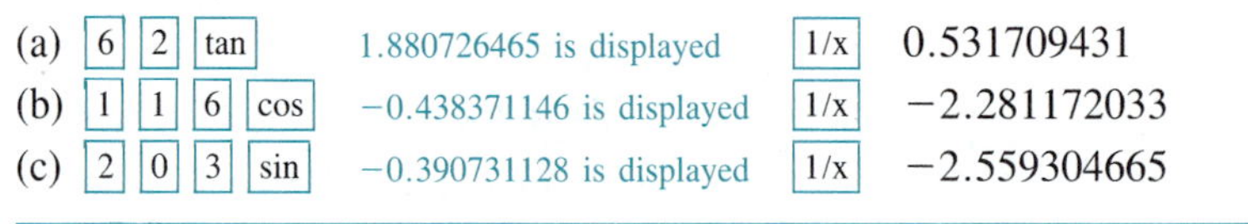

Problems 2.5

Readiness Check

I. Which of the following is equal to $\tan\theta$?

a. $1 - \cot\theta$ b. $\dfrac{\cos\theta}{\sin\theta}$

c. $\csc\theta - 1$ d. $\dfrac{\sin\theta}{\cos\theta}$

II. Which of the following is equal to $\sec^2\theta$?

a. $1 - \tan^2\theta$ b. $1 + \cot^2\theta$

c. $\tan^2\theta + 1$ d. $\csc^2\theta - 1$

III. Which of the following is $\sec\theta$ if the terminal side of θ passes through the point $(-5, -12)$?

a. $-\dfrac{13}{5}$ b. $\dfrac{12}{5}$ c. $-\dfrac{5}{13}$ d. $-\dfrac{13}{12}$

IV. Which of the following is undefined?

a. $\sec\pi$ b. $\tan(-\pi)$

c. $\csc\left(-\dfrac{\pi}{2}\right)$ d. $\cot\pi$

V. Which of the following is the reference angle to find $\tan\dfrac{5\pi}{6}$?

a. $-\dfrac{\pi}{6}$ b. $\dfrac{\pi}{6}$ c. $-\dfrac{5\pi}{6}$ d. π

In Problems 1–10 find tan θ, cot θ, sec θ, and csc θ when they are defined. Do not use Table 2 or a calculator.

1. $\theta = \dfrac{\pi}{4}$
2. $\theta = \dfrac{2\pi}{3}$
3. $\theta = \dfrac{7\pi}{6}$
4. $\theta = \dfrac{7\pi}{4}$
5. $\theta = -\dfrac{\pi}{2}$
6. $\theta = -\pi$
7. $\theta = 315°$
8. $\theta = 240°$
9. $\theta = 300°$
10. $\theta = 315°$

In Problems 11–20 one value from among sin θ, cos θ, tan θ, cot θ, sec θ, and csc θ is given together with the quadrant of θ. Find the values of the other five functions. Do not use a calculator.

11. $\sin\theta = \frac{1}{3}$, I
12. $\sin\theta = \frac{1}{4}$, II
13. $\tan\theta = 4$, I
14. $\tan\theta = 4$, III
15. $\sec\theta = -\frac{3}{2}$, II
16. $\sec\theta = -\frac{3}{2}$, III
17. $\cot\theta = -\frac{2}{5}$, II
18. $\cot\theta = -\frac{2}{5}$, IV
19. $\cos\theta = \frac{4}{5}$, I
20. $\cos\theta = \frac{4}{5}$, IV

In Problems 21–33 two values from among sin θ, cos θ, tan θ, cot θ, sec θ, and csc θ are given. Find (a) the quadrant of θ and (b) the other four values. Use a calculator only where indicated.

21. $\sin\theta = \frac{3}{5}$, $\cos\theta = \frac{4}{5}$
22. $\sin\theta = \frac{3}{5}$, $\cos\theta = -\frac{4}{5}$
23. $\sin\theta = -\frac{3}{5}$, $\cos\theta = \frac{4}{5}$
24. $\sin\theta = -\frac{3}{5}$, $\cos\theta = -\frac{4}{5}$
25. $\tan\theta = -\frac{8}{15}$, $\sec\theta = \frac{17}{15}$
26. $\tan\theta = \frac{8}{15}$, $\sec\theta = \frac{17}{15}$
27. $\tan\theta = \frac{8}{15}$, $\sec\theta = -\frac{17}{15}$
28. $\tan\theta = -\frac{8}{15}$, $\sec\theta = -\frac{17}{15}$
29. $\csc\theta = \frac{25}{24}$, $\cos\theta = -\frac{7}{25}$
30. $\csc\theta = -\frac{25}{24}$, $\cos\theta = -\frac{7}{25}$
31. $\cos\theta = 0.1432$, $\tan\theta = -6.9113$ [Round to 4 decimal places.]
32. $\cot\theta = -2.1309$, $\sec\theta = -1.1046$
33. $\csc\theta = -4.1722$, $\sec\theta = 1.0300$

In Problems 34–44 use a calculator to find each value. Be certain to use the correct mode.

34. tan 1
35. sec 3
36. csc 0.24
37. cot 0.9
38. sec 37°
39. csc 130°
40. cot 63.2°
41. tan 260°
42. sec (−1.6)
43. $\csc\left(\dfrac{1}{3.29}\right)$
44. $\cot\dfrac{2.08}{5.17}$

In Problems 45–55 use the results of Section 2.4 to prove each identity.

45. $\tan(\pi - \theta) = -\tan\theta$
46. $\tan(\pi + \theta) = \tan\theta$
47. $\sec(\pi - \theta) = -\sec\theta$
48. $\sec(\pi + \theta) = -\sec\theta$
49. $\csc(\pi - \theta) = \csc\theta$
50. $\csc(\pi + \theta) = -\csc\theta$
51. $\tan\left(\dfrac{\pi}{2} - \theta\right) = \cot\theta$ if $0 < \theta < \dfrac{\pi}{2}$
52. $\sec\left(\dfrac{\pi}{2} - \theta\right) = \csc\theta$ if $0 < \theta < \dfrac{\pi}{2}$
53. $\cot\left(\dfrac{\pi}{2} - \theta\right) = \tan\theta$ if $0 < \theta < \dfrac{\pi}{2}$
54. $\csc\left(\dfrac{\pi}{2} - \theta\right) = \sec\theta$ if $0 < \theta < \dfrac{\pi}{2}$
55. * $\tan\left(\theta + \dfrac{\pi}{2}\right) = -\cot\theta$
56. What values of θ in $\left[0, \dfrac{\pi}{2}\right)$ satisfy $\tan\theta > 1$?
57. What values of θ in $\left[0, \dfrac{\pi}{2}\right)$ satisfy $1 \le \sec\theta \le 2$?
58. What values of θ in $\left[\dfrac{\pi}{2}, \pi\right)$ satisfy $\csc\theta \ge \dfrac{2}{\sqrt{3}}$?
59. What values of θ in $\left(0, \dfrac{\pi}{2}\right]$ satisfy $\cot\theta > \sqrt{3}$?
60. Explain why the following problem does not have a unique solution. Find tan θ if $\sin\theta = \frac{5}{13}$ and $\csc\theta = \frac{13}{5}$.

Answers to Readiness Check

I. d II. c III. a IV. d V. b

2.6 Periodicity and Graphs of tan θ, cot θ, sec θ, and csc θ

Periodicity

Recall (p. 93) that $\sin(\theta + \pi) = -\sin\theta$ and $\cos(\theta + \pi) = -\cos\theta$. Then

$$\tan(\theta + \pi) = \frac{\sin(\theta + \pi)}{\cos(\theta + \pi)} = \frac{-\sin\theta}{-\cos\theta} = \frac{\sin\theta}{\cos\theta} = \tan\theta$$

and

$$\cot(\theta + \pi) = \frac{1}{\tan(\theta + \pi)} = \frac{1}{\tan \theta} = \cot \theta$$

It is true that $\tan(\theta + T) \neq \tan \theta$ if $|T| < \pi$, so

$\tan \theta$ and $\cot \theta$ are periodic of period π.

In addition, since $\cos \theta$ and $\sin \theta$ are periodic of period 2π,

$$\sec(\theta + 2\pi) = \frac{1}{\cos(\theta + 2\pi)} = \frac{1}{\cos \theta} = \sec \theta$$

$$\csc(\theta + 2\pi) = \frac{1}{\sin(\theta + 2\pi)} = \frac{1}{\sin \theta} = \csc \theta$$

That is,

$\sec \theta$ and $\csc \theta$ are periodic of period 2π.

Range and Graph of $\tan \theta$

Since $\tan \theta$ is periodic of period π, we first sketch its graph over an interval of length π. We choose an interval that does not include values for which $\tan \theta$ is not defined. That is, we avoid $\pi/2$, $-\pi/2$, $3\pi/2$, $-3\pi/2$, ... A natural interval is $(-\pi/2, \pi/2)$. Note that this is an open interval. The endpoints $-\pi/2$ and $\pi/2$ are not included.

What happens to $\tan \theta$ as θ increases from 0 to $\pi/2$ (but never reaches $\pi/2$)? Look at the ratio

$$\tan \theta = \frac{\sin \theta}{\cos \theta}$$

Refer to Table 1 on p. 91. As θ increases from 0 to $\pi/2$:

The numerator, $\sin \theta$, increases from 0 to 1.
The denominator, $\cos \theta$, decreases from 1 to 0.

That is, the numerator is getting bigger, and the denominator is getting smaller, so the ratio gets bigger. Thus, $\tan \theta$ increases as θ increases from 0 to $\pi/2$.

Suppose θ decreases from 0 to $-\pi/2$. Then

The numerator, $\sin \theta$, decreases from 0 to -1.
The denominator, $\cos \theta$, decreases from 1 to 0.

Here we see that the numerator is decreasing while remaining negative, and the denominator gets smaller and smaller while remaining positive.

Thus $\tan \theta$ decreases (increases in absolute value while remaining negative) as θ gets closer to $-\pi/2$ (with $-\pi/2 < \theta < 0$), so $\tan \theta$ increases as θ goes from $-\pi/2$ to 0. We have for $-\pi/2 < \theta < \pi/2$, $\tan \theta$ increases as θ

increases. In Table 1, we indicate what happens to tan θ as θ gets close to $\pi/2$ and $-\pi/2$.

Table 1

θ	θ (radians)†	tan θ	θ	θ (radians)	tan θ
45°	$\pi/4$	1	−45°	$-\pi/4$	−1
60°	$\pi/3$	$\sqrt{3} = 1.732$	−60°	$-\pi/3$	$-\sqrt{3} = -1.732$
75°	1.308996939	3.732	−75°	−1.308996939	−3.732
85°	1.483529864	11.43	−85°	−1.483529864	−11.43
87°	1.518436449	19.08	−87°	−1.518436449	−19.08
88°	1.535889742	28.64	−88°	−1.535889742	−28.64
89°	1.553343034	57.29	−89°	−1.553343034	−57.29
89.5°	1.562069681	114.59	−89.5°	−1.562069681	−114.59
89.9°	1.569050998	573	−89.9°	−1.569050998	−573
89.99°	1.570621794	5730	−89.99°	−1.570621794	−5730
89.999°	1.570778874	57,296	−89.999°	−1.570778874	−57,296
89.9999°	1.570794581	572,958	−89.9999°	−1.570794581	−572,958
89.99999°	1.570796152	5,729,578	−89.99999°	−1.570796152	−5,729,578

† As θ gets close to 90°, it gets close to $\pi/2 \approx 1.570796327$ radians.

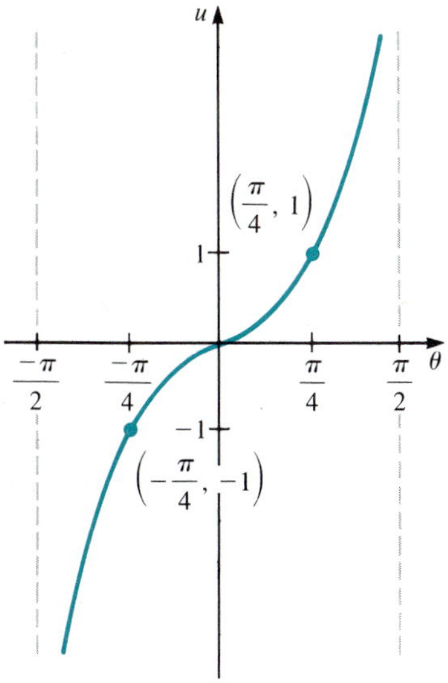

(a) $u = \tan\theta,\ \frac{-\pi}{2} < \theta < \frac{\pi}{2}$

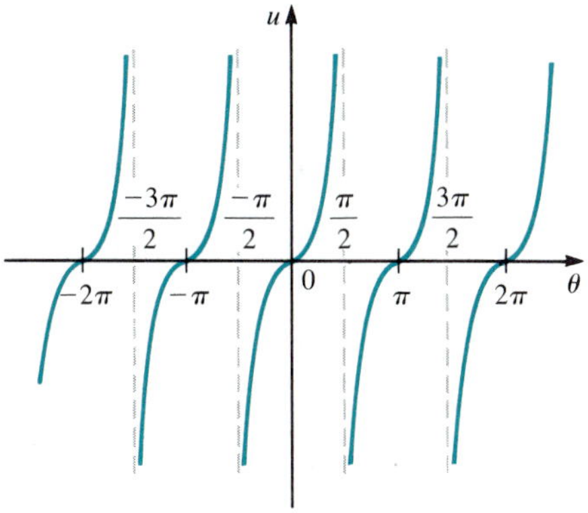

(b) $u = \tan\theta$

Figure 1 Graphs of $u = \tan\theta$.

NOTATION We write $\tan x \to \infty$ to indicate that $\tan x$ grows without bound. By writing $\tan x \to -\infty$, we mean that $\tan x$ remains negative and $|\tan x|$ grows without bound. In addition, the expression $x \to c^+$ indicates that x approaches the real number c from the right and $x \to c^-$ indicates that x approaches c from the left.

Using this notation we see that

$$\tan\theta \to \infty \qquad \text{as } \theta \to \frac{\pi}{2}^-$$

and

$$\tan\theta \to -\infty \qquad \text{as } \theta \to -\frac{\pi}{2}^+$$

That is,

The lines $\theta = \frac{\pi}{2}$ and $\theta = -\frac{\pi}{2}$ are vertical asymptotes of the graph of tan θ.

We put all of our information together to get the graph in Figure 1(a). We extend this graph to obtain the graph of $u = \tan\theta$ for all θ not an odd multiple of $\frac{\pi}{2}$.

In Figure 1, we see that tan θ takes on all real numbers. That is,

$$\text{range } \tan\theta \text{ is } (-\infty, \infty)$$

Graph and Range of cot θ

We could go through similar steps to obtain a graph of $u = \cot\theta$, but there is an easier way. Observe that since $\left(\theta - \frac{\pi}{2}\right) = -\left(\frac{\pi}{2} - \theta\right)$,

$$\cot\theta = \frac{\cos\theta}{\sin\theta} = \frac{\sin\left(\frac{\pi}{2} - \theta\right)}{\cos\left(\frac{\pi}{2} - \theta\right)} = \tan\left(\frac{\pi}{2} - \theta\right)$$

From eq. (3) on p. 104

$$= -\left(-\tan\left(\frac{\pi}{2} - \theta\right)\right) = -\tan\left(-\left(\frac{\pi}{2} - \theta\right)\right) = -\tan\left(\theta - \frac{\pi}{2}\right)$$

Thus the graph of $\cot\theta$ is the same as the graph of $-\tan\left(\theta - \frac{\pi}{2}\right)$. Then, from Section 1.4,

> The graph of $\tan\left(\theta - \frac{\pi}{2}\right)$ is the graph of $\tan\theta$ shifted $\frac{\pi}{2}$ units to the right.
>
> The graph of $-\tan\left(\theta - \frac{\pi}{2}\right)$ is the graph of $\tan\left(\theta - \frac{\pi}{2}\right)$ reflected about the horizontal axis (in this case the θ-axis).

To draw the graph of $\tan\left(\theta - \frac{\pi}{2}\right)$, observe that we have $-\frac{\pi}{2} < \theta - \frac{\pi}{2} < \frac{\pi}{2}$, so $0 < \theta < \pi$. This graph is given in Figure 2(b), and the graph

(a) $u = \tan\theta$, $-\frac{\pi}{2} < \theta < \frac{\pi}{2}$

(b) $u = \tan\left(\theta - \frac{\pi}{2}\right)$, $0 < \theta < \pi$

(c) $u = \cot\theta = -\tan\left(\theta - \frac{\pi}{2}\right)$, $0 < \theta < \pi$

Figure 2 To obtain the graph of $u = \cot\theta$, shift the graph of $\tan\theta$ $\frac{\pi}{2}$ units to the right and then reflect the resulting graph about the x-axis.

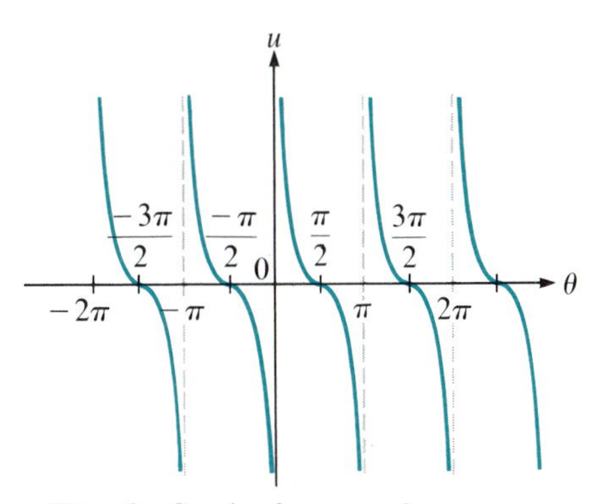

Figure 3 Graph of $u = \cot\theta$.

of cot θ for $0 < \theta < \pi$ is given in Figure 2(c). Note that the lines $\theta = 0$ and $\theta = \pi$ are vertical asymptotes of the graphs of cot θ. Note, too, that cot θ decreases as θ increases from 0 to π.

In Figure 3, we graph cot θ over a wider domain of numbers θ that are not multiples of π. We conclude from Figure 3 that

$$\text{range of } \cot\theta \text{ is } (-\infty, \infty)$$

The Graph and Range of sec θ

Since sec θ is periodic of period 2π, we compute values of sec θ in an interval of length 2π. We cannot avoid numbers at which sec θ is not defined because these numbers occur every π units: . . . $-\frac{3\pi}{2}, -\frac{\pi}{2}, \frac{\pi}{2}, \frac{3\pi}{2}$, ... We begin with the open interval $\left(-\frac{\pi}{2}, \frac{3\pi}{2}\right)$ because only one ''problem'' number, $\frac{\pi}{2}$, is in the interval.

Some representative values of sec θ are given in Table 2 on page 106. We divide our discussion into two parts:

Interval 1 $\theta \in \left(-\frac{\pi}{2}, \frac{\pi}{2}\right)$

Interval 2 $\theta \in \left(\frac{\pi}{2}, \frac{3\pi}{2}\right)$

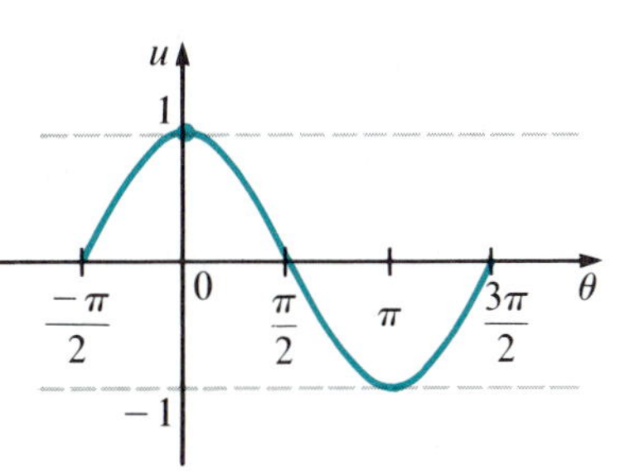

Figure 4 Graph of $u = \cos\theta$, $-\frac{\pi}{2} \le \theta \le \frac{3\pi}{2}$.

In Figure 4, we sketch the graph of cos θ for $-\frac{\pi}{2} \le \theta \le \frac{3\pi}{2}$. We see that

$$0 < \cos\theta \le 1 \text{ for } -\frac{\pi}{2} < \theta < \frac{\pi}{2}$$

and

$$-1 \le \cos\theta < 0 \text{ for } \frac{\pi}{2} < \theta < \frac{3\pi}{2}$$

Thus, since $\sec\theta = \frac{1}{\cos\theta}$, it follows that

$$\sec\theta \ge 1 \quad \text{for } -\frac{\pi}{2} < \theta < \frac{\pi}{2} \tag{1}$$

$$\sec\theta \le -1 \quad \text{for } \frac{\pi}{2} < \theta < \frac{3\pi}{2} \tag{2}$$

Now, $\cos\theta \to 0^+$ as $\theta \to \frac{\pi}{2}^-$. Thus, $\sec\theta = \frac{1}{\cos\theta} \to \infty$ as $\theta \to \frac{\pi}{2}^-$.

To convince you of this, look at the values in Table 2.

Table 2

θ (degrees)	θ (radians)	$\sec\theta = \dfrac{1}{\cos\theta}$
85°	1.483529864	11.47
89°	1.553343034	57.3
89.9°	1.569050998	573
89.99°	1.570621794	5730
89.9999°	1.570794581	572,958
89.99999°	1.570796152	5,729,578

Similarly, $\cos\theta$ increases from 0 to 1 as θ increases from $-\dfrac{\pi}{2}$ to 0, so $\sec\theta = \dfrac{1}{\cos\theta}$ decreases as θ increases from $-\dfrac{\pi}{2}$ to 0. Moreover, $\sec\theta \to \infty$ as $\theta \to -\dfrac{\pi}{2}^{+}$.

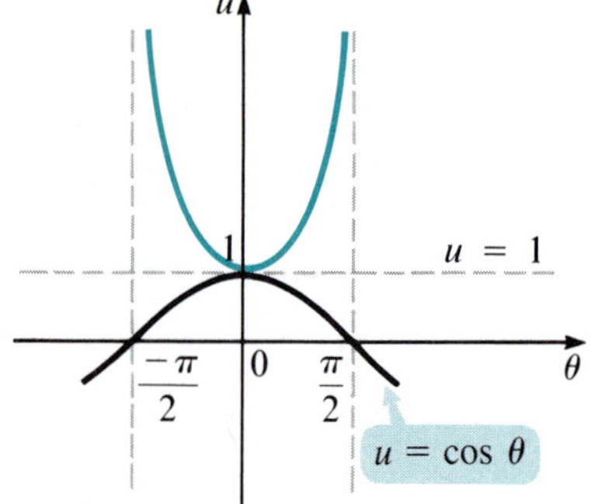

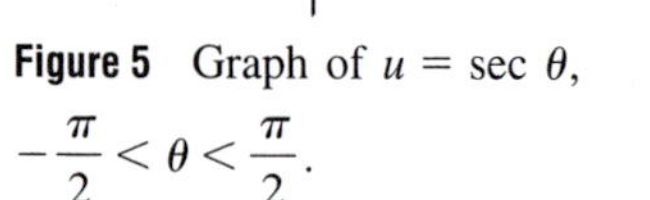
Figure 5 Graph of $u = \sec\theta$, $-\dfrac{\pi}{2} < \theta < \dfrac{\pi}{2}$.

Putting these facts together, we see that the lines $\theta = -\dfrac{\pi}{2}$ and $\theta = \dfrac{\pi}{2}$ are vertical asymptotes of the graph of $\sec\theta$. Further, $\sec\theta$ decreases as θ goes from $-\dfrac{\pi}{2}$ to 0 and increases as θ goes from 0 to $\dfrac{\pi}{2}$. Thus, it takes the minimum value 1 at $\theta = 0$. A graph of $\sec\theta$ in the interval $\left(-\dfrac{\pi}{2}, \dfrac{\pi}{2}\right)$ is sketched in Figure 5. In $\left(\dfrac{\pi}{2}, \dfrac{3\pi}{2}\right)$, the graph is similar except that now all the values are negative, and $\sec\theta$ achieves the maximum value of -1 at $\theta = \pi$.

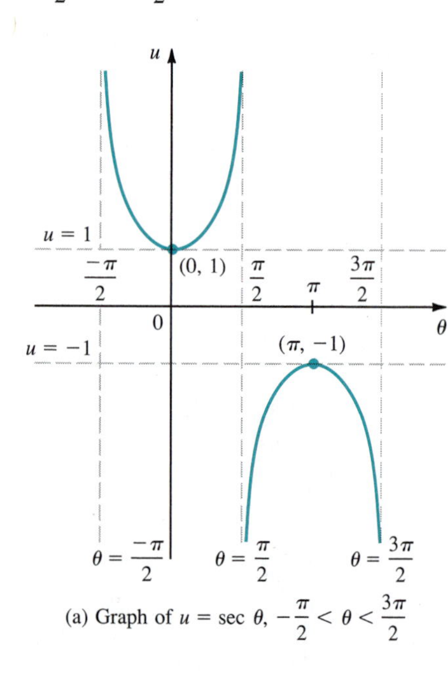

(a) Graph of $u = \sec\theta$, $-\dfrac{\pi}{2} < \theta < \dfrac{3\pi}{2}$

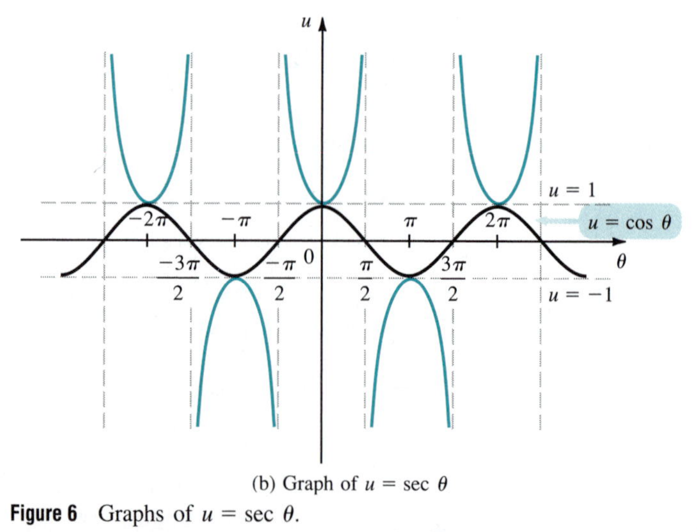

(b) Graph of $u = \sec\theta$

Figure 6 Graphs of $u = \sec\theta$.

In Figure 6(a), we sketch the graph of $u = \sec\theta$ for $-\frac{\pi}{2} < \theta < \frac{3\pi}{2}$, and in Figure 6(b) we extend the graph to other values of θ.

From Figure 6 we see that

$$\text{range of } \sec\theta \text{ is } \{u: |u| \geq 1\} = \{u: u \in (-\infty, -1] \text{ or } u \in [1, \infty)\}$$

Graph and Range of csc θ

We now obtain a graph of $u = \csc\theta$. We have

$$\sec\left(\theta - \frac{\pi}{2}\right) = \frac{1}{\cos\left(\theta - \frac{\pi}{2}\right)} = \frac{1}{\cos\left(-\left(\frac{\pi}{2} - \theta\right)\right)} \overset{\cos(-\theta) = \cos\theta}{=} \frac{1}{\cos\left(\frac{\pi}{2} - \theta\right)}$$

$$\overset{\cos\left(\frac{\pi}{2} - \theta\right) = \sin\theta}{=} \frac{1}{\sin\theta} = \csc\theta$$

Thus the graph of csc θ is the graph of sec θ shifted $\pi/2$ units to the right.

In Figure 7, we give the graph of $u = \csc\theta$. Note that the graph of csc θ has asymptotes at $0, \pm\pi, \pm 2\pi, \ldots$. This is no surprise since csc θ is not defined at these values.

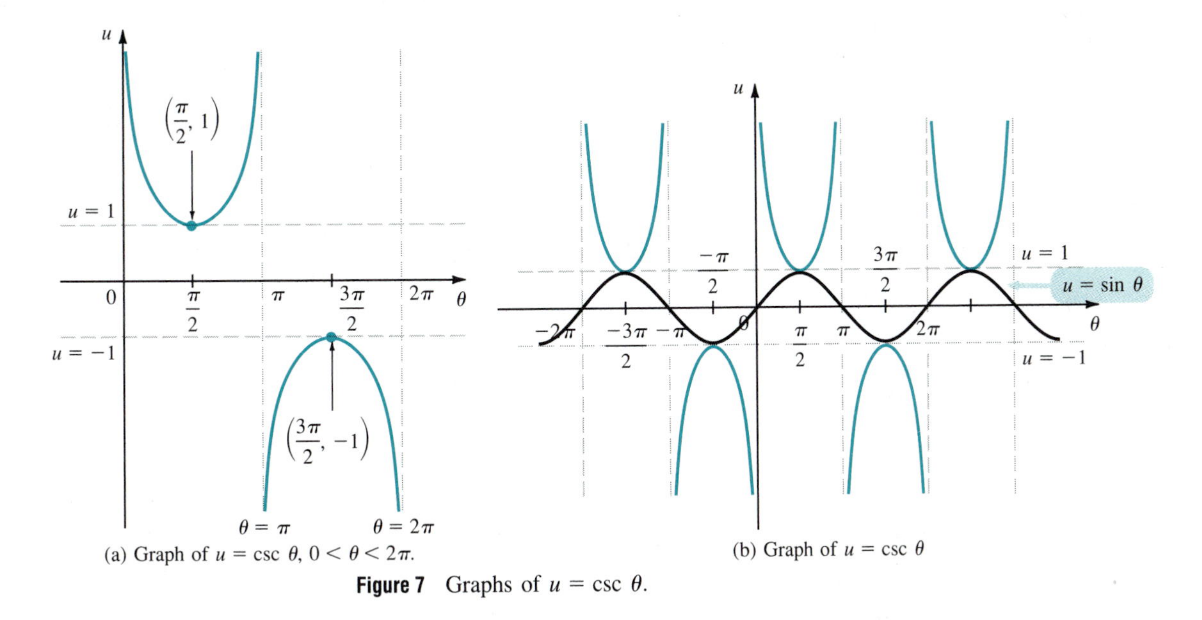

(a) Graph of $u = \csc\theta$, $0 < \theta < 2\pi$.

(b) Graph of $u = \csc\theta$

Figure 7 Graphs of $u = \csc\theta$.

Finally, from Figure 7 we see that

$$\text{range of } \csc\theta \text{ is } \{u: |u| \geq 1\} = \{u: u \in (-\infty, -1] \text{ or } u \in [1, \infty)\}$$

We close this section with Table 3, which summarizes some of the important properties of the six trigonometric functions.

Table 3

Function	Positive in Quadrant	Negative in Quadrant	Period	Domain (where defined)	Range (values taken)	Even or Odd
$\sin \theta$	I, II	III, IV	2π	all real numbers	$[-1, 1]$	odd: $\sin(-\theta) = -\sin \theta$
$\cos \theta$	I, IV	II, III	2π	all real numbers	$[-1, 1]$	even: $\cos(-\theta) = \cos \theta$
$\tan \theta$	I, III	II, IV	π	$\theta \neq \frac{\pi}{2} + n\pi$	$(-\infty, \infty)$	odd: $\tan(-\theta) = -\tan \theta$
$\cot \theta$	I, III	II, IV	π	$\theta \neq n\pi$	$(-\infty, \infty)$	odd: $\cot(-\theta) = -\cot \theta$
$\sec \theta$	I, IV	II, III	2π	$\theta \neq \frac{\pi}{2} + n\pi$	$(-\infty, -1]$ and $[1, \infty)$	even: $\sec(-\theta) = \sec \theta$
$\csc \theta$	I, II	III, IV	2π	$\theta \neq n\pi$	$(-\infty, -1]$ and $[1, \infty)$	odd: $\csc(-\theta) = -\csc \theta$

Problems 2.6

Readiness Check

I. Which of the following is an interval where $u = \cot \theta$ is increasing as θ increases?
  a. $(0, \pi/2]$
  b. $[\pi/2, \pi)$
  c. $(\pi, 3\pi/2]$
  d. There is no interval where $u = \cot \theta$ increases.

II. Which of the following is true?
  a. As $\theta \to \pi^-$, $\cot \to 0^+$.
  b. As $\theta \to (\pi/2)^+$, $\sec \theta \to +\infty$.
  c. As $\theta \to 0^-$, $\csc \theta \to -\infty$.
  d. As $\theta \to \pi^+$, $\tan \theta \to +\infty$.

III. Which of the following has a period of 2π?
  a. $\cot \theta$ b. $\csc \theta$ c. $\tan \theta$
  d. All trig functions have a period of 2π.

IV. Which of the following is equal to $\tan 7\pi/6$?
  a. $\tan \pi/6$ b. $\tan 11\pi/6$
  c. $\tan(-7\pi/6)$ d. $-\tan \pi/6$

V. Which of the following is true?
  a. $u = \tan \theta$ is a decreasing function.
  b. $u = \sec \theta$ has a vertical asymptote $\theta = -\pi$.
  c. $u = \cot \theta$ has a period of π.
  d. $u = \csc \theta$ passes through $(0, 0)$.

In Problems 1–10 find the given value, if it is defined. Do not use a calculator.

1. $\csc \frac{7\pi}{4}$ 2. $\cot \frac{7\pi}{4}$ 3. $\sec 11\pi$ 4. $\csc 11\pi$

5. $\cot 11\pi$ 6. $\csc \frac{11\pi}{3}$ 7. $\cot \frac{19\pi}{6}$ 8. $\sec 750°$

9. $\tan 750°$ 10. $\csc 1020°$

11. Find all real numbers θ such that $\tan \theta = 1$.
12. Find all real numbers θ such that $\cot \theta = -\sqrt{3}$.
13. Find all real numbers θ such that $\csc \theta = 1$.
14. Find all real numbers θ such that $\sec \theta = 2$.

In Problems 15–20 use the material in Section 1.4 to sketch the graph of the given function.

15. $u = \tan\left(\theta - \frac{\pi}{6}\right)$
16. $u = \tan \theta + 1$
17. $u = \sec \theta - 1$
18. $u = \sec\left(\theta + \frac{\pi}{4}\right)$
19. $u = \cot(\theta - \pi)$ [Hint: This is easier than it looks.]
20. $u = \csc \theta + 1$

Answers to Readiness Check

I. d II. c III. b IV. a V. c

2.7 Inverse Trigonometric Functions

The material in this section depends on the idea of an inverse function. This important topic is discussed in Section 1.6.

We summarize here some facts from Section 1.6. A function $f(x)$ is **one-to-one** on the interval $[a, b]$ if $x_1, x_2 \in [a, b]$ and $x_1 \neq x_2$ implies that $f(x_1) \neq f(x_2)$. A one-to-one function satisfies the **horizontal lines test:** If every line parallel to the x-axis intersects the graph of f at at most one point, then f is one-to-one. This is illustrated in Figure 4 on p. 46. If $f(x)$ is increasing or decreasing on $[a, b]$, then $f(x)$ is one-to-one on $[a, b]$.

The importance of one-to-one functions is that *every one-to-one function has an inverse*. That is, if $y = f(x)$ and f is one-to-one, then there exists a function $f^{-1}(y)$ such that $x = f^{-1}(y)$; for $x \in [a, b]$,

$$f^{-1}(f(x)) = x \qquad \text{and} \qquad f(f^{-1}(y)) = y$$

whenever $y = f(x)$ for some $x \in [a, b]$.

We now turn to the inverses of trigonometric functions.

So far in this chapter, we have used θ as the independent variable when we wrote the six trigonometric functions. In this section, we revert to the more common notations $y = f(x)$ or $x = f(y)$. That is, we write $\sin x$ or $\sin y$ instead of $\sin \theta$, $\cos x$ or $\cos y$ instead of $\cos \theta$, and so on.

The Inverse Sine Function

Let $y = \sin x$. Is f one-to-one? That is, can we solve for x as a function of y? To answer that question, let us give y a value, say $\frac{1}{2}$. Then we ask if there is a unique value of x such that $\sin x = \frac{1}{2}$. The answer is no. Do you see why? Did you answer $x = \dfrac{\pi}{6}$ or $x = 30°$? Of course, $\sin \dfrac{\pi}{6} = \dfrac{1}{2}$. But, also,

$$\sin\left(\pi - \frac{\pi}{6}\right) = \sin\frac{5\pi}{6} = \frac{1}{2} \qquad\qquad \sin\left(\frac{\pi}{6} + 2\pi\right) = \sin\frac{13\pi}{6} = \frac{1}{2}$$

and so on. We see that there are an infinite number of values of x (just keep adding 2π) such that $\sin x = \frac{1}{2}$. We cannot find a *unique* number x such that $\sin x = \frac{1}{2}$. In fact, if y_0 is any number in the interval $[-1, 1]$, then there are an *infinite* number of values of x for which $\sin x = y_0$. This is illustrated in Figure 1. At the marked points, $\sin x = y_0$. Simply put, the graph of $\sin x$

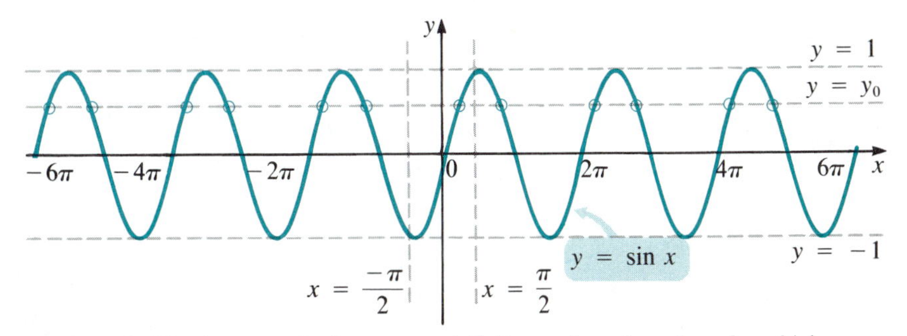

Figure 1 If $-1 \le y_0 \le 1$, there are an infinite number of x-values for which $\sin x = y_0$.

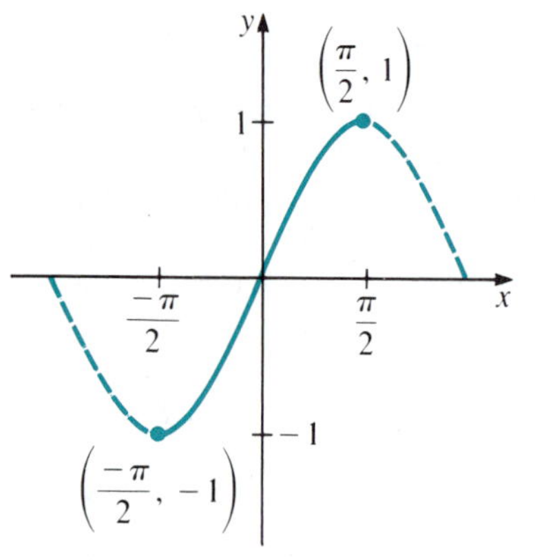

Figure 2 Graph of $f(x) = \sin x$,

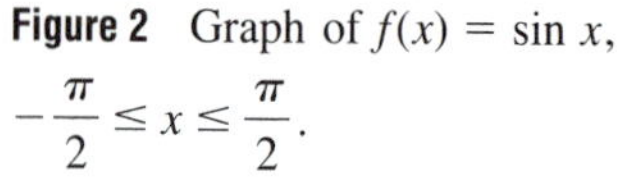

$-\frac{\pi}{2} \leq x \leq \frac{\pi}{2}$.

fails the horizontal lines test. Evidently, $\sin x$ is not a one-to-one function and, therefore, it does not have an inverse.

We can find an inverse for $\sin x$ if its domain is restricted in the following way: In Figure 2, we sketch the graph of $f(x) = \sin x$ for $-\frac{\pi}{2} \leq x \leq \frac{\pi}{2}$. We see that if x is restricted to this interval, then $\sin x$ is increasing and is therefore one-to-one. Thus, $f(x) = \sin x$ *has an inverse when x is restricted to the interval* $\left[-\frac{\pi}{2}, \frac{\pi}{2}\right]$.

NOTE We stress that the function $y = \sin x$ does *not* have an inverse. The function $y = \sin x$ with domain restricted to $\left[-\frac{\pi}{2}, \frac{\pi}{2}\right]$ *has* an inverse. The functions $f(x) = \sin x$, $x \in \left[-\frac{\pi}{2}, \frac{\pi}{2}\right]$ and $g(x) = \sin x$, $x \in \mathbb{R}$, are different functions because they have different domains.

When $f(x) = \sin x$ has restricted domain $\left[-\frac{\pi}{2}, \frac{\pi}{2}\right]$ and range $[-1, 1]$, its inverse function $f^{-1}(y) = \sin^{-1} y$ has domain $[-1, 1]$ and range $\left[-\frac{\pi}{2}, \frac{\pi}{2}\right]$. This means the following:

For any y in $[-1, 1]$, there is a unique x in $\left[-\frac{\pi}{2}, \frac{\pi}{2}\right]$ such that $\sin x = y$.

For example, if $y = \frac{1}{2}$, then $x = \frac{\pi}{6}$ is the only number in $\left[-\frac{\pi}{2}, \frac{\pi}{2}\right]$ such that $\sin x = \frac{1}{2}$. See Figure 3.

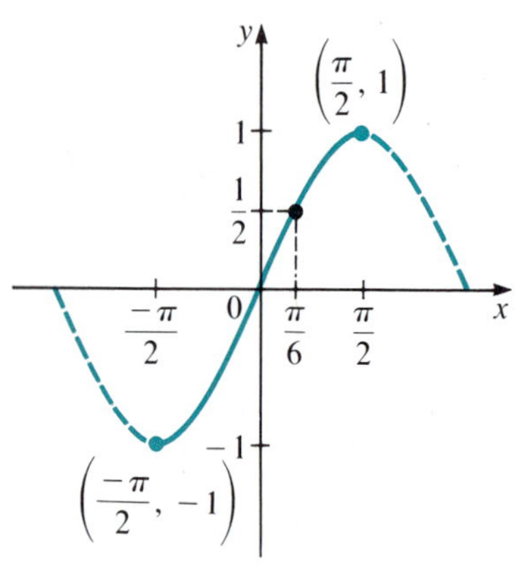

Figure 3 $\sin x$ is one-to-one if x is restricted to the interval $\left[-\frac{\pi}{2}, \frac{\pi}{2}\right]$.

We now give a formal definition of the inverse sine function. We reverse the roles of x and y in the formula $y = \sin x$ so that we can write the new function, the inverse sine function, as a function of x.

Definition of the Inverse Sine Function

The **inverse sine function** is the function that assigns to each number x in $[-1, 1]$ the unique number y in $\left[-\frac{\pi}{2}, \frac{\pi}{2}\right]$ such that $x = \sin y$. We write

$$y = \sin^{-1} x \tag{1}$$

That is,

$$y = \sin^{-1} x \qquad \text{if } x = \sin y \tag{2}$$

If y is measured in degrees, then $\sin^{-1} x$ is the degree measure of the angle whose sine is x.

It is important to keep in mind the domain and range of $\sin^{-1} x$.

Domain and Range of the Inverse Sine Function

$$\text{domain of } \sin^{-1} x \text{ is } [-1, 1] \tag{3}$$

$$\text{range of } \sin^{-1} x \text{ is } \left[-\frac{\pi}{2}, \frac{\pi}{2}\right] \tag{4}$$

From the definition of the inverse sine function, we obtain the following basic properties:

Basic Properties of $\sin^{-1} x$

$$\sin(\sin^{-1} x) = x \qquad \text{if } -1 \le x \le 1 \tag{5}$$

$$\sin^{-1}(\sin x) = x \qquad \text{if } -\frac{\pi}{2} \le x \le \frac{\pi}{2} \tag{6}$$

WARNING The -1 in $\sin^{-1} x$ does *not* mean $(\sin x)^{-1} = 1/(\sin x)$, which is equal to $\csc x$. ■

There are two other notations for the inverse sine function:

$$y = \text{arc} \sin x \tag{7}$$

is very common. In addition, some books write $\text{Sin}^{-1} x$ (with a capital S) to stress that $\sin x$ has many inverse functions (which depend on how we restrict the domains) and this one is the principal one. Another one has the range $\left[\frac{\pi}{2}, \frac{3\pi}{2}\right]$. This inverse exists because, from Figure 1, $\sin x$ is decreasing and thus is one-to-one in $\left[\frac{\pi}{2}, \frac{3\pi}{2}\right]$.

EXAMPLE 1 ***Computing Values of $\sin^{-1} x$***

Calculate $\sin^{-1} x$ for (a) $x = 0$, (b) $x = 1$, (c) $x = \frac{\sqrt{3}}{2}$, and (d) $x = -1$.

SOLUTION

(a) $\sin^{-1} 0 = 0$ since $\sin 0 = 0$ and $0 \in \left[-\frac{\pi}{2}, \frac{\pi}{2}\right]$. That is, the radian measure of the angle whose sine is 0 is 0

(b) $\sin^{-1} 1 = \frac{\pi}{2}$ since $\sin \frac{\pi}{2} = 1$ and $\frac{\pi}{2} \in \left[-\frac{\pi}{2}, \frac{\pi}{2}\right]$

(c) $\sin^{-1} \frac{\sqrt{3}}{2} = \frac{\pi}{3}$ since $\sin \frac{\pi}{3} = \frac{\sqrt{3}}{2}$ and $\frac{\pi}{3} \in \left[-\frac{\pi}{2}, \frac{\pi}{2}\right]$

(d) $\sin^{-1}(-1) = -\frac{\pi}{2}$ since $\sin\left(-\frac{\pi}{2}\right) = -1$ and $-\frac{\pi}{2} \in \left[-\frac{\pi}{2}, \frac{\pi}{2}\right]$

Values of $\sin^{-1} x$ can be given in radians or degrees. It all depends on the context. Here are the results of this example given in degrees:

(a) $\sin^{-1} 0 = 0°$
(b) $\sin^{-1} 1 = 90°$
(c) $\sin^{-1} \dfrac{\sqrt{3}}{2} = 60°$. That is, the degree measure of the angle whose sine is $\sqrt{3}/2$ is 60°
(d) $\sin^{-1} (-1) = -90°$

In Section 1.6, we saw that the graph of f^{-1} is the reflection about the line $y = x$ of the graph of f. That is, the graph of f^{-1} is the graph of f with the x- and y-axes reversed. In Figure 4(a), we give a graph of $y = \sin x$ for

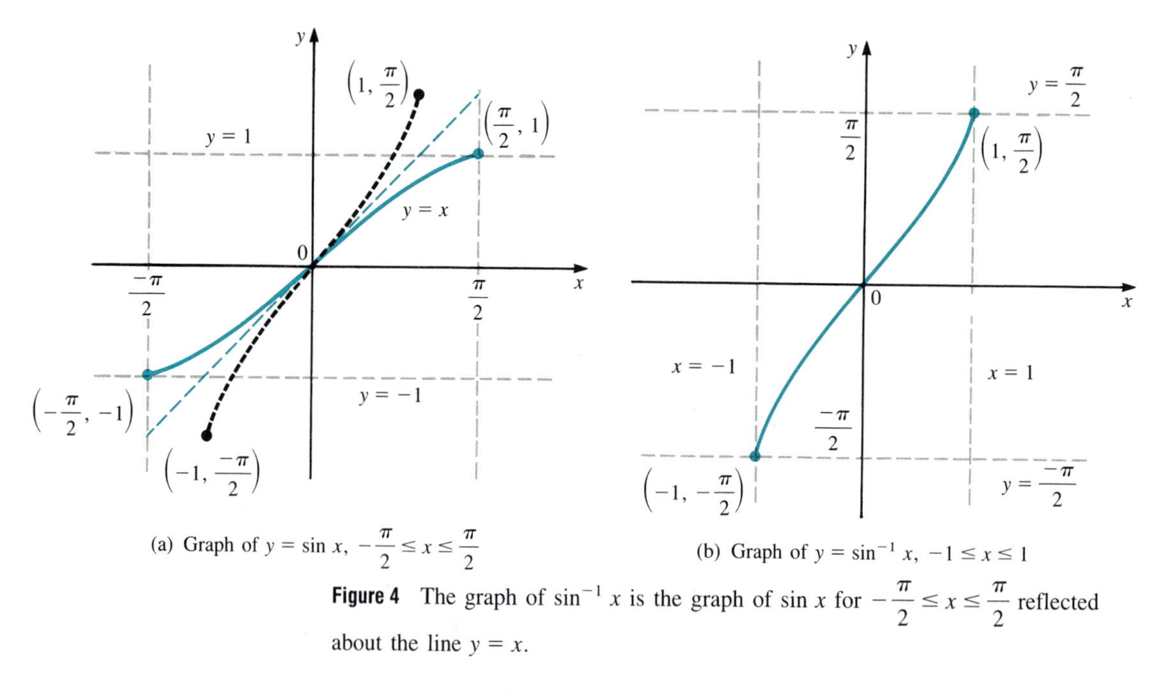

(a) Graph of $y = \sin x$, $-\dfrac{\pi}{2} \le x \le \dfrac{\pi}{2}$

(b) Graph of $y = \sin^{-1} x$, $-1 \le x \le 1$

Figure 4 The graph of $\sin^{-1} x$ is the graph of $\sin x$ for $-\dfrac{\pi}{2} \le x \le \dfrac{\pi}{2}$ reflected about the line $y = x$.

$-\dfrac{\pi}{2} \le x \le \dfrac{\pi}{2}$, and in Figure 4(b) we give the graph of $y = \sin^{-1} x$, obtained by interchanging the x- and y-axes in Figure 4(a).

EXAMPLE 2 *Computing a Cosine Value*

Compute $\cos (\sin^{-1} \frac{4}{7})$.

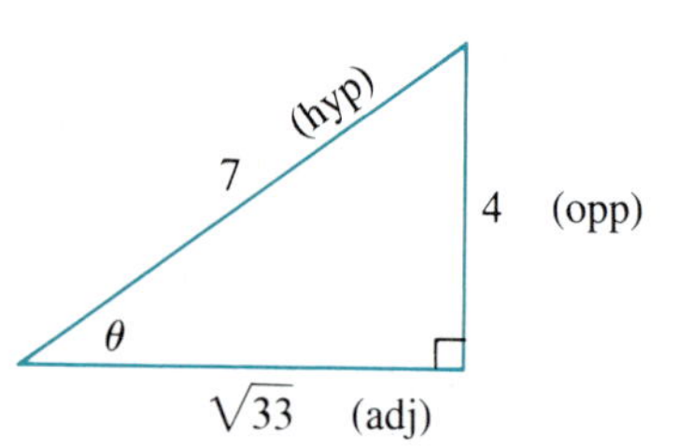

Figure 5 In this triangle $\sin \theta = \frac{4}{7}$.

SOLUTION Let $\theta = \sin^{-1} \frac{4}{7}$. Then $\sin \theta = \frac{4}{7}$ as in Figure 5. We compute

$$\text{adj}^2 = 7^2 - 4^2 = 49 - 16 = 33$$

so

$$\text{adj} = \sqrt{33}$$

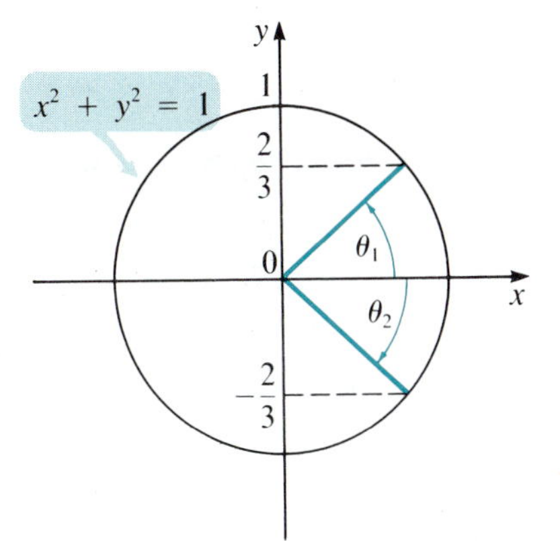

Figure 6 θ_1 is the reference angle for θ_2.

and

$$\cos\left(\sin^{-1}\frac{4}{7}\right) = \cos\theta = \frac{\sqrt{33}}{7}$$

Observe that $\cos\theta > 0$ because $-\frac{\pi}{2} < \sin^{-1}x < \frac{\pi}{2}$ and $\cos\theta > 0$ in $\left(-\frac{\pi}{2}, \frac{\pi}{2}\right)$. ■

EXAMPLE 3 ***Computing a Tangent Value***

Compute $\tan(\sin^{-1}(-\frac{2}{3}))$.

SOLUTION Let $\theta_1 = \sin^{-1}(\frac{2}{3})$ and $\theta_2 = \sin^{-1}(-\frac{2}{3})$. Then θ_1 is the reference angle for θ_2 (Figure 6). Since $-\frac{\pi}{2} < \theta_2 < 0$, θ_2 is in the fourth quadrant, $\tan\theta_2 < 0$, and, by the reference angle theorem (p. 97),

$$\tan\left(\sin^{-1}\left(-\frac{2}{3}\right)\right) = \tan\theta_2 = -\tan\theta_1 = -\tan\left(\sin^{-1}\left(\frac{2}{3}\right)\right) \tag{8}$$

Now, if $\theta_1 = \sin^{-1}\frac{2}{3}$, then $\sin\theta_1 = \frac{2}{3}$, the adjacent side is $\sqrt{3^2 - 2^2} = \sqrt{5}$, and, from Figure 7,

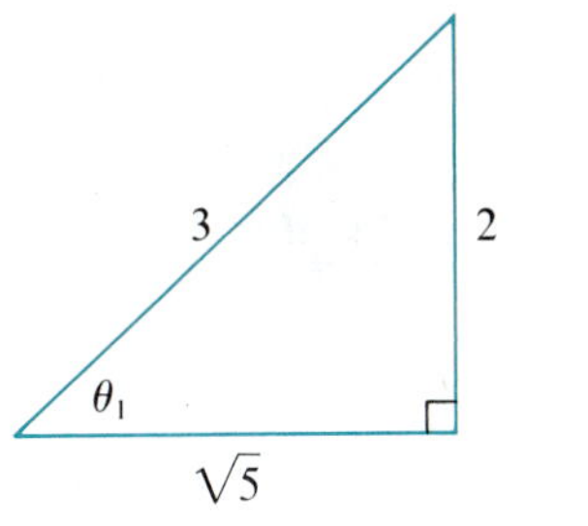

Figure 7 In this triangle $\sin\theta = \frac{2}{3}$.

$$\tan\theta_1 = \frac{2}{\sqrt{5}}$$

so

$$\tan\left(\sin^{-1}\left(-\frac{2}{3}\right)\right) = -\tan\left(\sin^{-1}\left(\frac{2}{3}\right)\right) = -\tan\theta_1 = -\frac{2}{\sqrt{5}}$$ ■

EXAMPLE 4 ***Computing a Secant Value***

Compute $\sec\left(\sin^{-1}\frac{x}{5}\right)$, $x > 0$.

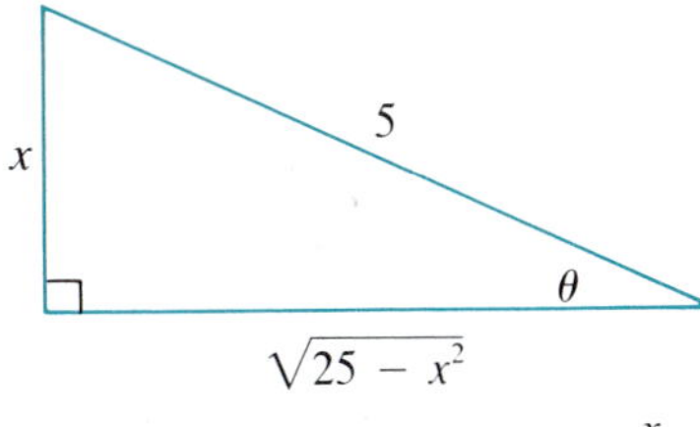

Figure 8 In this triangle $\sin\theta = \frac{x}{5}$.

SOLUTION Let $\theta = \sin^{-1}\frac{x}{5}$. Then $\sin\theta = \frac{x}{5}$. The adjacent side of the triangle in Figure 8 has length $\sqrt{5^2 - x^2} = \sqrt{25 - x^2}$, so

$$\sec\left(\sin^{-1}\frac{x}{5}\right) = \sec\theta = \frac{\text{hyp}}{\text{adj}} = \frac{5}{\sqrt{25 - x^2}}$$ ■

EXAMPLE 5 ***Showing That $\sin^{-1}(\sin x)$ Is Not Necessarily Equal to x***

Compute $\sin^{-1}(\sin 2\pi)$.

SOLUTION $\sin 2\pi = 0$, so $\sin^{-1}(\sin 2\pi) = \sin^{-1} 0 = 0$.

WARNING It is not true, in general, that

$$\sin^{-1}(\sin x) = x$$

We saw in Example 5 that $\sin^{-1}(\sin 2\pi) = 0 \neq 2\pi$. We repeat the correct statement:

$$\sin^{-1}(\sin x) = x \quad \text{if and only if} \quad -\frac{\pi}{2} \leq x \leq \frac{\pi}{2}$$ ■

$\sin^{-1} x$ on a Calculator

Every calculator that has a [sin] key also computes $\sin^{-1} x$.

To Obtain $\sin^{-1} x$ on a Calculator

Enter x and then press [INV] [sin] or [$\sin^{-1}$] or [2nd F] [sin]†

The result will be a number.

(a) between $-\frac{\pi}{2}$ and $\frac{\pi}{2}$ if the calculator is in radian mode,

(b) between -90 and 90 if the calculator is in degree mode.

EXAMPLE 6 ***Finding Values of $\sin^{-1} x$ on a Calculator***

Use a calculator to find $\sin^{-1} 0.7$ and $\sin^{-1}(-0.35)$.

SOLUTION

(a) In radian mode,

[.] [7] [INV] [sin] 0.775397496 is displayed $\quad \sin^{-1} 0.7 = 0.775397496$

[.] [3] [5] [+/−] [INV] [sin] −0.357571103 is displayed $\quad \sin^{-1}(-0.35) = -0.357571103$

(b) In degree mode,

[.] [7] [INV] [sin] 44.427004 is displayed $\quad \sin^{-1} 0.7 = 44.427004°$

[.] [3] [5] [+/−] [INV] [sin] −20.48731511 is displayed ■ $\quad \sin^{-1}(-0.35) = -20.48731511°$

EXAMPLE 7 ***Finding θ When sin θ and the Quadrant of θ Are Given***

Find an angle θ (in degrees) in the second quadrant such that $\sin \theta = 0.4$.

---

†On some calculators [INV] [sin] is pressed before the number x is entered.

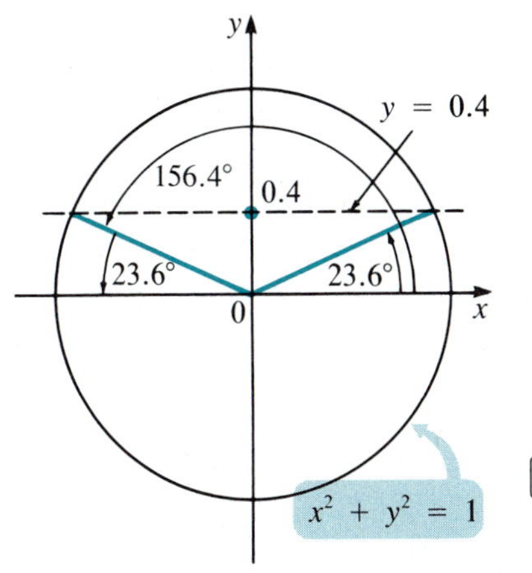

Figure 9 $\sin 23.6° = \sin 156.4° \approx 0.4$.

SOLUTION Since θ is in the second quadrant, its reference angle θ_r is in the first quadrant and (see p. 96) $\theta_r = 180° - \theta$. But $\sin\theta > 0$ in the second quadrant, so $\sin\theta_r = \sin\theta = 0.4$,

On a calculator
↓

$$\theta_r = \sin^{-1} 0.4 \approx 23.6°$$

and

$$\theta = 180° - 23.6° = 156.4°$$

The situation is illustrated in Figure 9. You should check on your calculator that $\sin 156.4° \approx 0.4$. ■

EXAMPLE 8 ***Finding θ When $\sin\theta$ and the Quadrant of θ Are Given***

Find an angle θ (in radians) in the third quadrant such that $\sin\theta = -0.85$.

SOLUTION Since θ is in the third quadrant, its reference angle is $\theta_r = \theta - \pi$. Then $\theta = \theta_r + \pi$,

On a calculator
↓

$$\theta_r = \sin^{-1} 0.85 \approx 1.0160$$

and

$$\theta = 1.0160 + \pi \approx 1.0160 + 3.1416 = 4.1576$$

Again, you should check that $\sin 4.1576 \approx -0.85$.

We summarize the results of Examples 7 and 8.

Quadrants for $\sin^{-1} x$

Let x be a number in the interval $(0, 1)$.

$\sin^{-1} x$ is in the first quadrant. **(9)**

$\pi - \sin^{-1} x \ (= 180° - \sin^{-1} x)$ is in the second quadrant. **(10)**

$\pi + \sin^{-1} x \ (= 180° + \sin^{-1} x)$ is in the third quadrant. **(11)**

$\sin^{-1}(-x) = -\sin^{-1} x$ is in the fourth quadrant. **(12)**

The Inverse Cosine Function

The graph of $y = \cos x$ is given in Figure 10. Evidently $\cos x$ is not one-to-

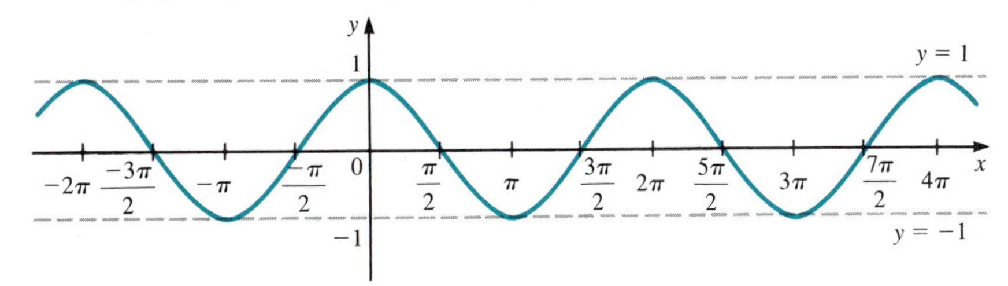

Figure 10 Graph of $y = \cos x$.

one. Note, however, that $\cos x$ decreases from 1 to -1 in the interval $[0, \pi]$. Thus, if $y = \cos x$ and x is restricted to lie in the interval $[0, \pi]$, then for every y in $[-1, 1]$ there is a unique x such that $\cos x = y$.

Definition of the Inverse Cosine Function

The **inverse cosine function** is the function that assigns to each number x in $[-1, 1]$ the unique number y in $[0, \pi]$ such that $x = \cos y$. We write $y = \cos^{-1} x$.

We stress that

Domain and Range of the Inverse Cosine Function

domain of $\cos^{-1} x$ is $[-1, 1]$ **(13)**

range of $\cos^{-1} x$ is $[0, \pi]$ **(14)**

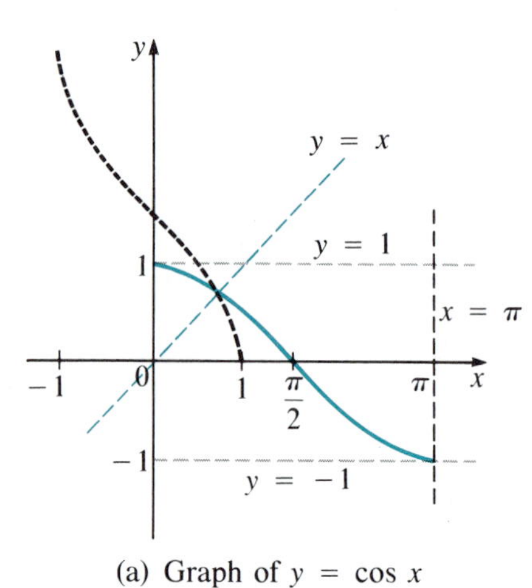

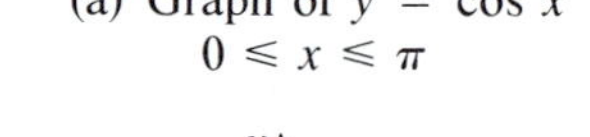
(a) Graph of $y = \cos x$
$0 \leqslant x \leqslant \pi$

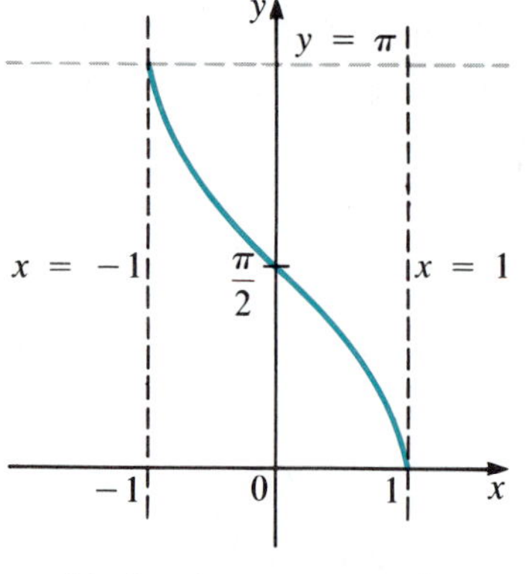

(b) Graph of $y = \cos^{-1} x$
$-1 \leqslant x \leqslant 1$

Figure 11 The graph of $\cos^{-1} x$ is the graph of $\cos x$ for $0 \leq x \leq \pi$ reflected about the line $y = x$.

NOTE $\cos^{-1} x$ is also written as arc cos x or $\text{Cos}^{-1} x$.

From the definition, we have

Basic Properties of $\cos^{-1} x$

$\cos(\cos^{-1} x) = x \quad$ if $-1 \leq x \leq 1$ **(15)**

$\cos^{-1}(\cos x) = x \quad$ if $0 \leq x \leq \pi$ **(16)**

EXAMPLE 9 ***Calculating Values of $\cos^{-1} x$***

Calculate $y = \cos^{-1} x$ for (a) $x = 0$, (b) $x = 1$, (c) $x = \dfrac{\sqrt{3}}{2}$, and (d) $x = -1$.

SOLUTION

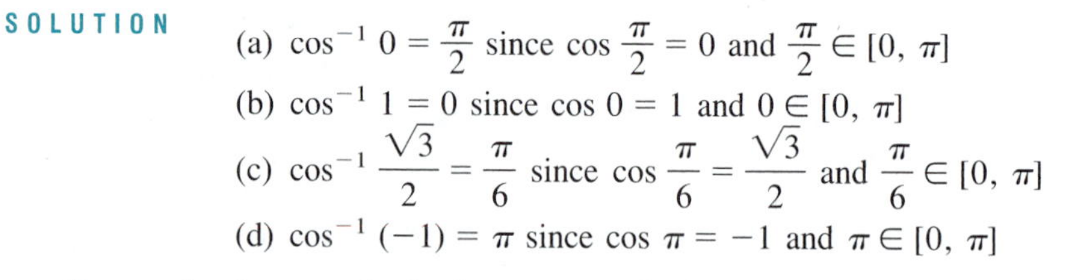

(a) $\cos^{-1} 0 = \frac{\pi}{2}$ since $\cos \frac{\pi}{2} = 0$ and $\frac{\pi}{2} \in [0, \pi]$

(b) $\cos^{-1} 1 = 0$ since $\cos 0 = 1$ and $0 \in [0, \pi]$

(c) $\cos^{-1} \dfrac{\sqrt{3}}{2} = \dfrac{\pi}{6}$ since $\cos \dfrac{\pi}{6} = \dfrac{\sqrt{3}}{2}$ and $\dfrac{\pi}{6} \in [0, \pi]$

(d) $\cos^{-1}(-1) = \pi$ since $\cos \pi = -1$ and $\pi \in [0, \pi]$

If we write these results in degrees, we obtain

(a) $\cos^{-1} 0 = 90°$ (b) $\cos^{-1} 1 = 0°$

(c) $\cos^{-1} \dfrac{\sqrt{3}}{2} = 30°$ (d) $\cos^{-1}(-1) = 180°$

In Figure 11(a), we give the graph of $y = \cos x$ for $0 \leq x \leq \pi$. In Figure 11(b), we reverse the roles of x and y to obtain the graph of $y = \cos^{-1} x$.

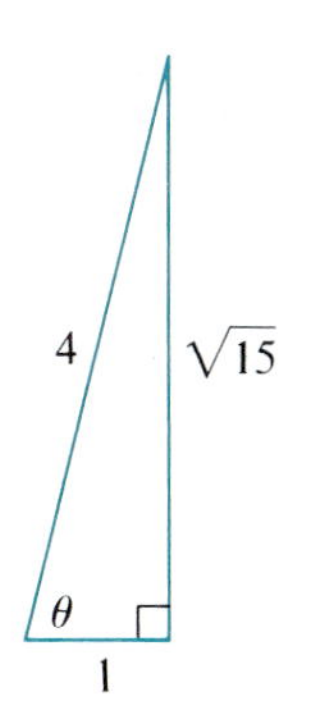

Figure 12 In this triangle $\cos \theta = \frac{1}{4}$.

EXAMPLE 10 *Computing a Tangent Value*

Compute $\tan \cos^{-1} \dfrac{1}{4}$.

SOLUTION We set $\theta = \cos^{-1} \dfrac{1}{4}$, so $\cos \theta = \dfrac{1}{4}$ as in Figure 12. The opposite side is $\sqrt{4^2 - 1^2} = \sqrt{15}$, so

$$\tan \cos^{-1} \frac{1}{4} = \tan \theta = \frac{\sqrt{15}}{1} = \sqrt{15} \quad \blacksquare$$

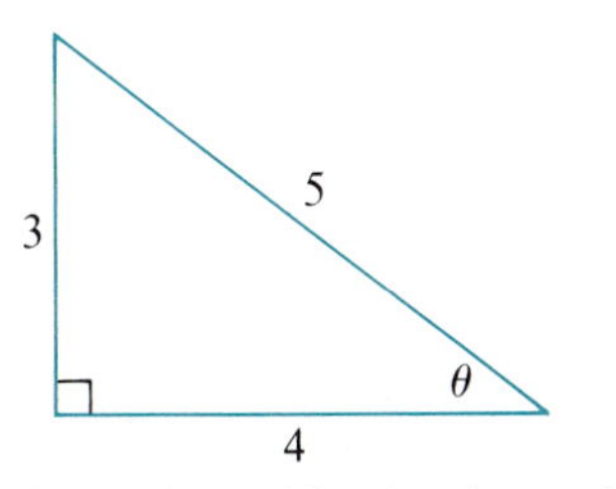

Figure 13 In this triangle $\cos \theta = \frac{4}{5}$.

EXAMPLE 11 *Computing a Tangent Value*

Compute $\tan\left(\cos^{-1}\left(-\dfrac{4}{5}\right)\right)$.

SOLUTION $\cos^{-1}\left(-\dfrac{4}{5}\right)$ is in $\left(\dfrac{\pi}{2}, \pi\right)$, the second quadrant, so $\tan\left(\cos^{-1}\left(-\dfrac{4}{5}\right)\right) < 0$ and, using the reasoning of Example 3,

$$\tan\left(\cos^{-1}\left(-\frac{4}{5}\right)\right) = -\tan\left(\cos^{-1}\left(\frac{4}{5}\right)\right) = -\frac{3}{4} \qquad \text{See Figure 13}$$

$\cos^{-1} x$ on a Calculator

To obtain $\cos^{-1} x$ on a calculator, enter x and then press [INV] [cos] or [$\cos^{-1}$] or [2nd F] [cos].† The result will be a number

(a) between 0 and π if the calculator is in radian mode, or
(b) between 0 and 180 if the calculator is in degree mode.

EXAMPLE 12 *Finding Values of $\cos^{-1} x$ on a Calculator (Radian Mode)*

Find $\cos^{-1} 0.7$ and $\cos^{-1}(-0.35)$.

SOLUTION In radian mode,

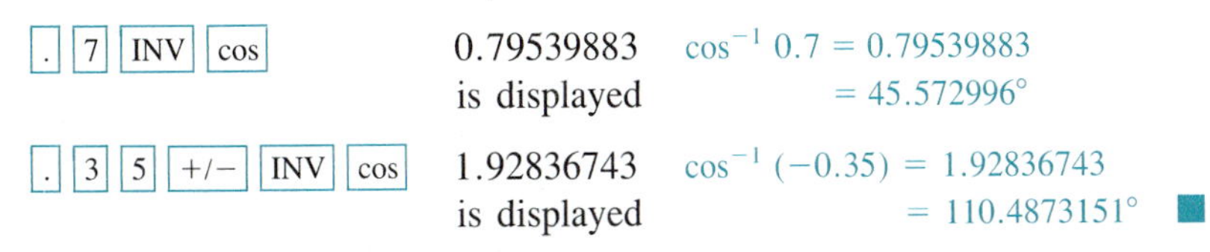

[.] [7] [INV] [cos] — 0.79539883 is displayed — $\cos^{-1} 0.7 = 0.79539883 = 45.572996°$

[.] [3] [5] [+/−] [INV] [cos] — 1.92836743 is displayed — $\cos^{-1}(-0.35) = 1.92836743 = 110.4873151°$ ■

EXAMPLE 13 *Finding Values of $\cos^{-1} x$ on a Calculator (Degree Mode)*

Find an angle θ (in degrees) in the third quadrant such that $\cos \theta = -0.82$.

† On some calculators [INV] [cos] is pressed *before* x is entered.

SOLUTION If θ is in the third quadrant and θ_r is its reference angle, then $\theta = \theta_r + 180°$ and

$$\theta = \cos^{-1}(0.82) + 180° \approx 34.915° + 180° = 214.915°$$

We have the following:

Quadrants for $\cos^{-1} x$

Let x be a number in the interval $(0, 1)$,

$\cos^{-1} x$ is in the first quadrant. **(17)**

$\cos^{-1}(-x)$ is in the second quadrant. **(18)**

$\pi + \cos^{-1} x$ $(= 180° + \cos^{-1} x)$ is in the third quadrant. **(19)**

$-\cos^{-1} x$ is in the fourth quadrant. **(20)**

One Further Note About Calculators

Calculators will tell you if you are making a serious error. For example, [2] [INV] [sin] or [2] [INV] [cos] will result in an error message since $\sin^{-1} 2$ and $\cos^{-1} 2$ are not defined.

Further, the inverse functions will bring you back where you started only if you start in $\left[-\frac{\pi}{2}, \frac{\pi}{2}\right]$ for $\sin^{-1} x$ or $[0, \pi]$ for $\cos^{-1} x$. For example, in degree mode,

[4] [1] [0] [sin] [INV] [sin] results in 50°, not 410°

because $410° = 360° + 50°$ is not in the interval $-90° \leq x \leq 90°$. On the other hand,

[5] [0] [sin] [INV] [sin] results in 50°

The Inverse Tangent Function

The graph of $y = \tan x$ is given in Figure 14. The function $\tan x$ takes all real

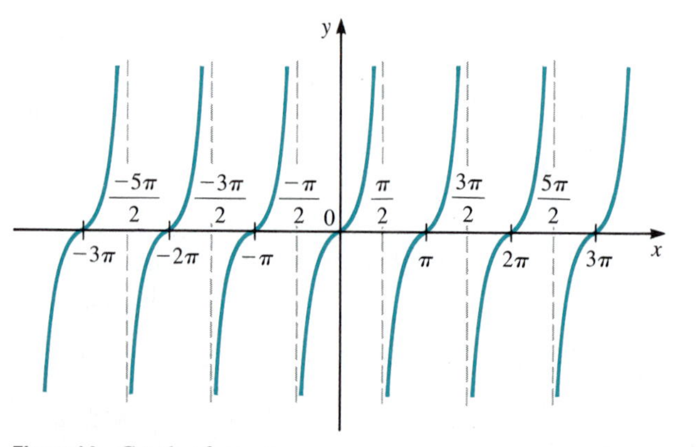

Figure 14 Graph of $y = \tan x$.

values; that is, its range is $\mathbb{R}$. To get a unique x for a given y, we restrict x to the interval $\left(-\frac{\pi}{2}, \frac{\pi}{2}\right)$.

Definition of the Inverse Tangent Function

The **inverse tangent function** is the function that assigns to each real number x the unique number y in $\left(-\frac{\pi}{2}, \frac{\pi}{2}\right)$ such that $x = \tan y$. We write $y = \tan^{-1} x$.

We have

Domain and Range of the Inverse Tangent Function

domain of $\tan^{-1} x$ is $(-\infty, \infty)$ **(21)**

range of $\tan^{-1} x$ is $\left(-\frac{\pi}{2}, \frac{\pi}{2}\right)$ **(22)**

We stress that $\frac{\pi}{2}$ is *not* in the range of $\tan^{-1} x$ because $\tan \frac{\pi}{2}$ is not defined.

NOTE $\tan^{-1} x$ is also written as arc tan x or $\text{Tan}^{-1} x$.

From the definition, we have

Basic Properties of $\tan^{-1} x$

$\tan(\tan^{-1} x) = x$ for all real x **(23)**

$\tan^{-1}(\tan x) = x$ if $-\frac{\pi}{2} < x < \frac{\pi}{2}$ **(24)**

EXAMPLE 14 *Computing Values of $\tan^{-1} x$*

Compute (in radians) (a) $\tan^{-1} 0$, (b) $\tan^{-1} 1$, (c) $\tan^{-1} \sqrt{3}$, (d) $\tan^{-1}\left(-\frac{1}{\sqrt{3}}\right)$.

SOLUTION

(a) $\tan^{-1} 0 = 0$ because $\tan 0 = 0$ and $0 \in \left(-\frac{\pi}{2}, \frac{\pi}{2}\right)$

(b) $\tan^{-1} 1 = \frac{\pi}{4}$ because $\tan \frac{\pi}{4} = 1$ and $\frac{\pi}{4} \in \left(-\frac{\pi}{2}, \frac{\pi}{2}\right)$

(c) $\tan^{-1} \sqrt{3} = \frac{\pi}{3}$ because $\tan \frac{\pi}{3} = \sqrt{3}$ and $\frac{\pi}{3} \in \left(-\frac{\pi}{2}, \frac{\pi}{2}\right)$

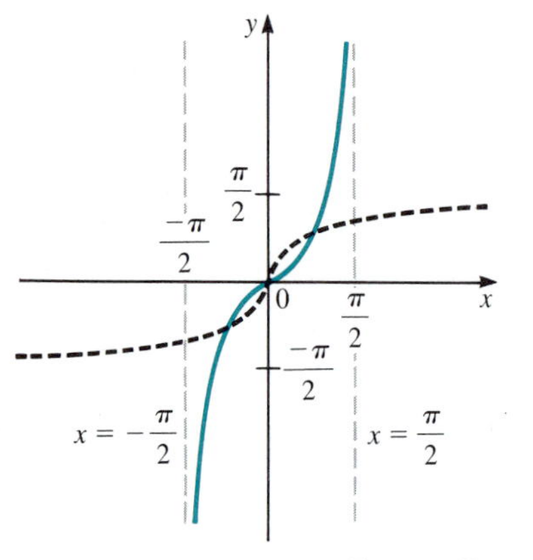

(a) Graph of $y = \tan x$, $-\frac{\pi}{2} < x < \frac{\pi}{2}$

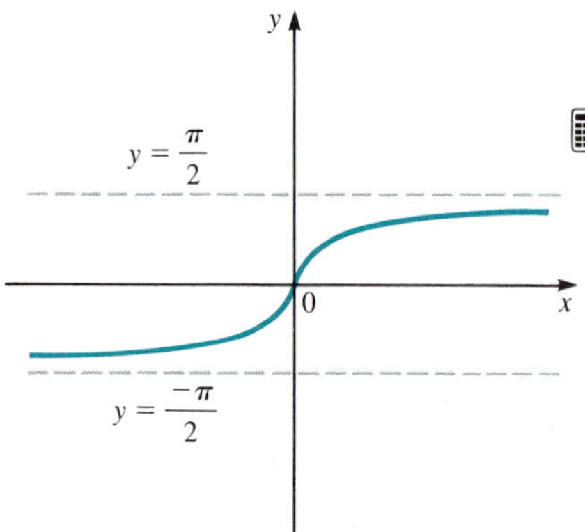

(b) Graph of $y = \tan^{-1} x$

Figure 15 The graph of $\tan^{-1} x$ is the graph of $\tan x$ for $-\frac{\pi}{2} < x < \frac{\pi}{2}$ reflected about the line $y = x$.

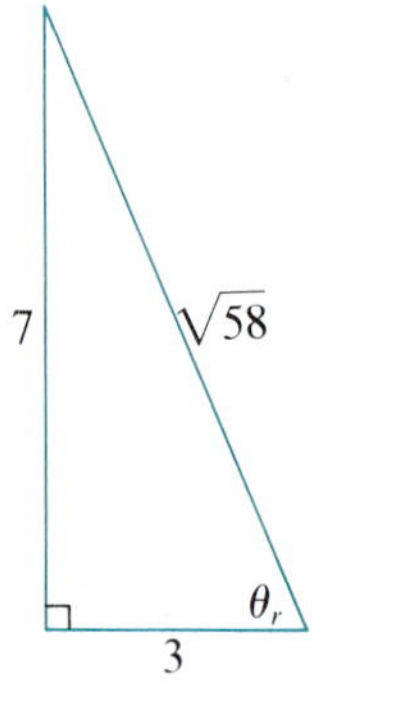

Figure 16 In this triangle $\tan \theta_r = \frac{7}{3}$.

(d) $\tan^{-1}\left(-\frac{1}{\sqrt{3}}\right) = -\frac{\pi}{6}$ because $\tan\left(-\frac{\pi}{6}\right) = -\tan\frac{\pi}{6} = -\frac{1}{\sqrt{3}}$ and $-\frac{\pi}{6} \in \left(-\frac{\pi}{2}, \frac{\pi}{2}\right)$

If we write these results in degrees, we obtain

(a) $\tan^{-1} 0 = 0°$ (b) $\tan^{-1} 1 = 45°$

(c) $\tan^{-1}\sqrt{3} = 60°$ (d) $\tan^{-1}\left(-\frac{1}{\sqrt{3}}\right) = -30°$

In Figure 15(a), we give the graph of $y = \tan x$ for $-\frac{\pi}{2} < x < \frac{\pi}{2}$. In Figure 15(b), we reverse the roles of x and y to obtain the graph of $y = \tan^{-1} x$.

EXAMPLE 15 *Computing a Sine Value*

Compute $\sin\left(\tan^{-1}\left(-\frac{7}{3}\right)\right)$.

SOLUTION $\theta_r = \tan^{-1}\left(\frac{7}{3}\right)$ is the reference angle to $\theta = \tan^{-1}\left(-\frac{7}{3}\right)$. θ is in the fourth quadrant since $-\frac{\pi}{2} < \theta < 0$, so $\sin\theta < 0$ and, from Figure 16,

$$\sin\left(\tan^{-1}\left(-\frac{7}{3}\right)\right) = -\sin\theta_r = -\frac{7}{\sqrt{58}}$$

$\tan^{-1} x$ on a Calculator

To obtain $\tan^{-1} x$ on a calculator, enter x and press [INV] [tan]† or [$\tan^{-1}$] or [2nd F] [tan]. The result will be a number

(a) between $-\frac{\pi}{2}$ and $\frac{\pi}{2}$ if the calculator is in radian mode, or

(b) between -90 and 90 if the calculator is in degree mode.

EXAMPLE 16 *Finding a Value of $\tan^{-1} x$ on a Calculator*

Find an angle θ (in degrees) in the second quadrant such that $\tan\theta = -2$.

---

†On some calculators the keys [INV] [tan] are pressed *before* x is entered.

SOLUTION If $\theta_r = \tan^{-1} 2$ is the reference angle for θ, then, since θ is in the second quadrant, $\theta_r = 180° - \theta$ or

$$\theta = 180° - \theta_r = 180° - \tan^{-1} 2 \approx 180° - 63.435° = 116.565°$$

Quadrants for $\tan^{-1} x$

Let x be a positive real number. Then

$\tan^{-1} x$ is in the first quadrant. (25)

$\pi - \tan^{-1} x$ ($= 180° - \tan^{-1} x$) is in the second quadrant. (26)

$\pi + \tan^{-1} x$ ($= 180° + \tan^{-1} x$) is in the third quadrant. (27)

$\tan^{-1}(-x) = -\tan^{-1} x$ is in the fourth quadrant. (28)

There are, of course, three other inverse trigonometric functions. These arise less frequently in applications, and we will not discuss them in this book.

Problems 2.7

Readiness Check

I. Which of the following is true?
  a. $\sin^{-1} x = 1/\sin x$ if $0 < x < \pi$
  b. $\sin^{-1}(\sin x) = x$ if $-\pi/2 \le x \le \pi/2$
  c. $\sin[\sin^{-1} x] = x$ if $0 \le x \le \pi$
  d. $\sin^{-1}(\sin x) = x$ if $-1 \le x \le 1$

II. Which of the following is true about $y = \cos^{-1} x$?
  a. Its domain is $[-\pi/2, \pi/2]$.
  b. Its domain is $[0, \infty)$.
  c. Its range is all real numbers.
  d. Its range is $[0, \pi]$.

III. Which of the following is true about $y = \sin^{-1} x$?
  a. It increases on the interval $[-1, 1]$.
  b. Its domain is $[-\pi/2, \pi/2]$.
  c. It passes through the point $(-1, 0)$.
  d. It is asymptotic to $y = 1$.

IV. Which of the following is true?
  a. The domain of $y = \sin^{-1} x$ is $[0, \pi]$.
  b. The range of $y = \tan^{-1} x$ is $(-\pi/2, \pi/2)$.
  c. The domain of $y = \cos^{-1} x$ is $[-\pi/2, \pi/2]$.
  d. The range of $y = \sin^{-1} x$ is $(0, \pi)$.

V. Which of the following has domain $(-\infty, \infty)$?
  a. $y = \sin^{-1} x$
  b. $y = \cos^{-1} x$
  c. $y = \tan^{-1} x$
  d. $y = \sin(\sin^{-1} x)$

In Problems 1–40 calculate the indicated value. Do not use a calculator.

1. $\cos^{-1} \frac{\sqrt{3}}{2}$
2. $\sin^{-1}\left(-\frac{\sqrt{3}}{2}\right)$
3. $\cos^{-1}\left(-\frac{1}{2}\right)$
4. $\tan^{-1}(-1)$
5. $\tan^{-1} \frac{1}{\sqrt{3}}$
6. $\tan^{-1}\left(-\frac{1}{\sqrt{3}}\right)$
7. $\tan^{-1}(-1)$
8. $\tan^{-1} \cos(-5\pi)$
9. $\tan^{-1}(\sin 30\pi)$
10. $\sin\left(\cos^{-1} \frac{3}{5}\right)$
11. $\cos\left[\sin^{-1}\left(-\frac{3}{5}\right)\right]$
12. $\tan\left(\sin^{-1} \frac{3}{5}\right)$

Answers to Readiness Check

I. b II. d III. a IV. b V. c

13. $\sin\left(\tan^{-1}\frac{3}{5}\right)$
14. $\sin\,[\tan^{-1}(-5)]$
* 15. $\tan(\sin^{-1} 5)$ [Hint: Watch out.]
16. $\sin(\tan^{-1} x)$
17. $\sin(\cos^{-1} x)$
18. $\tan(\sin^{-1} x)$
19. $\sec\left(\sin^{-1}\frac{3}{5}\right)$
20. $\csc\left(\cos^{-1}\frac{7}{9}\right)$
21. $\sin\left(\cos^{-1}\frac{9}{7}\right)$ [Hint: Watch out again.]
22. $\cot(\tan^{-1}(-0.112))$
23. $\cos(\cos^{-1} 10^{-55})$
24. $\sin^{-1}(\sin 3\pi)$
25. $\cos^{-1}\left(\cos\frac{5\pi}{6}\right)$
26. $\tan^{-1}\left(\tan\frac{3\pi}{4}\right)$
27. $\tan^{-1}(\tan 100\pi)$
28. $\sin^{-1}\left(\sin\frac{101\pi}{2}\right)$
29. $\sec(\tan^{-1} 10)$
30. $\csc\left(\sin^{-1}\frac{1}{2}\right)$
31. $\csc\left(\cos^{-1}\frac{4}{7}\right)$
32. $\cot(\tan^{-1} 4)$
33. $\cot(\tan^{-1}(-2))$
34. $\sin\left(\tan^{-1}\left(-\frac{1}{5}\right)\right)$
35. $\cos\left(\sin^{-1}\left(-\frac{5}{6}\right)\right)$
36. $\sin\left(\cos^{-1}\frac{x}{\sqrt{2}}\right)$
37. $\cos\left(\sin^{-1}\frac{x}{a}\right)$
38. $\sec\left(\tan^{-1}\frac{y}{x}\right)$
39. $\sec\left(\sin^{-1}\frac{x}{x+1}\right)$
40. $\sec\left(\tan^{-1}\frac{x}{\sqrt{x^2-4}}\right)$

In Problems 41–50 use a calculator to find each value. Give your answer in both degrees and radians and round to 4 decimal places.

41. $\cos^{-1}(0.31)$
42. $\sin^{-1}(-0.31)$
43. $\sin^{-1}(-0.42)$
44. $\cos^{-1}(0.89)$
45. $\tan^{-1} 2.7$
46. $\tan^{-1}(-5.8)$
47. $\tan^{-1}\left(\frac{1}{8}\right)$
48. $\tan^{-1}(-0.65)$
49. $\sin^{-1}\left(\frac{\pi}{4}\right)$
50. $\cos^{-1}\left(-\frac{\pi}{4}\right)$

In Problems 51–60 a trigonometric functional value and a quadrant containing θ are given. Find θ and express your answer in both degrees and radians. Round to 4 decimal places.

51. $\sin\theta = 0.2$; II
52. $\sin\theta = -0.2$; III
53. $\sin\theta = -0.2$; IV
54. $\cos\theta = 0.7$; IV
55. $\cos\theta = -\frac{2}{3}$; III
56. $\tan\theta = 10$; III
57. $\tan\theta = -\frac{1}{2}$; IV
58. $\sin\theta = \frac{1}{100}$; II
59. $\cos\theta = -\frac{4}{17}$; III
60. $\tan\theta = -0.356$; II
61. Find two numbers in $(4\pi, 6\pi)$ such that $\cos\theta = \frac{1}{2}$.
62. Find two numbers in $\left(\frac{7\pi}{2}, \frac{11\pi}{2}\right)$ such that $\sin\theta = 0.4$.
63. Find two numbers in $[10\pi, 12\pi]$ such that $\tan\theta = 7$.
64. Find one number in $\left(-\frac{5\pi}{2}, -\frac{3\pi}{2}\right)$ such that $\tan\theta = -\frac{1}{10}$.
65. Show that if $-1 \le x \le 1$, then $\sin(\cos^{-1} x) = \cos(\sin^{-1} x)$.
66. Show that as $x \to \infty$, $\tan^{-1} x \to \frac{\pi}{2}$.
67. Show that as $x \to -\infty$, $\tan^{-1} x \to -\frac{\pi}{2}$.
* 68. Show that $\sin^{-1} x + \cos^{-1} x = \frac{\pi}{2}$. [Hint: Treat the cases $x > 0$ and $x < 0$ separately. If $x > 0$, draw one right triangle with one acute angle $\theta_1 = \sin^{-1} x$ and another with acute angle $\theta_2 = \cos^{-1} x$. Show that the two triangles are congruent.]
69. Show that $\sin^{-1}(-x) = -\sin^{-1}(x)$. [Hint: Use the fact that $\sin(-x) = -\sin x$.]
70. Show that $\cos^{-1}(-x) = \pi - \cos^{-1}(x)$. [Hint: Use the fact that $\cos(\pi - x) = -\cos x$.]
71. Show that $\tan^{-1}(-x) = -\tan^{-1}(x)$. [Hint: Use the fact that $\tan(-x) = -\tan x$.]

2.8 Applications of Right Triangle Trigonometry

In the applications of trigonometry that we consider in this section, it is necessary to find all sides and angles of a right triangle. We can always do this in one of two situations. We can find all the sides and angles of a right triangle when (a) one acute angle and one side are given or (b) when two sides are given.

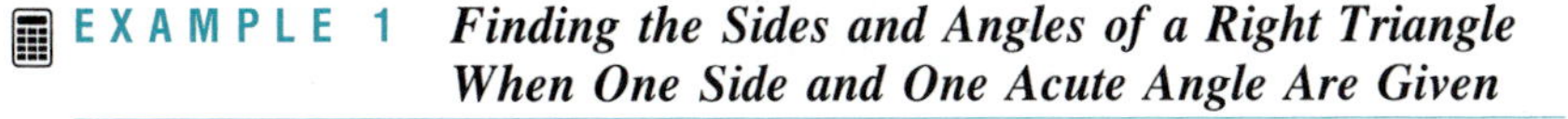

EXAMPLE 1 ***Finding the Sides and Angles of a Right Triangle When One Side and One Acute Angle Are Given***

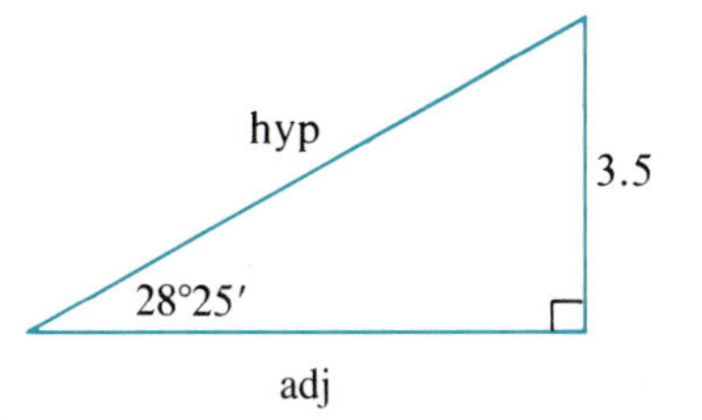

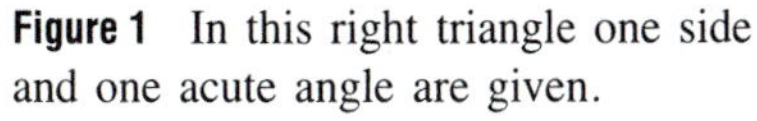

Figure 1 In this right triangle one side and one acute angle are given.

Find all sides and angles of the right triangle in Figure 1. In all steps round to 4 decimal places.

SOLUTION If we know two sides of a right triangle, we can compute the third side by the Pythagorean theorem. In this case, where one side and one acute angle are given, there are two ways to compute the length of a second side:

$$\text{(i)}\ \sin 28°25' = \frac{\text{opp}}{\text{hyp}} = \frac{3.5}{\text{hyp}}$$

$$\text{(ii)}\ \tan 28°25' = \frac{\text{opp}}{\text{adj}} = \frac{3.5}{\text{adj}}$$

In either case, we must convert 28°25′ to degrees. We have

$$25' = \left(\frac{25}{60}\right)^{\circ} \approx 0.4167° \quad \text{so} \quad 28°25' \approx 28.4167°$$

Then

$$\sin 28.4167° \approx 0.4759 = \frac{3.5}{\text{hyp}}$$

$$0.4759 \times \text{hyp} = 3.5$$

$$\text{hyp} = \frac{3.5}{0.4759} \approx 7.3545$$

To find adj, we use the Pythagorean theorem:

$$\text{adj}^2 + \text{opp}^2 = \text{hyp}^2$$

$$\text{adj}^2 + (3.5)^2 = (7.3545)^2$$

$$\text{adj}^2 + 12.25 = 54.0887$$

$$\text{adj}^2 = 41.8387$$

$$\text{adj} = \sqrt{41.8387} \approx 6.4683$$

We can check this by using approach (ii):

$$\tan 28.4167° \approx 0.5411 = \frac{3.5}{\text{adj}}$$

$$\text{adj} = \frac{3.5}{0.5411} \approx 6.4683$$

Finally, the other acute angle is $90° - 28°25' = 61°35' = 61\frac{35}{60}°$
$= 61.5833°$

NOTE From now on we will give all degree measurements in decimal notations except in navigation problems, where degrees and minutes are customarily given.

EXAMPLE 2 *Finding the Sides and Angles of a Right Triangle When Two Sides Are Given*

Find all sides and angles of the triangle in Figure 2.

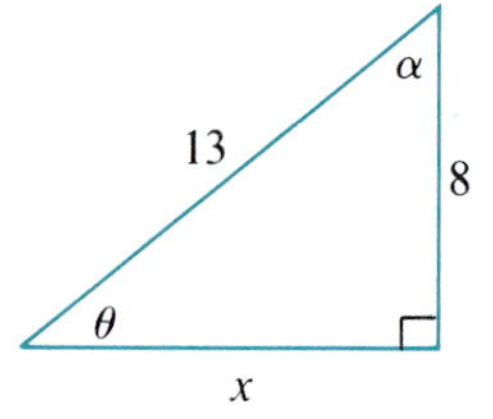

Figure 2 In this right triangle two sides are given.

SOLUTION Using the Pythagorean theorem, we see that

$$x = \sqrt{13^2 - 8^2} = \sqrt{169 - 64} = \sqrt{105}$$

What are the angles?

First, we observe that

$$\sin\theta = \frac{8}{13} = 0.615384615$$

Then

$$\theta = \sin^{-1} 0.615384615 \approx 37.98°$$

The angle α is $90° - 37.98° = 52.02°$. We can get α directly. We see that

$$\cos\alpha = \frac{8}{13} \text{ so } \alpha = \cos^{-1}\frac{8}{13} \approx \cos^{-1} 0.615384615 \approx 52.02°$$

Many applied problems can be solved by using right triangle trigonometry. We now provide a number of illustrations of this fact.

Measuring Heights and Distances: Angles of Elevation or Depression

In many situations, the trigonometric functions can be used to determine distances that are difficult to measure directly. Two such cases are illustrated in Figure 3.

In each case, an angle is formed by two lines: a horizontal line and a **line of sight.** If the angle is measured upward from the horizontal, as in Figure 3(a), then the angle is called the **angle of elevation.** If it is measured downward as in Figure 3(b), it is called an **angle of depression.**

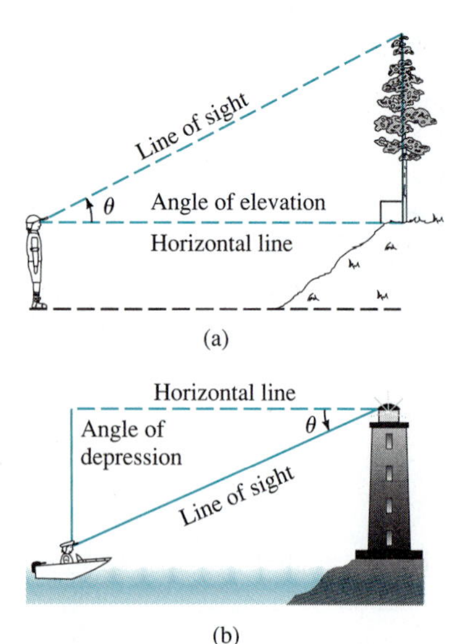

Figure 3 Illustrating angles of elevation and depression.

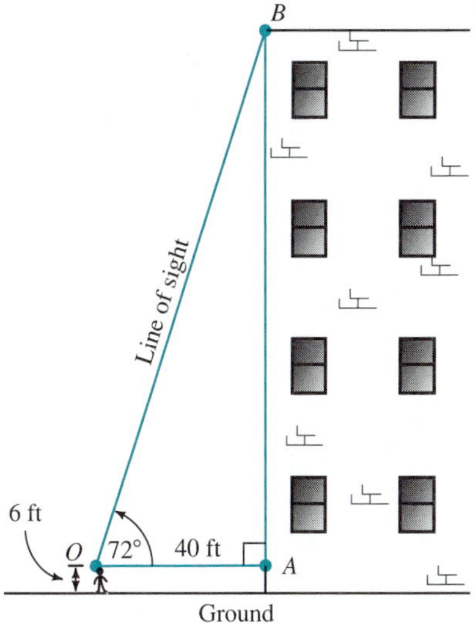

EXAMPLE 3 *Using an Angle of Elevation to Measure Height*

A man whose eye level is 6 ft above the ground stands 40 ft from a building. The angle of elevation from eye level to the top of the building is 72°.

(a) How tall is the building?
(b) What is the line-of-sight distance from the man's head to the top of the building?

SOLUTION

(a) From Figure 4, we see that the height of the building is 6 ft + $\overline{AB}$. But, in triangle AOB,

$$\tan 72^\circ = \frac{\overline{AB}}{\overline{OA}} = \frac{\overline{AB}}{40}$$

Calculator ↓

$$\overline{AB} = 40 \tan 72^\circ \approx 40(3.0776835) \approx 123.1 \text{ ft}$$

Thus, the height of the building is approximately 123.1 ft + 6 ft = 129.1 ft.

(b) The line-of-sight distance is $\overline{OB}$.

$$\cos 72^\circ = \frac{40}{\overline{OB}}$$

$$\overline{OB} = \frac{40}{\cos 72^\circ} \approx \frac{40}{0.309017} \approx 129.4 \text{ ft}$$

We could also obtain this answer from the Pythagorean theorem:

$$\overline{OB}^2 = \overline{OA}^2 + \overline{AB}^2$$
$$\overline{OB}^2 = 40^2 + (123.1)^2 = 1600 + 15153.61 = 16753.61$$
$$\overline{OB} = \sqrt{16,753.61} \approx 129.4 \text{ ft}$$

Figure 4 The angle of elevation from a spot 40 ft from a building is 72°.

WARNING When doing computations involving height or distance, you must usually figure the height of the observer into the calculation. ■

EXAMPLE 4 *Using an Angle of Depression to Measure Distance*

A woman standing on top of a cliff spots a boat in the sea, as in Figure 5. According to a nearby sign, the top of the cliff is 235 feet above the water level. Her eye level is 5 ft above the ground. If the angle of depression is 27°, how far is the boat from a point at sea level that is directly below the observer?

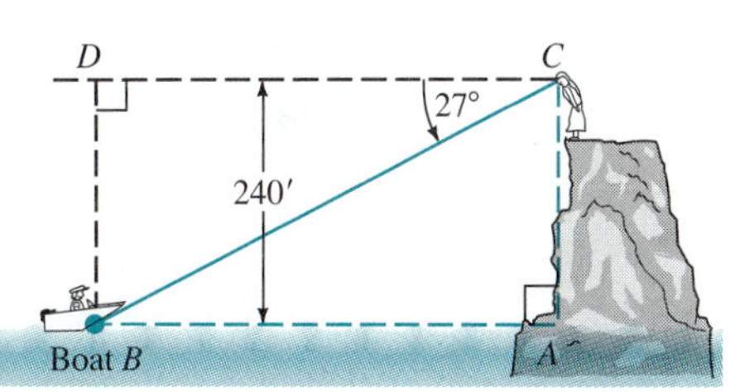

Figure 5 The angle of depression from a spot 235 ft above water level is 27°.

SOLUTION In Figure 5, the observer's eyes are 240 ft above the water level (235 ft for the cliff and 5 ft for the woman). Using triangle BCD, we compute

$$\tan 27^\circ = \frac{\overline{BD}}{\overline{DC}} = \frac{240}{\overline{DC}}$$

So

$$\overline{DC} = \frac{240}{\tan 27^\circ} \approx \frac{240}{0.5095} \approx 471 \text{ ft}$$

Thus, $\overline{AB} = 471$ ft is the answer.

Note that this may not be the same as the distance from the boat to the foot of the cliff if the cliff juts outward near the water level, as in Figure 5.

Surveying with a Transit

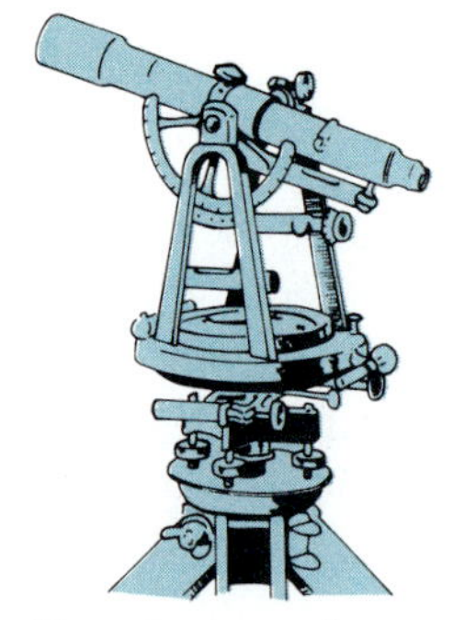

Figure 6 A transit.

In the construction of roads and bridges, distances must be calculated with considerable precision. Surveyors and engineers use an instrument of great versatility called a **transit** (see Figure 6). By the use of a telescope, a line of sight is well defined, and accurate sighting at long distances is possible. The transit contains a horizontal circle that is graduated every $10'$, $20'$, or $30'$, depending on the type of instrument. A transit also contains a device called a *vernier,* which can be used to measure fractional parts of the fixed scale (that is, parts of $10'$, $20'$, or $30'$). With the vernier, it is possible to lay off horizontal angles that are as small as $10''$. A transit also has a vertical circle and vernier for measuring vertical angles as well. Finally, each transit contains a level bubble that is sensitive enough to eliminate most uncertainty in measuring vertical angles.

The following example typifies the use of a transit by a civil engineer.

EXAMPLE 5

An engineer is hired to design a bridge to carry traffic across the river shown in Figure 7. Her first problem is to determine the distance from point A to point B. To do so, she starts at point A and measures a distance of 250 m in a direction at right angles to the segment AB. At point C, she uses her transit to find that angle ACB measures 49.4°. How long should the bridge be?

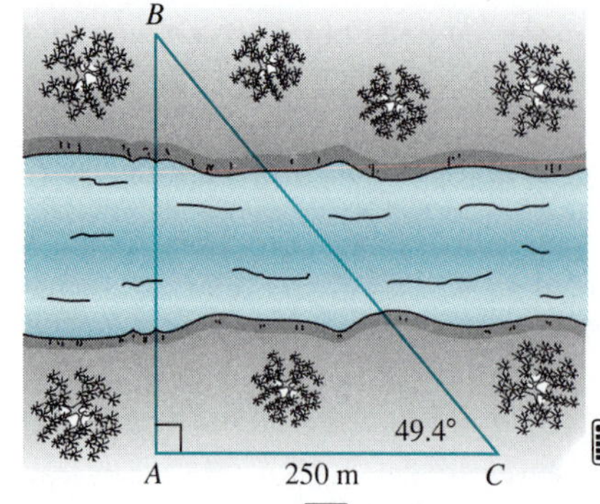

Figure 7 Given that $\overline{AC} = 250$ m and angle $C = 49.4°$ we must determine $\overline{AB}$.

SOLUTION From Figure 7, we find that

$$\tan 49.4^\circ = \frac{\overline{AB}}{\overline{AC}} = \frac{\overline{AB}}{250}$$

Then

$$\text{length of bridge} = \overline{AB} = 250 \tan 49.4 = 250(1.166720019)$$
$$\approx 291.7 \text{ m}$$

■

EXAMPLE 6

In order to measure the height of a mountain, a surveyor takes two sightings from a transit $1\frac{1}{2}$ m high. The sightings are taken 1200 m apart from the same ground elevation. The first measured angle of elevation is 51°, and the second is 38°. To the nearest meter, what is the height of the mountain (above ground elevation)?

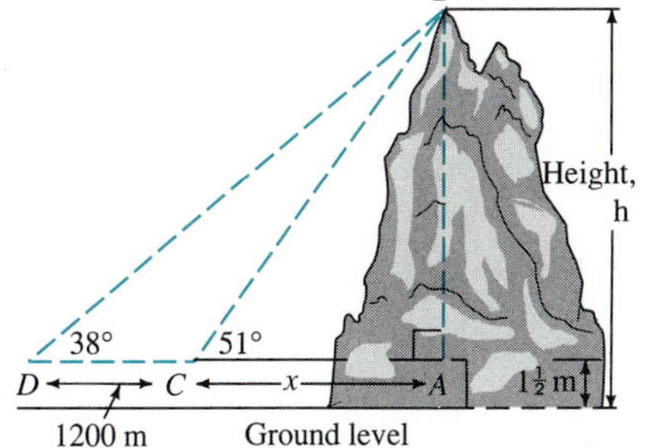

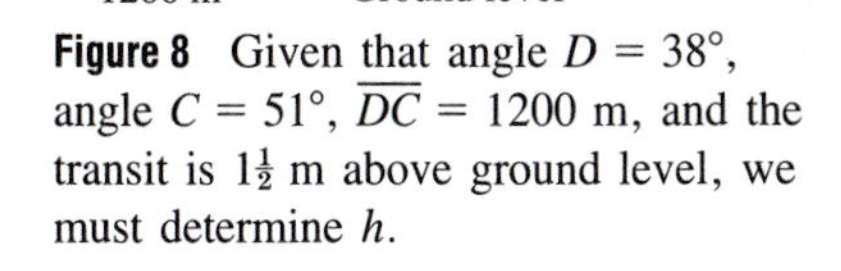

Figure 8 Given that angle $D = 38°$, angle $C = 51°$, $\overline{DC} = 1200$ m, and the transit is $1\frac{1}{2}$ m above ground level, we must determine h.

SOLUTION The situation is depicted in Figure 8. The height of the mountain is

$$\overline{AB} + 1\tfrac{1}{2}\text{ m} = h$$

From Figure 8, we find that

$$\tan 51° = \frac{\overline{AB}}{x} \qquad \text{and} \qquad \tan 38° = \frac{\overline{AB}}{x + 1200}$$

$$\overline{AB} = x \tan 51° \qquad \overline{AB} = (x + 1200) \tan 38°$$

$$\tan 51° \approx 1.235 \qquad \tan 38° \approx 0.7813$$

$$\overline{AB} \approx 1.235x \qquad \overline{AB} \approx (x + 1200)(0.7813) = 0.7813x + 937.56$$

Equating the two expressions for $\overline{AB}$, we have

$$\begin{aligned} 1.235x &= 0.7813x + 937.56 \\ (1.235 - 0.7813)x &= 937.56 \\ 0.4537x &= 937.56 \\ x &= \frac{937.56}{0.4537} \approx 2066.5 \end{aligned}$$

Thus

$$\overline{AB} = 1.235x = (1.235)(2066.5) \approx 2552.1 \text{ m}$$

and

$$\begin{aligned} \text{height of mountain} &= \overline{AB} + 1\tfrac{1}{2}\text{ m} = 2552.1 + 1.5 \\ &= 2553.6 \approx 2554 \text{ m} \end{aligned}$$

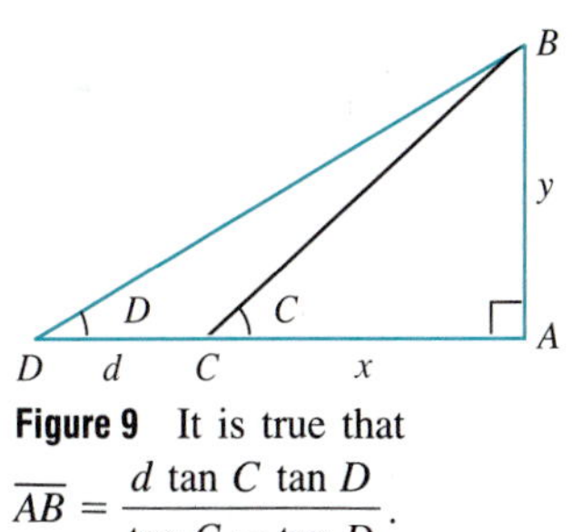

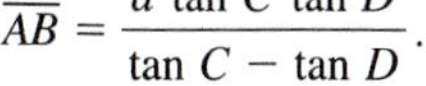

Figure 9 It is true that $\overline{AB} = \dfrac{d \tan C \tan D}{\tan C - \tan D}$.

NOTE In Example 6 we could have used the following fact: In the triangle of Figure 9,

$$y = \overline{AB} = \frac{d \tan C \tan D}{\tan C - \tan D} \qquad (1)$$

For example, if $d = 1200$, $D = 38°$, and $C = 51°$, (1) yields

$$\overline{AB} = \frac{1200 \tan 51° \tan 38°}{\tan 51° - \tan 38°} \approx 2552.3$$

(This answer is slightly different from the answer in Example 6 because here we didn't round until the end.)

It is not difficult to prove that (1) holds. We have, from Figure 9,

$$\tan C = \frac{y}{x} \qquad \text{so} \qquad y = x \tan C \qquad (2)$$

$$\tan D = \frac{y}{x + d} \qquad \text{so} \qquad y = (x + d) \tan D \qquad (3)$$

Equating the values for y in (2) and (3) gives us

$$x \tan C = (x + d) \tan D = x \tan D + d \tan D$$
$$x \tan C - x \tan D = d \tan D$$
$$x (\tan C - \tan D) = d \tan D$$
$$x = \frac{d \tan D}{\tan C - \tan D} \tag{4}$$

If we insert (4) into (2), we obtain

$$y = x \tan C = \tan C\left(\frac{d \tan D}{\tan C - \tan D}\right) = \frac{d \tan C \tan D}{\tan C - \tan D}$$

which is what we wanted to prove. ■

Navigation

In navigation, directions to and from a reference point are often given in terms of **bearings.** A bearing is an acute angle between a line of travel or line of sight and the north-south line. Bearings are usually given as so many degrees east of north, west of north, east of south, or west of south. If θ is the acute angle, then N θ° E is read as θ° east of north, and so on.

EXAMPLE 7 *Four Bearings*

The four bearings in Figure 10 are, respectively, (a) N30°E, (b) N55°W, (c) S60°E, and (d) S10°W.

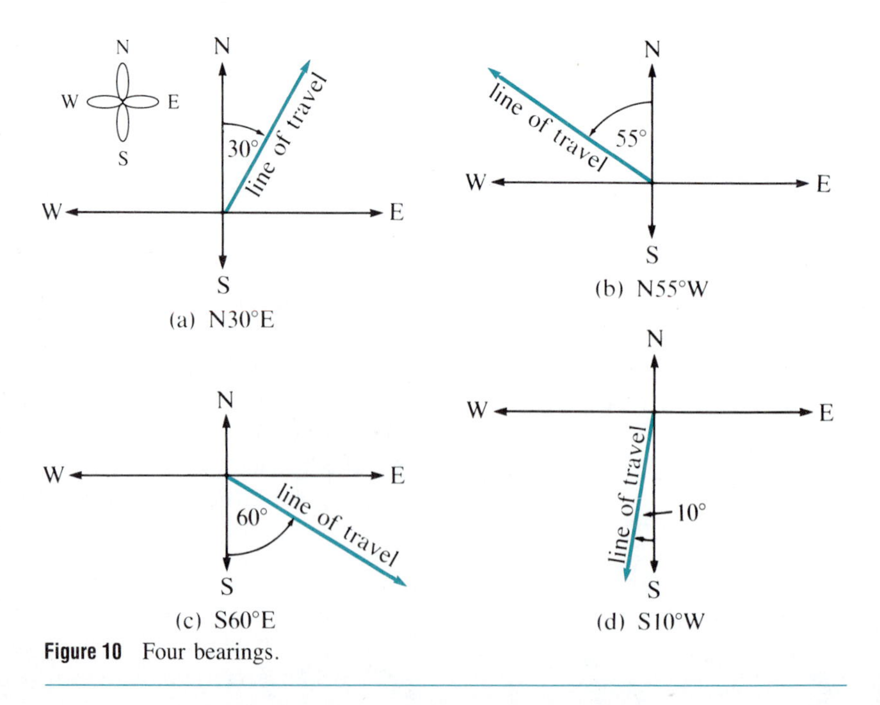

Figure 10 Four bearings.

NOTE In airplane navigation, bearings are often given in angles measured clockwise from the north. Thus, in this system, S60°E is a bearing of 120° and N55°W is a bearing of 305°. In this book, we will stick to the bearings as given in Example 7, but you should be aware that other ways are used to represent bearings.

EXAMPLE 8 *Determining Total Distance When Two Distances and Two Bearings Are Given*

A ship leaves port and travels 36 miles due west. It then changes course and sails 24 miles on a bearing of S33°W. How far is it from port at this point?

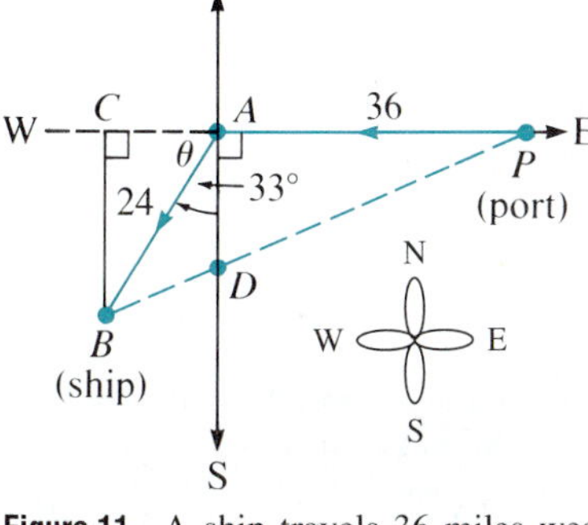

Figure 11 A ship travels 36 miles west from P to A and then travels 24 miles at a bearing of S33°W.

SOLUTION The situation is depicted in Figure 11. The ship is at point B and we must compute the distance $\overline{PB}$. In Section 3.2, we will learn the law of cosines, and this will enable us to compute $\overline{BP}$ in one step. Here we use several steps.

Method of attack

(1) We note that triangle PCB is a right triangle. Thus, if we know $\overline{PC}$ and $\overline{BC}$, we can compute $\overline{BP}$ by the Pythagorean theorem.
(2) $\overline{PC} = 36 + \overline{AC}$.
(3) We can find $\overline{AC}$ and $\overline{BC}$ if we can solve the right triangle ABC. We can do this if we know angle θ since one side, $\overline{AB}$, is known.

We observe that θ is the complement of the 33° bearing. Thus

$$\theta = 90° - 33° = 57°$$

Now

$$\cos\theta = \frac{\overline{AC}}{24} \quad \text{and} \quad \sin\theta = \frac{\overline{BC}}{24}$$

so

$$\overline{AC} = 24\cos 57° \approx 24(0.5446) \approx 13.07 \text{ miles}$$
$$\overline{BC} = 24\sin 57° \approx 24(0.8387) \approx 20.13 \text{ miles}$$

Then

$$\overline{PC} = 36 + \overline{AC} \approx 36 + 13.07 = 49.07 \text{ miles}$$

Finally,

$$\overline{BP}^2 = \overline{PC}^2 + \overline{BC}^2 \approx (49.07)^2 + (20.13)^2$$
$$\approx 2408 + 405 = 2813$$

Thus,

$$\text{distance from ship to port} = \overline{BP} \approx \sqrt{2813} \approx 53 \text{ miles}$$ ■

EXAMPLE 9 *Determining a Bearing*

An airplane is 255.2 kilometers due east of radar station R. A second radar station, Q, is 347.6 kilometers due north of radar station R. Find, to the nearest minute, the bearing of the plane from radar station Q.

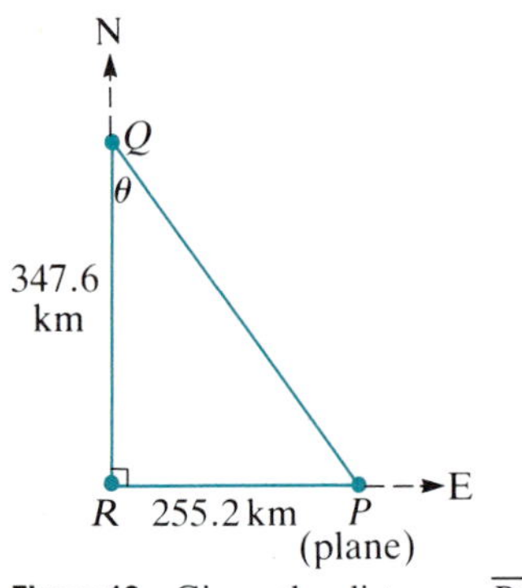

Figure 12 Given the distances $\overline{RQ}$ and $\overline{RP}$, we must determine θ.

SOLUTION From Figure 12, we see that the bearing of P from Q is $\theta°$ east of south. We have

$$\tan\theta = \frac{255.2}{347.6} \approx 0.734177215$$

$$\theta = \tan^{-1} 0.734177215 \approx 36.2853°$$

Now

$$0.2853° = [60(0.2853)]' \approx 17'$$

Thus,

$$\theta = 36°17'$$

That is,

the bearing of P from Q is S36°17′E

Problems 2.8

Readiness Check

I. Which of the following would yield a missing part on the right triangle in the figure below?

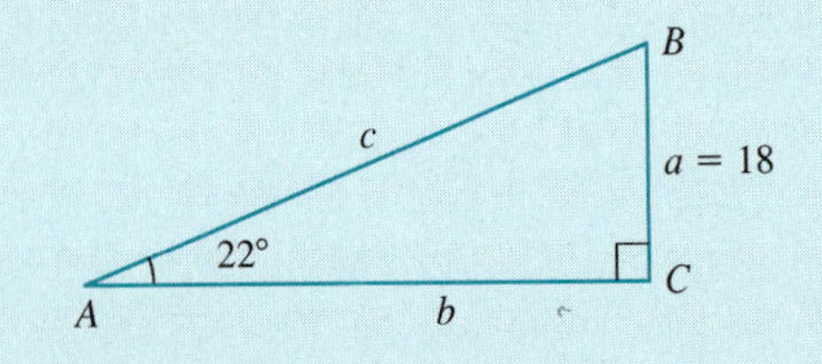

a. $B = (22° - 90°)$ b. $c = \dfrac{18}{\cos 22°}$

c. $b = \dfrac{18}{\tan 22°}$ d. $b = \dfrac{18}{\sin 22°}$

II. Which of the following is the angle of depression for a six-foot man standing on top of a 300-foot cliff to see a friend standing below 212 feet from the base of the cliff?

a. 55.3° b. 34.7° c. 54.8° d. 35.2°

III. Which of the following is the length of floor-length drapes Mary and John must buy if, when standing 72″ from the window, Mary, whose eye level is 62″, measures the angle of elevation when looking at the top of the window frame to be 15°?

a. 19″ b. 70″ c. 81″ d. 42″

IV. Which of the following is the height of a tree, to the nearest foot, that is 100 feet from a mouse if the angle of elevation for the mouse to see the tree top is 36°?

a. 73 feet b. 59 feet
c. 81 feet d. 137 feet

V. Which of the following equations could be used to solve the problem below?

A plane is flying at an altitude of 2000′. The pilot spots a boat on the water below at an angle of depression of 7°. What is the distance d from the plane to the boat?

a. $\sin 7° = 2000/d$
b. $\tan 7° = 2000/d$
c. $\cos 7° = 2000/d$
d. $\tan 83° = 2000/d$

Answers to Readiness Check

I. c II. a III. c IV. a V. a

In each problem, find the missing sides and angles of the given right triangle. Round each angle to two decimal places and round each side to 4 decimal places.

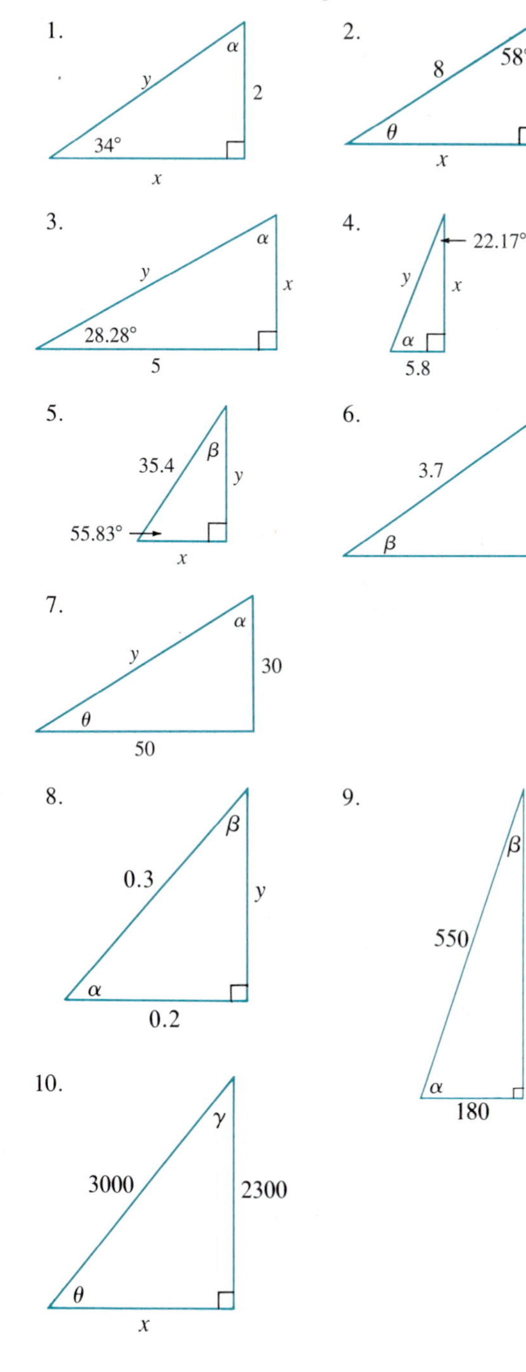

Each of the following problems requires the use of a calculator or a table of trigonometric functions. If a height or distance is asked for, give your answer to the nearest tenth. If an angle is requested, give your answer to the nearest tenth of a degree.

11. Find the height of the tree below.

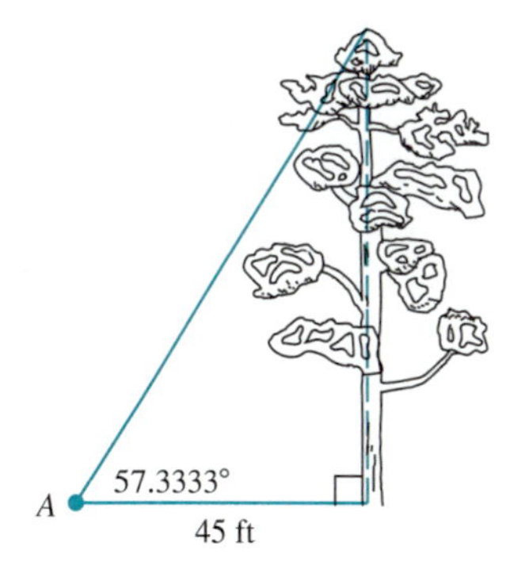

12. Find the distance from A to B across the river.

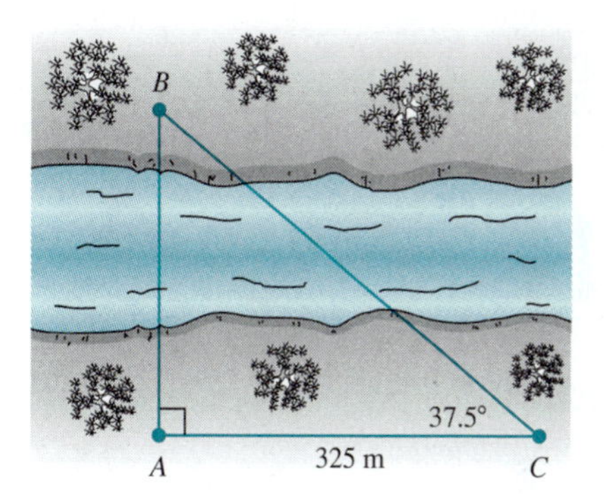

13. Find the distance across the pond.

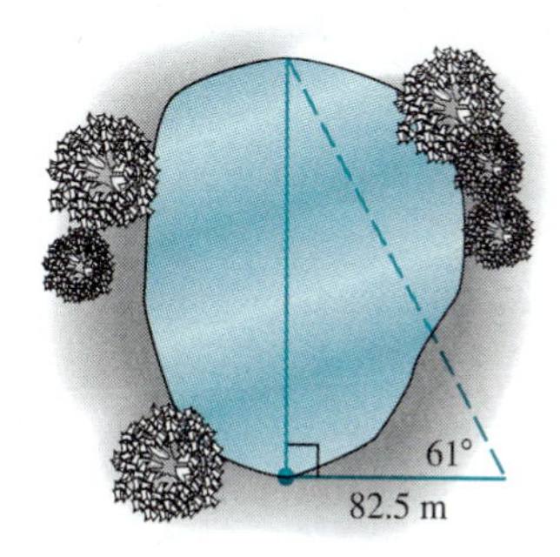

14. Find the height of the building below.

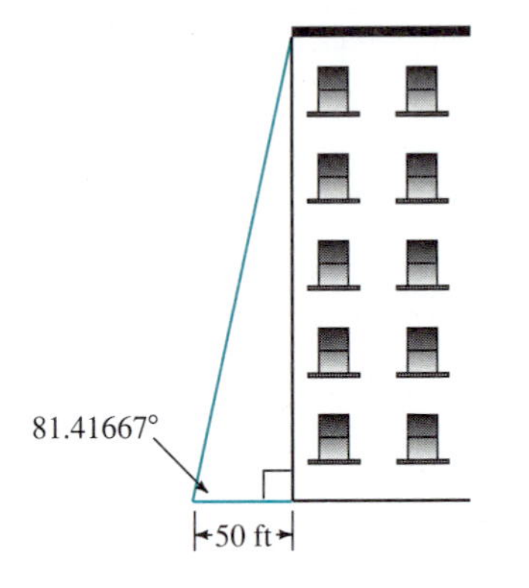

15. In Problem 11 find the line-of-sight distance from A to the top of the tree.
16. In Problem 12 find the line-of-sight distance from B to C.
17. In the figure below the angle of depression from the top of the lighthouse to a man in a ship is 23.5°. Find the distance $\overline{AB}$ from the man to the cliff.

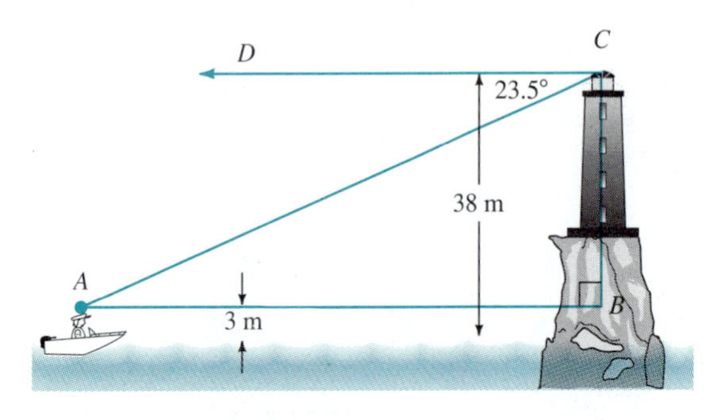

18. In Problem 17, find the line-of-sight distance, $\overline{AC}$, from the man to the top of the lighthouse.
19. A radio antenna is 50 ft high. The angle of elevation from the antenna to a landing airplane at an altitude of 420 ft is $6\frac{1}{3}°$. Find the line-of-sight distance from the top of the antenna to the plane.

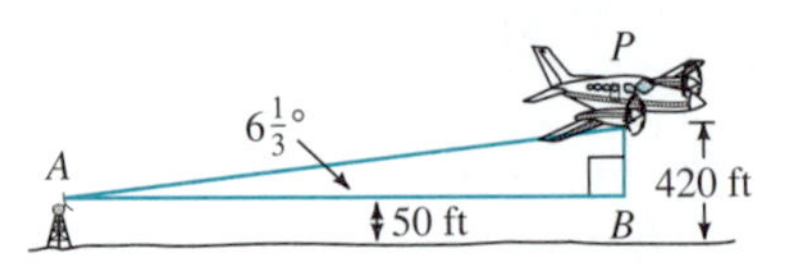

20. In Problem 19, find the horizontal distance, $\overline{AB}$, from the top of the antenna to a point directly below the plane.
21. A flagpole is 85 ft high. When the sun is 52° above the horizon, what is the length of the shadow cast by the flagpole? [Hint: The angle of elevation of the sun is 52°.]
22. In Problem 21, what is the angle of elevation of the sun if the length of the shadow cast by the flagpole is 100 ft?
23. Find the height of the Eiffel Tower in Paris, using the information below.

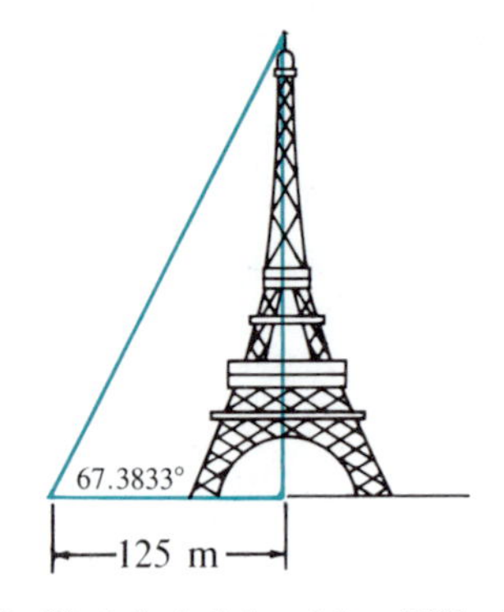

24. Find the height of the cliff in the figure. [Hint: Use equation (1).]

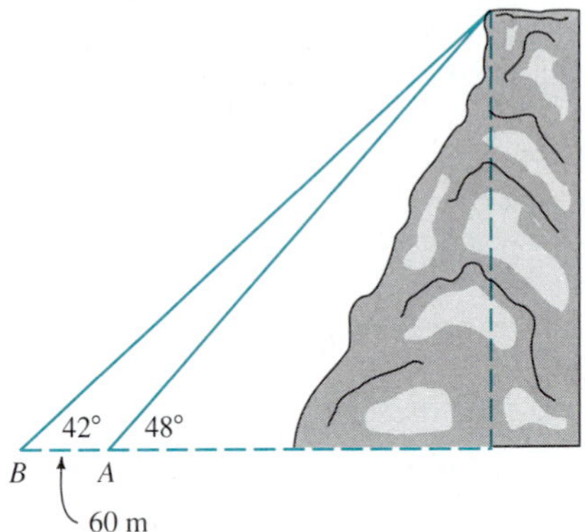

25. A man standing at the top of a 65-m lighthouse observes two boats. Using the data given below, determine the distance between the two boats.

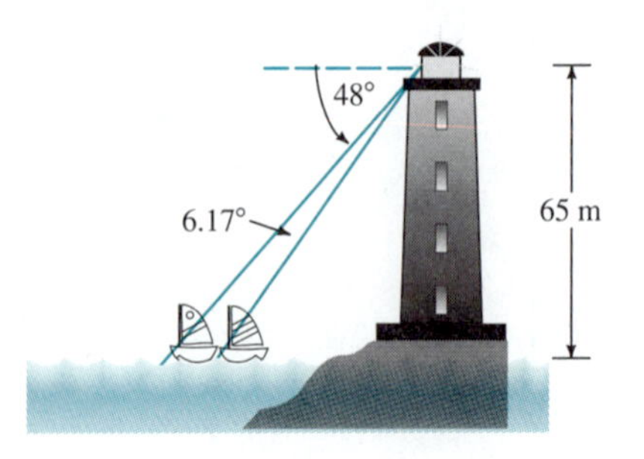

26. A 12-ft ladder leans against a cabin. The top of the ladder rests against the highest point on the cabin. If the foot of the ladder is 7 ft from the base of the cabin, how high is the cabin?
27. Find the length of the base of an isoceles triangle if each of the equal sides is 23 cm long and the angle between them is 28°.
* 28. A circle of radius 4 inches is inscribed in a regular hexagon. Find the length of one of its sides.
29. Find the angle between the diagonal and the longer side of a rectangle 5 cm by 8 cm.
* 30. A circle is inscribed in a regular pentagon whose perimeter is 25 cm. Find the radius of the circle.
31. An air traffic controller stands in a 120-ft-high control tower. Two planes wait in the same runway to take off. The angle of depression to the first plane is 14.6°, and the angle of depression to the second plane is 9.3°. Find the distance between the planes. Assume that the tower and the two planes lie on the same plane [no pun intended].
32. Answer the question of Problem 31 if the tower is 200 ft high.
* 33. A princess is imprisoned in a room of a castle that is surrounded by a moat. The angle of elevation from the edge of the moat to the bottom of the window in the princess's room is 71°. A knight equipped with a transit wishes to rescue the princess. He determines that the angle of elevation (from ground level) from a point 12 feet directly behind the moat and in line with the window is 52.7°. How long a ladder does he need in order to rescue the princess?
* 34. A satellite orbits the earth at a height of 520 miles. The angle θ to a point H on the horizon is 62.133°. Use these data to approximate the radius of the earth.

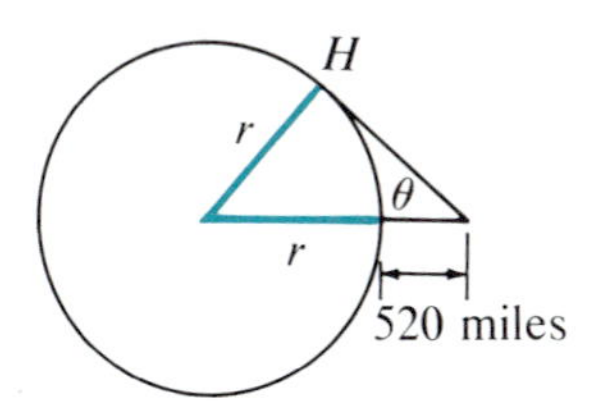

* 35. The diameter of the earth is 7926.4 miles. If you are flying in a plane at an altitude of 35,000 ft $\approx$ 6.6 miles, how far away is the horizon?
36. The sun is approximately 149.5 million km from the earth. From a point on the earth, the sun's diameter subtends an angle of 31′58″. What is the approximate diameter of the sun?

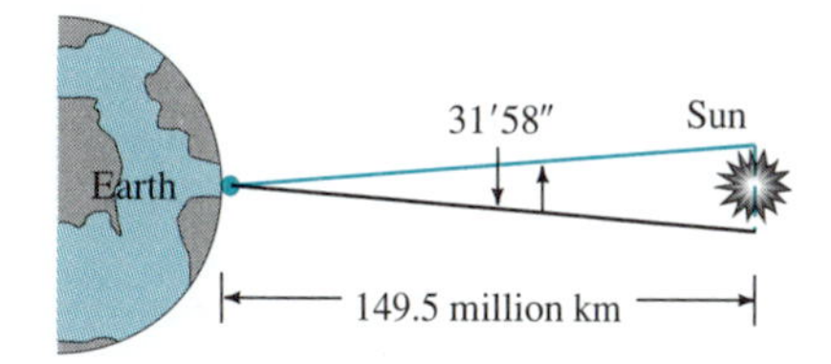

37. In December 1974, the American space probe *Pioneer 11* made its closest approach to Jupiter, which has a diameter of 88,000 miles. Photos showed that from the probe, the angle subtended by Jupiter's diameter was 57.7°. How close did the probe get to Jupiter?
38. A surveyor stands in a gully. Her eyes are 5 feet 2 inches above the ground, and she is standing 12 feet from one side of the gully. Using a transit, she estimates that the angle of elevation from her eyes to the top of that side of the gully is 17.37°. How deep is the gully?
39. The **grade** of a ramp or road is equal to the vertical distance the road rises divided by the horizontal length of the road. If a ramp has a 22% grade and is 12 m long, how high does it rise?

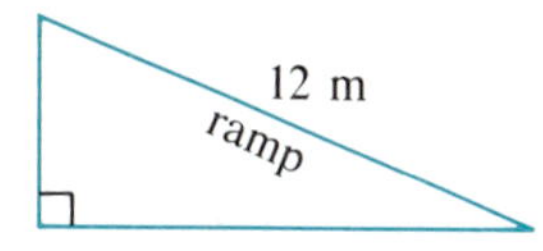

40. The exit ramp off a freeway has a constant 2.35% grade. Cars using this ramp arrive at an overpass that is 12.6 m above the freeway. What is the horizontal distance from the beginning of the exit ramp to a point on the freeway directly below the overpass?
* 41. A ship leaves port and travels 55 km due south. It then changes course and sails 41 km bearing N27°15′W. How far is the ship from port at this point?
* 42. A ship travels 16 miles/hr. It sails for $3\frac{1}{2}$ hours due north and then changes course and sails at a bearing of N61°25′E. How far is the ship from port 8 hours after leaving port?
43. A ship is 142 km due south of a naval base. A second ship is 216 km due east of the base. Find the bearing of the second ship from the first ship.

* 44. A ship started at point A and sailed 23.6 miles bearing N27°E. It then changed course and sailed 37.2 miles bearing S63°E to reach point B. How far is it from point A to point B?

* 45. In Problem 44, what is the bearing of the ship if it returns to point A by the most direct route?

* 46. A ship is sailing due east and at noon passes a lighthouse 8 miles due north. At 1 P.M. the bearing of the lighthouse is N23°W. Assuming that the ship is sailing at constant speed, find the speed of the ship in miles/hr.

47. A side view of the frame of one kind of symmetric roof is

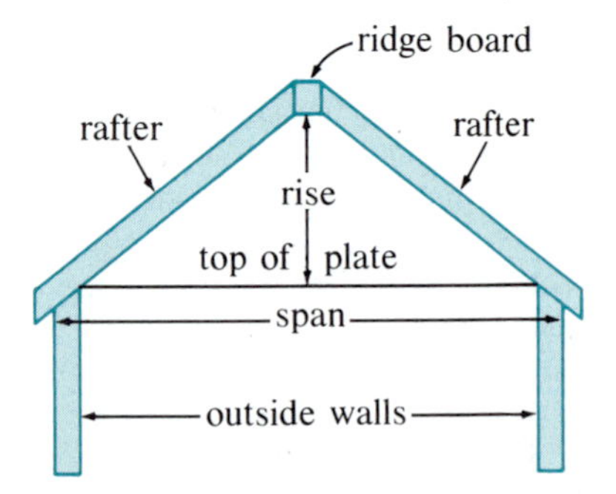

given above. The *span* of the roof is the horizontal distance between the outside walls while the *rise* is the vertical distance between the top of the plate and the center of the ridge board. The *pitch* of a roof is defined as

$$\text{pitch} = \frac{\text{rise}}{\text{span}}$$

Find the angle between a rafter and the top of the plate if the rise is 3 ft and the span is 7 ft.

48. Answer the question of Problem 47 if the rise is 12 ft and the span is 10 ft.

49. Find the pitch of a roof if the angle between the top of the plate and the rafter is 28.67°.

50. Answer the question of Problem 49 if the angle is 71.17°.

51. A *collar beam* is a horizontal beam that joins two rafters to add strength to the roof. In Problem 47 assume that the span is 20 feet and the center of a collar beam is 8 feet above the top of the plate. Find the length of the beam if the pitch is 0.5. [Hint: First draw a sketch like the one in Problem 47 and add in the collar beam.]

52. Answer the question of Problem 51 if the pitch is 1.2.

53. A truss† for a metal bridge has the measurements shown below. Find the length of the vertical bar AB.

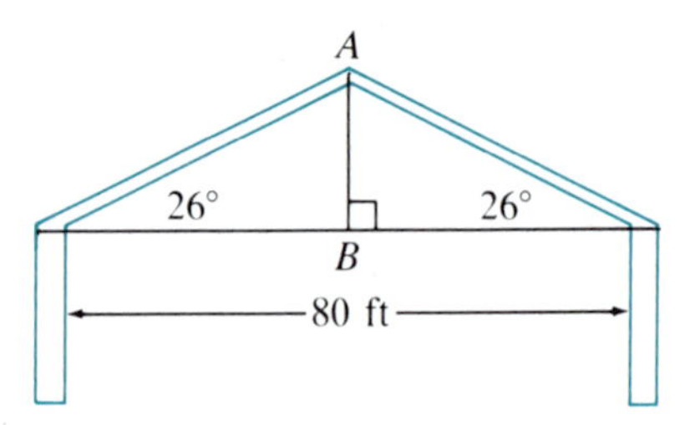

54. If the length of the vertical bar of the truss below is 2.5 m, find the angle of elevation θ.

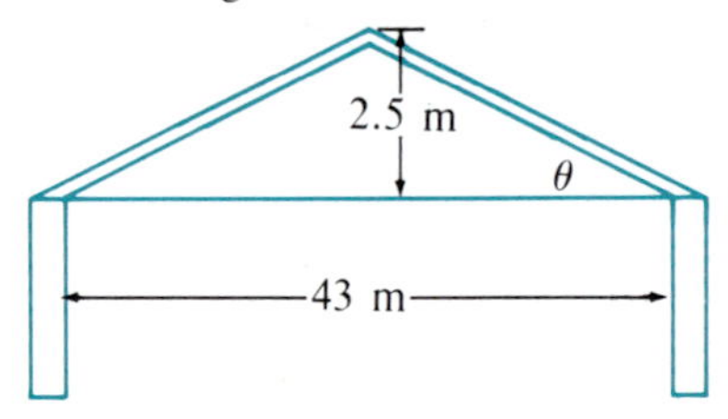

† A *truss* for a bridge is a collection of members such as bars or beams arranged, usually in the form of one or more triangles, in order to form a stable framework that gives support to the entire structure.

Summary Outline of Chapter 2

- An **angle** consists of two rays (an **initial** side and a **terminal** side) with a common initial point called the **vertex.** p. 55
- One **degree** (1°) is the measure of the angle obtained by rotating the terminal side of the angle $\frac{1}{360}$ of a complete revolution in the positive (counterclockwise) direction. p. 56
- The **complement** of $\theta = 90° - \theta$; the **supplement** of $\theta = 180° - \theta$. p. 57
- 1 **minute** (1′) = $\frac{1}{60}$ of a degree; 1 **second** (1″) = $\frac{1}{60}$ of a minute. p. 59

- The **radian measure** of an angle in standard position (vertex at the origin, initial side along the positive x-axis) is the length of the arc of the unit circle cut off by the sides of the angle. p. 61

- **Conversion Formulas:**

$$\theta \text{ (degrees)} = \frac{180}{\pi}\,\theta \text{ (radians)}; \qquad \theta \text{ (radians)} = \frac{\pi}{180}\,\theta \text{ (degrees)}$$ p. 61

- If a central angle θ subtends (cuts off) an arc of length L in a circle of radius r, then the radian measure of θ is given by $\theta = \dfrac{L}{r}$. p. 64

- If an object is moving on a circle at n rpm, then the **angular speed** $= \omega = 2\pi n$ radians/minute. p. 65

- The **linear speed** $= v = 2\pi nr$ units/minute where r is the radius of the circle. p. 65

- **Right Triangle Definitions of the Trigonometric Functions:** Let adj = side adjacent to the angle θ in a right triangle, opp = opposite side, and hyp = the hypotenuse.

$$\mathbf{sin}\ \boldsymbol{\theta} = \frac{\text{opp}}{\text{hyp}} \qquad \mathbf{cos}\ \boldsymbol{\theta} = \frac{\text{adj}}{\text{hyp}} \qquad \mathbf{tan}\ \boldsymbol{\theta} = \frac{\text{opp}}{\text{adj}}$$ p. 69

$$\mathbf{csc}\ \boldsymbol{\theta} = \frac{\text{hyp}}{\text{opp}} \qquad \mathbf{sec}\ \boldsymbol{\theta} = \frac{\text{hyp}}{\text{adj}} \qquad \mathbf{cot}\ \boldsymbol{\theta} = \frac{\text{adj}}{\text{opp}}$$

- **Circular Definitions of the Sine and Cosine Functions:**

  If the terminal side of an angle θ in standard position intersects the circle of radius r centered at the origin at the point (x, y), then $\mathbf{cos}\ \boldsymbol{\theta} = \dfrac{x}{r}$ and $\mathbf{sin}\ \boldsymbol{\theta} = \dfrac{y}{r}$. If $r = 1$ (the unit circle), then $\cos\theta = x$ and $\sin\theta = y$. pp. 77, 80

- The **reference angle** of θ (in standard position) is the positive acute angle, θ_r, that the terminal side of θ makes with the x-axis. p. 96

- **Reference Angle Theorem:** $\sin\theta = \pm\sin\theta_r$ and $\cos\theta = \pm\cos\theta_r$. p. 97

- **Identities Involving sin $\boldsymbol{\theta}$ and cos $\boldsymbol{\theta}$** $\left(\text{given in radian measure: } 2\pi = 360°,\ \pi = 180°,\ \dfrac{\pi}{2} = 90°\right)$:

$\sin^2\theta + \cos^2\theta = 1$		p. 78
$\cos(\theta + 2\pi) = \cos\theta$	$\sin(\theta + 2\pi) = \sin\theta$	p. 89
$\cos(-\theta) = \cos\theta$	$\sin(-\theta) = -\sin\theta$	p. 92
$\cos(\theta + \pi) = -\cos\theta$	$\sin(\theta + \pi) = -\sin\theta$	p. 93
$\cos(\theta - \pi) = -\cos\theta$	$\sin(\theta - \pi) = -\sin\theta$	p. 93
$\cos(\pi - \theta) = -\cos\theta$	$\sin(\pi - \theta) = \sin\theta$	p. 94
$\cos\left(\dfrac{\pi}{2} - \theta\right) = \sin\theta$	$\sin\left(\dfrac{\pi}{2} - \theta\right) = \cos\theta$	p. 94

- The zeros of $\sin\theta$ are $0, \pm\pi, \pm 2\pi, \ldots$. p. 99

- The zeros of $\cos\theta$ are $\pm\dfrac{\pi}{2}, \pm\dfrac{3\pi}{2}, \pm\dfrac{5\pi}{2}, \ldots$. p. 99

- **Circular Definitions of Other Trigonometric Functions:**

  If x, y, and r are as in the definitions of $\sin\theta$ and $\cos\theta$,

  $\tan\theta = \frac{y}{x}$ $\quad$ $\cot\theta = \frac{x}{y}$ $\quad$ $\sec\theta = \frac{1}{x}$ $\quad$ $\csc\theta = \frac{1}{y}$ p. 101

- $\tan\theta = \frac{\sin\theta}{\cos\theta}$ $\quad$ $\cot\theta = \frac{\cos\theta}{\sin\theta} = \frac{1}{\tan\theta}$ $\quad$ $\sec\theta = \frac{1}{\cos\theta}$ $\quad$ $\csc\theta = \frac{1}{\sin\theta}$ p. 101

- $\tan\theta$ and $\sec\theta$ are undefined when $\theta = \pm\frac{\pi}{2}, \pm\frac{3\pi}{2}, \pm\frac{5\pi}{2}, \ldots$. p. 103

- $\cot\theta$ and $\csc\theta$ are undefined when $\theta = 0, \pm\pi, \pm 2\pi, \ldots$. p. 103

- **Identities Involving $\tan\theta$, $\cot\theta$, $\sec\theta$, and $\csc\theta$**

$\tan(-\theta) = -\tan\theta$	$\cot(-\theta) = -\cot\theta$	p. 104
$\sec(-\theta) = \sec\theta$	$\csc(-\theta) = -\csc\theta$	p. 104
$\tan(\theta + \pi) = \tan\theta$	$\cot(\theta + \pi) = \cot\theta$	pp. 111, 112
$\sec(\theta + 2\pi) = \sec\theta$	$\csc(\theta + 2\pi) = \csc\theta$	p. 112
$\tan^2\theta + 1 = \sec^2\theta$	$\cot^2\theta + 1 = \csc^2\theta$	pp. 106, 107

- **Three Inverse Functions**

Function	Domain	Range	Also called	
$\sin^{-1} x$	$[-1, 1]$	$\left[-\frac{\pi}{2}, \frac{\pi}{2}\right]$	arc sin x	p. 120
$\cos^{-1} x$	$[-1, 1]$	$[0, \pi]$	arc cos x	p. 126
$\tan^{-1} x$	$(-\infty, \infty)$	$\left(-\frac{\pi}{2}, \frac{\pi}{2}\right)$	arc tan x	p. 129

Review Exercises for Chapter 2

In Exercises 1–5 determine the quadrant of each angle.

1. 114° 2. 2 3. 5.924 4. 0.63 5. 287°
6. Find the complement of 62°41′.
7. Find the supplement of 43°42′.
8. Express 41.572° in degrees, minutes, and seconds. Round to the nearest second.
9. Write 127°28′43″ as a decimal. Round to 4 decimal places.
10. Convert from radians to degrees:
    (a) $\frac{3\pi}{8}$ (b) $\frac{7\pi}{6}$
    (c) 2.37
11. Convert from degrees to radians:
    (a) 165°
    (b) $-67\frac{1}{2}°$
    (c) 112°
12. In a circle of radius 3, find the length of an arc subtended by an angle of
    (a) $\frac{\pi}{4}$ (b) $\frac{7\pi}{6}$ (c) 105°
13. A weight tied to the end of a rope 3 m long is swung in a circle at a rate of 35 rpm.
    (a) What is the angular speed of the weight?
    (b) What is its linear speed?
    (c) How far does the weight travel in 12 seconds?
14. The planet Mercury revolves about the sun once every 88 days. Its average distance from the sun is 36 million miles.
    (a) What is the angle swept out by a line joining Mercury and the sun in 11 days?
    (b) What is the linear speed of Mercury around the sun?
    (c) How far does Mercury travel around the sun in one of its years?

15. How far is Peoria, Illinois (latitude 40°42′) from the North Pole? [See Problem 97 on p. 68.]

In Exercises 16–27 find the values of sin θ, cos θ, tan θ, cot θ, sec θ, and csc θ whenever each is defined. Do not use a calculator.

16. $\theta = \frac{\pi}{4}$ 17. $\theta = 30°$ 18. $\theta = \frac{\pi}{2}$
19. $\theta = 0$ 20. $\theta = 180°$ 21. $\theta = -\frac{\pi}{6}$
22. $\theta = 4\pi$ 23. $\theta = 21\pi$ 24. $\theta = -135°$
25. $\theta = 270°$ 26. $\theta = \frac{7\pi}{6}$ 27. $\theta = 315°$

In Exercises 28–35 use a calculator to determine sin θ, cos θ, tan θ, cot θ, sec θ, and csc θ. Round to 4 decimal places.

28. $\theta = 1.62$ 29. $\theta = 3.85$ 30. $\theta = 84.26°$
31. $\theta = 217°$ 32. $\theta = 5.82$ 33. $\theta = -1.76$
34. $\theta = 28°15'$ 35. $\theta = 71°28'09''$

In Exercises 36–44 sin θ, cos θ, tan θ, cot θ, sec θ, or csc θ is given together with the quadrant of θ. Find the values of the other five functions. Do not use a calculator.

36. $\sin\theta = \frac{2}{3}$, I 37. $\tan\theta = 4$, I
38. $\tan\theta = \frac{1}{4}$, III 39. $\cos\theta = -\frac{1}{2}$, III
40. $\sin\theta = -0.7$, III 41. $\sec\theta = 3$, IV
42. $\csc\theta = 2.5$, II 43. $\cot\theta = -6$, II
44. $\cot\theta = -6$, IV
45. Given that $\cos 1.35 = 0.2190$, find two angles α and β in $(0, 2\pi)$ such that $\sin\alpha = \sin\beta = 0.2190$.
46. In Exercise 45, find two numbers γ and δ in $(0, 2\pi)$ such that $\sin\gamma = \sin\delta = -0.2190$.
47. Find the four smallest positive numbers such that $\cos\theta = \frac{\sqrt{3}}{2}$.
48. What values of θ in $[0, \pi]$ satisfy $\sin\theta > \frac{1}{2}$?
49. Find tan θ, cot θ, sec θ, and csc θ if $\sin\theta = \frac{3}{5}$ and $\cos\theta = -\frac{4}{5}$.
50. Find the four smallest positive numbers that satisfy $\tan\theta = -1$.
51. Sketch the graph of $f(\theta) = \cos\left(\theta - \frac{\pi}{3}\right)$.
52. Sketch the graph of $f(\theta) = \tan\theta - 2$.
53. Find $\sin\theta$ if $\tan\theta = \frac{x}{3}$, $x > 0$.
54. Find $\sec\theta$ if $\sin\theta = \frac{x}{\sqrt{x^2+5}}$, $x > 0$.

In Exercises 55–66 calculate the indicated value. Do not use a calculator.

55. $\sin^{-1}\frac{1}{2}$ 56. $\cos^{-1}\left(-\frac{\sqrt{3}}{2}\right)$
57. $\tan^{-1} 1$ 58. $\tan^{-1}(-\sqrt{3})$
59. $\cos\left(\sin^{-1}\frac{5}{8}\right)$ 60. $\tan\left(\sin^{-1}\frac{1}{3}\right)$
61. $\sin\left(\tan^{-1}\frac{1}{3}\right)$ 62. $\sec\left(\tan^{-1} 4\right)$
63. $\cot\left(\sin^{-1}\left(-\frac{3}{8}\right)\right)$ 64. $\csc\left(\cos^{-1}\frac{3}{4}\right)$
65. $\sin\left(\cos^{-1}\frac{x}{4}\right)$ 66. $\sec\left(\tan^{-1}\frac{x+1}{x+2}\right)$

In Exercises 67–69 find the missing sides and angles of the given right triangle. Round each angle to the nearest tenth of a degree and round each side to four decimal places.

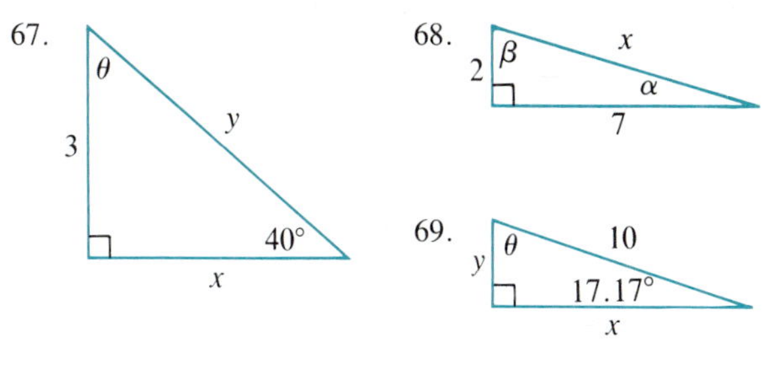

Exercises 70–78 require the use of a calculator.

70. Find the height of the building.

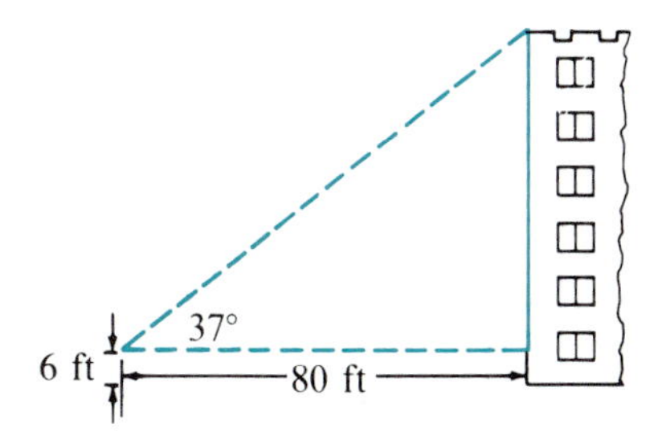

71. Find the distance from A to B across a river.

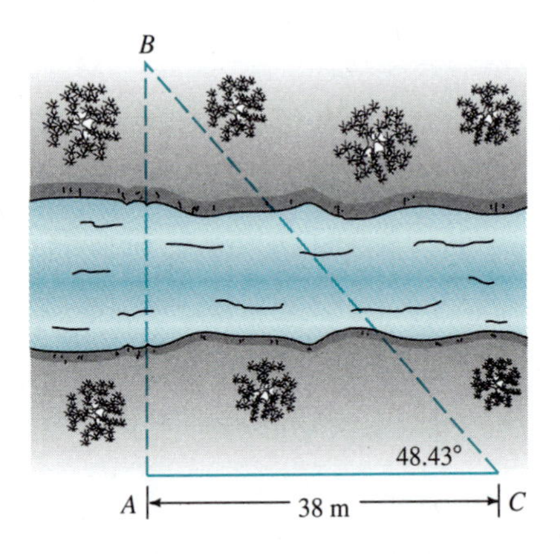

72. The angle of depression from the top of a lighthouse, 23 m above sea level, to the bow of a ship is 16.73°. How far is the ship from the base of the lighthouse?
73. A billboard is 42 ft high. What is the length of the shadow cast by the billboard when the sun is 36° above the horizon?
74. What is the height of the mountain in the accompanying figure?

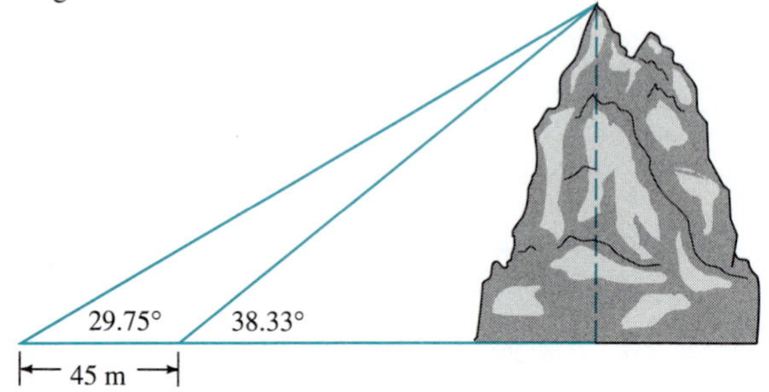

75. The diameter of the earth is 7926.4 miles. If you are flying in a plane at an altitude of 30,000 ft, how far away is the horizon?
76. A ship leaves port and travels 124 km due east. It then changes course and sails 180 km bearing N31°20′W. How far is the ship from port?
77. In Exercise 76, what is the bearing of the ship if it returns to port by the most direct route?
78. Find the pitch of a roof if the angle between the top of the plate and the rafter is 48.53°.

Chapter 3

Trigonometric Identities and Graphs

In Chapter 2, we introduced the six trigonometric functions and discussed some of their properties. In this chapter, we discuss trigonometric identities and graphs that are useful both in the study of calculus and in the analysis of a variety of phenomena in the physical, biological, and social sciences and in business.

3.1 Basic Trigonometric Identities

We begin by repeating an important definition. Let f and g denote functions.

The equation

$$f(x) = g(x) \tag{1}$$

is called an **identity** if $f(x) = g(x)$ at every x in both the domain of f and the domain of g. That is, $f(x) = g(x)$ wherever both values are defined.

EXAMPLE 1 *Three Identities*

The following are identities:

(a) $x^2 + x - 12 = (x + 4)(x - 3)$

(b) $\sin^2 x + \cos^2 x = 1$ — Equation (2) in Section 2.3

(c) $\dfrac{x}{x} = 1$ — Both functions equal 1 when $x \neq 0$

The function $\dfrac{x}{x}$ is not defined at $x = 0$, so we exclude this value.

The identity $\sin^2 x + \cos^2 x = 1$ is called a **trigonometric identity** because it is an identity involving trigonometric functions. Note that $f(x) = \sin^2 x + \cos^2 x$ is defined for every real number x. The same holds for $g(x) = 1$. Thus $f(x)$ and $g(x)$ are defined and equal for every real number x. We will not discuss the domains of functions further here, but you should always be aware of them.

We list below some of the trigonometric identities obtained in Chapter 2. We write the independent variable as x instead of θ.

Basic Trigonometric Identities

	Identity	Source	
Ratio identities	$\tan x = \dfrac{\sin x}{\cos x}$	Section 2.5	**(2)**
	$\cot x = \dfrac{\cos x}{\sin x}$	Section 2.5	**(3)**
Reciprocal identities	$\sec x = \dfrac{1}{\cos x}$	Section 2.5	**(4)**
	$\csc x = \dfrac{1}{\sin x}$	Section 2.5	**(5)**
Circular or Pythagorean identities	$\sin^2 x + \cos^2 x = 1$	Equation (2) in Section 2.3	**(6)**
	$\tan^2 x + 1 = \sec^2 x$	Equation (8) in Section 2.5	**(7)**
	$1 + \cot^2 x = \csc^2 x$	Equation (9) in Section 2.5	**(8)**
Even-odd identities	$\cos(-x) = \cos x$	Equation (6) in Section 2.4	**(9)**
	$\sin(-x) = -\sin x$	Equation (7) in Section 2.4	**(10)**
	$\tan(-x) = -\tan x$	Equation (3) in Section 2.5	**(11)**

Many other identities can be obtained from these ten identities. In fact, some of the identities listed can be obtained from others. For example, we can start with $\sin^2 x + \cos^2 x = 1$ and divide by $\cos^2 x$ to obtain (7). Also, we can use (2), (9), and (10) to obtain (11):

$$\tan(-x) \overset{(2)}{=} \frac{\sin(-x)}{\cos(-x)} \overset{(9)\text{ and }(10)}{=} -\frac{\sin x}{\cos x} = -\tan x$$

which is identity (11).

You should *know* identities (2)–(11) because they are fundamental in working with the trigonometric functions. In the rest of this section we will find other identities using the basic ones.

EXAMPLE 2 *Using Identities to Simplify a Trigonometric Function*

Simplify the function

$$y = (\tan x + \cot x) \sin x$$

SOLUTION

$$
\begin{aligned}
y &= (\tan x + \cot x)\sin x \\
&= \left(\frac{\sin x}{\cos x} + \frac{\cos x}{\sin x}\right)\sin x && \text{(2) and (3)} \\
&= \frac{\sin x}{\cos x}\cdot \sin x + \frac{\cos x}{\sin x}\cdot \sin x && \text{Distributive law} \\
&= \frac{\sin^2 x}{\cos x} + \cos x \\
&= \frac{\sin^2 x}{\cos x} + \cos x\,\frac{\cos x}{\cos x} && \text{Multiply } \cos x \text{ by } 1 = \frac{\cos x}{\cos x} \\
&= \frac{\sin^2 x}{\cos x} + \frac{\cos^2 x}{\cos x} = \frac{\sin^2 x + \cos^2 x}{\cos x} \\
&\overset{(6)}{=} \frac{1}{\cos x} \overset{(4)}{=} \sec x
\end{aligned}
$$

Thus $(\tan x + \cot x)\sin x = \sec x$.

It often helps, as here, to write everything in terms of sines and cosines.

EXAMPLE 3 ***Verifying a Trigonometric Identity***

Show that

$$\sin \theta = \frac{\tan \theta \cos \theta - \sin^2 \theta}{1 - \sin \theta} \tag{12}$$

SOLUTION

$$
\begin{aligned}
\frac{\tan \theta \cos \theta - \sin^2 \theta}{1 - \sin \theta} &\overset{\text{Identity (2)}}{=} \frac{\dfrac{\sin \theta}{\cos \theta}\cdot \cos \theta - \sin^2 \theta}{1 - \sin \theta} \\
&= \frac{\sin \theta - \sin^2 \theta}{1 - \sin \theta} \overset{\text{Factor}}{=} \frac{\sin \theta\,\cancel{(1 - \sin \theta)}}{\cancel{1 - \sin \theta}} = \sin \theta
\end{aligned}
$$

NOTE Equation (12) is an identity even though $\sin \theta$ is defined for all real θ, whereas $\dfrac{\tan \theta \cos \theta - \sin^2 \theta}{1 - \sin \theta}$ is defined only when $\tan \theta$ exists and $1 - \sin \theta \neq 0$. Equation (12) certainly holds when $\tan \theta$ exists and $\sin \theta \neq 1$; that is, when $\theta \neq \dfrac{\pi}{2} + 2n\pi$, $n = 0, \pm 1, \pm 2, \ldots$.

USEFUL HINT Often, expressions have the terms $1 - \cos x$, $1 + \cos x$, $1 - \sin x$, or $1 + \sin x$ in the denominator. These terms can be simplified by noting that

$$(1 + \cos \theta)(1 - \cos \theta) = 1 - \cos^2 \theta = \sin^2 \theta \tag{13}$$

$$(1 + \sin \theta)(1 - \sin \theta) = 1 - \sin^2 \theta = \cos^2 \theta \tag{14}$$

The process described above is called **conjugation.**

EXAMPLE 4 *Using Conjugation to Verify a Trigonometric Identity*

Show that

$$\frac{1 - \cos u}{\sin u} = \frac{\sin u}{1 + \cos u}$$

SOLUTION We start with the right side:

$$\frac{\sin u}{1 + \cos u} = \frac{(\sin u)(1 - \cos u)}{(1 + \cos u)(1 - \cos u)}$$

Multiply numerator and denominator by $1 - \cos u$ and then use (13)

$$= \frac{(\sin u)(1 - \cos u)}{1 - \cos^2 u} = \frac{(\sin u)(1 - \cos u)}{\sin^2 u}$$

$$= \frac{\cancel{(\sin u)}(1 - \cos u)}{\cancel{\sin u} \cdot \sin u} = \frac{1 - \cos u}{\sin u} \quad \blacksquare$$

EXAMPLE 5 *Verifying a Trigonometric Identity*

Show that

$$\frac{1 + \sin t}{1 - \sin t} = (\sec t + \tan t)^2$$

SOLUTION We start with the right-hand side:

$$(\sec t + \tan t)^2 = \left(\frac{1}{\cos t} + \frac{\sin t}{\cos t}\right)^2 = \left(\frac{1 + \sin t}{\cos t}\right)^2 = \frac{(1 + \sin t)^2}{\cos^2 t}$$

Identity (6)

$$= \frac{(1 + \sin t)^2}{1 - \sin^2 t} = \frac{\cancel{(1 + \sin t)}(1 + \sin t)}{\cancel{(1 + \sin t)}(1 - \sin t)} = \frac{1 + \sin t}{1 - \sin t}$$

WARNING It is sometimes necessary to work with one side of a supposed identity at a time to establish the identity. If you work with both sides at once, you can get into trouble. For example, it is not true that

$$\sin \theta = -\sin \theta \tag{15}$$

We can provide a *false* proof of (15) by squaring both sides:

$$(\sin\,\theta)^2 = (-\sin\,\theta)^2$$
$$\sin^2\theta = \sin^2\theta$$

But this, of course, *is* true, so (15) is true. Right? Wrong! The problem is that the operation of squaring is not reversible. That is, $x^2 = y^2$ does not imply that $x = y$ $[(-2)^2 = 2^2$, but $-2 \neq 2]$. ■

EXAMPLE 6 *A Technique Useful in Calculus*

Write the expression $\sqrt{a^2 + x^2}$ in terms of trigonometric functions by making the substitution $x = a\tan\theta$, $a > 0$, $-\dfrac{\pi}{2} < \theta < \dfrac{\pi}{2}$.

SOLUTION

$$a^2 + x^2 = a^2 + (a\tan\theta)^2 = a^2 + a^2\tan^2\theta$$

From (7)
↓

$$= a^2(1 + \tan^2\theta) = a^2\sec^2\theta$$

Thus

$$\sqrt{a^2 + x^2} = \sqrt{a^2\sec^2\theta} = |a\sec\theta|$$

But $a > 0$ and, for $-\dfrac{\pi}{2} < \theta < \dfrac{\pi}{2}$, $\sec\theta > 0$. Hence,

$$\sqrt{a^2 + x^2} = |a\sec\theta| = a\sec\theta$$

This problem can also be solved by using Figure 1, since, in the figure, $\sec\theta = \dfrac{\sqrt{a^2 + x^2}}{a}$. This type of substitution is very useful in solving certain types of problems in integral calculus.

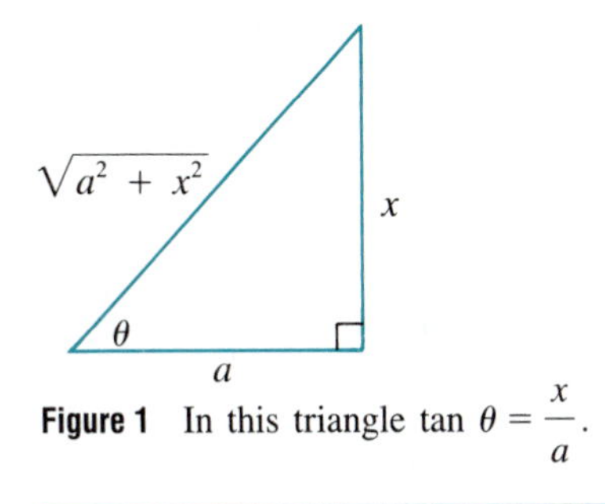

Figure 1 In this triangle $\tan\theta = \dfrac{x}{a}$.

Problems 3.1

Readiness Check

I. Which of the following is called a circular identity?

a. $\sin^2 x - \cos^2 x = \cos 2x$
b. $\tan^2 x = \sec^2 x + 1$
c. $1 + \cot^2 x = \csc^2 x$
d. $x^2 - y^2 = r^2$

II. Which of the following is equal to $(\sin x + \cos x)^2$?

a. $\sin^2 x + \cos^2 x$ b. $1 + 2 \sin x \cos x$
c. 1 d. $\sin^2 x + \tan x - 1$

III. Which of the following is an even identity?

a. $1 = \sec^2 x + \tan^2 x$ b. $\tan x = \dfrac{\sin x}{\cos x}$
c. $\cos(-x) = \cos x$ d. $\sin(-x) = -\sin x$

IV. Which of the following is a trigonometric identity?

a. $\tan x \cdot \dfrac{\sin x}{\cos x} = 1$
b. $1 + \sin^2 x = \cos^2 x$
c. $\sin(-x) = \sin x$
d. $\csc x \cdot \sin x = 1$

V. Which of the following is not a trigonometric identity?

a. $\tan\theta \cot\theta = 1$
b. $\sec\theta \csc\theta = 1$
c. $\csc^2\theta - 1 = \cot^2\theta$
d. $\cos^2\theta - 1 = -\sin^2\theta$

In Problems 1–10 write each expression in terms of sines and cosines and simplify.

1. $\cos x \tan x$
2. $\sin x \cot x$
3. $\cos\theta \cot\theta$
4. $\sin\theta \tan\theta$
5. $\dfrac{1 + \sec t}{1 - \tan t}$
6. $\dfrac{\sec^2 t + \sin t}{1 - \cot t}$
7. $\dfrac{\tan u + \sec u}{\tan u - \sec u}$
8. $\dfrac{\sec v \cos v}{\csc v \sin v}$
9. $\dfrac{1 + \tan s}{1 - \tan s}$
10. $\dfrac{1 - \cot s}{\sin s - \cos s}$

In Problems 11–15 write each expression in terms of $\cos x$ only and simplify.

11. $\sin^2 x$
12. $\sin x \tan x$
13. $\dfrac{1 + \sec x}{1 - \sec x}$
14. $\tan^2 x$
15. $(1 - \sin x)^2 + 2 \sin x$

In Problems 16–20 write each expression in terms of $\sin x$ only and simplify.

16. $\cos^2 x$
17. $\cos x \cot x$
18. $\dfrac{1 + \csc x}{1 - \csc x}$
19. $\tan^2 x$
20. $(1 - \cos x)^2 + 2 \cos x$

In Problems 21–80 show that each equation is an identity.

21. $\sin\theta \csc\theta = 1$
22. $\cos x \sec x = 1$
23. $\tan u \cot u = 1$
24. $\sin\alpha \cos\alpha \csc\alpha \sec\alpha = 1$
25. $\dfrac{\sec\beta}{\csc\beta} = \tan\beta$
26. $\sin\theta(\tan\theta + \cot\theta) = \sec\theta$
27. $\cos\gamma(\tan\gamma + \cot\gamma) = \csc\gamma$
28. $\dfrac{1}{\tan x + \cot x} = \sin x \cos x$
29. $(\cos x - \sin x)^2 = 1 - 2 \cos x \sin x$
30. $(\cos x + \sin x)^2 = 1 + 2 \cos x \sin x$
31. $\tan\theta + \cot\theta = \sec\theta \csc\theta$
32. $\dfrac{\cos\alpha}{1 + \sin\alpha} + \tan\alpha = \sec\alpha$
33. $1 - 2\cos^2\beta = 2\sin^2\beta - 1$
34. $\sec^2\gamma + \csc^2\gamma = \tan^2\gamma + \cot^2\gamma + 2$
35. $\dfrac{\tan x - \sin x}{\tan x + \sin x} = \dfrac{\sec x - 1}{\sec x + 1}$
36. $\dfrac{1 - \cos^2\theta}{\cos^2\theta} = \tan^2\theta$
37. $\cos x \sec x - \sin^2 x = \cos^2 x$
38. $\cos^2 u - \sin u \csc u = -\sin^2 u$
39. $(1 - \cos^2\theta)(1 + \cot^2\theta) = 1$
40. $\dfrac{\csc\alpha - \sin\alpha}{\cos^2\alpha} = \csc\alpha$
41. $\dfrac{\sec\beta - \cos\beta}{\sin^2\beta} = \sec\beta$
42. $\dfrac{\sin\gamma \sec^2\gamma - \sin\gamma}{\cos\gamma} = \tan^3\gamma$

Answers to Readiness Check

I. c II. b III. c IV. d V. b

43. $\dfrac{2}{\sin x + 1} - \dfrac{2}{\sin x - 1} = 4 \sec^2 x$

44. $\dfrac{\cot^2 \theta - 1}{1 - \tan^2 \theta} = \cot^2 \theta$

45. $\dfrac{\tan \alpha}{\sec \alpha + 1} = \dfrac{1}{\cot \alpha + \csc \alpha}$

46. $\dfrac{1}{1 - \sin \beta} + \dfrac{1}{1 + \sin \beta} = 2 \sec^2 \beta$

47. $\dfrac{1}{1 - \cos \gamma} + \dfrac{1}{1 + \cos \gamma} = 2 \csc^2 \gamma$

48. $\dfrac{\cos^2 x + 3 \cos x + 2}{\sin^2 x} = \dfrac{2 + \cos x}{1 - \cos x}$

49. $\cos^4 \theta + \sin^2 \theta = \cos^2 \theta + \sin^4 \theta$

50. $\dfrac{\cos^3 \alpha - \sin^3 \alpha}{\cos \alpha - \sin \alpha} = 1 + \cos \alpha \sin \alpha$

51. $\dfrac{1}{\tan \beta + \cot \beta} = \sin \beta \cos \beta$

52. $\sin (-u) \sec (-u) = -\tan u$

53. $\dfrac{1 - \sin v}{1 + \sin v} = (\tan v - \sec v)^2$

54. $\sin^4 x - \cos^4 x = \sin^2 x - \cos^2 x$

55. $\dfrac{\sec \theta \sin \theta}{\tan \theta + \cot \theta} = \sin^2 \theta$

56. $\dfrac{\csc u \cos u}{\tan u + \cot u} = \cos^2 u$

57. $(1 - \sin^2 v)(1 + \tan^2 v) = 1$

58. $\dfrac{\cos \theta}{1 + \sin \theta} + \dfrac{1 + \sin \theta}{\cos \theta} = 2 \sec \theta$

59. $\dfrac{\cot \alpha}{1 + \sin \alpha} = \dfrac{1 - \sin \alpha}{\sin \alpha \cos \alpha}$

60. $\dfrac{\cos^2 \beta}{\sin^4 \beta + \sin^2 \beta \cos^2 \beta} = \cot^2 \beta$

* 61. $\dfrac{\cos \gamma - \sin \gamma + 1}{\cos \gamma + \sin \gamma - 1} = \dfrac{\cos \gamma + 1}{\sin \gamma}$

* 62. $\dfrac{1 + \sin x + \cos x}{1 + \cos x - \sin x} = \sec x + \tan x$

* 63. $\dfrac{1 + \sin x + \cos x}{1 + \sin x - \cos x} = \csc x + \cot x$

* 64. $\dfrac{\sin^3 \theta + \cos^3 \theta}{1 - \sin \theta \cos \theta} = \sin \theta + \cos \theta$

65. $\dfrac{\sin^3 \theta - \cos^3 \theta}{1 + \sin \theta \cos \theta} = \sin \theta - \cos \theta$

66. $\sqrt{\dfrac{1 - \sin \theta}{1 + \sin \theta}} = \dfrac{1 - \sin \theta}{|\cos \theta|}$

67. $\sqrt{\dfrac{1 + \cos \alpha}{1 - \cos \alpha}} = \dfrac{1 + \cos \alpha}{|\sin \alpha|}$

68. $a^2 \sin^2 \beta + a^2 \cos^2 \beta = a^2$

69. $(a \cos \gamma + b \sin \gamma)^2 + (b \cos \gamma - a \sin \gamma)^2 = a^2 + b^2$

70. $a^2 \sec^2 x - a^2 = a^2 \tan^2 x$

71. $a^2 \csc^2 x - a^2 = a^2 \cot^2 x$

72. $(a \tan u + a)^2 + (a \tan u - a)^2 = 2a^2 \sec^2 u$

73. $(a \cot u + a)^2 + (a \cot u - a)^2 = 2a^2 \csc^2 u$

74. $\ln |\sec \theta| = -\ln |\cos \theta|$†

75. $\ln |\csc \theta| = -\ln |\sin \theta|$

76. $\ln |\tan \alpha| = \ln |\sin \alpha| - \ln |\cos \alpha|$

$\left[\text{Hint: } \ln \dfrac{x}{y} = \ln x - \ln y\right]$

77. $\ln |\cot \beta| = \ln |\cos \beta| - \ln |\sin \beta|$

78. $\ln |\sin^2 x + \cos^2 x| = 0$

79. $\ln |\sin x| + \ln |\csc x| = 0$

80. $\ln |\tan x| + \ln (1 + \cot^2 x) = \ln |\tan x + \cot x|$

In Problems 81–92 show that each equation is *not* an identity by finding a value of x for which the equation is false.

81. $\sin x = \sqrt{\sin^2 x}$

82. $\tan x = \sqrt{\tan^2 x}$

83. $\sin x = \sqrt{1 - \cos^2 x}$

84. $\sec x = \sqrt{1 + \tan^2 x}$

85. $\cot x = \sqrt{\csc^2 x - 1}$

86. $\sin x + \cos x = \csc x + \sec x$

87. $\sqrt{\sin^2 x + \cos^2 x} = \sin x + \cos x$

88. $\sin^2 x + \cos^2 x = (\sin x + \cos x)^2$

89. $\ln \dfrac{1}{\cos x} = \dfrac{1}{\ln \cos x}$

90. $\ln \sin^2 x + \ln \cos^2 x = \ln 1 = 0$

91. $\sin (\sec x) = 1$

92. $\tan (\cot x) = 1$

93. For what values of x is it true that $\sin x = \sqrt{1 - \cos^2 x}$?

94. For what values of x is it true that $\cos x = \sqrt{1 - \sin^2 x}$?

Problems Useful in Calculus

In Problems 95–110 an expression in x is given. By using the given substitution, write the expression as a trigonometric function of θ for θ in the indicated range. Assume that $a > 0$ and that n is a positive integer.

95. $\sqrt{a^2 - x^2}$; $x = a \sin \theta$; $-\dfrac{\pi}{2} \le \theta \le \dfrac{\pi}{2}$

96. $\sqrt{2 - x^2}$; $x = \sqrt{2} \cos \theta$; $0 \le \theta \le \pi$

97. $\sqrt{4 + x^2}$; $x = 2 \tan \theta$; $-\dfrac{\pi}{2} < \theta < \dfrac{\pi}{2}$

† Problems 74–80 use material on natural logarithms in Section 7.4. Omit them if you have not seen natural logarithms before.

98. $(a^2 + x^2)^{3/2}$; $x = a \tan \theta$; $-\dfrac{\pi}{2} < \theta < \dfrac{\pi}{2}$

99. $\dfrac{2}{\sqrt{9 - x^2}}$; $x = 3 \sin \theta$; $-\dfrac{\pi}{2} < \theta < \dfrac{\pi}{2}$

100. $x\sqrt{1 - x^2}$; $x = \cos \theta$; $0 \le \theta \le \dfrac{\pi}{2}$

101. $\sqrt{x^2 - a^2}$; $x = a \sec \theta$; $0 < \theta < \dfrac{\pi}{2}$

102. $x^2\sqrt{x^2 - 10}$; $x = \sqrt{10} \sec \theta$; $0 < \theta < \dfrac{\pi}{2}$

103. $\dfrac{1}{(x^2 - 4)^{3/2}}$; $x = 2 \sec \theta$; $-\dfrac{\pi}{2} < \theta < \dfrac{\pi}{2}$

104. $\dfrac{1}{(x^2 + 16)^{5/2}}$; $x = 4 \tan \theta$; $-\dfrac{\pi}{2} < \theta < \dfrac{\pi}{2}$

105. $\dfrac{x^2}{\sqrt{3 - x^2}}$; $x = \sqrt{3} \cos \theta$; $0 < \theta < \pi$

106. $\dfrac{x^3}{(x^2 - a^2)^{3/2}}$; $x = a \sec \theta$; $0 < \theta < \dfrac{\pi}{2}$

107. $\dfrac{x^n}{(a^2 - x^2)^{n/2}}$; $x = a \sin \theta$; $-\dfrac{\pi}{2} < \theta < \dfrac{\pi}{2}$

108. $x^n\sqrt{a^2 + x^2}$; $x = a \tan \theta$; $-\dfrac{\pi}{2} < \theta < \dfrac{\pi}{2}$

109. $\dfrac{\sqrt{x^2 - a^2}}{x^n}$; $x = a \sec \theta$; $0 < \theta < \dfrac{\pi}{2}$

110. $\sqrt{a + x}\,\sqrt{a - x}$; $x = a \sin \theta$; $-\dfrac{\pi}{2} \le \theta \le \dfrac{\pi}{2}$

3.2 The Sum and Difference Identities

In this section, we prove six important identities. Five of them follow from the first one.

Identity for Cosine of a Difference

For any real numbers α and β

$$\cos(\alpha - \beta) = \cos\alpha \cos\beta + \sin\alpha \sin\beta \qquad (1)$$

Proof

We assume that $\alpha > \beta$. This doesn't make any difference since $\cos(\alpha - \beta) = \cos(-(\beta - \alpha)) = \cos(\beta - \alpha)$, so we will get the same result if $\beta > \alpha$. In Figure 1(a), we place α and β in standard position and draw the unit circle. The terminal side of β intersects the unit circle at the point $A = (\cos\beta, \sin\beta)$ (see page 77). The terminal side of α intersects the circle at $B = (\cos\alpha, \sin\alpha)$. In Figure 1(b), we depict the angle $\alpha - \beta$ in standard position. The terminal side of $\alpha - \beta$ intersects the unit circle at $C = (\cos(\alpha - \beta), \sin(\alpha - \beta))$.

Observe that the line segment AB in Figure 1(a) has the same length as the line segment DC in Figure 1(b) since DC is obtained by rotating AB β radians (or degrees) clockwise. That is,

$$\overline{AB} = \overline{DC} \quad \text{so} \quad \overline{AB}^2 = \overline{DC}^2$$

Using the distance formula (p. 10), we have

$$\begin{aligned}\overline{AB}^2 &= (\cos\alpha - \cos\beta)^2 + (\sin\alpha - \sin\beta)^2 \\ &= \cos^2\alpha - 2\cos\alpha\cos\beta + \cos^2\beta + \sin^2\alpha - 2\sin\alpha\sin\beta + \sin^2\beta\end{aligned}$$

Figure 1 The angle in (b) is obtained by rotating the angle in (a) β degrees or radians clockwise.

Rearrange terms
↓
$$= \cos^2 \alpha + \sin^2 \alpha + \cos^2 \beta + \sin^2 \beta - 2\,(\cos \alpha \cos \beta + \sin \alpha \sin \beta)$$

Basic identity (6) in Section 3.1
↓
$$= 1 + 1 - 2\,(\cos \alpha \cos \beta + \sin \alpha \sin \beta)$$

or

$$\overline{AB}^2 = 2 - 2\,(\cos \alpha \cos \beta + \sin \alpha \sin \beta) \qquad (2)$$

Next, we compute

$$\begin{aligned}\overline{DC}^2 &= (\cos(\alpha - \beta) - 1)^2 + (\sin(\alpha - \beta) - 0)^2 \\ &= \cos^2(\alpha - \beta) - 2\cos(\alpha - \beta) + 1 + \sin^2(\alpha - \beta) \\ &= \cos^2(\alpha - \beta) + \sin^2(\alpha - \beta) + 1 - 2\cos(\alpha - \beta)\end{aligned}$$

Identity (6) in Section 3.1
↓
$$= 1 + 1 - 2\cos(\alpha - \beta)$$

or

$$\overline{DC}^2 = 2 - 2\cos(\alpha - \beta) \qquad (3)$$

Equating (2) and (3), we have

$$\begin{aligned}2 - 2\cos(\alpha - \beta) &= 2 - 2\,(\cos \alpha \cos \beta + \sin \alpha \sin \beta) \\ -2\cos(\alpha - \beta) &= -2\,(\cos \alpha \cos \beta + \sin \alpha \sin \beta) && \text{Subtract 2} \\ \cos(\alpha - \beta) &= \cos \alpha \cos \beta + \sin \alpha \sin \beta && \text{Divide by } -2 \quad \blacksquare\end{aligned}$$

NOTE A very elegant "proof without words" of (1) is suggested in Problem 74.

EXAMPLE 1 *Using Identity (1) to Compute a Value of cos θ*

Compute $\cos \frac{\pi}{12}$ exactly.

SOLUTION We first observe that† $\frac{\pi}{12} = \frac{\pi}{3} - \frac{\pi}{4}$. Thus,

$$\begin{aligned}\cos \frac{\pi}{12} = \cos\left(\frac{\pi}{3} - \frac{\pi}{4}\right) &= \cos \frac{\pi}{3} \cos \frac{\pi}{4} + \sin \frac{\pi}{3} \sin \frac{\pi}{4} \\ &= \frac{1}{2} \cdot \frac{1}{\sqrt{2}} + \frac{\sqrt{3}}{2} \cdot \frac{1}{\sqrt{2}} = \frac{1 + \sqrt{3}}{2\sqrt{2}}\end{aligned}$$

The value $\cos \frac{\pi}{12} = \frac{1 + \sqrt{3}}{2\sqrt{2}}$ is exact. A calculator can give you only an approximate value. Using a calculator, we can obtain $\cos \frac{\pi}{12} = 0.9659258263$ which is correct to 10 decimal places, but it is not exact.

† We write $\frac{\pi}{12}$ as $\frac{\pi}{3} - \frac{\pi}{4}$ because we know the values of $\sin \frac{\pi}{4}$, $\cos \frac{\pi}{4}$, $\sin \frac{\pi}{3}$, and $\cos \frac{\pi}{3}$.

EXAMPLE 2 *Using Identity (1) to Prove a Different Trigonometric Identity*

Prove the identity

$$\cos(\pi - \theta) = -\cos\theta$$

SOLUTION

$$\cos(\pi - \theta) = \cos\pi\cos\theta + \sin\pi\sin\theta = (-1)\cos\theta + 0\sin\theta$$
$$= -\cos\theta$$

NOTE This proof is easier than the ones given on p. 94.

The five other identities we discuss in this section all follow from (1).

Identity for Cosine of a Sum

$$\cos(\alpha + \beta) = \cos\alpha\cos\beta - \sin\alpha\sin\beta \tag{4}$$

Proof

From (1)

$$\cos(\alpha + \beta) = \cos(\alpha - (-\beta)) = \cos\alpha\cos(-\beta) + \sin\alpha\sin(-\beta)$$

$\cos(-\beta) = \cos\beta$
$\sin(-\beta) = -\sin\beta$

$$= \cos\alpha\cos\beta + \sin\alpha(-\sin\beta)$$
$$= \cos\alpha\cos\beta - \sin\alpha\sin\beta \quad ■$$

EXAMPLE 3 *Using Identity (4) to Compute a Value of cos θ*

Compute cos 165° exactly.

SOLUTION

$$\cos 165° = \cos(120° + 45°) = \cos 120°\cos 45° - \sin 120°\sin 45°$$

$$= -\frac{1}{2}\cdot\frac{\sqrt{2}}{2} - \frac{\sqrt{3}}{2}\cdot\frac{\sqrt{2}}{2} \qquad \cos 120° = -\cos 60° = \frac{-1}{2}$$

$$= \frac{-\sqrt{2} - \sqrt{6}}{4} \qquad \sin 120° = \sin 60° = \frac{\sqrt{3}}{2} \quad ■$$

EXAMPLE 4 *Using Identity (1) to Prove Two Trigonometric Identities*

Show that

$$\cos\left(\frac{\pi}{2}-\theta\right)=\cos\,(90^\circ-\theta)=\sin\,\theta \tag{5}$$

and

$$\sin\left(\frac{\pi}{2}-\theta\right)=\sin\,(90^\circ-\theta)=\cos\,\theta \tag{6}$$

NOTE We proved identities (5) and (6) in Section 2.2 (p. 71) only under the condition that θ is an acute angle. We now prove it for all values of θ.

SOLUTION

$$\text{(a)}\ \cos\left(\frac{\pi}{2}-\theta\right)=\cos\frac{\pi}{2}\cos\theta+\sin\frac{\pi}{2}\sin\theta$$
$$=0\cdot\cos\theta+1\cdot\sin\theta=\sin\theta$$

$$\text{(b)}\ \cos\theta=\cos\left(\frac{\pi}{2}-\left(\frac{\pi}{2}-\theta\right)\right)$$
$$=\cos\frac{\pi}{2}\cos\left(\frac{\pi}{2}-\theta\right)+\sin\frac{\pi}{2}\sin\left(\frac{\pi}{2}-\theta\right)$$
$$=0\cdot\cos\left(\frac{\pi}{2}-\theta\right)+1\cdot\sin\left(\frac{\pi}{2}-\theta\right)=\sin\left(\frac{\pi}{2}-\theta\right)$$

We list below the other four identities of this section.

Identity for Sine of a Sum

$$\sin\,(\alpha+\beta)=\sin\alpha\cos\beta+\cos\alpha\sin\beta \tag{7}$$

Identity for Sine of a Difference

$$\sin\,(\alpha-\beta)=\sin\alpha\cos\beta-\cos\alpha\sin\beta \tag{8}$$

Identity for Tangent of a Sum

$$\tan\,(\alpha+\beta)=\frac{\tan\alpha+\tan\beta}{1-\tan\alpha\tan\beta} \tag{9}$$

Identity for Tangent of a Difference

$$\tan\,(\alpha-\beta)=\frac{\tan\alpha-\tan\beta}{1+\tan\alpha\tan\beta} \tag{10}$$

We prove (7) and (9) and leave the proofs of (8) and (10) as exercises (see Problems 72 and 73).

Proof of (7)

From (5)

$$\sin(\alpha+\beta) = \cos\left(\frac{\pi}{2}-(\alpha+\beta)\right) = \cos\left(\left(\frac{\pi}{2}-\alpha\right)-\beta\right)$$

From (1)

$$= \cos\left(\frac{\pi}{2}-\alpha\right)\cos\beta + \sin\left(\frac{\pi}{2}-\alpha\right)\sin\beta$$

From (5) and (6)

$$= \sin\alpha\cos\beta + \cos\alpha\sin\beta \quad \blacksquare$$

Proof of (9)

From (7) and (4)

$$\tan(\alpha+\beta) = \frac{\sin(\alpha+\beta)}{\cos(\alpha+\beta)} = \frac{\sin\alpha\cos\beta+\cos\alpha\sin\beta}{\cos\alpha\cos\beta-\sin\alpha\sin\beta}$$

We now divide numerator and denominator by $\cos\alpha\cos\beta$

$$= \frac{\dfrac{\sin\alpha\cos\beta}{\cos\alpha\cos\beta}+\dfrac{\cos\alpha\sin\beta}{\cos\alpha\cos\beta}}{\dfrac{\cos\alpha\cos\beta}{\cos\alpha\cos\beta}-\dfrac{\sin\alpha}{\cos\alpha}\cdot\dfrac{\sin\beta}{\cos\beta}} = \frac{\tan\alpha+\tan\beta}{1-\tan\alpha\tan\beta} \quad \blacksquare$$

EXAMPLE 5 *Using Identity (7) to Compute a Value of sin θ*

Compute $\sin\frac{5\pi}{12}$ exactly.

SOLUTION

$$\sin\frac{5\pi}{12} = \sin\left(\frac{\pi}{4}+\frac{\pi}{6}\right) = \sin\frac{\pi}{4}\cos\frac{\pi}{6}+\cos\frac{\pi}{4}\sin\frac{\pi}{6}$$

$$= \frac{\sqrt{2}}{2}\frac{\sqrt{3}}{2}+\frac{\sqrt{2}}{2}\frac{1}{2} = \frac{\sqrt{6}+\sqrt{2}}{4} \quad \blacksquare$$

EXAMPLE 6 *Using Identity (9) to Compute a Value of tan θ*

Compute $\tan 195°$ exactly.

SOLUTION

$$\tan 195° = \tan(135°+60°) = \frac{\tan 135° + \tan 60°}{1-\tan 135°\tan 60°}$$

$$= \frac{-1+\sqrt{3}}{1-(-1)(\sqrt{3})} = \frac{\sqrt{3}-1}{\sqrt{3}+1} = \frac{\sqrt{3}-1}{\sqrt{3}+1}\cdot\frac{\sqrt{3}-1}{\sqrt{3}-1} = 2-\sqrt{3}$$

$\tan 135° = -\tan 45° = -1$ $\blacksquare$

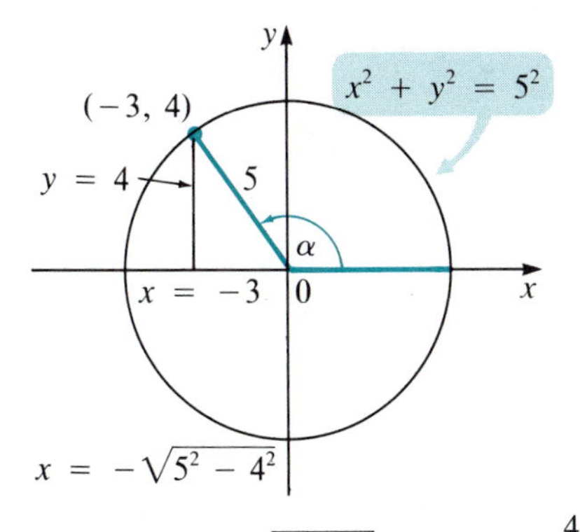

Figure 2 $x = -\sqrt{5^2 - 4^2}$ and $\sin \alpha = \dfrac{4}{5}$

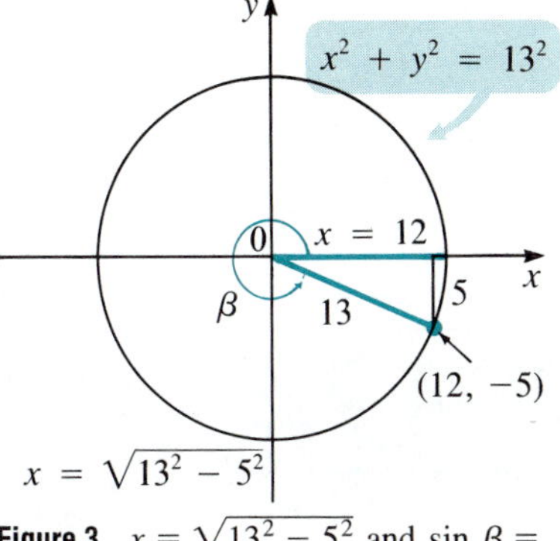

Figure 3 $x = \sqrt{13^2 - 5^2}$ and $\sin \beta = -\dfrac{5}{13}$

EXAMPLE 7 *Finding sin $(\alpha - \beta)$ When sin α, sin β, and the Quadrants of α and β Are Given*

Find $\sin(\alpha - \beta)$ if $\sin \alpha = \frac{4}{5}$ and α is in the second quadrant while $\sin \beta = -\frac{5}{13}$ and β is in the fourth quadrant.

SOLUTION In order to use identity (8), we must compute $\cos \alpha$ and $\cos \beta$. We have, from Figures 2 and 3,

$$\cos \alpha = \frac{x}{r} = -\frac{3}{5} \quad \text{and} \quad \cos \beta = \frac{x}{r} = \frac{12}{13}$$

Thus,

$$\begin{aligned} \sin(\alpha - \beta) &= \sin \alpha \cos \beta - \cos \alpha \sin \beta \\ &= \frac{4}{5} \cdot \frac{12}{13} - \left(-\frac{3}{5}\right)\left(-\frac{5}{13}\right) = \frac{48}{65} - \frac{15}{65} = \frac{33}{65} \end{aligned}$$

■

EXAMPLE 8 *An Identity Useful in Calculus*

Let $f(x) = \sin x$. Prove that

$$\begin{aligned} \frac{f(x+h) - f(x)}{h} &= \frac{\sin(x+h) - \sin x}{h} \\ &= \cos x \left(\frac{\sin h}{h}\right) + \sin x \left(\frac{\cos h - 1}{h}\right) \end{aligned}$$

SOLUTION

$$\begin{aligned} \frac{\sin(x+h) - \sin x}{h} &\overset{\text{Using (7)}}{=} \frac{(\sin x \cos h + \cos x \sin h) - \sin x}{h} \\ &= \frac{\cos x \sin h}{h} + \frac{\sin x \cos h - \sin x}{h} \\ &= \cos x \left(\frac{\sin h}{h}\right) + \sin x \left(\frac{\cos h - 1}{h}\right) \end{aligned}$$

■

EXAMPLE 9 *Using the Identity for the Sine of a Sum to Evaluate a Function*

Compute $\sin\left(\cos^{-1} x + \sin^{-1} \dfrac{x}{x+1}\right)$, $0 < x < 1$.

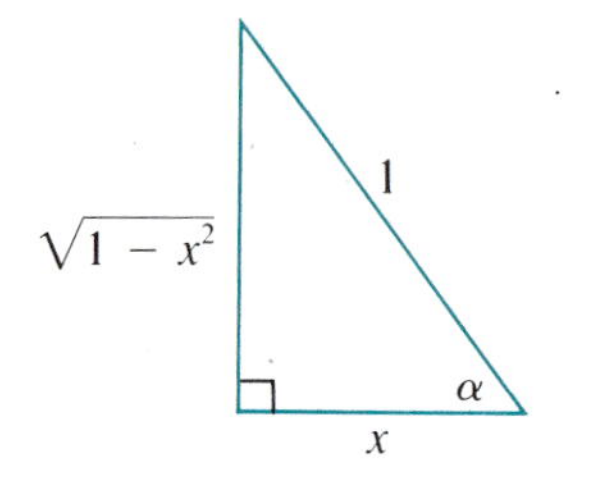

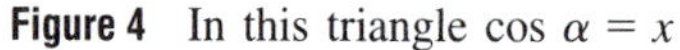

Figure 4 In this triangle $\cos \alpha = x$

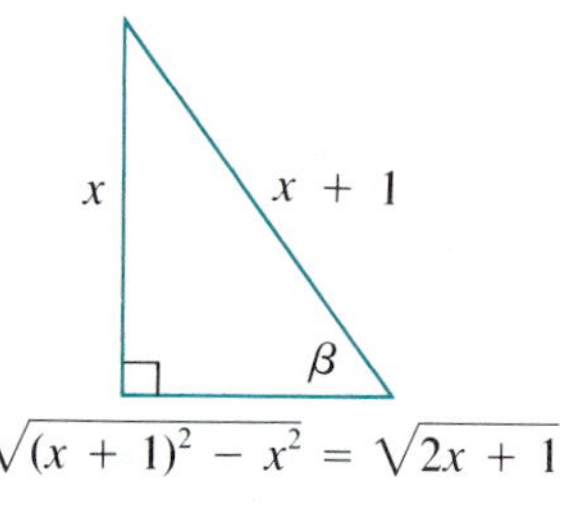

Figure 5 In this triangle $\sin \beta = \dfrac{x}{x+1}$

SOLUTION Let $\alpha = \cos^{-1} x$ and $\beta = \sin^{-1} \dfrac{x}{x+1}$. Since $0 < x < 1$, both α and β are in the first quadrant, $\cos \alpha = x$, and $\sin \beta = \dfrac{x}{x+1}$. From Figures 4 and 5, we see that

$$\sin \alpha = \sqrt{1-x^2} \qquad \text{and} \qquad \cos \beta = \frac{\sqrt{2x+1}}{x+1}$$

Thus,

$$\begin{aligned}\sin\left(\cos^{-1} x + \sin^{-1} \frac{x}{x+1}\right) &= \sin(\alpha + \beta) = \sin \alpha \cos \beta + \cos \alpha \sin \beta \\ &= \frac{\sqrt{1-x^2}\sqrt{2x+1}}{x+1} + x\frac{x}{x+1} \\ &= \frac{\sqrt{1-x^2}\sqrt{2x+1} + x^2}{x+1}\end{aligned}$$

Problems 3.2

Readiness Check

I. Which of the following is equal to $\cos \dfrac{\pi}{6}$?

a. $\sin \dfrac{\pi}{2} \cos \dfrac{\pi}{3} - \cos \dfrac{\pi}{2} \sin \dfrac{\pi}{3}$

b. $\sin \dfrac{\pi}{2} \cos \dfrac{\pi}{3} + \cos \dfrac{\pi}{2} \sin \dfrac{\pi}{3}$

c. $\sin \dfrac{\pi}{2} \sin \dfrac{\pi}{3} - \cos \dfrac{\pi}{2} \cos \dfrac{\pi}{3}$

d. $\cos \dfrac{\pi}{2} \cos \dfrac{\pi}{3} + \sin \dfrac{\pi}{2} \sin \dfrac{\pi}{3}$

II. Which of the following is the exact value of $\sin 75°$?

a. $\dfrac{\sqrt{2}+\sqrt{6}}{4}$ b. $\dfrac{\sqrt{2}-\sqrt{6}}{4}$

c. $2 - \dfrac{\sqrt{6}}{2}$ d. $2 + \dfrac{\sqrt{6}}{2}$

III. Which of the following is true about $\tan(\alpha - \beta)$ if $\sin \alpha = -12/13$, $\cos \beta = -3/5$ and α and β are both in the third quadrant?

a. $\tan(\alpha - \beta)$ is positive.
b. $\tan(\alpha - \beta)$ is negative.
c. $\tan(\alpha - \beta)$ is zero.
d. $\tan(\alpha - \beta)$ is undefined.

IV. Which of the following is true about $(\alpha + \beta)$ if $\sin \alpha = 4/5$ in quadrant II and $\cos \beta = -12/13$ in quadrant III?

a. $\sin(\alpha + \beta) = -33/65$
b. $(\alpha + \beta)$ is in quadrant I.
c. $\cos(\alpha + \beta) = -16/65$
d. $\tan(\alpha + \beta) = 63/16$

In Problems 1–10 compute each value exactly.

1. $\cos \dfrac{7\pi}{12}$ 2. $\sin \dfrac{7\pi}{12}$ 3. $\tan \dfrac{7\pi}{12}$
4. $\sin 15°$ 5. $\tan 15°$ 6. $\cos 375°$
7. $\sin \dfrac{13\pi}{12}$ 8. $\tan \dfrac{25\pi}{12}$ 9. $\sin 345°$
10. $\tan 285°$

Answers to Readiness Check

I. d II. a III. a $[\tan(\alpha - \beta) = \frac{16}{63}]$ IV. a

In Problems 11–22 write each expression as $\sin\theta$, $\cos\theta$, or $\tan\theta$ for a single value of θ.

11. $\cos 22^\circ \cos 47^\circ - \sin 22^\circ \sin 47^\circ$
[Hint: $22^\circ + 47^\circ = 69^\circ$]
12. $\cos 1 \cos 2 + \sin 1 \sin 2$
13. $\sin 1.5 \cos 2.3 - \cos 1.5 \sin 2.3$
14. $\sin 1.5 \cos 2.3 + \cos 1.5 \sin 2.3$
15. $\cos 3.4 \cos 2.6 + \sin 3.4 \sin 2.6$
16. $\cos 57^\circ \cos 82^\circ + \sin 57^\circ \sin 82^\circ$
17. $\dfrac{\tan 1 + \tan \frac{1}{2}}{1 - \tan 1 \tan \frac{1}{2}}$
18. $\dfrac{\tan \frac{1}{2} - \tan 1}{1 + \tan \frac{1}{2} \tan 1}$
19. $\sin x \cos 2x + \cos x \sin 2x$
20. $\cos\theta \cos 3\theta + \sin\theta \sin 3\theta$
21. $\dfrac{\tan\alpha - \tan 4\alpha}{1 + \tan\alpha \tan 4\alpha}$
* 22. $2 \sin\theta \cos\theta$

In Problems 23–28 two values and the quadrants of α and β are given. (a) Determine $\sin(\alpha+\beta)$, $\sin(\alpha-\beta)$, $\cos(\alpha+\beta)$, $\cos(\alpha-\beta)$, $\tan(\alpha+\beta)$, and $\tan(\alpha-\beta)$. (b) Find the quadrants of $\alpha+\beta$ and $\alpha-\beta$.

23. $\cos\alpha = \frac{4}{5}$, I; $\sin\beta = \frac{12}{13}$, I
24. $\sin\alpha = \frac{3}{5}$, II; $\cos\beta = \frac{12}{13}$, IV
25. $\sin\alpha = -\frac{3}{5}$, IV; $\cos\beta = -\frac{12}{13}$, II
26. $\sin\alpha = \frac{5}{13}$, II; $\tan\beta = -\frac{4}{3}$, II
27. $\sin\alpha = \frac{5}{13}$, I; $\tan\beta = \frac{4}{3}$, III
28. $\sin\alpha = -\frac{5}{13}$, IV; $\tan\beta = \frac{3}{4}$, III

In Problems 29–54 verify the given identity using the identities in this section.

29. $\sin(x + 2\pi) = \sin x$
30. $\cos(x + 2\pi) = \cos x$
31. $\tan(x + \pi) = \tan x$
32. $\sin(x + \pi) = -\sin x$
33. $\cos(x + \pi) = -\cos x$
34. $\sin(\pi - x) = \sin x$
35. $\tan(\pi - x) = -\tan x$
36. $\sin 2x = 2 \sin x \cos x$
37. $\cos 2x = \cos^2 x - \sin^2 x$
38. $\cos 2x = 1 - 2\sin^2 x$
39. $\cos 2x = 2\cos^2 x - 1$
40. $\sin\left(\dfrac{\pi}{2} + x\right) = \cos x$
41. $\cos\left(\dfrac{\pi}{2} + x\right) = -\sin x$
42. $\sin\left(x + \dfrac{\pi}{4}\right) = \dfrac{1}{\sqrt{2}}(\sin x + \cos x)$
43. $\cos\left(x + \dfrac{\pi}{4}\right) = \dfrac{1}{\sqrt{2}}(\cos x - \sin x)$
44. $\sin\left(\dfrac{3\pi}{2} - x\right) = -\cos x$
45. $\cos\left(\dfrac{3\pi}{2} - x\right) = -\sin x$
46. $\tan\left(x + \dfrac{\pi}{4}\right) = \dfrac{1 + \tan x}{1 - \tan x}$
47. $\tan\left(x - \dfrac{\pi}{4}\right) = \dfrac{\tan x - 1}{\tan x + 1}$
* 48. $\sin^2 x - \sin^2 y = \sin(x+y)\sin(x-y)$
* 49. $\cos^2 x - \sin^2 y = \cos(x+y)\cos(x-y)$
50. $\sin(x+y) + \sin(x-y) = 2\sin x \cos y$
51. $\cos(x+y) - \cos(x-y) = -2\sin x \sin y$
52. $\cos(n\pi + x) = (-1)^n \cos x$, n is a positive integer
53. $\sin(n\pi + x) = (-1)^n \sin x$, n is a positive integer
* 54. $\sec(x+y) = \dfrac{\cos(y-x)}{\cos^2 y - \sin^2 x}$

In Problems 55–63 compute each value without a calculator.

55. $\sin\left(\cos^{-1}\dfrac{1}{2} + \sin^{-1} 0\right)$
56. $\sin\left(\cos^{-1}\dfrac{1}{2} + \sin^{-1} 1\right)$
57. $\cos\left(\cos^{-1}\dfrac{1}{2} + \sin^{-1}\dfrac{\sqrt{2}}{2}\right)$
58. $\cos\left(\sin^{-1}\left(-\dfrac{\sqrt{3}}{2}\right) + \cos^{-1}\left(-\dfrac{1}{2}\right)\right)$
59. $\tan\left(\sin^{-1}\dfrac{1}{2} + \cos^{-1}\dfrac{1}{2}\right)$
60. $\sin(\tan^{-1} 1 + \tan^{-1} 2)$
61. $\tan\left(\cos^{-1}\dfrac{3}{4} + \sin^{-1}\dfrac{4}{5}\right)$
62. $\tan\left(\cos^{-1}\left(-\dfrac{3}{7}\right) - \sin^{-1}\dfrac{5}{8}\right)$
63. $\sin\left(\tan^{-1} 3 - \tan^{-1}\dfrac{1}{3}\right)$
64. Evaluate $\sin(\cos^{-1} x - \cos^{-1} x^2)$, $0 < x < 1$.
65. Evaluate $\cos(\tan^{-1} x - \tan^{-1}(x+2))$, $x > 0$.
66. Evaluate $\tan\left(\sin^{-1}\dfrac{1}{x} + \cos^{-1}(x-1)\right)$, $1 < x < 2$.
67. Evaluate $\sin(\sin^{-1} 2x + \cos^{-1} x)$, $-\dfrac{1}{2} < x < 0$.
68. Show that for $-1 \le x \le 1$,

$$\cos(\sin^{-1} x + \cos^{-1} x) = 0$$

69. Show that for $-1 \leq x \leq 1$,

$$\sin(\cos^{-1} x + \sin^{-1} x) = 1$$

* 70. Write $\sin 3x$ in terms of $\sin x$ and $\cos x$.

* 71. Write $\cos 4x$ in terms of $\sin x$ and $\cos x$.

72. Prove identity (8).
[Hint: $\sin(\alpha - \beta) = \sin(\alpha + (-\beta))$]

73. Prove identity (10). [Hint: Use (9) and the fact that $\tan(-\theta) = -\tan\theta$.]

74. (a) In the figure below explain why area of triangle BCA = area of triangle BCD + area of triangle DCA

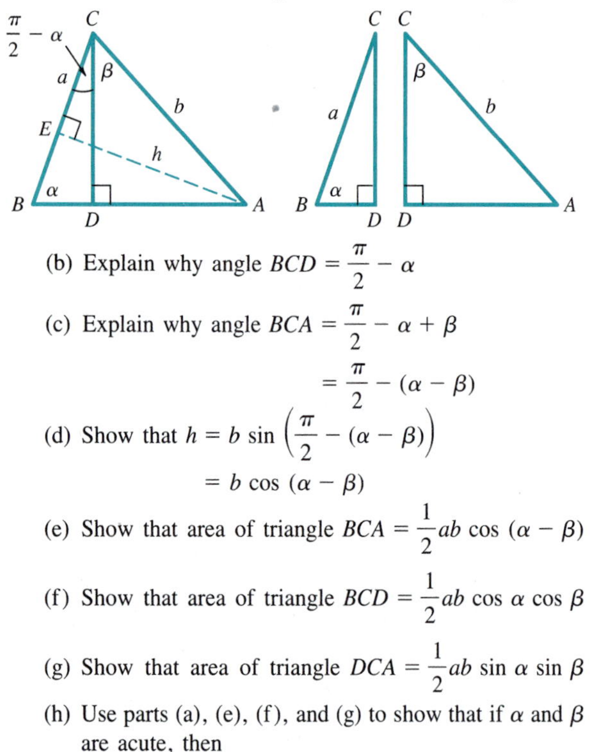

(b) Explain why angle $BCD = \dfrac{\pi}{2} - \alpha$

(c) Explain why angle $BCA = \dfrac{\pi}{2} - \alpha + \beta$

$$= \frac{\pi}{2} - (\alpha - \beta)$$

(d) Show that $h = b \sin\left(\dfrac{\pi}{2} - (\alpha - \beta)\right)$

$$= b\cos(\alpha - \beta)$$

(e) Show that area of triangle $BCA = \dfrac{1}{2}ab\cos(\alpha - \beta)$

(f) Show that area of triangle $BCD = \dfrac{1}{2}ab\cos\alpha\cos\beta$

(g) Show that area of triangle $DCA = \dfrac{1}{2}ab\sin\alpha\sin\beta$

(h) Use parts (a), (e), (f), and (g) to show that if α and β are acute, then

$$\cos(\alpha - \beta) = \cos\alpha\cos\beta + \sin\alpha\sin\beta$$

* 75. Follow the steps of Problem 74 to explain the following "proof without words."†

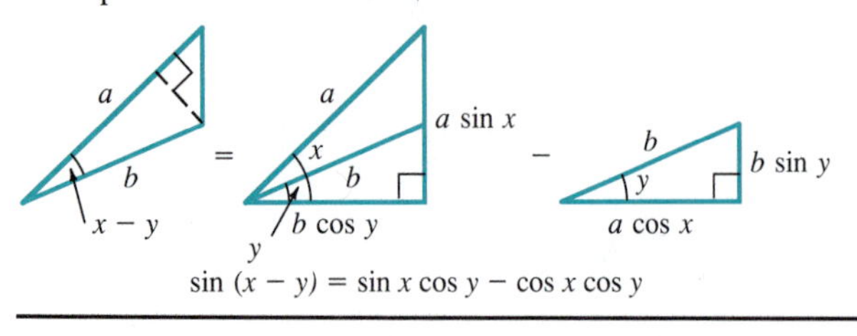

$\sin(x - y) = \sin x \cos y - \cos x \cos y$

* 76. Suppose that a mass is attached to a spring as in the figure below. If the mass is pulled out a distance of x_0 meters

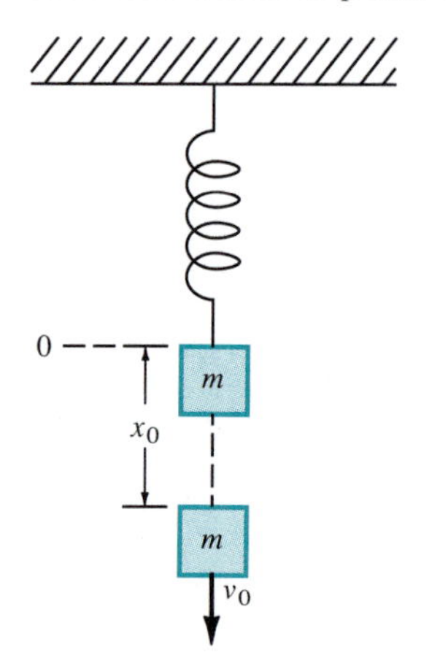

below its equilibrium position and given an initial velocity of v_0 meters/sec, then, ignoring friction and air resistance, the mass will vibrate according to the formula

$$x(t) = A\sin(\omega_0 t + \phi)$$

where $x(t)$ denotes the position of the mass at time t (in meters above (if $x(t) < 0$) or below (if $x(t) > 0$) the equilibrium position denoted by 0. Here $A = \sqrt{x_0^2 + \left(\dfrac{v_0}{\omega_0}\right)^2}$ and $\phi = \tan^{-1}\left(\dfrac{v_0/\omega_0}{x_0}\right)$. Show that $x(t)$ can be written as

$$x(t) = x_0 \cos\omega_0 t + \frac{v_0}{\omega_0}\sin\omega_0 t$$

We will say more about this kind of motion in Sections 3.6 and 3.8.

† The proofs in Problems 74 and 75 are adapted from Sidney H. Kung, "Proof without Words: Area and Difference Formulas," *Mathematics Magazine* 62(5), Dec. 1989, p. 317.

3.3 The Double-Angle and Half-Angle Identities

In this section, we derive formulas for $\sin 2x$, $\cos 2x$, $\sin \frac{1}{2}x$, $\cos \frac{1}{2}x$, and related functions.

Double-Angle Identities

For every real number x,

$$\sin 2x = 2 \sin x \cos x \tag{1}$$

$$\cos 2x = \cos^2 x - \sin^2 x \tag{2}$$

$$\cos 2x = 2 \cos^2 x - 1 \tag{3}$$

$$\cos 2x = 1 - 2 \sin^2 x \tag{4}$$

Proof

We prove (1) and leave the proofs of the others as exercises (see Problems 29–31).

According to the identity for sine of a sum (equation (7) in Section 3.2),

$$\sin (\alpha + \beta) = \sin \alpha \cos \beta + \cos \alpha \sin \beta \tag{5}$$

Setting $\alpha = x$ and $\beta = x$ in (5), we have

$$\begin{aligned} \sin 2x &= \sin (x + x) = \sin x \cos x + \cos x \sin x \\ &= 2 \sin x \cos x \quad \blacksquare \end{aligned}$$

EXAMPLE 1 *Verifying Identities (1) and (2) Using Known Values*

As a check of identities (1) and (2), observe that

$$\begin{aligned} \sin \frac{\pi}{3} &= \sin 2\left(\frac{\pi}{6}\right) = 2 \sin \frac{\pi}{6} \cos \frac{\pi}{6} = 2 \cdot \frac{1}{2} \cdot \frac{\sqrt{3}}{2} = \frac{\sqrt{3}}{2} \\ \cos \frac{\pi}{3} &= \cos 2\left(\frac{\pi}{6}\right) = \cos^2 \frac{\pi}{6} - \sin^2 \frac{\pi}{6} = \left(\frac{\sqrt{3}}{2}\right)^2 - \left(\frac{1}{2}\right)^2 \\ &= \frac{3}{4} - \frac{1}{4} = \frac{1}{2} \quad \blacksquare \end{aligned}$$

EXAMPLE 2 *Using the Double Angle Identities*

Find $\sin 2\theta$ and $\cos 2\theta$ if θ is in the second quadrant and $\sin \theta = \frac{2}{3}$.

SOLUTION First we compute $\cos\theta$, which is negative because cosine is negative in the second quadrant. We have

$$\cos^2\theta = 1 - \sin^2\theta = 1 - \frac{4}{9} = \frac{5}{9}$$

so

$$\cos\theta = -\sqrt{\frac{5}{9}} = -\frac{\sqrt{5}}{3}$$

Then

$$\sin 2\theta = 2\sin\theta\cos\theta = 2\left(\frac{2}{3}\right)\left(-\frac{\sqrt{5}}{3}\right) = \frac{-4\sqrt{5}}{9}$$

$$\cos 2\theta = \cos^2\theta - \sin^2\theta = \frac{5}{9} - \frac{4}{9} = \frac{1}{9}$$

Note that 2θ is in the fourth quadrant because $\sin 2\theta < 0$ and $\cos 2\theta > 0$. ■

EXAMPLE 3 *Using the Double Angle Identities to Rewrite sin 3x*

Express $\sin 3x$ in terms of powers of $\sin x$.

SOLUTION

Identity (7) on p. 159 ↓

$$\sin 3x = \sin(2x + x) = \sin 2x\cos x + \cos 2x\sin x$$

Identities (1) and (2) ↓

$$= (2\sin x\cos x)\cos x + (\cos^2 x - \sin^2 x)\sin x$$

$$= 2\sin x\cos^2 x + \cos^2 x\sin x - \sin^3 x$$

$$= 3\sin x\cos^2 x - \sin^3 x$$

$$= 3\sin x(1 - \sin^2 x) - \sin^3 x \qquad \cos^2 x = 1 - \sin^2 x$$

$$= 3\sin x - 3\sin^3 x - \sin^3 x$$

or

$$\sin 3x = 3\sin x - 4\sin^3 x \quad ■ \tag{6}$$

EXAMPLE 4 *Proving An Identity*

Prove the **double-angle identity for tangents.**

Double-Angle Identity for Tangents

$$\tan 2x = \frac{2\tan x}{1 - \tan^2 x} \tag{7}$$

SOLUTION

$$\tan 2x = \frac{\sin 2x}{\cos 2x} = \frac{2 \sin x \cos x}{\cos^2 x - \sin^2 x}$$

Divide numerator and denominator by $\cos^2 x$

$$= \frac{\dfrac{2 \sin x \cos x}{\cos^2 x}}{\dfrac{\cos^2 x}{\cos^2 x} - \dfrac{\sin^2 x}{\cos^2 x}} = \frac{\dfrac{2 \sin x}{\cos x}}{1 - \left(\dfrac{\sin x}{\cos x}\right)^2} = \frac{2 \tan x}{1 - \tan^2 x}$$

Before continuing, we warn you against a common error.

WARNING Do *not* cancel 2's in the following expressions:

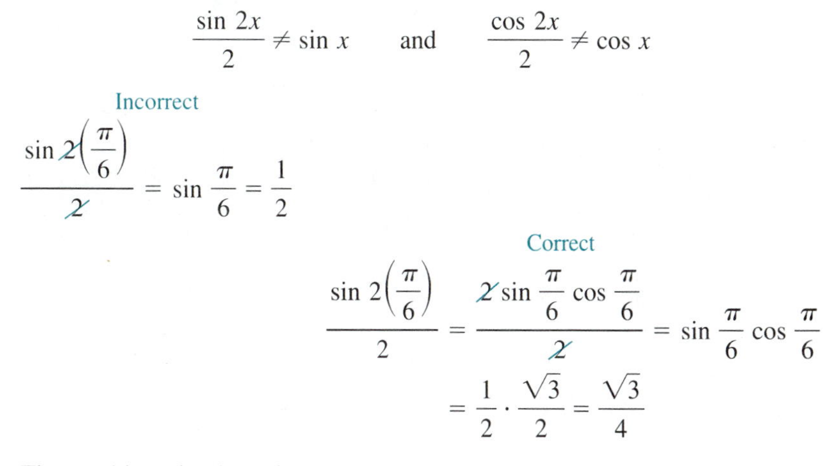

The problem is that $\sin 2x \neq 2 \sin x$. The correct formula is $\sin 2x = 2 \sin x \cos x$. ■

In calculus and other areas in which trigonometry is applied, it is frequently useful to rewrite powers of $\sin x$ and $\cos x$. The following identities are useful:

Three Identities Useful in Calculus

$$\sin^2 x = \frac{1 - \cos 2x}{2} \qquad (8)$$

$$\cos^2 x = \frac{1 + \cos 2x}{2} \qquad (9)$$

$$\tan^2 x = \frac{1 - \cos 2x}{1 + \cos 2x} \qquad (10)$$

Proof

Formula (8) follows from (4):

$$1 - 2\sin^2 x = \cos 2x$$
$$-2\sin^2 x = \cos 2x - 1$$
$$\sin^2 x = \frac{1 - \cos 2x}{2}$$

Formula (9) follows from (3). To obtain (10), observe that

$$\tan^2 x = \frac{\sin^2 x}{\cos^2 x} = \frac{\dfrac{1 - \cos 2x}{2}}{\dfrac{1 + \cos 2x}{2}} = \frac{1 - \cos 2x}{1 + \cos 2x} \quad ■$$

EXAMPLE 5 *Rewriting $\sin^4 x$ Without Any Exponents*

Write $\sin^4 \theta$ without any exponents.

SOLUTION $\sin^4 \theta = (\sin^2 \theta)^2 = \left(\dfrac{1 - \cos 2\theta}{2}\right)^2$

$$= \frac{1}{4}(1 - 2\cos 2\theta + \cos^2 2\theta)$$

But

From (9) with $x = 2\theta$

$$\cos^2 2\theta = \frac{1 + \cos 4\theta}{2} = \frac{1}{2} + \frac{1}{2}\cos 4\theta$$

Thus,

$$\sin^4 \theta = \frac{1}{4}\Bigl(1 - 2\cos 2\theta + \overbrace{\frac{1}{2} + \frac{1}{2}\cos 4\theta}^{\cos^2 2\theta}\Bigr)$$
$$= \frac{1}{4} - \frac{1}{2}\cos 2\theta + \frac{1}{8} + \frac{\cos 4\theta}{8}$$
$$= \frac{3}{8} - \frac{1}{2}\cos 2\theta + \frac{1}{8}\cos 4\theta$$

Half-Angle Identities

For every real number x,

$$\sin \frac{x}{2} = \pm\sqrt{\frac{1 - \cos x}{2}} \qquad \textbf{(11)}$$

$$\cos \frac{x}{2} = \pm\sqrt{\frac{1 + \cos x}{2}} \qquad \textbf{(12)}$$

$$\tan \frac{x}{2} = \pm\sqrt{\frac{1 - \cos x}{1 + \cos x}} \qquad \textbf{(13)}$$

The sign depends on the quadrant of $\dfrac{x}{2}$.

$$\tan \frac{x}{2} = \frac{1 - \cos x}{\sin x} = \frac{\sin x}{1 + \cos x} \qquad (14)$$

Proof

We prove (11). The proofs of (12) and (13) are similar. See Problem 55 for a suggested proof of (14). From (8), using $\frac{x}{2}$ in place of x,

$$\sin^2 \frac{x}{2} = \frac{1 - \cos x}{2}$$

Thus

$$\sqrt{\sin^2 \frac{x}{2}} = \left|\sin \frac{x}{2}\right| = \sqrt{\frac{1 - \cos x}{2}}$$

or

$$\sin \frac{x}{2} = \pm\sqrt{\frac{1 - \cos x}{2}} \quad \blacksquare$$

EXAMPLE 6 *Using the Half-Angle Identities*

Compute $\sin \frac{\pi}{8}$, $\cos \frac{\pi}{8}$, and $\tan \frac{\pi}{8}$ exactly.

SOLUTION $\frac{\pi}{8} = \frac{1}{2}\left(\frac{\pi}{4}\right)$ and $\cos \frac{\pi}{4} = \frac{1}{\sqrt{2}} = \frac{\sqrt{2}}{2}$; so, with $\frac{x}{2} = \frac{\pi}{8}$ and $x = \frac{\pi}{4}$,

$$\sin \frac{\pi}{8} = \sin \frac{1}{2}\left(\frac{\pi}{4}\right) = \sqrt{\frac{1 - \frac{\sqrt{2}}{2}}{2}} = \sqrt{\frac{2 - \sqrt{2}}{4}} = \frac{1}{2}\sqrt{2 - \sqrt{2}}$$

$$\cos \frac{\pi}{8} = \cos \frac{1}{2}\left(\frac{\pi}{4}\right) = \sqrt{\frac{1 + \frac{\sqrt{2}}{2}}{2}} = \frac{1}{2}\sqrt{2 + \sqrt{2}}$$

$$\tan \frac{\pi}{8} = \tan \frac{1}{2}\left(\frac{\pi}{4}\right) = \frac{\sin \frac{\pi}{8}}{\cos \frac{\pi}{8}} = \sqrt{\frac{2 - \sqrt{2}}{2 + \sqrt{2}}}$$

Note that we took the positive square roots because $\frac{\pi}{8}$ is in the first quadrant.

EXAMPLE 7 *Using the Half-Angle Identities*

If $\sin \theta = -\dfrac{3}{5}$ and θ is in the third quadrant, find the exact value of $\sin \dfrac{\theta}{2}$, $\cos \dfrac{\theta}{2}$, and $\tan \dfrac{\theta}{2}$.

SOLUTION $\cos^2 \theta = 1 - \sin^2 \theta = 1 - \dfrac{9}{25} = \dfrac{16}{25}$, so $\cos \theta = \pm\dfrac{4}{5}$.

Since θ is in the third quadrant, $\cos \theta < 0$, so $\cos \theta = -\dfrac{4}{5}$. Then

$$\sin \frac{\theta}{2} = \pm\sqrt{\frac{1 - \left(-\frac{4}{5}\right)}{2}} = \pm\sqrt{\frac{\frac{9}{5}}{2}} = \pm\sqrt{\frac{9}{10}}$$
$$= \pm\frac{3}{\sqrt{10}}$$

Since θ is in the third quadrant, $\pi < \theta < \dfrac{3\pi}{2}$, so $\dfrac{\pi}{2} < \dfrac{\theta}{2} < \dfrac{3\pi}{4}$. That is, $\dfrac{\theta}{2}$ is in the second quadrant. In the second quadrant, $\sin \dfrac{\theta}{2} > 0$, $\cos \dfrac{\theta}{2} < 0$, and $\tan \dfrac{\theta}{2} < 0$. Thus, $\sin \dfrac{\theta}{2} = \dfrac{3}{\sqrt{10}}$,

$\cos \dfrac{\theta}{2} < 0$ ↘

$$\cos \frac{\theta}{2} = -\sqrt{\frac{1 + \left(-\frac{4}{5}\right)}{2}} = -\sqrt{\frac{1}{10}} = -\frac{1}{\sqrt{10}}$$

$\tan \dfrac{\theta}{2} < 0$ ↘

$$\tan \frac{\theta}{2} = -\sqrt{\frac{1 - \left(-\frac{4}{5}\right)}{1 + \left(-\frac{4}{5}\right)}} = -\sqrt{\frac{\frac{9}{10}}{\frac{1}{10}}} = -3$$

Problems 3.3

Readiness Check

I. Which of the following is equal to $\cos 10x$?

a. $1 - 2\sin^2 5x$ b. $\sqrt{\dfrac{1 + \cos 5x}{2}}$

c. $\cos^2 20x - \sin^2 20x$ d. $\sqrt{\dfrac{1 + \cos 20x}{2}}$

II. Which is true if θ is in quadrant III?

a. $\sin \dfrac{\theta}{2} < 0$ b. $\dfrac{\theta}{2}$ is in quadrant I or IV.

c. $\cos \dfrac{\theta}{2} > 0$ d. $\cos \dfrac{\theta}{2} < 0$

III. Which of the following is $\sin \frac{\theta}{2}$ if $\sin \theta = -2/3$ and θ is in quadrant III?

a. $-\sqrt{\frac{3-\sqrt{5}}{\sqrt{6}}}$ b. $-\sqrt{\frac{3+\sqrt{5}}{6}}$

c. $\sqrt{\frac{3-\sqrt{5}}{6}}$ d. $\sqrt{\frac{3+\sqrt{5}}{6}}$

IV. Which of the following is $\sin 2x$ if $\sin x = \frac{1}{4}$ and x is in quadrant II?

a. $\frac{1}{2}$ b. $-\frac{\sqrt{15}}{8}$ c. $\frac{\sqrt{15}}{8}$ d. $-\frac{1}{2}$

In Problems 1–12 compute each value exactly.

1. $\sin 15°$
2. $\cos 15°$
3. $\tan 15°$
4. $\sin \frac{\pi}{24}$ [Hint: Use the result of Problem 2.]
5. $\cos \frac{\pi}{24}$
6. $\tan \frac{\pi}{24}$
7. $\sin \frac{3\pi}{8}$
8. $\cos \frac{7\pi}{8}$
9. $\tan \frac{11\pi}{12}$
10. $\sin \left(-\frac{7\pi}{12}\right)$
11. $\sec \frac{\pi}{8}$
12. $\cot \frac{3\pi}{8}$

In Problems 13–22 find $\sin 2\theta$, $\cos 2\theta$, $\tan 2\theta$, $\sin \frac{\theta}{2}$, $\cos \frac{\theta}{2}$, and $\tan \frac{\theta}{2}$ using the information given.

13. $\cos \theta = \frac{4}{5}$, θ in first quadrant.
14. $\cos \theta = \frac{4}{5}$, θ in fourth quadrant.
15. $\tan \theta = \frac{5}{12}$, θ in third quadrant.
16. $\tan \theta = \frac{5}{12}$, θ in first quadrant.
17. $\sin \theta = -\frac{3}{7}$, θ in third quadrant.
18. $\cos \theta = -\frac{1}{3}$, θ in third quadrant.
19. $\cos \theta = -\frac{2}{3}$, θ in second quadrant.
20. $\tan \theta = -5$, θ in second quadrant.
21. $\sec \theta = 4$, θ in first quadrant.
22. $\tan \theta = -\frac{4}{3}$, θ in fourth quadrant.
23. Express $\cos 3x$ in terms of powers of $\cos x$.
24. Express $\cos 4x$ in terms of powers of $\cos x$.
25. Express $\sin 4x$ in terms of powers of $\sin x$ and $\cos x$.
26. Express $\cos \frac{x}{4}$ in terms of powers of $\cos x$.
27. Write $\sin^3 \theta \cos^3 \theta$ without any exponents.
$\left[\text{Hint: } \sin^3 \theta \cos^3 \theta = \sin \theta \sin^2 \theta \cos \theta \cos^2 \theta \text{ and } \sin^2 2\theta = \frac{1-\cos 4\theta}{2}\right]$
28. Write $\sin^2 \theta \cos^4 \theta$ without any exponents.
29. Show that $\cos 2x = \cos^2 x - \sin^2 x$. [Hint: Use the identity for the cosine of a sum.]
30. Show that $\cos 2x = 2\cos^2 x - 1$. [Hint: Use the result of Problem 29 and the fact that $\sin^2 x + \cos^2 x = 1$.]
31. Show that $\cos 2x = 1 - 2\sin^2 x$.

In Problems 32–51 prove each identity.

32. $\sec 2x = \frac{1}{2\cos^2 x - 1}$
33. $\csc 2x = \frac{\sec x \csc x}{2}$
* 34. $\tan x = \csc 2x - \cot 2x$
35. $\sin 2x = \frac{2\tan x}{\sec^2 x}$
36. $\cos 2x = \frac{\csc^2 x - 2}{\csc^2 x}$
37. $\tan x = \frac{\sin 2x}{1 + \cos 2x}$
38. $\frac{\sin 2x}{\sin x} - \frac{\cos 2x}{\cos x} = \sec x$
39. $\tan 2x = \frac{\sin 4x}{1 + \cos 4x}$
40. $\sec 2x = \frac{\sec^2 x}{2 - \sec^2 x}$
41. $\cot 2x = \frac{\cot^2 x - 1}{2\cot x}$

Answers to Readiness Check

I. a II. d III. d IV. b

42. $\sec \frac{x}{2} = \frac{\pm\sqrt{2(1 + \cos x)}}{1 + \cos x}$

43. $\csc \frac{x}{2} = \frac{\pm\sqrt{2(1 + \cos x)}}{\sin x}$

44. $\cot \frac{x}{2} = \frac{\sin x}{1 - \cos x}$

45. $\csc 2x = \frac{1}{2}(\tan x + \cot x)$

46. $\frac{\sin 4x}{\sin 2x} = 2 \cos 2x$

47. $\tan x = \frac{1}{2} (\sin 2x)(1 + \tan^2 x)$

48. $\frac{1 - \tan x}{1 + \tan x} = \frac{1 - \sin 2x}{\cos 2x}$

49. $\cot x = \frac{\cos^2 \frac{x}{2} - \sin^2 \frac{x}{2}}{\sin x}$

50. $\frac{\sin^2 x}{2 \cos^2 \frac{x}{2}} = 1 - \cos x$

51. $\sec x = \frac{1 + \tan^2 \frac{x}{2}}{1 - \tan^2 \frac{x}{2}}$

* 52. Early in the morning a tree casts a shadow of 20 ft. Later in the day, when the angle of inclination of the sun is twice as large, the shadow is 8 ft. How tall is the tree?

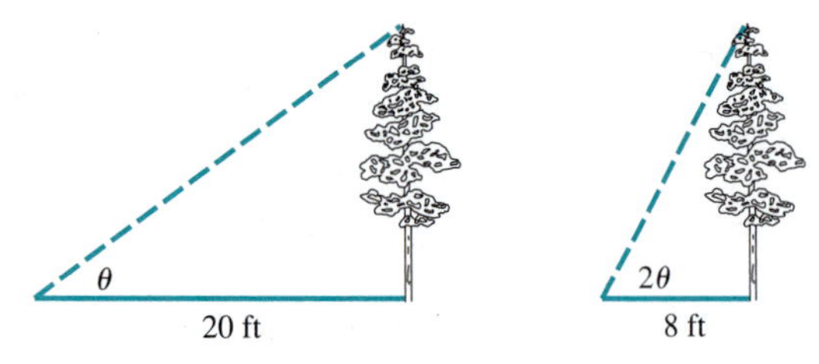

53. The area of a triangle is $\frac{1}{2}$ base × height. Find the area of the isosceles triangle in the figure below.

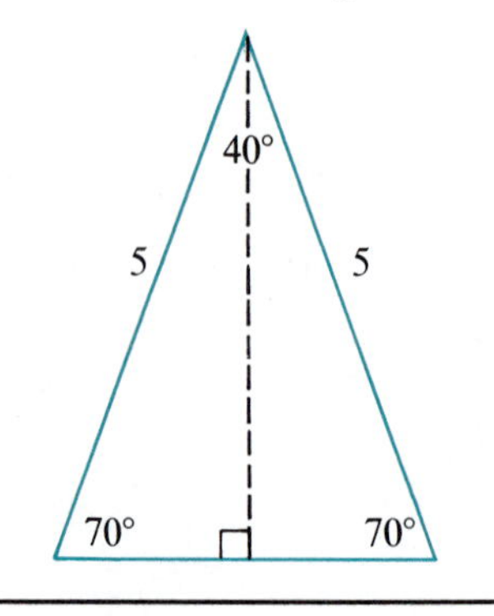

54. Find a formula for the area of an isosceles triangle with two sides equal to s when the angle between the two equal sides is denoted by θ. [Hint: See the figure below.]

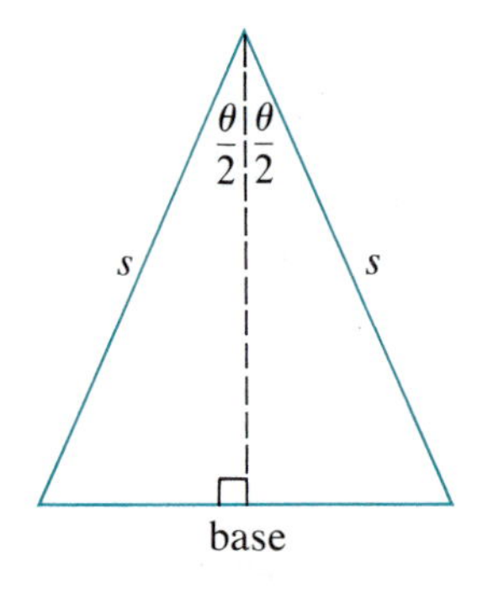

* 55. Show that $\tan \frac{x}{2} = \frac{1 - \cos x}{\sin x} = \frac{\sin x}{1 + \cos x}$. [Hint: Square both sides of (13) and multiply numerator and denominator by $1 - \cos x$.]

56. **A problem that arises in calculus.** Let $u = \tan \frac{x}{2}$. Show that

$$\sin x = \frac{2u}{1 + u^2} \quad \text{and} \quad \cos x = \frac{1 - u^2}{1 + u^2}$$

57. Oil is pumped through a large pipe into smaller ones, as in the figure below. If the angle between the two smaller pipes is ϕ, show that
distance along the pipes from A to C = $\overline{AB} + \overline{BC} = \frac{1}{2} h \tan \frac{\phi}{4} + l$

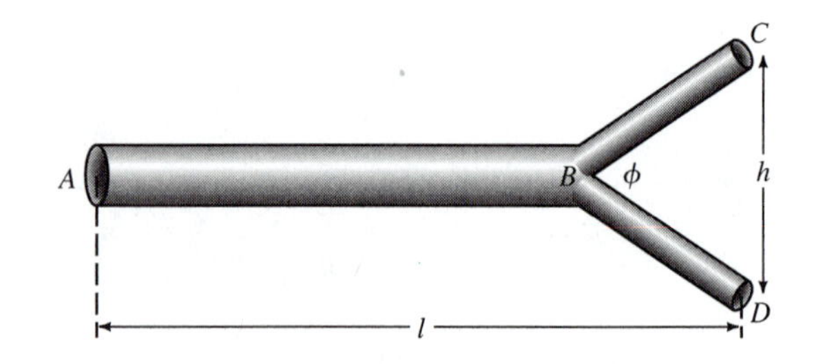

58. In Problem 57, compute $\overline{AB} + \overline{BC}$ if h = 100 ft, $l = 5000$ ft and $\phi = 30°$.

3.4 Other Trigonometric Identities

There are other identities that arise in applications. We give several of these in this section.

Product-to-Sum Identities

$$\sin \alpha \sin \beta = \frac{1}{2}[\cos (\alpha - \beta) - \cos (\alpha + \beta)] \quad (1)$$

$$\sin \alpha \cos \beta = \frac{1}{2}[\sin (\alpha + \beta) + \sin (\alpha - \beta)] \quad (2)$$

$$\cos \alpha \sin \beta = \frac{1}{2}[\sin (\alpha + \beta) - \sin (\alpha - \beta)] \quad (3)$$

$$\cos \alpha \cos \beta = \frac{1}{2}[\cos (\alpha + \beta) + \cos (\alpha - \beta)] \quad (4)$$

Each identity can be verified by using the identities for $\sin (\alpha + \beta)$, $\sin (\alpha - \beta)$, $\cos (\alpha + \beta)$, and $\cos (\alpha - \beta)$. We prove (1) here and leave the other proofs as exercises (see Problems 37–39).

Proof of (1)

$$\cos (\alpha - \beta) = \cos \alpha \cos \beta + \sin \alpha \sin \beta$$
$$\cos (\alpha + \beta) = \cos \alpha \cos \beta - \sin \alpha \sin \beta$$

$$\cos (\alpha - \beta) - \cos (\alpha + \beta) = 2 \sin \alpha \sin \beta \qquad \text{Subtract}$$

so

$$\frac{1}{2}[\cos (\alpha - \beta) - \cos (\alpha + \beta)] = \sin \alpha \sin \beta \quad \blacksquare$$

EXAMPLE 1 ***Writing a Product as a Sum***

Express $\sin 40° \cos 60°$ as a sum.

SOLUTION Using (2), we have

$$\sin 40° \cos 60° = \frac{1}{2}[\sin (40° + 60°) + \sin (40° - 60°)]$$

$$= \frac{1}{2}[\sin 100° + \sin (-20°)] = \frac{1}{2}[\sin 100° - \sin 20°] \qquad \sin(-\theta) = -\sin \theta$$

EXAMPLE 2 *Writing a Product as a Sum*

Express $\cos 2\theta \cos 3\theta$ as a sum.

SOLUTION From (4), we have

$$\cos 2\theta \cos 3\theta = \frac{1}{2}[\cos(2\theta + 3\theta) + \cos(2\theta - 3\theta)]$$

$$= \frac{1}{2}[\cos 5\theta + \cos(-\theta)] \underset{\cos(-\theta) = \cos\theta}{=} \frac{1}{2}[\cos 5\theta + \cos\theta]$$

Sum-to-Product Identities

$$\sin\alpha + \sin\beta = 2\sin\frac{\alpha+\beta}{2}\cos\frac{\alpha-\beta}{2} \qquad (5)$$

$$\sin\alpha - \sin\beta = 2\cos\frac{\alpha+\beta}{2}\sin\frac{\alpha-\beta}{2} \qquad (6)$$

$$\cos\alpha + \cos\beta = 2\cos\frac{\alpha+\beta}{2}\cos\frac{\alpha-\beta}{2} \qquad (7)$$

$$\cos\alpha - \cos\beta = 2\sin\frac{\alpha+\beta}{2}\sin\frac{\beta-\alpha}{2} \qquad (8)$$

NOTE In (8), the last factor is $\sin\frac{\beta-\alpha}{2}$ rather than $\sin\frac{\alpha-\beta}{2}$.

We prove (5) and leave the others as exercises (see Problems 40–42).

Proof of (5)

We rewrite (2) by multiplying both sides by 2 and using x and y instead of α and β:

$$\sin(x + y) + \sin(x - y) = 2\sin x \cos y \qquad (9)$$

In (9), set $\alpha = x + y$ and $\beta = x - y$. Then

$$\alpha + \beta = (x + y) + (x - y) = 2x \quad \text{and} \quad x = \frac{\alpha+\beta}{2}$$

$$\alpha - \beta = (x + y) - (x - y) = 2y \quad \text{so} \quad y = \frac{\alpha-\beta}{2}$$

Thus, inserting $x + y = \alpha$, $x - y = \beta$, $x = \frac{\alpha+\beta}{2}$, and $y = \frac{\alpha-\beta}{2}$ into (9) yields

$$\sin\alpha + \sin\beta = 2\sin\frac{\alpha+\beta}{2}\cos\frac{\alpha-\beta}{2} \quad ■$$

EXAMPLE 3 *Writing a Sum as a Product*

Express cos 25° + cos 40° as a product.

SOLUTION

From (7)

$$\cos 25^\circ + \cos 40^\circ = 2\cos\frac{25^\circ + 40^\circ}{2}\cos\frac{25^\circ - 40^\circ}{2}$$
$$= 2\cos\frac{65^\circ}{2}\cos\left(-\frac{15^\circ}{2}\right)$$
$$= 2\cos 32.5^\circ \cos(-7.5^\circ)$$

$\cos(-\theta) = \cos\theta$

$$= 2\cos 32.5^\circ \cos 7.5^\circ \quad \blacksquare$$

EXAMPLE 4 *Writing a Difference as a Product*

Express sin $3x$ − sin $4x$ as a product.

SOLUTION

From (6)

$$\sin 3x - \sin 4x = 2\cos\frac{3x + 4x}{2}\sin\frac{3x - 4x}{2}$$

$\sin(-x) = -\sin x$

$$= 2\cos\frac{7x}{2}\sin\left(-\frac{x}{2}\right) = -2\cos\frac{7x}{2}\sin\frac{x}{2} \quad \blacksquare$$

EXAMPLE 5 *Using the Sum-to-Product Identities to Prove a New Trigonometric Identity*

Prove that

$$\frac{\sin\alpha + \sin\beta}{\cos\alpha + \cos\beta} = \tan\frac{\alpha + \beta}{2}$$

SOLUTION

(5) and (7)

$$\frac{\sin\alpha + \sin\beta}{\cos\alpha + \cos\beta} = \frac{2\sin\dfrac{\alpha + \beta}{2}\cos\dfrac{\alpha - \beta}{2}}{2\cos\dfrac{\alpha + \beta}{2}\cos\dfrac{\alpha - \beta}{2}} = \tan\frac{\alpha + \beta}{2} \quad \blacksquare$$

EXAMPLE 6 *Writing cos θ + 2 sin θ as a Single Cosine Function*

Write $\cos\theta + 2\sin\theta$ as a single cosine function.

SOLUTION We wish to write $\cos\theta + 2\sin\theta = A\cos(\theta - \delta)$ for some positive constants A and δ. But

$$\begin{aligned} A\cos(\theta - \delta) &= A(\cos\theta\cos\delta + \sin\theta\sin\delta) \\ &= (A\cos\delta)\cos\theta + (A\sin\delta)\sin\theta \\ &= \cos\theta + 2\sin\theta \end{aligned}$$

Thus, equating coefficients of $\cos\theta$ and $\sin\theta$, we have

$$A\cos\delta = 1 \qquad A\sin\delta = 2$$

$$\cos\delta = \frac{1}{A} \qquad \sin\delta = \frac{2}{A}$$

and

$$1 = \cos^2\delta + \sin^2\delta = \frac{1}{A^2} + \frac{4}{A^2} = \frac{5}{A^2}$$

so

$$A^2 = 5 \quad \text{and} \quad A = \sqrt{5}$$

Then $\cos\delta = \dfrac{1}{\sqrt{5}}$, $\sin\delta = \dfrac{2}{\sqrt{5}}$, δ is in the first quadrant (since $\sin\delta > 0$ and $\cos\delta > 0$) and $\delta = \cos^{-1}\dfrac{1}{\sqrt{5}} \approx \cos^{-1} 0.4472 \approx 1.1$ ($\approx 63.4°$).

Thus

$$\cos\theta + 2\sin\theta \approx \sqrt{5}\cos(\theta - 1.1)$$

We now generalize the result of Example 6.

The following identities are very useful in graphing certain trigonometric functions.

Harmonic Identities

$$a\cos\omega t + b\sin\omega t = A\cos(\omega t - \delta) \qquad \textbf{(10)}$$

$$a\cos\omega t + b\sin\omega t = A\sin(\omega t + \phi) \qquad \textbf{(11)}$$

Here $A = \sqrt{a^2 + b^2}$, δ is the unique number such that $\cos\delta = \dfrac{a}{\sqrt{a^2 + b^2}} = \dfrac{a}{A}$ and $\sin\delta = \dfrac{b}{A}$, so $\tan\delta = \dfrac{b}{a}$. ϕ is the unique number such that $\cos\phi = \dfrac{b}{A}$ and $\sin\phi = \dfrac{a}{A}$, so $\tan\phi = \dfrac{a}{b}$.

NOTE In (10), A is called the **amplitude** of the function $A\cos(\omega t - \delta)$, $\dfrac{\omega}{2\pi}$ is called the **frequency** of the function, and $\dfrac{|\delta|}{\omega}$ is called the **phase shift.** We will say more about these notions in Sections 3.6 and 3.8.

We prove (10) and leave (11) as an exercise (see Problem 43).

Proof of (10)

We first note that

$$\left(\frac{a}{\sqrt{a^2+b^2}}\right)^2 + \left(\frac{b}{\sqrt{a^2+b^2}}\right)^2 = \frac{a^2}{a^2+b^2} + \frac{b^2}{a^2+b^2} = \frac{a^2+b^2}{a^2+b^2} = 1$$

so

$$\left(\frac{a}{\sqrt{a^2+b^2}}, \frac{b}{\sqrt{a^2+b^2}}\right)$$

is a point on the unit circle. If δ is chosen as in Figure 1, then

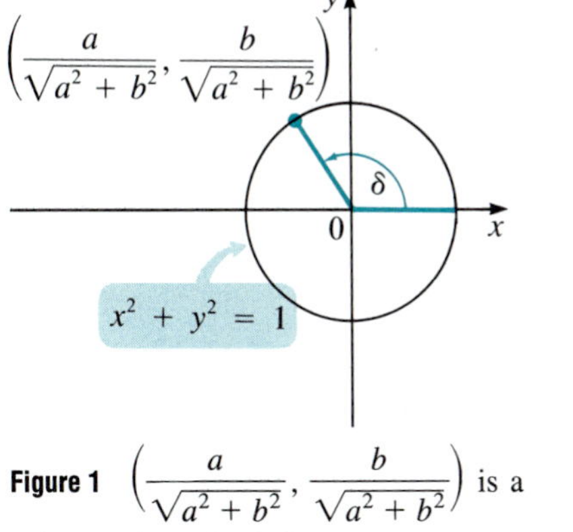

Figure 1 $\left(\frac{a}{\sqrt{a^2+b^2}}, \frac{b}{\sqrt{a^2+b^2}}\right)$ is a point on the unit circle.

$$\cos\delta = \frac{a}{\sqrt{a^2+b^2}} = \frac{a}{A} \quad \text{and} \quad \sin\delta = \frac{b}{\sqrt{a^2+b^2}} = \frac{b}{A}$$

Then

Equation (1) in Section 3.2
↓

$$\begin{aligned} A\cos(\omega t - \delta) &= A[\cos\omega t\cos\delta + \sin\omega t\sin\delta] \\ &= A\cos\delta\cos\omega t + A\sin\delta\sin\omega t \\ &= A\left(\frac{a}{A}\right)\cos\omega t + A\left(\frac{b}{A}\right)\sin\omega t \\ &= a\cos\omega t + b\sin\omega t \end{aligned}$$ ■

EXAMPLE 7 ***Using a Harmonic Identity to Write $3\cos 2t - 7\sin 2t$ as a Single Sine Function***

Write $3\cos 2t - 7\sin 2t$ as a sine function.

SOLUTION Here $a = 3$, $b = -7$, and $A = \sqrt{3^2 + (-7)^2} = \sqrt{58}$. Also, $(\cos\phi, \sin\phi) = \left(\frac{b}{A}, \frac{a}{A}\right) = \left(\frac{-7}{\sqrt{58}}, \frac{3}{\sqrt{58}}\right)$. Then ϕ is in the second quadrant (negative x-coordinate, positive y-coordinate), so from equation (18) on p. 128,

$$\phi = \cos^{-1}\left(\frac{-7}{\sqrt{58}}\right) \approx 2.74 \quad (\approx 157°)$$

and

$$3\cos 2t - 7\sin 2t \approx \sqrt{58}\sin(2t + 2.74)$$

EXAMPLE 8 ***Writing $-\cos \pi t - 4 \sin \pi t$ as a Single Cosine Function***

Write $-\cos \pi t - 4 \sin \pi t$ as a cosine function.

SOLUTION Here $a = -1$, $b = -4$, and $A = \sqrt{(-1)^2 + (-4)^2} = \sqrt{17}$. Then $(\cos \delta, \sin \delta) = \left(\frac{-1}{\sqrt{17}}, \frac{-4}{\sqrt{17}}\right)$, δ is in the third quadrant, and, from equation (11) on p. 125,

$$\delta = \pi + \sin^{-1} \frac{4}{\sqrt{17}} \approx 3.1416 + 1.3258$$
$$= 4.4674 \quad (\approx 256^\circ)$$

Thus,

$$-\cos \pi t - 4 \sin \pi t \approx \sqrt{17} \cos (\pi t - 4.4674)$$

Problems 3.4

Readiness Check

I. Which of the following is equal to $\cos 3 + \cos 1/3$?
a. $\cos 1$ b. $\cos \frac{10}{3}$
c. $2 \cos \frac{5}{3} \cos \frac{4}{3}$ d. undefined

II. Which of the following is $\sin (\theta + \pi) \sin (\theta - \pi)$?
a. $\frac{1}{2} (\sin \theta \cos \theta)$ b. $\frac{1}{2} (\sin \theta + \cos \pi)$
c. $\frac{1}{2} (1 - \cos 2\theta)$ d. $\frac{1}{2} (\sin 2\theta - 1)$

III. Which of the following is equal to $\cos y \sin (2x + y)$?
a. $\frac{1}{2} [\sin (2x + 2y) + \sin 2x]$
b. $\frac{1}{2} [\sin (2x + 2y) - \sin 2x]$
c. $\sin (x + y) + \sin x$
d. $\sin (x + y) - \sin x$

In Problems 1–14 write each product as a sum.

1. $\sin 30^\circ \cos 50^\circ$
2. $\cos 60^\circ \cos 200^\circ$
3. $\sin 54^\circ \cos 46^\circ$
4. $\sin \frac{\pi}{3} \cos \frac{\pi}{4}$
5. $\cos \frac{\pi}{4} \sin \frac{\pi}{7}$
6. $\sin 1 \cos 2$
7. $\cos \frac{2}{3} \cos \frac{3}{2}$
8. $\cos 3 \cos 4$
9. $\cos 3\alpha \sin 4\alpha$
10. $\frac{1}{3} \cos 3\beta \sin 5\beta$
11. $3 \sin (x + 1) \cos (x - 1)$
12. $2 \cos x^2 \sin 2x^2$
13. * $\cos t \cos 2t \cos 4t$ [Hint: Apply identity (4) twice.]
14. * $\sin x \cos 4x \sin 7x$

In Problems 15–24 write each sum as a product.

15. $\cos 15^\circ + \cos 35^\circ$
16. $\sin 40^\circ + \sin 50^\circ$
17. $\cos 3 + \cos 5$
18. $\cos \frac{1}{3} - \cos 2$
19. $\sin \frac{\pi}{8} - \sin \frac{\pi}{12}$
20. $\cos \frac{3\pi}{4} + \cos \frac{5\pi}{6}$
21. $\sin 5x - \sin 2x$
22. $\cos 3t + \cos 5t$
23. $\sin (-2\theta) - \sin (-5\theta)$
24. $\cos (-4\alpha) + \cos (-\alpha)$

In Problems 25–36 write each sum as a single cosine function and a single sine function. Use a value of δ or ϕ in the interval $(0, 2\pi)$.

25. $2 \cos t + 3 \sin t$
26. $-2 \cos t + 3 \sin t$

Answers to Readiness Check

I. c II. c III. a

27. $-2 \cos t - 3 \sin t$
28. $2 \cos t - 3 \sin t$
29. $4 \cos 2\theta + 3 \sin 2\theta$
30. $-3 \cos 2\theta - 4 \sin 2\theta$
31. $-5 \cos \frac{\alpha}{2} + 12 \sin \frac{\alpha}{2}$
32. $12 \cos \frac{\alpha}{2} - 5 \sin \frac{\alpha}{2}$
33. $\frac{1}{2} \cos 4\beta - \frac{1}{4} \sin 4\beta$
34. $3 \cos \pi x + 2 \sin \pi x$
35. $-7 \cos \frac{\pi}{2}x - 3 \sin \frac{\pi}{2}x$
36. $x_0 \cos \omega_0 t + \frac{v_0}{\omega_0} \sin \omega_0 t$
37. Prove identity (2). [Hint: Add the identities for $\sin(\alpha + \beta)$ and $\sin(\alpha - \beta)$.]
38. Prove identity (3). [Hint: Use (2) and the fact that $\sin(-\theta) = -\sin \theta$.]
39. Prove identity (4).
40. Prove identity (6). [Hint: Substitute $-\beta$ for β in (5).]
41. Prove identity (7). [Hint: Use identity (4).]
42. Prove identity (8). [Hint: Use identity (1) and the fact that $\sin(-\theta) = -\sin \theta$.]
43. Prove identity (11). [Hint: Use the sum identity for sines.]

In Problems 44–54 prove each identity.

44. $\frac{\sin 3x + \sin x}{2 \sin 2x} = \cos x$
45. $\frac{\cos 3x + \cos x}{2 \cos 2x} = \cos x$
46. $\frac{\sin 6x + \sin 10x}{\cos 6x + \cos 10x} = \tan 8x$
47. $\frac{\cos 10x - \cos 6x}{\sin 10x - \sin 6x} = -\tan 8x$
48. $\frac{\cos \alpha - \cos \beta}{\cos \alpha + \cos \beta} = -\tan\left(\frac{\alpha + \beta}{2}\right) \tan\left(\frac{\alpha - \beta}{2}\right)$
49. $\tan \frac{\alpha + \beta}{2} = \frac{\sin \alpha + \sin \beta}{\cos \alpha + \cos \beta}$
50. $\frac{\sin 3x - \sin x}{\sin 3x + \sin x} = \frac{1 - \tan^2 x}{2}$
51. $\frac{\cos 2\theta + \cos 4\theta}{\sin 2\theta + \sin 4\theta} = \cot 3\theta$
52. $\frac{\sin 7t - \sin 5t}{\sin 7t + \sin 5t} = \frac{\tan t}{\tan 6t}$
53. $\frac{\sin \alpha + \sin \beta}{\sin \alpha - \sin \beta} = \frac{\tan\left(\frac{\alpha + \beta}{2}\right)}{\tan\left(\frac{\alpha - \beta}{2}\right)}$
54. $\sin 2t + \sin 4t + \sin 6t = 4 \sin 3t \cos t \cos 2t$
55. Use identity (5) to show that $\sin \theta + \sin(\theta + \pi) = 0$.
56. Use identity (7) to show that $\cos \theta + \cos(\theta + \pi) = 0$.
57. * (a) Verify on a calculator that $\sin 10° \sin 50° \sin 70° = \frac{1}{8}$.
    (b) Prove that this is true using the identities of this section.

    [Hint: First show that $\sin 10° \sin 50° = \frac{1}{2} \cos 40° - \frac{1}{4}$. Then use identity (3). Explain why $\sin 110° - \sin 70° = 0$.]
58. ** (a) Verify on a calculator that

$$\sin 6° \sin 42° \sin 66° \sin 78° = \frac{1}{16}.\dagger$$

    (b) Show that this is true by using the identities of this section.
59. If middle C is struck on a keyboard with an amplitude of 2, the resulting sound wave has the equation $y = 2 \sin 528\pi t$ (see Example 3 on page 206). If a note one octave below middle C is struck with amplitude 2, the resulting sound wave has the equation $y = 2 \sin 264\pi t$. Suppose the two notes are struck together, each with amplitude 2. Express the equation of the resulting sound wave as a product. [Hint: The equation of the resulting sound wave is the sum of the equations of the sound waves that produced it].

† For an interesting extension of the ideas behind Problems 57 and 58, see Z. Usiskin, "Products of Sines," *The Two Year College Math. Journal* 10 (1979), pp. 334–340 and Steven Golovich, "Products of Sines and Cosines," *Mathematics Magazine* 60(2), 1987, pp. 105–113.

3.5 Trigonometric Equations

A **trigonometric equation** is an equation involving one or more of the six basic trigonometric functions. Unless otherwise requested, we will give all solutions in radians.

EXAMPLE 1 *Solving a Trigonometric Equation*

Solve the equation $\sin x = \frac{1}{2}$.

SOLUTION We know that $\sin \frac{\pi}{6} = \frac{1}{2}$, so one solution is $x = \frac{\pi}{6}$. However, as $\sin \frac{\pi}{6} = \sin\left(\frac{\pi}{6} \pm 2\pi\right) = \sin\left(\frac{\pi}{6} \pm 4\pi\right)$, etc, there are an infinite number of solutions. $\frac{\pi}{6}$ is the reference angle for $\frac{5\pi}{6}$, and we have

$$\sin \frac{\pi}{6} = \sin\left(\pi - \frac{\pi}{6}\right) = \sin \frac{5\pi}{6} = \frac{1}{2}$$

So $\frac{5\pi}{6}$ is another solution. In addition,

$$\sin\left(\frac{\pi}{6} + 2n\pi\right) = \sin\left(\frac{5\pi}{6} + 2n\pi\right) = \frac{1}{2} \qquad \text{for } n = 0, \pm 1, \pm 2, \ldots$$

So all the solutions are

$$x = \frac{\pi}{6} + 2n\pi \quad \text{and} \quad x = \frac{5\pi}{6} + 2n\pi, \qquad n = 0, \pm 1, \pm 2, \ldots$$

We write out the first few:

$$\frac{\pi}{6}, \frac{5\pi}{6}, -\frac{11\pi}{6}, -\frac{7\pi}{6}, \frac{13\pi}{6}, \frac{17\pi}{6}, -\frac{23\pi}{6}, -\frac{19\pi}{6}, \ldots$$

NOTE There is an important difference between a trigonometric identity and a trigonometric equation. The identity is true for *all* x (or θ) for which all given functions are defined. The equation is true only for some values of x. Thus, $\sin^2 x + \cos^2 x = 1$ is a trigonometric identity, whereas $\sin x = \frac{1}{2}$ and $\sin \theta = \cos \theta$ are trigonometric equations.

EXAMPLE 2 *Solving a Trigonometric Equation*

Find all solutions to the equation $\sin \theta = \cos \theta$.

SOLUTION If $\sin\theta = \cos\theta$, then $\cos\theta \neq 0$ (explain why). Thus we can divide by $\cos\theta$:

$$\sin\theta = \cos\theta$$
$$\frac{\sin\theta}{\cos\theta} = \frac{\cos\theta}{\cos\theta}$$
$$\tan\theta = 1$$

One solution is $\theta = \frac{\pi}{4}$ since $\tan\frac{\pi}{4} = 1$. But $\tan\theta$ is periodic of period π, so the following are solutions:

$$\theta = \frac{\pi}{4} + n\pi, \quad n = 0, \pm 1, \pm 2, \ldots$$
$$= \frac{\pi}{4}, \frac{5\pi}{4}, -\frac{3\pi}{4}, \frac{9\pi}{4}, -\frac{7\pi}{4}, \ldots$$

EXAMPLE 3 *Solving a Quadratic Trigonometric Equation*

Find all solutions to $4\sin^2\theta + 3\cos\theta - 2 = 0$ in $[0, 2\pi]$.

SOLUTION Since $\sin^2\theta = 1 - \cos^2\theta$, we can write everything in terms of $\cos\theta$:

$$4\overbrace{(1 - \cos^2\theta)}^{=\sin^2\theta} + 3\cos\theta - 2 = 0$$
$$4 - 4\cos^2\theta + 3\cos\theta - 2 = 0$$
$$4\cos^2\theta - 3\cos\theta - 2 = 0$$

Set $x = \cos\theta$:

$$4x^2 - 3x - 2 = 0$$
$$x = \frac{3 \pm \sqrt{(-3)^2 - 4(4)(-2)}}{8} = \frac{3 \pm \sqrt{41}}{8} \quad \text{Use the quadratic formula}$$
$$x = \frac{3 + \sqrt{41}}{8} \approx 1.1754 \quad \text{and} \quad x = \frac{3 - \sqrt{41}}{8} \approx -0.4254$$

Thus, $\cos\theta = 1.1754$ or $\cos\theta = -0.4254$. But $|\cos\theta| \leq 1$, so the first possibility is ruled out. We therefore have

$$\cos\theta \approx -0.4254$$

and one solution is

$$\theta_1 \approx \cos^{-1}(-0.4254) \approx 2.0102$$

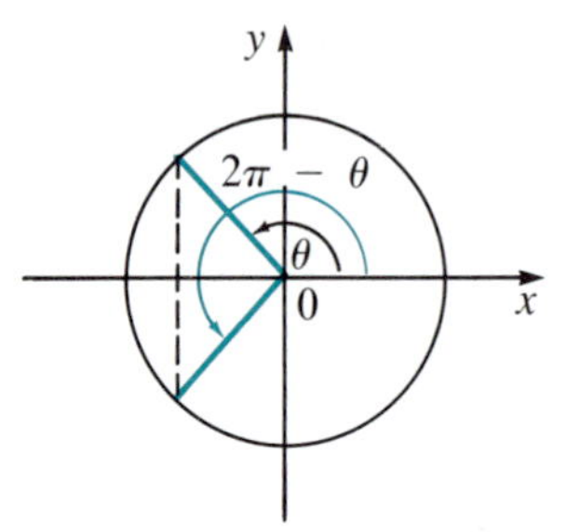

Figure 1 There are two values of θ for which $\cos\theta = -0.4254$.

This value is in the second quadrant. There is another solution in the third quadrant since $\cos\theta < 0$ in quadrants II and III (see Figure 1). Since $\cos(2\pi - \theta) = \cos\theta$, a second solution is

$$\theta_2 = 2\pi - \theta_1 \approx 6.2832 - 2.0102 = 4.2730$$

The two solutions are 2.0102 ($\approx 115°$) and 4.273 ($\approx 245°$).

Check We check the first answer:

$$\begin{aligned}4\sin^2 2.0102 + 3\cos 2.0102 - 2 &= 4(0.9050)^2 + 3(-0.4254) - 2\\ &= 4(0.8190) - 1.2762 - 2\\ &= 3.2760 - 1.2762 - 2 = -0.0002\end{aligned}$$

Our answer is essentially correct. We just have some roundoff error. ■

EXAMPLE 4 *A Trigonometric Equation with No Solution*

Show that there is no real number x such that

$$\sin x \cos x = \frac{3}{4}$$

SOLUTION We use the fact that $\sin 2x = 2\sin x\cos x$. If $\sin x\cos x = \frac{3}{4}$, then $\sin 2x = 2\sin x\cos x = 2\left(\frac{3}{4}\right) = \frac{3}{2} = 1.5 > 1$. This is impossible because $|\sin 2x| \le 1$. Thus, there is no real number x such that $\sin x\cos x = \frac{3}{4}$. ■

EXAMPLE 5 *Solving a Quadratic Trigonometric Equation*

Find all solutions to $\sec^2\theta + 2\tan\theta = 4$ in the interval $\left(-\frac{\pi}{2}, \frac{\pi}{2}\right)$.

SOLUTION We first convert everything to $\tan\theta$ by using the identity $\sec^2\theta = 1 + \tan^2\theta$:

$$\begin{aligned}\sec^2\theta + 2\tan\theta &= 4\\ 1 + \tan^2\theta + 2\tan\theta - 4 &= 0\\ \tan^2\theta + 2\tan\theta - 3 &= 0\\ x^2 + 2x - 3 &= 0 \qquad x = \tan\theta\\ (x+3)(x-1) &= 0\\ x = -3 \quad \text{and} &\quad x = 1\end{aligned}$$

Thus, in $\left(-\frac{\pi}{2}, \frac{\pi}{2}\right)$ the solutions are

$$\theta_1 = \tan^{-1}(-3) \approx -1.2490 \quad (\approx -72^\circ)$$

and

$$\theta_2 = \tan^{-1} 1 = \frac{\pi}{4} \quad (= 45^\circ)$$

Fermat's Principle and Snell's Law

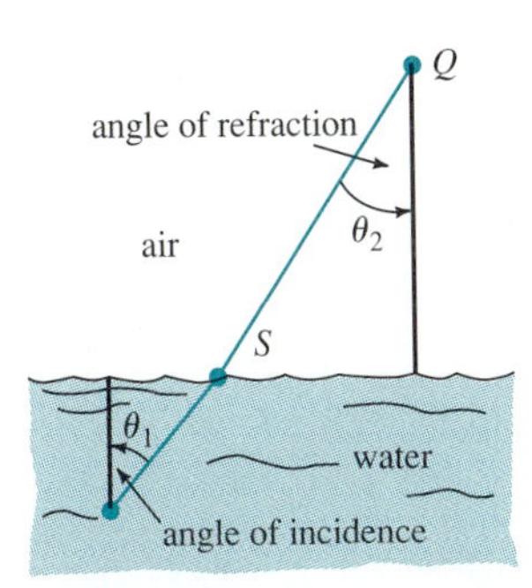

Figure 2 Light travels from P to S in water and from S to Q in air. The light refracts at S.

According to **Fermat's principle,** *when light travels through various media, it takes the path that requires the least time*. Suppose that light is traveling from a point P beneath the surface of water to a point Q in the air, above the surface (see Figure 2). In each medium (water and air), the light ray, according to Fermat's principle, will travel in a straight line. However, at the point S, where the light leaves the water, the light ray will bend or **refract** because the velocity of light in water is different from the velocity in air. The acute angle at P between the light ray and a vertical line is called the **angle of incidence** and is denoted in Figure 2 by θ_1. The acute angle at Q between the ray and a vertical line is called the **angle of refraction** and is denoted here by θ_2. Let v_1 and v_2 denote the velocity of light in the first medium (water) and the second medium (air), respectively. The following physical law can be derived, using calculus, from Fermat's principle.

Snell's Law†

$$\frac{\sin \theta_1}{\sin \theta_2} = \frac{v_1}{v_2}$$

That is,

$$\frac{\text{sine of angle of incidence}}{\text{sine of angle of refraction}} = \frac{\text{velocity of light in first medium}}{\text{velocity of light in second medium}}$$

The number $\eta_{12} = \frac{v_1}{v_2}$ is called the **index of refraction.**

EXAMPLE 6 *Computing the Angle of Refraction*

The velocity of light in air is approximately 300,000 km/sec.‡ The velocity of light in water is approximately 225,400 km/sec. Find the angle of refraction in Figure 2 if the angle of incidence is 35°. Round to the nearest minute.

† Willebrord Snell (1591–1626) was a Dutch physicist.

‡ More precisely, the speed of light in a vacuum is 299,792.4 km/sec. In air at 1 atmosphere of pressure and a temperature of 20°C, the speed is 299,702.5 km/sec. For our computations here and in the problems, the value of 300,000 km/sec or 186,000 miles/sec is an acceptable approximation.

SOLUTION Here $v_1 = 225{,}400$ and $v_2 = 300{,}000$, so the index of refraction is $\eta_{12} = \frac{v_1}{v_2} = \frac{225{,}400}{300{,}000} \approx 0.7513$. Thus

$$\frac{\sin 35°}{\sin \theta_2} = 0.7513$$

$$\sin \theta_2 \approx \frac{\sin 35°}{0.7513} \approx \frac{0.573576}{0.7513} \approx 0.76345$$

Thus,

$$\text{angle of refraction} = \theta_2 \approx \sin^{-1} 0.76345 \approx 49.769°$$

But $0.769° = 0.769 \times 60$ minutes $\approx 46'$. Thus

$$\text{angle of refraction} \approx 49°46'$$

Problems 3.5

Readiness Check

I. Which of the following has a solution?
a. $\csc x = -\frac{1}{2}$ b. $\sec x = \frac{2}{3}$
c. $4 \sin x = 5$ d. $5 \cos x = -4$

II. Which of the following has no solution?
a. $\sin \theta = -1$ b. $\sec \theta = 0$
c. $\cos \theta = 0$ d. $\csc \theta = 1$

III. Which of the following is one of the factors needed to solve $6 \cos^2 x - \cos x = 1$?
a. $(6 \cos x - 1)$ b. $(3 \cos x + 1)$
c. $(3 \cos x - 1)$ d. $(2 \cos x + 1)$

IV. Which of the following should be done first in solving $\sin 2x = -\sin x$ for x?
a. Replace $\sin 2x$ with $2 \sin x \cos x$.
b. Add $\sin 2x$ and $\sin x$ to get $\sin 3x$.
c. Factor out 2 on the left side of the equation.
d. Let $\sin 2x = 0$.

In Problems 1–18 find all solutions to each equation. Where the use of a calculator is indicated, round to 4 decimal places. Otherwise, do not use a calculator.

1. $\cos x = 0$
2. $\sin \theta = \frac{1}{2}$
3. $\cos \theta = -\frac{1}{2}$
4. $\tan 2\alpha = 1$
5. $\sqrt{3} \tan 4\alpha + 1 = 0$
6. $2 \sin 4t = \sqrt{2}$
7. $\sin x - \cos x = 0$
8. $\sec \beta = 2$
9. $\csc \omega = -2$
10. $\tan 5\theta = 0$
11. $\cot 3x = 0$
12. $\cos \alpha + \sqrt{2} \sin \alpha = 0$
13. $\sin \theta = \frac{2}{3}$
14. $\tan x = 6$
15. $\cos \alpha = 0.55$
16. $2 \sin \beta - 3 \cos \beta = 0$
17. $\sec \theta - \csc \theta = 0$
18. * $\tan x = \cot x$
19. Show that the equation $\tan x + \cot x = 0$ has no solution.
20. Show that the equation $\cos x + \sec x = 1$ has no solution.

For each equation in Problems 21–52 find all solutions in the interval $[0, 2\pi]$, if any. A calculator will be necessary in some (but not all) problems. When it is needed, round to 4 decimal places.

21. $4 \cos^2 \theta = 3$
22. $\cos^2 \gamma = 2$
23. $\sin^3 x + \sin x = 0$
24. $\cos^3 x - \cos x = 0$
25. $\sin^3 \beta + 2 \sin \beta = 0$
26. $4 \cos^4 2\theta - 1 = 0$

Answers to Readiness Check

I. d II. b III. b IV. a

27. $2 \cos 3x = 1$

28. $4 \sin^4 \left(\frac{\alpha}{2}\right) = 1$

29. $2 \sin^2 x - \cos x - 1 = 0$

30. $2 \sin^2 x - \sin x - 1 = 0$

31. $2 \cos^2 x + 3 \cos x + 1 = 0$

32. $\sin^2 x - \sqrt{2} \sin x + \frac{1}{2} = 0$

33. $2 \cos^2 \theta = \cos \theta$

34. $4 \cos^2 \beta - 4 \cos \beta + 1 = 0$

35. $6 \cos^2 \theta + 7 \cos \theta - 5 = 0$

36. $2 \sin^2 x - \cos x - 1 = 0$

37. $\sin^2 u - 5 \sin u + 6 = 0$

38. $\cos^2 \theta + \sin \theta = \frac{5}{4}$

39. $\sin^2 x - \frac{1}{2} \cos x = 0.76$

40. $2 \cos^2 \alpha + 3 \cos \alpha - 4 = 0$

41. $2 \cos^2 \theta + 3 \cos \theta + 4 = 0$

42. $\tan^2 \beta - 2 \tan \beta + 1 = 0$

43. $\tan^2 \theta + \sec \theta - 4 = 0$

44. $\tan^2 \theta + \sec \theta - 2 = 0$

* 45. $2 \sec^2 x - 5 \tan x - 3 = 0$

46. $\sin 2\theta = \cos \theta$

47. $\sin x = \sin \frac{x}{2}$

48. $3 \cos x = \cos \frac{x}{2}$

49. $\tan^2 \theta + \sec^2 \theta = 3$

50. $\cot^2 u + \csc^2 u = 3$

* 51. $4 \tan 2\beta + 5 \cot \beta = 0$

52. $\sin \theta(\sec^2 \theta - \csc^2 \theta) = 0$ [Hint: It is a mistake to begin by dividing both sides of the equation by $\sin \theta$.]

* 53. Find the largest positive number c such that the equation $\sin \theta \cos \theta = c$ has solutions.

54. The speed of light in ethyl alcohol is approximately 220,400 km/sec. A light ray leaves a point in air and strikes a point in a jar of ethyl alcohol with an angle of refraction of 38°. To the nearest minute, what is the angle of incidence?

55. In Problem 54 find the angle of refraction if the angle of incidence is 30°.

56. A light ray leaves a point in water with an angle of incidence of 27°. It strikes a point in dense flint glass at an angle of refraction of 21°20′. What is the velocity of light in dense flint glass?

57. According to **Poiseuille's† law,** the resistance R of a blood vessel of length l and radius r is given by $R = \alpha l/r^4$, where α is a constant of proportionality determined by the viscosity of blood. Blood is transported in the body by means of arteries, veins, arterioles, and capillaries. This procedure is ideally carried out in such a way as to minimize the energy required to transport the blood from the heart to the organs and back again. With calculus,‡ it can be shown that if blood moves from one vessel of radius r_1 to a vessel of radius r_2, with $r_1 > r_2$, then the branching angle θ that minimizes the resistance of the total flow satisfies the equation

$$\cos \theta = \left(\frac{r_2}{r_1}\right)^4$$

Note that θ does not depend on the length of either blood vessel. If $r_1 = 2r_2$, compute the optimal branching angle θ.

58. In Problem 57 compute the ratio $\frac{r_2}{r_1}$ if the optimal branching angle θ is 57°.

* 59. In a simple electric circuit, an EMF of $10 \sin \frac{\pi}{4} t$ volts is connected to a resistor [see Example 1 on page 204]. For what percentage of the time is the voltage equal to at least 7 volts?

* 60. If the EMF of Problem 59 is $35 \sin \frac{\pi}{10} t$ volts, for what percentage of the time is the voltage no greater than 15 volts?

* 61. A bacteria population varies according to the equation $P(t) = 10{,}000 + 2000 \cos \frac{\pi}{3} t$ where t is measured in hours.
   (a) What is the population when $t = 0$?
   (b) For what percentage of the time is the population between 8500 and 11,000 bacteria?

†Jean Louis Poiseuille (1799–1869) was a French physiologist.

‡See S. Grossman, *Calculus*, 5th ed., Saunders College Publishing, Philadelphia, 1993, p. 221.

62. The graph of the function $f(x) = e^{-x/3} \cos 2x$ is given in the figure below, for $x \geq 0$.
The points at which the curve turns are solutions to the equation $-2 \sin 2x - \frac{1}{3} \cos 2x = 0$. [This can be proved using calculus.] Find all solutions to this equation for $0 \leq x \leq 0$ and determine the value of $f(x)$ at each of these solutions. Give each answer to four decimal place accuracy.

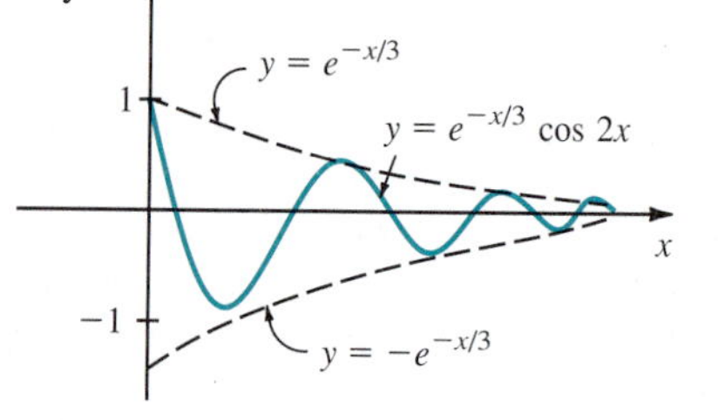

Graphing Calculator Problems

Find all solutions in $[0, 2\pi]$ to each equation in Problems 63–70 on a graphing calculator. Give each answer to two decimal place accuracy. Before starting, read Appendix A.

63. $2 \sin^2 x - 5 \sin x - 2 = 0$
64. $\sin^3 x + \sin^2 x + \sin x - 1 = 0$
65. $\sin x + \cos 3x - \tan 2x = 0$
66. $\cos^4 x - \tan x + 3 = x$
67. $\sin x = \dfrac{\cos 2x}{3 - \sin 3x}$
68. $\sec x + 5 = \tan^2 x$

3.6 Graphs Involving Sine and Cosine Functions I: Graphing $A \cos(\omega t - \delta)$ and $A \sin(\omega t - \delta)$

Functions that are combinations of sine and cosine functions come up frequently in applications. We will see several examples of this when we discuss harmonic motion in Section 3.8. In this section, we discuss graphs involving the sine and cosine functions.

In graphing a sine or cosine function, three properties are needed: amplitude, period, and phase shift. We deal with them one at a time. However, we first remind you of the graphs of $y = \sin t$ and $y = \cos t$, which are given in Figure 1.

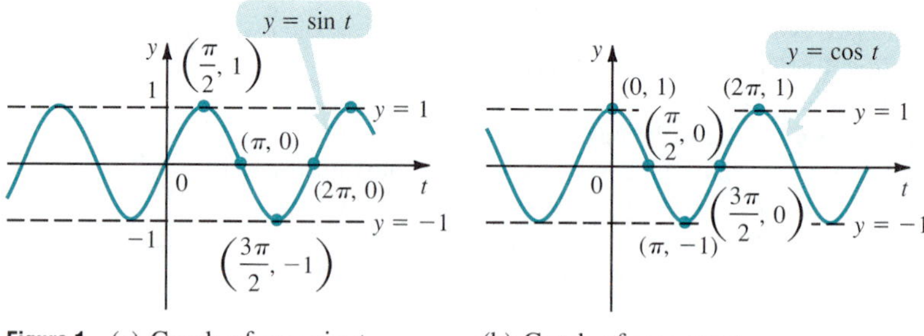

Figure 1 (a) Graph of $y = \sin t$. (b) Graph of $y = \cos t$.

I. Amplitude

EXAMPLE 1 ***The Graph of a Sine Function with an Amplitude of 3***

Sketch the graph of $y = 3 \sin t$.

SOLUTION Since $-1 \le \sin t \le 1$, we have $-3 \le 3 \sin t \le 3$. Now, $3 \sin t$ takes values at each t that are 3 times as large as the values taken by $\sin t$. Otherwise the graphs are the same. The graphs of $\sin t$ and $3 \sin t$ are given in Figure 2.

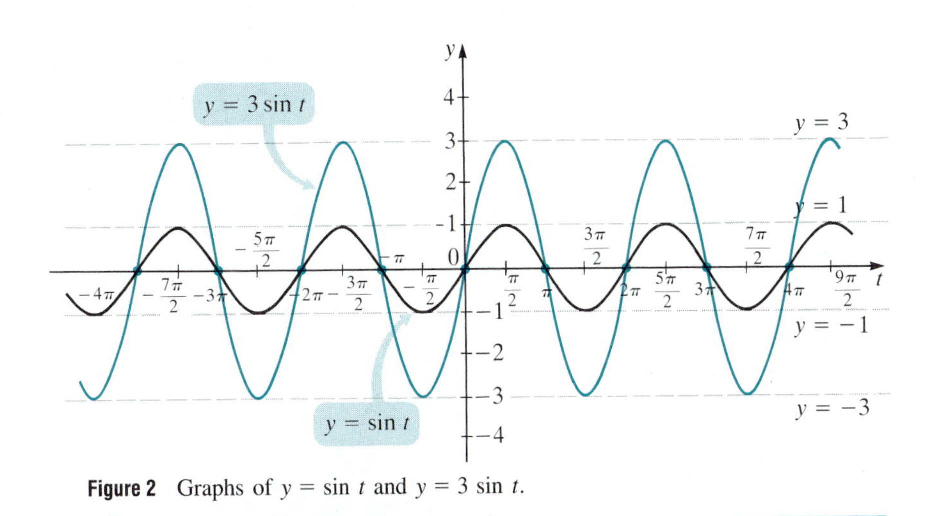

Figure 2 Graphs of $y = \sin t$ and $y = 3 \sin t$.

For the function $y = A \sin t$ or $y = A \cos t$, the number $|A| > 0$ is called the **amplitude** of the function. We have

Amplitude

$$-|A| \le A \sin t \le |A| \quad \text{and} \quad -|A| \le A \cos t \le |A| \tag{1}$$

EXAMPLE 2 ***The Graph of a Cosine Function with an Amplitude of $\frac{1}{2}$***

Sketch the graph of $y = -\dfrac{1}{2} \cos t$.

SOLUTION The graph of $\dfrac{1}{2} \cos t$ is just like the graph of $\cos t$ except that each value of y is half as large. That is, the amplitude is $\dfrac{1}{2}$. To obtain the graph of $-\dfrac{1}{2} \cos t$, we reflect the graph of $\dfrac{1}{2} \cos t$ about the t-axis (see

Section 1.4, p. 34). The graphs of $\cos t$, $\frac{1}{2}\cos t$, and $-\frac{1}{2}\cos t$ are given in Figure 3.

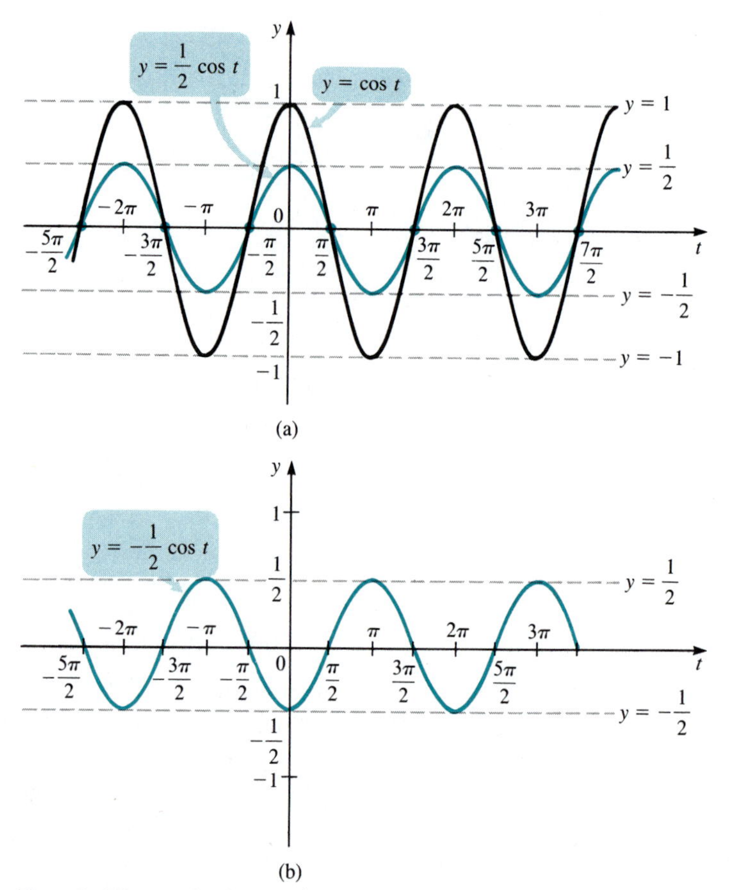

Figure 3 The graph of $y = -\frac{1}{2}\cos t$ is the graph of $y = \frac{1}{2}\cos t$ reflected about the t-axis.

II. Period

EXAMPLE 3 *The Graph of a Sine Function with a Period of $\frac{\pi}{2}$*

Sketch the graph of $y = \sin 4t$.

SOLUTION We first note that $\sin 4\left(t + \frac{\pi}{2}\right) = \sin\left(4t + 4 \cdot \frac{\pi}{2}\right) = \sin(4t + 2\pi) = \sin 4t$. Thus, $\sin 4t$ is periodic of period $\frac{\pi}{2}$. Put another way, as t increases from 0 to $\frac{\pi}{2}$, $4t$ increases from 0 to 2π. Thus, since $\sin t$

is periodic of period 2π, $\sin 4t$ is periodic of period $\frac{\pi}{2}$. In Table 1, we give some values of $\sin 4t$ for $0 \leq t \leq \frac{\pi}{2}$.

Table 1

t	0	$\frac{\pi}{24}$	$\frac{\pi}{16}$	$\frac{\pi}{12}$	$\frac{\pi}{8}$	$\frac{3\pi}{16}$	$\frac{\pi}{4}$	$\frac{5\pi}{16}$	$\frac{3\pi}{8}$	$\frac{7\pi}{16}$	$\frac{\pi}{2}$
$4t$	0	$\frac{\pi}{6}$	$\frac{\pi}{4}$	$\frac{\pi}{3}$	$\frac{\pi}{2}$	$\frac{3\pi}{4}$	π	$\frac{5\pi}{4}$	$\frac{3\pi}{2}$	$\frac{7\pi}{4}$	2π
$\sin 4t$	0	$\frac{1}{2}$	$\frac{1}{\sqrt{2}}$	$\frac{\sqrt{3}}{2}$	1	$\frac{1}{\sqrt{2}}$	0	$-\frac{1}{\sqrt{2}}$	-1	$-\frac{1}{\sqrt{2}}$	0

Sin $4t$ takes the same values as $\sin t$, but it takes them sooner. Put another way, $\sin 4t$ goes through one complete cycle every $\frac{\pi}{2}$ units. Loosely speaking, the scale is ''squeezed down.'' In Figure 4, we provide a graph of $y = \sin 4t$.

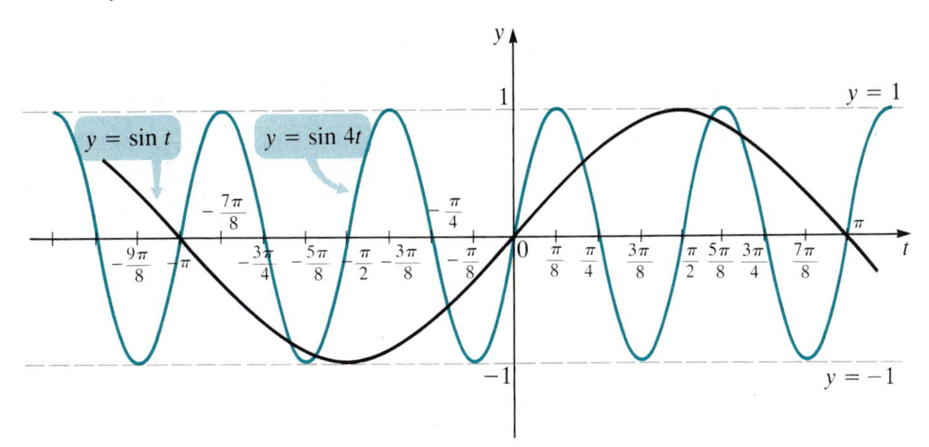

Figure 4 Graph of $\sin 4t$ (blue) and graph of $\sin t$ (black).

In Example 3, we found that the period of $\sin 4t$ is $\frac{\pi}{2} = \frac{2\pi}{4}$. We generalize this result.

Let ω (the Greek omega) be a positive real number. Then

Period of $A \sin \omega t$ and $A \cos \omega t$

$$A \sin \omega t \text{ and } A \cos \omega t \text{ are periodic of period } \frac{2\pi}{\omega}. \quad (2)$$

Proof

$$y = A \sin \omega t \overset{\text{sin } x \text{ is periodic of period } 2\pi}{=} A \sin(\omega t + 2\pi) = A \sin \omega\left(t + \frac{2\pi}{\omega}\right)$$

Thus $A \sin \omega t$ is periodic of period $\frac{2\pi}{\omega}$. A similar proof shows that $A \cos \omega t$ is periodic of period $\frac{2\pi}{\omega}$. ■

EXAMPLE 4 *Determining the Period of a Cosine Function*

What is the period of $y = \sqrt{23} \cos \frac{1}{3}t$?

SOLUTION $\omega = \frac{1}{3}$ so

$$\text{period} = \frac{2\pi}{\omega} = \frac{2\pi}{\left(\frac{1}{3}\right)} = 6\pi$$

Before continuing, we make some observations that will make it easier to graph $y = A \cos \omega t$ or $y = A \sin \omega t$. We refer again to the graphs of $y = \sin t$ and $y = \cos t$. Because each function is periodic of period 2π, we draw each graph for $0 \le t \le 2\pi$ (in Figure 5).

Since the period of $\sin t$ and $\cos t$ is 2π, we calculate that

$$\frac{1}{4}\text{period} = \frac{1}{4}(2\pi) = \frac{\pi}{2} \qquad \frac{1}{2}\text{period} = \pi$$

$$\frac{3}{4}\text{period} = \frac{3\pi}{2} \qquad 1 \text{ period} = 2\pi$$

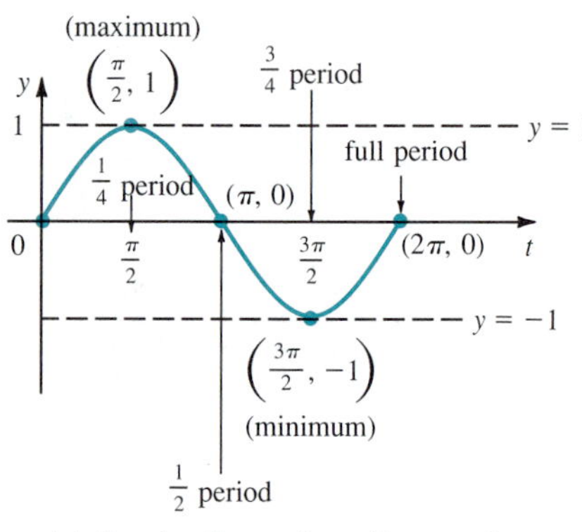

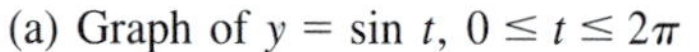

(a) Graph of $y = \sin t$, $0 \le t \le 2\pi$

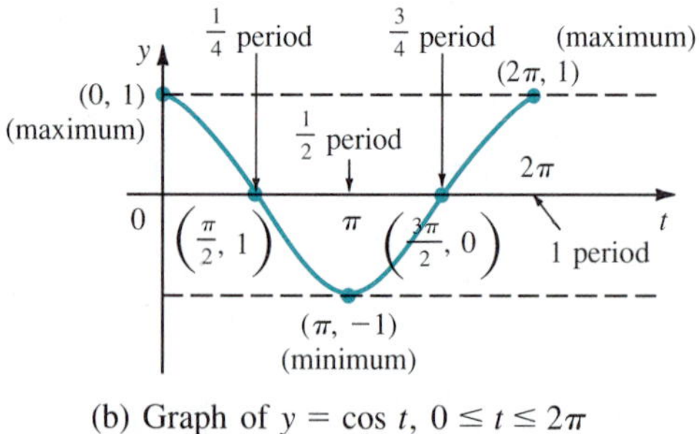

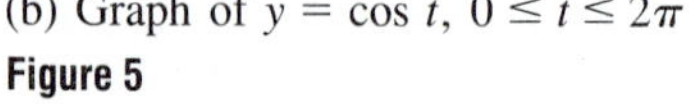

(b) Graph of $y = \cos t$, $0 \le t \le 2\pi$

Figure 5

From Figure 5, we see that

$\sin x$ is zero at 0 and at $\frac{1}{2}$ period.
$\sin x$ takes the maximum value of 1 at $\frac{1}{4}$ period.
$\sin x$ takes the minimum value of -1 at $\frac{3}{4}$ period.

Similarly,

$\cos x$ is zero at $\frac{1}{4}$ period and $\frac{3}{4}$ period.
$\cos x$ takes the maximum value of 1 at 0.
$\cos x$ takes the minimum value of -1 at $\frac{1}{2}$ period.
We can generalize these results.

Zeros, Maximum, and Minimum Values of $A \sin \omega t$ and $A \cos \omega t$

Suppose that $A > 0$ and $\omega > 0$. Then $A \sin \omega t$ and $A \cos \omega t$ are periodic of period $\dfrac{2\pi}{\omega}$. The values

$$0 \qquad \frac{1}{4} \text{ period} = \frac{\pi}{2\omega} \qquad \frac{1}{2} \text{ period} = \frac{\pi}{\omega} \qquad \frac{3}{4} \text{ period} = \frac{3\pi}{2\omega}$$

are called the four **vital points** and we have

$A \sin \omega t$ is zero at 0 and $\dfrac{\pi}{\omega}$.

$A \sin \omega t$ takes its maximum value of A at $\dfrac{\pi}{2\omega}$.

$A \sin \omega t$ takes its minimum value of $-A$ at $\dfrac{3\pi}{2\omega}$.

$A \cos \omega t$ is zero at $\dfrac{\pi}{2\omega}$ and $\dfrac{3\pi}{2\omega}$.

$A \cos \omega t$ takes its maximum value of A at 0.

$A \cos \omega t$ takes its minimum value of $-A$ at $\dfrac{\pi}{\omega}$.

EXAMPLE 5 ***Using the Vital Points to Help Sketch the Graph of a Cosine Function***

Sketch the graph of $y = \sqrt{23} \cos \frac{1}{3}t$.

SOLUTION We saw in Example 4 that the period is 6π. The amplitude is $\sqrt{23} \approx 4.8$. Then the four vital points are 0, $\dfrac{6\pi}{4} = \dfrac{3\pi}{2}$, $\dfrac{6\pi}{2} = 3\pi$, and $\dfrac{3\pi}{2(\frac{1}{3})} = \dfrac{9\pi}{2}$. We have

$\sqrt{23} \cos \dfrac{1}{3}t$ is zero at $\dfrac{3\pi}{2}$ and $\dfrac{9\pi}{2}$.

$\sqrt{23} \cos \dfrac{1}{3}t$ takes its maximum value of $\sqrt{23}$ at 0 and 6π.

$\sqrt{23} \cos \dfrac{1}{3}t$ takes its minimum value of $-\sqrt{23}$ at 3π.

Putting this all together, we obtain the graph in Figure 6(a), which we extend periodically to the graph in Figure 6(b).

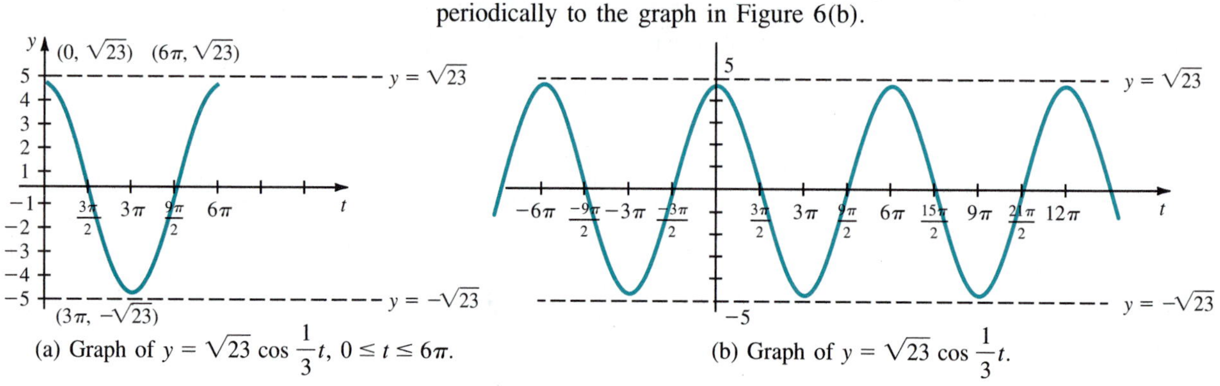

(a) Graph of $y = \sqrt{23}\cos\frac{1}{3}t,\ 0 \le t \le 6\pi$.

(b) Graph of $y = \sqrt{23}\cos\frac{1}{3}t$.

Figure 6

III. Phase Shift

EXAMPLE 6 ***The Graph of a Cosine Function with a Phase Shift of $\frac{1}{2}$***

Sketch the graph of $y = \cos\left(\pi t - \dfrac{\pi}{2}\right)$.

SOLUTION The function $y = \cos \pi t$ is periodic of period $\dfrac{2\pi}{\omega} = \dfrac{2\pi}{\pi} = 2$. From the results of Section 1.4 (p. 33), the graph of $\cos\left(\pi t - \dfrac{\pi}{2}\right) = \cos \pi\left(t - \dfrac{1}{2}\right)$ is the graph of $\cos \pi t$ shifted $\dfrac{1}{2}$ unit to the right. In Figure 7, we give the graph of $\cos \pi t$ (black) and the graph of $\cos \pi\left(t - \dfrac{1}{2}\right)$ (in blue).

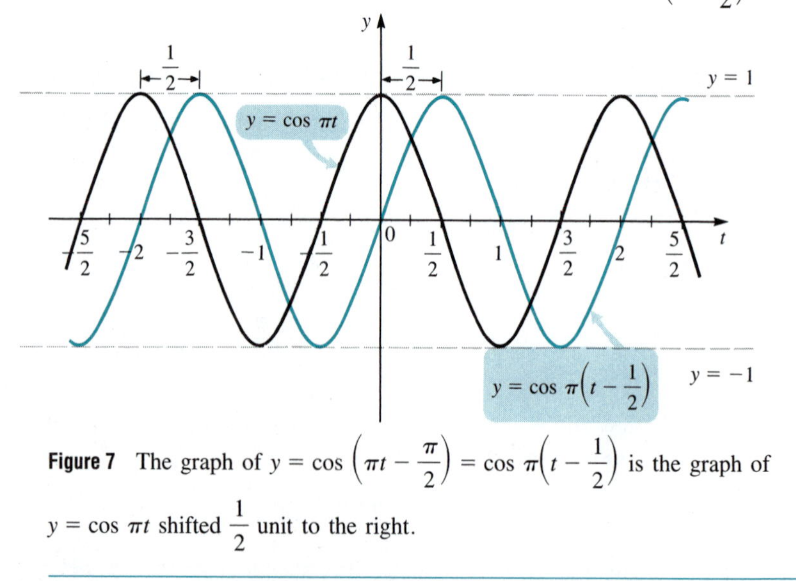

Figure 7 The graph of $y = \cos\left(\pi t - \dfrac{\pi}{2}\right) = \cos \pi\left(t - \dfrac{1}{2}\right)$ is the graph of $y = \cos \pi t$ shifted $\dfrac{1}{2}$ unit to the right.

In the graphs of $A \cos(\omega t - \delta)$ and $A \sin(\omega t - \delta)$ for $\omega > 0$, the number $\dfrac{|\delta|}{\omega}$ is called a **phase shift.** The graph of $A \cos(\omega t - \delta) = A \cos \omega\left(t - \dfrac{\delta}{\omega}\right)$ is the graph of $A \cos \omega t$ shifted $\dfrac{\delta}{\omega}$ units to the right if $\delta > 0$ and $\dfrac{|\delta|}{\omega}$ units to the left if $\delta < 0$.

NOTE

$$\cos\left(t - \frac{\pi}{2}\right) = \cos t \cos \frac{\pi}{2} + \sin t \sin \frac{\pi}{2} = \sin t$$

Thus, as we saw in Section 2.4,

The graph of $\sin t$ is the graph of $\cos t$ shifted $\dfrac{\pi}{2}$ units to the right.

We summarize the results of this section:

To Sketch the Function $y = A \cos(\omega t - \delta)$, $\omega > 0$

Step 1 Sketch the function $y = \cos \omega t$ with **period** $\dfrac{2\pi}{\omega}$. To do so, use the values at the four vital points $0, \dfrac{\pi}{2\omega}, \dfrac{\pi}{\omega}$, and $\dfrac{3\pi}{2\omega}$.

Step 2 Sketch the function $y = |A| \cos \omega t$, where $|A|$ is the **amplitude.** This graph has the same vital points as $\cos \omega t$. Simply change the maximum ($|A|$) and minimum ($-|A|$) values.

Step 3 Sketch the function

$$y = |A| \cos(\omega t - \delta) = |A| \cos\left(\omega\left(t - \frac{\delta}{\omega}\right)\right)$$

with **phase shift** $\dfrac{|\delta|}{\omega}$. To do so,

shift the graph of $|A| \cos \omega t$ $\dfrac{\delta}{\omega}$ units to the right if $\delta > 0$

or

shift the graph of $|A| \cos \omega t$ $\dfrac{|\delta|}{\omega}$ units to the left if $\delta < 0$

Step 4 If $A > 0$, you are done. If $A < 0$, then $A = -|A|$, and you can sketch the graph of $A \cos(\omega t - \delta) = -|A| \cos(\omega t - \delta)$ by reflecting the graph of $|A| \cos(\omega t - \delta)$ about the t-axis.

To sketch the function $y = A \sin(\omega t - \delta)$, follow the same four steps, replacing cos by sin in each step.

Problems 3.6

Readiness Check

I. Which of the following is the amplitude of $y = -\frac{2}{3}\cos(5t + 4)$?

a. $-\dfrac{2}{3}$ b. $\dfrac{2\pi}{5}$ c. $\dfrac{4}{5}$ d. $\dfrac{2}{3}$

II. Which of the following is graphed below?

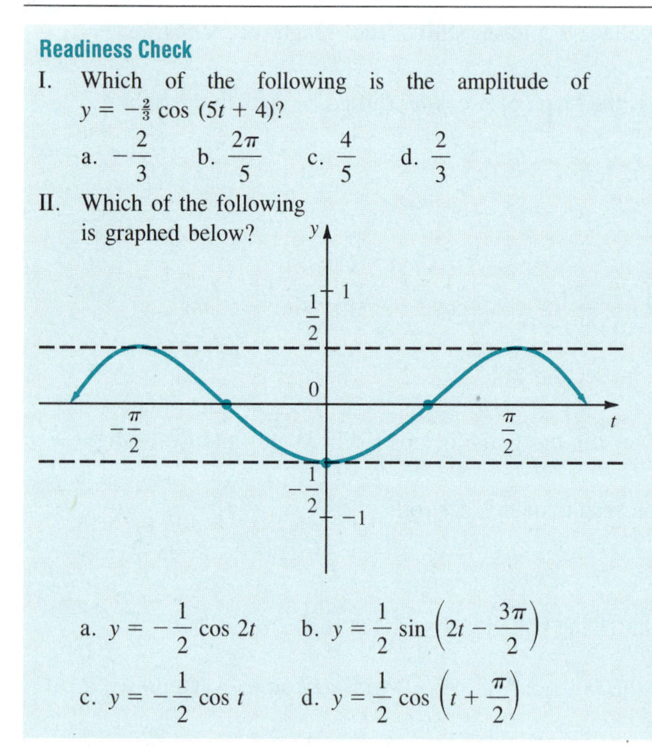

a. $y = -\dfrac{1}{2}\cos 2t$ b. $y = \dfrac{1}{2}\sin\left(2t - \dfrac{3\pi}{2}\right)$

c. $y = -\dfrac{1}{2}\cos t$ d. $y = \dfrac{1}{2}\cos\left(t + \dfrac{\pi}{2}\right)$

III. Which of the following has a period of $\pi/5$?

a. $y = 5\sin(10t + 2)$

b. $y = -5\cos(5t - 3)$

c. $y = -5\sin\left(\dfrac{t}{5} - \pi\right)$

d. $y = 5\cos\left(\dfrac{t}{10} + 2\pi\right)$

IV. Which of the following is the phase shift for the graph of $y = 2/3\cos(\pi t/4 + \pi/2)$?

a. $\dfrac{\pi}{2}$ b. $\dfrac{1}{2}$ c. 2 d. $\dfrac{1}{4}$

V. Which of the following is an x-value in $[1/3, 7/3]$ at which a maximum occurs on the graph of $y = 4\sin(\pi t - \pi/3)$?

a. $\dfrac{4}{3}$ b. $\dfrac{5}{6}$ c. $\dfrac{11}{6}$ d. $\dfrac{7}{3}$

In Problems 1–12 a function taking the form $y = A\cos\omega t$ or $y = A\sin\omega t$ is sketched. Determine each function and find its period.

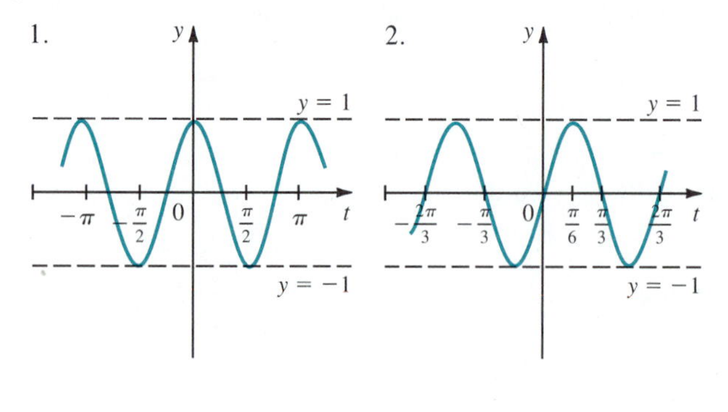

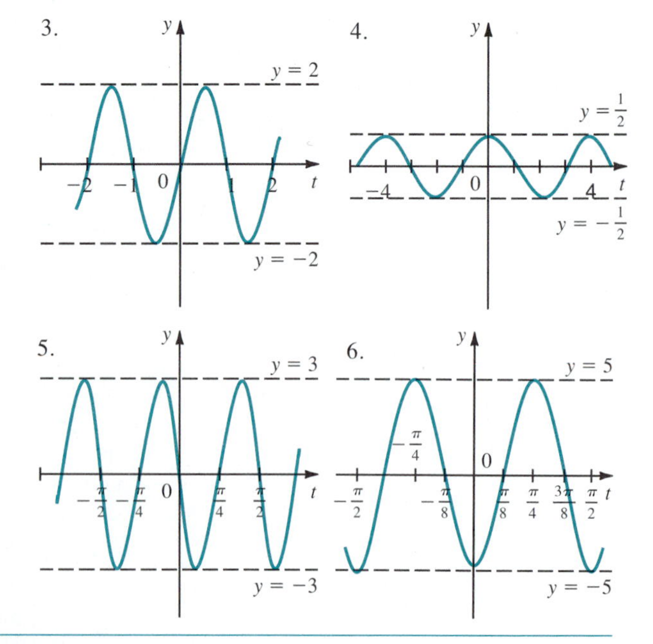

Answers to Readiness Check

I. d II. a III. a IV. c V. b

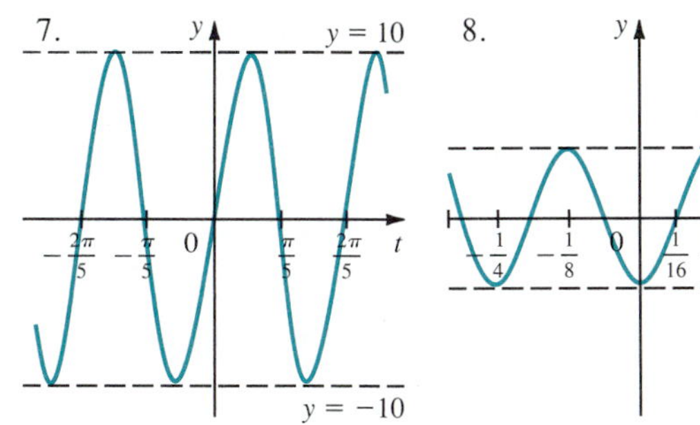

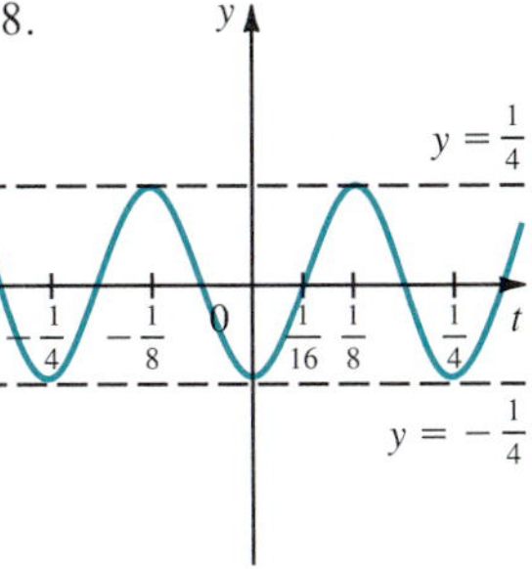

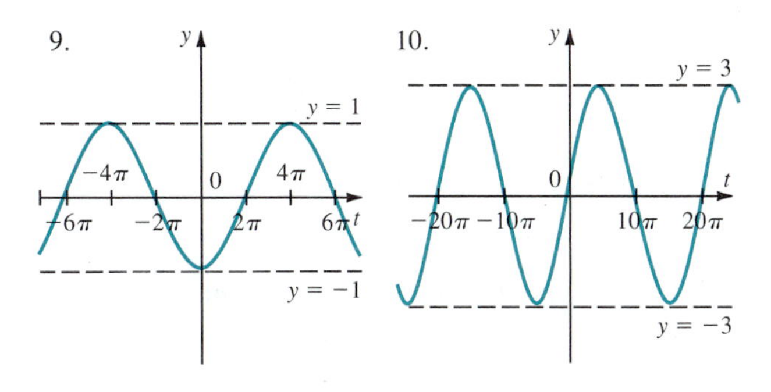

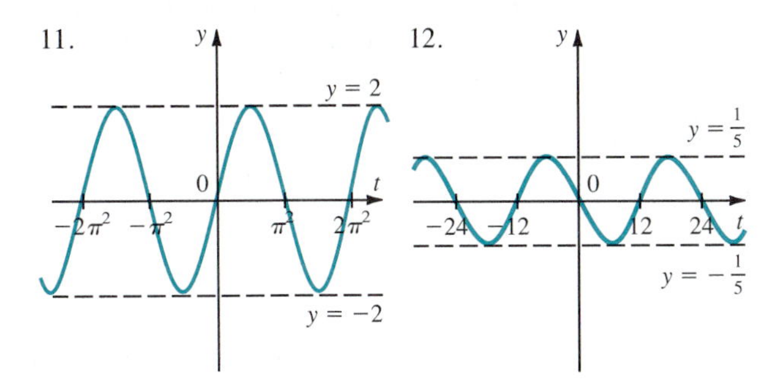

In Problems 13–26 find the amplitude, period, and phase shift of each function, and sketch the graph.

13. $2 \sin t$
14. $4 \cos t$
15. $-2 \sin t$
16. $-4 \cos t$
17. $2 \sin 2t$
18. $5 \cos 3t$
19. $\dfrac{1}{2} \sin \dfrac{1}{2}t$
20. $-\dfrac{1}{2} \sin 2t$
21. $-\sin \pi t$
22. $\sqrt{2} \cos \dfrac{\pi}{2}t$
23. $\cos(t - \pi)$
24. $\sin\left(2t + \dfrac{\pi}{2}\right)$
25. $\cos\left(\dfrac{t}{2} + \dfrac{\pi}{4}\right)$
26. $-\cos \dfrac{\pi}{2}(t - 1)$

In Problems 27–32 a function taking the form $y = A \sin(\omega t - \delta)$ is sketched. Determine A, ω, and δ.

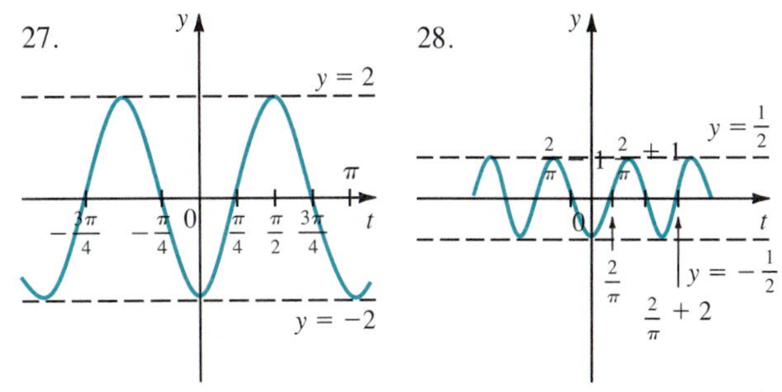

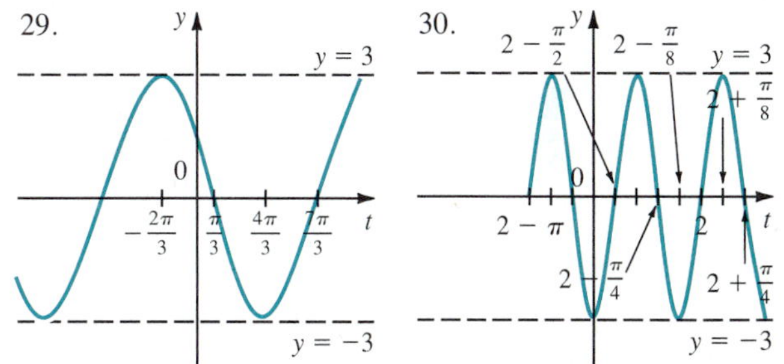

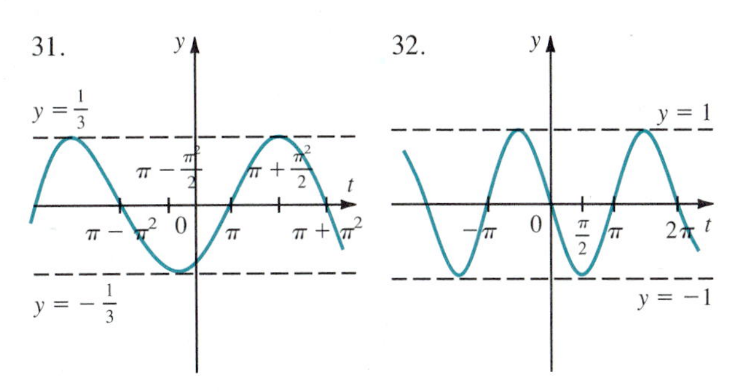

Graphing Calculator Problems

In Problems 33–42 obtain the graph of each function on a graphing calculator. Pay attention to the scales on the x- and y-axes. If you haven't already done so, read the material on graphing in Appendix A. In particular, see Example 5.

33. $y = 3 \cos 2.6x$

34. $y = -6.4 \sin \dfrac{x}{7}$

35. $y = 2.3 \cos\left(\dfrac{x}{4} - 5\right)$

36. $y = -1.8 \cos\left(2x + \dfrac{\pi}{5}\right)$

37. $y = 0.02 \sin\left(\frac{x}{137} + 0.15\pi\right)$
38. $y = 2385 \sin(12.6x - 1.27)$
39. $y = \frac{1}{50} \cos \pi^3 x$
40. $y = -\frac{27}{8} \sin\left(\frac{x}{\pi^5} - \pi^4\right)$
41. $y = 4 \sin(2x + 0.78)$
42. $y = 3.4 \cos\left(\frac{x}{5} - \frac{\pi}{20}\right)$

3.7 Graphs Involving Sine and Cosine Functions II: Sums of Functions and Vertical Shifts

In this section we discuss the graphs of some more complicated functions involving sines and cosines.

Vertical Shifts

In Section 1.4 (p. 32) we saw that the graph of $y = f(x) + c$ is obtained by shifting the graph of $f(x)$ c units up if $c > 0$ and $|c|$ units down if $c < 0$. We can apply that technique to graphs of trigonometric functions.

EXAMPLE 1 ***Obtaining a Sine Graph by Shifting Up***

Sketch the graph of $y = \sin 2t + 3$.

SOLUTION The graph of $\sin 2t + 3$ is the graph of $\sin 2t$ shifted 3 units up. The graphs of $\sin 2t$ and $\sin 2t + 3$ are given in Figure 1.

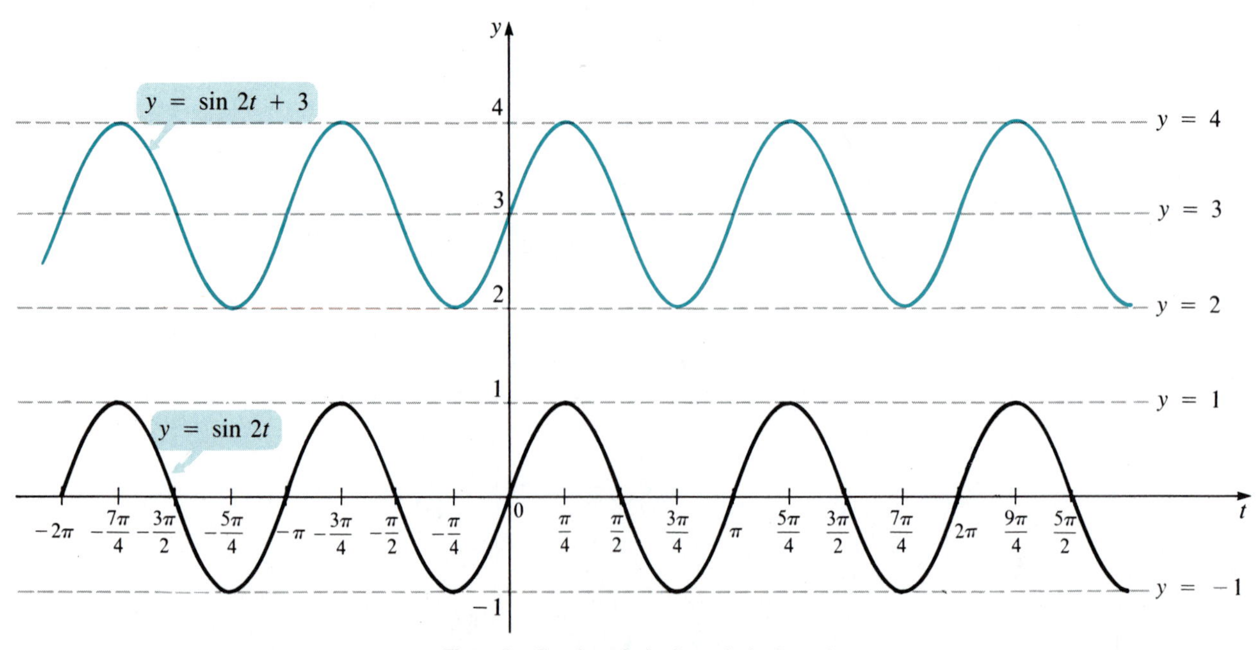

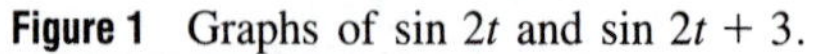
Figure 1 Graphs of $\sin 2t$ and $\sin 2t + 3$.

Graphing $a \cos \omega t + b \sin \omega t$: Using the Harmonic Identities

We repeat from the previous section facts about the functions $A \cos(\omega t - \delta)$ and $A \sin(\omega t - \delta)$.

Amplitude, Period, and Phase Shift

In the graphs of $y = A \cos(\omega t - \delta)$ and $y = A \sin(\omega t - \delta)$, with $\omega > 0$,

$|A|$ is the amplitude,

$\dfrac{2\pi}{\omega}$ is the period,

$\dfrac{|\delta|}{\omega}$ is the phase shift (to the right if $\delta > 0$ and to the left if $\delta < 0$).

We can use this information to obtain the graphs of more complicated functions. Recall the **harmonic identities** (equations (10) and (11) on p. 176).

$$a \cos \omega t + b \sin \omega t = A \cos(\omega t - \delta) = A \sin(\omega t + \phi) \qquad \textbf{(1)}$$

where $A = \sqrt{a^2 + b^2}$, $\cos \delta = \sin \phi = \dfrac{a}{A}$, and $\sin \delta = \cos \phi = \dfrac{b}{A}$.

EXAMPLE 2 ***Graphing a Function of the Form $a \cos \omega t + b \sin \omega t$***

Sketch the graph of $y = \cos t + 2 \sin t$.

SOLUTION In Example 6 in Section 3.4 (p. 175), we found that

$$\cos t + 2 \sin t \approx \sqrt{5} \cos(t - 1.1)$$

The graph of $\sqrt{5} \cos t$ has amplitude $\sqrt{5}$ and the graph of $\sqrt{5} \cos(t - 1.1)$ is the graph of $\sqrt{5} \cos t$ shifted 1.1 units to the right. This graph is given in Figure 2.

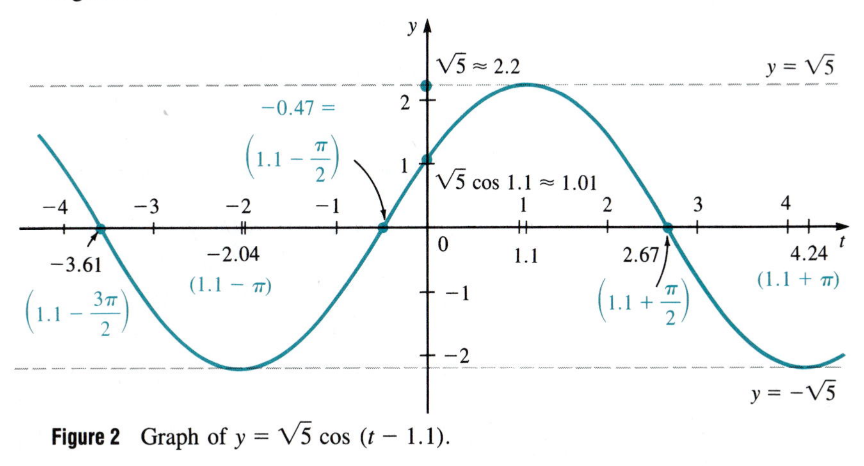

Figure 2 Graph of $y = \sqrt{5} \cos(t - 1.1)$.

EXAMPLE 3 *Graphing a Function of the Form $a \cos \omega t + b \sin \omega t$*

Sketch the graph of $y = 3 \cos 2t - 7 \sin 2t$.

SOLUTION In Example 7 in Section 3.4 (p. 177), we found that

$$3 \cos 2t - 7 \sin 2t \approx \sqrt{58} \sin (2t + 2.74)$$

Now

$$\sqrt{58} \sin 2t \text{ has amplitude } \sqrt{58} \approx 7.6 \text{ and period } \frac{2\pi}{2} = \pi$$

The graph of $\sqrt{58} \sin (2t + 2.74) = \sqrt{58} \sin 2(t + 1.37) = \sqrt{58} \sin 2(t - (-1.37))$ is the graph of $\sqrt{58} \sin 2t$ shifted 1.37 units to the left. It is sketched in Figure 3.

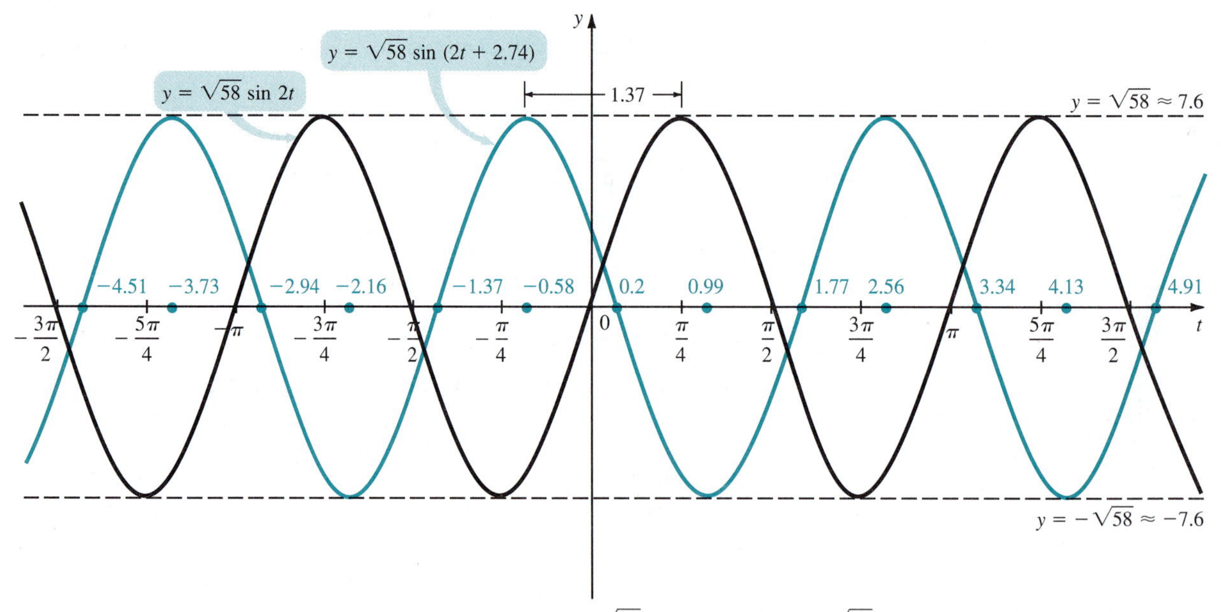

Figure 3 Graphs of $\sqrt{58} \sin 2t$ (black) and $\sqrt{58} \sin (2t + 2.74)$ (blue).

Graphing Sums of Functions by Adding y-Values

None of the techniques discussed so far will enable us easily to graph $a \sin \alpha x + b \cos \beta x$ if $\alpha \neq \beta$. In this case, the easiest thing to do is to graph each function separately on the same axes and then add the y-values to obtain the graph of the sum.

EXAMPLE 4 *Obtaining a Graph by Adding y-Values*

Sketch the graph of $4 \sin t + 2 \cos 2t$.

SOLUTION In Figure 4, we first sketch $4 \sin t$ (the grey line) and $2 \cos 2t$ (the black line). The blue curve is the graph of $4 \sin t + 2 \cos 2t$, obtained by

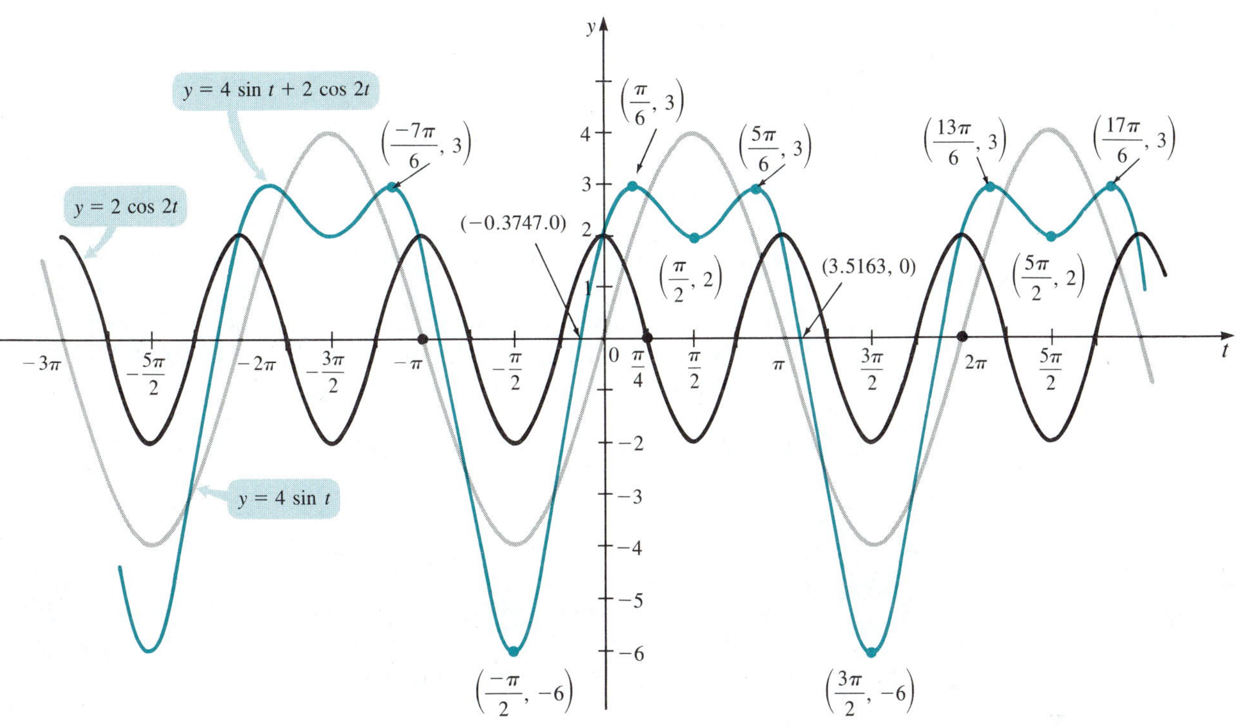

Figure 4 The blue curve is the graph of $y = 4 \sin t + 2 \cos 2t$.

adding the y-values at each point. Note, for example, the following values:

$$4 \sin \frac{\pi}{6} + 2 \cos\left(2 \cdot \frac{\pi}{6}\right) = 4 \sin \frac{\pi}{6} + 2 \cos \frac{\pi}{3}$$

$$= 4 \cdot \frac{1}{2} + 2 \cdot \frac{1}{2} = 2 + 1 = 3$$

$$4 \sin \frac{\pi}{3} + 2 \cos\left(2 \cdot \frac{\pi}{3}\right) = 4 \sin \frac{\pi}{3} + 2 \cos \frac{2\pi}{3}$$

$$= 4\left(\frac{\sqrt{3}}{2}\right) + 2\left(-\frac{1}{2}\right) = 2\sqrt{3} - 1$$

$$\approx 3.46 - 1 = 2.46$$

It is evident from Example 4 that obtaining a graph by adding y-values is not easy. Fortunately, many computer software packages exist for generating highly accurate graphs of virtually any function you can write down. More recently, several brands of hand-held calculators with graphing capabilities have become available. A description of how to obtain graphs on these calculators is given in Appendix A.

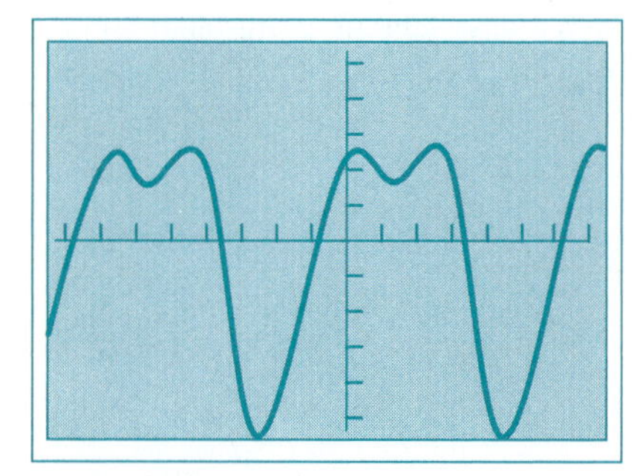

Figure 5 Graph of $y = 4 \sin t + 2 \cos 2t$ obtained on a graphing calculator for $-7 \le t \le 7$ and $-6 \le y \le 6$.

We can use a software program or a graphing calculator to obtain the graph of Example 4: $y = 4 \sin t + 2 \cos 2t$. If we use the range of values $-7 \le t \le 7$ and $-6 \le y \le 6$, we obtain a graph like the one in Figure 5. See Example 5 in Appendix A for more details.

Fourier Series (Optional)

Many important problems in physics, biology, and economics involve periodic phenomena. At the beginning of the nineteenth century, the French physicist Jean Baptiste Joseph Fourier (1768–1830) discovered that most periodic functions could be written as an infinite sum of functions of the form $A_n \cos n\omega t$ or $B_n \sin n\omega t$. Such sums are called **Fourier series.** The mathematics of Fourier series is beyond the scope of this book. However, we give one example illustrating the great power of the process of taking sums of trigonometric functions.

EXAMPLE 5 *A Fourier Series for the Sawtooth Function*

Let

$$f(x) = \begin{cases} x + 1, & -1 \le x \le 0 \\ -x + 1, & 0 \le x \le 1 \end{cases}$$

This function is sketched in Figure 6(a). If we extend this graph periodically in both directions to the graph of a function that is periodic of period 2, we obtain the graph of the **sawtooth function.** See Figure 6(b).

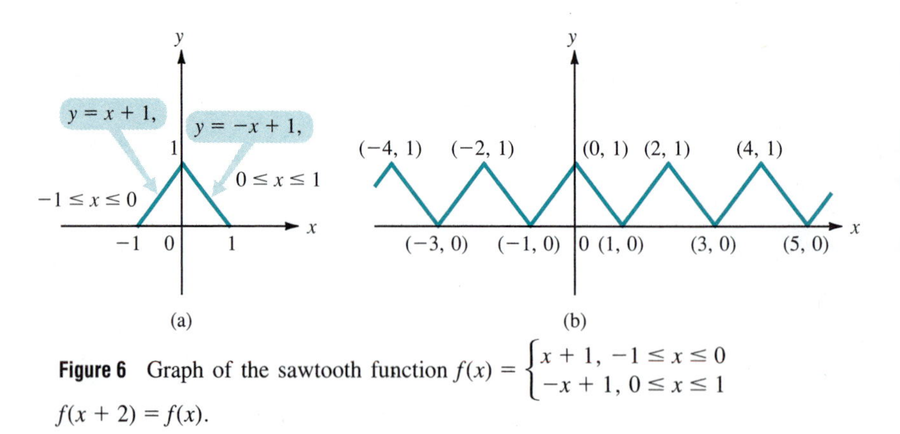

Figure 6 Graph of the sawtooth function $f(x) = \begin{cases} x + 1, & -1 \le x \le 0 \\ -x + 1, & 0 \le x \le 1 \end{cases}$ $f(x + 2) = f(x)$.

The sawtooth function arises in a number of physical applications.

It can be shown that the sawtooth function can be written as the following Fourier series:

$$f(x) = \frac{1}{2} + \frac{4}{\pi^2}\left(\cos \pi x + \frac{1}{3^2} \cos 3\pi x + \frac{1}{5^2} \cos 5\pi x + \frac{1}{7^2} \cos 7\pi x + \cdots\right) \quad \textbf{(2)}$$

We let S_n denote the sum of $\frac{1}{2}$ and the first n terms in (2). Then

$$S_1 = \frac{1}{2} + \frac{4}{\pi^2} \cos \pi x$$

$$S_2 = \frac{1}{2} + \frac{4}{\pi^2} \left(\cos \pi x + \frac{1}{9} \cos 3\pi x\right)$$

$$S_3 = \frac{1}{2} + \frac{4}{\pi^2} \left(\cos \pi x + \frac{1}{9} \cos 3\pi x + \frac{1}{25} \cos 5\pi x\right)$$

$$S_4 = \frac{1}{2} + \frac{4}{\pi^2} \left(\cos \pi x + \frac{1}{9} \cos 3\pi x + \frac{1}{25} \cos 5\pi x + \frac{1}{49} \cos 7\pi x\right)$$

and so on. In Figure 7, we sketch the graphs of S_1, S_2, S_3, and S_4 for $-1.2 \le x \le 1.2$. It is remarkable that even using as few as four terms in the sum (2), we obtain a graph that very closely resembles the graph of the sawtooth function.

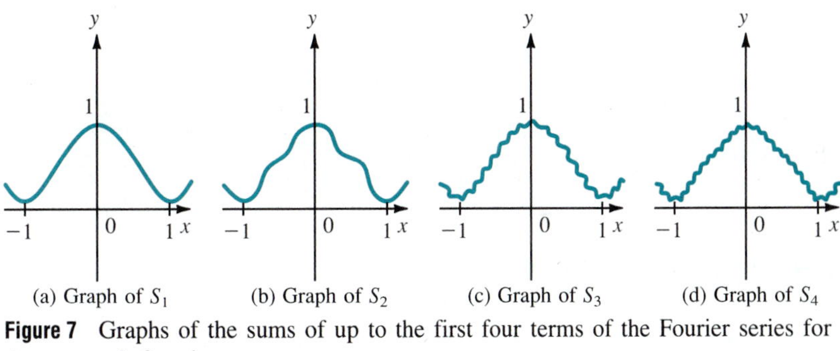

Figure 7 Graphs of the sums of up to the first four terms of the Fourier series for the sawtooth function.

In Problems 35–38 you are asked to use a calculator or a computer to graph the first n terms of the Fourier series for other important functions.

Problems 3.7

Readiness Check

I. When graphing the function $f(t) = 2 \sin t - 3 \cos 2t$ by adding the y-values at each point, which of the following are the two y-values to be added to find the y-value at $t = 3\pi/2$?
a. $-1, 1$ b. $-2, 0$ c. $-2, 3$ d. $0, -3$

II. Which of the following is true about the graph of $y = -3 \cos (x/2 + \pi/2)$?
a. Its amplitude is -3. b. Its period is π.
c. Its phase shift is $\pi/4$.
d. It is the reflection of $y = 3 \cos (x/2 + \pi/2)$ about the x-axis.

III. $4 \cos 2\theta + 3 \sin 2\theta = 5 \cos (2\theta - \delta)$. Then $\delta \approx$ ____
a. 0.9273 b. 0.6435
c. −0.6435 d. −0.9273

Answers to Readiness Check

I. c II. d III. b

In Problems 1–19 sketch the graph of each function.

1. $\sin t + 2$
2. $3 \sin \dfrac{t}{2} - 1$
3. $2 \cos\left(t - \dfrac{\pi}{6}\right) + 1$
4. $\dfrac{1}{2} \sin \pi(t-1) - 1$
5. $\sin t + \cos t$
6. $\sin t - \cos t$
7. $\cos t - \sin t$
8. $2 \sin t - 3 \cos t$
9. $-2 \sin t - 3 \cos t$
10. $2 \sin t + 3 \cos t$
11. $-2 \sin t + 3 \cos t$
12. $4 \cos 2t + 3 \sin 2t$
13. $4 \cos 2t - 3 \sin 2t$
14. $-4 \cos 2t - 3 \sin 2t$
15. $-4 \cos 2t + 3 \sin 2t$
16. $3 \cos \pi t + 2 \sin \pi t$
17. $-\cos \dfrac{\pi}{2} t + 4 \sin \dfrac{\pi}{2} t$
18. $5 \cos \dfrac{t}{4} - 12 \sin \dfrac{t}{4}$
19. $12 \cos \dfrac{\pi}{4} t + 5 \sin \dfrac{\pi}{4} t$

Sketch the graphs in Problems 20–26 by adding the y-values.

20. $\sin t + \cos 2t$
21. $\sin t - \cos 2t$
22. $-\sin t + \cos 2t$
23. $-\sin t - \cos 2t$
24. $2 \sin 3t + 5 \cos 2t$
25. $\cos \dfrac{\pi}{2} t - \sin \pi t$
26. $2 \sin \dfrac{\pi}{4} t - \cos \pi t$

Graphing Calculator Problems

In Problems 27–34 obtain each graph of the given function on a graphing calculator. The range of x-values should include at least one full period of the function. Before starting, read Example 5 in Appendix A.

27. $3 \sin 2x - 5 \cos 3x$
28. $2.6 \sin \dfrac{8}{3} x - \dfrac{1}{2} \cos \dfrac{x}{7}$
29. $12 \cos \dfrac{x}{2} + 8 \cos \dfrac{2x}{3}$
30. $-4 \sin \dfrac{\pi}{5} x + 7 \sin \dfrac{\pi}{3} x$
31. $3 \sin 2x - \cos x + 5 \sin 4x$
32. $\cos \dfrac{x}{2} + 2 \sin \dfrac{2x}{5} - 3 \cos \dfrac{7x}{9}$
33. $\sin x + \dfrac{\sin 2x}{2} + \dfrac{\sin 3x}{3}$
34. $\cos \pi x + 2 \cos \dfrac{\pi}{2} x + 3 \cos \dfrac{\pi}{3} x$

In Problems 35–38 a periodic function is sketched and the Fourier series for that function is given. In each case sketch the graphs of (a) the first term in the Fourier series, (b) the sum of the first two terms, S_2, (c) S_3, and, (d) S_4.

35. $f(x) = x, \ -\pi < x < \pi; \ f(x + 2\pi) = f(x)$

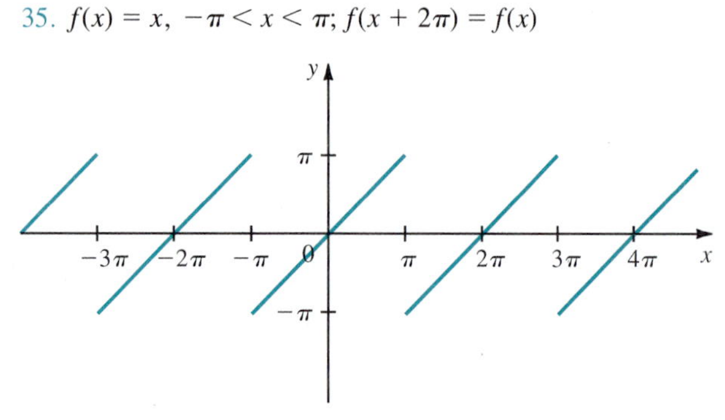

$f(x) = 2(\sin x - \frac{1}{2} \sin 2x + \frac{1}{3} \sin 3x - \frac{1}{4} \sin 4x + \cdots)$

36. $f(x) = x^2, \ -1 < x < 1; \ f(x + 2) = f(x)$

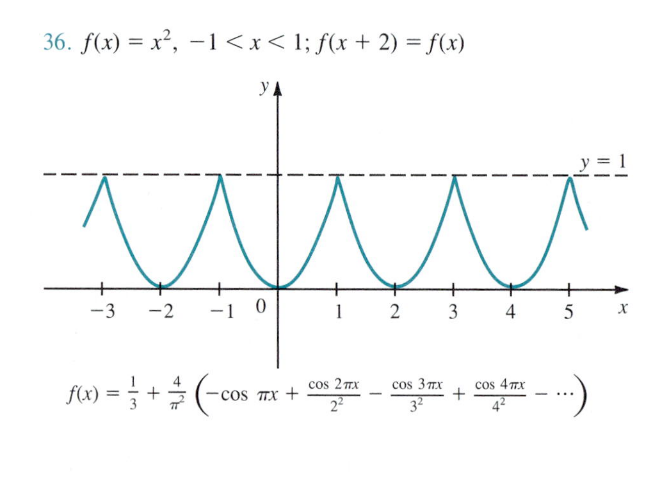

$f(x) = \frac{1}{3} + \frac{4}{\pi^2}\left(-\cos \pi x + \frac{\cos 2\pi x}{2^2} - \frac{\cos 3\pi x}{3^2} + \frac{\cos 4\pi x}{4^2} - \cdots\right)$

37. $f(x) = \begin{cases} 1, & 0 \le x \le 1 \\ -1, & -1 < x < 0 \end{cases} \quad f(x + 2) = f(x)$

$f(x) = \frac{4}{\pi}(\sin \pi x + \frac{1}{3} \sin 3\pi x + \frac{1}{5} \sin 5\pi x + \cdots)$

38. $f(x) = \begin{cases} 3 \sin 2x, \; 0 < x < \dfrac{\pi}{2} \\ 0, \; -\dfrac{\pi}{2} < x < 0 \end{cases}$ $\quad f(x + \pi) = f(x)$

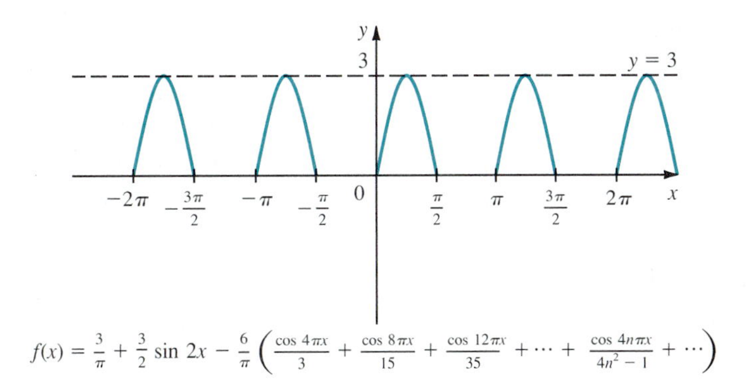

3.8 Harmonic Motion (Optional)

In physics, biology, and economics, many quantities are periodic. Examples include the vibration or oscillation of a pendulum or a spring, periodic fluctuations in the population of a species, and periodic fluctuations in a business cycle. Many of these quantities can be described by harmonic functions.

A **harmonic function** is a function that can be written in the form

$$g(t) = a \cos \omega t + b \sin \omega t \tag{1}$$

Recall from Section 3.4 (p. 176) that (1) can be written in the forms

$$a \cos \omega t + b \sin \omega t = A \cos (\omega t - \delta) \tag{2}$$

$$a \cos \omega t + b \sin \omega t = A \sin (\omega t + \phi) \tag{3}$$

where $A = \sqrt{a^2 + b^2}$, $(\cos \delta, \sin \delta) = \left(\dfrac{a}{A}, \dfrac{b}{A}\right)$, and $(\cos \phi, \sin \phi) = \left(\dfrac{b}{A}, \dfrac{a}{A}\right)$.

In (2) or (3), the period is $\dfrac{2\pi}{\omega}$. The **natural frequency** f of the function is the number of complete periods per unit time. Since $y = A \cos (\omega t - \delta)$ or $y = A \sin (\omega t + \phi)$ returns to the same y-value in one period equal to $\dfrac{2\pi}{\omega}$ time units, we have

Natural Frequency of a Function

$$\text{frequency} = f = \frac{\omega}{2\pi} \tag{4}$$

Units of frequency are cycles/sec (also called **hertz**). In this section, we will show how harmonic functions can arise in applications.

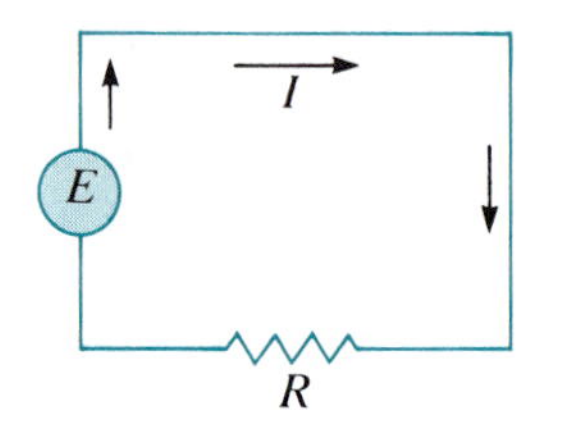

Figure 1 An electric circuit with a resistor.

A Simple Electric Circuit

In an electric circuit, such as the one depicted in Figure 1, an **electromotive force** E (volts), usually a battery or generator, drives an electric charge Q (coulombs) and produces a current I (amperes). In the circuit of Figure 1, a resistor of resistance R (ohms) is a component of the circuit that opposes the current, dissipating the energy in the form of heat. It produces a drop in the voltage given by **Ohm's law:**

$$E = RI \tag{5}$$

The electromotive force (EMF) may be **direct** or **alternating.** A direct EMF is given by a constant voltage. An alternating EMF is usually given as a sine function:

$$E = E_0 \sin \omega_0 t, \ \mathrm{E}_0 > 0$$

Since $-1 \le \sin \omega_0 t \le 1$, we see that

$$-E_0 \le E \le E_0$$

Thus, E_0 is the maximum voltage, and $-E_0$ is the minimum voltage.

EXAMPLE 1 *Computing EMF, Frequency, and Maximum Current in an Electric Circuit*

Suppose that an EMF of $E = 10 \sin \dfrac{\pi}{4} t$ volts is connected in the circuit of Figure 1 to a resistance of 5 ohms.

(a) What is the period of the EMF?
(b) What is the frequency?
(c) What is the maximum current in the system?

SOLUTION

(a) Period $= \dfrac{2\pi}{\omega} = \dfrac{2\pi}{\left(\dfrac{\pi}{4}\right)} = \dfrac{8\pi}{\pi} = 8$

(b) Frequency $= \dfrac{\omega}{2\pi} = \dfrac{1}{8}$ cycle/sec

(c) From (5),

$$I = \frac{E}{R} = \frac{10 \sin \dfrac{\pi}{4} t}{5} = 2 \sin \frac{\pi}{4} t \text{ amperes}$$

The maximum current is 2 amperes.

The next example uses the exponential function e^x, discussed in Section 7.2. If you haven't seen this function before, skip the example and Problems 5, 6, and 7 on pages 206 and 207.

EXAMPLE 2 *A Predator-Prey Model*

In many ecosystems some species are predators (eaters), whereas others are prey (the eaten). As prey become more numerous, so do predators, for the simple reason that there is more food. On the other hand, as predators become more numerous, prey become less so because they are more likely to be killed and devoured. It is not difficult to see that predator and prey populations fluctuate in size in relation to their own size and the size of the other species.

Much current research deals with modeling such phenomena. In one simple model, the population of each of two species (one predator, one prey) is the product of an exponential function and a harmonic function. Let $x(t)$ denote the population of the predator species, and $y(t)$ the population of the prey species. If the initial predator population is 500, whereas that of the prey species is 1000, then, according to one such model,

$$x(t) = e^t(500 \cos t + 1000 \sin t)$$
$$y(t) = e^t(1000 \cos t - 500 \sin t)$$

where t is measured in years. When is the prey species extinct?

SOLUTION The prey species is extinct when $y(t) = 0$. That is, when

$$y(t) = e^t(1000 \cos t - 500 \sin t) = 0 \tag{6}$$

But $e^t > 0$ for all t, so (6) can hold only when

$$1000 \cos t - 500 \sin t = 0$$
$$1000 \cos t = 500 \sin t$$
$$2 = \frac{1000}{500} = \frac{\sin t}{\cos t} = \tan t$$
$$t = \tan^{-1} 2 \approx 1.107 \text{ years} \approx 1 \text{ year } 1.29 \text{ months}$$

Sound Waves

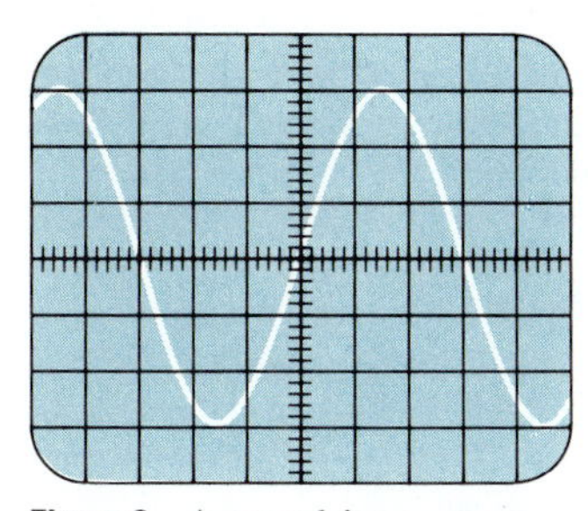

Figure 2 A sound image on an oscilloscope.

Vibrations, such as those created by plucking a violin string or striking a wooden tube, cause sound waves, which may or may not be audible to the human ear. However, sound waves can be converted to an image on an oscilloscope, such as the one depicted in Figure 2. Often, sound waves are sinusoidal and can therefore be written in the form

$$y = A \sin \omega t \tag{7}$$

Here we assume that there is no phase shift [that is, $\phi = 0$ in equation (3)]. The amplitude A in (7) is related to the loudness of the sound, which is measured in **decibels.**

EXAMPLE 3 *Finding an Equation for a Sound Wave*

Middle C is struck on a piano with an amplitude of $A = 2$. The frequency of middle C is 264 cycles/sec. Write an equation for the resulting sound wave.

SOLUTION With $A = 2$, equation (7) is

$$y = 2 \sin \omega t$$

But

$$\text{frequency} = \frac{\omega}{2\pi} = 264$$

so

$$\omega = 264(2\pi) = 528\pi$$

and

$$y = 2 \sin 528\pi t$$

is the equation of the sound wave.

Problems 3.8

Readiness Check

I. Which of the following is true of the graph of a harmonic function?
  a. It passes through the origin.
  b. As $x \to +\infty$, $y \to +\infty$
  c. It is a sine or cosine curve.
  d. It is asymptotic to the positive x-axis.

II. Which of the following determines the period of the graph of the harmonic motion equation $y = A \sin(\omega t + \phi)$?
  a. A b. ω c. t d. ϕ

III. Which of the following are the times t when a buoy, whose position y is given by $y = 5 \cos \pi t/6$, will be in equilibrium position (at sea level) in the interval [0, 12]?
  a. 0, 6, 12 b. 3, 9
  c. $0, \frac{2}{3}, \frac{4}{3}$ d. $\frac{1}{3}, 1$

In Problems 1–4 an EMF E and a resistance R are given. (a) Determine the natural frequency of the current. (b) Find the maximum current.

1. $E = 20 \sin 2\pi t$ volts, $R = 10$ ohms
2. $E = 5 \sin 6\pi t$ volts, $R = \frac{1}{2}$ ohm
3. $E = 50 \sin \frac{\pi}{2} t$ volts, $R = 1$ ohm
4. $E = 35 \sin 3t$ volts, $R = 7$ ohms

In Problems 5–7 (a) determine when the prey species ($y(t)$) will be extinct; (b) what is the predator population ($x(t)$) at the time the prey species becomes extinct?

5. $x(t) = e^{t/4}(200 \cos 2t + 500 \sin 2t)$
   $y(t) = e^{t/4}(1000 \cos 2t - 100 \sin 2t)$
6. $x(t) = e^{t/2}\left(400 \cos \frac{t}{4} + 1000 \sin \frac{t}{4}\right)$
   $y(t) = e^{t/2}\left(2000 \cos \frac{t}{4} - 50 \sin \frac{t}{4}\right)$

Answers to Readiness Check

I. c II. b III. b

7. $x(t) = e^{t/10}\left(1000 \cos \dfrac{\pi}{2}t + 1500 \sin \dfrac{\pi}{2}t\right)$

   $y(t) = e^{t/10}\left(400 \cos \dfrac{\pi}{2}t - 600 \sin \dfrac{\pi}{2}t\right)$

8. An anchored buoy bobs on ocean waves. At a given time it is observed that its up-down motion is harmonic. The difference between its highest and lowest positions is 6 feet, and it returns every 15 seconds to the same position in the cycle. Write an equation that gives the motion of the buoy as a function of time.
9. A wire coil is immersed in an electric field and produces the voltage $E = 50 \sin 200\pi t$.
   (a) What is the natural frequency of the coil?
   (b) What is the voltage after $t = 0.002$ sec?
10. The frequency of a note one octave below middle C is 132 cycles/sec. Find the equation of a sound wave struck at that frequency with an amplitude of 5.
11. Answer the question of Problem 10 for a note struck two octaves above middle C ($f = 1056$ cycles/sec).
12. Sound is audible to the human ear in the range 20 to 20,000 cycles/sec. Find equations for the sound wave at each extreme of the **audible range.** Assume that $A = 1$.

In Problems 13 and 14 the sound wave pattern on an oscilloscope is reproduced. Write the equation of each sound wave.

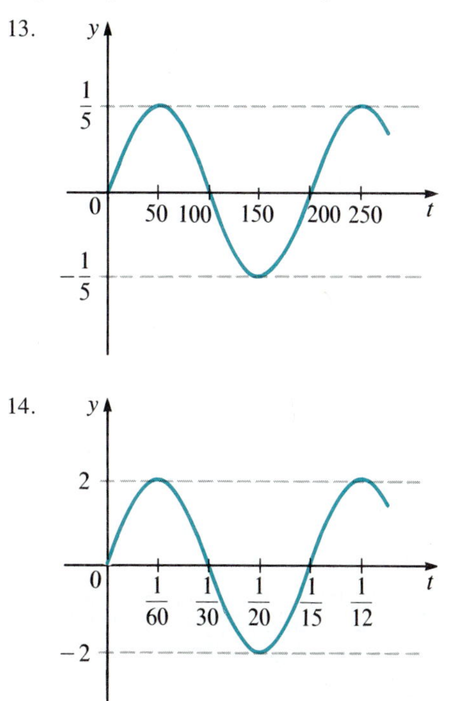

Summary Outline of Chapter 3

- **Trigonometric Identities** [Note: The identities marked (C) are especially useful in calculus.]

(C) • **Identities That Follow from the Definitions**

$\tan x = \dfrac{\sin x}{\cos x}$ $\quad \cot x = \dfrac{\cos x}{\sin x} = \dfrac{1}{\tan x}$ pp. 101, 150

$\sec x = \dfrac{1}{\cos x}$ $\quad \csc x = \dfrac{1}{\sin x}$ pp. 101, 150

(C) • **Circular or Pythagorean Identities**

$\sin^2 x + \cos^2 x = 1$ pp. 78, 150

$1 + \tan^2 x = \sec^2 x$ pp. 106, 150

$1 + \cot^2 x = \csc^2 x$ pp. 107, 150

(C) • **Even-Odd Identities**

$\cos(-x) = \cos x$ $\quad \sin(-x) = -\sin x$ pp. 92, 150

$\tan(-x) = -\tan x$ $\quad \cot(-x) = -\cot x$ pp. 104, 150

$\sec(-x) = \sec x$ $\quad \csc(-x) = -\csc x$ pp. 104, 150

- **Reduction Identities**

$\sin(x + 2\pi) = \sin x$ — p. 89

$\cos(x + 2\pi) = \cos x$ — p. 89

$\tan(x + \pi) = \tan x$ — p. 111

$\sin\left(\frac{\pi}{2} - x\right) = \cos x$ — p. 94

$\cos\left(\frac{\pi}{2} - x\right) = \sin x$ — p. 94

$\cos(x + \pi) = -\cos x$ — p. 93

$\sin(x + \pi) = -\sin x$ — p. 93

$\cos(\pi - x) = -\cos x$ — p. 94

$\sin(\pi - x) = \sin x$ — p. 94

- **Sum and Difference Identities**

$\cos(\alpha + \beta) = \cos\alpha\cos\beta - \sin\alpha\sin\beta$ — p. 158

$\cos(\alpha - \beta) = \cos\alpha\cos\beta + \sin\alpha\sin\beta$ — p. 156

$\sin(\alpha + \beta) = \sin\alpha\cos\beta + \cos\alpha\sin\beta$ — p. 159

$\sin(\alpha - \beta) = \sin\alpha\cos\beta - \cos\alpha\sin\beta$ — p. 159

$\tan(\alpha + \beta) = \dfrac{\tan\alpha + \tan\beta}{1 - \tan\alpha\tan\beta}$ — p. 159

$\tan(\alpha - \beta) = \dfrac{\tan\alpha - \tan\beta}{1 + \tan\alpha\tan\beta}$ — p. 159

- **Double-Angle Identities**

$\cos 2x = \cos^2 x - \sin^2 x = 2\cos^2 x - 1 = 1 - 2\sin^2 x$ — p. 165

$\sin 2x = 2\sin x\cos x \qquad \tan 2x = \dfrac{2\tan x}{1 - \tan^2 x}$ — pp. 165, 166

- **Half-Angle Identities**

$\sin\dfrac{x}{2} = \pm\sqrt{\dfrac{1 - \cos x}{2}} \qquad \cos\dfrac{x}{2} = \pm\sqrt{\dfrac{1 + \cos x}{2}}$ — p. 168

$\tan\dfrac{x}{2} = \pm\sqrt{\dfrac{1 - \cos x}{1 + \cos x}} = \dfrac{1 - \cos x}{\sin x} = \dfrac{\sin x}{1 + \cos x}$ — pp. 168, 169

- **Identities that Follow From the Half-Angle Identities**

$\sin^2 x = \dfrac{1 - \cos 2x}{2} \qquad \cos^2 x = \dfrac{1 + \cos 2x}{2}$ — p. 167

- **Product-to-Sum Identities**

$\sin\alpha\sin\beta = \frac{1}{2}[\cos(\alpha - \beta) - \cos(\alpha + \beta)]$ — p. 173

$\sin\alpha\cos\beta = \frac{1}{2}[\sin(\alpha + \beta) + \sin(\alpha - \beta)]$ — p. 173

$\cos\alpha\sin\beta = \frac{1}{2}[\sin(\alpha + \beta) - \sin(\alpha - \beta)]$ — p. 173

$\cos\alpha\cos\beta = \frac{1}{2}[\cos(\alpha + \beta) + \cos(\alpha - \beta)]$ — p. 173

- **Sum-to-Product Identities**

$\sin \alpha + \sin \beta = 2 \sin \dfrac{\alpha + \beta}{2} \cos \dfrac{\alpha - \beta}{2}$ p. 174

$\sin \alpha - \sin \beta = 2 \cos \dfrac{\alpha + \beta}{2} \sin \dfrac{\alpha - \beta}{2}$ p. 174

$\cos \alpha + \cos \beta = 2 \cos \dfrac{\alpha + \beta}{2} \cos \dfrac{\alpha - \beta}{2}$ p. 174

$\cos \alpha - \cos \beta = 2 \sin \dfrac{\alpha + \beta}{2} \sin \dfrac{\beta - \alpha}{2}$ p. 174

- **Harmonic Identities**

$a \cos \omega t + b \sin \omega t = \sqrt{a^2 + b^2} \cos (\omega t - \delta)$ where $(\cos \delta, \sin \delta) = (a, b)$ p. 176

$a \cos \omega t + b \sin \omega t = \sqrt{a^2 + b^2} \sin (\omega t + \phi)$ where $(\cos \phi, \sin \phi) = (b, a)$ p. 176

- In the graphs of $y = A \cos (\omega t - \delta)$ and $y = A \sin (\omega t - \delta)$, with $\omega > 0$

$|A|$ is the **amplitude** pp. 176, 187

$\dfrac{2\pi}{\omega}$ is the **period** pp. 176, 189

$\dfrac{|\delta|}{\omega}$ is the **phase shift** (to the right if $\delta > 0$ and to the left if $\delta < 0$) pp. 176, 193

Review Exercises for Chapter 3

1. Write $\tan x \sec x$ in terms of $\sin x$ and $\cos x$.
2. Write $\dfrac{\sec^2 \theta}{\cos^2 \theta} + \cot^2 \theta$ in terms of $\sin \theta$ and $\cos \theta$.
3. Write $\dfrac{1 - \sec x}{1 + \sec x} \csc^2 x$ in terms of $\cos x$ only.

In Exercises 4–8 show that each equation is an identity.

4. $\cos \theta \sec \theta = 1$
5. $\dfrac{\csc \alpha}{\sec \alpha} = \cot \alpha$
6. $\dfrac{2}{\tan x + \cot x} = \sin 2x$
7. $\sin \theta \csc \theta - \cos^2 \theta = \sin^2 \theta$
8. $\dfrac{\cos \beta}{\sin \beta \sec^2 \beta - \sin \beta} = \cot^3 \beta$
9. For what values of x is it true that $\sec x = \sqrt{1 + \tan^2 x}$?
10. Write $\sqrt{9 - x^2}$ as a function of θ for $-\dfrac{\pi}{2} \leq \theta \leq \dfrac{\pi}{2}$ by making the substitution $x = 3 \sin \theta$.
11. Write $\sqrt{16 + x^2}$ as a function of θ for $-\dfrac{\pi}{2} < \theta < \dfrac{\pi}{2}$ by making the substitution $x = 4 \tan \theta$.
12. Compute $\sin \dfrac{7\pi}{8}$ exactly.
13. Compute $\tan \dfrac{5\pi}{12}$ exactly.
14. Compute $\cos 22\frac{1}{2}°$ exactly.
15. Compute $\tan \dfrac{3\pi}{8}$ exactly.
16. Write $\cos 3 \cos 2.5 + \sin 3 \sin 2.5$ as $\cos \theta$ for a single value of θ.
17. Write $\sin 37° \cos 41° - \cos 37° \sin 41°$ as $\sin \theta$ for a single value of θ.

18. Write $\dfrac{\tan 2 + \tan 5}{1 - \tan 2 \tan 5}$ as $\tan \theta$ for a single value of θ.
19. Verify that $\cos^2 \alpha - \sin^2 \beta = \cos(\alpha - \beta)\cos(\alpha + \beta)$.
20. Suppose that $\sin \alpha = \frac{5}{13}$, α is in the first quadrant, $\cos \beta = -\frac{1}{4}$, and β is in the second quadrant.
    (a) Determine $\sin(\alpha + \beta)$, $\sin(\alpha - \beta)$, $\cos(\alpha + \beta)$, $\cos(\alpha - \beta)$, $\tan(\alpha + \beta)$, and $\tan(\alpha - \beta)$.
    (b) Find the quadrants of $\alpha + \beta$ and $\alpha - \beta$.
21. Answer the questions in Exercise 20 if $\cos \alpha = -\frac{2}{3}$, α is in the third quadrant, $\cot \beta = -2$, and β is in the second quadrant.
22. Compute $\cos(\sin^{-1} \frac{2}{3} + \tan^{-1} \frac{3}{7})$ without a calculator.
23. Compute $\tan(\sin^{-1}(-\frac{2}{3}) - \cos^{-1} \frac{3}{7})$ without a calculator.
24. Evaluate $\sin(\cos^{-1} x - \sin^{-1}(x + 1))$.

In Exercises 25–27 find $\sin 2\theta$, $\cos 2\theta$, $\tan 2\theta$, $\sin \dfrac{\theta}{2}$, $\cos \dfrac{\theta}{2}$, and $\tan \dfrac{\theta}{2}$ using the information given.

25. $\sin \theta = \dfrac{3}{5}$ and θ is in the first quadrant.
26. $\tan \theta = -2$ and θ is in the second quadrant.
27. $\cos \theta = -\dfrac{3}{4}$ and θ is in the third quadrant.
28. Prove that $\dfrac{2 - \sec^2 x}{\sec^2 x} = \cos 2x$.
29. Prove that $\dfrac{2 \cot \theta}{\cot^2 \theta - 1} = \tan 2\theta$.
30. Prove that $\dfrac{1}{1 - \cos x} = \dfrac{2 \cos^2 \dfrac{x}{2}}{\sin^2 x}$.

In Exercises 31–33 write each product as a sum.

31. $\sin 55° \cos 15°$
32. $\cos \frac{1}{2} \cos 4$
33. $\sin 2x \sin 5x$

In Exercises 34–36 write each sum as a product.

34. $\cos 1 - \cos 4$
35. $\sin 20° + \sin 50°$
36. $\cos 4\theta + \cos 6\theta$

In Exercises 37–39 write each sum in the form $A \cos(\omega t - \delta)$ and $A \sin(\omega t + \phi)$.

37. $5 \cos t - 3 \sin t$
38. $3 \cos \dfrac{x}{2} + 4 \sin \dfrac{x}{2}$
39. $-6 \cos 3\theta - 8 \sin 3\theta$
40. Prove that $\dfrac{\sin 2x + \sin 4x}{\cos 2x + \cos 4x} = \cot 3x$.
41. Prove that $\dfrac{\sin \alpha - \sin \beta}{\sin \alpha + \sin \beta} = \dfrac{\tan\left(\dfrac{\alpha - \beta}{2}\right)}{\tan\left(\dfrac{\alpha + \beta}{2}\right)}$.

In Exercises 42–54 find all solutions to each equation in the interval $[0, 2\pi]$. When necessary, round to 4 decimal places.

42. $\cos \theta = 0$
43. $\sin \theta = -\dfrac{\sqrt{3}}{2}$
44. $\tan x = -3$
45. $\cot 3\alpha = -1$
46. $\csc \beta - \sec \beta = 0$
47. $\sin \theta = \dfrac{3}{4}$
48. $4 \cos^2 \theta = 1$
49. $2 \sin x + \tan x = 0$
50. $\sin^2 \theta + 2 \cos \theta - 1 = 0$
51. $\cos^2 \theta - 3 \cos \theta - 4 = 0$
52. $\cot^2 \beta - 2 \cot \beta + 1 = 0$
53. $\cos x = \cos \dfrac{x}{2}$
54. $4 \cot 2\theta + 5 \tan \theta = 0$

In Exercises 55–61 sketch the graph of the given function.

55. $2 \sin 3t$
56. $-4 \sin \dfrac{\pi}{2} t$
57. $3 \cos\left(2t - \dfrac{\pi}{6}\right)$
58. $-2 \cos t + \sin t$
59. $4 \cos \dfrac{t}{2} + 3 \sin \dfrac{t}{2}$
60. $\cos 2t - 2 \sin t$
61. $3 \sin \dfrac{\pi}{2} t + 2 \cos 2\pi t$

Chapter 4

Further Applications of Trigonometry

4.1 The Law of Sines

In Section 2.8, we saw how to find the sides and acute angles of a right triangle if either one acute angle and one side of the triangle or two sides of the triangle are given. A triangle that is not a right triangle is called an **oblique triangle.** In this section and in Section 4.2, we will solve oblique triangles.

An oblique triangle is sketched in Figure 1. Using a common convention, we label the vertices of the angles with capital letters A, B, and C and the lengths of the sides opposite these angles by a, b, and c. The angle A between sides b and c is called the **included angle** of sides b and c.

In solving a triangle, we seek three sides and three angles. We can find these if we know the length of one side and two other parts (sides or angles). This leads to four possible cases.

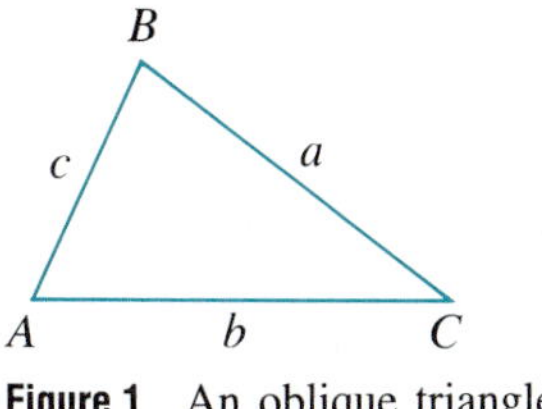

Figure 1 An oblique triangle.

Solving Oblique Triangles

We can solve an oblique triangle if we are given

Statement	*Notation*	*Picture*
1. Two angles and one side	ASA or AAS	
2. Two sides and an angle opposite one of them	SSA	
3. Two sides and the included angle	SAS	
4. Three sides	SSS	

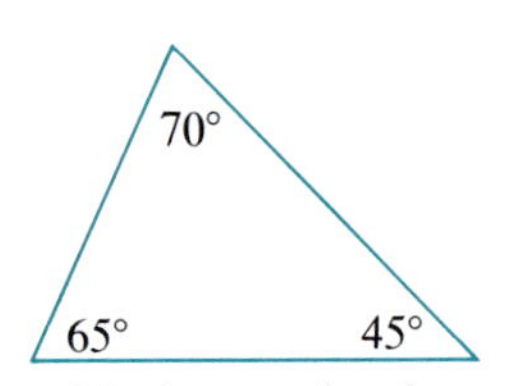

(a) Acute triangle

In this section, we use the law of sines to solve triangles in the first two cases. In Section 4.2, we use the law of cosines to solve triangles in cases 3 and 4.

Before stating the law of sines, we make an observation: Since the sum of the angles in a triangle is 180°, we can divide oblique triangles into two categories:

acute triangles (all three angles are acute)
obtuse triangles (one angle is obtuse, the other two are acute)

These triangles are illustrated in Figure 2.

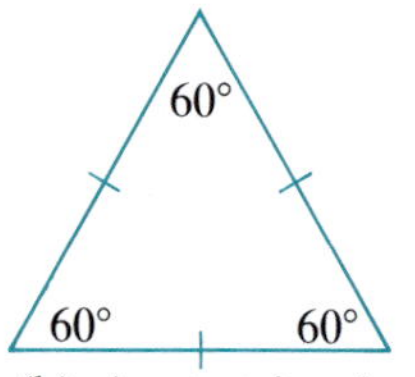

(b) Acute triangle (equilateral)

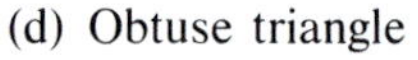

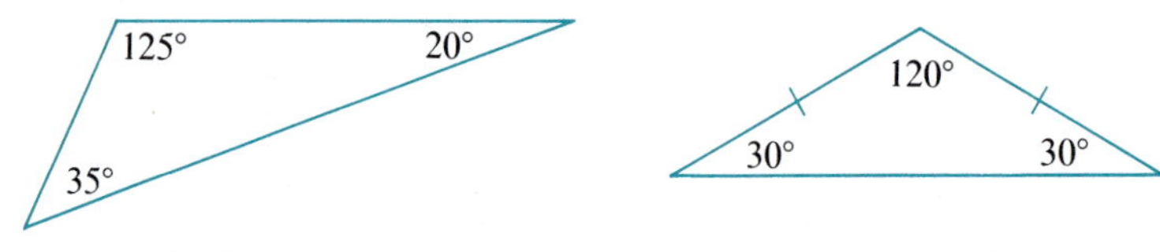

(c) Acute triangle (isosceles)

(d) Obtuse triangle

(e) Obtuse triangle (isosceles)

Figure 2 Five triangles

Law of Sines

For an oblique triangle with angles A, B, and C and opposite sides a, b, and c, respectively,

$$\frac{\sin A}{a} = \frac{\sin B}{b} = \frac{\sin C}{c} \tag{1}$$

Proof

There are two cases: the triangle is acute or obtuse. In either case, we draw the altitude h from vertex A as in Figure 3.

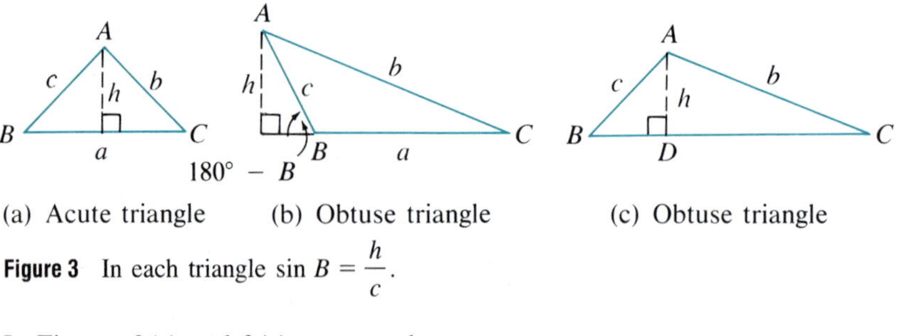

(a) Acute triangle (b) Obtuse triangle (c) Obtuse triangle

Figure 3 In each triangle $\sin B = \dfrac{h}{c}$.

In Figures 3(a) and 3(c), we see that

$$\sin B = \frac{h}{c} \quad \text{or} \quad h = c \sin B$$

$$\sin C = \frac{h}{b} \quad \text{or} \quad h = b \sin C$$

so

$$c \sin B = b \sin C$$

$$\frac{\sin B}{b} = \frac{\sin C}{c} \quad \text{Divide both sides by } bc$$

In Figure 3(b)

$$\sin(180^\circ - B) = \frac{h}{c} \quad \text{or} \quad h = c \sin(180^\circ - B) \overset{\text{Equation (15), p. 94}}{=} c \sin B$$

Similarly,

$$\sin C = \frac{h}{b} \qquad \text{or} \qquad h = b \sin C$$

So, as before,

$$c \sin B = b \sin C$$

and

$$\frac{\sin B}{b} = \frac{\sin C}{c}$$

If, in Figure 3, we draw the altitude from vertex C, then the same proof shows that

$$\frac{\sin A}{a} = \frac{\sin B}{b}$$

Since $\frac{\sin B}{b} = \frac{\sin C}{c}$, the proof is complete. ■

USEFUL GEOMETRIC FACT In any triangle, the longest side is opposite the largest angle and the shortest side is opposite the smallest angle.

A NOTE ON ROUNDING In this chapter, when the desired accuracy is not specified, we will adopt the following convention:

1. In computations involving sides, round to five significant figures.
2. In computations involving angles, round to the nearest tenth of a degree or the nearest minute.

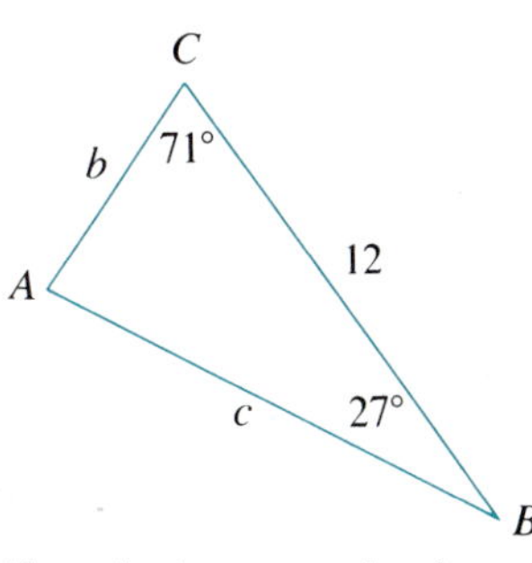

Figure 4 An acute triangle.

EXAMPLE 1 *Solving an Acute Triangle When Two Angles and the Included Side Are Given (ASA)*

Two angles of a triangle are 27° and 71°, and the included side has length 12. Find the remaining angle and the lengths of the other two sides.

SOLUTION The triangle is sketched in Figure 4. First, we compute

$$A + 27^\circ + 71^\circ = 180^\circ$$

so

$$A = 180° - 27° - 71° = 180° - 98° = 82°$$

Next, from the law of sines,

$$\frac{\sin A}{a} = \frac{\sin B}{b} = \frac{\sin C}{c}$$

$$\frac{\sin 82°}{12} = \frac{\sin 27°}{b} = \frac{\sin 71°}{c}$$

We solve for b and c: Since $\dfrac{\sin 82°}{12} = \dfrac{\sin 27°}{b}$,

$$b \sin 82° = 12 \sin 27°$$

That is,

$$b = \frac{12 \sin 27°}{\sin 82°} \approx \frac{12(0.45399)}{0.99027} \approx 5.5014 \quad \text{5 significant figures}$$

Similarly,

$$c = \frac{12 \sin 71°}{\sin 82°} \approx \frac{12(0.94552)}{0.99027} \approx 11.458 \quad \text{5 significant figures}$$

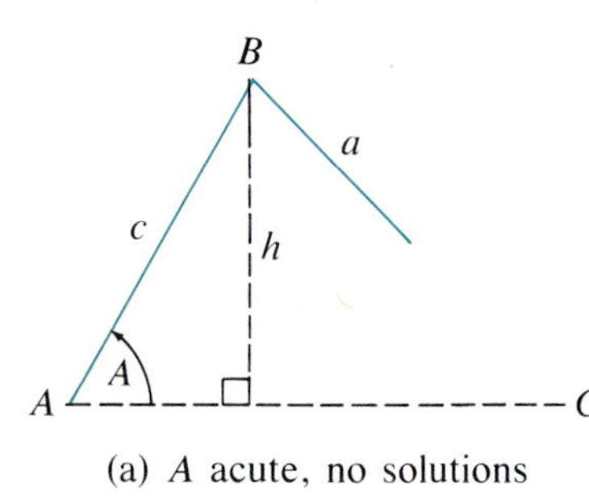

(a) A acute, no solutions

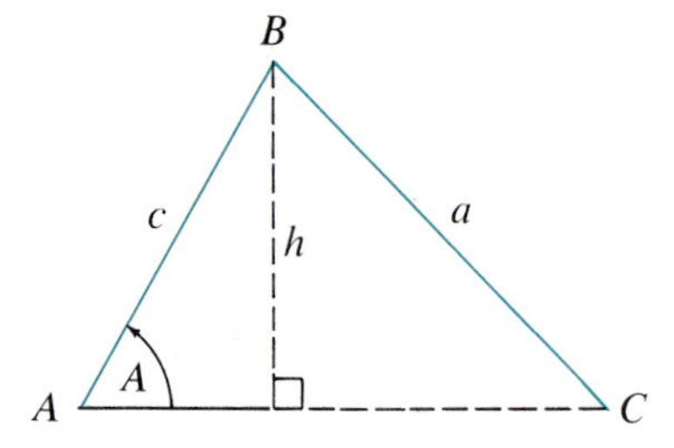

(b) A acute, one solution

Two triangles may be congruent if corresponding parts are equal. There are four cases: ASA, AAS, SAS, and SSS. Thus, as in Example 1, two angles and any side completely determine a triangle. Also, the triangle is determined if two sides and the included angle or three sides are given. However, two triangles that have two sides and an opposite angle in common (SSA) are not necessarily congruent.

Ambiguous Case

The case SSA is called the **ambiguous case** because there may be no, one, or two triangles having the two given sides a and c and opposite angle A. This is illustrated in Figure 5.

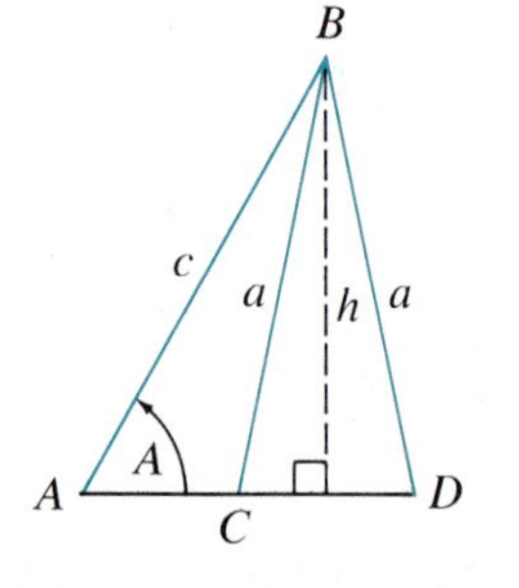

(c) A acute, two solutions

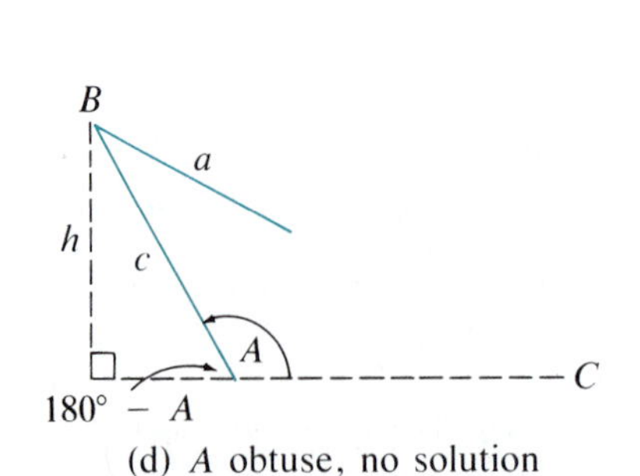

(d) A obtuse, no solution

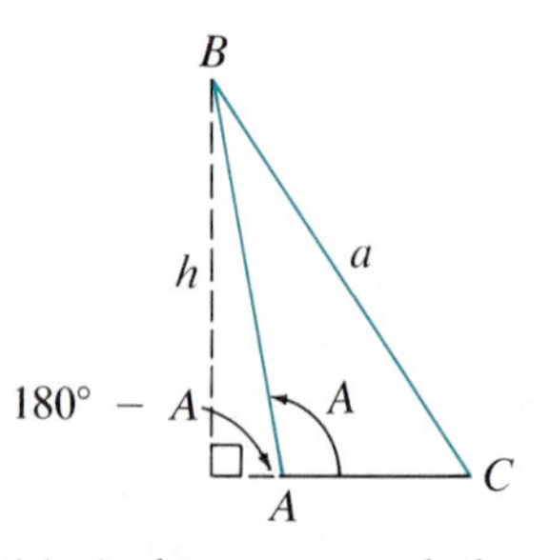

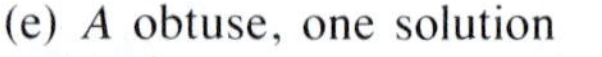
(e) A obtuse, one solution

Figure 5 When two sides and an opposite angle are given, there are three possibilities: no triangle contains them, exactly one does, or two do.

The proof of the following theorem is suggested by Examples 2 and 3.

Theorem

Let c and a be two positive numbers, and let A be an acute or obtuse angle.

(i) If A is acute and $a < c \sin A$, then there is no triangle with two sides having lengths a and c in which the measure of the angle opposite the side of length a is equal to the measure of A.

(ii) If A is acute and $a = c \sin A$ or if $a \geq c$, then there is one such triangle.

(iii) If A is acute and $c \sin A < a < c$, then there are two such triangles.

(iv) If A is obtuse and $a \leq c$, then there is no such triangle.

(v) If A is obtuse and $a > c$, then there is one such triangle.

Note that since $\sin A = h/c$, $h = c \sin A$ is the length of the altitude (height) from B.

EXAMPLE 2 *The Ambiguous (SSA) Case: No Solution*

Find all triangles with $a = 2$, $c = 4$, and $A = 35°$.

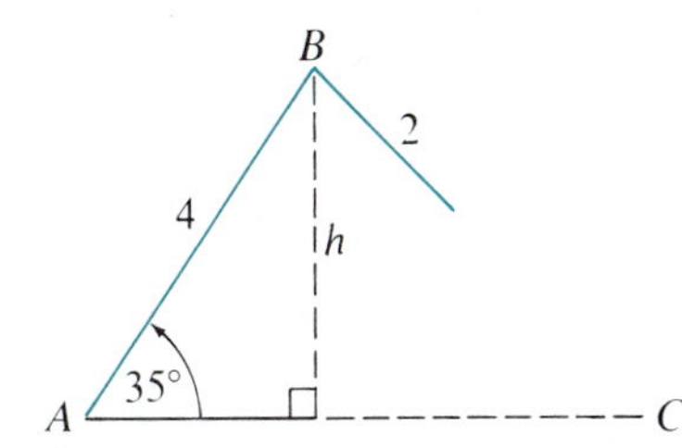

Figure 6 No triangle has sides of 4 and 2 with an angle of 35° opposite the side of length 2.

SOLUTION As in Figure 6, $\sin 35° = \frac{h}{4}$, so $h = 4 \sin 35° \approx 4(0.57358) \approx 2.2943$. But $2 < 2.2943$, so there is no solution.

Suppose we apply the law of sines to obtain the angle C in Figure 6. We have

$$\frac{\sin C}{4} = \frac{\sin 35°}{2}$$

$$\sin C = \frac{4}{2} \sin 35° = 2 \sin 35° \approx 1.1472$$

But since $0 < \sin C < 1$, this is impossible. So we conclude, again, that there is no triangle that meets the given conditions. ■

EXAMPLE 3 *The Ambiguous (SSA) Case: Two Solutions*

Find all triangles with $a = 8$, $c = 10$, and $A = 48°$.

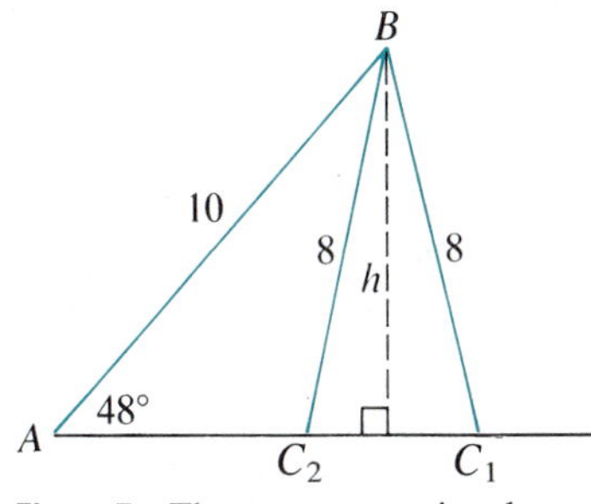

Figure 7 There are two triangles with sides of 8 and 10 such that the angle opposite the side of length 8 is 48°.

SOLUTION Here $h = c \sin A = 10 \sin 48° \approx 7.43 < a < c$, so there are two such triangles. From Figure 7, we find that

$$\frac{\sin C_1}{10} = \frac{\sin 48°}{8}$$

$$\sin C_1 = \frac{10 \sin 48°}{8} \approx 0.928931032$$

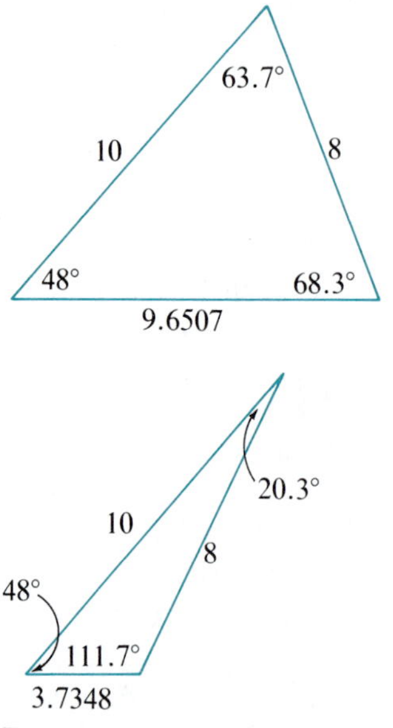

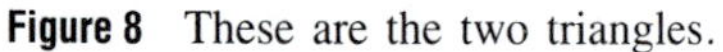

Figure 8 These are the two triangles.

so

$$C_1 = \sin^{-1} 0.928931032 \approx 68.3°$$

Another angle with the same sine value is

$$C_2 = 180° - C_1 \approx 180° - 68.3° = 111.7°$$

Note that

$$48° + 68.3° = 116.3° < 180°$$

and

$$48° + 111.7° = 159.7° < 180°$$

so both values of C are possible and there are two triangles.

Triangle 1 $C = 68.3°$ and $B = 180° - 48° - 68.3° = 63.7°$, so

$$\frac{\sin B}{b} = \frac{\sin 63.7°}{b} = \frac{\sin 48°}{8}$$

$$b = \frac{8 \sin 63.7°}{\sin 48°} \approx 9.6507$$

Triangle 2 $C = 111.7°$ and $B = 180° - 48° - 111.7° = 20.3°$, so

$$\frac{\sin 20.3°}{b} = \frac{\sin 48°}{8}$$

$$b = \frac{8 \sin 20.3°}{\sin 48°} \approx 3.7348$$

The two triangles are sketched in Figure 8.

We have shown you examples of no solutions or two solutions in the ambiguous case. In Problem 17, you are asked to show that there is exactly one triangle with $a = 6$, $c = 5$, and $A = 110°$.

Finding the Area of a Triangle

We can use the proof of the law of sines to compute the area of any triangle. Look at Figure 9. Recall that the area of a triangle is given by the formula

$$\text{area of a triangle} = \frac{1}{2}\ \text{base} \times \text{height}$$

In the proof of the law of sines we found that for each triangle in Figure 9, $h = c \sin B$ so

$$\text{area} = \frac{1}{2}\ \text{base} \times \text{height} = \frac{1}{2}ah = \frac{1}{2}ac \sin B$$

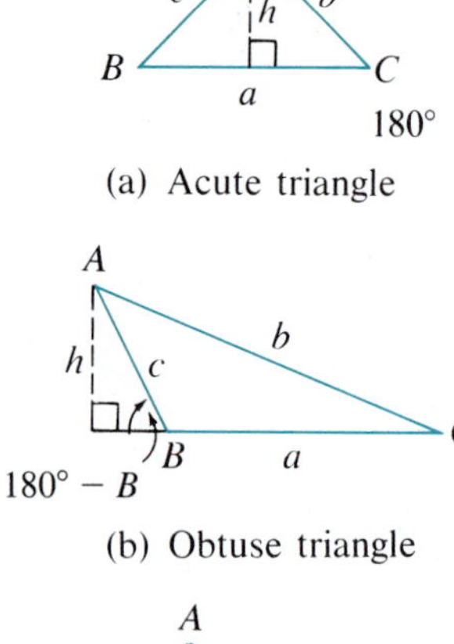

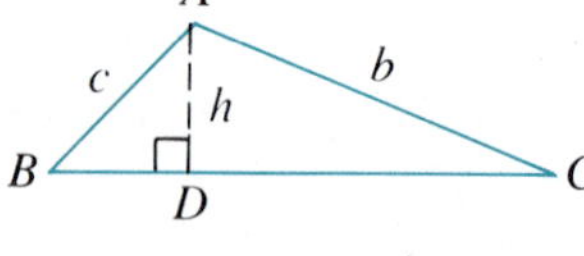

(c) Obtuse triangle

Figure 9 In each triangle $\sin B = \dfrac{h}{c}$.

Using similar reasoning we can show that

$$\text{area} = \frac{1}{2}ab \sin C = \frac{1}{2}bc \sin A$$

That is,

Area of a Triangle

The area of any triangle is equal to one half the product of the lengths of two sides and the sine of the angle between them. That is,

$$\begin{aligned}\text{Area of triangle } ABC &= \frac{1}{2}ab \sin C \\ &= \frac{1}{2}ac \sin B = \frac{1}{2}bc \sin A\end{aligned} \qquad (2)$$

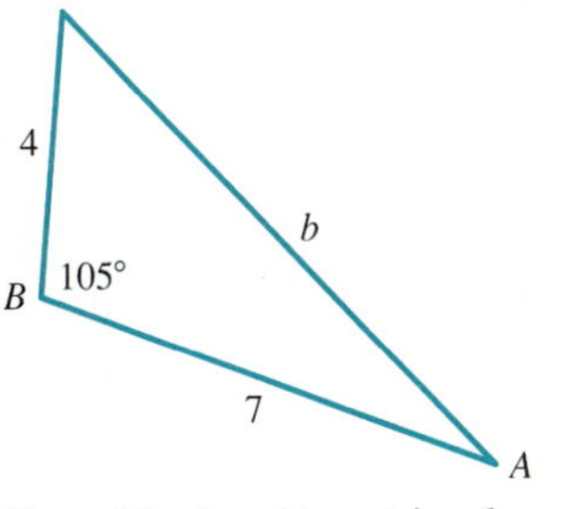

Figure 10 An obtuse triangle

EXAMPLE 4 *Calculating the Area of a Triangle*

Compute the area of a triangle with side of 4 and 7 and an included angle of 105°.

SOLUTION The triangle is sketched in Figure 10. From (2), we have

$$\text{area} = \frac{1}{2}ac \sin B = \frac{1}{2}(4)(7) \sin 105° \approx 14(0.96593) \approx 13.523$$

Problems 4.1

Readiness Check

I. Which of the following is given in the ambiguous case?
a. AAS b. SSA c. SSS d. AAA

II. Which of the following is a missing part of the triangle where $C = 86°$, $b = 17$, and $c = 19$?
a. $a = 10.6$ b. $A = 30.8°$
c. $B = 53.2°$ d. $A = 90°$

III. In which of the following would there be no solution?
a. $a = 37$, $b = 19$, $c = 40$
b. $A = 63°$, $B = 45°$, $b = 23$
c. $A = 25°$, $c = 17$, $a = 7.2$
d. $C = 35°$, $b = 22.1$, $c = 11.3$

IV. Which of the following would yield two triangles?
a. $A = 37°$, $a = 18$, $b = 17$
b. $B = 53°$, $a = 25$, $b = 15$
c. $C = 47°$, $b = 10$, $c = 8$
d. $A = 64°$, $a = 19$, $c = 18$

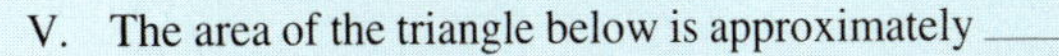
V. The area of the triangle below is approximately ______.
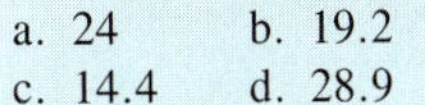
a. 24 b. 19.2
c. 14.4 d. 28.9

6
37°
8

Answers to Readiness Check

I. b II. b III. d IV. c V. c

In Problems 1–8 solve each oblique triangle.

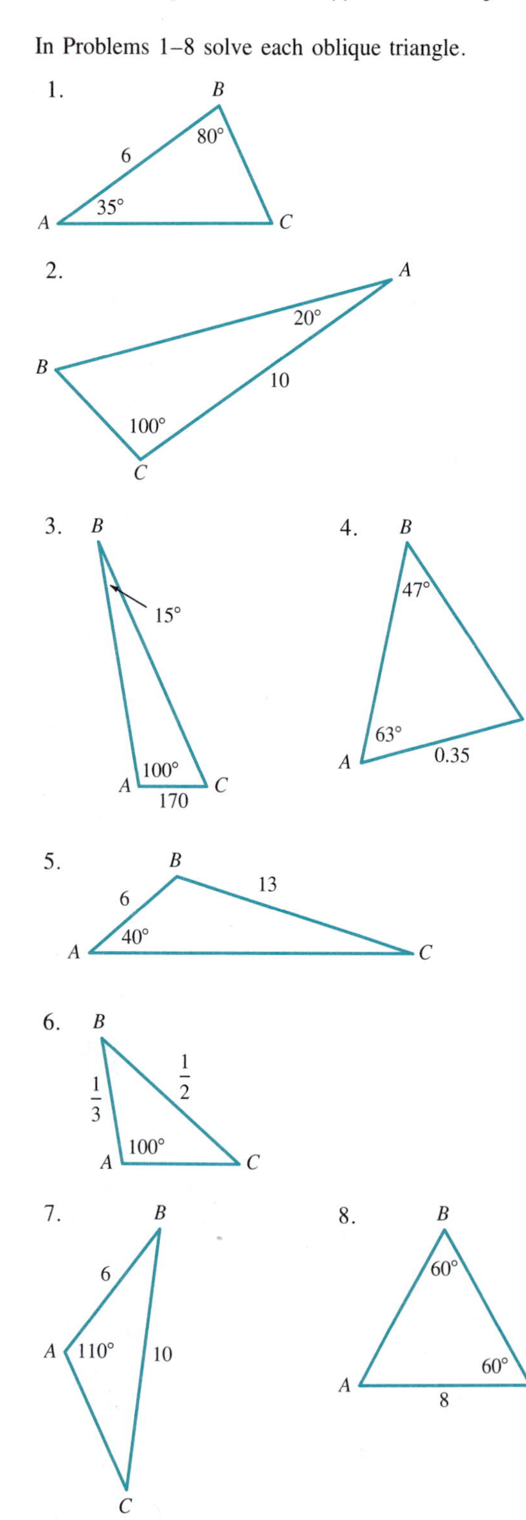

In Problems 9–16 two angles and one side of a triangle are given. Find the remaining angle and sides.

9. $a = 15$, $C = 65°$, $B = 55°$
10. $c = 4$, $A = 67.2°$, $B = 42.3°$
11. $c = 150$, $C = 105°$, $A = 10°$
12. $a = 2500$, $A = 46.2°$, $B = 41.5°$
13. $b = 255{,}000$, $A = 38.2°$, $C = 59.7°$
14. $a = 0.315$, $B = 57.4°$, $C = 95.8°$
15. $a = 3.17$, $A = 8°$, $B = 20°$
16. $b = 4 \times 10^7$, $N = 9°$, $C = 15°$
17. Show that there is exactly one triangle with sides $a = 6$, $c = 5$, and opposite angle $A = 110°$.

In Problems 18–28 find the measurements of all triangles (if any) that have the given sides and opposite angle.

18. $a = 10$, $b = 6$, $A = 58°$
19. $a = 6$, $b = 10$, $A = 58°$
20. $a = 9$, $b = 10$, $A = 58°$
21. $a = 30$, $b = 40$, $B = 27°$
22. $a = 30$, $b = 20$, $B = 27°$
23. $a = 30$, $b = 20$, $B = 27°$
24. $b = 2$, $c = 1$, $C = 102°$
25. $b = 2$, $c = 3$, $C = 102°$
26. $a = 100$, $c = 150$, $A = 15.6°$
27. $a = 100$, $c = 150$, $A = 46.8°$
28. $a = 200$, $c = 150$, $A = 46.8°$
29. Let $a = 10$, $b = x$, and $A = 62°$. For what values of x will there be
    (a) no triangle with sides a and b and opposite angle A?
    (b) one such triangle?
    (c) two such triangles?
30. In triangle ABC let $a = 37$ and $b = 58$. For what values of A will there be
    (a) no triangle with sides a and b and opposite angle A?
    (b) one such triangle?
    (c) two such triangles?

In Problems 31–36 compute the area of each triangle. In some cases it will first be necessary to use the law of sines.

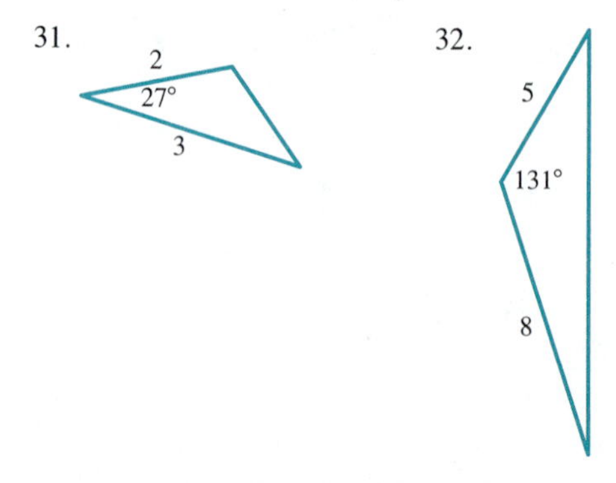

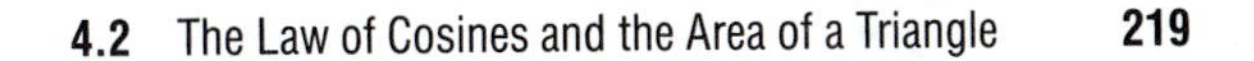

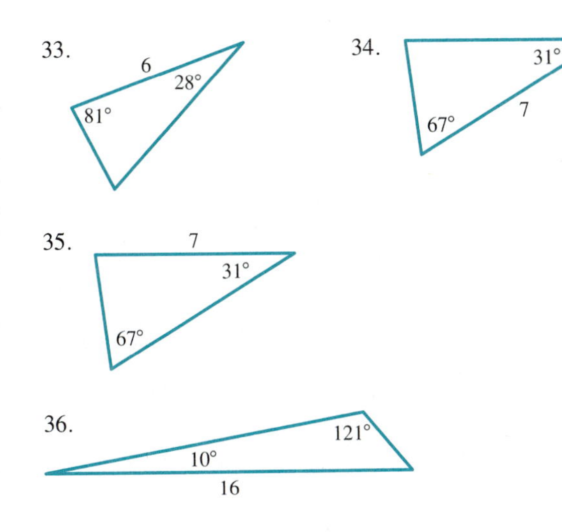

* 37. Prove the law of sines in the case where C is a right angle.

38. Show that for any triangle ABC,

$$\frac{b-c}{c} = \frac{\sin B - \sin C}{\sin C}$$

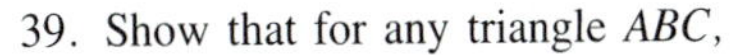

39. Show that for any triangle ABC,
 (i) $a \geq b \sin A$
 (ii) $b \geq a \sin B$
 (iii) $c \geq b \sin C$

** 40. Prove the **law of tangents.** In any triangle ABC,

$$\frac{b-a}{b+a} = \frac{\tan\left(\frac{B-A}{2}\right)}{\tan\left(\frac{B+A}{2}\right)}, \quad \frac{a-c}{a+c} = \frac{\tan\left(\frac{A-C}{2}\right)}{\tan\left(\frac{A+C}{2}\right)}$$

$$\frac{c-b}{c+b} = \frac{\tan\left(\frac{C-B}{2}\right)}{\tan\left(\frac{C+B}{2}\right)}$$

[Hint: Use the law of sines to show that $\frac{b-a}{b+a} = \frac{\sin B - \sin A}{\sin B + \sin A}$ and then use appropriate sum-to-product identities from Chapter 3.]

4.2 The Law of Cosines and the Area of a Triangle

The law of cosines enables us to solve oblique triangles of the type SAS or SSS.

Law of Cosines

Let ABC be a triangle with sides a, b, and c. Then

$$a^2 = b^2 + c^2 - 2bc \cos A \tag{1}$$

$$b^2 = a^2 + c^2 - 2ac \cos B \tag{2}$$

$$c^2 = a^2 + b^2 - 2ab \cos C \tag{3}$$

That is, the square of the length of any one side of a triangle equals the sum of the squares of the lengths of the other two sides minus twice the product of the lengths of the other two sides and the cosine of the angle between them.

The law of cosines can be written in terms of angles. For example, if we solve equation (1) for $\cos A$, we obtain

$$\cos A = \frac{b^2 + c^2 - a^2}{2bc} \tag{4}$$

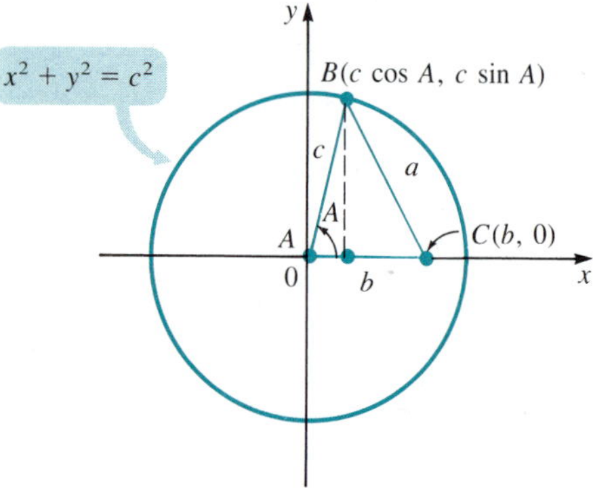

(a) Acute triangle

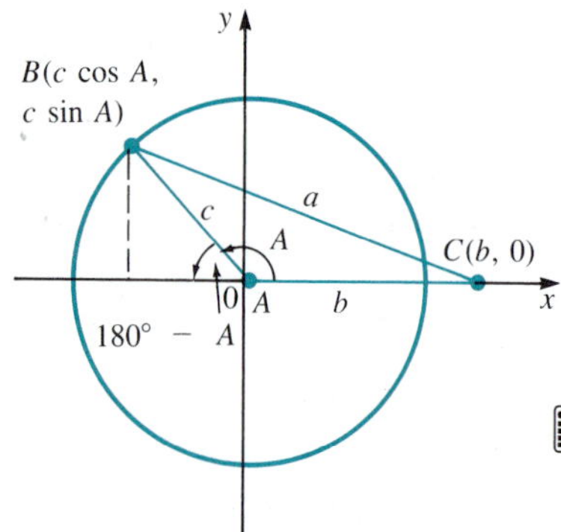

(b) Obtuse triangle

Figure 1 If the coordinates of B are (x, y), then, by definition, $\cos A = \dfrac{x}{c}$ and $\sin A = \dfrac{y}{c}$.

Proof

We prove (1) only. The proofs of (2) and (3) are identical. Then (4) follows from (1).

There are three cases to consider:

(i) $A = 90°$.
(ii) A is acute.
(iii) A is obtuse.

If $A = 90°$, then $\cos A = 0$, ABC is a right triangle with hypotenuse a, and (1) is simply the Pythagorean theorem. To prove (1) when A is acute or obtuse, we place the triangle with A in standard position and the side AC along the positive x-axis (see Figure 1). We draw circles of radius c that are centered at the origin. By the alternative definitions of the sine and cosine functions on p. 80, the point B in either figure has the coordinates $(c \cos A, c \sin A)$.

Then, from the distance formula (see p. 10),

$$
\begin{aligned}
a^2 = \overline{BC}^2 &= (c \cos A - b)^2 + (c \sin A - 0)^2 \\
&= c^2 \cos^2 A - 2bc \cos A + b^2 + c^2 \sin^2 A \\
&= b^2 + c^2(\sin^2 A + \cos^2 A) - 2bc \cos A && \text{Rearrange terms} \\
&= b^2 + c^2 - 2bc \cos A && \sin^2 A + \cos^2 A = 1
\end{aligned}
$$

EXAMPLE 1 *Solving a Triangle When Two Sides and the Included Angle Are Given*

Solve the triangle in Figure 2.

SOLUTION From (1),

$$
\begin{aligned}
a^2 &= b^2 + c^2 - 2bc \cos A \\
a^2 &= 5^2 + 8^2 - 2(5)(8) \cos 41° \\
&= 25 + 64 - 80(0.75470958) \\
&= 89 - 60.37676642 \approx 28.623
\end{aligned}
$$

Thus

$$a \approx \sqrt{28.623} \approx 5.3500$$

We now know all three sides. To obtain a second angle, we use the law of sines:

$$\frac{\sin B}{5} = \frac{\sin A}{a}$$

$$\sin B = \frac{5 \sin A}{a} = \frac{5 \sin 41°}{5.3500} \approx 0.61314$$

Then

$$
\begin{aligned}
B &= \sin^{-1} 0.61314 \quad &\text{or} \quad B &= 180° - \sin^{-1} 0.61314 \\
B &\approx 37.8° \quad &\text{or} \quad B &\approx 180° - 37.8° = 142.2°
\end{aligned}
$$

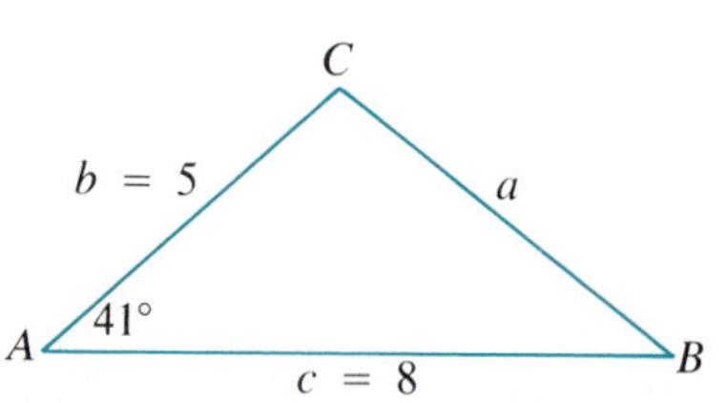

Figure 2 In this triangle sides b and c and angle A are given.

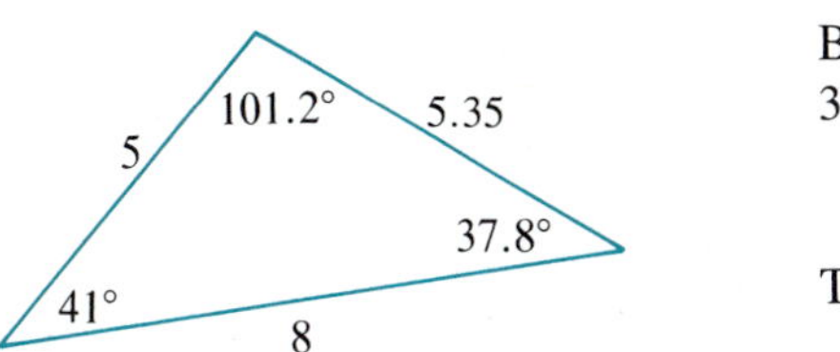

Figure 3 The (obtuse) triangle solved.

But 142.2° is impossible since $142.2° + 41° = 183.2° > 180°$. Thus $B = 37.8°$ and

$$C = 180° - 41° - 37.8° = 101.2°$$

The complete measurements are shown in Figure 3.

Alternative Approach We can modify formula (4) to obtain B:

$$\cos B = \frac{a^2 + c^2 - b^2}{2ac} \approx \frac{28.623 + 64 - 25}{2(5.35)(8)} = \frac{67.623}{85.6} \approx 0.7900$$

and

$$B \approx \cos^{-1} 0.79 \approx 37.8°$$

NOTE Using the alternative approach serves as a check of our computations. ■

EXAMPLE 2 *Solving a Triangle When All Three Sides Are Given*

Solve the triangle in Figure 4.

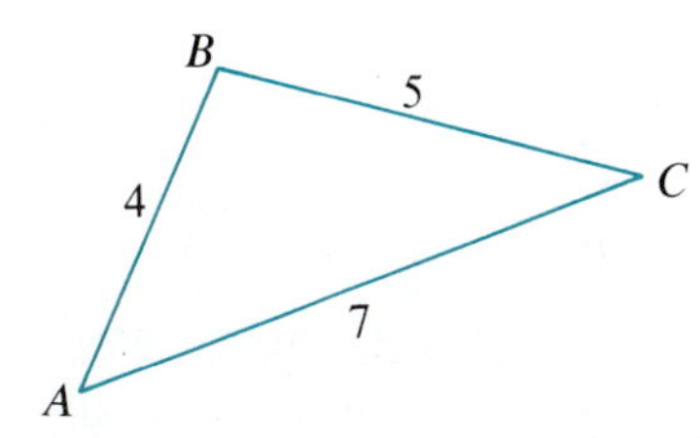

Figure 4 In this triangle the three sides are given.

SOLUTION We use the law of cosines (formula (4)) to obtain angles A and B.

$$\cos A = \frac{b^2 + c^2 - a^2}{2bc} = \frac{7^2 + 4^2 - 5^2}{2(7)(4)} = \frac{49 + 16 - 25}{56}$$

$$= \frac{40}{56} = 0.714285714$$

and

$$A = \cos^{-1} 0.714285714 \approx 44.4°$$

$$\cos B = \frac{a^2 + c^2 - b^2}{2ac} = \frac{5^2 + 4^2 - 7^2}{2(5)(4)} = \frac{25 + 16 - 49}{40}$$

$$= -\frac{8}{40} = -0.2$$

and

$$B = \cos^{-1} (-0.2) \approx 101.5°$$

Then

$$C = 180° - A - B = 180° - 44.4° - 101.5° = 34.1°$$

Two ways to check.

Method 1 Use (4):

$$\cos C = \frac{a^2 + b^2 - c^2}{2ab} = \frac{5^2 + 7^2 - 4^2}{2(5)(7)} = \frac{58}{70} = 0.828571428$$

and

$$C = \cos^{-1} 0.828571428 \approx 34.048°$$

This rounds to 34.0°. The discrepancy is due to roundoff error.

Method 2 Use the law of sines: If our computations are correct, then

$$\frac{\sin A}{a} = \frac{\sin B}{b} = \frac{\sin C}{c}$$

$$\frac{\sin 44.4^\circ}{5} = \frac{\sin 101.5^\circ}{7} = \frac{\sin 34.1^\circ}{4}$$

$$0.13993 = 0.13999 = 0.14016$$

This is close enough. Again, the discrepancy is due to roundoff error.

WARNING When asked to solve an oblique triangle, first make certain that there is a triangle to solve. ■

EXAMPLE 3 *An Impossible Triangle*

Solve the triangle with sides 4, 5, and 12.

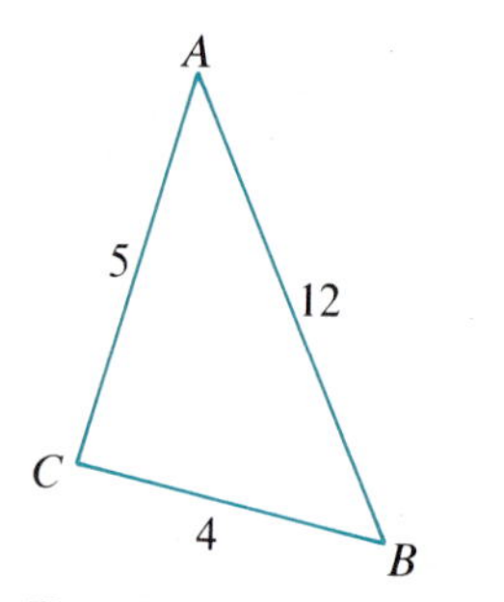

Figure 5 An impossible triangle.

SOLUTION The "triangle" is sketched in Figure 5. But there is no such triangle. Do you see why? From geometry, the shortest distance between two points is a straight line. Thus

$$\overline{AB} = \text{shortest distance from } A \text{ to } B < \overline{AC} + \overline{CB}$$

But $\overline{AB} = 12$, whereas $\overline{AC} + \overline{CB} = 5 + 4 = 9$. Thus the "triangle" in Figure 5 doesn't exist, so it makes no sense to try to solve it. If we blindly apply formula (4), we obtain

$$\cos A = \frac{b^2 + c^2 - a^2}{2bc} = \frac{5^2 + 12^2 - 4^2}{2(5)(12)} = \frac{153}{120} > 1$$

which is, of course, impossible.

We summarize our result:

The length of one side of a triangle is always less than the sum of the lengths of the other two sides. Do not try to solve impossible triangles where this condition is violated.

Heron's Formula and the Area of a Triangle

We can use the law of cosines to prove a remarkable formula that gives the area of a triangle in terms of the lengths of its sides. This formula is known as **Heron's formula** in honor of the Greek mathematician Heron of Alexandria (born ca. 75 B.C.). However, the Persian mathematician al-Biruni (who first proved the law of sines) wrote that Heron's formula was proved earlier by perhaps the greatest mathematician of all time, Archimedes of Syracuse.

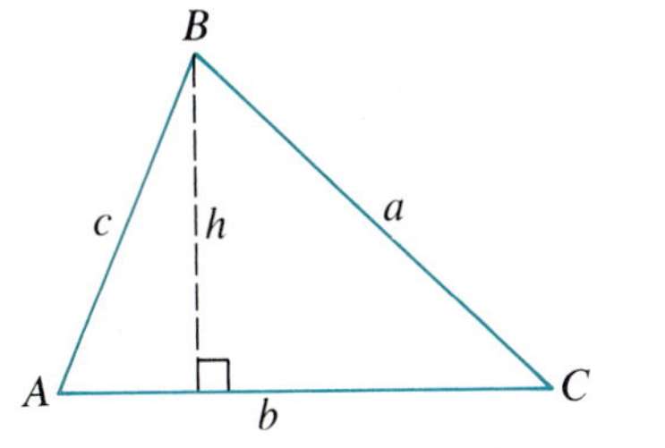

Figure 6 The area of this triangle equals $\frac{1}{2}bc \sin A$.

Consider the triangle in Figure 6. From equation (2) on p. 217, we know that

$$\text{area of triangle} = \frac{1}{2}bc \sin A \tag{5}$$

We can use this fact to derive a way to compute the area knowing only the lengths of the three sides of the triangle.

We define s to be half the perimeter of the triangle:

$$s = \frac{1}{2}(a + b + c)$$

Heron's Formula

$$\text{Area of triangle } ABC = \sqrt{s(s - a)(s - b)(s - c)} \tag{6}$$

where $s = \frac{1}{2}(a + b + c)$

A proof of Heron's formula is given after an example.†

EXAMPLE 4 ***Using Heron's Formula to Find the Area of a Triangle***

Find the area of the triangle with sides 4, 7, and 9.

SOLUTION Here $s = \frac{1}{2}(4 + 7 + 9) = \frac{1}{2}(20) = 10$, so

$$\text{Area} = \sqrt{10(10 - 4)(10 - 7)(10 - 9)} = \sqrt{10(6)(3)(1)}$$
$$= \sqrt{180} \approx 13.416$$

Proof of Heron's Formula (Optional)

We start with (5).

$$\text{Area} = \frac{1}{2}bc \sin A = \sqrt{\frac{1}{4}b^2c^2 \sin^2 A}$$
$$= \sqrt{\frac{1}{4}b^2c^2(1 - \cos^2 A)} = \sqrt{\frac{1}{4}b^2c^2(1 + \cos A)(1 - \cos A)}$$

or

$$\text{Area} = \sqrt{\left[\frac{1}{2}bc(1 + \cos A)\right]\left[\frac{1}{2}bc(1 - \cos A)\right]} \tag{7}$$

† For an interesting discussion of how a student might derive Heron's formula, see Roger C. Alperin, "Heron's Area Formula," *The College Mathematics Journal* (18)(2), 1987.

From (4)

$$\cos A = \frac{b^2 + c^2 - a^2}{2bc}$$

Then

$$\begin{aligned}
\frac{1}{2}bc(1 + \cos A) &= \frac{1}{2}bc\left(1 + \frac{b^2 + c^2 - a^2}{2bc}\right) \\
&= \frac{1}{2}bc\left(\frac{2bc + b^2 + c^2 - a^2}{2bc}\right) \\
&= \frac{2bc + b^2 + c^2 - a^2}{4} = \frac{(b^2 + 2bc + c^2) - a^2}{4} \\
&= \frac{(b + c)^2 - a^2}{4} \\
&= \left(\frac{b + c + a}{2}\right)\left(\frac{b + c - a}{2}\right)
\end{aligned}$$

Similarly

$$\begin{aligned}
\frac{1}{2}bc(1 - \cos A) &= \frac{1}{2}bc\left(1 - \frac{b^2 + c^2 - a^2}{2bc}\right) \\
&= \frac{1}{2}bc\left(\frac{2bc - b^2 - c^2 + a^2}{2bc}\right) \\
&= \frac{a^2 - (b^2 + c^2 - 2bc)}{4} = \frac{a^2 - (b - c)^2}{4} \\
&= \left(\frac{a - (b - c)}{2}\right)\left(\frac{a + (b - c)}{2}\right) \\
&= \left(\frac{a - b + c}{2}\right)\left(\frac{a + b - c}{2}\right)
\end{aligned}$$

Thus

$$\text{area} = \sqrt{\left(\frac{b + c + a}{2}\right)\left(\frac{b + c - a}{2}\right)\left(\frac{a - b + c}{2}\right)\left(\frac{a + b - c}{2}\right)} \qquad \textbf{(8)}$$

If $s = \dfrac{a + b + c}{2}$, then

$$s - a = \frac{b + c - a}{2}, \quad s - b = \frac{a - b + c}{2}, \quad \text{and} \quad s - c = \frac{a + b - c}{2}$$

Inserting these values into (8), we conclude that

$$\text{area} = \sqrt{s(s - a)(s - b)(s - c)} \quad ■$$

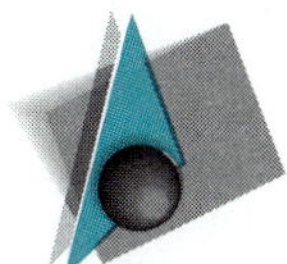

FOCUS ON
The Law of Cosines

The law of cosines was first stated by Euclid in Book II of his *Elements*, but was not proved until 1579 by François Viète (1540–1603). Viète is considered the finest French mathematician of the sixteenth century. In his 1579 work *Canon mathematicus seu ad triangula,* Viète proved the following version of the law of cosines:

$$\frac{2ab}{a^2 + b^2 - c^2} = \frac{1}{\sin(90° - C)}$$

Viète did many other things as well. In his most famous work *In Artem analyticam isagoge,* published in 1591, Viète developed much of the algebraic notation we use today. For example, he wrote *Aq* (*A quadratam* or *A* squared) to denote A^2. Before Viète, mathematicians used different letters to denote different powers of an unknown quantity. In another of his works, Viète described a method for successively approximating the roots of a polynomial. This method was in wide use until about 1680, when Newton's method was found to be superior.

Problems 4.2

Readiness Check

I. Which of the following is true about the law of cosines?
 a. It can be used to solve a triangle when given 3 angles.
 b. It can be used to solve a triangle when an angle, a side adjacent, and the side opposite the angle are given.
 c. The Pythagorean Theorem is a special case of the law of cosines.
 d. It expresses the relationship of a side of a triangle to a function of an adjacent angle.

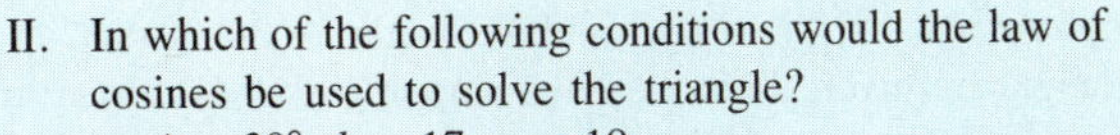

II. In which of the following conditions would the law of cosines be used to solve the triangle?
 a. $A = 30°$, $b = 17$, $a = 19$
 b. $C = 125°$, $a = 16$, $b = 64$
 c. $A = 64°$, $B = 33°$, $C = 83°$
 d. $A = 25°$, $B = 100°$, $b = 31$

III. Which of the following would replace s in Heron's formula for the area of the triangle with $a = 22$, $b = 18$, and $c = 14$?
 a. 108 b. 54 c. 27 d. 13.5

In Problems 1–8 solve each triangle, if possible.

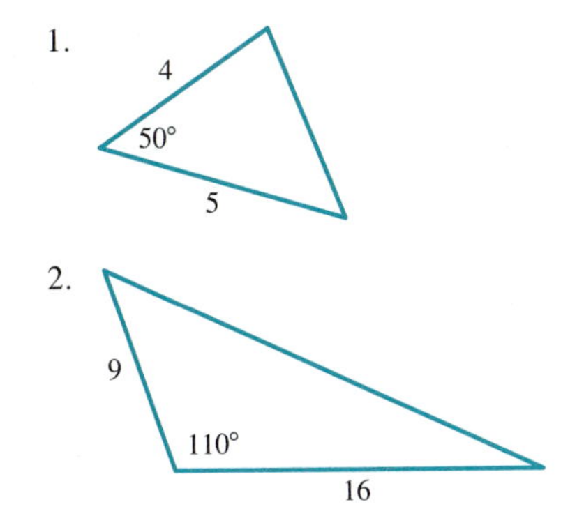

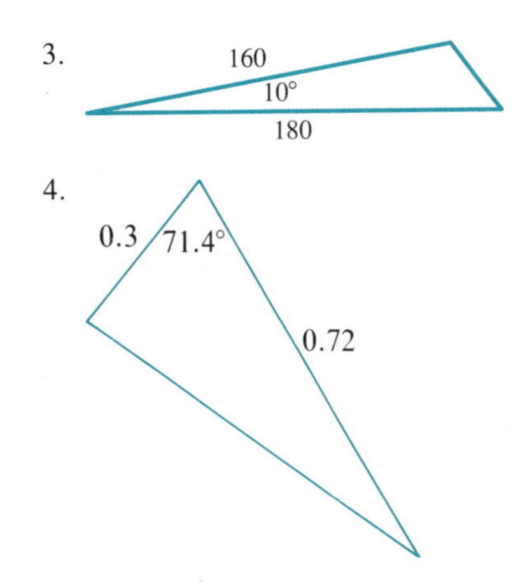

Answers to Readiness Check

I. c II. b III. c

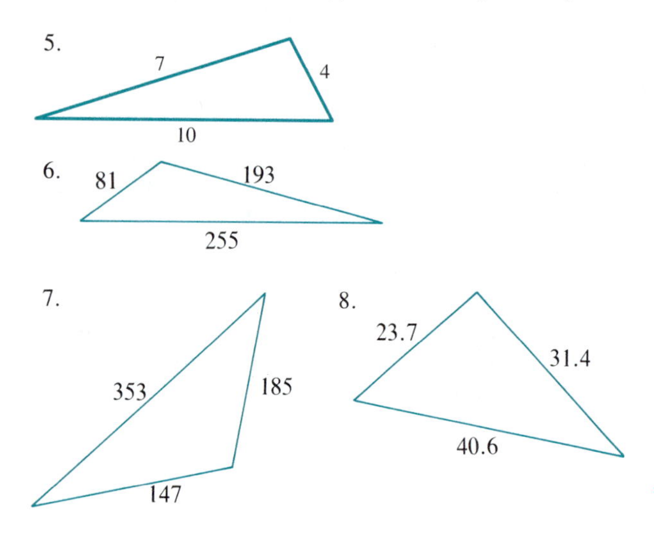

In Problems 9–18 two sides and one angle, or three sides, of a triangle are given. Find the remaining angles and, if missing, the remaining side. Note those triangles that do not exist.

9. $A = 42°$, $c = 4$, $b = 5$
10. $b = 14$, $c = 20$, $A = 59°$
11. $C = 38°$, $a = 120$, $b = 170$
12. $a = 15$, $c = 27$, $B = 109°$
13. $b = 1251$, $a = 4238$, $C = 152°$
14. $a = 8$, $b = 10$, $c = 12$
15. $a = 11$, $b = 14$, $c = 20$
16. $a = 1$, $b = 2$, $c = 3$
17. $a = 15$, $b = 8$, $c = 8$
18. $a = 38$, $b = 51$, $c = 38$

In Problems 19–26 find the area of the triangle whose sides have the given lengths.

19. 5, 12, 13
20. 25, 30, 47
21. 8, 13, 21
22. 215, 302, 417
23. 0.71, 0.94, 1.23
24. 2341, 5172, 2409
25. 3127, 4183, 2801
26. $\sqrt{3}$, $\sqrt{5}$, $\sqrt{7}$
27. Prove that for any triangle ABC,

$$a^2 + b^2 + c^2 = 2(ab \cos C + bc \cos A + ac \cos B)$$

28. Prove that for any triangle ABC,

$$c^2 - a^2 = b(c \cos A - a \cos C)$$

29. Prove that for any triangle ABC,

$$\frac{a^2 + b^2 + c^2}{2abc} = \frac{\cos A}{a} + \frac{\cos B}{b} + \frac{\cos C}{c}$$

30. Find a formula for the height h, as in Figure 6, in terms of a, b, and c.
31. An equilateral triangle has an area of 25 cm^2. Use Heron's formula to find, to the nearest tenth of a centimeter, the length of one of its sides.
32. The two equal sides of an isosceles triangle have length l. The third side has length 5. Use Heron's formula to write the area of the triangle in terms of l.

* 33. Prove **Newton's formula:**

$$\frac{a - b}{c} = \frac{\sin \frac{1}{2}(A - B)}{\cos \frac{1}{2}C}$$

* 34. Show that

$$\text{Area of triangle } ABC = \frac{c^2}{2}\frac{\sin A \sin B}{\sin (A + B)}$$

* 35. The following "wordless" proof of the law of cosines appeared on p. 259 of the October 1988 issue of *Mathematics Magazine*. It was sent in by Timothy A. Sipka at Anderson University. Explain how the proof works.

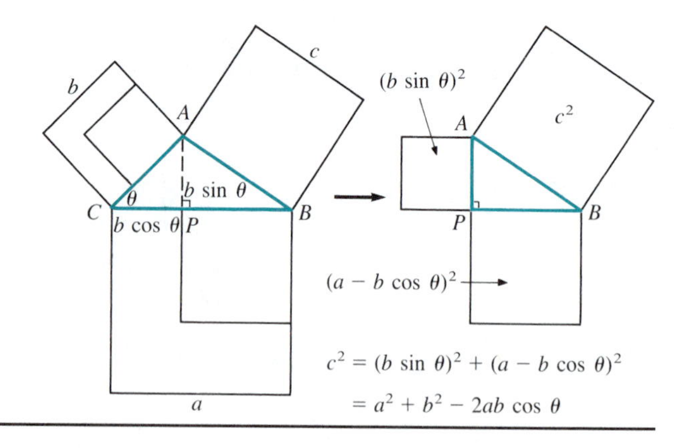

4.3 Applications of the Laws of Sines and Cosines

In Section 2.8, we gave a number of applications of right triangle trigonometry. There we were limited to problems involving right triangles. Using the laws of sines and cosines, we can remove that limitation.

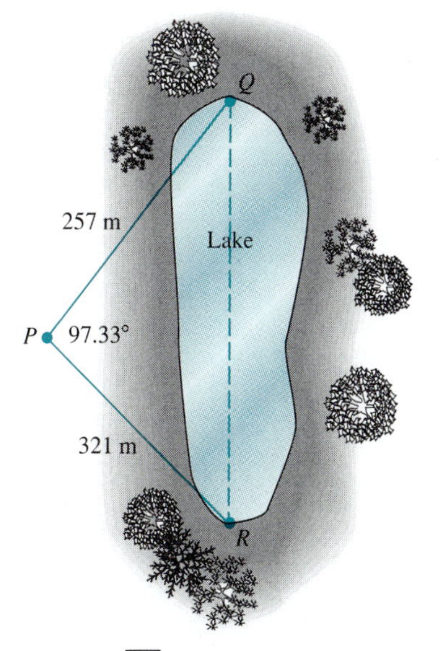

Figure 1 $\overline{RQ}$ is the (unknown) distance across the lake.

Measuring Distances and Heights

EXAMPLE 1 *Determining the Distance Across a Lake*

Two markers lie at opposite ends of a lake (see Figure 1). From a reference point P, the distance to the marker labeled Q is 257 m, whereas the distance to the marker labeled R is 321 m. A surveyor using a transit determines that the measure of the angle QPR is 97.33°. What is the distance, to the nearest meter, between the markers?

SOLUTION We apply the law of cosines to triangle QPR. We have

$$\overline{QR}^2 = \overline{PQ}^2 + \overline{PR}^2 - 2(\overline{PQ})(\overline{PR}) \cos P$$

$$\overline{QR}^2 = 257^2 + 321^2 - 2(257)(321) \cos 97.33°$$

$$\approx 66{,}049 + 103{,}041 - 164{,}994(-0.12758)$$

$$\approx 169{,}090 + 21{,}050.6 = 190{,}140.6$$

and

$$\overline{QR} \approx \sqrt{190{,}140.6} \approx 436 \text{ m}$$

In Example 1, we saw that cos 97.33° ≈ −0.1276. When we use the law of cosines it helps to remember that the cosine of an obtuse angle is negative.

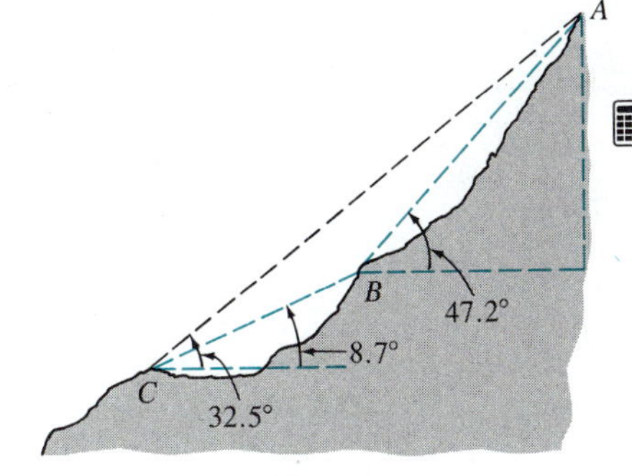

(a)

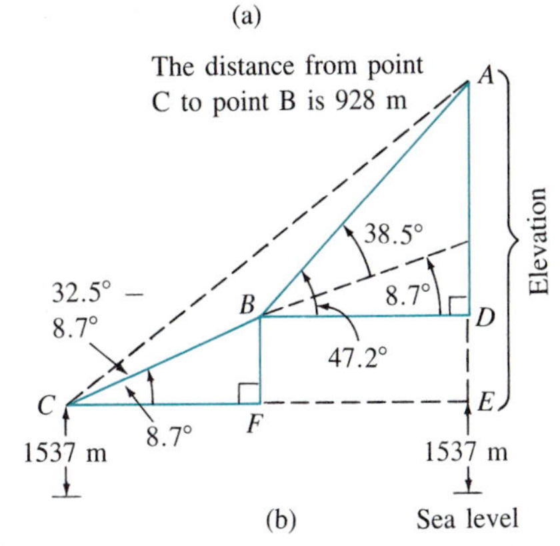

(b)

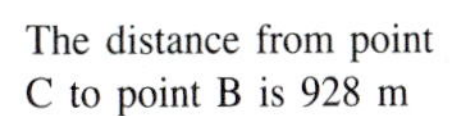
The distance from point C to point B is 928 m

Figure 2 The elevation at A equals $\overline{AD} + \overline{DE} + 1537$.

EXAMPLE 2 *Determining Elevation*

In hilly or mountainous terrain, it is often difficult to determine heights. There are two problems. First, it may be impossible to find an expanse of level ground from which measurements can be taken. Second, sighting may be limited. That is, points of reference such as mountain peaks may be visible only from certain places. One such situation is illustrated in Figure 2(a).

A geologist took measurements at two points, B and C, 928 m apart. The angle of elevation from B to the top of a mountain at A is 47.2°. The angle of elevation from C to B is 8.7°, and the angle of elevation from C to A is 32.5°. It is known that the elevation at C is 1537 m. To the nearest meter, what is the elevation at A?

SOLUTION In Figure 2(b), we depict the information given. The solution is long, and we will do it in stages. We seek the distance $\overline{AE} = \overline{AD} + \overline{DE}$. $\overline{AD}$ is one side of the right triangle ADB. We can find $\overline{AD}$ if we know $\overline{AB}$. Also $\overline{DE} = \overline{BF}$, and we can determine $\overline{BF}$ since it is one side of the right triangle BCF, whose hypotenuse $\overline{CB}$ is known.

Step 1 Determine $\overline{AB}$ We need to solve the triangle CBA. We already know one angle and one side:

$$\measuredangle ACB^{\dagger} = 32.5° - 8.7° = 23.8°$$

$$\overline{CB} = 928$$

† $\measuredangle ACB$ denotes the measure of the angle ACB.

We need one more angle in order to use the law of sines. Look at the point B. There are four angles that meet at B, and their measures must add up to 360°.

$$\measuredangle ABD = 47.2° \quad \text{Given}$$
$$\measuredangle DBF = 90° \quad \text{A right angle}$$
$$\measuredangle FBC = 81.3° \quad \measuredangle FBC \text{ and } \measuredangle BCF \text{ are complements, so } \measuredangle FBC = 90° - 8.7°$$
$$\measuredangle CBA = x$$

Thus,

$$47.2° + 90° + 81.3° + x = 360°$$
$$218.5° + x = 360°$$
$$\measuredangle CBA = x = 360° - 218.5° = 141.5°$$

Then

$$\measuredangle BAC = 180° - 23.8° - 141.5° = 180° - 165.3° = 14.7°$$

From the law of sines,

$$\frac{\sin \measuredangle ACB}{\overline{AB}} = \frac{\sin \measuredangle BAC}{\overline{CB}}$$
$$\frac{\sin 23.8°}{\overline{AB}} = \frac{\sin 14.7°}{928}$$
$$\overline{AB} \approx \frac{928 \sin 23.8°}{\sin 14.7°} \approx 1475.8 \text{ m}$$

Step 2 Determine $\overline{AD}$

$$\sin 47.2° = \frac{\overline{AD}}{\overline{AB}}$$
$$\overline{AD} = \overline{AB} \sin 47.2° \approx 1475.8 \sin 47.2° \approx 1083 \text{ m}$$

Step 3 Determine $\overline{DE} = \overline{BF}$

$$\sin 8.7° = \frac{\overline{BF}}{\overline{CB}}$$
$$\overline{DE} = \overline{BF} = \overline{CB} \sin 8.7° \approx 928 \sin 8.7° \approx 140 \text{ m}$$

Finally,

$$\text{elevation at } A = 1537 \text{ m} + \overline{DE} + \overline{AD}$$
$$\approx 1537 \text{ m} + 140 \text{ m} + 1083 \text{ m} = 2760 \text{ m}$$ ■

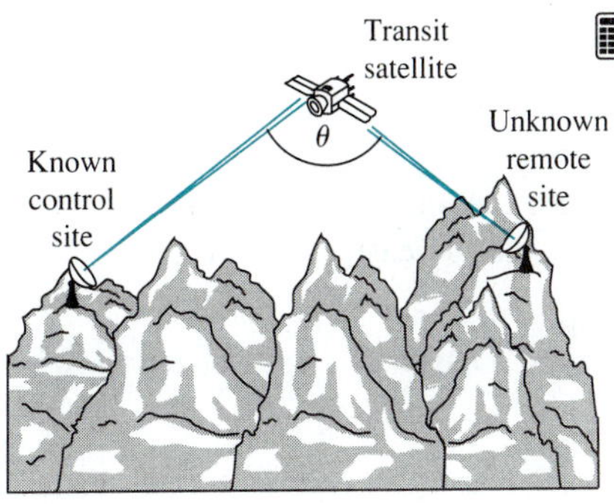

Figure 3 Using a transit

EXAMPLE 3 *Measuring Distances with a Transit Satellite*

In 1964 the U.S. Navy, in cooperation with the Applied Physics Laboratory of Johns Hopkins University, developed the *U.S. Navy Navigation Satellite* or *Transit*. The system consists of a satellite and two ground antennas. One of the antennas is installed at a known control site and the other is set up at a remote site that is to be surveyed (see Figure 3). The satellite orbits the earth (around the poles) in about 1 hr 47 min at an altitude of about 1000 km above the earth's surface. It broadcasts a steady stream of digital data that enables an observer at the control site to determine both the distance from the satellite to the control site as well as the angle θ between the line from the transit to the control site and the line from the transit to the remote site.

In one measurement, it was determined that the distance from the transit to the control site was 2235 km while the angle θ was $102°38'$. Compute, to the nearest kilometer, the horizontal distance between the control site and the remote site.

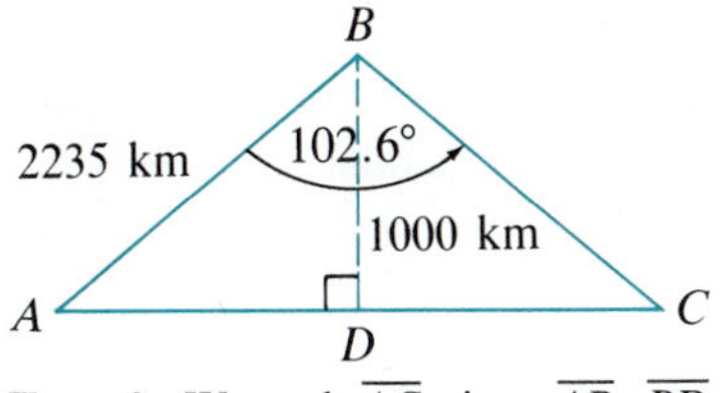

Figure 4 We seek $\overline{AC}$ given $\overline{AB}$, $\overline{BD}$, and angle B.

SOLUTION We need to determine the distance $\overline{AC}$ in Figure 4. Since we already know one side and one angle of triangle ABC, we can determine $\overline{AC}$ if we know one other angle. We use the fact that the satellite is 1000 km above sea level. Triangle ABD is a right triangle, so

$$\sin A = \frac{\overline{BD}}{\overline{AB}} = \frac{1000}{2235} = 0.447427293$$

and

$$A = \sin^{-1} 0.447427293 \approx 26.6°$$

Also

$$B = 102°38' = 102\tfrac{38}{60}° \approx 102.6°$$

Then

$$C \approx 180° - 26.6° - 102.6° = 50.8°$$

We now can use the law of sines:

$$\frac{\sin C}{\overline{AB}} = \frac{\sin B}{\overline{AC}}$$

$$\overline{AC} = \frac{\overline{AB} \sin B}{\sin C} = \frac{2235 \sin 102.6°}{\sin 50.8°} \approx 2815 \text{ km}$$

For more information about the transiting satellites (the U.S. Navy now operates five of them), see *Electronic Surveying in Practice* by S. H. Laurila, Wiley, New York, 1983, p. 302.

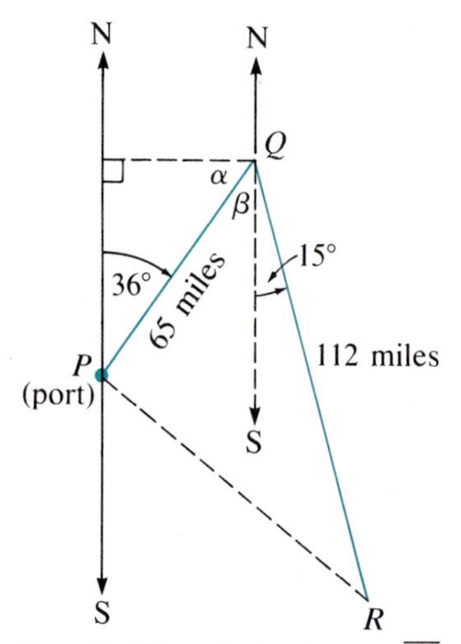

Figure 5 We seek the distance $\overline{PR}$.

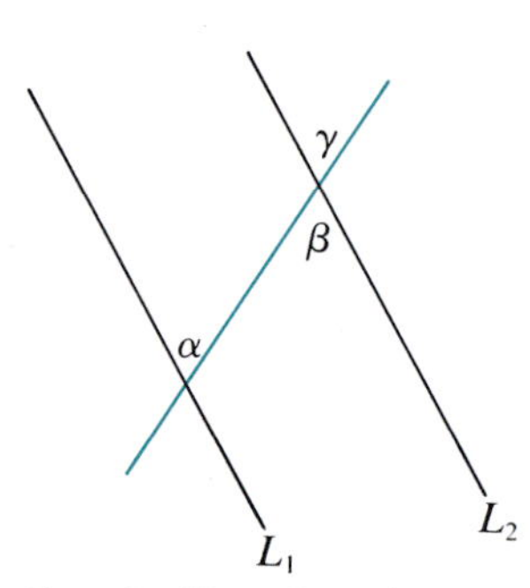

Figure 6 Since L_1 and L_2 are parallel, the alternate interior angles α and β are equal.

Navigation

Recall from Section 2.8 (see p. 138) that a **bearing** is an acute angle between a line of travel or line of sight and the north-south line. Bearings are given as so many degrees east of north, west of north, east of south, or west of south. If θ is the acute angle, then Nθ°E is read as θ° east of north, and so on.

EXAMPLE 4 *Determining Distance Traveled After a Change of Course*

A ship leaves port and travels 65 miles on a bearing of N36°E. It then changes course and sails 112 miles on a bearing of S15°E. How far is it from port at this point?

SOLUTION We depict the information in the problem in Figure 5. After 65 miles, the ship has reached point Q. To clarify the picture, we draw in a new north-south line at Q in order to depict the new bearing. We seek the distance from P to R. We can find it by using the law of cosines if we know the measure of angle PQR. This takes a bit of work.

Using the notation in Figure 5, we see that α is the complement of 36° and β is the complement of α. Therefore $\beta = 36°$. This also follows from a fact of geometry. In Figure 6, the lines L_1 and L_2 are parallel. The angles α and β are called **alternate interior angles.** In high school geometry, the following result is proved:

> Alternate interior angles formed by a transversal and two parallel lines have equal measure.

Also, the angles β and γ, called **vertical angles,** are equal. Thus, in Figure 6, $\alpha = \beta = \gamma$. In either case, we see that $\beta = 36°$. Thus

$$\measuredangle PQR = 36° + 15° = 51°$$

Now we can use the law of cosines:

$$\overline{PR}^2 = \overline{PQ}^2 + \overline{QR}^2 - 2(\overline{PQ})(\overline{QR}) \cos (\measuredangle PQR)$$
$$\overline{PR}^2 = 65^2 + 112^2 - 2(65)(112) \cos 51°$$
$$\overline{PR}^2 = 4225 + 12{,}544 - (14{,}560)(0.629320391) \approx 7606$$

and

$$\overline{PR} = \sqrt{7606} \approx 87 \text{ miles}$$ ■

EXAMPLE 5 *Determining Distance and Bearing*

An airplane is 421 km from radar station R at a bearing of N23°25′W.† A second radar station Q is 210 km from R at a bearing S57°44′E. Find, to the nearest kilometer and minute, the distance and bearing from Q to the airplane.

---

†Recall from p. 139 that bearings may also be given as degrees (0° to 360°) measured clockwise from north. Thus, a bearing of N23°25′W is sometimes referred to as a bearing of 336°35′, and S57°44′E would be 122°16′.

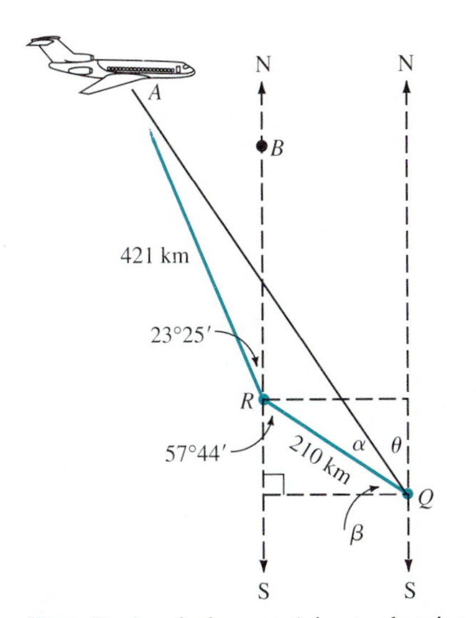

Figure 7 An airplane at A is at a bearing of N23°25′W from radar station R. The radar station Q is at a bearing of S57°44′E.

SOLUTION The data of the problem are depicted in Figure 7. We added point B as a reference point for angles. The airplane is at point A. From the figure, we see that the plane is at a distance $\overline{AQ}$ from Q at a bearing of $\theta°$ west of north.

To use the law of cosines to find $\overline{AQ}$, we need to find the measure of angle QRA. But

$$\measuredangle QRA = \measuredangle QRB + \measuredangle BRA$$

$$\measuredangle QRB \text{ is the supplement of } 57°44'$$

so

$$\begin{aligned}\measuredangle QRA &= (180° - 57°44') + 23°25' \\ &= 122°16' + 23°25' = 145°41' \\ &= 145\tfrac{41}{60}° \approx 145.68°\end{aligned}$$

Thus, from the law of cosines,

$$\begin{aligned}\overline{AQ}^2 &= \overline{AR}^2 + \overline{RQ}^2 - 2(\overline{AR})(\overline{RQ}) \cos 145.68° \\ &= 421^2 + 210^2 - 2(421)(210) \cos 145.68° \\ &= 177{,}241 + 44{,}100 - 176{,}820(-0.825901536) \\ &\approx 367{,}377\end{aligned}$$

so, to the nearest kilometer, $\overline{AQ} = 606$ km

To find θ, we note that $\alpha + \beta + \theta = 90°$

β is the complement of 57°44′, so

$$\beta = 90° - 57°44' = 32°16'$$

We obtain α from the law of sines:

$$\frac{\sin \alpha}{\overline{AR}} = \frac{\sin \measuredangle QRA}{\overline{AQ}}$$

$$\sin \alpha = \frac{\overline{AR} \sin \measuredangle QRA}{\overline{AQ}} \approx \frac{421}{606} \sin 145.68°$$

$$\approx 0.3916928$$

so

$$\alpha \approx \sin^{-1} 0.3916928 = 23.0599°$$

But

$$0.0599° = (0.0599) \times 60' \approx 4'$$

so

$$\alpha = 23°4'$$

Finally,

$$\begin{aligned}\theta &= 90° - \alpha - \beta = 90° - 23°4' - 32°16' \\ &= 90° - 55°20' = 34°40'\end{aligned}$$

Thus, the airplane is 606 km from the radar station Q at a bearing of N34°40′W.

Problems like the one solved in Example 5 require many steps. There are no shortcuts. In each case, your task will be simplified if you first draw an appropriate picture and then determine exactly what is being asked for in the problem.

Problems 4.3

Readiness Check

I. Which of the following bearings would be 20° below the negative x-axis?
 a. S20°E
 b. N20°W
 c. S20°W
 d. S70°W

II. Which of the following would be used to solve the problem below?

 On a map city A is 4.3 inches due south of city B. City C is 5.2 inches from city A and 1.4 inches from city B. What is the angle formed at city B?
 a. Law of cosines
 b. Law of sines
 c. Pythagorean theorem
 d. Heron's formula

III. Which of the following is true of a pilot flying at a bearing of N30°E at a rate of 550 mph from an airport if the airport is thought to be at the origin?
 a. He will be flying into quadrant I.
 b. He will be making a 30° angle with the positive x-axis.
 c. After 2 hours he will be 1100 miles due east of the airport.
 d. He will be making an angle of 150° with the positive y-axis.

In the following problems, unless otherwise requested, give all distances to the nearest tenth of a unit and all angles to the nearest tenth of a degree—except in navigation problems, where all angles should be given to the nearest minute.

1. In surveying a tract of land, a surveyor found it impractical to measure the distance $\overline{AB}$ because of thick brush lying

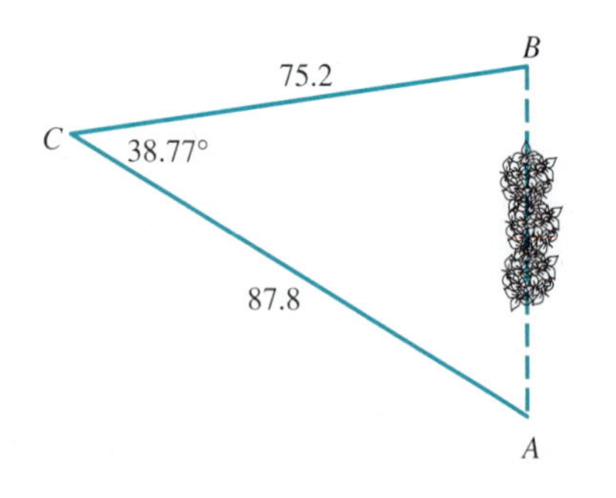

between A and B. The surveyor did measure the distances $\overline{CB} = 75.2$ ft and $\overline{AC} = 87.8$ ft. Moreover, she determined that $\measuredangle ACB = 38.77°$. Find $\overline{AB}$.

2. A surveyor wishes to determine the distance from point A to a point B on the other side of a river. He obtained the measurements given below. Find $\overline{AB}$.

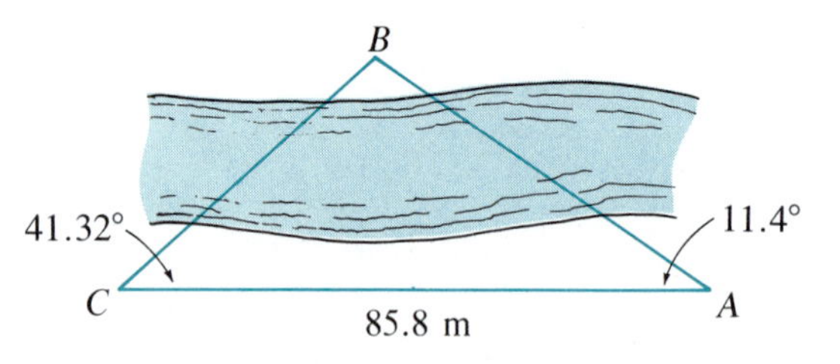

3. A hill has a constant slope of 24°. The angle of elevation from a point 345 feet from the base of the slope to the top of the hill is 18.33°. How long is the hill?

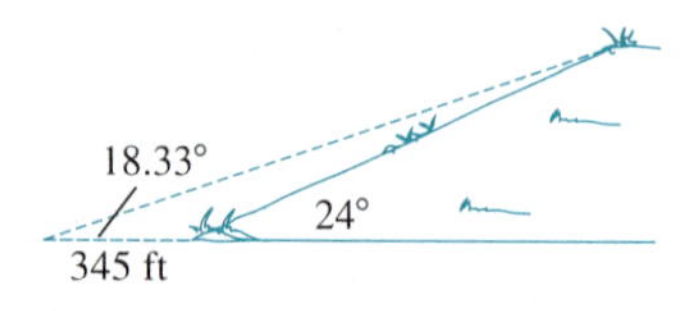

Answers to Readiness Check

I. d II. a III. a

4. A car approaches a large city at 65 miles per hour. At one moment, a passenger with a transit determines that the angle of elevation to the top of a downtown building is 2.67°. Five minutes later, the angle is 4.5°. (a) How tall is the building? (b) How far was the car from the base of the building at the time of the first sighting?
5. The chemical symbol for boron trifluoride is BF_3. One molecule of BF_3 consists of 1 boron and 3 fluorine atoms and takes the shape of a triangle as in the figure below.

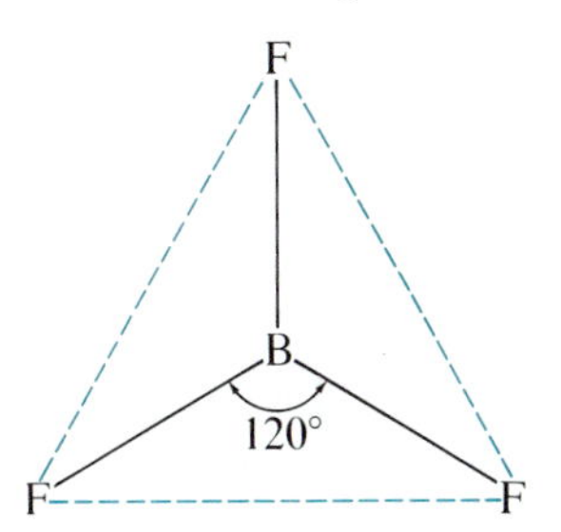

The angle FBF, called the **bond angle,** is approximately 120°. The distance from the boron atom to each fluorine atom is approximately 1.29 angstroms (Å) [1 Å = 10^{-10} m]. Find the distance between 2 fluorine atoms to the nearest hundredth of an angstrom.
6. In a water molecule (H_2O), the ratio of the distance between the two hydrogen atoms and the distance from the oxygen atom to 1 hydrogen atom is approximately 1.58. Find the bond angle for water (the angle HOH) to the nearest tenth of a degree.

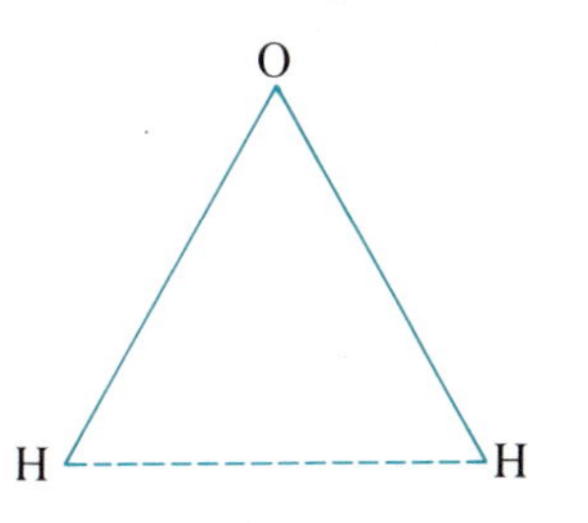

7. In hydrogen sulfide, H_2S, the bond angle is 92°. Find the ratio of distance $\overline{HH}$ to distance $\overline{HS}$.

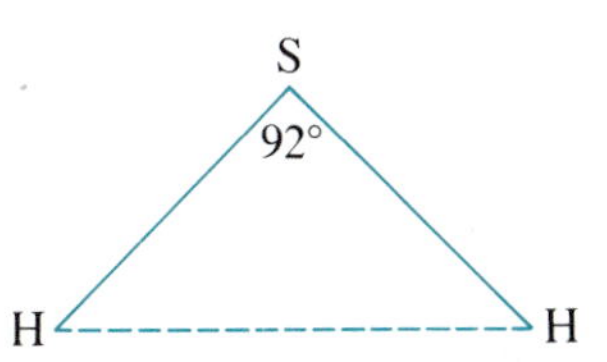

8. An airplane flies above the earth. An observer at point A estimates that the angle of elevation to the plane is 83°. A second observer at point B, 846 m from point A, estimates that the angle of elevation is 88°. Approximately how high is the plane?

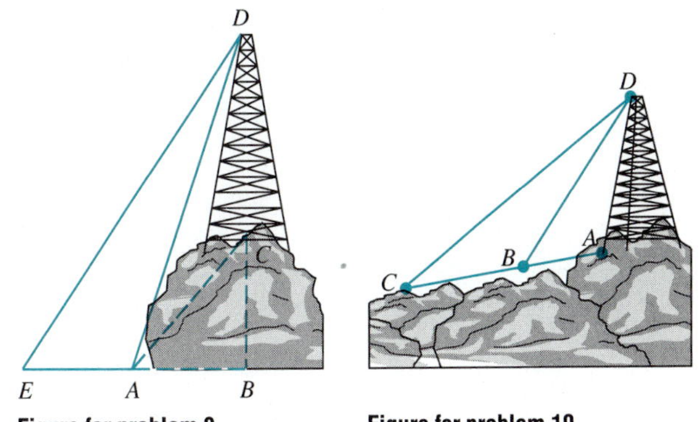

Figure for problem 9 **Figure for problem 10**

9. A transmission tower stands atop a hill. The angle of inclination BAC to the top of the hill is 44.58° and the angle of elevation BAD to the top of the tower is 59.8°. The horizontal line $\overline{AE}$ is measured to be 55 meters. If the angle of inclination, BED, to the top of the tower is 46.42°, determine (a) the height, $\overline{BC}$, of the hill and (b) the height, $\overline{CD}$, of the tower.
10. A tower stands in hilly terrain as in the figure. The following measurements are obtained. $\overline{AB} = 45$ ft, $\overline{BC} = 68$ ft, angle $ABD = 50.25°$, and angle $BCD = 30.75°$. Find $\overline{AD}$, the height of the tower.
11. In Example 3, it was determined that the distance from the transit to the control site was 1784 km while the angle θ was 94.43°. Determine, to the nearest kilometer, the horizontal distance between the control site and the remote site.
12. In the mountainous terrain of Example 2, the angle of inclination at point B (in Figure 2(a)) is 61.23°, the distance from B to C is 651 m, the angle of elevation from C to B is 6.85°, the angle of elevation from C to A is 32.5°, and the elevation at C is 2156 m. To the nearest meter, what is the elevation at A?
13. The leaning tower of Pisa leans at an angle of 5.5° from the vertical and is approximately 55.6 m long. If the sun is shining on the tower from a direction opposite to that in which the tower leans, at an angle of 52°, how long is the shadow cast by the tower?

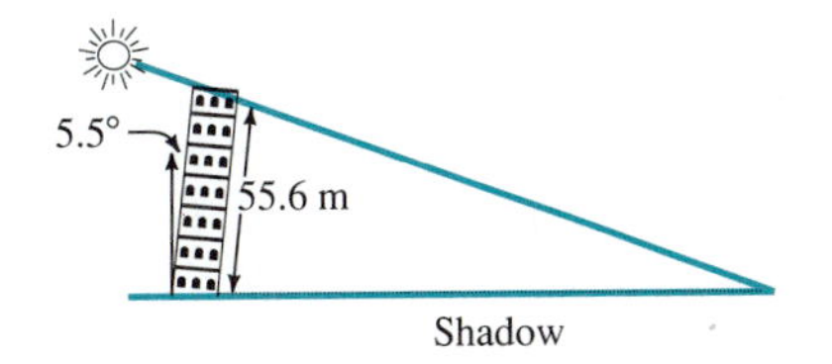

14. A tunnel is to be blasted through the center of a mountain. The measurements in the figure below were taken. How long will the tunnel be?

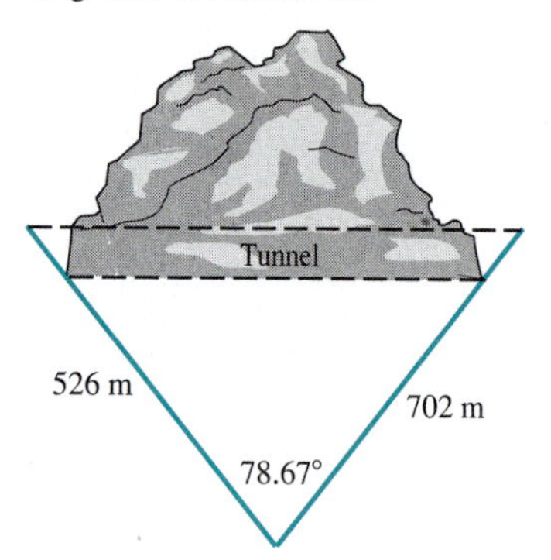

15. A tree is blown out of a vertical position by a high wind. A 12-ft rope is tied to the trunk 7 ft along the trunk (from the ground). When the rope is pulled taut, it reaches a point on the ground 10.6 ft from the base of the tree. Determine the acute angle between the tree and the ground.
16. A ship leaves port and travels 137 km on a bearing of S47°W. It then changes course and sails 89 km on a bearing of N9°W. How far is it from port at this point?
17. In Problem 16 determine the bearing at which the ship must sail in order to get back to port.
18. In Problem 16 the ship sails an additional 214 km at a bearing of N51°E. How far is it from port now?
19. * In Problem 18 determine the bearing at which the ship must sail in order to return to port.
20. A ship travels at a constant speed of 12 mph. It travels $2\frac{1}{2}$ hours at a bearing of S37°20′E and then travels 4 hours at a bearing of S87°15′W. How long will it take to return directly to port?
21. Two streets meet at an angle of 104°. A third street intersects the first two to form a triangular lot. Two sides of the lot, formed from the first two streets, are 85 ft and 63 ft. (a) What is the length of the third side of the lot? (b) What is the area of the lot?

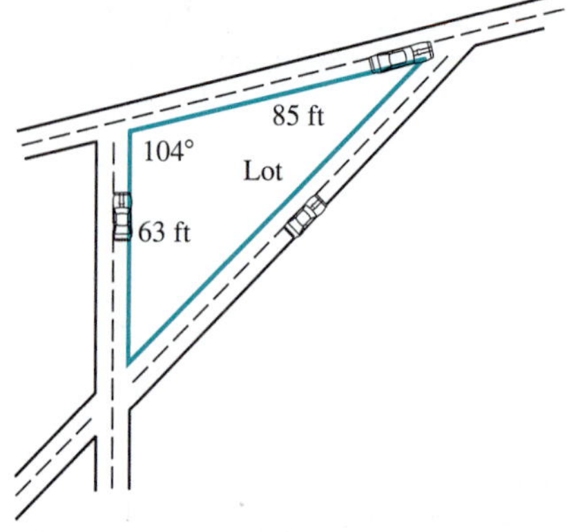

22. To an observer on Earth at a particular time, the angle between lines joining Venus to the Earth and the Earth to the Sun is 29°. The distance from Venus to the Sun is approximately 67 million miles, whereas the distance from the Earth to the Sun is approximately 93 million miles. Estimate, to the nearest million miles, the distance from the Earth to Venus at this time. [Hint: Explain why there are two possible answers and give both of them.]

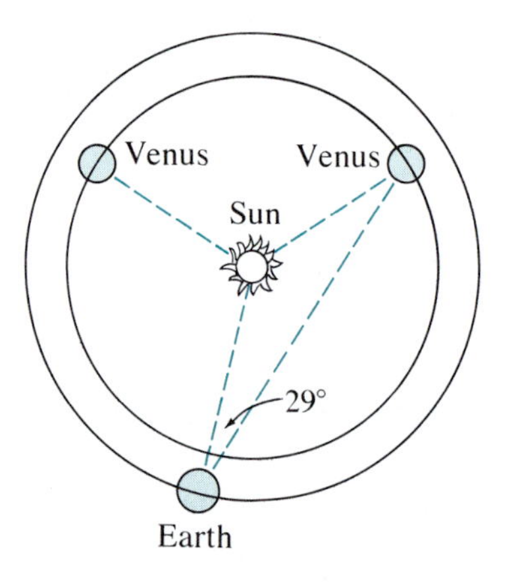

23. At a time when Venus is precisely 100 million miles from Earth, what is the angle between the straight lines joining the Earth to the Sun and to Venus?
24. The distance from Mars to the Sun varies between 128.5 and 155 million miles. At a certain time, Mars is 142 million miles from the Sun and the angle MES is 126°. Estimate the distance between the Earth and Mars at that time.

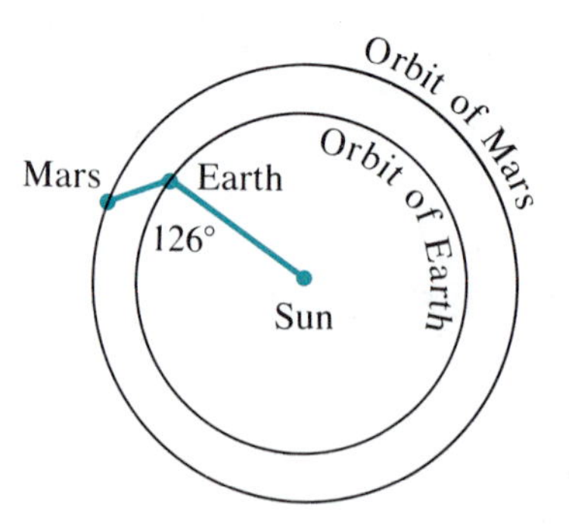

25. Two lighthouses 58 km apart receive a distress signal from a ship. They determine that the angle between the lines joining the ship and the first lighthouse and the line between the two lighthouses is 98°. The angle between the line joining the ship and the second lighthouse and the line joining the two lighthouses is 79°. Which lighthouse is closer to the ship, and how far away is it?

26. A 125-ft tower sits on a cliff. The angle of depression from the top of the tower to a nearby ship is 29°, whereas the angle of depression from the top of the cliff to the ship is 21°. (a) How high is the cliff? (b) How far is the ship from a point directly below the edge of the cliff?

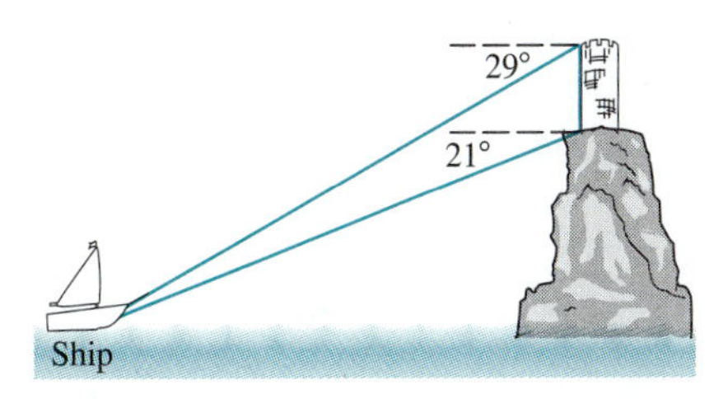

27. The markers P and Q lie in flat terrain 6 km apart. The angles of elevation to a passing airplane are 57° at P and 71° at Q. The airplane is to the west of P and Q and in the same vertical plane. How high is the plane?
28. A ship travels due north parallel to the shoreline at a speed of 17 km/hr. The bearing to a radar station at noon is S47°W. At 12:20 P.M. the bearing is S43°20′W. What is the distance from the ship to the radar station at 12:20 P.M.?
29. In Problem 28 what is the bearing from the ship to the radar station at 1:00 P.M., assuming that the shoreline continues to be straight and the ship continues in the same direction and at the same speed?
30. An airplane is 214 miles from radar station A, which is at a bearing of N14°40′E from the plane. A second radar station B is 85 miles from A at a bearing of S61°26′W from A. Find the distance and bearing from B to the airplane.
31. A naval ship is 286 km from a lighthouse at a bearing of S36°51′W from the lighthouse. A naval base is 406 km from the lighthouse at a bearing of N7°22′W from the lighthouse. At what bearing must the ship sail to return to base?
32. In Problem 31 determine how long it will take the ship to return to base if it travels at a speed of 28 km/hr.
33. A hiker walks 8000 m on a course of S81°33′E. She then changes direction and hikes 5000 m on a course of N32°7′W. How far is the hiker from her starting point, and on what course must she travel to return to the starting point?
* 34. The following cartoon appeared in 1987 in many local newspapers.
    (a) Explain why the problem stated in the cartoon cannot be uniquely solved. That is, explain why the distance from A to C could take many different values. [Hint: Draw a picture.]
    (b) What is the smallest possible distance from A to C?
    (c) What additional information is needed in order to find a unique answer to the question posed?

4.4 Vectors

In many applications of mathematics, scientists are concerned with entities that have both magnitude (length) and direction. Examples include the notions of force, velocity, acceleration, and momentum. These quantities are called **vectors.**

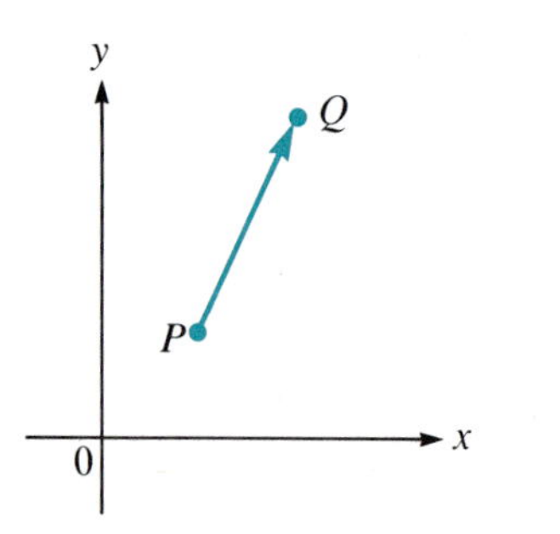

Figure 1 A directed line segment from P to Q.

The modern study of vectors began essentially with the work of the great Irish mathematician Sir William Rowan Hamilton (1805–1865) who worked with what he called quaternions.† After Hamilton's death, his work on quaternions was supplanted by the more adaptable work on vector analysis by the American mathematician and physicist Josiah Willard Gibbs (1839–1903) and the general treatment of ordered n-tuples by the German mathematician Hermann Grassmann (1809–1877).

Throughout Hamilton's life and for the remainder of the nineteenth century, there was considerable debate over the usefulness of quaternions and vectors. At the end of the century, the great British physicist Lord Kelvin wrote that quaternions, "although beautifully ingenious, have been an unmixed evil to those who have touched them in any way. . . . Vectors . . . have never been of the slightest use to any creature."

But Kelvin was wrong. Today nearly all branches of classical and modern physics are represented using the language of vectors. Vectors are also used with increasing frequency in the social and biological sciences. Quaternions, too, have recently been used in physics — in particle theory and other areas.

In order to define a vector geometrically, we first define a directed line segment.

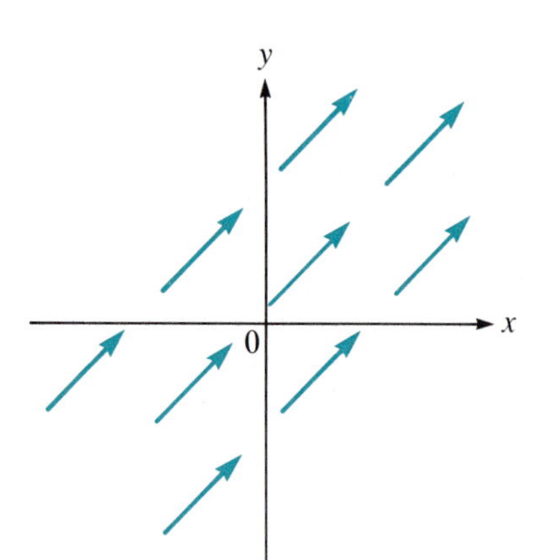

Figure 2 Equivalent directed line segments.

Let P and Q be two different points in the plane. Then the **directed line segment** from P to Q, denoted by $\overrightarrow{PQ}$, is the straight-line segment that extends from P to Q (see Figure 1). Note that the directed line segments $\overrightarrow{PQ}$ and $\overrightarrow{QP}$ are different because they point in opposite directions.

The point P in the directed line segment $\overrightarrow{PQ}$ is called the **initial point** of the segment, and the point Q is called the **terminal point.** The two important properties of a directed line segment are its magnitude (length) and its direction. If two directed line segments $\overrightarrow{PQ}$ and $\overrightarrow{RS}$ in the plane have the same magnitude and direction, we say that they are **equivalent,** no matter where they are located with respect to the origin. The directed line segments in Figure 2 are all equivalent.

Geometric Definition of a Vector

The set of all directed line segments equivalent to a given directed line segment is called a **vector.** Any directed line segment in that set is called a **representation** of the vector. Note that the directed line segments in Figure 2 are all representations of the same vector.

NOTATION We will denote vectors by lowercase boldface letters such as **v**, **w**, **a**, **b**.

From the definition we see that a given vector **v** can be represented in many different ways. In fact, let $\overrightarrow{PQ}$ be a representation of **v**. Then without changing magnitude or direction, we can move $\overrightarrow{PQ}$ in a parallel way so that its initial point is shifted to the origin. We then obtain the directed line segment $\overrightarrow{OR}$, which is another representation of the vector **v** (see Figure 3). The vector

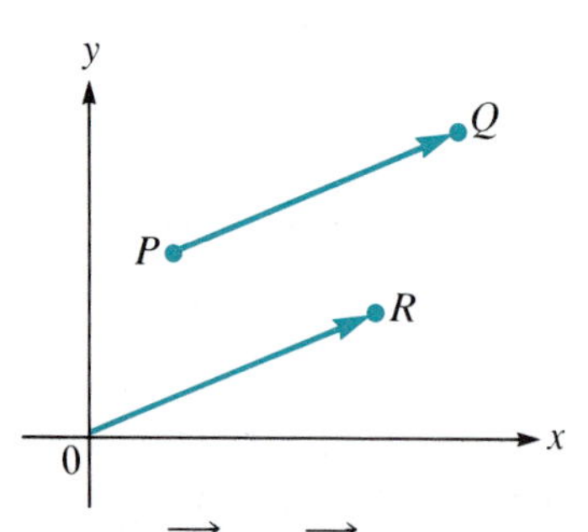

Figure 3 $\overrightarrow{OR}$ and $\overrightarrow{PQ}$ are equivalent. They represent the same vector.

†See the biographical sketch of Hamilton on p. 244.

$\overrightarrow{OR}$ is called the **standard representation** of the vector.

Now suppose that R has the Cartesian coordinates (a, b). Then we can describe the directed line segment $\overrightarrow{OR}$ by the coordinates (a, b). That is, $\overrightarrow{OR}$ is the directed line segment with initial point $(0, 0)$ and terminal point (a, b). Since one representation of a vector is as good as another, we can write the vector $\mathbf{v}$ as (a, b). We see that a vector has a *unique* representation as a directed line segment with initial point at the origin, and this (standard) representation is completely determined by the terminal point (a, b). Therefore, we can work with either directed line segments or points (a, b) that are thought of as terminal points of vectors with initial points at $(0, 0)$.

The vector $(0, 0)$ is called the **zero vector.**

Magnitude of a Vector

The **magnitude** or **length** of the vector (a, b) is defined by

$$|\mathbf{v}| = \text{magnitude of } \mathbf{v} = \sqrt{a^2 + b^2} \qquad \textbf{(1)}$$

This definition makes sense. We see, in Figure 4, that if $\mathbf{v} = \overrightarrow{OR} = (a, b)$, then $|\overrightarrow{OR}| = \sqrt{a^2 + b^2}$. This follows from the Pythagorean theorem.

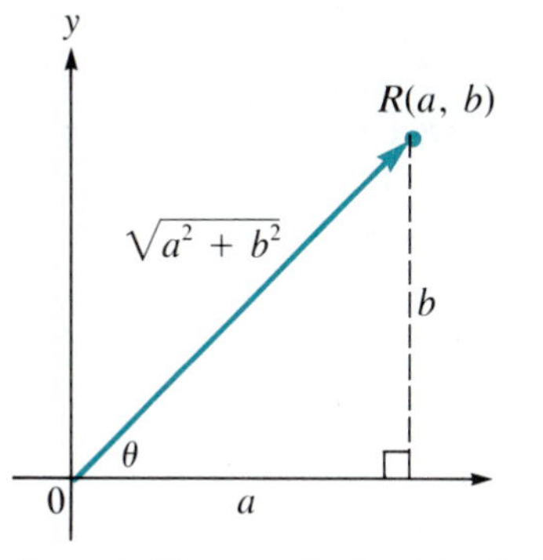

Figure 4 The magnitude of (a, b) is $\sqrt{a^2 + b^2}$.

NOTE $|\mathbf{v}|$ is a nonnegative real number, not a vector.

EXAMPLE 1 *Computing the Magnitudes of Five Vectors*

Calculate the magnitudes of the vectors (a) $(2, 2)$, (b) $(2, 2\sqrt{3})$, (c) $(-2\sqrt{3}, 2)$, (d) $(-3, -3)$, (e) $(6, -6)$.

(a) $|\mathbf{v}| = \sqrt{2^2 + 2^2} = \sqrt{8} = 2\sqrt{2}$

(b) $|\mathbf{v}| = \sqrt{2^2 + (2\sqrt{3})^2} = \sqrt{4 + 12} = \sqrt{16} = 4$

(c) $|\mathbf{v}| = \sqrt{(-2\sqrt{3})^2 + 2^2} = 4$

(d) $|\mathbf{v}| = \sqrt{(-3)^2 + (-3)^2} = \sqrt{18} = 3\sqrt{2}$

(e) $|\mathbf{v}| = \sqrt{6^2 + (-6)^2} = \sqrt{72} = 6\sqrt{2}$

Direction of a Vector

The **direction** of a vector (a, b) is defined as the angle θ, in standard position, that the vector makes with the positive x-axis. By convention the angle is measured as we rotate the positive x-axis counterclockwise to the vector; we choose θ such that $0 \le \theta < 2\pi$ $(0 \le \theta < 360°)$. From the definition of $\tan^{-1} x$ in Section 2.7, we know that $0 < \tan^{-1} \frac{b}{a} < \frac{\pi}{2}$ if $\frac{b}{a} > 0$ and $-\frac{\pi}{2} < \tan^{-1} \frac{b}{a} < 0$ if $\frac{b}{a} < 0$. Thus,

$$\text{direction of } \mathbf{v} = \theta = \tan^{-1} \frac{b}{a} \text{ if } (a, b) \text{ is in the first quadrant} \quad (2)$$

$$\text{direction of } \mathbf{v} = \theta = \pi + \tan^{-1} \frac{b}{a} \text{ if } (a, b) \text{ is in the second quadrant} \quad (3)$$

$$\text{direction of } \mathbf{v} = \theta = \pi + \tan^{-1} \frac{b}{a} \text{ if } (a, b) \text{ is in the third quadrant} \quad (4)$$

$$\text{direction of } \mathbf{v} = \theta = 2\pi + \tan^{-1} \frac{b}{a} \text{ if } (a, b) \text{ is in the fourth quadrant} \quad (5)$$

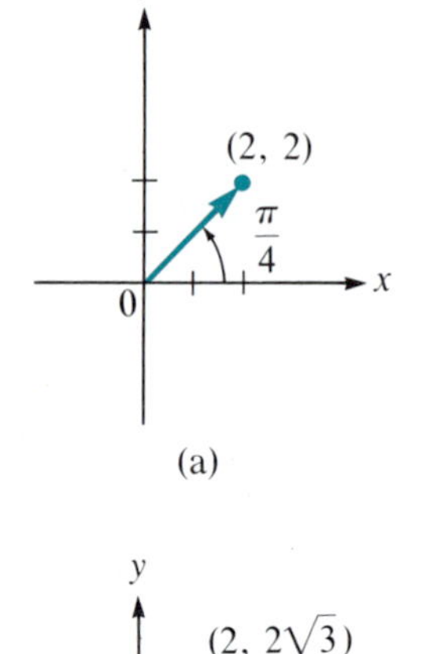

Some comments on this definition:

(i) The zero vector has a magnitude of 0. Since the initial and terminal points coincide, we say that the *zero vector has no direction*.
(ii) If the initial point of **v** is not at the origin, then the direction of **v** is defined as the direction of the vector equivalent to **v** whose initial point is at the origin.
(iii) It follows from (ii) that *parallel vectors have the same direction*.
(iv) This definition works only for vectors in the plane. Different ideas must be developed to discuss the direction of a vector in space.

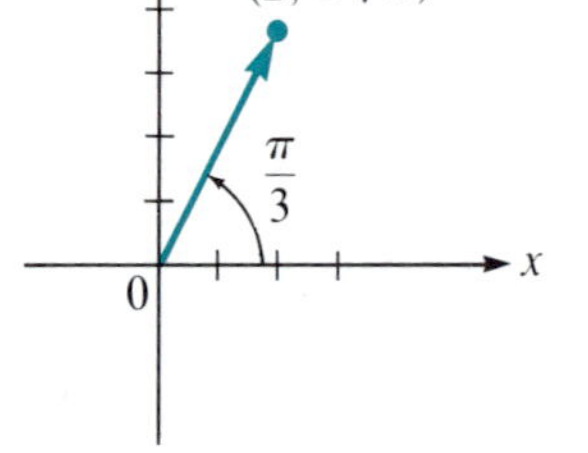

EXAMPLE 2 ***Computing the Directions of Five Vectors***

Calculate the directions of the vectors in Example 1.

SOLUTION We depict these 5 vectors in Figure 5.

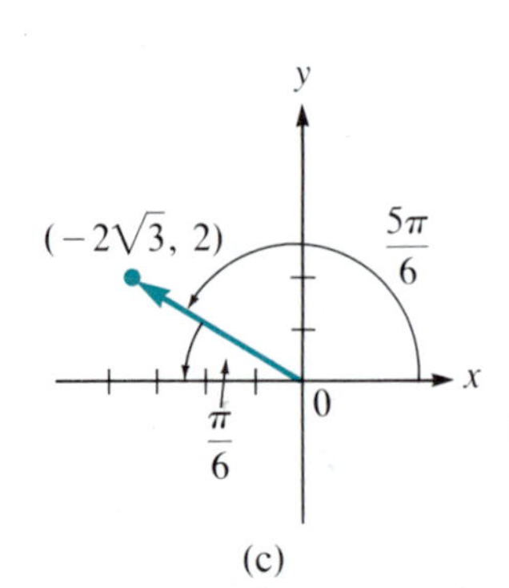

Figure 5 Five vectors in the plane.

(a) Here **v** is in the first quadrant, and since $\tan\theta = \frac{2}{2} = 1$, $\theta = \frac{\pi}{4}$.

(b) Here $\theta = \tan^{-1}\frac{2\sqrt{3}}{2} = \tan^{-1}\sqrt{3} = \frac{\pi}{3}$.

(c) Here **v** is in the second quadrant, and $\tan^{-1}\left(\frac{2}{-2\sqrt{3}}\right) = \tan^{-1}(-1/\sqrt{3}) = -\frac{\pi}{6}$. Thus, from (3),

$$\theta = \pi + \tan^{-1}(-1/\sqrt{3}) = \pi - \frac{\pi}{6} = \frac{5\pi}{6}$$

(d) **v** is in the third quadrant, and $\tan^{-1}\left(\frac{-3}{-3}\right) = \tan^{-1} 1 = \frac{\pi}{4}$, so, from (4), $\theta = \pi + \frac{\pi}{4} = \frac{5\pi}{4}$.

(e) **v** is in the fourth quadrant, and $\tan^{-1}\left(-\frac{6}{6}\right) = \tan^{-1}(-1) = -\frac{\pi}{4}$, so, from (5), $\theta = 2\pi - \frac{\pi}{4} = \frac{7\pi}{4}$. ■

EXAMPLE 3 *Computing the Magnitudes and Directions of Two Vectors*

Compute the magnitudes and directions (in degrees) of the vectors (a) $(2, -3)$ and (b) $(-7, 4)$.

SOLUTION

(a) $|(2, -3)| = \sqrt{2^2 + (-3)^2} = \sqrt{4+9} = \sqrt{13}$

$$\tan^{-1}\left(-\frac{3}{2}\right) = \tan^{-1}(-1.5) \approx -56.3°$$

Since $(2, -3)$ is in the fourth quadrant,

From (5)
↓
$$\theta = 360° - 56.3° = 303.7°$$

(b) $|(-7, 4)| = \sqrt{(-7)^2 + 4^2} = \sqrt{49+16} = \sqrt{65}$

$$\tan^{-1}\left(\frac{4}{-7}\right) = \tan^{-1}(-0.571428571) \approx -29.7°$$

So, since $(-7, 4)$ is in the second quadrant,

From (3)
↓
$$\theta = 180° - 29.7° = 150.3°$$

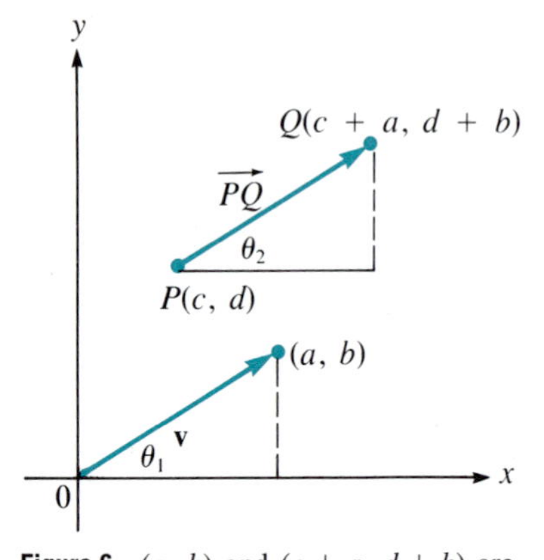

Figure 6 (a, b) and $(c + a, d + b)$ are representations of the same vector because they have the same magnitude and direction.

As in Figure 6, let $\mathbf{v} = (a, b)$, $P = (c, d)$, and $Q = (c + a, d + b)$. Then

$$\mathbf{v} \text{ and } \overrightarrow{PQ} \text{ are representations of the same vector}$$

To see this, we compute

$$|\mathbf{v}| = \sqrt{a^2 + b^2}$$
$$|\overrightarrow{PQ}| = \sqrt{[(c + a) - c]^2 + [(d + b) - d]^2} = \sqrt{a^2 + b^2}$$

Thus $\mathbf{v}$ and $\overrightarrow{PQ}$ have the same magnitude. If θ_1 = direction of $\mathbf{v}$ and θ_2 = direction of $\overrightarrow{PQ}$, then, from Figure 6,

$$\tan\theta_1 = \frac{b}{a} \qquad \text{and} \qquad \tan\theta_2 = \frac{(d + b) - d}{(c + a) - c} = \frac{b}{a}$$

When both θ_1 and θ_2 are in the first quadrant as in Figure 6, and $\theta_1 = \theta_2$, then $\mathbf{v}$ and $\overrightarrow{PQ}$ have the same direction. It is also true, although we do not prove it here, that $\theta_1 = \theta_2$ in other quadrants as well. Since $\mathbf{v}$ and $\overrightarrow{PQ}$ have the same magnitude and direction, they are equivalent.

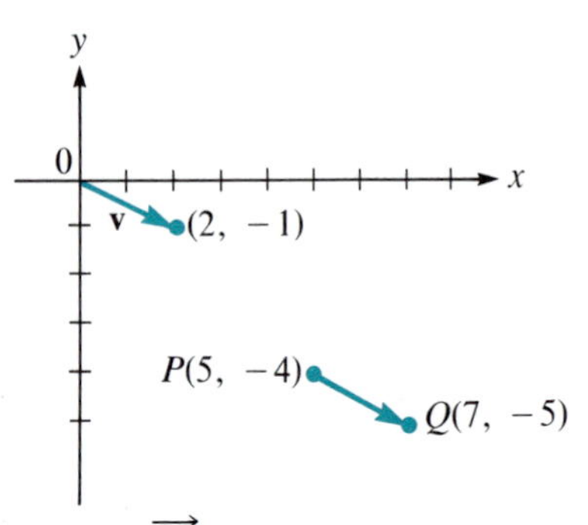

Figure 7 $\overrightarrow{PQ}$ and $\mathbf{v}$ are representations of the same vector.

EXAMPLE 4 ***Finding a Representation of a Vector with a Given Initial Point***

Find a representation of the vector $(2, -1)$ whose initial point is the point $P = (5, -4)$.

SOLUTION Let $Q = (5 + 2, -4 - 1) = (7, -5)$. Then $\overrightarrow{PQ}$ is one of many possible representations of the vector $(2, -1)$. This is illustrated in Figure 7.

Addition, Subtraction, and Scalar Multiplication of Vectors

When dealing with vectors, we will refer to a real number as a **scalar.** Let $\mathbf{u} = (a_1, b_1)$ and $\mathbf{v} = (a_2, b_2)$ be two vectors in the plane, and let α be a scalar. Then we define

Algebra of Vectors

Sum of Two Vectors

$$\mathbf{u} + \mathbf{v} = (a_1 + a_2, b_1 + b_2)$$

Multiplication of a Vector by a Scalar

$$\alpha\mathbf{u} = (\alpha a_1, \alpha b_1)$$

Negative of a Vector

$$-\mathbf{v} = (-1)\mathbf{v} = (-a_2, -b_2)$$

Difference of Two Vectors

$$\mathbf{u} - \mathbf{v} = \mathbf{u} + (-\mathbf{v}) = (a_1 - a_2, b_1 - b_2)$$

The vector $-\mathbf{v}$ is said to have the **opposite direction** as $\mathbf{v}$. This means that

$$|\text{direction of } (-\mathbf{v}) - \text{direction of } \mathbf{v}| = \pi$$

To add two vectors, we add their corresponding components, and to multiply a vector by a scalar, we multiply each of its components by that scalar.

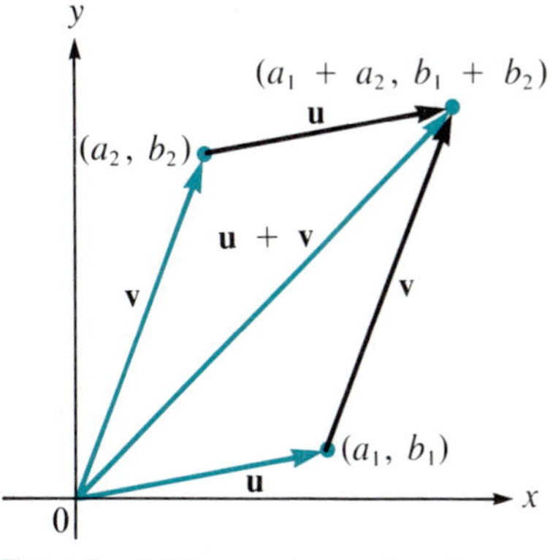

Figure 8 Adding vectors using the parallelogram rule.

EXAMPLE 5 *Adding Vectors and Multiplying Them by Scalars*

Let $\mathbf{u} = (1, 3)$ and $\mathbf{v} = (-2, 4)$. Calculate (a) $\mathbf{u} + \mathbf{v}$, (b) $3\mathbf{u}$, (c) $-\mathbf{v}$, (d) $\mathbf{u} - \mathbf{v}$, and (e) $-3\mathbf{u} + 5\mathbf{v}$.

SOLUTION

(a) $\mathbf{u} + \mathbf{v} = (1 + (-2), 3 + 4) = (-1, 7)$

(b) $3\mathbf{u} = 3(1, 3) = (3, 9)$

(c) $-\mathbf{v} = (-1)(-2, 4) = (2, -4)$

(d) $\mathbf{u} - \mathbf{v} = \mathbf{u} + (-\mathbf{v}) = (1 + 2, 3 - 4) = (3, -1)$

(e) $-3\mathbf{u} + 5\mathbf{v} = (-3, -9) + (-10, 20) = (-13, 11)$

Parallelogram Rule for Adding Vectors

Now suppose we add the vectors $\mathbf{u} = (a_1, b_1)$ and $\mathbf{v} = (a_2, b_2)$, as in Figure 8. From the figure, we see that the vector $\mathbf{u} + \mathbf{v} = (a_1 + a_2, b_1 + b_2)$ can be obtained by shifting the representation of the vector $\mathbf{v}$ so that its initial point coincides with the terminal point (a_1, b_1) of the vector $\mathbf{u}$. We can obtain the vector $\mathbf{u} + \mathbf{v}$ by drawing a parallelogram with one vertex at the origin and sides $\mathbf{u}$ and $\mathbf{v}$. Then $\mathbf{u} + \mathbf{v}$ is the vector that points from the origin along the diagonal of the parallelogram.

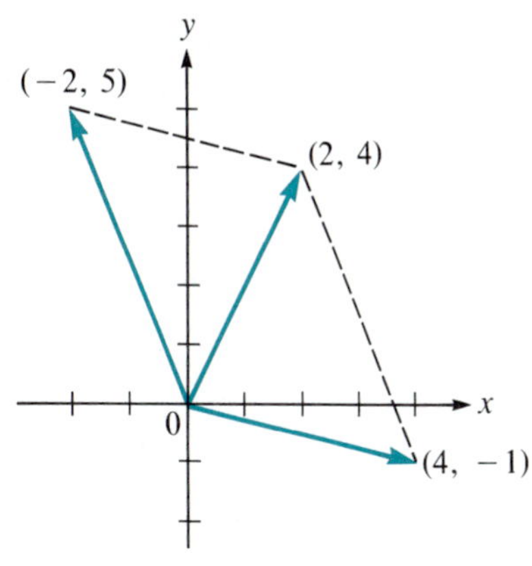

Figure 9 (2, 4) is the sum of (4, −1) and (−2, 5).

EXAMPLE 6 *Adding Two Vectors Graphically*

Depict graphically the sum $(4, -1) + (-2, 5)$.

SOLUTION In Figure 9, we draw the two vectors and then form the parallelogram with the two vectors as adjacent sides. The diagonal of the parallelogram is the required sum.† It is a vector in the first quadrant. We compute $(4, -1) + (-2, 5) = (4 - 2, -1 + 5) = (2, 4)$.

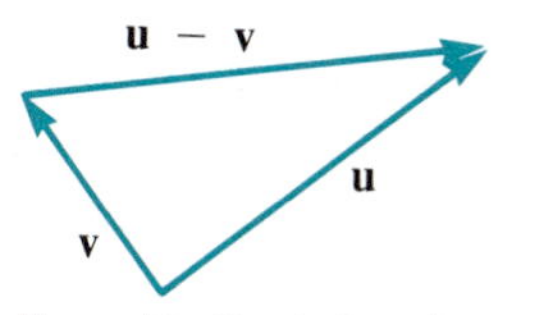

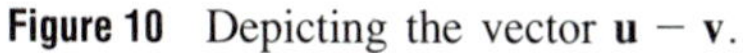

Figure 10 Depicting the vector $\mathbf{u} - \mathbf{v}$.

We can also obtain a geometric representation of the vector $\mathbf{u} - \mathbf{v}$. Since $\mathbf{u} = \mathbf{u} - \mathbf{v} + \mathbf{v}$, the vector $\mathbf{u} - \mathbf{v}$ is the vector that must be added to $\mathbf{v}$ to obtain $\mathbf{u}$. This is illustrated in Figure 10.

There are two special vectors in the plane that allow us to represent other vectors in a convenient way. We will denote the vector (1, 0) by the vector

† A parallelogram has *two* diagonals. The diagonal used here is the one that does *not* join the terminal points of **u** and **v**.

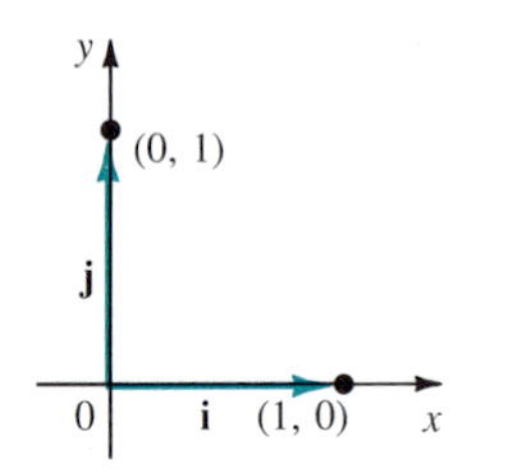

Figure 11 The vectors **i** and **j**.

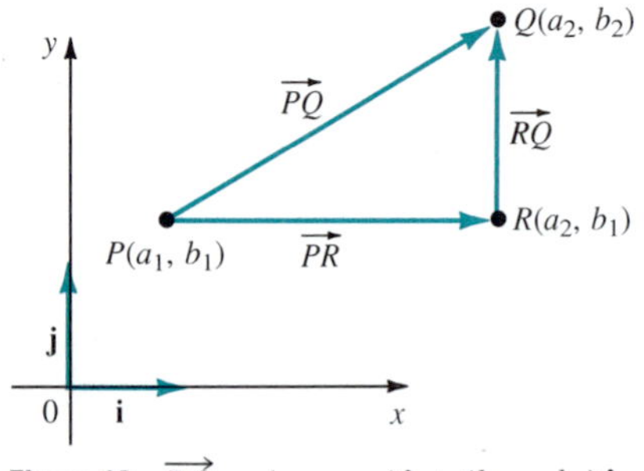

Figure 12 $\overrightarrow{PQ} = (a_2 - a_1)\mathbf{i} + (b_2 - b_1)\mathbf{j}$.

symbol **i** and the vector (0, 1) by the vector symbol **j** (see Figure 11).† If (a, b) denotes any other vector in $\mathbb{R}^2$, then since $(a, b) = a(1, 0) + b(0, 1)$, we may write

$$\mathbf{v} = (a, b) = a\mathbf{i} + b\mathbf{j} \tag{6}$$

Moreover, any vector in $\mathbb{R}^2$ can be represented in a unique way in the form $a\mathbf{i} + b\mathbf{j}$ since the representation of (a, b) as a point in the plane is unique. (Put another way, a point in the xy-plane has one and only one x-coordinate and one and only one y-coordinate.) Thus, all the properties of vectors we have seen apply with this new representation as well.

When the vector **v** is written in the form $\mathbf{v} = a\mathbf{i} + b\mathbf{j}$, we say that **v** is **resolved into its horizontal and vertical components** since a is the horizontal component of **v** and b is its vertical component.

Now suppose that a vector **v** can be represented by the directed line segment $\overrightarrow{PQ}$, where $P = (a_1, b_1)$ and $Q = (a_2, b_2)$ (see Figure 12). If we label the point (a_2, b_1) as R, then we immediately see that

$$\mathbf{v} = \overrightarrow{PQ} = \overrightarrow{PR} + \overrightarrow{RQ} \tag{7}$$

If $a_2 \geq a_1$, then the length of $\overrightarrow{PR}$ is $a_2 - a_1$, and since $\overrightarrow{PR}$ has the same direction as **i** (since they are parallel), we can write

$$\overrightarrow{PR} = (a_2 - a_1)\mathbf{i} \tag{8}$$

If $a_2 < a_1$, then the length of $\overrightarrow{PR}$ is $a_1 - a_2$, but then $\overrightarrow{PR}$ has the same direction as $-\mathbf{i}$ so $\overrightarrow{PR} = (a_1 - a_2)(-\mathbf{i}) = (a_2 - a_1)\mathbf{i}$ again. Similarly,

$$\overrightarrow{RQ} = (b_2 - b_1)\mathbf{j} \tag{9}$$

and we may write [using (7), (8), and (9)]

$$\mathbf{v} = (a_2 - a_1)\mathbf{i} + (b_2 - b_1)\mathbf{j} \tag{10}$$

EXAMPLE 7 *Resolving a Vector Into its Horizontal and Vertical Components*

Resolve the vector represented by the directed line segment from $(-2, 3)$ to $(1, 5)$ into its vertical and horizontal components.

SOLUTION Using (10), we have

$$\mathbf{v} = (a_2 - a_1)\mathbf{i} + (b_2 - b_1)\mathbf{j} = [1 - (-2)]\mathbf{i} + (5 - 3)\mathbf{j} = 3\mathbf{i} + 2\mathbf{j}$$

We now define a kind of vector that is very useful in certain types of applications.

Unit Vector

A **unit vector u** is a vector that has length 1.

† The symbols **i** and **j** were first used by Hamilton who defined a quarternion as a quantity of the form $a + b\mathbf{i} + c\mathbf{j} + d\mathbf{k}$, where a was the "scalar part" and $b\mathbf{i} + c\mathbf{j} + d\mathbf{k}$ the "vector part."

EXAMPLE 8 *A Unit Vector*

The vector $\mathbf{u} = (1/2)\mathbf{i} + (\sqrt{3}/2)\mathbf{j}$ is a unit vector since

$$|\mathbf{u}| = \sqrt{\left(\frac{1}{2}\right)^2 + \left(\frac{\sqrt{3}}{2}\right)^2} = \sqrt{\frac{1}{4} + \frac{3}{4}} = 1$$

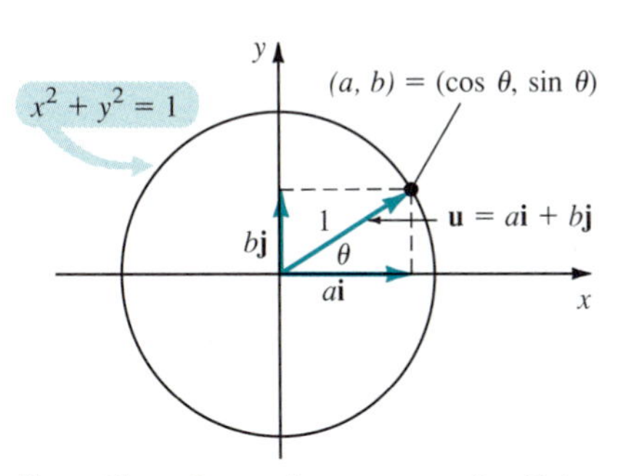

Figure 13 $\mathbf{u}$ is a unit vector, so (a, b) is a point on the unit circle.

Let $\mathbf{u} = a\mathbf{i} + b\mathbf{j}$ be a unit vector. Then $|\mathbf{u}| = \sqrt{a^2 + b^2} = 1$, so $a^2 + b^2 = 1$ and $\mathbf{u}$ is a point on the unit circle (see Figure 13). If θ is the direction of $\mathbf{u}$, then we immediately see (from the definition of $\cos\theta$ and $\sin\theta$ on p. 77) that $a = \cos\theta$ and $b = \sin\theta$. Thus any unit vector $\mathbf{u}$ can be written in the form

$$\mathbf{u} = (\cos\theta)\mathbf{i} + (\sin\theta)\mathbf{j} \tag{11}$$

where θ is the direction of $\mathbf{u}$.

EXAMPLE 9 *Writing a Unit Vector as $\cos\theta\mathbf{i} + \sin\theta\mathbf{j}$*

The unit vector $\mathbf{u} = (1/2)\mathbf{i} + (\sqrt{3}/2)\mathbf{j}$ of Example 8 can be written in the form (11) with $\theta = \cos^{-1}(1/2) = \pi/3$. Note that since $\cos\theta = 1/2$ and $\sin\theta = \sqrt{3}/2$, θ is in the first quadrant. We need this fact to conclude that $\theta = \pi/3$. It is also true that $\cos 5\pi/3 = 1/2$, but $5\pi/3$ is in the fourth quadrant. We have $\mathbf{u} = \cos\frac{\pi}{3}\mathbf{i} + \sin\frac{\pi}{3}\mathbf{j}$.

Finally (see Problem 41):

> Let $\mathbf{v}$ be any nonzero vector. Then $\mathbf{u} = \mathbf{v}/|\mathbf{v}|$ is the unit vector having the same direction as $\mathbf{v}$.

EXAMPLE 10 *Finding a Unit Vector Having the Same Direction as a Given Vector*

Find the unit vector having the same direction as $\mathbf{v} = 2\mathbf{i} - 3\mathbf{j}$.

SOLUTION Here $|\mathbf{v}| = \sqrt{4 + 9} = \sqrt{13}$, so $\mathbf{u} = \mathbf{v}/|\mathbf{v}| = (2/\sqrt{13})\mathbf{i} - (3/\sqrt{13})\mathbf{j}$ is the required unit vector. ■

EXAMPLE 11 *Finding a Vector Having a Given Magnitude and Direction*

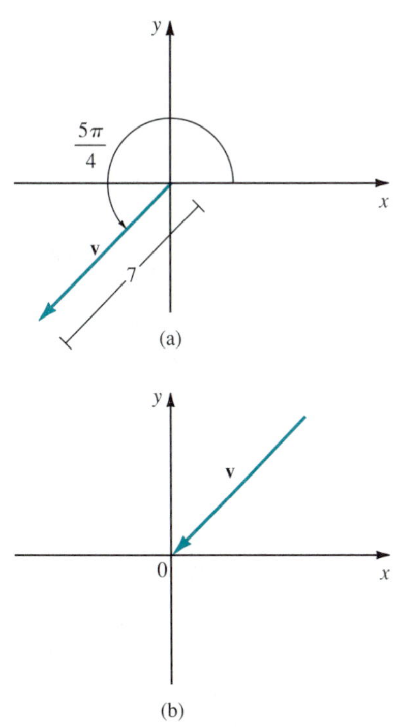

Figure 14 Two representations of the vector having direction $5\pi/4$ and magnitude 7

Find the vector $\mathbf{v}$ whose direction is $5\pi/4$ and whose magnitude is 7.

SOLUTION A unit vector $\mathbf{u}$ with direction $5\pi/4$ is given by

$$\mathbf{u} = \left(\cos\frac{5\pi}{4}\right)\mathbf{i} + \left(\sin\frac{5\pi}{4}\right)\mathbf{j} = -\frac{1}{\sqrt{2}}\mathbf{i} - \frac{1}{\sqrt{2}}\mathbf{j}$$

Then $\mathbf{v} = 7\mathbf{u} = -(7/\sqrt{2})\mathbf{i} - (7/\sqrt{2})\mathbf{j}$. This vector is sketched in Figure 14a. In Figure 14b we have translated $\mathbf{v}$ so that it points toward the origin. This representation of $\mathbf{v}$ is sometimes useful.

We conclude this section with a summary of properties of vectors, given in Table 1.

Table 1
Some Properties of Vectors

Object	Intuitive Definition	Expression in Terms of Components if $\mathbf{u} = u_1\mathbf{i} + u_2\mathbf{j} = (u_1, u_2)$ and $\mathbf{v} = v_1\mathbf{i} + v_2\mathbf{j} = (v_1, v_2)$
Vector **v**	Magnitude and direction ↗**v**	$v_1\mathbf{i} + v_2\mathbf{j}$ or (v_1, v_2)
$\lvert\mathbf{v}\rvert$	Magnitude or length of **v**	$\sqrt{v_1^2 + v_2^2}$
$\alpha\mathbf{v}$	↗**v** ↗α**v** (Here $\alpha = 2$)	$\alpha v_1\mathbf{i} + \alpha v_2\mathbf{j}$ or $(\alpha v_1, \alpha v_2)$
$-\mathbf{v}$	↗**v** ↙$-$**v**	$-v_1\mathbf{i} - v_2\mathbf{j}$ or $(-v_1, -v_2)$ or $-(v_1, v_2)$
$\mathbf{u} + \mathbf{v}$	**u** + **v**, **v**, **u** (triangle diagram)	$(u_1 + v_1)\mathbf{i} + (u_2 + v_2)\mathbf{j}$ or $(u_1 + v_1, u_2 + v_2)$
$\mathbf{u} - \mathbf{v}$	**v**, **u** − **v**, **u** (triangle diagram)	$(u_1 - v_1)\mathbf{i} + (u_2 - v_2)\mathbf{j}$ or $(u_1 - v_1, u_2 - v_2)$

FOCUS ON
Sir William Rowan Hamilton 1805–1865

Born in Dublin in 1805, where he spent most of his life, William Rowan Hamilton was without question Ireland's greatest mathematician. Hamilton's father (an attorney) and mother died when he was a small boy. His uncle, a linguist, took over the boy's education. By his fifth birthday, Hamilton could read English, Hebrew, Latin, and Greek. By his thirteenth birthday he had mastered not only the languages of continental Europe, but also Sanscrit, Chinese, Persian, Arabic, Malay, Hindi, Bengali, and several others as well. Hamilton liked to write poetry, both as a child and as an adult, and his friends included the great English poets Samuel Taylor Coleridge and William Wordsworth. Hamilton's poetry was considered so bad, however, that it is fortunate that he developed other interests—especially in mathematics.

Although he enjoyed mathematics as a young boy, Hamilton's interest was greatly enhanced by a chance meeting at the age of 15 with Zerah Colburn, the American lightning calculator. Shortly afterward, Hamilton began to read important mathematical books of the time. In 1823, at the age of 18, he discovered an error in Simon Laplace's *Mécanique céleste* and wrote an impressive paper on the subject. A year later he entered Trinity College in Dublin.

Sir William Rowan Hamilton *(The Granger Collection)*

Hamilton's university career was astonishing. At the age of 21, while still an undergraduate, he had so impressed the faculty that he was appointed Royal Astronomer of Ireland and Professor of Astronomy at the University. Shortly thereafter, he wrote what is now considered a classical work on optics. Using only mathematical theory, he predicted conical refraction in certain types of crystals. Later, this theory was confirmed by physicists. Largely because of this work, Hamilton was knighted in 1835.

Hamilton's first great purely mathematical paper appeared in 1833. In this work he described an algebraic way to manipulate pairs of real numbers. This work gives rules that are used today to add, subtract, multiply, and divide complex numbers. At first, however, Hamilton was unable to devise a multiplication for triples or n-tuples of numbers for $n > 2$. For 10 years he pondered this problem, and it is said that he solved it in an inspiration while walking on the Brougham Bridge in Dublin in 1843. The key was to discard the familiar commutative property of multiplication. The new objects he created were called *quaternions,* which were the precursors of what we now call vectors.

For the rest of his life, Hamilton spent most of his time developing the algebra of quaternions. He felt that they would have revolutionary significance in mathematical physics. His monumental work on this subject, *Treatise on Quaternions,* was published in 1853. Thereafter, he worked on an enlarged work, *Elements of Quaternions.* Although Hamilton died in 1865 before his *Elements* was completed, the work was published by his son in 1866.

Students of mathematics and physics know Hamilton in a variety of other contexts. In mathematical physics, for example, one encounters the Hamiltonian function, which often represents the total energy in a system and the Hamilton-Jacobi differential equations of dynamics. In matrix theory, the Cayley-Hamilton theorem states that every matrix satisfies its own characteristic equation.

Despite the great work he was doing, Hamilton's final years were a torment to him. His wife was a semi-invalid and he was plagued by alcoholism. It is therefore gratifying to point out that during these last years, the newly formed American National Academy of Sciences elected Sir William Rowan Hamilton to be its first foreign associate.

For a fascinating account of Hamilton's discoveries, read the article "Hamilton, Rodrigues, and the Quaternion Scandal—What went wrong with one of the major mathematical discoveries of the nineteenth century" in *Mathematics Magazine* 62(5) December, 1989, 291–308.

Problems 4.4

Readiness Check

I. A *vector* is ______.
   a. two points in the xy-plane
   b. a line segment between two points
   c. a directed line segment from one point to another
   d. a collection of equivalent directed line segments

II. If $P = (3, -4)$ and $Q = (8, 6)$, the vector $\overrightarrow{PQ}$ has length ______.
   a. $|3| + |-4|$ b. $(3)^2 + (-4)^2$
   c. $(3-8)^2 + (-4-6)^2$ d. $\sqrt{(8-3)^2 + (6-(-4))^2}$

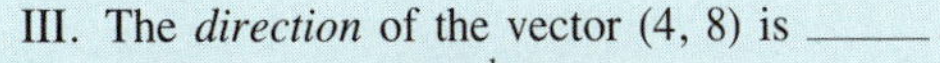

III. The *direction* of the vector (4, 8) is ______.
   a. π b. $\tan^{-1}(8-4)$
   c. $(\frac{8}{4})\pi$ d. $\tan^{-1}(\frac{8}{4})$

IV. If $\mathbf{u} = (3, 4)$ and $\mathbf{v} = (5, 8)$, then $\mathbf{u} + \mathbf{v} =$ ______.
   a. (7, 13)   b. (8, 12)
   c. (2, 4)   d. (15, 32)

V. If $\mathbf{u} = (4, 3)$ then the unit vector with the same direction as $\mathbf{u}$ is ______.
   a. (0.4, 0.3)   b. (0.8, 0.6)
   c. $(\frac{4}{5}, \frac{3}{5})$ d. $(\frac{4}{7}, \frac{3}{7})$

In Problems 1–10 find the magnitude and direction of the given vector. Then write the vector in the form $a\mathbf{i} + b\mathbf{j}$.

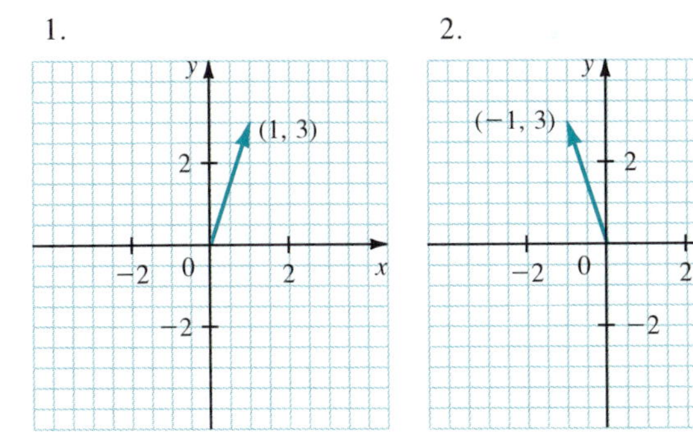

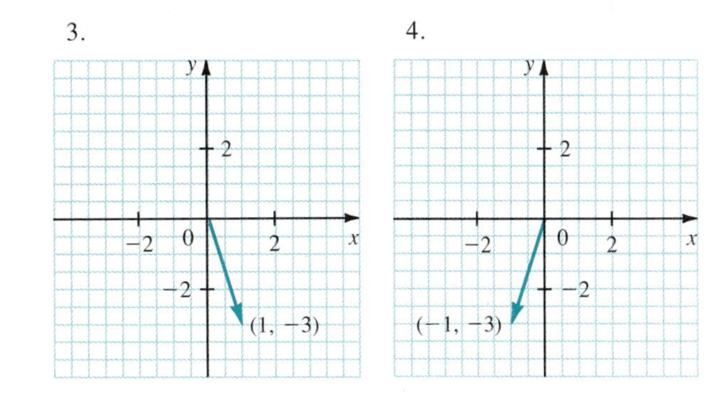

Answers to Readiness Check

I. d   II. d   III. d   IV. b   V. b, c

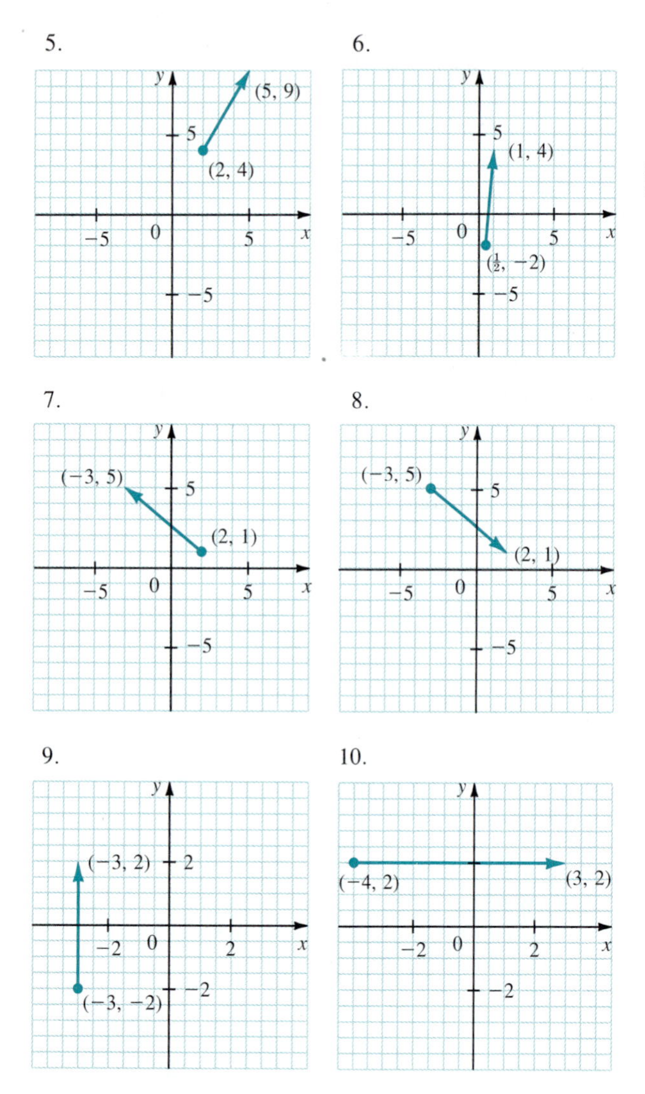

In Problems 11–16 a vector $\mathbf{v}$ and a point P are given. Find a point Q such that the directed line segment $\overrightarrow{PQ}$ is a representation of $\mathbf{v}$. Sketch $\mathbf{v}$ and $\overrightarrow{PQ}$.

11. $\mathbf{v} = (2, 5)$; $P = (1, -2)$
12. $\mathbf{v} = (5, 8)$; $P = (3, 8)$
13. $\mathbf{v} = (-3, 7)$; $P = (7, -3)$
14. $\mathbf{v} = -\mathbf{i} - 7\mathbf{j}$; $P = (0, 1)$
15. $\mathbf{v} = 5\mathbf{i} - 3\mathbf{j}$; $P = (-7, -2)$
16. $\mathbf{v} = e\mathbf{i} + \pi\mathbf{j}$; $P = (\pi, \sqrt{2})$

In Problems 17–28 find the magnitude and direction of the given vector.

17. $\mathbf{v} = (4, 4)$
18. $\mathbf{v} = (-4, 4)$
19. $\mathbf{v} = (4, -4)$
20. $\mathbf{v} = (-4, -4)$
21. $\mathbf{v} = (\sqrt{3}, 1)$
22. $\mathbf{v} = (1, \sqrt{3})$
23. $\mathbf{v} = (-1, \sqrt{3})$
24. $\mathbf{v} = (1, -\sqrt{3})$
25. $\mathbf{v} = (-1, -\sqrt{3})$
26. $\mathbf{v} = (1, 2)$
27. $\mathbf{v} = (-5, 8)$
28. $\mathbf{v} = (11, -14)$

In Problems 29–36 write in the form $a\mathbf{i} + b\mathbf{j}$ the vector $\mathbf{v}$ that is represented by $\overrightarrow{PQ}$. Sketch $\overrightarrow{PQ}$ and $\mathbf{v}$.

29. $P = (1, 2)$; $Q = (1, 3)$
30. $P = (2, 4)$; $Q = (-7, 4)$
31. $P = (5, 2)$; $Q = (-1, 3)$
32. $P = (8, -2)$; $Q = (-3, -3)$
33. $P = (7, -1)$; $Q = (-2, 4)$
34. $P = (3, -6)$; $Q = (8, 0)$
35. $P = (-3, -8)$; $Q = (-8, -3)$
36. $P = (2, 4)$; $Q = (-4, -2)$
37. Let $\mathbf{u} = (2, 3)$ and $\mathbf{v} = (-5, 4)$. Find the following:
    (a) $3\mathbf{u}$ (b) $\mathbf{u} + \mathbf{v}$
    (c) $\mathbf{v} - \mathbf{u}$ (d) $2\mathbf{u} - 7\mathbf{v}$
38. Let $\mathbf{u} = 2\mathbf{i} - 3\mathbf{j}$ and $\mathbf{v} = -4\mathbf{i} + 6\mathbf{j}$. Find the following:
    (a) $\mathbf{u} + \mathbf{v}$ (b) $\mathbf{u} - \mathbf{v}$
    (c) $3\mathbf{u}$ (d) $-7\mathbf{v}$
    (e) $8\mathbf{u} - 3\mathbf{v}$ (f) $4\mathbf{v} - 6\mathbf{u}$
39. Show that the vectors $\mathbf{i}$ and $\mathbf{j}$ are unit vectors.
40. Show that the vector $(1/\sqrt{2})\mathbf{i} + (1/\sqrt{2})\mathbf{j}$ is a unit vector.
41. Show that if $\mathbf{v} = a\mathbf{i} + b\mathbf{j} \neq \mathbf{0}$, then $\mathbf{u} = (a/\sqrt{a^2 + b^2})\mathbf{i} + (b/\sqrt{a^2 + b^2})\mathbf{j}$ is a unit vector having the same direction as $\mathbf{v}$.

In Problems 42–47 find a unit vector having the same direction as the given vector.

42. $\mathbf{v} = 2\mathbf{i} + 3\mathbf{j}$
43. $\mathbf{v} = \mathbf{i} - \mathbf{j}$
44. $\mathbf{v} = (3, 4)$
45. $\mathbf{v} = (3, -4)$
46. $\mathbf{v} = -3\mathbf{i} + 4\mathbf{j}$
47. $\mathbf{v} = (a, a)$, $a \neq 0$
48. If $\mathbf{v} = a\mathbf{i} + b\mathbf{j} \neq \mathbf{0}$, show that $a/\sqrt{a^2 + b^2} = \cos\theta$ and $b/\sqrt{a^2 + b^2} = \sin\theta$, where θ is the direction of $\mathbf{v}$.
49. For $\mathbf{v} = 2\mathbf{i} - 3\mathbf{j}$, find $\sin\theta$ and $\cos\theta$.
50. For $\mathbf{v} = -3\mathbf{i} + 8\mathbf{j}$, find $\sin\theta$ and $\cos\theta$.

A vector $\mathbf{v}$ has a direction opposite to that of a vector $\mathbf{u}$ if $\mathbf{v} = -\mathbf{u}$. In Problems 51–56 find a unit vector $\mathbf{u}$ that has a direction opposite the direction of the given vector $\mathbf{v}$.

51. $\mathbf{v} = \mathbf{i} + \mathbf{j}$
52. $\mathbf{v} = 2\mathbf{i} - 3\mathbf{j}$
53. $\mathbf{v} = (-3, 4)$
54. $\mathbf{v} = (-2, 3)$
55. $\mathbf{v} = -3\mathbf{i} - 4\mathbf{j}$
56. $\mathbf{v} = (8, -3)$
57. Let $\mathbf{u} = 2\mathbf{i} - 3\mathbf{j}$ and $\mathbf{v} = -\mathbf{i} + 2\mathbf{j}$. Find a unit vector having the same direction as the following:
    (a) $\mathbf{u} + \mathbf{v}$ (b) $2\mathbf{u} - 3\mathbf{v}$ (c) $3\mathbf{u} + 8\mathbf{v}$

58. Let $P = (c, d)$ and $Q = (c + a, d + b)$. Show that the magnitude of $\overrightarrow{PQ}$ is $\sqrt{a^2 + b^2}$.
59. Show that the direction of $\overrightarrow{PQ}$ in Problem 58 is the same as the direction of the vector (a, b). [Hint: If $R = (a, b)$, show that the line passing through the points P and Q is parallel to the line passing through the points 0 and R.]

In Problems 60–67 find a vector $\mathbf{v}$ having the given magnitude and direction. [Hint: See Example 11.]

60. $|\mathbf{v}| = 3;\ \theta = \pi/6$
61. $|\mathbf{v}| = 8;\ \theta = \pi/3$
62. $|\mathbf{v}| = 7;\ \theta = \pi$
63. $|\mathbf{v}| = 4;\ \theta = \pi/2$
64. $|\mathbf{v}| = 1;\ \theta = \pi/4$
65. $|\mathbf{v}| = 6;\ \theta = 2\pi/3$
66. $|\mathbf{v}| = 8;\ \theta = 3\pi/2$
67. $|\mathbf{v}| = 6;\ \theta = 11\pi/6$

4.5 Applications of Vectors [Optional]

In this section, we present three interesting applications of vectors.

Force Vectors

According to Newton's first law of motion, the law of inertia, an object in a state of rest or uniform motion in a straight line stays in that state unless acted on by an outside force. Loosely speaking, **force** is that thing that causes a change in acceleration of an object. More precisely, according to Newton's second law of motion,

$$F = ma \tag{1}$$

where F is the force, m is the mass of the object, and a is the acceleration of the object. There are two systems of units for measuring force:

In the metric system, mass is measured in kilograms (kg), acceleration is measured in meters per second per second (m/sec^2), force is measured in newtons (N).

In the English system, mass is measured in slugs, acceleration is measured in feet per second per second (ft/sec^2), force is measured in pounds (lb).

The **weight** W of an object is the gravitational force exerted on it by the earth. The acceleration due to the earth's gravity is denoted by g. We use the constants

$$g = 9.8 \text{ m/sec}^2 \qquad \text{and} \qquad g = 32 \text{ ft/sec}^2$$

Then,

Units of Weight

In metric units,

$$W = mg \text{ is measured in newtons} \tag{2}$$

In English units,

$$W = mg \text{ is measured in pounds} \tag{3}$$

In most applications, we are concerned not only with the magnitude of a force but also its direction. Thus, force and acceleration are considered as vector quantities and are denoted by **F** and **a**. By (1), we have

$$|\mathbf{F}| = m|\mathbf{a}| \tag{4}$$

If more than one force is applied to an object, then we define the **resultant** force applied to the object to be the *vector sum* of these forces. We can think of the resultant as the *net* applied force.

EXAMPLE 1 *Computing Forces in Two Directions*

A car weighing 4500 lb rolls down a ramp (or inclined plane) that makes an angle of 23° with the horizontal ground.

(a) What is the magnitude of the force pushing the car down the ramp?
(b) What is the magnitude of the force pushing the car against the ramp?

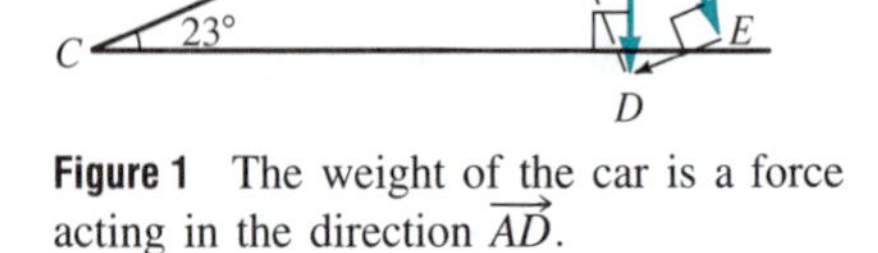

Figure 1 The weight of the car is a force acting in the direction $\overrightarrow{AD}$.

SOLUTION The situation is depicted in Figure 1. The weight of the car is the force $\overrightarrow{AD}$ that acts downward. $\overrightarrow{AD}$ is the resultant of two forces: the force $\overrightarrow{AB}$ that pushes the car down the ramp, and the force $\overrightarrow{AE}$ that pushes the car against the ramp. These two force vectors are always perpendicular. We have

$$\overrightarrow{AD} = \overrightarrow{AB} + \overrightarrow{AE}$$

The problem asks us to find $|\overrightarrow{AB}|$ and $|\overrightarrow{AE}|$. Since $\overrightarrow{AB}$ and $\overrightarrow{ED}$ are parallel and have the same length, they are representations of the same vector. Hence,

$$|\overrightarrow{AB}| = |\overrightarrow{ED}|$$

We observe that

$$\measuredangle DAE = 23°$$

because both $\measuredangle DAE$ and $\measuredangle DCB$ are complements of the same angle ($\measuredangle DAB$) and $\measuredangle DCB = 23°$. Then

$$\sin \measuredangle DAE = \frac{|\overrightarrow{ED}|}{|\overrightarrow{AD}|}$$

$$\sin 23° = \frac{|\overrightarrow{ED}|}{|\overrightarrow{AD}|}$$

$$|\overrightarrow{ED}| = |\overrightarrow{AD}| \sin 23°$$

But $|\overrightarrow{AD}|$ is the magnitude of the gravitational force (weight), so $|\overrightarrow{AD}| = 4500$ lb and

$$|\overrightarrow{AB}| = |\overrightarrow{ED}| = 4500 \sin 23° = 4500(0.390731128) \approx 1758.3 \text{ lb}$$

Similarly,

$$|\overrightarrow{AE}| = |\overrightarrow{AD}| \cos 23° = 4500(0.920504853) \approx 4142.3 \text{ lb}$$

Check: By the Pythagorean theorem,

$$
\begin{aligned}
|\overrightarrow{AD}|^2 &= |\overrightarrow{AE}|^2 + |\overrightarrow{ED}|^2 \\
4500^2 &= 4142.3^2 + 1758.3^2 \\
20{,}250{,}000 &= 17{,}158{,}649.29 + 3{,}091{,}618.89 \\
&= 20{,}250{,}268.18
\end{aligned}
$$

This is close enough. We lost some accuracy in rounding. ■

EXAMPLE 2 *Determining the Maximum Height at Which a Force is Effective*

A ramp 25 m long is used to pull a 50-kg object up to the top of a loading dock. If the force on the incline is not to exceed 380 N, what is the maximum height of the dock?

SOLUTION Look at Figure 2. We first compute the force due to gravity (the weight), $\mathbf{F}_w$. By (2),

$$|\mathbf{F}_w| = W = mg = (50)(9.8) = 490 \text{ N}$$

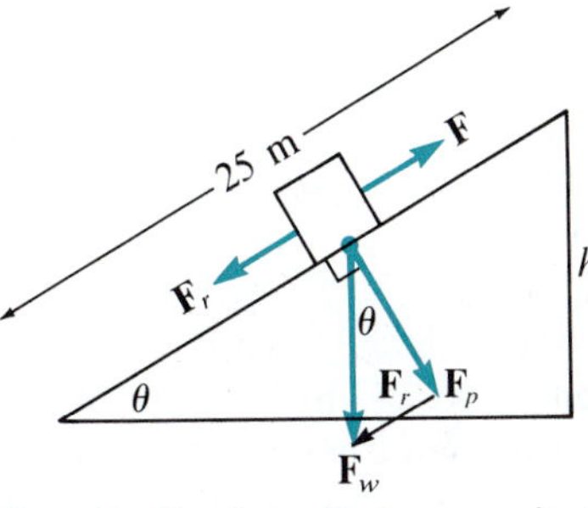

Figure 2 The force $\mathbf{F}_w$ has magnitude $mg = 490$ N and acts downward.

Let $\mathbf{F}_r$ be the force that pushes the object down the ramp. The force, $\mathbf{F}$, needed to overcome $\mathbf{F}_r$ is $-\mathbf{F}_r$ (equal magnitude but opposite direction). Let θ denote the angle the ramp makes with the ground. If $\mathbf{F}_p$ denotes the force of the object against the ramp, then by the same reasoning as in Example 1, the angle between $\mathbf{F}_p$ and $\mathbf{F}_w$ is also θ. We have

$$\sin\theta = \frac{|\mathbf{F}_r|}{|\mathbf{F}_w|}$$

$$|\mathbf{F}| = |-\mathbf{F}_r| = |\mathbf{F}_r| = |\mathbf{F}_w| \sin\theta = 490 \sin\theta$$

Now we are told that $|\mathbf{F}|$ cannot exceed 380 N. Thus we seek the θ that leads to $|\mathbf{F}| = 380$. We have

$$380 = 490 \sin\theta$$

$$\sin\theta = \frac{380}{490} \approx 0.77551$$

But from Figure 2,

$$\sin\theta = \frac{h}{25}$$

$$h = 25 \sin\theta = 25(0.77551) \approx 19.4 \text{ m}$$

Note that if $h < 19.4$, then less force is needed to pull the object up the ramp. For example, if $h = 15$ m, then

$$\sin\theta = \frac{h}{25} = \frac{15}{25} = 0.6$$

and

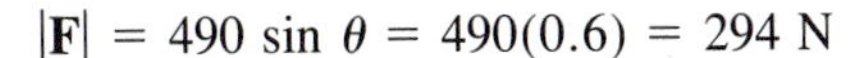

$$|\mathbf{F}| = 490 \sin\theta = 490(0.6) = 294 \text{ N}$$

If $h > 19.4$, then more force than the maximum allowed is needed. For example, if $h = 21$ m, then

$$\sin\theta = \frac{21}{25} = 0.84$$

and

$$|\mathbf{F}| = 490 \sin\theta = 490(0.84) = 411.6 \text{ N}$$

is required to pull the object up. This exceeds the available force of 380 N.

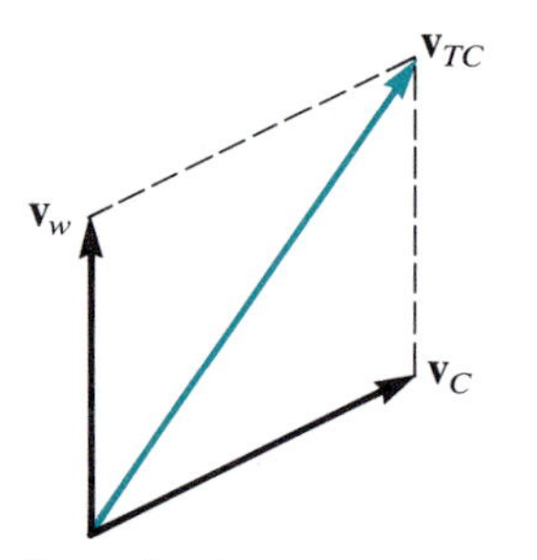

Figure 3 The true course vector $\mathbf{v}_{TC}$ is the resultant of the course vector $\mathbf{v}_C$ and the wind vector $\mathbf{v}_w$.

Velocity Vectors and Navigation

Suppose an object moves with constant speed s in a constant direction θ. Then the vector $\mathbf{v}$ whose magnitude is s and whose direction is θ is called the **velocity** vector of the object.

Often there are two or more velocity vectors appearing in a single problem. One example is given by a flying plane. We are interested in both the velocity vector of the plane's motion and the velocity vector of the wind's motion. The true velocity of the plane is the resultant of these two vectors. A second example is a boat moving in a river where both the velocity of the boat and the velocity of the current are important.

The **airspeed** and **course** of a plane are the speed and bearing at which the plane would be flying if there were no wind. The airspeed is the speed generated by the plane's engine. The **ground speed** and **true course** of the plane are the speed and bearing of the plane that result from the action of the wind. Look at Figure 3. If $\mathbf{v}_C$ denotes the velocity vector of the plane without wind and $\mathbf{v}_w$ denotes the wind's velocity vector, then the ground speed and true course are the magnitude and direction of the vector $\mathbf{v}_{TC}$, which is the resultant of $\mathbf{v}_w$ and $\mathbf{v}_C$.

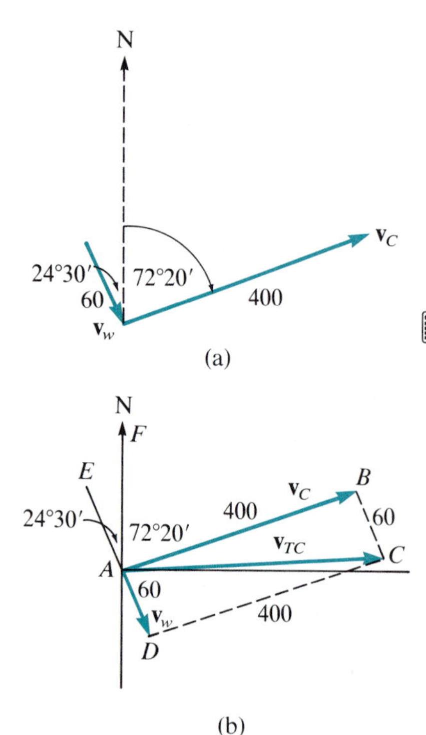

Figure 4 $\mathbf{v}_{TC}$ is the sum of the vectors $\mathbf{v}_w$ and $\mathbf{v}_C$.

EXAMPLE 3 *Finding the True Course and Ground Speed in the Presence of Wind*

A pilot heads her plane at a bearing of N72°20′E at 400 mph. A wind is blowing from N24°30′W at 60 mph. Find the ground speed and true course of her plane.

SOLUTION The situation is depicted in Figure 4(a). Note that the wind is coming *from* a northwesterly direction, so it blows toward the southeast. In Figure 4(b), we move the wind velocity vector $\mathbf{v}_w$ so that it starts at the same point as the plane's velocity vector $\mathbf{v}_C$. The true course velocity vector $\mathbf{v}_{TC}$ is the resultant of $\mathbf{v}_C$ and $\mathbf{v}_w$. In Figure 4(b), the angles EAB and ABC are alternate interior angles (see p. 230), so

$$\measuredangle ABC = \measuredangle EAB = 24°30' + 72°20' = 96°50' \approx 96.83°$$

Also, $|\overrightarrow{BC}| = |\mathbf{v}_w| = 60$. Then, from the law of cosines applied to triangle

ABC, we have

$$|\mathbf{v}_{TC}|^2 = |\mathbf{v}_C|^2 + |\overrightarrow{BC}|^2 - 2|\mathbf{v}_C||\overrightarrow{BC}| \cos(\measuredangle ABC)$$
$$= 400^2 + 60^2 - 2(400)(60)\cos 96.83° \approx 169{,}308$$
$$\text{Ground speed} = |\mathbf{v}_{TC}| \approx \sqrt{169{,}308} \approx 411.5 \text{ mph}$$

To compute the true course, we need to find $\measuredangle BAC$. From the law of sines,

$$\frac{\sin \measuredangle BAC}{|\overrightarrow{BC}|} = \frac{\sin \measuredangle ABC}{|\mathbf{v}_{TC}|}$$
$$\sin \measuredangle BAC = \frac{|\overrightarrow{BC}|}{|\mathbf{v}_{TC}|} \sin \measuredangle ABC$$
$$\approx \frac{60}{411.5} \sin 96.83° \approx 0.14477$$

and

$$\measuredangle BAC \approx \sin^{-1} 0.14477 \approx 8.324° \approx 8°19'$$

Then

$$\measuredangle FAC = 72°20' + 8°19' = 80°39'$$

and

$$\text{True course} = \text{N}80°39'\text{E} \quad \blacksquare$$

EXAMPLE 4 ***Determining Direction and Airspeed Necessary to Achieve a Desired Ground Speed and Bearing***

The wind is blowing from S42°E at a speed of 85 km/hr. An aviator wishes to achieve a ground speed of 650 km/hr with a bearing of S63°W. In what direction must he head (that is, what is his course), and at what airspeed should he fly?

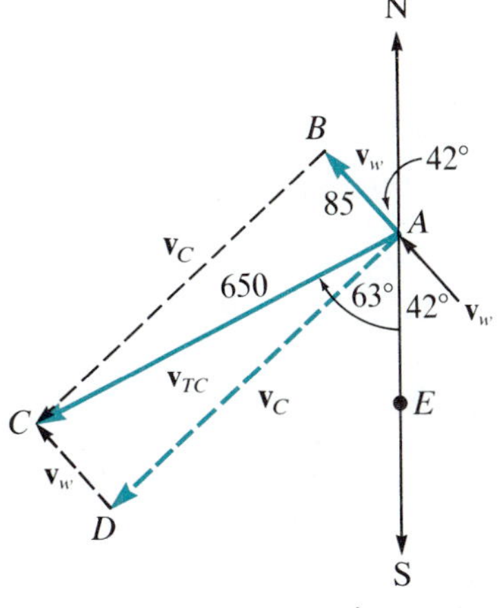

Figure 5 $\mathbf{v}_w$ and $\mathbf{v}_{TC}$ are given. $\mathbf{v}_C = \mathbf{v}_{TC} - \mathbf{v}_w$.

SOLUTION Here $\mathbf{v}_{TC}$ and $\mathbf{v}_w$ are given and we must find $\mathbf{v}_C$. The situation is sketched in Figure 5. As before, $\mathbf{v}_w$ has been drawn so that the vectors $\mathbf{v}_w$ and $\mathbf{v}_{TC}$ have the same initial point. One representation of $\mathbf{v}_C$ is drawn so that

$$\mathbf{v}_{TC} = \mathbf{v}_w + \mathbf{v}_C$$

In Figure 5, we see that $\overrightarrow{BC} = |\mathbf{v}_C|$ and

$$\measuredangle BAC = 180° - 42° - 63° = 180° - 105° = 75°$$

So, from the law of cosines applied to triangle ABC,

$$|\mathbf{v}_C|^2 = |\mathbf{v}_w|^2 + |\mathbf{v}_{TC}|^2 - 2|\mathbf{v}_w||\mathbf{v}_{TC}| \cos 75°$$
$$= 85^2 + 650^2 - 2(85)(650)\cos 75° \approx 401{,}125.5$$

and

$$\text{Airspeed} \approx \sqrt{401{,}125.5} \approx 633.3 \text{ km/hr}$$

To compute the course, we need to subtract $\measuredangle CAD$ from 63°. But $\measuredangle BCA$ and $\measuredangle CAD$ are alternate interior angles and are therefore equal. We use the law of sines to find $\measuredangle BCA$:

$$\frac{\sin \measuredangle BCA}{|\mathbf{v}_w|} = \frac{\sin \measuredangle BAC}{|\mathbf{v}_C|}$$

$$\sin \measuredangle BCA = \frac{|\mathbf{v}_w|}{|\mathbf{v}_C|} \sin \measuredangle BAC$$

$$\approx \frac{85}{633.3} \sin 75° \approx 0.12964$$

and

$$\measuredangle CAD = \measuredangle BCA \approx \sin^{-1} 0.12964 \approx 7.45° \approx 7°27'$$

$$\measuredangle EAD \approx 63° - 7°27' = 55°33'$$

Thus

$$\text{Course at take-off} = \text{S}55°33'\text{W}$$

Geometric Applications of Vectors

Vectors can be used to solve problems in geometry.

EXAMPLE 5 ***Using Vectors to Prove a Geometric Fact***

Show that the midpoints of the sides of a quadrilateral are the vertices of a parallelogram.

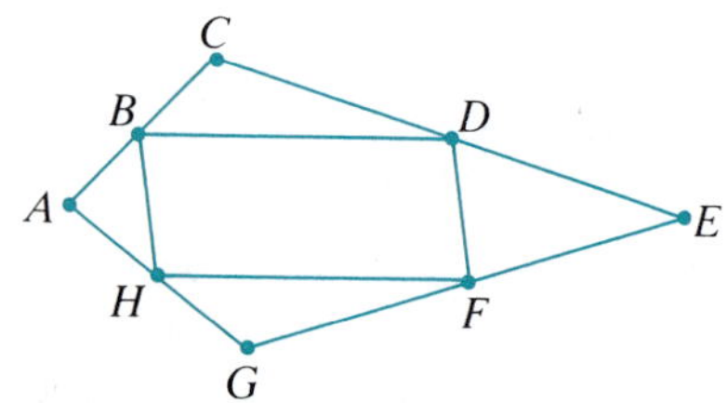

Figure 6 $ACEG$ is a quadrilateral. B, D, F, and H mark the midpoints of its sides.

SOLUTION The situation is sketched in Figure 6. Using the fact that B, D, F, and H are midpoints, we have

$$\overrightarrow{BD} = \overrightarrow{BC} + \overrightarrow{CD} = \tfrac{1}{2}\overrightarrow{AC} + \tfrac{1}{2}\overrightarrow{CE} = \tfrac{1}{2}(\overrightarrow{AC} + \overrightarrow{CE}) = \tfrac{1}{2}\overrightarrow{AE}$$

and

$$\overrightarrow{HF} = \overrightarrow{HG} + \overrightarrow{GF} = \tfrac{1}{2}\overrightarrow{AG} + \tfrac{1}{2}\overrightarrow{GE} = \tfrac{1}{2}(\overrightarrow{AG} + \overrightarrow{GE}) = \tfrac{1}{2}\overrightarrow{AE}$$

Thus $\overrightarrow{BD} = \overrightarrow{HF}$. Similarly, $\overrightarrow{HB} = \overrightarrow{FD}$, so $BDFH$ is a parallelogram.

Problems 4.5

Readiness Check

I. Which of the following is the resultant vector of vectors $\mathbf{v} = (7, 1)$ and $\mathbf{u} = (-2, 9)$?
   a. (9, 10) b. (9, 8)
   c. (5, 10) d. (5, −10)

II. Which of the following describes the difference between the airspeed and the ground speed of a plane?
   a. The airspeed and the ground speed are the same.
   b. The airspeed is always less than the ground speed.
   c. The airspeed includes the speed of the wind, and the ground speed does not.
   d. The ground speed includes the speed of the wind, and the airspeed does not.

III. Which of the following is true of a boat having a bearing of S34°E traveling 24 mph if a current of 3 mph flows due north?
   a. The boat will have a more westerly true course.
   b. The current will slow the boat.
   c. The boat will have a more southerly true course.
   d. The current will have no effect on the boat.

Answers to Readiness Check

I. c II. d III. b (the true speed of the boat will be $\sqrt{24^2 + 3^2 - 2(3)(24)\cos 34°} \approx$ 21.6 mph.)

1. A block weighing 122 lb is sliding down a ramp that makes an angle of 31° with the ground.

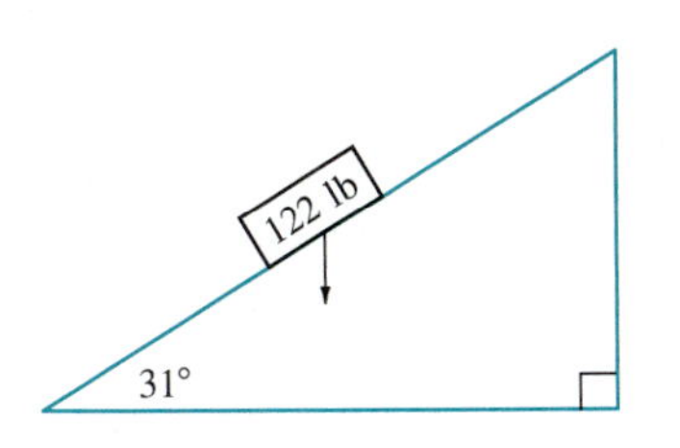

   (a) What is the magnitude of the force pushing the block down the ramp?
   (b) What is the force of the block against the ramp?
2. Answer the questions in Problem 1 if the block has a mass of 35 kg.
3. It takes a force of 85 lb to pull a 120-lb object up a ramp. What angle does the ramp make with the ground?
4. It takes a force of 636 N to pull a 87-kg mass up a ramp. What angle does the ramp make with the ground?
5. One end of a rope is attached to a box weighing 160 lb. If the box is pulled up, the **tension** in the rope is defined as the magnitude of the force needed to do the pulling. If $\theta = 58°$, what is the tension in the rope?

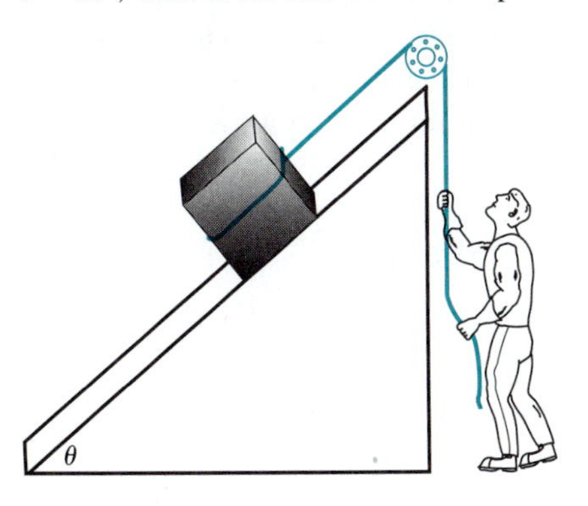

6. In Problem 5 how high is the end of the ramp if the man pulls with a force of 95 lb?
7. Find the angle θ in Problem 5 if the tension is 620 N and the magnitude of the force pushing the box against the ramp is 437 N.
8. If $\theta = 29°$ in Problem 5 and the man pulls with a force of 410 N, what is the mass of the box?
9. A girl pushes a lawn mower with a force of 26 lb (see the figure in the next column). The handle makes an angle of 42° with the horizontal. What are the vertical and horizontal components of the force?
10. Answer the questions in Problem 9 if she pushes with a force of 112 N and the angle is 36°.

11. If, in Problem 9, the vertical and horizontal components of the force are 15 lb and 19.5 lb, find (a) the force she exerts on the mower and (b) the angle the handle makes with the horizontal.
12. A section of road has a 4% grade. That is, it rises 4 ft for each 100 ft of road. A car weighing 3174 lb is parked on the hill. What is the force that tends to make the car roll downhill?

13. On a hill with a 3.2% grade, a force of 2,850 N is needed to prevent a truck from rolling down. What is the mass of the truck?
14. A force of 2845 N is needed to prevent a 6400-kg van from rolling down a hill. What is the grade of the hill?
15. A wind is blowing from N23°E at 45 mph. Find the ground speed and true course of a plane if the plane takes off on the course N42°35′W at 430 mph.
16. The pilot of the plane in Problem 15 wants to fly at a ground speed of 430 mph and true course of N42°35′W. What must be his airspeed and course at take-off?
17. In Problem 15 what must be the airspeed and course at take-off if a ground speed of 510 mph and a true course of S33°10′E is desired?
18. A wind is blowing from N82°20′W at 105 km/hr. If a plane takes off bearing N28°45′E and achieves an airspeed of 316 km/hr, determine the true course and ground speed.
19. In Problem 18 what must be the airspeed and course at take-off if a ground speed of 340 km/hr and a true course of N5°W is desired?
20. A plane takes off bearing S71°E at an airspeed of 211 mph. The true course is S64°52′E, and the ground speed is 223.4 mph. From which direction is the wind blowing, and what is the wind speed?
21. A river that is 1465 ft wide flows due south at a rate of 255 ft/min. A boat attempts to cross the river in 5 minutes. What must be the bearing and speed of the boat if it is to reach a point directly across from the point at which it starts?

22. The distance across the river from point A to point B is 640 m. The current flows at 3.4 km/hr. A boat heading due east crosses the river along line AB in 8 minutes. What is the boat's velocity?

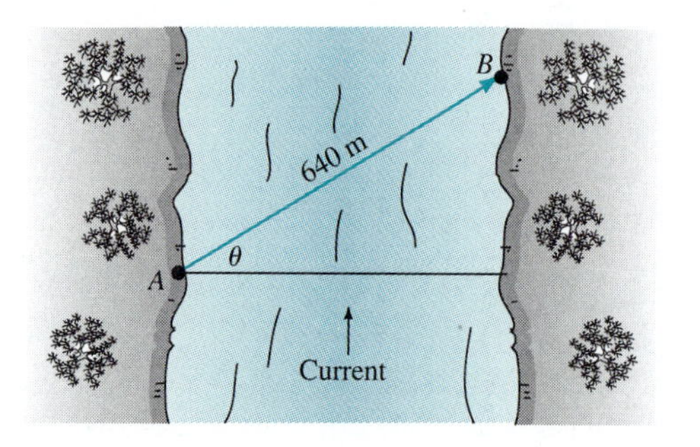

23. Using the information from Problem 22, determine the angle θ, called the **drift angle.**
24. Show that the diagonals of a parallelogram bisect each other.
25. Use vector methods to show that the angles opposite the equal sides in an isosceles triangle are equal.
26. Show that in a trapezoid the line segment that joins the midpoints of the two sides that are not parallel is parallel to the parallel sides and has a length equal to the average of the lengths of the parallel sides.

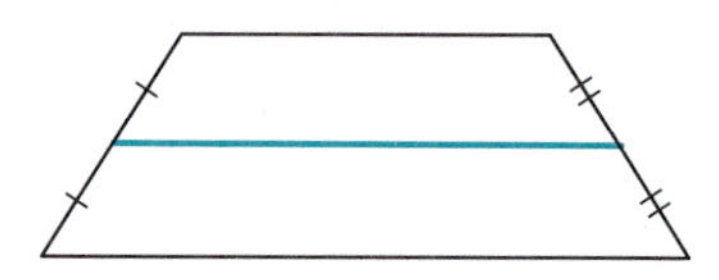

* 27. Prove that the medians of a triangle intersect at a fixed point that is two thirds the distance from each vertex to the opposite side.

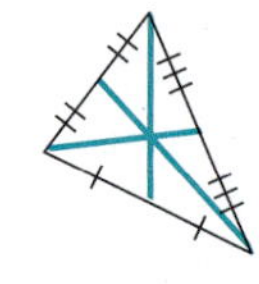

28. Use the result of Problem 27 to find the point of intersection of the medians of the triangle with vertices at (3, 4), (2, −1), and (−3, 2).

4.6 The Dot Product of Two Vectors

In Section 4.4 we showed how a vector could be multiplied by a scalar but not how two vectors could be multiplied. Actually, there are several ways to define the product of two vectors, and in this section we will discuss one of them.

Definition of the Dot Product

Let $\mathbf{u} = (a_1, b_1) = a_1\mathbf{i} + b_1\mathbf{j}$ and $\mathbf{v} = (a_2, b_2) = a_2\mathbf{i} + b_2\mathbf{j}$. Then the **dot product** of **u** and **v**, denoted by $\mathbf{u} \cdot \mathbf{v}$, is defined by

$$\mathbf{u} \cdot \mathbf{v} = a_1a_2 + b_1b_2 \tag{1}$$

NOTE We stress that $\mathbf{u} \cdot \mathbf{v}$ is a scalar (a real number). It is not another vector.

EXAMPLE 1 ***Computing a Dot Product***

If $\mathbf{u} = (1, 3)$ and $\mathbf{v} = (4, -7)$, then

$$\mathbf{u} \cdot \mathbf{v} = 1(4) + 3(-7) = 4 - 21 = -17$$

The following theorem follows from (1) and the definition of $|\mathbf{v}|$ on page 237.

Theorem: Five Properties of the Dot Product

For any vectors **u**, **v**, **w**, and scalar a,

(i) $\mathbf{u} \cdot \mathbf{v} = \mathbf{v} \cdot \mathbf{u}$
(ii) $(\mathbf{u} + \mathbf{v}) \cdot \mathbf{w} = \mathbf{u} \cdot \mathbf{w} + \mathbf{v} \cdot \mathbf{w}$
(iii) $(a\mathbf{u}) \cdot \mathbf{v} = a(\mathbf{u} \cdot \mathbf{v})$
(iv) $\mathbf{u} \cdot \mathbf{u} \geq 0$ and $\mathbf{u} \cdot \mathbf{u} = 0$ if and only if $\mathbf{u} = \mathbf{0}$
(v) $|\mathbf{u}| = \sqrt{\mathbf{u} \cdot \mathbf{u}}$

The dot product is useful in a wide variety of applications. An interesting one follows.

Definition of the Angle Between Two Vectors

Let **u** and **v** be two nonzero vectors. Then the **angle** ϕ between **u** and **v** is defined to be the smallest angle† between the representations of **u** and **v** that have the origin as their initial points. If $\mathbf{u} = \alpha\mathbf{v}$ for some scalar α, then we define $\phi = 0$ if $\alpha > 0$ and $\phi = \pi$ if $\alpha < 0$.

Theorem

Let **u** and **v** be two nonzero vectors. Then if ϕ is the angle between them,

$$\cos \phi = \frac{\mathbf{u} \cdot \mathbf{v}}{|\mathbf{u}||\mathbf{v}|} \tag{2}$$

Proof

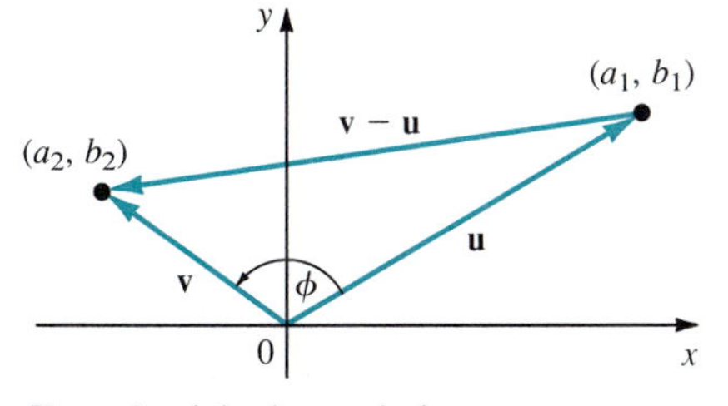

Figure 1 ϕ is the angle between **u** and **v**.

We place **u** and **v** in standard position as in Figure 1. Then, from the law of cosines

$$|\mathbf{v} - \mathbf{u}|^2 = |\mathbf{v}|^2 + |\mathbf{u}|^2 - 2|\mathbf{u}||\mathbf{v}| \cos \phi \tag{3}$$

Using properties (ii) and (v) of the dot product, we have

$$|\mathbf{v} - \mathbf{u}|^2 = (\mathbf{v} - \mathbf{u}) \cdot (\mathbf{v} - \mathbf{u}) = \mathbf{v} \cdot \mathbf{v} - 2\mathbf{u} \cdot \mathbf{v} + \mathbf{u} \cdot \mathbf{u} = |\mathbf{v}|^2 - 2\mathbf{u} \cdot \mathbf{v} + |\mathbf{u}|^2$$

If we insert this last equation into (3), we obtain

$$|\mathbf{v}|^2 - 2\mathbf{u} \cdot \mathbf{v} + |\mathbf{u}|^2 = |\mathbf{v}|^2 + |\mathbf{u}|^2 - 2|\mathbf{u}||\mathbf{v}| \cos \phi$$

Then

$$-2\mathbf{u} \cdot \mathbf{v} = -2|\mathbf{u}||\mathbf{v}| \cos \phi$$

or

$$\cos \phi = \frac{\mathbf{u} \cdot \mathbf{v}}{|\mathbf{u}||\mathbf{v}|} \quad ■$$

NOTE Using this theorem we could *define* the dot product $\mathbf{u} \cdot \mathbf{v}$ by

$$\mathbf{u} \cdot \mathbf{v} = |\mathbf{u}||\mathbf{v}| \cos \phi \tag{4}$$

† The smallest angle will be in the interval $[0, \pi]$.

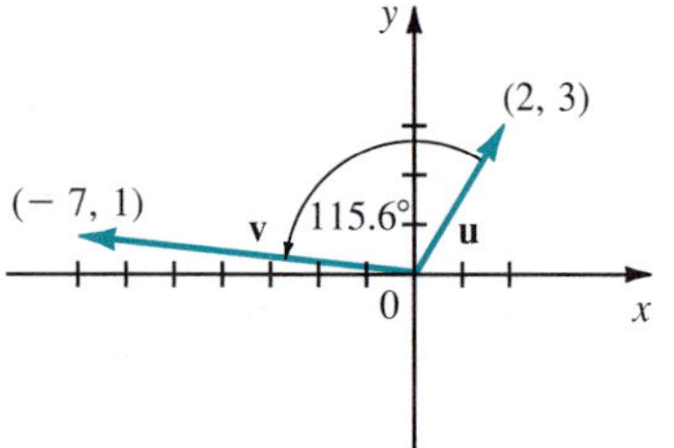
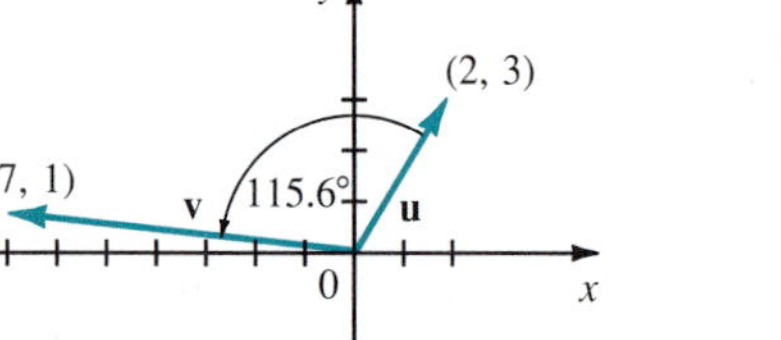

Figure 2 The angle between **u** and **v** is approximately 115.6°.

EXAMPLE 2 *Computing the Angle Between Two Vectors*

Find the angle between the vectors $\mathbf{u} = 2\mathbf{i} + 3\mathbf{j}$ and $\mathbf{v} = -7\mathbf{i} + \mathbf{j}$.

SOLUTION $\mathbf{u} \cdot \mathbf{v} = -14 + 3 = -11$, $|\mathbf{u}| = \sqrt{2^2 + 3^2} = \sqrt{13}$, and $|\mathbf{v}| = \sqrt{(-7)^2 + 1^2} = \sqrt{50}$, so

$$\cos \phi = \frac{\mathbf{u} \cdot \mathbf{v}}{|\mathbf{u}||\mathbf{v}|} = \frac{-11}{\sqrt{13}\sqrt{50}} = \frac{-11}{\sqrt{650}} \approx -0.431455.$$

Then

$$\phi \approx \cos^{-1}(-0.431455) \approx 115.6°.$$

The two vectors are sketched in Figure 2.

Parallel Vectors

Two nonzero vectors **u** and **v** are **parallel** if the angle between them is 0 or π.

Figure 3 **u** and **v** are parallel.

EXAMPLE 3 *Two Parallel Vectors*

Show that the vectors $\mathbf{u} = (2, -3)$ and $\mathbf{v} = (-4, 6)$ are parallel.

SOLUTION

$$\cos \phi = \frac{\mathbf{u} \cdot \mathbf{v}}{|\mathbf{u}||\mathbf{v}|} = \frac{-8 - 18}{\sqrt{13}\sqrt{52}} = \frac{-26}{\sqrt{13}(2\sqrt{13})} = \frac{-26}{2(13)} = -1$$

so $\phi = \cos^{-1}(-1) = \pi$.

The two vectors are sketched in Figure 3.

There is a much easier way to determine whether two vectors are parallel. If $\mathbf{v} = a\mathbf{u}$ for some constant a then,

$$\cos \phi = \frac{\mathbf{u} \cdot \mathbf{v}}{|\mathbf{u}||\mathbf{v}|} = \frac{\mathbf{u} \cdot a\mathbf{u}}{|\mathbf{u}||a\mathbf{u}|} = \frac{a(\mathbf{u} \cdot \mathbf{u})}{|\mathbf{u}||a||\mathbf{u}|} = \frac{a|\mathbf{u}|^2}{|a||\mathbf{u}|^2} = \frac{a}{|a|} = \begin{cases} 1, & \text{if } a > 0 \\ -1, & \text{if } a < 0 \end{cases}$$

Property (v) ↓ Property (iii) ↓

NOTE $|a\mathbf{u}| = \sqrt{(a\mathbf{u}) \cdot (a\mathbf{u})} = \sqrt{a^2(\mathbf{u} \cdot \mathbf{u})} = \sqrt{a^2}\sqrt{\mathbf{u} \cdot \mathbf{u}} = |a||\mathbf{u}|$

Thus $\cos \phi = 1$ and $\phi = 0$ or $\cos \phi = -1$ and $\phi = \pi$. In either case, **u** and **v** are parallel. Conversely, it can be shown (see Problems 43–45) that if $\mathbf{u} \neq \mathbf{0}$ and **v** are parallel, then $\mathbf{v} = a\mathbf{u}$ for some constant a. Thus we have the following result.

Theorem

Two nonzero vectors are parallel if and only if one is a scalar multiple of the other. If $\mathbf{u} \neq \mathbf{0}$, then $\mathbf{v} = a\mathbf{u}$ for some nonzero constant a if and only if $\mathbf{u}$ and $\mathbf{v}$ are parallel.

EXAMPLE 4 ***Two Parallel Vectors***

The vectors $\mathbf{u} = (2, -3)$ and $\mathbf{v} = (-4, 6)$ are parallel because $\mathbf{v} = -2\mathbf{u}$.

Orthogonal Vectors

The nonzero vectors $\mathbf{u}$ and $\mathbf{v}$ are called **orthogonal** (or **perpendicular**) if the angle between them is $\pi/2$.

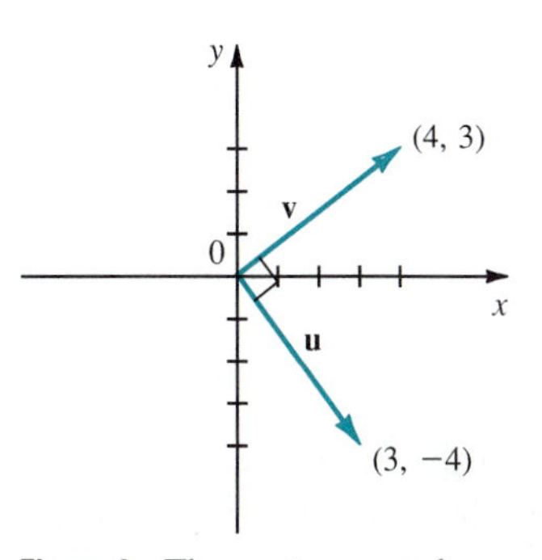

Figure 4 The vectors $\mathbf{u}$ and $\mathbf{v}$ are orthogonal.

EXAMPLE 5 ***Two Orthogonal Vectors***

Show that the vectors $\mathbf{u} = 3\mathbf{i} - 4\mathbf{j}$ and $\mathbf{v} = 4\mathbf{i} + 3\mathbf{j}$ are orthogonal.

SOLUTION $\mathbf{u} \cdot \mathbf{v} = 3 \cdot 4 - 4 \cdot 3 = 0$. This implies that

$$\cos \phi = (\mathbf{u} \cdot \mathbf{v})/(|\mathbf{u}||\mathbf{v}|) = 0.$$

Since ϕ is in the interval $[0, \pi]$, $\phi = \cos^{-1} 0 = \pi/2$. The vectors are sketched in Figure 4.

Theorem

The nonzero vectors $\mathbf{u}$ and $\mathbf{v}$ are orthogonal if and only if $\mathbf{u} \cdot \mathbf{v} = 0$.

Proof

Suppose that $\mathbf{u}$ and $\mathbf{v}$ are orthogonal. Then $\phi = \pi/2$ and, from (4),

$$\mathbf{u} \cdot \mathbf{v} = |\mathbf{u}||\mathbf{v}| \cos \frac{\pi}{2} = |\mathbf{u}||\mathbf{v}|(0) = 0$$

Conversely, if $\mathbf{u} \cdot \mathbf{v} = 0$, then

$$\cos \phi = \frac{\mathbf{u} \cdot \mathbf{v}}{|\mathbf{u}||\mathbf{v}|} = \frac{0}{|\mathbf{u}||\mathbf{v}|} = 0 \quad \text{so } \phi = \cos^{-1} 0 = \frac{\pi}{2}$$

and the vectors are orthogonal. ■

EXAMPLE 6 ***Two Orthogonal Vectors***

The vectors $\mathbf{u} = 3\mathbf{i} - 4\mathbf{j}$ and $\mathbf{v} = 4\mathbf{i} + 3\mathbf{j}$ are orthogonal because

$$\mathbf{u} \cdot \mathbf{v} = 3(4) + (-4)(3) = 12 - 12 = 0$$

NOTE The condition $\mathbf{u} \cdot \mathbf{v} = 0$ is often given as the *definition* of orthogonal vectors.

A number of interesting problems involve the notion of the *projection* of one vector onto another. Before defining this term, we make an observation. Let **u** and **v** be nonzero vectors and define **w** by

$$\mathbf{w} = \mathbf{u} - \frac{\mathbf{u} \cdot \mathbf{v}}{|\mathbf{v}|^2}\mathbf{v}$$

Then

$$\mathbf{w} \cdot \mathbf{v} = \left(\mathbf{u} - \frac{(\mathbf{u} \cdot \mathbf{v})\mathbf{v}}{|\mathbf{v}|^2}\right) \cdot \mathbf{v} = \mathbf{u} \cdot \mathbf{v} - \frac{(\mathbf{u} \cdot \mathbf{v})(\mathbf{v} \cdot \mathbf{v})}{|\mathbf{v}|^2}$$

$$= \mathbf{u} \cdot \mathbf{v} - \frac{(\mathbf{u} \cdot \mathbf{v})|\mathbf{v}|^2}{|\mathbf{v}|^2} = \mathbf{u} \cdot \mathbf{v} - \mathbf{u} \cdot \mathbf{v} = 0$$

That is, **w** is orthogonal to **v**.

The vectors **u**, **v**, and **w** are illustrated in Figure 5.

This fact enables us to make an important definition.

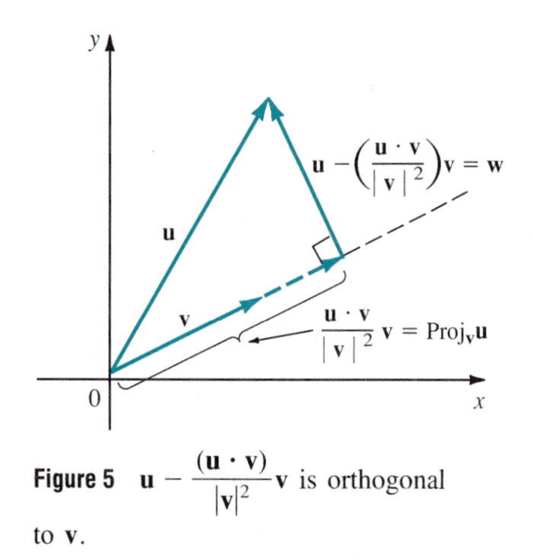

Figure 5 $\mathbf{u} - \frac{(\mathbf{u} \cdot \mathbf{v})}{|\mathbf{v}|^2}\mathbf{v}$ is orthogonal to **v**.

Projection and Component

Let **u** and **v** be nonzero vectors. Then the **projection of u onto v** is a vector, denoted by $\mathrm{Proj}_{\mathbf{v}}\,\mathbf{u}$, which is defined by

$$\mathrm{Proj}_{\mathbf{v}}\,\mathbf{u} = \frac{\mathbf{u} \cdot \mathbf{v}}{\mathbf{v} \cdot \mathbf{v}}\mathbf{v} = \frac{\mathbf{u} \cdot \mathbf{v}}{|\mathbf{v}|^2}\mathbf{v} = \left(\frac{\mathbf{u} \cdot \mathbf{v}}{|\mathbf{v}|}\right)\frac{\mathbf{v}}{|\mathbf{v}|} \quad (5)$$

The **component** *of* **u** *in the direction* **v** is $\dfrac{\mathbf{u} \cdot \mathbf{v}}{|\mathbf{v}|}$. (6)

Note that $\dfrac{\mathbf{v}}{|\mathbf{v}|}$ is a unit vector in the direction of **v**.

Here are some facts about projections.

1. From Figure 5 and the fact that $\cos\phi = (\mathbf{u} \cdot \mathbf{v})/(|\mathbf{u}||\mathbf{v}|)$, we find that **v** and $\mathrm{Proj}_{\mathbf{v}}\,\mathbf{u}$ have
   (i) the same direction if $\mathbf{u} \cdot \mathbf{v} > 0$ and
   (ii) opposite directions if $\mathbf{u} \cdot \mathbf{v} < 0$.
2. $\mathrm{Proj}_{\mathbf{v}}\,\mathbf{u}$ can be thought of as the "**v**-component" of the vector **u**. We will see an illustration of this in our discussions of force and work.
3. If **u** and **v** are orthogonal, then $\mathbf{u} \cdot \mathbf{v} = 0$ so that $\mathrm{Proj}_{\mathbf{v}}\,\mathbf{u} = \mathbf{0}$.

EXAMPLE 7 *Calculating a Projection*

Let $\mathbf{u} = 2\mathbf{i} + 3\mathbf{j}$ and $\mathbf{v} = \mathbf{i} + \mathbf{j}$. Calculate $\mathrm{Proj}_{\mathbf{v}}\,\mathbf{u}$.

SOLUTION $\mathrm{Proj}_{\mathbf{v}}\,\mathbf{u} = \dfrac{(\mathbf{u} \cdot \mathbf{v})\mathbf{v}}{|\mathbf{v}|^2} = \dfrac{5}{(\sqrt{2})^2}\mathbf{v} = \dfrac{5}{2}\mathbf{i} + \dfrac{5}{2}\mathbf{j}$. This is illustrated in Figure 6.

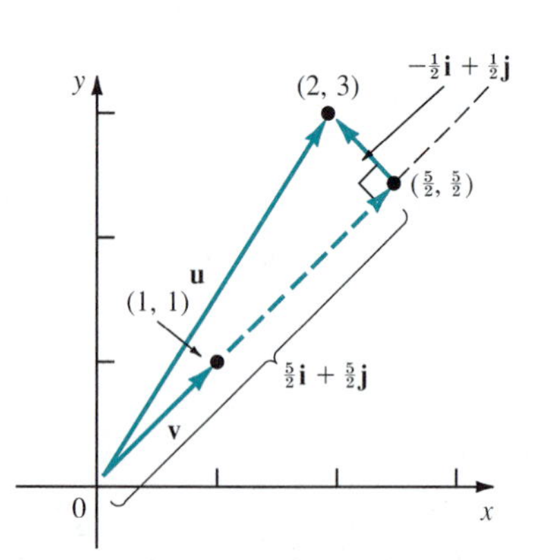

Figure 6 $\mathrm{Proj}_{\mathbf{v}}\,\mathbf{u} = \frac{5}{2}\mathbf{i} + \frac{5}{2}\mathbf{j}$. **v** and $\mathrm{Proj}_{\mathbf{v}}\,\mathbf{u}$ have the same direction.

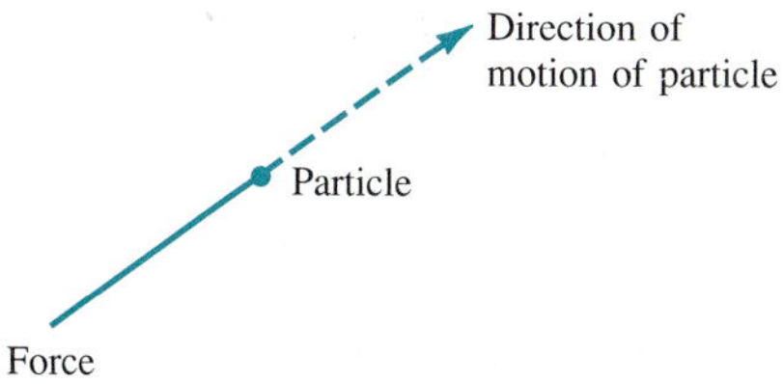

Direction of motion of particle

Particle

Force

Figure 7 The direction of a force.

There are many applications of projections. We give one of them here.

Work

Consider a particle acted on by a force.† In the simplest case, the force is constant, and the particle moves in a straight line in the direction of the force (see Figure 7). In this situation, we define the *work W done by the force on the particle* as the product of the magnitude of the force F and the distance s through which the particle travels:

The Work Done by a Constant Force Over a Distance s

$$W = Fs \tag{7}$$

One unit of work is the work done by a unit force in moving a body a unit distance in the direction of the force. In the metric system, the unit of work is 1 newton-meter (N · m), called 1 *joule* (J). In the English system, the unit of work is the *foot-pound* (ft · lb).

EXAMPLE 8 ***The Work Done by a Constant Force (in English Units)***

How much work is done in lifting a 25-lb weight 5 ft off the ground?

SOLUTION $W = Fs = 25 \text{ lb} \times 5 \text{ ft} = 125 \text{ ft} \cdot \text{lb}$ ■

EXAMPLE 9 ***The Work Done by a Constant Force (in Metric Units)***

A block of mass 10.0 kg is raised 5 m off the ground. How much work is done?

SOLUTION From Newton's second law, force = (mass) × (acceleration). The acceleration here is opposing acceleration due to gravity and is therefore equal to 9.8 m/sec^2. We have $F = ma = (10 \text{ kg})(9.8 \text{ m/sec}^2) = 98$ N. Therefore, $W = Fs = (98) \times 5 \text{ m} = 490$ J.

In formula (7) it is assumed that the force is applied in the same direction as the direction of motion. However, this is not always the case. In general, we may define

$$W = (\text{component of } \mathbf{F} \text{ in the direction of motion}) \times (\text{distance moved}) \tag{8}$$

If the object moves from P to Q, then the distance moved is $|\overrightarrow{PQ}|$. The vector **d**, one of whose representations is $\overrightarrow{PQ}$, is called a **displacement vector.** Then from equation (6)

$$\text{component of } \mathbf{F} \text{ in direction of motion} = \frac{\mathbf{F} \cdot \mathbf{d}}{|\mathbf{d}|} \tag{9}$$

†See page 247 for a discussion of force.

Finally, combining (8) and (9), we obtain

$$W = \frac{\mathbf{F} \cdot \mathbf{d}}{|\mathbf{d}|}|\mathbf{d}| = \mathbf{F} \cdot \mathbf{d} \tag{10}$$

That is,

> The work done is the dot product of the force **F** and the displacement vector **d**. Note that if **F** acts in the direction **d** and if ϕ denotes the angle (which is zero) between **F** and **d**, then $\mathbf{F} \cdot \mathbf{d} = |\mathbf{F}||\mathbf{d}| \cos \phi = |\mathbf{F}||\mathbf{d}| \cos 0 = |\mathbf{F}||\mathbf{d}|$, which is formula (7).

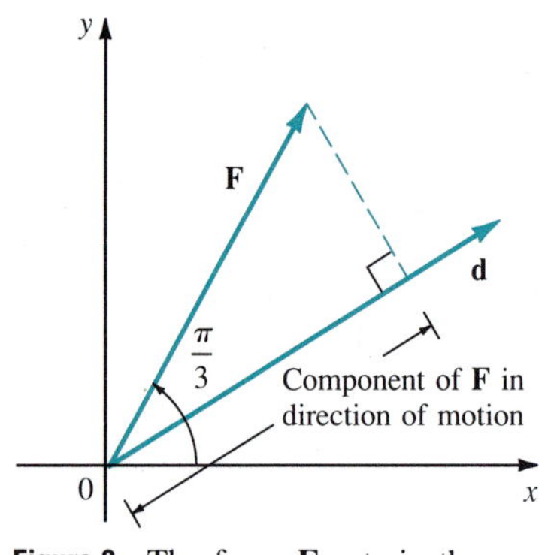

Figure 8 The force **F** acts in the direction $\pi/3$.

EXAMPLE 10 *Computing Work*

A force of 4 N has the direction $\pi/3$. What is the work done in moving an object from the point (1, 2) to the point (5, 4), where distances are measured in meters?

SOLUTION A unit vector with direction $\pi/3$ is given by $\mathbf{u} = (\cos \pi/3)\mathbf{i} + (\sin \pi/3)\mathbf{j} = (1/2)\mathbf{i} + (\sqrt{3}/2)\mathbf{j}$. Thus $\mathbf{F} = 4\mathbf{u} = 2\mathbf{i} + 2\sqrt{3}\mathbf{j}$. The displacement vector **d** is given by $(5 - 1)\mathbf{i} + (4 - 2)\mathbf{j} = 4\mathbf{i} + 2\mathbf{j}$. Thus

$$W = \mathbf{F} \cdot \mathbf{d} = (2\mathbf{i} + 2\sqrt{3}\mathbf{j}) \cdot (4\mathbf{i} + 2\mathbf{j}) = (8 + 4\sqrt{3}) \approx 14.93 \text{ J}$$

The component of **F** in the direction of motion is sketched in Figure 8. ■

EXAMPLE 11 *Computing the Work Done When a Bicycle Is Pushed Up a Hill*

A woman pushes a 55-lb bicycle up a hill that makes a 30° angle with the horizontal. How much work is done in pushing the bicycle 150 feet?

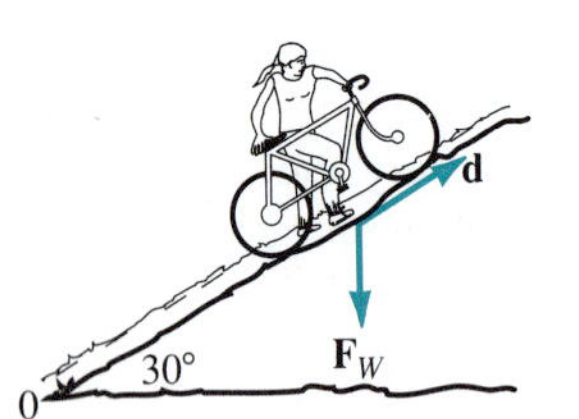

Figure 9 Pushing a bicycle up a 30° slope.

SOLUTION The downward force caused by gravity is, as in Figure 9, denoted by $\mathbf{F}_w$. Since weight is a force acting downward, its direction is $-\mathbf{j}$ and

$$\mathbf{F}_w = -55\mathbf{j}$$

In order to push the bicycle up, the woman must overcome this force. A unit vector in the direction of motion is

$$\mathbf{u} = \cos 30°\mathbf{i} + \sin 30°\mathbf{j} = \frac{\sqrt{3}}{2}\mathbf{i} + \frac{1}{2}\mathbf{j}$$

and, since the bicycle moves 150 feet, the displacement vector is

$$\mathbf{d} = 150\mathbf{u} = 75\sqrt{3}\mathbf{i} + 75\mathbf{j}$$

Now, from (10),

the work done by gravity as the bicycle moves $= \mathbf{F}_w \cdot \mathbf{d}$

$$= (75\sqrt{3}\mathbf{i} + 75\mathbf{j}) \cdot (-55\mathbf{j}) = (75)(-55) = -4125 \text{ ft} \cdot \text{lb}$$

Thus, in order to overcome gravity, the woman must do 4125 ft · lb of work.

Problems 4.6

Readiness Check

I. $\mathbf{i} \cdot \mathbf{j} =$ ____.
  a. 1 b. $\sqrt{(0-1)^2 + (1-0)^2}$
  c. 0 d. $\mathbf{i} + \mathbf{j}$

II. $(3, 4) \cdot (3, 2) =$ ____
  a. $(3 + 3)(4 + 2) = 36$ b. $(3)(3) + (4)(2) = 17$
  c. $(3 - 3)(2 - 4) = 0$ d. $(3)(3) - (4)(2) = 1$

III. The cosine of the angle between $\mathbf{i} + \mathbf{j}$ and $\mathbf{i} - \mathbf{j}$ is ____.
  a. $0\mathbf{i} + 0\mathbf{j}$ b. 0
  c. $\sqrt{2}$ d. $1/\sqrt{2 + 0}$

IV. The vectors $2\mathbf{i} - 12\mathbf{j}$ and $3\mathbf{i} + (1/2)\mathbf{j}$ are ____.
  a. neither parallel nor orthogonal
  b. parallel
  c. orthogonal
  d. identical

V. $\text{Proj}_{\mathbf{w}}\, \mathbf{u} =$ ____.
  a. $\dfrac{\mathbf{u} \cdot \mathbf{w}}{|\mathbf{w}|}$ b. $\dfrac{\mathbf{w}}{|\mathbf{w}|}$
  c. $\dfrac{\mathbf{u} \cdot \mathbf{w}}{|\mathbf{w}|} \dfrac{\mathbf{w}}{|\mathbf{w}|}$ d. $\dfrac{\mathbf{u} \cdot \mathbf{w}}{|\mathbf{u}|} \dfrac{\mathbf{u}}{|\mathbf{u}|}$

In Problems 1–10 calculate both the dot product of the two vectors and the cosine of the angle between them.

1. $\mathbf{u} = \mathbf{i} + \mathbf{j}$; $\mathbf{v} = \mathbf{i} - \mathbf{j}$
2. $\mathbf{u} = 3\mathbf{i}$; $\mathbf{v} = -7\mathbf{j}$
3. $\mathbf{u} = -5\mathbf{i}$; $\mathbf{v} = 18\mathbf{j}$
4. $\mathbf{u} = \alpha\mathbf{i}$; $\mathbf{v} = \beta\mathbf{j}$; α, β real and nonzero
5. $\mathbf{u} = 2\mathbf{i} + 5\mathbf{j}$; $\mathbf{v} = 5\mathbf{i} + 2\mathbf{j}$
6. $\mathbf{u} = 2\mathbf{i} + 5\mathbf{j}$; $\mathbf{v} = 5\mathbf{i} - 3\mathbf{j}$
7. $\mathbf{u} = -3\mathbf{i} + 4\mathbf{j}$; $\mathbf{v} = -2\mathbf{i} - 7\mathbf{j}$
8. $\mathbf{u} = 4\mathbf{i} + 5\mathbf{j}$; $\mathbf{v} = 7\mathbf{i} - 4\mathbf{j}$
9. $\mathbf{u} = 11\mathbf{i} - 8\mathbf{j}$; $\mathbf{v} = 4\mathbf{i} - 7\mathbf{j}$
10. $\mathbf{u} = -13\mathbf{i} + 8\mathbf{j}$; $\mathbf{v} = 2\mathbf{i} + 11\mathbf{j}$
11. Show that for any nonzero real numbers α and β, the vectors $\mathbf{u} = \alpha\mathbf{i} + \beta\mathbf{j}$ and $\mathbf{v} = \beta\mathbf{i} - \alpha\mathbf{j}$ are orthogonal.
12. Let $\mathbf{u}$, $\mathbf{v}$, and $\mathbf{w}$ denote three arbitrary vectors. Explain why the product $\mathbf{u} \cdot \mathbf{v} \cdot \mathbf{w}$ is *not defined*.

In Problems 13–20 determine whether the given vectors are orthogonal, parallel, or neither. Then sketch each pair.

13. $\mathbf{u} = 3\mathbf{i} + 5\mathbf{j}$; $\mathbf{v} = -6\mathbf{i} - 10\mathbf{j}$
14. $\mathbf{u} = 2\mathbf{i} + 3\mathbf{j}$; $\mathbf{v} = 6\mathbf{i} - 4\mathbf{j}$
15. $\mathbf{u} = 2\mathbf{i} + 3\mathbf{j}$; $\mathbf{v} = 6\mathbf{i} + 4\mathbf{j}$
16. $\mathbf{u} = 2\mathbf{i} + 3\mathbf{j}$; $\mathbf{v} = -6\mathbf{i} + 4\mathbf{j}$
17. $\mathbf{u} = 7\mathbf{i}$; $\mathbf{v} = -23\mathbf{j}$
18. $\mathbf{u} = 2\mathbf{i} - 6\mathbf{j}$; $\mathbf{v} = -\mathbf{i} + 3\mathbf{j}$
19. $\mathbf{u} = \mathbf{i} + \mathbf{j}$; $\mathbf{v} = \alpha\mathbf{i} + \alpha\mathbf{j}$; α real
20. $\mathbf{u} = -2\mathbf{i} + 3\mathbf{j}$; $\mathbf{v} = -\mathbf{i} + 2\mathbf{j}$
21. Let $\mathbf{u} = 3\mathbf{i} + 4\mathbf{j}$ and $\mathbf{v} = \mathbf{i} + \alpha\mathbf{j}$.
    (a) Determine α such that $\mathbf{u}$ and $\mathbf{v}$ are orthogonal.
    (b) Determine α such that $\mathbf{u}$ and $\mathbf{v}$ are parallel.
    (c) Determine α such that the angle between $\mathbf{u}$ and $\mathbf{v}$ is $\pi/4$.
    (d) Determine α such that the angle between $\mathbf{u}$ and $\mathbf{v}$ is $\pi/3$.
22. Let $\mathbf{u} = -2\mathbf{i} + 5\mathbf{j}$ and $\mathbf{v} = \alpha\mathbf{i} - 2\mathbf{j}$.
    (a) Determine α such that $\mathbf{u}$ and $\mathbf{v}$ are orthogonal.
    (b) Determine α such that $\mathbf{u}$ and $\mathbf{v}$ are parallel.
    (c) Determine α such that the angle between $\mathbf{u}$ and $\mathbf{v}$ is $2\pi/3$.
    (d) Determine α such that the angle between $\mathbf{u}$ and $\mathbf{v}$ is $\pi/3$.
23. In Problem 21 show that there is no value of α for which $\mathbf{u}$ and $\mathbf{v}$ have opposite directions.
24. In Problem 22 show that there is no value of α for which $\mathbf{u}$ and $\mathbf{v}$ have the same direction.

In Problems 25–38 calculate $\text{Proj}_{\mathbf{v}}\, \mathbf{u}$.

25. $\mathbf{u} = 3\mathbf{i}$; $\mathbf{v} = \mathbf{i} + \mathbf{j}$
26. $\mathbf{u} = -5\mathbf{j}$; $\mathbf{v} = \mathbf{i} + \mathbf{j}$
27. $\mathbf{u} = 2\mathbf{i} + \mathbf{j}$; $\mathbf{v} = \mathbf{i} - 2\mathbf{j}$
28. $\mathbf{u} = 2\mathbf{i} + 3\mathbf{j}$; $\mathbf{v} = 4\mathbf{i} + \mathbf{j}$
29. $\mathbf{u} = \mathbf{i} + \mathbf{j}$; $\mathbf{v} = 2\mathbf{i} - 3\mathbf{j}$
30. $\mathbf{u} = \mathbf{i} + \mathbf{j}$; $\mathbf{v} = 2\mathbf{i} + 3\mathbf{j}$
31. $\mathbf{u} = 4\mathbf{i} + 5\mathbf{j}$; $\mathbf{v} = 2\mathbf{i} + 4\mathbf{j}$
32. $\mathbf{u} = 4\mathbf{i} + 5\mathbf{j}$; $\mathbf{v} = 2\mathbf{i} - 4\mathbf{j}$
33. $\mathbf{u} = -4\mathbf{i} + 5\mathbf{j}$; $\mathbf{v} = 2\mathbf{i} - 4\mathbf{j}$
34. $\mathbf{u} = -4\mathbf{i} - 5\mathbf{j}$; $\mathbf{v} = -2\mathbf{i} - 4\mathbf{j}$
35. $\mathbf{u} = \alpha\mathbf{i} + \beta\mathbf{j}$; $\mathbf{v} = \mathbf{i} + \mathbf{j}$
36. $\mathbf{u} = \mathbf{i} + \mathbf{j}$; $\mathbf{v} = \alpha\mathbf{i} + \beta\mathbf{j}$
37. $\mathbf{u} = \alpha\mathbf{i} - \beta\mathbf{j}$; $\mathbf{v} = \mathbf{i} + \mathbf{j}$
38. $\mathbf{u} = \alpha\mathbf{i} - \beta\mathbf{j}$; $\mathbf{v} = -\mathbf{i} + \mathbf{j}$
39. Let $\mathbf{u} = a_1\mathbf{i} + b_1\mathbf{j}$ and $\mathbf{v} = a_2\mathbf{i} + b_2\mathbf{j}$. Give a condition on a_1, b_1, a_2, and b_2 that will ensure that $\mathbf{v}$ and $\text{Proj}_{\mathbf{v}}\, \mathbf{u}$ have the same direction.

Answers to Readiness Check

I. c II. b III. b IV. c V. c

40. In Problem 39 give a condition that will ensure that $\mathbf{v}$ and $\text{Proj}_{\mathbf{v}}\,\mathbf{u}$ have opposite directions.
41. Let $P = (2, 3)$, $Q = (5, 7)$, $R = (2, -3)$, and $S = (1, 2)$. Calculate $\text{Proj}_{\overrightarrow{PQ}}\,\overrightarrow{RS}$ and $\text{Proj}_{\overrightarrow{RS}}\,\overrightarrow{PQ}$.
42. Let $P = (-1, 3)$, $Q = (2, 4)$, $R = (-6, -2)$, and $S = (3, 0)$. Calculate $\text{Proj}_{\overrightarrow{PQ}}\,\overrightarrow{RS}$ and $\text{Proj}_{\overrightarrow{RS}}\,\overrightarrow{PQ}$.
43. Show that if $\mathbf{u}$ and $\mathbf{v}$ are unit vectors and the angle between them is 0, then $\mathbf{u} = \mathbf{v}$.
44. Show that if $\mathbf{u}$ and $\mathbf{v}$ are unit vectors and the angle between them is π, then $\mathbf{v} = -\mathbf{u}$. [Hint: If $\mathbf{u} = \cos\theta\mathbf{i} + \sin\theta\mathbf{j}$, then $\mathbf{v} = \cos(\theta + \pi)\mathbf{i} + \sin(\theta + \pi)\mathbf{j}$.]
45. Show that if $\mathbf{u} \neq \mathbf{0}$ and $\mathbf{v} \neq \mathbf{0}$ are parallel, then there is a constant a such that $\mathbf{v} = a\mathbf{u}$. $\left[\text{Hint: } \frac{\mathbf{u}}{|\mathbf{u}|} \text{ and } \frac{\mathbf{v}}{|\mathbf{v}|} \text{ are unit vectors. Use the results of Problems 43 and 44.}\right]$

In Problems 46–54 find the work done when the force with given magnitude and direction moves an object from P to Q. All distances are measured in meters. (Note that work can be negative.)

46. $|\mathbf{F}| = 3$ N; $\theta = 0$; $P = (2, 3)$; $Q = (1, 7)$
47. $|\mathbf{F}| = 2$ N; $\theta = \pi/2$; $P = (5, 7)$; $Q = (1, 1)$
48. $|\mathbf{F}| = 6$ N; $\theta = \pi/4$; $P = (2, 3)$; $Q = (-1, 4)$
49. $|\mathbf{F}| = 4$ N; $\theta = \pi/6$; $P = (-1, 2)$; $Q = (3, 4)$
50. $|\mathbf{F}| = 7$ N; $\theta = 2\pi/3$; $P = (4, -3)$; $Q = (1, 0)$
51. $|\mathbf{F}| = 3$ N; $\theta = 3\pi/4$; $P = (2, 1)$; $Q = (1, 2)$
52. $|\mathbf{F}| = 6$ N; $\theta = \pi$; $P = (3, -8)$; $Q = (5, 10)$
53. $|\mathbf{F}| = 4$ N; θ is the direction of $2\mathbf{i} + 3\mathbf{j}$; $P = (2, 0)$; $Q = (-1, 3)$
54. $|\mathbf{F}| = 5$ N; θ is the direction of $-3\mathbf{i} + 2\mathbf{j}$; $P = (1, 3)$; $Q = (4, -6)$
55. Two tugboats are towing a barge. Tugboat 1 pulls with a force of 500 N at an angle of 20° with the horizontal. Tugboat 2 pulls with a force of x newtons at an angle of 30°. The barge moves horizontally (i.e., $\theta = 0$). Find x. [Hint: See the discussion of resultants on page 248.]

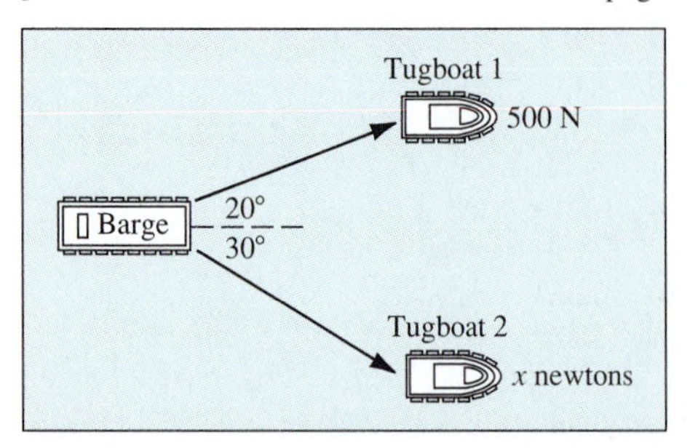

56. Answer the question of Problem 55 if the angles are 50° and 75°, respectively, and all other data remain the same.
57. In Problem 55 how much work is done by each tugboat in moving the barge a distance of 750 m?
58. In Problem 56 how much work is done by each tugboat in moving the barge a distance of 2 km?
59. A block weighing 122 lb is sliding down a ramp that makes an angle of 31° with the ground.

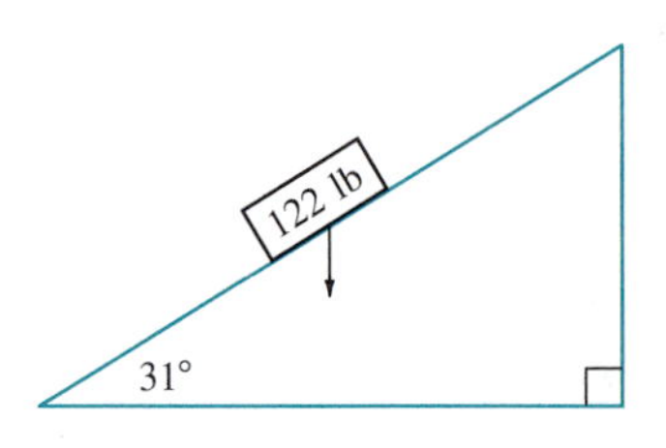

    (a) What is the magnitude of the force pushing the block down the ramp? [Hint: See Example 2 on page 249.]
    (b) How much work is done if the block slides a total of 25 ft?
60. One end of a rope is attached to a box weighing 160 lb. If the box is pulled up, the **tension** in the rope is defined as the magnitude of the force needed to do the pulling?

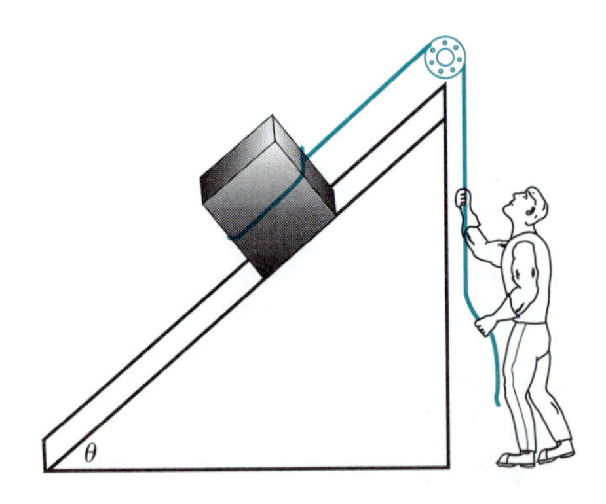

    (a) If $\theta = 58°$, what is the tension in the rope?
    (b) How much work is done in pulling the box 12 ft?
61. A girl pushes a lawn mower with a force of 26 lb. The handle makes an angle of 42° with the horizontal. How much work does she do when she pushes the lawn mower 80 ft?

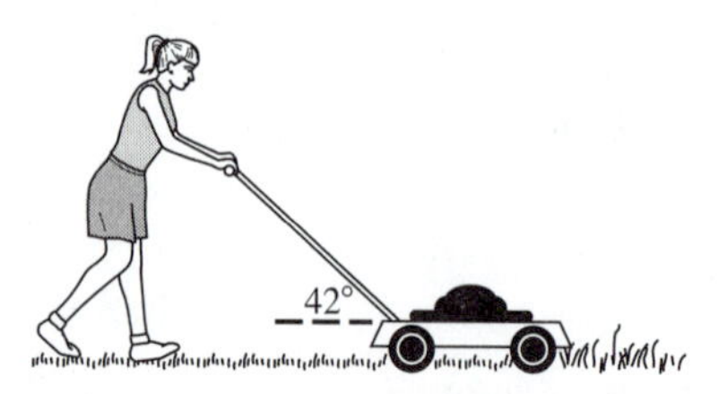

62. Answer the questions in Problem 61 if she pushes with a force of 112 N and the angle is 36°.
63. A section of road has a 4% grade. That is, it rises 4 ft for each 100 ft of road. A car weighing 3174 lb is parked on the hill.
    (a) What is the force that tends to make the car roll downhill?
    (b) How much work does the car do if it rolls 300 ft down the hill?

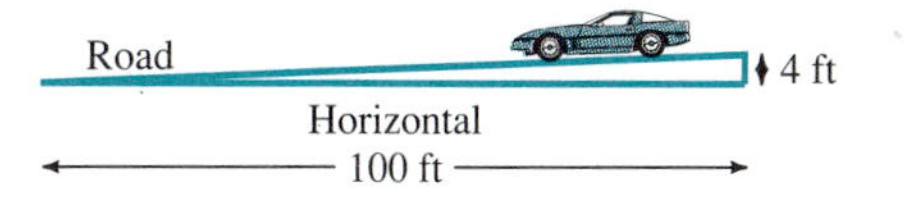

4.7 Polar Coordinates

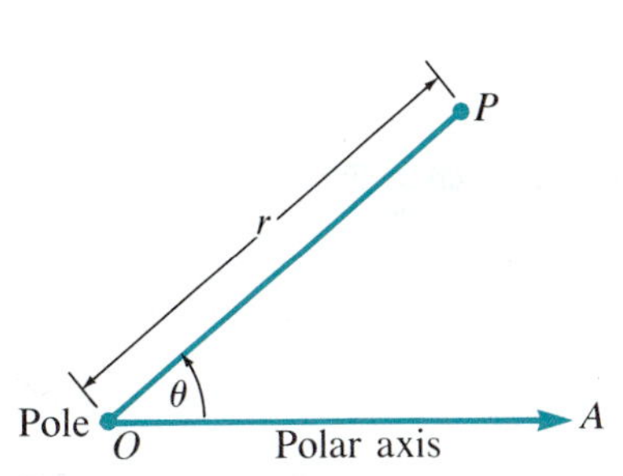

Figure 1 Representing a point in polar coordinates.

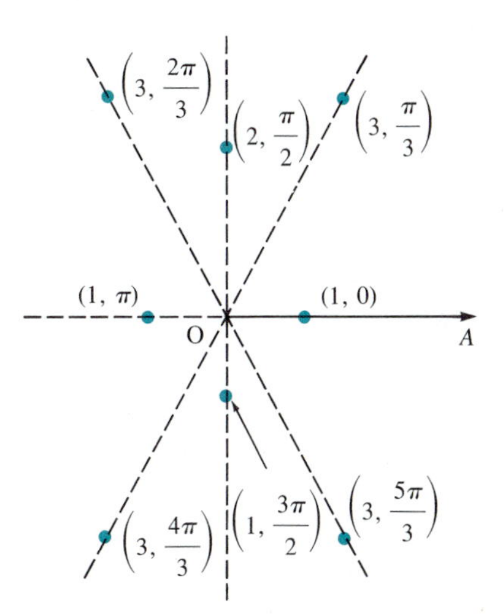

Figure 2 Eight points in polar coordinates.

In Section 1.2, we introduced the Cartesian plane (the xy-plane), and up to this moment we have represented points in the plane by their x- and y-coordinates. There are many other ways to represent points in the plane, the most important of which is called the **polar coordinate system.**

We begin by choosing a fixed point, which we label O, and a ray (half-line) that extends in one direction from O, which we label OA. The fixed point is called the **pole** (or **origin**), and the ray OA is called the **polar axis.** In Figure 1, the polar axis is drawn as a horizontal ray that extends infinitely to the right (just like the positive x-axis). This representation is a matter of convention, although it would be correct to have the polar axis extend in any direction.

If P (not the origin) is any other point in the plane, let r denote the distance between O and P, and let θ represent the angle (in radians) between OA and OP, measured counterclockwise from OA to OP. Then every point in the plane, except the pole, can be represented by a pair of numbers (r, θ), where $r > 0$ and $0 \leq \theta < 2\pi$. The pole can be represented as $(0, \theta)$ for any number θ. The representation $P = (r, \theta)$ is called the **polar representation** of the point, and r and θ are called the **polar coordinates** of P. Some typical points are sketched in Figure 2.

Suppose we are given two numbers r and θ with $r \neq 0$. We draw the point $P = (r, \theta)$ in the plane as follows:

Case 1: $r > 0$ To locate the point $P = (r, \theta)$, we first rotate the polar axis through an angle of $|\theta|$ in the counterclockwise direction if θ is positive and in the clockwise direction if θ is negative. Let us call this new ray OB. Then if $r > 0$, the point P is placed on the ray OB, r units from the pole.

Case 2: $r < 0$ Draw the ray OB as in Case 1. Then, extend the ray OB backward through the pole. The point P is then located $|r|$ units from the pole along this extended ray.

EXAMPLE 1 *Plotting Four Points in Polar Coordinates*

Plot the following points: (a) $\left(2, \frac{8\pi}{3}\right)$, (b) $\left(-1, \frac{\pi}{4}\right)$, (c) $\left(3, -22\frac{1}{2}\pi\right)$, (d) $\left(-2, -\frac{\pi}{2}\right)$.

SOLUTION The four points are plotted in Figure 3.

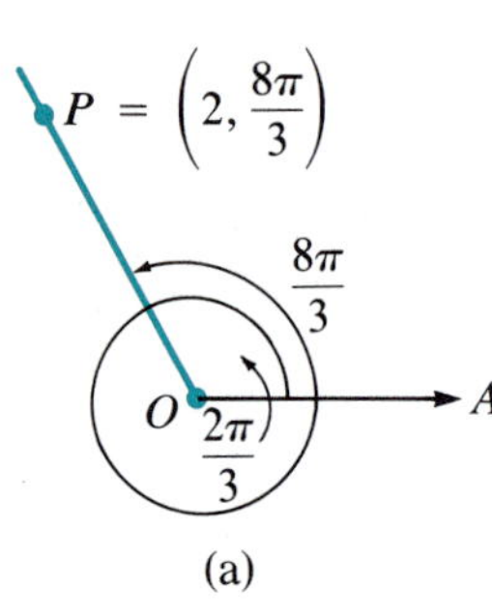

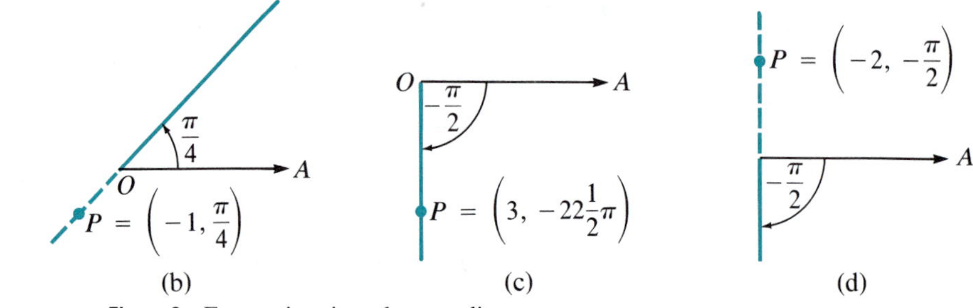

Figure 3 Four points in polar coordinates.

In (a), since $\frac{8\pi}{3} = 2\pi + \frac{2\pi}{3}$, the line OB makes an angle of $\frac{2\pi}{3}$ with OA and the point P lies 2 units along this line.

In (b), we rotate $\frac{\pi}{4}$ units and then extend OB backward since $r < 0$. The point P is located 1 unit along this extended line.

In (c), we rotate in a clockwise direction and find that the line OB makes an angle of $-\frac{\pi}{2}$ with the polar axis since $-22\frac{1}{2}\pi = -11(2\pi) - \frac{\pi}{2}$. The point P is located 3 units along OB.

In (d), we rotate OA in a clockwise direction $-\frac{\pi}{2}$ radians and then extend OB backward (and therefore upward) since $r < 0$. The point P is then located 2 units along this line.

We immediately notice that if we do not restrict r and θ, then there are many (in fact, infinitely many) representations for each point in the plane. For example, in Figure 3(d) we see that $\left(-2, -\frac{\pi}{2}\right) = \left(2, \frac{\pi}{2}\right) = \left(2, \frac{\pi}{2} + 2n\pi\right)$ for any integer n. In general,

The Representations of a Point in Polar Coordinates

$$P = (r, \theta) = (-r, \theta + \pi) = (r, \theta + 2n\pi) \qquad (1)$$

for $n = 0, \pm 1, \pm 2, \ldots$

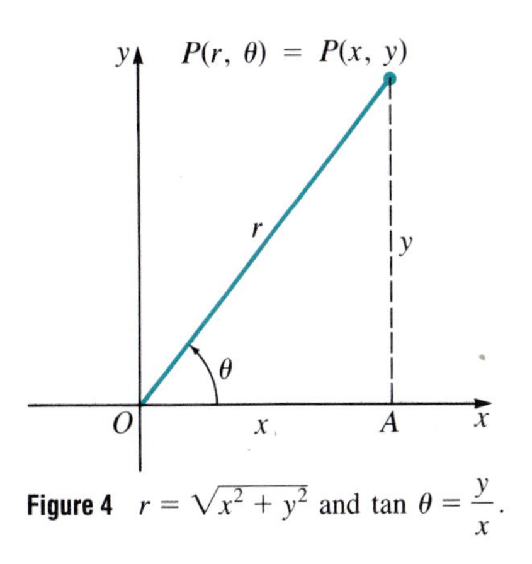

Figure 4 $r = \sqrt{x^2 + y^2}$ and $\tan\theta = \dfrac{y}{x}$.

(see Problem 36). Actually, these ordered pairs are not equal. The equal sign indicates that the points they represent are the same. Therefore, with this understanding, we will write $(r_1, \theta_1) = (r_2, \theta_2)$ in the rest of this chapter when the points they represent are the same, even though r_1 may not be equal to r_2, and θ_1 may not be equal to θ_2. However, to avoid difficulties, if a point $P \neq O$ is given, then we can write $P = (r, \theta)$ in polar coordinates in a unique way if we specify that $r > 0$ and $0 \le \theta < 2\pi$. If P is the pole O, then $P = (0, \theta)$ for any real number θ, and there is no unique representation unless we specify a value for θ.

What is the relationship between polar and rectangular (Cartesian) coordinates? To find it, we place the pole at the origin in the xy-plane with the polar axis along the x-axis as in Figure 4. Let $P = (x, y) = (r, \theta)$ be a point in the plane. Then as is evident from Figure 4, we have the following:

If $P = (r, \theta)$ is given in polar coordinates and if (x, y) is the representation of the point in rectangular coordinates, then

$$\cos\theta = \frac{x}{r} \qquad \text{and} \qquad \sin\theta = \frac{y}{r}$$

or

Converting from Polar to Rectangular Coordinates

$$x = r\cos\theta \qquad \text{and} \qquad y = r\sin\theta \tag{2}$$

Let (x, y) be the rectangular representation of a point and let (r, θ) be the polar representation with $r > 0$ and $0 \le \theta < 2\pi$. Then, from the Pythagorean theorem, we see that $x^2 + y^2 = r^2$, so, since $r > 0$, $r = \sqrt{x^2 + y^2}$. In Figure 4 we see that $\tan\theta = y/x$ and we have the following.

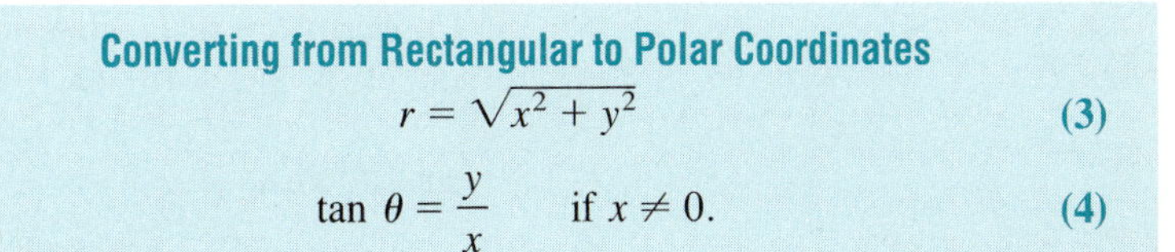

Converting from Rectangular to Polar Coordinates

$$r = \sqrt{x^2 + y^2} \tag{3}$$

$$\tan\theta = \frac{y}{x} \qquad \text{if } x \neq 0. \tag{4}$$

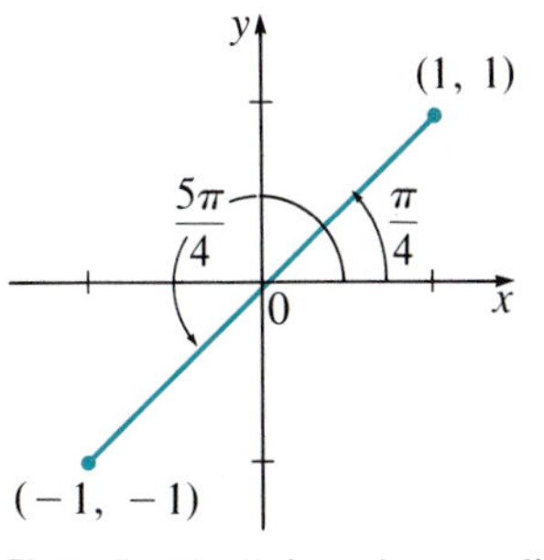

Figure 5 (1, 1) in polar coordinates is $\left(\sqrt{2}, \dfrac{\pi}{4}\right)$. $(-1, -1)$ in polar coordinates is $\left(\sqrt{2}, \dfrac{5\pi}{4}\right)$.

Before citing examples, we note that these conversion formulas have been illustrated only for $r > 0$ and $0 \le \theta < \pi/2$. The fact that $0 \le \theta < \pi/2$ in Figure 4 is irrelevant since, using the alternative definition of $\sin x$ and $\cos x$ on page 80, we see that $x/r = \cos\theta$ and $y/r = \sin\theta$ for any real number θ.

WARNING Formula (4) does *not* determine θ uniquely. The signs of x and y must be taken into account. For example, for $(1, 1)$ and $(-1, -1)$, $\tan\theta = 1$. But in the first case, $\theta = \pi/4$, and in the second case $\theta = 5\pi/4$ (see Figure 5).

With these formulas, we can always convert from polar to rectangular coordinates in a unique way. To convert from rectangular to polar coordinates in a unique way, we must specify that $r > 0$ and $0 \le \theta < 2\pi$. ■

EXAMPLE 2 *Converting from Polar to Rectangular Coordinates*

Convert from polar to rectangular coordinates: (a) $\left(3, \frac{\pi}{6}\right)$, (b) $\left(4, \frac{2\pi}{3}\right)$, (c) $\left(-6, \frac{\pi}{4}\right)$, (d) $(2, 0)$, (e) $(1, -\pi)$.

SOLUTION

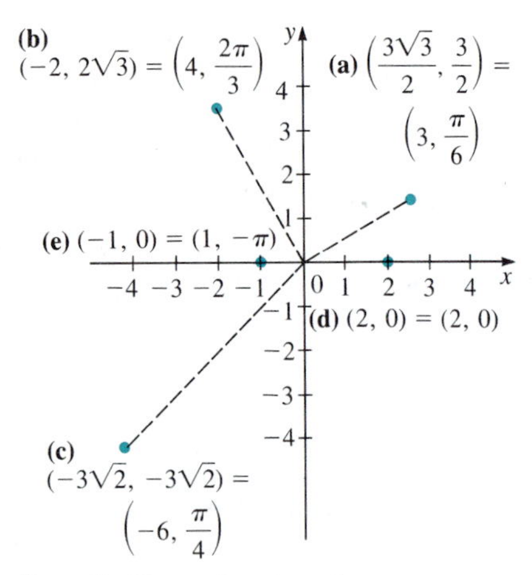

Figure 6 Five points in polar and rectangular coordinates.

(a) $x = 3\cos\frac{\pi}{6}$ and $y = 3\sin\frac{\pi}{6}$ so $(x, y) = \left(\frac{3\sqrt{3}}{2}, \frac{3}{2}\right)$.

(b) $x = 4\cos\frac{2\pi}{3} = -4\cos\frac{\pi}{3} = -\frac{4}{2} = -2$ and

$y = 4\sin\frac{2\pi}{3} = 4\sin\frac{\pi}{3} = \frac{4\sqrt{3}}{2} = 2\sqrt{3}$, so $(x, y) = (-2, 2\sqrt{3})$.

(c) $\left(-6, \frac{\pi}{4}\right) = \left(6, \frac{\pi}{4} + \pi\right) = \left(6, \frac{5\pi}{4}\right)$ [from (1)], so we have $x = 6\cos\frac{5\pi}{4} = 6\left(-\frac{\sqrt{2}}{2}\right) = -3\sqrt{2}$, $y = 6\sin\frac{5\pi}{4} = -3\sqrt{2}$, and $(x, y) = (-3\sqrt{2}, -3\sqrt{2})$.

(d) $x = 2\cos 0 = 2$ and $y = 2\sin 0 = 0$, so $(x, y) = (2, 0)$.

(e) Since $(1, -\pi) = (1, -\pi + 2\pi) = (1, \pi)$, we have $x = \cos\pi = -1$, $y = \sin\pi = 0$, and $(x, y) = (-1, 0)$.

These five points are sketched in Figure 6. ■

EXAMPLE 3 *Converting from Rectangular to Polar Coordinates*

Convert from rectangular to polar coordinates (with $r > 0$ and $0 \le \theta < 2\pi$): (a) $(1, \sqrt{3})$ (b) $(2, -2)$ (c) $(-2\sqrt{3}, -2)$

SOLUTION

(a) $r = \sqrt{x^2 + y^2} = \sqrt{1^2 + \sqrt{3}^2} = \sqrt{1 + 3} = 2$ and $\tan\theta = \frac{\sqrt{3}}{1} = \sqrt{3}$. Since $(1, \sqrt{3})$ is in the first quadrant, we have $\theta = \tan^{-1}\sqrt{3} = \frac{\pi}{3}$ so that $(r, \theta) = \left(2, \frac{\pi}{3}\right)$.

(b) $r = \sqrt{(2)^2 + (-2)^2} = \sqrt{8} = 2\sqrt{2}$ and $\tan\theta = -\frac{2}{2} = -1$. Since $(2, -2)$ is in the fourth quadrant, $\theta = \frac{7\pi}{4}$ and $(r, \theta) = \left(2\sqrt{2}, \frac{7\pi}{4}\right)$.

(c) $r = \sqrt{(-2\sqrt{3})^2 + (-2)^2} = \sqrt{12 + 4} = 4$ and $\tan\theta = \frac{-2}{-2\sqrt{3}} =$

$\dfrac{1}{\sqrt{3}}$. Since $(-2\sqrt{3}, -2)$ is in the third quadrant, we have $\theta = \dfrac{7\pi}{6}$ and $(r, \theta) = \left(4, \dfrac{7\pi}{6}\right)$.

These three points are sketched in Figure 7.

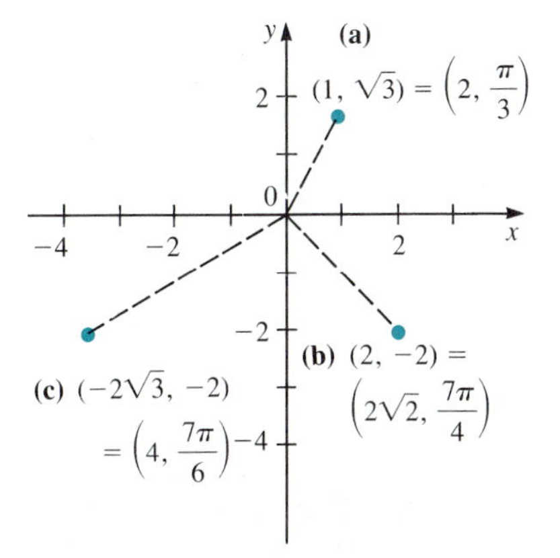

Figure 7 Three points in polar and rectangular coordinates.

Problems 4.7

Readiness Check

I. Which of the following is true?

a. There are many pairs of rectangular coordinates and only one pair of polar coordinates for a given point.

b. There are many pairs of polar coordinates and only one pair of rectangular coordinates for a given point.

c. In rectangular form a point is located by rotating through an angle θ and measuring a distance r on its terminal side.

d. In polar form a point is located by a horizontal distance from the origin followed by a vertical distance.

II. Which of the following names the same point as the rectangular coordinates $(-1, 0)$?

a. $(1, 2\pi)$ b. $(-1, -\pi)$

c. $(-1, 3\pi)$ d. $(1, \pi)$

III. Which of the following is true of the point named by the polar coordinates $(-3, -3\pi/2)$?

a. It is on the negative y-axis.

b. It is on the positive y-axis.

c. It is in quadrant III.

d. It is in quadrant II.

IV. Which of the following polar coordinates represents the same point as the polar coordinates $(5, \pi/3)$?

a. $\left(5, \dfrac{4\pi}{3}\right)$ b. $\left(5, -\dfrac{4\pi}{3}\right)$

c. $\left(-5, \dfrac{2\pi}{3}\right)$ d. $\left(-5, -\dfrac{2\pi}{3}\right)$

V. Which of the following polar coordinates does *not* give the same point as the rectangular point $(3, -3)$?

a. $\left(3\sqrt{2}, -\dfrac{\pi}{4}\right)$ b. $\left(-3\sqrt{2}, -\dfrac{\pi}{4}\right)$

c. $\left(3\sqrt{2}, \dfrac{7\pi}{4}\right)$ d. $\left(-3\sqrt{2}, \dfrac{3\pi}{4}\right)$

Answers to Readiness Check

I. b II. d III. a IV. d V. b

In Problems 1–18 a point is given in polar coordinates. Write the point in rectangular coordinates and plot it in the xy-plane, showing both representations.

1. $(2, 0)$
2. $\left(3, \frac{\pi}{4}\right)$
3. $(-5, 0)$
4. $\left(-4, \frac{\pi}{4}\right)$
5. $\left(6, \frac{7\pi}{6}\right)$
6. $\left(-3, -\frac{\pi}{6}\right)$
7. $(5, \pi)$
8. $(5, -\pi)$
9. $(-5, -\pi)$
10. $\left(-3, \frac{\pi}{4}\right)$
11. $\left(3, \frac{11\pi}{4}\right)$
12. $\left(3, \frac{5\pi}{4}\right)$
13. $\left(-3, -\frac{\pi}{4}\right)$
14. $\left(3, \frac{7\pi}{4}\right)$
15. $\left(1, \frac{3\pi}{2}\right)$
16. $\left(1, -\frac{\pi}{2}\right)$
17. $\left(-1, \frac{\pi}{2}\right)$
18. $\left(-1, -\frac{3\pi}{2}\right)$

In Problems 19–35 a point is given in rectangular coordinates. Write the point in polar coordinates with $r > 0$ and $0 \le \theta < 2\pi$. Then plot the point with both representations in the xy-plane.

19. $(2, 0)$
20. $(7, 0)$
21. $(-3, 0)$
22. $(2, 2)$
23. $(-2, 2)$
24. $(2, -2)$
25. $(-2, -2)$
26. $(0, 1)$ [Hint: Draw a sketch first.]
27. $(0, -1)$
28. $(2, 2\sqrt{3})$
29. $(-2, 2\sqrt{3})$
30. $(2, -2\sqrt{3})$
31. $(-2, -2\sqrt{3})$
32. $(2\sqrt{3}, 2)$
33. $(2\sqrt{3}, -2)$
34. $(-2\sqrt{3}, 2)$
35. $(-2\sqrt{3}, -2)$
36. Show that $(-r, \theta)$ and $(r, \theta + \pi)$ represent the same point. [Hint: Draw a sketch.]

4.8 Graphing in Polar Coordinates

In rectangular coordinates, we defined the graph of the equation $y = f(x)$ as the set of points (x, y) whose coordinates satisfy the equation. In polar coordinates, however, we must be careful since each point in the plane has infinitely many representations.

The **graph** of an equation written in polar coordinates r and θ consists of those points P having at least one representation $P = (r, \theta)$ whose coordinates satisfy the equation.

EXAMPLE 1 *A Circle in Polar Coordinates*

Find the graph of the polar equation $r = 1$.

SOLUTION The set of points for which $r = 1$ is the set of points 1 unit from the pole, which is the definition of the unit circle. Thus the graph of $r = 1$ is the unit circle. This curve is sketched in Figure 1, together with the curves $r = 2$ (the circle of radius 2), $r = 3$, and $r = \frac{1}{2}$.

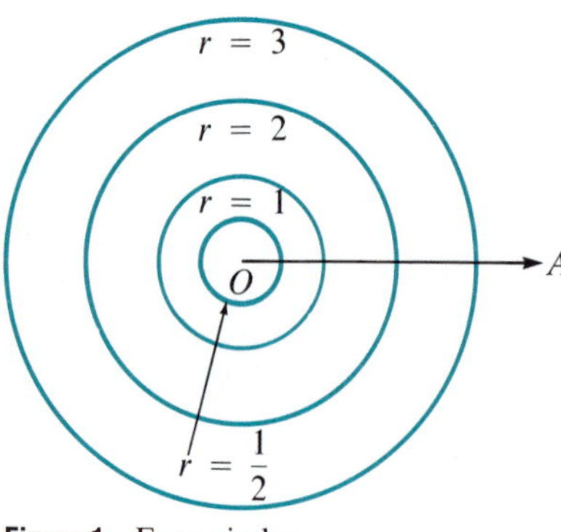

Figure 1 Four circles

We now consider the graphs of some more general curves in polar coordinates. To aid us in obtaining sketches of these curves, we cite three rules that are often useful.

Rules of Symmetry

i. If in a polar equation θ can be replaced by $-\theta$ without changing the equation, then the polar graph is symmetric about the polar axis (see Figure 2(a)).

ii. If θ can be replaced by $\pi - \theta$ without changing the equation, then the polar graph is symmetric about the line $\theta = \pi/2$ (see Figure 2(b)).
iii. If r can be replaced by $-r$, or, equivalently, if θ can be replaced by $\theta + \pi$, without changing the equation, then the polar graph is symmetric about the pole (see Figure 2(c)).

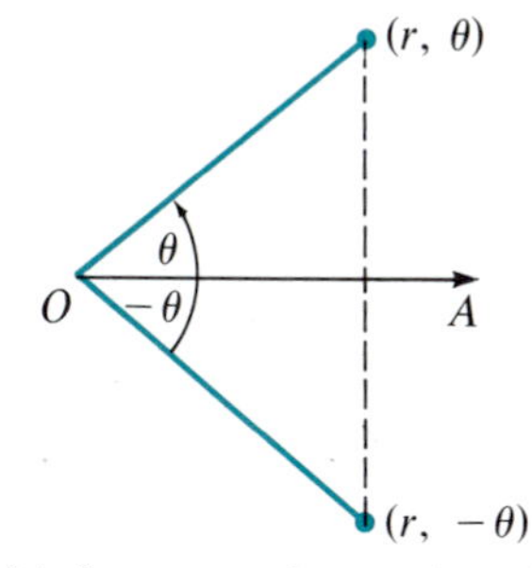

(a) Symmetry about polar axis
$f(r, \theta) = f(r, -\theta)$

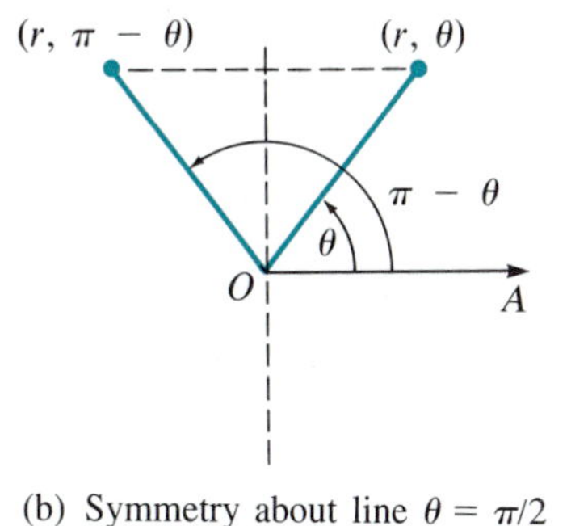

(b) Symmetry about line $\theta = \pi/2$
$f(r, \pi - \theta) = f(r, \theta)$

(c) Symmetry about pole
$f(-r, \theta) = f(r, \theta) = f(r, \theta + \pi)$

Figure 2 Three kinds of symmetry.

EXAMPLE 2 *Sketching a Cardioid*

Sketch the curve $r = 1 + \sin \theta$.

SOLUTION Since $\sin(\pi - \theta) = \sin \theta$, the curve is symmetric about the line $\theta = \dfrac{\pi}{2}$. We therefore need only consider values of θ in $\left[0, \dfrac{\pi}{2}\right]$ and $\left[\dfrac{3\pi}{2}, 2\pi\right]$. Typical values of the function are given in Table 1. We use these values in Figure 3 and then use symmetry to reflect the curve about the line $\theta = \dfrac{\pi}{2}$. The heart-shaped curve we have sketched is called a **cardioid,** from the Greek *kardia,* meaning "heart."

Table 1

θ	0	$\frac{\pi}{6}$	$\frac{\pi}{4}$	$\frac{\pi}{3}$	$\frac{\pi}{2}$	$\frac{3\pi}{2}$	$\frac{5\pi}{3}$	$\frac{7\pi}{4}$	$\frac{11\pi}{6}$
$r = 1 + \sin \theta$	1	$\frac{3}{2}$	$1 + \frac{\sqrt{2}}{2}$	$1 + \frac{\sqrt{3}}{2}$	2	0	$1 - \frac{\sqrt{3}}{2}$	$1 - \frac{\sqrt{2}}{2}$	$\frac{1}{2}$

Figure 3 The cardioid $r = 1 + \sin \theta$.

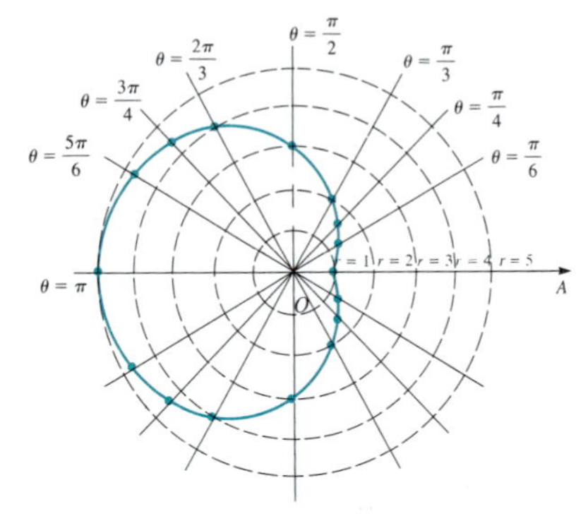

Figure 4 The limaçon $r = 3 - 2\cos\theta$.

EXAMPLE 3 *Sketching a Limaçon*

Sketch the curve $r = 3 - 2\cos\theta$.

SOLUTION The curve is symmetric about the polar axis (since $\cos(-\theta) = \cos\theta$), and we therefore calculate values of r for θ in $[0, \pi]$ (Table 2). We then obtain the graph sketched in Figure 4. The curve sketched in Figure 4 is called a **limaçon.**†

Table 2

θ	0	$\frac{\pi}{6}$	$\frac{\pi}{4}$	$\frac{\pi}{3}$	$\frac{\pi}{2}$	$\frac{2\pi}{3}$	$\frac{3\pi}{4}$	$\frac{5\pi}{6}$	π
$r = 3 - 2\cos\theta$	1	$3 - \sqrt{3}$	$3 - \sqrt{2}$	2	3	4	$3 + \sqrt{2}$	$3 + \sqrt{3}$	5

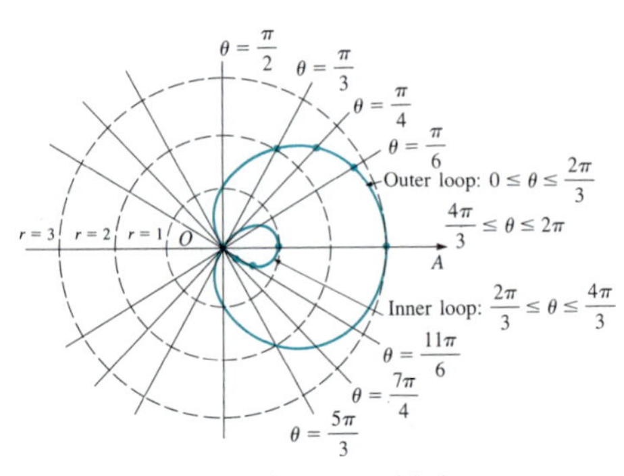

Figure 5 The limaçon with loop $r = 1 + 2\cos\theta$.

EXAMPLE 4 *Sketching a Limaçon with Loop*

Sketch the curve $r = 1 + 2\cos\theta$.

SOLUTION This curve is also symmetric about the polar axis, and we therefore tabulate the function for θ in $[0, \pi]$ (Table 3).

Using the values in Table 3 together with the symmetry around the polar axis, we obtain the graph in Figure 5. This curve is also called a limaçon or a **limaçon with loop.**

Table 3

θ	0	$\frac{\pi}{6}$	$\frac{\pi}{4}$	$\frac{\pi}{3}$	$\frac{\pi}{2}$	$\frac{2\pi}{3}$	$\frac{3\pi}{4}$	$\frac{5\pi}{6}$	π
$r = 1 + 2\cos\theta$	3	$1 + \sqrt{3}$	$1 + \sqrt{2}$	2	1	0	$1 - \sqrt{2}$	$1 - \sqrt{3}$	-1

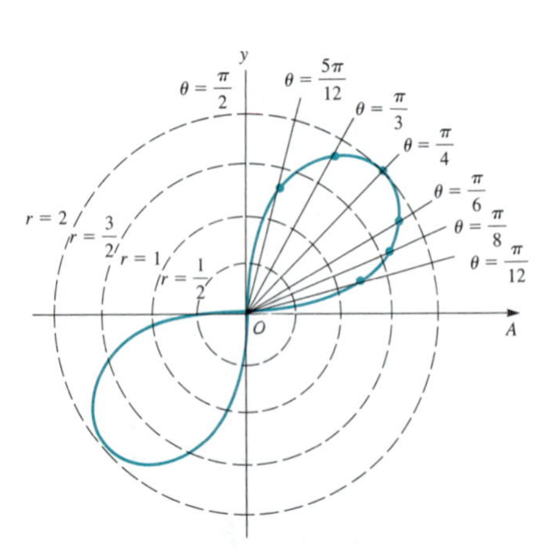

Figure 6 The lemniscate $r^2 = 4\sin 2\theta$.

EXAMPLE 5 *Sketching a Lemniscate*

Sketch the curve $r^2 = 4\sin 2\theta$.

SOLUTION Two things should be immediately evident from the equation. First, since $(-r)^2 = r^2$, the graph is symmetric about the pole. Second, since $r^2 \geq 0$, the function is defined only for values of θ such that $\sin 2\theta \geq 0$. If θ is restricted to the interval $[0, 2\pi]$, then $\sin 2\theta \geq 0$ if and only if θ is in $\left[0, \frac{\pi}{2}\right]$ or $\left[\pi, \frac{3\pi}{2}\right]$. Then using Table 4 and the symmetry about the pole, we obtain the graph sketched in Figure 6. This curve is called a **lemniscate.**‡

† From the Latin word *limax*, meaning "snail." The curve is also referred to as *Pascal's limaçon* since it was discovered by Etienne Pascal (1588–1640), father of the famous French mathematician Blaise Pascal (1623–1662).

‡ From the Greek *lemniskos* and the Latin *lemniscus*, meaning "knotted ribbon." Actually, this curve was originally called the *lemniscate* of *Bernoulli*, named after the Swiss mathematician Jacques Bernoulli (1654–1705), who described the curve in the *Acta Eruditorum*, 1694.

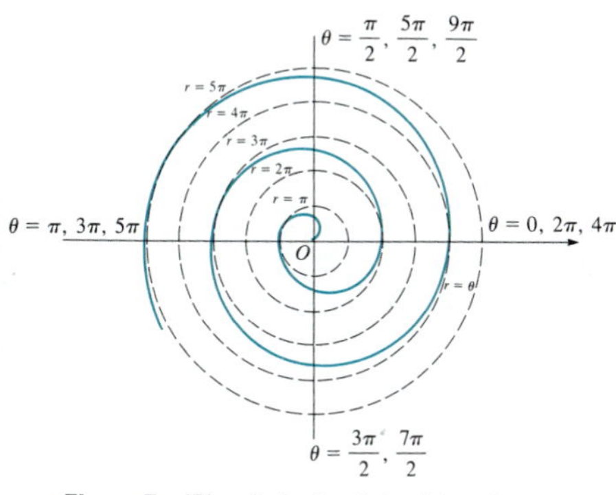

Figure 7 The Spiral of Archimedes $r = \theta$, $r \geq 0$.

Table 4

θ	0	$\frac{\pi}{12}$	$\frac{\pi}{8}$	$\frac{\pi}{6}$	$\frac{\pi}{4}$	$\frac{\pi}{3}$	$\frac{5\pi}{12}$	$\frac{\pi}{2}$
$r = \sqrt{4 \sin 2\theta}$	0	$\sqrt{2}$	$2^{3/4}$	$\sqrt{2} \cdot 3^{1/4}$	2	$\sqrt{2} \cdot 3^{1/4}$	$\sqrt{2}$	0

EXAMPLE 6 *Sketching a Spiral of Archimedes*

Sketch the curve $r = \theta$, $\theta \geq 0$.

SOLUTION This curve has no symmetry. To sketch its graph, we note that as we move around the pole, r increases as θ increases. The graph is sketched in Figure 7, and the curve depicted is called the **Spiral of Archimedes.**† It intersects the polar axis when $\theta = 2n\pi$, n an integer, and it intersects the line $\theta = \frac{\pi}{2}$ when $\theta = \frac{(2n+1)}{2}\pi$.

Problems 4.8

Readiness Check

I. Which of the following is the polar form of the equation $y = 3$?

a. $\theta = 3r$ b. $r = \frac{3}{\sin \theta}$

c. $r = \frac{3}{\cos \theta}$ d. $\frac{3}{r} = \cos \theta$

II. Which of the following is true of the graph of $r = 3 + \sin \theta$?

a. It is symmetric with respect to the polar axis.
b. It is symmetric with respect to the line $\theta = \pi/2$.
c. It is symmetric about the pole.
d. It has no symmetry.

III. Which of the following is symmetric with respect to the polar axis only?

a. $r = 2 \sin \theta$ b. $r = 3 \cos \theta$

c. $r = 2$ d. $\theta = \frac{\pi}{3}$

IV. Which of the following is true about the graph of $\theta = 5\pi/6$?

a. It is symmetric about the line $\theta = \frac{\pi}{2}$.
b. It is symmetric about the polar axis.
c. It is symmetric about the y-axis.
d. Its graph is a straight line in quadrants II and IV.

In Problems 1–65 sketch the graph of the given equation, indicating any symmetry about the polar axis, the line $\theta = \pi/2$, and/or the pole.

1. $r = 5$ 2. $r = 7$

3. $r = -4$ 4. $\theta = \frac{3\pi}{8}$

5. $\theta = -\frac{\pi}{6}$ 6. $\theta = \frac{13\pi}{5}$

7. $r = 5 \sin \theta$ 8. $r = 5 \cos \theta$

9. $r = -5 \sin \theta$ 10. $r = -5 \cos \theta$

11. $r = 5 \cos \theta + 5 \sin \theta$

12. $r = -5 \cos \theta + 5 \sin \theta$

Answers to Readiness Check

I. b II. b III. b IV. d

† Archimedes devoted a great deal of study to the spiral that can be described by the polar equation $r = a\theta$. He used it initially in his attempt to solve the ancient problem of trisecting the angle. Later he calculated part of its area by his method of exhaustion. He described his work on this subject in one of his most important works, entitled *On Spirals*.

13. $r = 5\cos\theta - 5\sin\theta$
14. $r = -5\cos\theta - 5\sin\theta$
15. $r = 2 + 2\sin\theta$
16. $r = 2 + 2\cos\theta$
17. $r = 2 - 2\sin\theta$
18. $r = 2 - 2\cos\theta$
19. $r = -2 + 2\sin\theta$
20. $r = -2 + 2\cos\theta$
21. $r = -2 - 2\sin\theta$
22. $r = -2 - 2\cos\theta$
23. $r = 1 + 3\sin\theta$
24. $r = 3 + \sin\theta$
25. $r = -2 + 4\cos\theta$
26. $r = -4 + 2\cos\theta$
27. $r = -3 - 4\cos\theta$
28. $r = -4 - 3\cos\theta$
29. $r = -3 - 4\sin\theta$
30. $r = -4 - 3\sin\theta$
31. $r = 4 - 3\cos\theta$
32. $r = 3 - 4\cos\theta$
33. $r = 4 + 3\sin\theta$
34. $r = 3 + 4\sin\theta$
35. $r = 3\sin 2\theta$ (This curve is called a **four-leafed rose.**)
36. $r = -3\sin 2\theta$
37. $r = 3\cos 2\theta$
38. $r = -3\cos 2\theta$
39. $r = 5\sin 3\theta$ (This curve is called a **three-leafed rose.**)
40. $r = 5\cos 3\theta$
41. $r = -5\sin 3\theta$
42. $r = -5\cos 3\theta$
43. $r = 2\cos 4\theta$ (This curve is called an **eight-leafed rose.**)
44. $r = 2\sin 4\theta$
45. $r = -2\cos 4\theta$
46. $r = 3\theta,\ \theta \geq 0$ (46–48 are all spirals of Archimedes.)
47. $r = -5\theta,\ \theta \geq 0$
48. $r = \theta/2,\ \theta \geq 0$
49. $r = e^{\theta}$ (This curve is called a **logarithmic spiral** since $\ln r = \theta$.)†
50. $r = e^{\theta/2}$
51. $r = e^{3\theta}$
52. $r^2 = \cos 2\theta$
53. $r^2 = \sin 2\theta$
54. $r^2 = -\cos 2\theta$
55. $r^2 = -\sin 2\theta$
56. $r^2 = 4\sin 2\theta$
57. $r^2 = -25\cos 2\theta$
58. $r^2 = -25\sin 2\theta$
59. $r = \sin\theta\tan\theta$ (This curve is called a **cissoid.**)
60. $r = 2 - 3\sec\theta$ (This curve is called a **conchoid.**)
61. $r = 4 + 3\csc\theta$
62. $r^2 = \theta$ (This curve is called a **parabolic spiral.**)
63. $(r + 1)^2 = 3\theta$
64. $r = |\sin\theta|$
65. $r = |\cos\theta|$
66. * By translating into rectangular coordinates and completing the square, show that for any real number a the graph of the equation $r = a\cos\theta$ is a circle in the xy-plane with center at $\left(\frac{a}{2}, 0\right)$ and radius $\left|\frac{a}{2}\right|$.
67. * Show that the graph of the equation $r = b\sin\theta$ in the xy-plane is a circle with radius $\left|\frac{b}{2}\right|$ centered at $\left(0, \frac{b}{2}\right)$.
68. * Show that the graph of the equation $r = a\cos\theta + b\sin\theta$ in the xy-plane is a circle centered at $\left(\frac{a}{2}, \frac{b}{2}\right)$ with radius $\frac{\sqrt{a^2 + b^2}}{2}$.
69. Find the polar equation of the circle centered at $(-2, \frac{3}{2})$ with radius $\frac{5}{2}$.
70. Show that the polar equation $r\cos\theta = a$ is the equation of a vertical line for any real number a.
71. Show that the polar equation $r\sin\theta = a$ is the equation of a horizontal line for any real number a.
72. Show that the polar equation $r(a\cos\theta + b\sin\theta) = c$ with a, b, c real numbers is the equation of a straight line, if $a^2 + b^2 \neq 0$.
73. Sketch the graphs:
    (a) $r\sin\theta = 3$
    (b) $r\cos\theta = -2$
    (c) $3r\cos\theta = 8$
    (d) $r(2\sin\theta - 3\cos\theta) = 4$
    (e) $r(-5\sin\theta + 10\cos\theta) = 20$

† Problems 49–51 make use of material in Chapter 7.

Summary Outline of Chapter 4

A, B, C denote angles, and a, b, c denote sides of a triangle.

- **Law of Sines** $\dfrac{\sin A}{a} = \dfrac{\sin B}{b} = \dfrac{\sin C}{c}$ p. 212
- **Law of Cosines** $a^2 = b^2 + c^2 - 2bc\cos A$ p. 219

$$\cos A = \frac{b^2 + c^2 - a^2}{2bc}$$

- **Heron's Formula** p. 223

$$\text{Area of triangle } ABC = \sqrt{s(s-a)(s-b)(s-c)}$$

where $s = \frac{1}{2}(a + b + c)$

- A **directed line segment** from P to Q, denoted by $\overrightarrow{PQ}$, is the straight-line segment that extends from P to Q. p. 236
- A **vector** $\mathbf{v}$ is a collection of equivalent directed line segments. p. 236
  If one representation starts at (0, 0) and extends to (a, b), then the vector can be denoted by $\mathbf{v} = (a, b)$.
- The **magnitude** of $\mathbf{v} = (a, b)$ is $|\mathbf{v}| = \sqrt{a^2 + b^2}$. p. 237
- The **direction** of $\mathbf{v}$ is the angle θ in standard position that the vector makes with the positive x-axis. p. 238
  By definition, $0 \le \theta < 2\pi \quad (0° \le \theta° < 360°)$
- **Vector Algebra** p. 240
  addition: $(a_1, b_1) + (a_2, b_2) = (a_1 + a_2, b_1 + b_2)$
  scalar multiplication: $\alpha(a, b) = (\alpha a, \alpha b)$
- In $\mathbb{R}^2$, let $\mathbf{i} = (1, 0)$ and $\mathbf{j} = (0, 1)$. Then $\mathbf{v} = (a, b)$ can be written p. 242

$$\mathbf{v} = a\mathbf{i} + b\mathbf{j}$$

- A **unit vector u** in $\mathbb{R}^2$ is a vector that satisfies $|\mathbf{u}| = 1$. In $\mathbb{R}^2$, a unit vector can be written p. 243

$$\mathbf{u} = (\cos\theta)\mathbf{i} + (\sin\theta)\mathbf{j}$$

where θ is the direction of $\mathbf{u}$.

- Let $\mathbf{u} = (a_1, b_1)$ and $\mathbf{v} = (a_2, b_2)$. Then the **dot product** of $\mathbf{u}$ and $\mathbf{v}$, written $\mathbf{u} \cdot \mathbf{v}$, is given by p. 254

$$\mathbf{u} \cdot \mathbf{v} = a_1a_2 + b_1b_2$$

- **Properties of the Dot Product in $\mathbb{R}^2$** For any vectors $\mathbf{u}$, $\mathbf{v}$, $\mathbf{w}$, and scalar α, p. 255
  (i) $\mathbf{u} \cdot \mathbf{v} = \mathbf{v} \cdot \mathbf{u}$
  (ii) $(\mathbf{u} + \mathbf{v}) \cdot \mathbf{w} = \mathbf{u} \cdot \mathbf{w} + \mathbf{v} \cdot \mathbf{w}$
  (iii) $(\alpha\mathbf{u}) \cdot \mathbf{v} = \alpha(\mathbf{u} \cdot \mathbf{v})$
  (iv) $\mathbf{u} \cdot \mathbf{u} \ge 0$ and $\mathbf{u} \cdot \mathbf{u} = 0$ if and only if $\mathbf{u} = \mathbf{0}$
  (v) $|\mathbf{u}|^2 = \mathbf{u} \cdot \mathbf{u}$.
- The angle ϕ between two vectors $\mathbf{u}$ and $\mathbf{v}$ in $\mathbb{R}^2$ is the unique number in $[0, \pi]$ that satisfies p. 255

$$\cos\phi = \frac{\mathbf{u} \cdot \mathbf{v}}{|\mathbf{u}||\mathbf{v}|}$$

- Two vectors in $\mathbb{R}^2$ are parallel if the angle between them is 0 or π. p. 256
- Two vectors in $\mathbb{R}^2$ are **orthogonal** if the angle between them is $\frac{\pi}{2}$. p. 257
- Let $\mathbf{u}$ and $\mathbf{v}$ be two nonzero vectors in $\mathbb{R}^2$. Then the **projection** of $\mathbf{u}$ on $\mathbf{v}$ is a vector, denoted by $\text{Proj}_{\mathbf{v}}\, \mathbf{u}$, which is defined by p. 258

$$\text{Proj}_{\mathbf{v}}\, \mathbf{u} = \frac{\mathbf{u} \cdot \mathbf{v}}{|\mathbf{v}|^2}\mathbf{v}$$

- The vector $\frac{\mathbf{u} \cdot \mathbf{v}}{|\mathbf{v}|}$ is called the **component** of $\mathbf{u}$ in the direction $\mathbf{v}$. p. 258

- **Polar Coordinates** p. 263

  $P = (r, \theta)$ where
  r is the distance from the origin to the point
  θ is the angle from the polar axis to OP
  We generally have $r > 0$ and $0 \leq \theta < 2\pi$.

- **Conversion Formulas** p. 265

From polar to rectangular coordinates	From rectangular to polar coordinates
$x = r\cos\theta \qquad y = r\sin\theta$	$r = \sqrt{x^2 + y^2}$ and $\tan\theta = \dfrac{y}{x}, \quad x \neq 0$

- $(-r, \theta) = (r, \theta + \pi)$ p. 264

- **Graph in Polar Coordinates** The **graph** of an equation written in polar coordinates r and θ consists of those points P having at least one representation $P = (r, \theta)$ whose coordinates satisfy the equation. p. 268

- **Rules of symmetry in polar coordinates** pp. 268–269

  (i) If in a polar equation θ can be replaced by $-\theta$ without changing the equation, then the polar graph is symmetric about the polar axis.
  (ii) If θ can be replaced by $\pi - \theta$ without changing the equation, then the polar graph is symmetric about the line $\theta = \pi/2$.
  (iii) If r can be replaced by $-r$, or equivalently, if θ can be replaced by $\theta + \pi$, without changing the equation, then the polar graph is symmetric about the pole.

Review Exercises for Chapter 4

In Exercises 1–8 solve each oblique triangle.

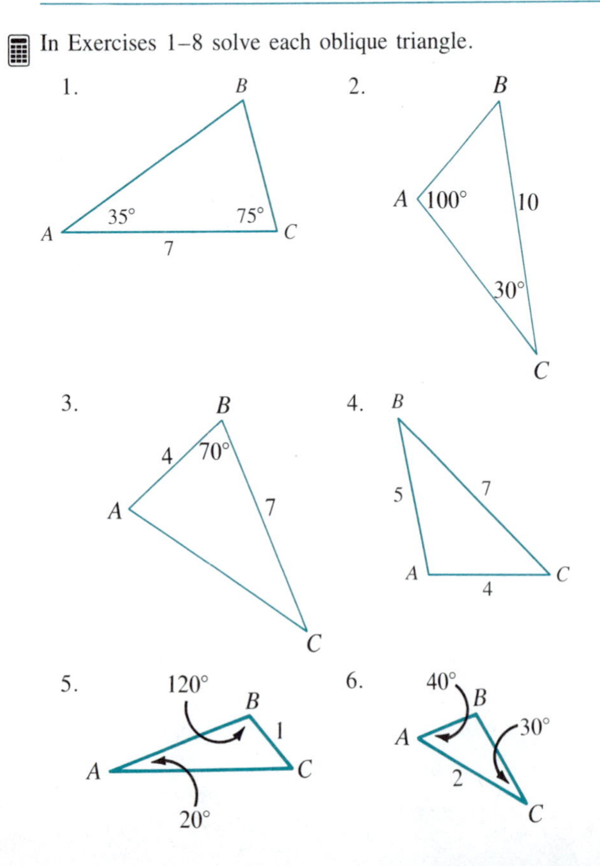

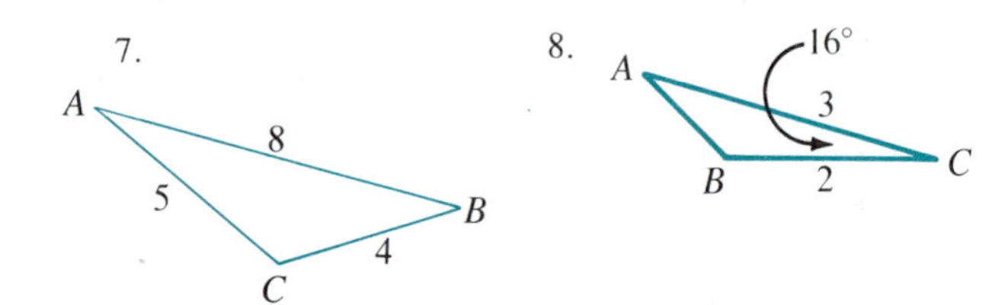

In Exercises 9–14 three angles or sides of a triangle ABC are given. Find the remaining angles or sides.

9. $a = 12$, $B = 40°$, $C = 95°$
10. $c = 9$, $A = 42.3°$, $B = 65.5°$
11. $a = 5$, $b = 4$, $C = 47°$
12. $a = \frac{1}{2}$, $c = \frac{3}{4}$, $B = 102°$
13. $a = 7$, $b = 2$, $c = 8$
14. $a = 425$, $b = 362$, $c = 519$

In Exercises 15–20 find the measurements of all triangles (if any) that have the given sides and opposite angle.

15. $a = 20$, $b = 30$, $A = 37°$
16. $a = 20$, $b = 30$, $A = 57°$
17. $a = 30$, $b = 20$, $A = 47°$
18. $a = 23.4$, $b = 17.8$, $B = 77°$
19. $a = 3$, $b = 5$, $A = 107°$
20. $a = 0.7$, $b = 0.4$, $B = 121°$

In Exercises 21 and 22 find the area of the triangle whose sides have the given lengths.

21. 5, 6, 7 22. 238, 195, 306

23. Find the distance from A to B across the pond in the figure below.

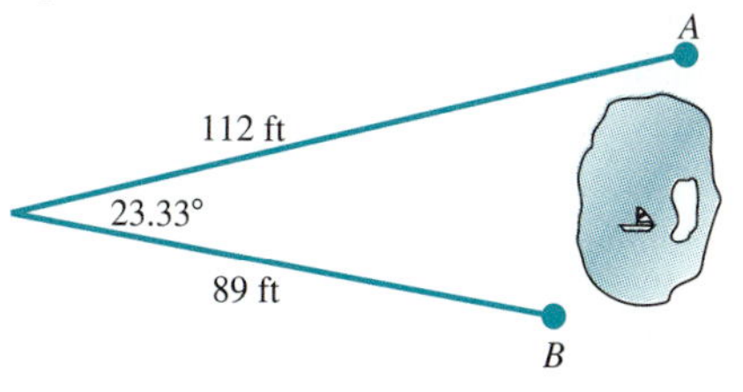

24. Find the distance from A to B across the river in the figure below.

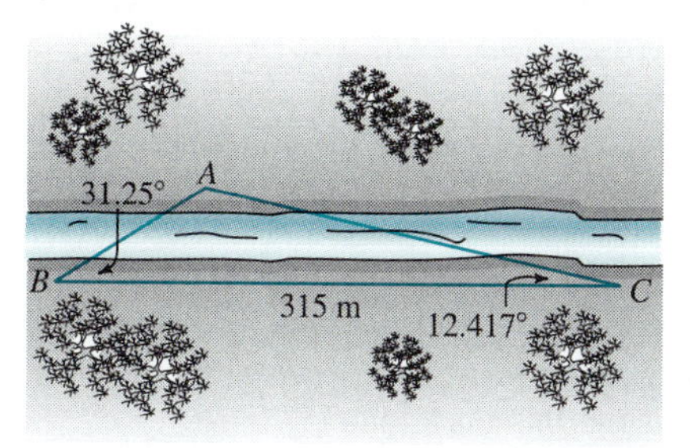

25. Find the elevation at the top of the mountain (point A) in the figure below if it is known that the elevation at C is 2144 m.

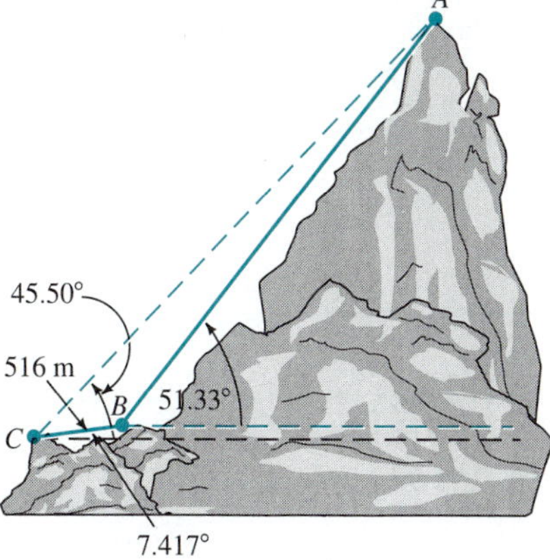

26. A transit satellite orbits 1000 km above the earth's surface. In one measurement, it is determined that the distance from the transit to a known radio control site is 3115 km. The angle formed by the line from the transit to the known control site and the line from the transit to an unknown remote site is 97°26′. Determine, to the nearest kilometer, the horizontal distance between the control site and the remote site.

27. A ship leaves port and travels 88 km on a bearing of N25°W. It then changes course and sails 102 km at a bearing of S82°W. How far is it from port at this point?

28. In Exercise 27 determine the bearing at which the ship must sail in order to get back to port.

29. An airplane is 186 miles from radar station P, which is at a bearing of S16°15′E from the plane. A second radar station Q is 94 miles from P at a bearing of N87°30′W from P. Find the distance and bearing from Q to the airplane.

In Exercises 30–33 find the magnitude and direction of the given vector.

30. $\mathbf{v} = (3, 3)$
31. $\mathbf{v} = (-3, 3)$
32. $\mathbf{v} = (2, -2\sqrt{3})$
33. $\mathbf{v} = (\sqrt{3}, 1)$

In Exercises 34 and 35 write the vector $\mathbf{v}$ that is represented by $\overrightarrow{PQ}$ in the form (a, b).

34. $P = (2, 3)$; $Q = (4, 5)$
35. $P = (1, -2)$; $Q = (7, 12)$
36. Let $\mathbf{u} = (2, 1)$ and $\mathbf{v} = (-3, 4)$. Find (a) $5\mathbf{u}$, (b) $\mathbf{u} - \mathbf{v}$, (c) $-8\mathbf{u} + 5\mathbf{v}$
37. Let $\mathbf{u} = (-4, 1)$ and $\mathbf{v} = (-3, -4)$. Find (a) $-3\mathbf{v}$, (b) $\mathbf{u} + \mathbf{v}$, (c) $3\mathbf{u} - 6\mathbf{v}$

In Exercises 38–43 find a unit vector having the same direction as the given vector.

38. $\mathbf{v} = -\mathbf{i} + \mathbf{j}$
39. $\mathbf{v} = 2\mathbf{i} + 5\mathbf{j}$
40. $\mathbf{v} = -7\mathbf{i} + 3\mathbf{j}$
41. $\mathbf{v} = 3\mathbf{i} + 4\mathbf{j}$
42. $\mathbf{v} = -2\mathbf{i} - 2\mathbf{j}$
43. $\mathbf{v} = a\mathbf{i} - a\mathbf{j}$
44. For $\mathbf{v} = 4\mathbf{i} - 7\mathbf{j}$, find $\sin\theta$ and $\cos\theta$, where θ is the direction of $\mathbf{v}$.
45. Find a unit vector with direction opposite to that of $\mathbf{v} = 5\mathbf{i} + 2\mathbf{j}$.
46. Find two unit vectors orthogonal to $\mathbf{v} = \mathbf{i} - \mathbf{j}$.
47. Find a unit vector with direction opposite to that of $\mathbf{v} = 10\mathbf{i} - 7\mathbf{j}$.

In Exercises 48–51 find a vector $\mathbf{v}$ having the given magnitude and direction.

48. $|\mathbf{v}| = 2$; $\theta = \pi/3$
49. $|\mathbf{v}| = 1$; $\theta = \pi/2$
50. $|\mathbf{v}| = 4$; $\theta = \pi$
51. $|\mathbf{v}| = 7$; $\theta = 5\pi/6$

In Exercises 52–55 calculate the dot product of the two vectors and the cosine of the angle between them.

52. $\mathbf{u} = \mathbf{i} - \mathbf{j}$; $\mathbf{v} = \mathbf{i} + 2\mathbf{j}$
53. $\mathbf{u} = -4\mathbf{i}$; $\mathbf{v} = 11\mathbf{j}$
54. $\mathbf{u} = 4\mathbf{i} - 7\mathbf{j}$; $\mathbf{v} = 5\mathbf{i} + 6\mathbf{j}$
55. $\mathbf{u} = -\mathbf{i} - 2\mathbf{j}$; $\mathbf{v} = 4\mathbf{i} + 5\mathbf{j}$

In Exercises 56–61 determine whether the given vectors are orthogonal, parallel, or neither. Then sketch each pair.

56. $\mathbf{u} = 2\mathbf{i} - 6\mathbf{j}$; $\mathbf{v} = -\mathbf{i} + 3\mathbf{j}$
57. $\mathbf{u} = 4\mathbf{i} - 5\mathbf{j}$; $\mathbf{v} = 5\mathbf{i} - 4\mathbf{j}$
58. $\mathbf{u} = 4\mathbf{i} - 5\mathbf{j}$; $\mathbf{v} = -5\mathbf{i} + 4\mathbf{j}$
59. $\mathbf{u} = -7\mathbf{i} - 7\mathbf{j}$; $\mathbf{v} = \mathbf{i} + \mathbf{j}$
60. $\mathbf{u} = -7\mathbf{i} - 7\mathbf{j}$; $\mathbf{v} = -\mathbf{i} + \mathbf{j}$
61. $\mathbf{u} = -7\mathbf{i} - 7\mathbf{j}$; $\mathbf{v} = -\mathbf{i} - \mathbf{j}$
62. Let $\mathbf{u} = 2\mathbf{i} + 3\mathbf{j}$ and $\mathbf{v} = 4\mathbf{i} + \alpha\mathbf{j}$.
    (a) Determine α such that $\mathbf{u}$ and $\mathbf{v}$ are orthogonal.
    (b) Determine α such that $\mathbf{u}$ and $\mathbf{v}$ are parallel.
    (c) Determine α such that the angle between $\mathbf{u}$ and $\mathbf{v}$ is $\pi/4$.
    (d) Determine α such that the angle between $\mathbf{u}$ and $\mathbf{v}$ is $\pi/6$.

In Exercises 63–68 calculate $\text{Proj}_{\mathbf{v}}\, \mathbf{u}$.

63. $\mathbf{u} = 14\mathbf{i}$; $\mathbf{v} = \mathbf{i} + \mathbf{j}$
64. $\mathbf{u} = 14\mathbf{i}$; $\mathbf{v} = \mathbf{i} - \mathbf{j}$
65. $\mathbf{u} = 3\mathbf{i} - 2\mathbf{j}$; $\mathbf{v} = 3\mathbf{i} + 2\mathbf{j}$
66. $\mathbf{u} = 3\mathbf{i} + 2\mathbf{j}$; $\mathbf{v} = \mathbf{i} - 3\mathbf{j}$
67. $\mathbf{u} = 2\mathbf{i} - 5\mathbf{j}$; $\mathbf{v} = -3\mathbf{i} - 7\mathbf{j}$
68. $\mathbf{u} = 4\mathbf{i} - 5\mathbf{j}$; $\mathbf{v} = -3\mathbf{i} - \mathbf{j}$

In Exercises 69–71 find the work done when the force with given magnitude and direction moves an object from P to Q. All distances are measured in meters.

69. $|\mathbf{F}| = 2$ N; $\theta = \pi/4$; $P = (1, 6)$; $Q = (2, 4)$
70. $|\mathbf{F}| = 3$ N; $\theta = \pi/2$; $P = (3, -5)$; $Q = (2, 7)$
71. $|\mathbf{F}| = 11$ N; $\theta = \pi/6$; $P = (-1, -2)$; $Q = (-7, -4)$

Use a calculator to solve Problems 72–79.

72. From a point A on level ground, the angles of elevation to the top D and bottom B of a flagpole situated on top of a hill are measured as 47.9° and 39.75°. Find the height of the hill if the flagpole is 115 feet tall.

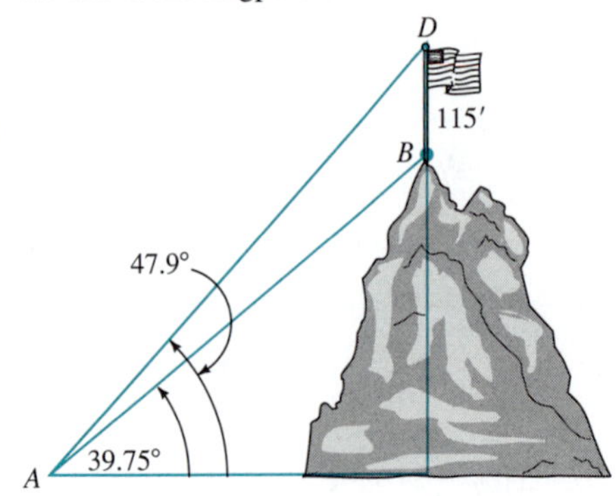

73. A slide that is 45 feet long has a height of 12 feet. If the force required to pull an object up the slide is 80 pounds, find the weight of the object.
74. A 325-pound black bear is resting at the top of a slide, whose angle to the ground is 18.58°.
    (a) Find the force tending to slide the bear down the slide.
    (b) Find the force of the bear on the slide.
    (c) How much work does the bear do when it slides 8 ft down the slide?
75. From a point on a road, the angle of elevation to the top of a power pole is 39.25°. By moving 40 feet closer to the pole, the angle of elevation is now 48.62°. Find the height of the power pole.
76. A submarine leaves dock X and sails on course N32°48′W for 145 km. The boat then changes its course to N83°46′E and sails for 112 km. At this point, the boat again changes its course, this time to N42°35′E, and sails for 190 km. How far is the boat from its starting point (dock X)?
77. In Exercise 76 what is the bearing from dock X to the submarine at this final point?
78. An airplane takes off on course N23°40′W at 475 mph. If the wind is blowing from N16°37′E at 60 mph, find the plane's true course and ground speed.
79. In Exercise 78 find the course and airspeed at which the plane should take off if the pilot wants the final course and ground speed to be N23°40′E and 475 mph.

In Exercises 80–84 convert from polar to rectangular coordinates.

80. $(2, 0)$ 81. $\left(3, \dfrac{\pi}{6}\right)$ 82. $\left(7, \dfrac{3\pi}{2}\right)$
83. $\left(4, \dfrac{2\pi}{3}\right)$ 84. $\left(1, \dfrac{5\pi}{3}\right)$

In Exercises 85–90 convert from rectangular to polar coordinates with $r > 0$ and $0 \le \theta < 2\pi$.

85. $(4, 0)$ 86. $(\sqrt{3}, 1)$ 87. $(\sqrt{3}, -1)$
88. $(6, 6)$ 89. $(-6, -6)$ 90. $(-6, 6)$

In Exercises 91–100 sketch the graph of the given equation, indicating any symmetry about the polar axis, the line $\theta = \dfrac{\pi}{2}$, and/or the pole.

91. $r = 8$ 92. $\theta = \dfrac{\pi}{3}$
93. $r = 2\cos\theta$ 94. $r = 3 - 3\sin\theta$
95. $r = 3 - 2\sin\theta$ 96. $r = 2 - 3\sin\theta$
97. $r = 5\cos 2\theta$ 98. $r^2 = 4\sin 2\theta$
99. $r = 3\sin 4\theta$ 100. $r = 3\theta,\ \theta \ge 0$
101. Find in rectangular coordinates an equation of the circle $r = 4\sin\theta - 6\cos\theta$.
102. Find a polar equation of the circle centered at $(\frac{5}{2}, -6)$ with radius $\frac{13}{2}$.

Chapter 5

Conic Sections and Parametric Equations

5.1 Introduction to the Conic Sections

The period from about 300 to 200 B.C. is known as the Golden Age of Greek mathematics because three of the world's greatest mathematicians lived during that period. The first of these, Euclid, invented much of the geometry that is studied in high schools today. The second, Archimedes, is considered by many to be the finest mathematician of any era. The third of the great Greek mathematicians, Apollonius of Perga, made many of the discoveries that are part of what we now call *analytic geometry*. The exact dates of Apollonius's life are not known, but he is believed to have lived from approximately 260 to 190 B.C.

Like most of the mathematicians of his day, Apollonius was an *applied* mathematician. He studied certain kinds of curves because they arose in practical ways. For example, he applied his work to the analysis of planetary motion and is considered to be the founder of Greek mathematical astronomy.

One of Apollonius's most important discoveries was that four different types of curves are obtained if a right circular cone is cut by a plane. These curves are circles, ellipses, parabolas, and hyperbolas. Because of the way they are formed, they are called **conic sections.** In Figure 1, we illustrate how the four curves can be obtained.

(a) Circle (b) Ellipse (c) Parabola (d) Hyperbola

Figure 1 Four cross sections of a right circular cone.

It turns out that each of the four curves can be written in the form

$$Ax^2 + Bxy + Cy^2 + Dx + Ey + F = 0 \qquad \textbf{(1)}$$

for certain constants A, B, C, D, E, and F.

We saw special cases of equation (1) earlier in this book. In Section 1.2, we discussed circles. In Example 7 on p. 12, we saw that the equation:

$$x^2 + y^2 - 6x + 2y - 17 = 0$$

is an equation of the circle of radius $\sqrt{27}$ centered at $(3, -1)$. This is equation (1) with $A = C = 1$, $B = 0$, $D = -6$, $E = 2$, and $F = -17$.

In Example 1 in Section 1.4 (p. 28) we saw that the graph of the squaring function $y = x^2$ is a parabola (see Figure 1 on p. 29). The equation $y = x^2$ can be written in the form (1) with $A = 1$, $B = C = D = F = 0$ and $E = -1$.

In the rest of this chapter, we will discuss equations that can be written in the form (1) or some equivalent form. We also will show how these equations, and the curves they represent, arise in applications.

5.2 The Ellipse and Translation of Axes

Definition of an Ellipse

An **ellipse** is the set of points (x, y) such that the sum of the distances from (x, y) to two given points is fixed. Each of the two points is called a **focus** of the ellipse.

NOTE The plural of focus is foci. Thus we speak about the two foci of an ellipse.

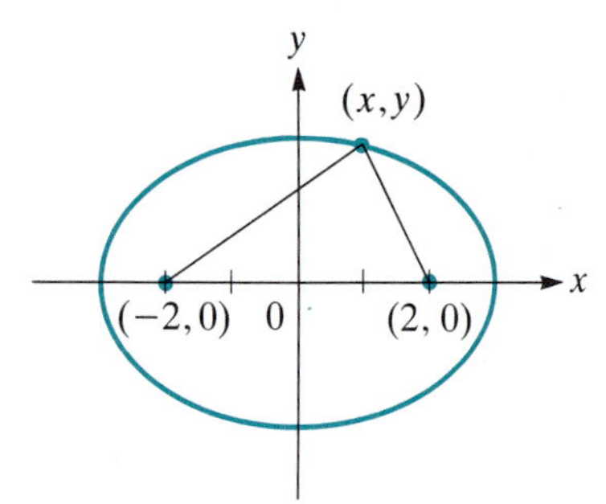

Figure 1 An ellipse with foci at $(-2, 0)$ and $(2, 0)$.

EXAMPLE 1 ***Finding an Equation of an Ellipse with Given Foci***

Find an equation of the ellipse with foci at $(-2, 0)$ and $(2, 0)$ such that the sum of the distances to the foci is 6.

SOLUTION As in Figure 1, let (x, y) denote a point on the ellipse. Then

$$[\text{distance from } (x, y) \text{ to } (2, 0)] + [\text{distance from } (x, y) \text{ to } (-2, 0)] = 6$$

or, from the distance formula (equation (1) on p. 10),

$$\sqrt{(x-2)^2 + (y-0)^2} + \sqrt{(x+2)^2 + (y-0)^2} = 6 \quad \text{or}$$

$$\sqrt{(x-2)^2 + y^2} = 6 - \sqrt{(x+2)^2 + y^2} \qquad (1)$$

$$(x-2)^2 + y^2 = 36 - 12\sqrt{(x+2)^2 + y^2} + (x+2)^2 + y^2 \quad \text{We squared both sides}$$

$$x^2 - 4x + 4 + y^2 = 36 - 12\sqrt{(x+2)^2 + y^2} + x^2 + 4x + 4 + y^2$$

$$-4x = 36 - 12\sqrt{(x+2)^2 + y^2} + 4x$$

We subtracted $x^2 + y^2 + 4$ from both sides

$$36 + 8x = 12\sqrt{(x+2)^2 + y^2}$$

We rearranged terms

$$3 + \frac{2}{3}x = \sqrt{(x+2)^2 + y^2} \quad \text{We divided both sides by 12}$$

$$9 + 4x + \frac{4}{9}x^2 = (x+2)^2 + y^2 = x^2 + 4x + 4 + y^2 \quad \text{We squared again}$$

$$\frac{5}{9}x^2 + y^2 = 5 \quad \text{We combined terms}$$

$$\frac{x^2}{9} + \frac{y^2}{5} = 1 \quad \text{We divided by 5} \tag{2}$$

This is the **standard equation** of the ellipse with foci at $(-2, 0)$ and $(2, 0)$ and sum of distances equal to 6.

Suppose now that the foci, F_1 and F_2, of an ellipse are the points $(-c, 0)$ and $(c, 0)$. (See Figure 2.) If $P = (x, y)$ is on the ellipse, then by the definition of the ellipse, the distance from P to the first focus plus the distance from P to the second focus equals $2a$; that is,

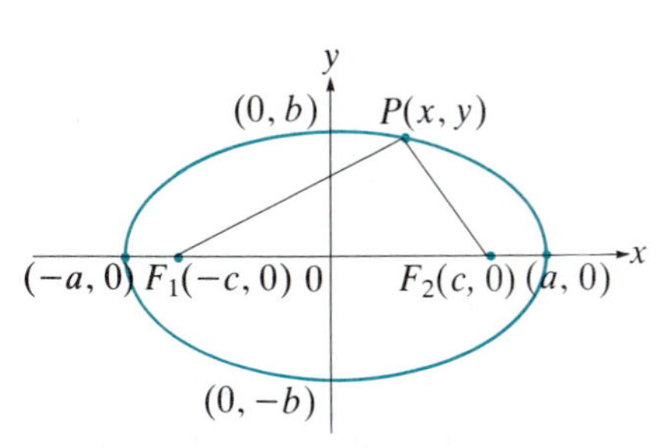

Figure 2 An ellipse with foci $(c, 0)$ and $(-c, 0)$ such that the sum of the distances from a point to each of the two foci is $2a$.

$$\overline{F_1P} + \overline{F_2P} = 2a \tag{3}$$

Since a straight line is the shortest distance between two points, we have

$$\overline{F_1F_2} = 2c < \overline{F_1P} + \overline{F_2P} = 2a$$

That is,

$$2a > 2c \quad \text{or} \quad a > c$$

($2c$ is the distance between the foci and $2a$ is the given sum of the distances.) Then

$$\sqrt{(x + c)^2 + y^2} + \sqrt{(x - c)^2 + y^2} = 2a \quad \text{From (3)}$$

or

$$\sqrt{(x + c)^2 + y^2} = 2a - \sqrt{(x - c)^2 + y^2}$$

$$(x + c)^2 + y^2 = 4a^2 - 4a\sqrt{(x - c)^2 + y^2} + (x - c)^2 + y^2 \quad \text{We squared both sides}$$

$$x^2 + 2xc + c^2 + y^2 = 4a^2 - 4a\sqrt{(x - c)^2 + y^2} + x^2 - 2xc + c^2 + y^2$$

$$4xc = 4a^2 - 4a\sqrt{(x - c)^2 + y^2} \quad \text{We subtracted } x^2 + c^2 + y^2 \text{ from both sides and simplified}$$

$$\sqrt{(x - c)^2 + y^2} = a - \frac{c}{a}x \quad \text{We divided by } 4a \text{ and rearranged terms}$$

$$(x - c)^2 + y^2 = a^2 - 2cx + \frac{c^2}{a^2}x^2 \quad \text{We squared again}$$

Then

$$x^2 - 2xc + c^2 + y^2 = a^2 - 2xc + \frac{c^2}{a^2}x^2$$

$$x^2\left(1 - \frac{c^2}{a^2}\right) + y^2 = a^2 - c^2$$

or

$$x^2\left(\frac{a^2 - c^2}{a^2}\right) + y^2 = a^2 - c^2$$

and after dividing both sides by $a^2 - c^2$, which is positive since $a > c$, we have

$$\frac{x^2}{a^2} + \frac{y^2}{a^2 - c^2} = 1$$

Finally, we define the positive number b by $b^2 = a^2 - c^2$ to obtain the standard equation of the ellipse.

Standard Equation of an Ellipse

$$\frac{x^2}{a^2} + \frac{y^2}{b^2} = 1 \quad \textbf{(4)}$$

Here, since b^2 is defined by $b^2 = a^2 - c^2$, we have

$$a^2 = b^2 + c^2$$

where $(c, 0)$ and $(-c, 0)$ are the foci. Note that this ellipse is symmetric about both the x- and y-axes. Since $a > b$, the line segment from $(-a, 0)$ to $(a, 0)$ is called the **major axis,** and the line segment from $(0, -b)$ to $(0, b)$ is called the **minor axis.** The point $(0, 0)$, which is the intersection of the axes, is called the **center** of the ellipse. The points $(a, 0)$ and $(-a, 0)$ are called the **vertices** of the ellipse. In general, the vertices of an ellipse are the endpoints of the major axis.

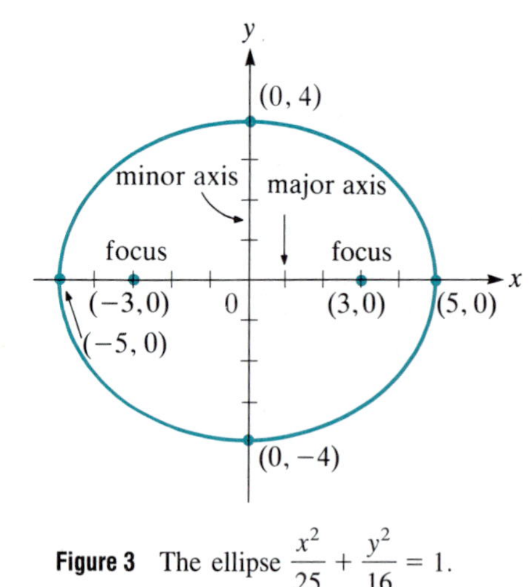

Figure 3 The ellipse $\frac{x^2}{25} + \frac{y^2}{16} = 1$.

EXAMPLE 2 *Finding the Standard Equation of an Ellipse*

Find the equation of the ellipse with foci at $(-3, 0)$ and $(3, 0)$ and with $a = 5$.

SOLUTION $c = 3$ and $a = 5$ so that $b^2 = a^2 - c^2 = 16$, and we obtain

$$\frac{x^2}{25} + \frac{y^2}{16} = 1$$

The ellipse is sketched in Figure 3.

We can reverse the roles of x and y in the preceding discussion. Suppose that the foci are at $(0, c)$ and $(0, -c)$ on the y-axis. Then if the fixed sum of the distances is given as $2b$, we obtain, using similar reasoning,

$$\frac{x^2}{a^2} + \frac{y^2}{b^2} = 1$$

where $a^2 = b^2 - c^2$. Now the major axis is on the y-axis, the minor axis is on the x-axis, and the vertices are $(0, b)$ and $(0, -b)$. In general, if the ellipse is given by (4), then we have the equations given in Table 1.

The last two entries in Table 1 are called **degenerate ellipses.** Degenerate ellipses have graphs containing one point or no point. For example, the equa-

Table 1 ***Standard Ellipses***

Equation	Description	Picture
$\frac{x^2}{a^2} + \frac{y^2}{b^2} = 1,\ a > b$	Ellipse with major axis on x-axis; $a^2 = b^2 + c^2$	
$\frac{x^2}{a^2} + \frac{y^2}{b^2} = 1,\ b > a$	Ellipse with major axis on y-axis; $b^2 = a^2 + c^2$	
$\frac{x^2}{a^2} + \frac{y^2}{b^2} = 1,\ b = a$	Circle with radius a $(= b)$	
$\frac{x^2}{a^2} + \frac{y^2}{b^2} = 0$	Degenerate ellipse; single point $(0, 0)$	
$\frac{x^2}{a^2} + \frac{y^2}{b^2} = -1$	Degenerate ellipse; graph is empty	

tion $\frac{x^2}{4} + \frac{y^2}{9} = 0$ is satisfied only by the point $(0, 0)$. Similarly, no point lies on the graph of $\frac{x^2}{2} + \frac{y^2}{3} = -1$. The equations $\frac{x^2}{4} + \frac{y^2}{9} = 0$ and $\frac{x^2}{2} + \frac{y^2}{3} = -1$ are similar to the standard equation of an ellipse (with 0 or -1 in place of 1). That is one reason why they are called degenerate *ellipses*.

We note, too, that when $a = b$ in equation (4), we obtain

$$\frac{x^2}{a^2} + \frac{y^2}{a^2} = 1 \quad \text{or} \quad x^2 + y^2 = a^2$$

This is an equation of the circle centered at $(0, 0)$ with radius a (see equation (2) on p. 12). Thus a circle can be thought of as a special kind of ellipse. We discussed circles in Section 1.2, so we will say no more about them here.

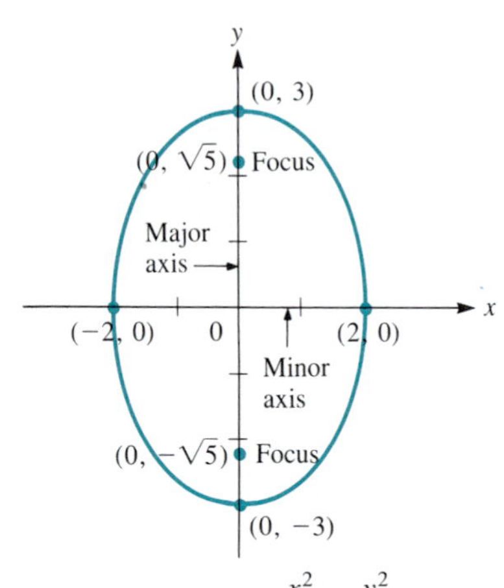

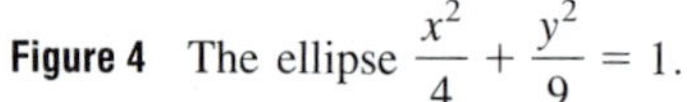

Figure 4 The ellipse $\dfrac{x^2}{4} + \dfrac{y^2}{9} = 1$.

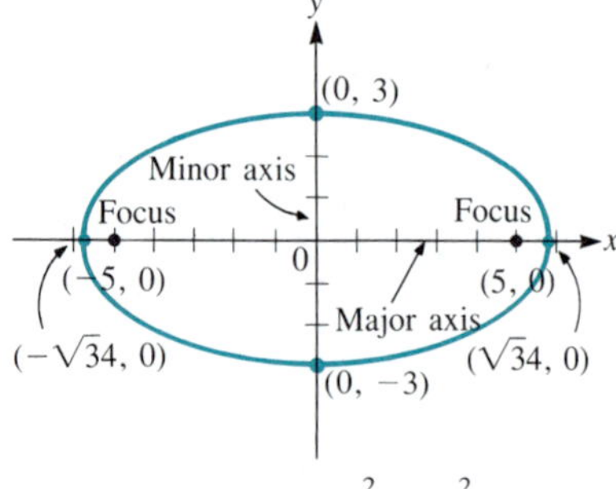

Figure 5 The ellipse $\dfrac{x^2}{34} + \dfrac{y^2}{9} = 1$.

EXAMPLE 3 ***Showing That a Second-Degree Curve Is the Equation of an Ellipse***

Discuss the curve $9x^2 + 4y^2 = 36$.

SOLUTION Dividing both sides by 36, we obtain

$$\frac{x^2}{4} + \frac{y^2}{9} = 1$$

Here $a = 2$, $b = 3$, and the major axis is on the y-axis. Since $c^2 = 9 - 4 = 5$, the foci are at $(0, \sqrt{5})$ and $(0, -\sqrt{5})$. This curve is sketched in Figure 4. ■

EXAMPLE 4 ***Finding the Equation of an Ellipse Given Its Foci and Minor Axis***

Find the equation of the ellipse with foci at $(-5, 0)$ and $(5, 0)$ and whose minor axis is the line segment extending from $(0, -3)$ to $(0, 3)$.

SOLUTION We have $c = 5$ and $b = 3$ so that

$$a^2 = b^2 + c^2 = 34$$

and the equation of the ellipse is

$$\frac{x^2}{34} + \frac{y^2}{9} = 1$$

It is sketched in Figure 5.

Definition of Eccentricity

The **eccentricity** e of an ellipse is defined by

$$e = \frac{c}{a} \quad \text{if} \quad a \geq b \tag{5}$$

and

$$e = \frac{c}{b} \quad \text{if} \quad b \geq a \tag{6}$$

NOTE If $a \geq b$, then the length of the major axis is $2a$; if $b \geq a$, the length of the major axis is $2b$. In both cases the distance between the foci is $2c$. Since $2c/2a = c/a$ and $2c/2b = c/b$, we have

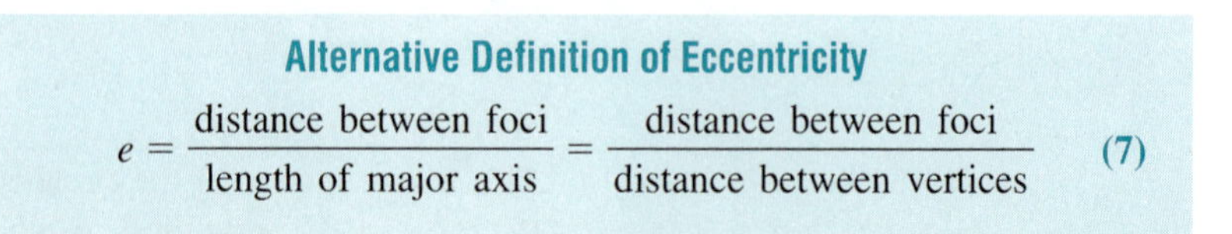

Alternative Definition of Eccentricity

$$e = \frac{\text{distance between foci}}{\text{length of major axis}} = \frac{\text{distance between foci}}{\text{distance between vertices}} \tag{7}$$

The eccentricity of an ellipse is a measure of the shape of the ellipse and is always a number in the interval $[0, 1]$. If $e = 0$, then the ellipse is a circle, since in that case $c = 0$ so that $a^2 = b^2$, and the foci now coincide and are the center of the circle. As e approaches 1, the ellipse becomes progressively flatter and approaches the major axis: the straight-line segment from $(-a, 0)$ to $(a, 0)$ if $a > b$, and from $(0, -b)$ to $(0, b)$ if $b > a$. In general,

The larger the eccentricity, the flatter the ellipse.

In Example 2, $e = 3/5 = 0.6$. In Example 3, $c^2 = b^2 - a^2 = 5$ so that $e = \sqrt{5}/3 \approx 0.74536$. In Example 4, $e = 5/\sqrt{34} \approx 0.85749$.

In Figure 6, we draw four different ellipses showing how the ellipses get flatter as the eccentricity increases.

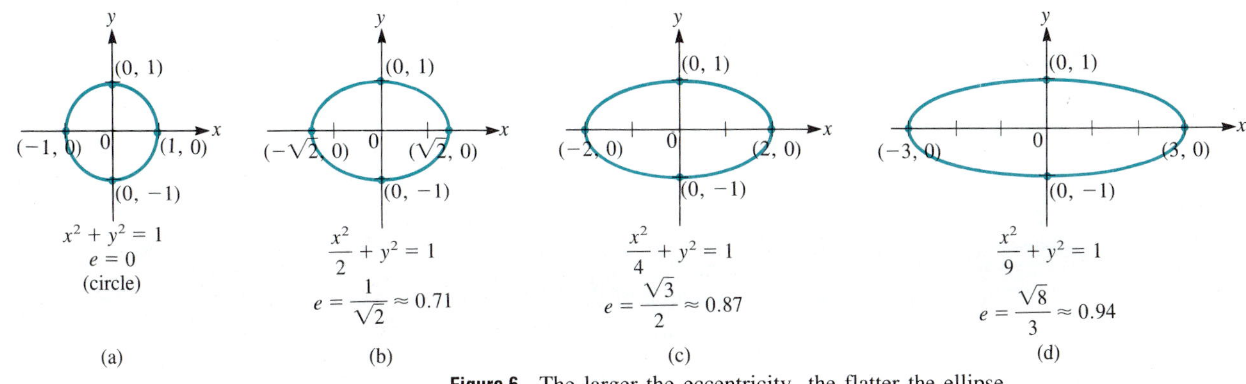

Figure 6 The larger the eccentricity, the flatter the ellipse.

Translation of Axes

We now turn to a different question. What happens if the center of the ellipse $(x^2/a^2) + (y^2/b^2) = 1$ is shifted from $(0, 0)$ to a new point (x_0, y_0) while the major and minor axes remain parallel to the x-axis and y-axis? Consider the equation

$$\frac{(x - x_0)^2}{a^2} + \frac{(y - y_0)^2}{b^2} = 1 \tag{8}$$

If we define two new variables by

$$x' = x - x_0 \quad \text{and} \quad y' = y - y_0 \tag{9}$$

then (8) becomes

$$\frac{(x')^2}{a^2} + \frac{(y')^2}{b^2} = 1 \tag{10}$$

This equation is the equation of an ellipse centered at the origin in the new coordinate system (x', y'). But $(x', y') = (0, 0)$ implies $(x - x_0, y - y_0) = (0, 0)$, or $x = x_0$ and $y = y_0$. That is, equation (8) is the equation of the ''shifted'' ellipse. See Figure 7. We have performed what is called a **translation of axes.** That is, we moved (or **translated**) the x- and y-axes to new positions so that they intersect at the point (x_0, y_0). If the original ellipse had its foci at $(-c, 0)$ and $(c, 0)$, then the translated ellipse has its foci at $(-c + x_0, y_0)$ and $(c + x_0, y_0)$. If the original ellipse had its foci at $(0, -c)$ and $(0, c)$, then the translated ellipse has its foci at $(x_0, -c + y_0)$, $(x_0, c + y_0)$.

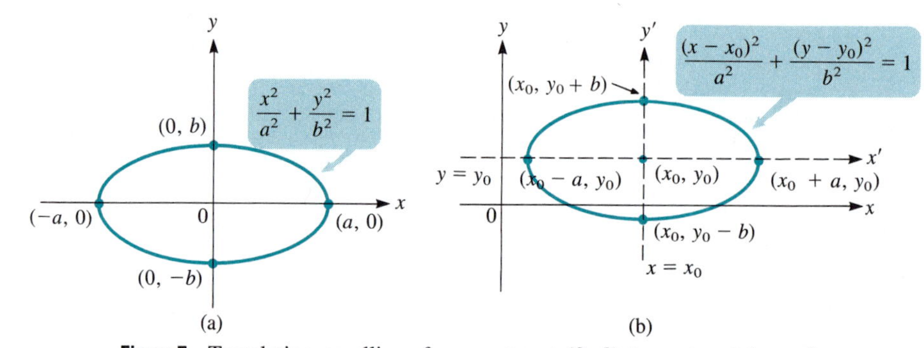

Figure 7 Translating an ellipse from center at $(0, 0)$ to center at (x_0, y_0).

EXAMPLE 5 *Finding the Equation of a Translated Ellipse*

Find the equation of the ellipse centered at $(4, -2)$ with foci at $(1, -2)$ and $(7, -2)$ and minor axis joining the points $(4, 0)$ and $(4, -4)$.

SOLUTION The points are sketched in Figure 8a. Here $c = (7 - 1)/2 = 3$, $b = [0 - (-4)]/2 = 2$, and $a^2 = b^2 + c^2 = 13$. Also, $(x_0, y_0) = (4, -2)$ so $x - x_0 = x - 4$ and $y - y_0 = y - (-2) = y + 2$. Thus, from (8), we have

$$\frac{(x - 4)^2}{13} + \frac{(y + 2)^2}{4} = 1$$

This ellipse is sketched in Figure 8b. Its major axis is on the line $y = -2$, and its minor axis is on the line $x = 4$. Its eccentricity is $3/\sqrt{13} \approx 0.83205$.

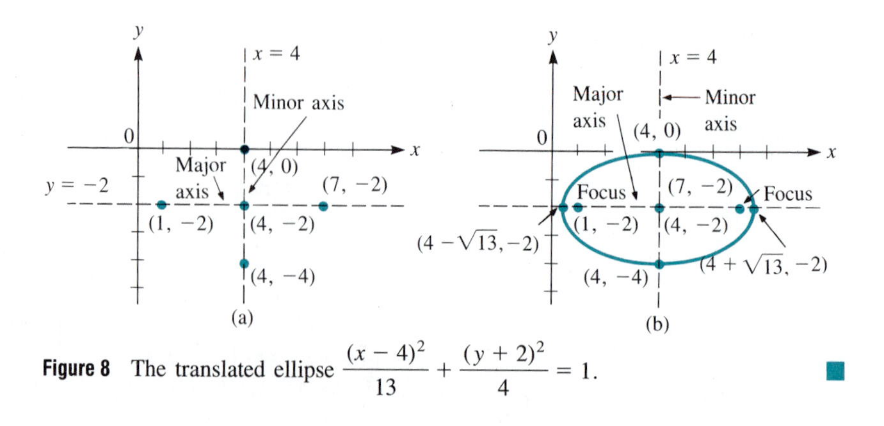

Figure 8 The translated ellipse $\frac{(x - 4)^2}{13} + \frac{(y + 2)^2}{4} = 1$. ■

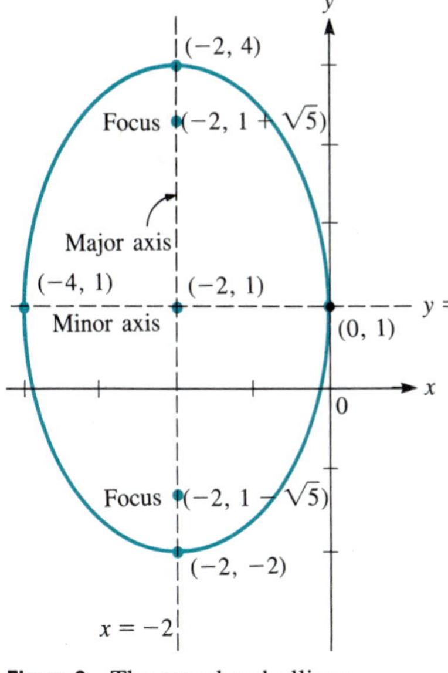

Figure 9 The translated ellipse $\frac{(x+2)^2}{4} + \frac{(y-1)^2}{9} = 1$.

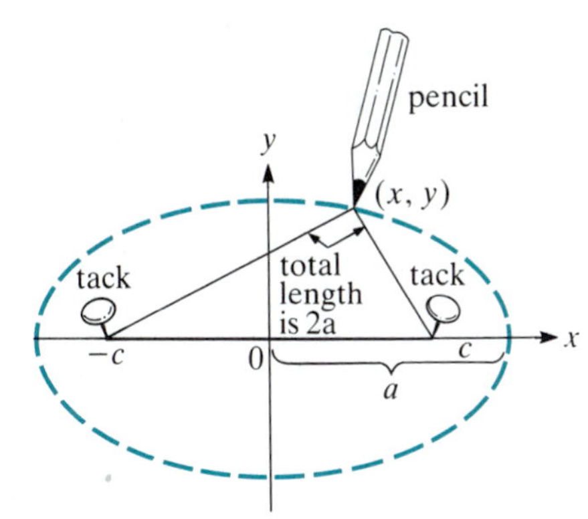

Figure 10 Drawing an ellipse.

EXAMPLE 6 ***Showing That a Second-Degree Equation Is the Equation of a Translated Ellipse by Completing Two Squares***

Discuss the curve given by

$$9x^2 + 36x + 4y^2 - 8y + 4 = 0$$

SOLUTION We write this expression as

$$9(x^2 + 4x) + 4(y^2 - 2y) = -4$$

Then after completing the squares, we obtain

$$9(\underbrace{x^2 + 4x + 4}_{\text{Added}}) \underbrace{- 36}_{\text{Subtracted } 9 \cdot 4} + 4(\underbrace{y^2 - 2y + 1}_{\text{Added}}) \underbrace{- 4}_{\text{Subtracted } 4 \cdot 1} = -4$$

or

$$9(x + 2)^2 + 4(y - 1)^2 = 36$$

and, dividing both sides by 36,

$$\frac{(x+2)^2}{4} + \frac{(y-1)^2}{9} = 1$$

Since $x + 2 = x - x_0$ and $y - 1 = y - y_0$, $x_0 = -2$ and $y_0 = 1$. Thus this is the equation of an ellipse centered at $(-2, 1)$. Here $a = 2$, $b = 3$, and $c = \sqrt{b^2 - a^2} = \sqrt{5}$. The foci are therefore at $(-2, -\sqrt{5} + 1)$ and $(-2, \sqrt{5} + 1)$. The major axis is on the line $x = -2$, and the minor axis is on the line $y = 1$. The eccentricity is $\sqrt{5}/3 \approx 0.74536$. The ellipse is sketched in Figure 9.

How to Draw an Ellipse

Pick two positive numbers a and c with $a > c$. Place two tacks $2c$ units apart as in Figure 10, and attach a string of length $2a$ to the tacks. Then proceed as in the figure to obtain the ellipse whose equation is $\frac{x^2}{a^2} + \frac{y^2}{a^2 - c^2} = 1$.

FOCUS ON
The Ellipse in the Real World

Ellipses are all around you. To see one, take a circular glass of water (or any other liquid) and tilt it. The surface of the liquid now forms an ellipse. As another common example, hold a spherical ball in front of a light. The ball will cast an elliptical shadow (see Figure 11).

Ellipses are very important in astronomy. In 1609, the German astronomer Johannes Kepler (1571–1630) discovered that each planet follows an elliptical path with the sun at one focus. Except for Mercury and Pluto, the nearest and farthest planets

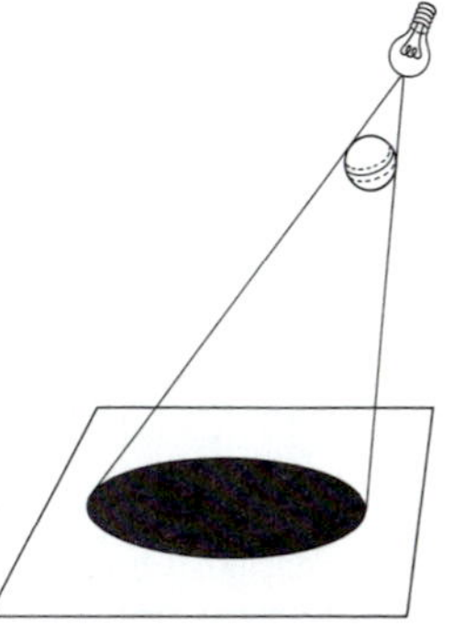

Figure 11 A ball casts an elliptical shadow.

from the sun, the orbits are nearly circular; that is, the eccentricities are close to 0. The table below gives the eccentricities of the nine planets.

Planet	Eccentricity, e
Mercury	0.2056
Venus	0.0068
Earth	0.0167
Mars	0.0934
Jupiter	0.0484
Saturn	0.0543
Uranus	0.0460
Neptune	0.0082
Pluto	0.2481

EXAMPLE 7

The comet Katounek orbits the sun in a very flat elliptical path. If 1 AU (astronomical unit) denotes the distance between the earth and the sun (1 AU $\approx$ 93 million miles), then the length of the minor axis of Katounek's orbit is 44 AU, while the length of the major axis is 3600 AU. Find the eccentricity of Katounek's orbit.

SOLUTION If the major axis is on the x-axis and the minor axis is on the y-axis, then

$$2a = 3600 \text{ AU}, 2b = 44 \text{ AU}, \qquad a = 1800 \text{ AU}, b = 22 \text{ AU}$$

and

$$c = \sqrt{a^2 - b^2} = \sqrt{(1800)^2 - (22)^2} = \sqrt{3{,}239{,}516} = 1799.865551$$

Then

$$e = \frac{c}{a} = \frac{1799.865551}{1800} \approx 0.999925$$

In Figure 12, we depict the very flat orbit of Katounek and use the orbit of Pluto as a reference.

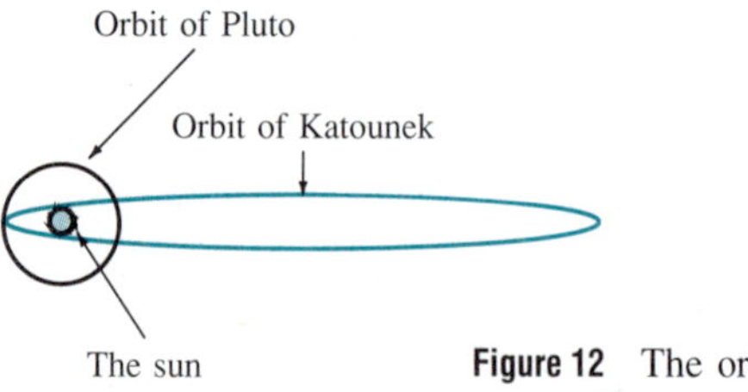

Figure 12 The orbit of Katounek.

The ellipse is used to design a more effective bicycle gear for racers. See Figure 13. The gear is designed to respond to the natural strengths and weaknesses of the racer's legs. At the top and bottom of the powerstroke, where the legs have the least leverage, the gear offers less resistance, but as the gear rotates, the resistance increases. This allows the legs to apply more power where it is most naturally available.

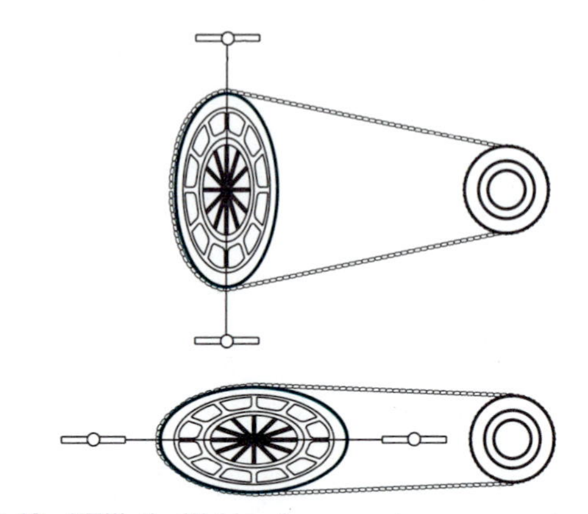

Figure 13 "Elliptical" bicycle gears give racers a significant advantage.

The Reflected-Wave Property of Ellipses and the Whispering-Gallery Effect

Let P be a point on an ellipse that is not a vertex (see Figure 14).

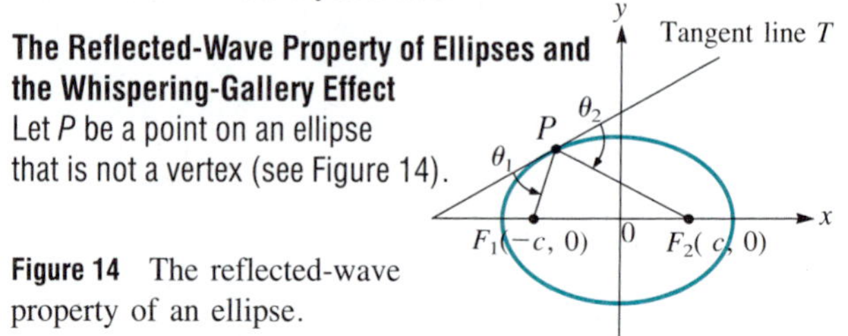

Figure 14 The reflected-wave property of an ellipse.

We draw the tangent line T that passes through P and draw lines joining P to each of the two foci of the ellipse. Let θ_1 denote the angle between T and PF_1, and let θ_2 denote the angle between T and PF_2. Then $\theta_1 = \theta_2$. This property is called the **reflected-wave property** of ellipses.

The reflected-wave property has been used to good effect by both Renaissance and modern architects. If the upper half of the ellipse in Figure 14 is rotated around the x-axis, then it will form a dome. In any room with an elliptical domed ceiling, a sound made at one focus will be reflected to the other focus where it will be heard very clearly. This phenomenon is known as the **whispering-gallery effect.** Some of the most famous rooms that exhibit the whispering-gallery effect include St. Paul's Cathedral in London (designed by the most famous of British architects, Sir Christopher Wren), the Caryatids room in the Louvre Museum in Paris, and the National Statuary Hall at the U.S. Capitol (the original House of Representatives) in Washington, D.C.

For more details on these and other interesting applications, see the excellent article, "The Standup Conic Presents: The Ellipse and Applications" by Lee Whitt in *The UMAP Journal,* Vol. 4, No. 2, 1983, pp. 157–186.

Problems 5.2

Readiness Check

I. Of the following, ______ is *not* an equation for an ellipse.

a. $x^2 + y^2 = 36$ b. $\left(\frac{x}{3}\right)^2 + \left(\frac{y}{4}\right)^2 = 1$

c. $16x^2 + 9y = 144$ d. $16x^2 + \frac{y^2}{9} = 12$

II. Among the ellipses satisfying the following equations, the one satisfying ______ is *not* a translation of the others.

a. $\frac{x^2}{25} + \frac{y^2}{16} = 1$

b. $25(x-1)^2 + 16(y-3)^2 = 400$

c. $\left(\frac{x}{5}\right)^2 + \frac{(y+3)^2}{16} = 1$

d. $16(x+7)^2 + 25(y-8)^2 = 16 \cdot 25$

III. The ellipse with foci $(-6, 0)$ and $(6, 0)$ and vertices $(-10, 0)$ and $(10, 0)$ can be drawn by attaching the ends of a string, ______ units in length, to thumbtacks at the foci and then tracing the outline with a pencil that moves in such a way as to keep the string taut.

a. 24 b. 20 c. 16 d. 12

IV. The ellipse with foci at $(-3, 0)$ and $(3, 0)$ and vertices $(-5, 0)$ and $(5, 0)$ has equation ______.

a. $\left(\frac{x}{5}\right)^2 + \left(\frac{y}{4}\right)^2 = 1$ b. $\left(\frac{x}{3}\right)^2 + \left(\frac{y}{4}\right)^2 = 1$

c. $\left(\frac{x}{3}\right)^2 + \left(\frac{y}{5}\right)^2 = 1$ d. $\left(\frac{x}{4}\right)^2 + \left(\frac{y}{3}\right)^2 = 1$

V. Which of the following is graphed below?

a. $x^2 + 9y^2 + 8x - 36y + 43 = 0$

b. $x^2 + 9y^2 - 8x + 36y + 43 = 0$

c. $9x^2 + y^2 - 72x + 4y - 139 = 0$

d. $9x^2 + y^2 + 72x - 4y - 139 = 0$

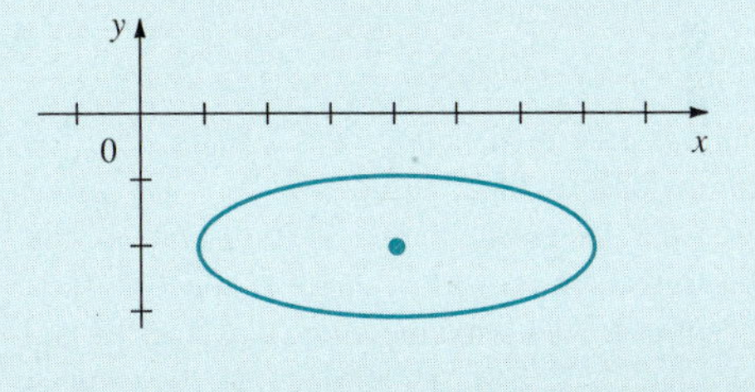

In Problems 1–18 the equation of an ellipse (or circle) is given. Find its center, foci, vertices, major and minor axes, and eccentricity. Then sketch it.

1. $\frac{x^2}{16} + \frac{y^2}{25} = 1$ 2. $\frac{x^2}{25} + \frac{y^2}{16} = 1$

3. $x^2 + \frac{y^2}{9} = 1$ 4. $\frac{x^2}{9} + y^2 = 1$

5. $x^2 + 4y^2 = 16$ 6. $4x^2 + y^2 = 16$

7. $\frac{(x-1)^2}{16} + \frac{(y+3)^2}{25} = 1$

8. $\frac{(x+3)^2}{25} + \frac{(y-1)^2}{16} = 1$

9. $2x^2 + 2y^2 = 2$ 10. $4x^2 + y^2 = 9$

11. $x^2 + 4y^2 = 9$

12. $4(x-3)^2 + (y-7)^2 = 9$

13. $4x^2 + 8x + y^2 + 6y = 3$

14. $x^2 + 6x + 4y^2 + 8y = 3$

15. $4x^2 + 8x + y^2 - 6y = 3$

16. $x^2 + 2x + y^2 + 2y = 7$

17. $3x^2 + 12x + 8y^2 - 4y = 20$

18. $2x^2 - 3x + 4y^2 + 5y = 37$

19. Find the equation of an ellipse with foci at $(0, 4)$ and $(0, -4)$ and vertices at $(0, 5)$ and $(0, -5)$.

20. Find the equation of the ellipse with vertices at $(2, 0)$ and $(-2, 0)$ and eccentricity 0.8.

* 21. Find the equation of an ellipse with center at $(-1, 4)$ that is a translation of the ellipse of Problem 19.

* 22. Find the equation of the "ellipse" centered at the origin that is symmetric with respect to both the x- and y-axes and that passes through the points $(1, 2)$ and $(-1, -4)$.

23. Show that the graph of the equation $x^2 + 2x + 2y^2 + 12y = c$ is:
(a) An ellipse if $c > -19$.
(b) A single point if $c = -19$.
(c) Empty if $c < -19$.

* 24. Find conditions on the numbers a, b, and c in order that the graph of the equation $x^2 + ax + 2y^2 + by = c$ be (a) an ellipse, (b) a single point, (c) empty.

Answers to Readiness Check

I. c II. b III. b IV. a V. b

Use a calculator to solve Problems 25–29.

25. The orbit of Halley's Comet is an ellipse with major axis approximately 36.2 AU and minor axis approximately 9.1 AU. Find the eccentricity of the orbit.

* 26. The major axis of the earth's orbit is approximately 185.5 million miles, and the eccentricity of the orbit is 0.0167. Find the largest and smallest distances between the earth and the sun. (The closest and farthest positions of the planet from the sun are called the **perihelion** and **aphelion,** respectively.) [Hint: Remember that the sun is at one focus of the orbit.]

27. A body orbits around the sun with major axis $2a$ measured in astronomical units (AU). Let T, measured in years, denote the period of the orbit. Then, according to **Kepler's third law,**

$$T^2 = a^3$$

If an asteroid has an orbital period of 8.4 years, find the length of the major axis of its orbit.

28. Kepler showed that the perihelion (closest) distance of the planet's orbit to the sun is $a(1 - e)$ and its aphelion (farthest) distance is $a(1 + e)$. Find these two distances for the asteroid of Problem 27 if the eccentricity of the orbit is 0.058.

29. The roof of a six-lane highway tunnel is constructed in the form of elliptical arches. Each of the six car lanes is 14 feet wide. Using the measurements in the figure, determine the vertical clearance in each lane; that is, determine how tall a truck can drive through without hitting any part of the roof. [Hint: First find the equation of the ellipse.]

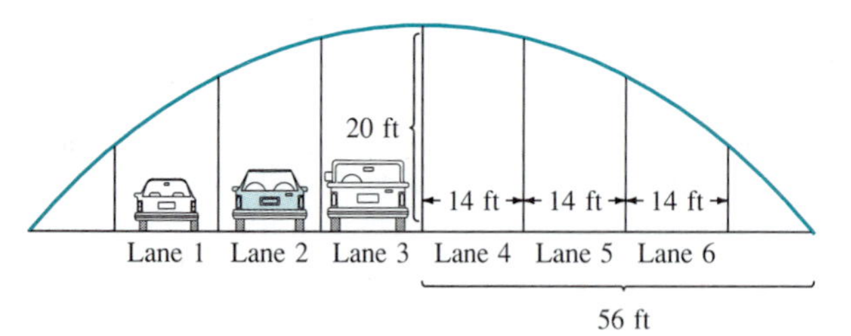

Eight ellipses are sketched in the right-hand column. Match each sketch with the equations given in Problems 30–37. Find the unmarked foci and vertices of each ellipse.

30. $(x - 3)^2 + 16(y - 2)^2 = 16$
31. $4x^2 + 25y^2 = 100$
32. $25x^2 + 21y^2 = 525$
33. $9(x + 2)^2 + 25(y - 1)^2 = 225$
34. $16(x + 2)^2 + 4(y + 1)^2 = 64$
35. $13x^2 + 4y^2 = 52$
36. $9x^2 + 16y^2 = 144$
37. $9(x - 3)^2 + 4(y - 1)^2 = 36$

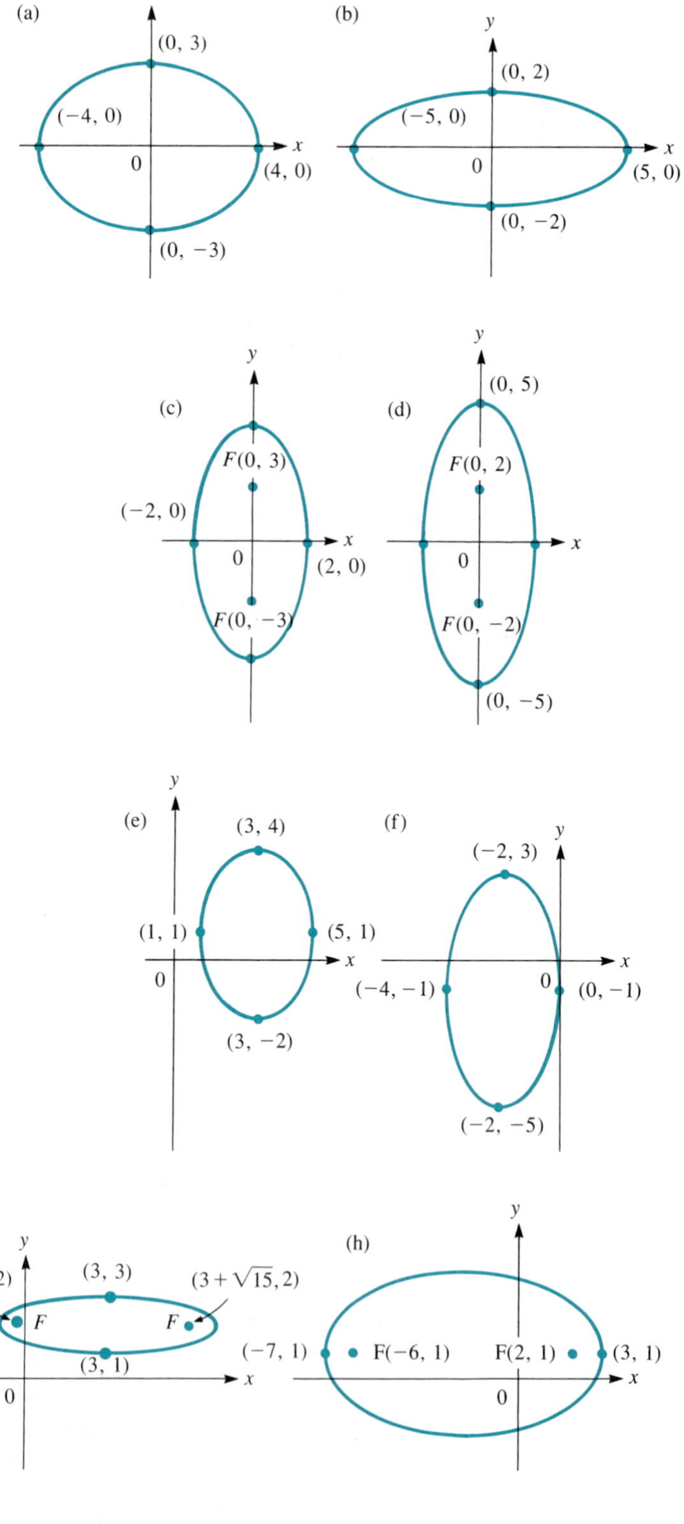

Graphing Calculator Problems

In Problems 38–47 obtain the graph of each ellipse on a graphing calculator. Before starting, read the material in Appendix A dealing with graphing conic sections. In particular, read Example 11 in that appendix.

38. $\dfrac{x^2}{10} + \dfrac{y^2}{37} = 1$

39. $\dfrac{x^2}{50} + \dfrac{y^2}{23} = 1$

40. $3x^2 + 8y^2 = 5$

41. $17x^2 + 10y^2 = 85$

42. $\dfrac{(x+3)^2}{20} + \dfrac{(y+1)^2}{30} = 1$

43. $\dfrac{3(x+4)^2}{7} + \dfrac{5(y+7)^2}{12} = 1$

44. $x^2 + 4x + y^2 + 10y = 18$

45. $x^2 - 7x + y^2 + 9y + 3 = 0$

46. $2x^2 + 3y^2 - 6y = 2$

* 47. $14x^2 + 23x + 27y^2 - 33y = 57$

5.3 The Parabola

In Figure 1 on page 29 we graphed the parabola $y = x^2$. In this section, we discuss parabolas in more generality. We begin with a definition.

Definition of a Parabola

A **parabola** is the set of points (x, y) equidistant from a fixed point and a fixed line that does not contain the fixed point. The fixed point is called the **focus,** and the fixed line is called the **directrix.**

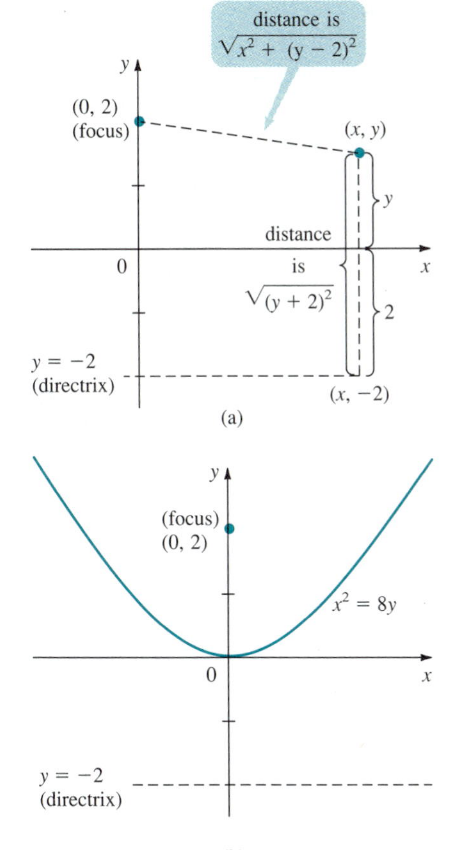

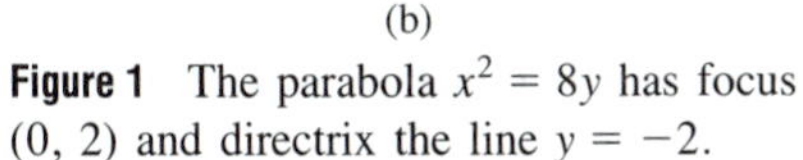

Figure 1 The parabola $x^2 = 8y$ has focus $(0, 2)$ and directrix the line $y = -2$.

EXAMPLE 1 ***Finding an Equation of a Parabola with Given Focus and Directrix***

Find an equation of the parabola whose focus is the point $(0, 2)$ and whose directrix is the line $y = -2$.

SOLUTION As in Figure 1(a), if (x, y) is a point on the parabola, then the distance from $(0, 2)$ to (x, y) is, from the distance formula (p. 10), equal to $\sqrt{(x-0)^2 + (y-2)^2}$. The distance between (x, y) and the line $y = -2$ is defined as the shortest distance from the point to the line. This is obtained by "dropping a perpendicular" from (x, y) to the line. Since the line $y = -2$ is horizontal, the perpendicular line will be vertical and will intersect $y = -2$ at the point $(x, -2)$. The distance between (x, y) and $(x, -2)$ is $\sqrt{(x-x)^2 + (y+2)^2} = \sqrt{(y+2)^2}$. Setting these two distances equal and squaring, we obtain

$$x^2 + (y-2)^2 = (y+2)^2$$

$$x^2 + y^2 - 4y + 4 = y^2 + 4y + 4$$

$$x^2 - 4y = 4y \qquad \text{Subtract } y^2 + 4 \text{ from both sides}$$

$$x^2 = 8y$$

This is an equation of the parabola we sought. Its vertex is at the origin. The parabola is sketched in Figure 1(b).

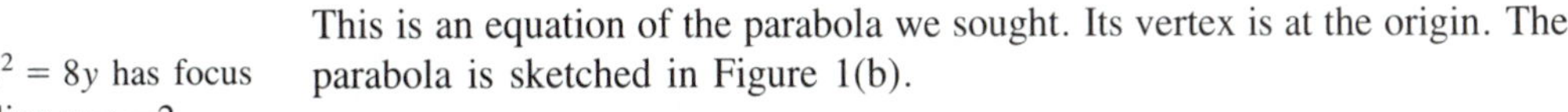

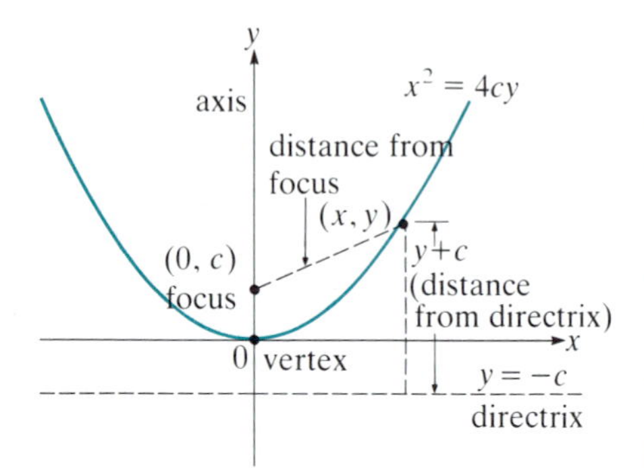

Figure 2 The parabola with focus $(0, c)$ and directrix the line $y = -c$.

We now calculate a more general equation of a parabola. We place the axes so that the focus is the point $(0, c)$ and the directrix is the line $y = -c$. (See Figure 2.) If $P = (x, y)$ is a point on the parabola, then, as in Example 1, we obtain (with c instead of 2)

$$\sqrt{x^2 + (y - c)^2} = \sqrt{(y + c)^2}$$

or squaring,

$$x^2 + (y - c)^2 = (y + c)^2$$

which reduces to

The Standard Equation of a Parabola

The standard equation of a parabola with vertex at the origin, focus at $(0, c)$, and directrix the line $y = -c$ is

$$x^2 = 4cy \tag{1}$$

The parabola given by (1) is symmetric about the y-axis. This line is called the **axis** of the parabola. Note that the axis contains the focus and is perpendicular to the directrix. The point at which the axis and the parabola intersect is called the **vertex.** The vertex is equidistant from the focus and the directrix.

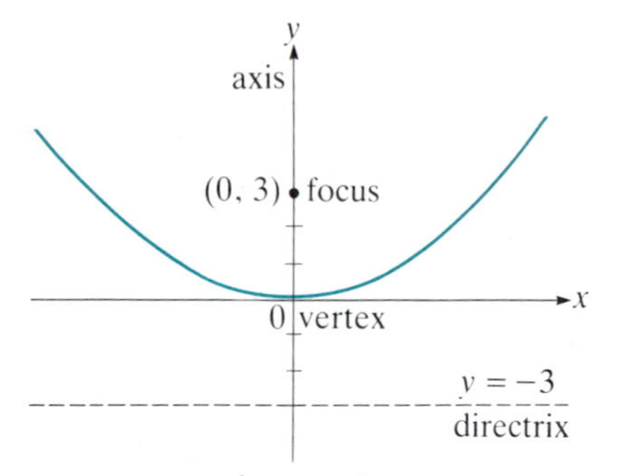

Figure 3 The parabola $x^2 = 12y$ has focus $(0, 3)$ and directrix the line $y = -3$.

EXAMPLE 2 *Sketching a Parabola*

Describe the parabola given by $x^2 = 12y$.

SOLUTION Here, as in equation (1), $4c = 12$, so that $c = 3$, the focus is the point $(0, 3)$, and the directrix is the line $y = -3$. The axis of the parabola is the y-axis, and the vertex is the origin. The curve is sketched in Figure 3. ■

EXAMPLE 3 *A Parabola That Opens Downward*

Describe the parabola given by $x^2 = -8y$.

SOLUTION Here $4c = -8$ so that $c = -2$, and the focus is $(0, -2)$, the directrix is the line $y = 2$, and the curve opens downward, as shown in Figure 4.

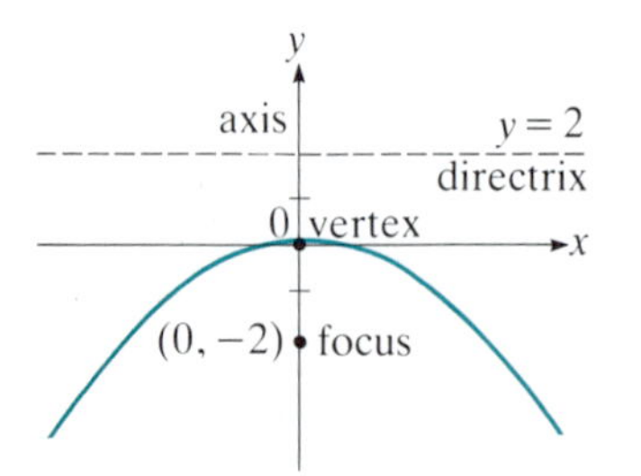

Figure 4 The parabola $x^2 = -8y$ opens downward.

In general,

Determining Whether a Parabola Opens Upward or Downward

The parabola described by $x^2 = 4cy$ opens upward if $c > 0$ and opens downward if $c < 0$.

As with the ellipse, we can exchange the role of x and y. We then obtain the following:

The Standard Equation of a Parabola

The standard equation of the parabola with vertex at the origin, focus at $(c, 0)$, and directrix the line $x = -c$ is

$$y^2 = 4cx \tag{2}$$

If $c > 0$, the parabola opens to the right; and if $c < 0$, the parabola opens to the left.

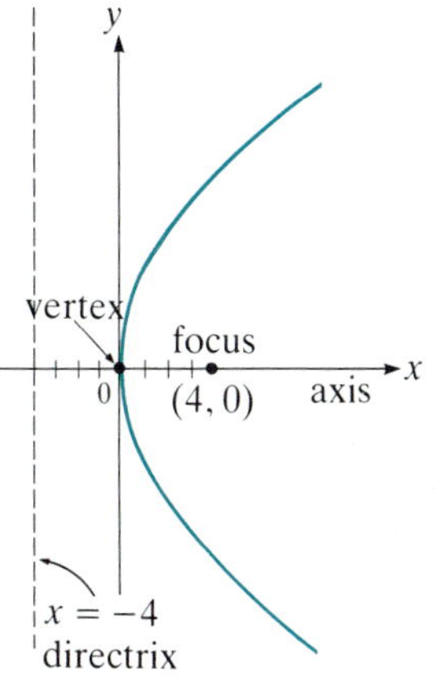

Figure 5 The parabola $y^2 = 16x$ opens to the right.

EXAMPLE 4 *A Parabola That Opens to the Right*

Describe the parabola $y^2 = 16x$.

SOLUTION Here $4c = 16$ so that $c = 4$, and the focus is $(4, 0)$, the directrix is the line $x = -4$, the axis is the x-axis (since the parabola is symmetric about the x-axis), and the vertex is the origin. The curve is sketched in Figure 5. Note that it opens to the right.

In Table 1 below we list the **standard parabolas.**

Table 1
Standard Parabolas with Vertex at the Origin

Equation	Description	Picture
$x^2 = 4cy$	Focus: $(0, c)$ Directrix: $y = -c$ Axis: y-axis Vertex: $(0, 0)$ Curve opens upward if $c > 0$ and downward if $c < 0$	
$y^2 = 4cx$	Focus: $(c, 0)$ Directrix: $x = -c$ Axis: x-axis Vertex: $(0, 0)$ Curve opens to the right if $c > 0$ and to the left if $c < 0$	
$x^2 = 0$	Degenerate parabola; graph is y-axis (one line)	
$y^2 = 0$	Degenerate parabola; graph is x-axis (one line)	
$x^2 = 1$	Degenerate parabola; graph consists of the two lines $x = 1$ and $x = -1$	
$y^2 = 1$	Degenerate parabola; graph consists of the two lines $y = 1$ and $y = -1$	

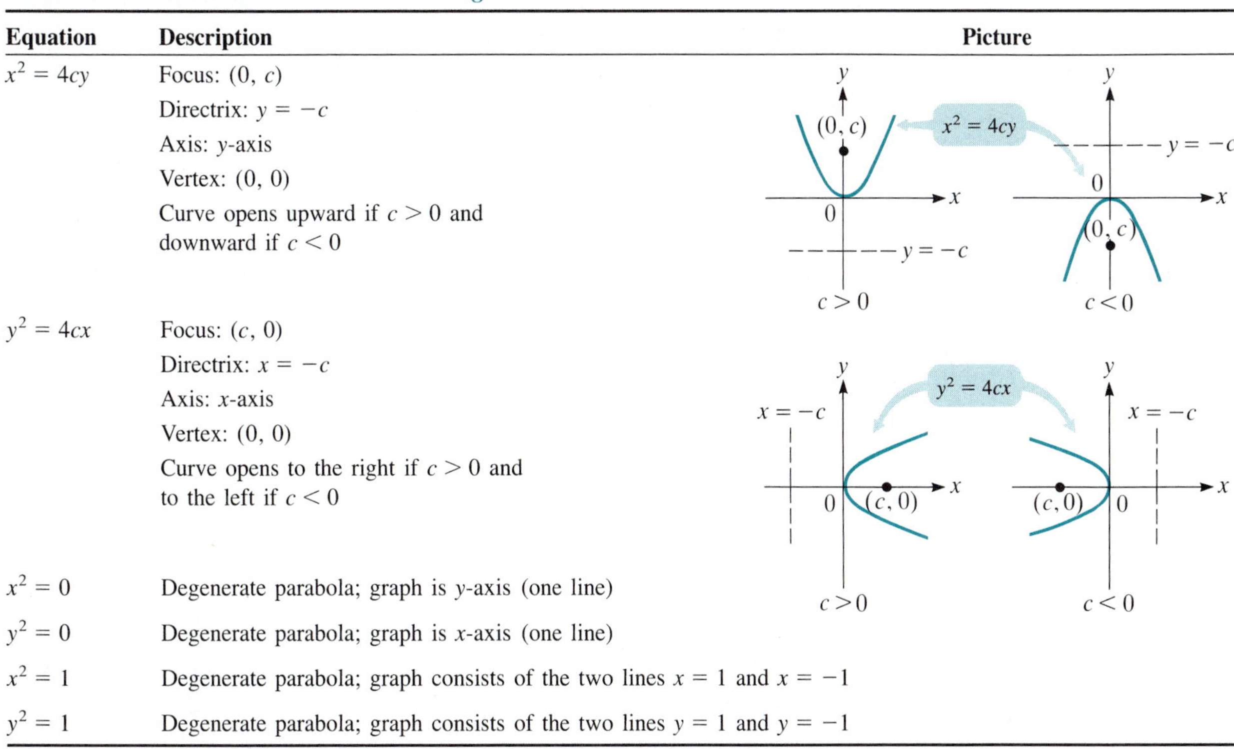

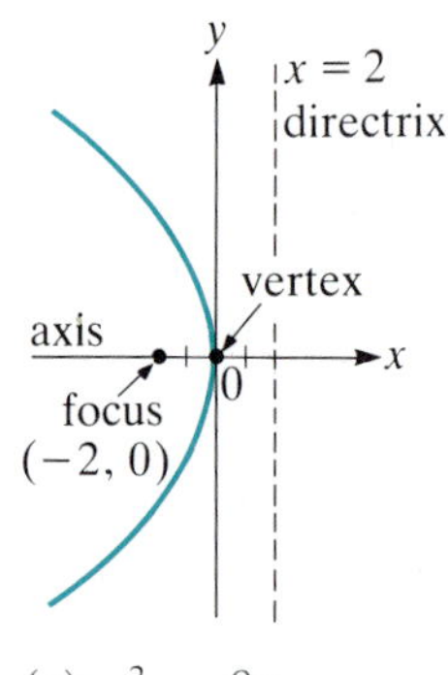

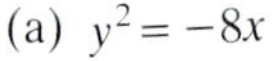
(a) $y^2 = -8x$

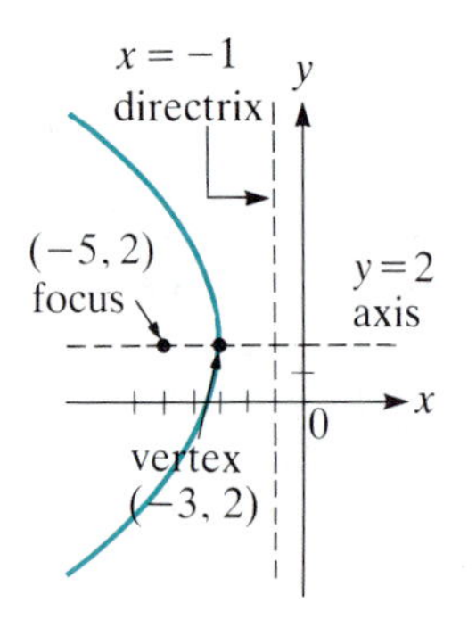

(b) $(y-2)^2 = -8(x+3)$

Figure 6 The parabola $(y - 2)^2 = -8(x + 3)$ is obtained by translating the parabola $y^2 = -8x$.

All the parabolas we have drawn so far have had their vertices at the origin. Other parabolas can be obtained by a simple translation of axes. The parabolas

$$(x - x_0)^2 = 4c(y - y_0) \tag{3}$$

and

$$(y - y_0)^2 = 4c(x - x_0) \tag{4}$$

have vertices at the point (x_0, y_0).

EXAMPLE 5 *A Translated Parabola*

Describe the parabola $(y - 2)^2 = -8(x + 3)$.

SOLUTION This parabola has its vertex at $(-3, 2)$. It is obtained by shifting the parabola $y^2 = -8x$ three units to the left and two units up (see Section 1.4). Since $4c = -8$, $c = -2$ and the focus and directrix of $y^2 = -8x$ are $(-2, 0)$ and $x = 2$. Hence after translation, the focus and directrix of $(y - 2)^2 = -8(x + 3)$ are $(-5, 2)$ and $x = -1$. The axis of this curve is $y = 2$. The two parabolas are sketched in Figure 6. ■

EXAMPLE 6 *Showing That a Second-Degree Equation Is the Equation of a Translated Parabola by First Completing the Square*

Describe the curve $x^2 - 4x + 2y + 10 = 0$.

SOLUTION We first complete the square:

$$x^2 - 4x + 2y + 10 = (x - 2)^2 - 4 + 2y + 10 = 0$$
$$(x - 2)^2 = -2y - 6$$
$$(x - 2)^2 = -2(y + 3)$$

This expression is the equation of a parabola with vertex at $(2, -3)$. Since $4c = -2$, $c = -\frac{1}{2}$, and the focus is $(2, -3 - \frac{1}{2}) = (2, -\frac{7}{2})$. The directrix is the line $y = -3 - (-\frac{1}{2}) = -\frac{5}{2}$, and the axis is the line $x = 2$. The curve is sketched in Figure 7.

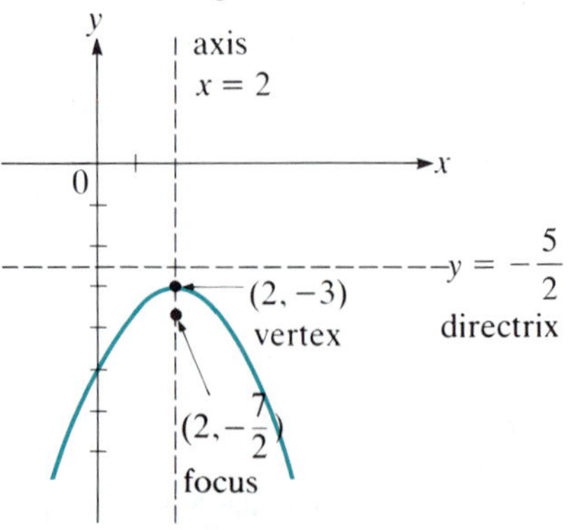

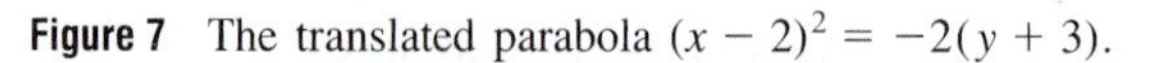
Figure 7 The translated parabola $(x - 2)^2 = -2(y + 3)$.

Translated Parabolas

The parabola $(x - x_0)^2 = 4c(y - y_0)$ has its vertex at (x_0, y_0) and opens upward if $c > 0$ or downward if $c < 0$.

The parabola $(y - y_0)^2 = 4c(x - x_0)$ has its vertex at (x_0, y_0) and opens to the right if $c > 0$ or to the left if $c < 0$.

The equation

$$Ax^2 + Dx + Ey + F = 0$$

can be written in the form (3) by first dividing by A and then completing the square.

The equation

$$Cy^2 + Dx + Ey + F = 0$$

can be written in the form (4) by first dividing by C and then completing the square.

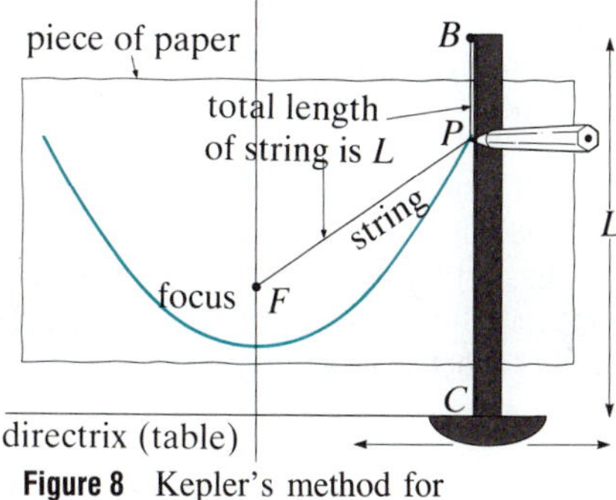

Figure 8 Kepler's method for drawing a parabola.

How to Draw a Parabola

Much of what we now know about the parabola was discovered by the great German physicist, astronomer, and mathematician Johannes Kepler (1571–1630). Kepler was the first to use the term *focus* (Latin for ''hearthside'') in the context of this section. He constructed parabolas using a table, a piece of string, a pencil, and the seventeenth-century version of a T-square. Place a piece of paper along a wall above the edge of a horizontal table as in Figure 8. The table's edge is the directrix of our parabola.

If the T-square has length L, then choose a string of length L. Pin one end to the focus F, and pin the other end to the top of the T-square. Slide a pencil along the string as in Figure 8, keeping the string taut. Then, as the T-square is moved along the side of the table, the pencil will trace out a parabola. The reason for this is that $\overline{FP} + \overline{PB} = \overline{CP} + \overline{PB} = L$ so that $\overline{FP} = \overline{CP}$. That is, the distance from a point P on the parabola to the focus equals the distance from the point to the directrix.

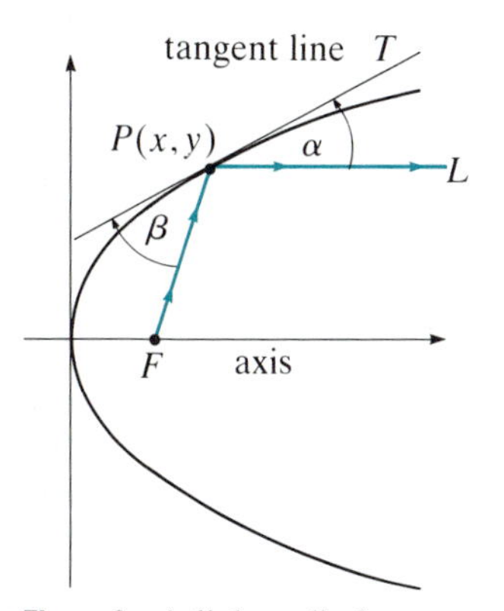

Figure 9 A light radio beam emanating from F is reflected parallel to the axis along the line L.

The Reflective Property of a Parabola

Consider the parabola with focus F sketched in Figure 9. Let $P(x, y)$ be a point on the parabola, T the line tangent to the parabola at P, and L the line passing through P that is parallel to the axis. Finally, let α denote the angle between T and L, and let β denote the angle between T and PF. Then $\alpha = \beta$. This means that a beam that starts at the focus F will be reflected off the parabola parallel to the axis. This **reflective property of the parabola,** as it is called, is useful in a wide variety of applications, as we will soon see.

The opposite effect is also seen: A beam coming in parallel to the axis will be reflected back to the focus. This can be illustrated in Figure 9 by reversing the arrows.

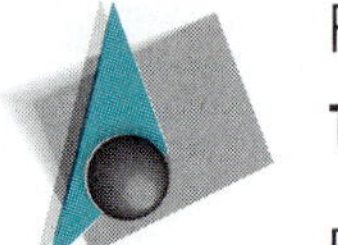

FOCUS ON
The Parabola in the Real World

Parabolas are all around us. The automobile headlight has the parabolic shape obtained by rotating a parabola around its axis. All the light emanating from a bulb placed at the focus is reflected parallel to the parabola's axis, that is, parallel to the ground. This follows from the reflective property just discussed (see Figure 10).

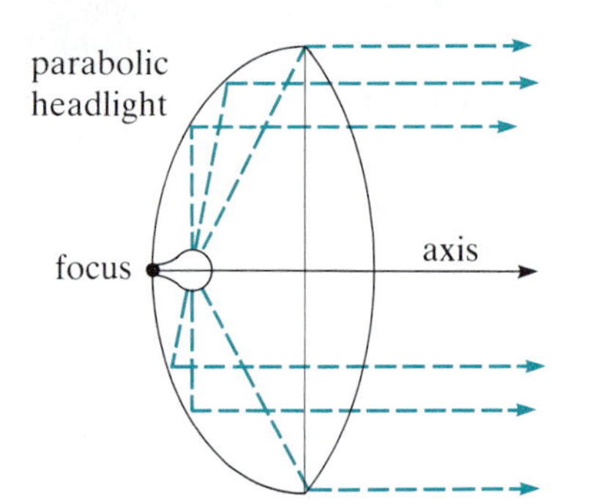

Figure 10 A parabolic headlight. Light rays from a bulb at the focus are reflected parallel to the axis.

Parabolic reflectors can be found in communications systems, electronic surveillance systems, radar systems, and telescopes. In radar, an electromagnetic beam is bounced off a target, and a parabolic reflector is used to collect and concentrate the deflected beam for signal processing. This works because the reflector reflects all returning signals back to the focus. In Figure 11, we show a parabolic reflector used on earth to track space probes.

Figure 11 Credit: Photo Researchers, Inc.

When a projectile is shot into the air, its path takes the shape of a parabola. This famous phenomenon was discovered by Galileo (1564–1642). Renaissance scientists were fascinated by this fact. The great artist Leonardo da Vinci drew the path of exploding mortar shells toward the end of the fifteenth century (Figure 12).

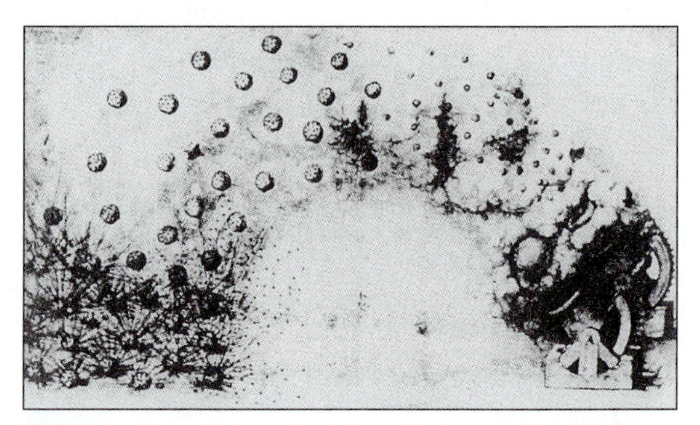

Figure 12 Cannon in action. The flight of exploding mortar shells, drawn by Leonardo da Vinci.

Sound waves can be transmitted effectively using parabolic reflectors. This is illustrated in Figure 13. One such "double reflector" device can be found at the Exploratorium in San Francisco. In this model, the reflectors have diameters of 8 feet and are placed facing each other about 50 feet apart.

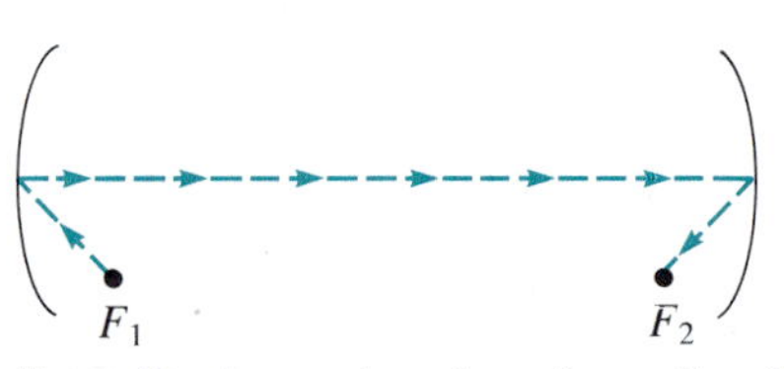

Figure 13 A sound made at focus F_1 will be reflected twice and clearly heard at focus F_2.

Parabolic reflectors also have been used in warfare and other unfortunate circumstances. The most famous example comes from Archimedes of Syracuse (287–212 B.C.), who seemed to have dabbled in everything. According to the Greek historian Plutarch, Syracuse was besieged by Romans led by their great general Marcellus. Archimedes helped save the city by designing, among other things, "burning" mirrors. These were parabolic mirrors capable of concentrating the rays of the sun onto attacking ships. Archimedes' clever devices also included a catapult that used the principle of the lever to hurl huge boulders. Because of Archimedes, Syracuse was able to hold out for nearly three years.

The story of the siege of Syracuse had an unhappy ending. After Marcellus failed to take Syracuse by a frontal siege, the city fell after a circuitous attack. A Roman soldier was sent by Marcellus to bring Archimedes to him. According to one version of the story, Archimedes had drawn a diagram in the sand and asked the soldier to stand away from it. This angered the soldier, who killed Archimedes with his spear.

The Greeks were fascinated by the parabola and sometimes suggested applications that were, to put it mildly, very unpleasant. The Greek Diocles, who lived in the second century B.C., wrote a book entitled *Burning Mirrors*. In his work, Diocles suggested that if victims were to be sacrificed in front of large crowds, parabolic mirrors could be used to provide a visible burning spot on the victim's body. It is not clear whether this idea was ever put into practice.

Parabolas are used in civil engineering. Bridges are built with twin parabolic cables that, ideally, will support a uniform horizontal load (see Figure 14).

These and many other applications of parabolas can be found in the fascinating paper, "The Standup Conic Presents: The Parabola and Applications," by Lee Whitt in *The UMAP Journal*, Vol. 3, No. 3, 1983, pp. 285–316.

Figure 14 In the ideal case, the main cable of a suspension bridge is parabolic. Credit: B.A. Lang Sr. Photo Researchers, Inc.

Problems 5.3

Readiness Check

I. The graph of $\frac{x}{-4} = \left(\frac{y}{3}\right)^2$ is a parabola opening ____.

a. to the right
b. to the left
c. upward
d. downward

II. The set of points $\{(a, b): 4a = -b^2\}$ is a ____.

a. vertical line
b. horizontal line
c. single point
d. parabola

III. Answer True or False to each of the following assertions about the parabola satisfying

$$y = -(x - 1)^2.$$

a. The focus is $(0, 0)$.
b. The vertex is $(0, 0)$.
c. The vertex is $(1, 0)$.
d. The focus is $(1, 4)$.
e. The focus is $(1, -\frac{1}{4})$.
f. The directrix passes through the vertex.
g. The directrix passes through the focus.
h. The directrix is perpendicular to the axis of the parabola.

In Problems 1–18 the equation of a parabola is given. Find its focus, directrix, axis, and vertex. Determine whether it opens up, down, right, or left. Then sketch it.

1. $x^2 = 16y$
2. $y^2 = 16x$
3. $x^2 = -16y$
4. $y^2 = -16x$
5. $2x^2 = 3y$
6. $2y^2 = 3x$
7. $4x^2 = -9y$
8. $7y^2 = -20x$
9. $(x - 1)^2 = -16(y + 3)$
10. $(y - 1)^2 = -16(x - 3)$
11. $x^2 + 4y = 9$
12. $(x + 1)^2 + 25y = 50$
13. $x^2 + 2x + y + 1 = 0$
14. $x + y - y^2 = 4$
15. $x^2 + 4x + y = 0$
16. $y^2 + 4y + x = 0$
17. $x^2 + 4x - y = 0$
18. $y^2 + 4y - x = 0$
19. Find the equation of the parabola with focus $(0, 4)$ and directrix the line $y = -4$.
20. Find the equation of the parabola with focus $(-3, 0)$ and directrix the line $x = 3$.

Answers to Readiness Check

I. b II. d III. a. False b. False c. True d. False e. True f. False g. False h. True

21. Find the equation of the parabola obtained when the parabola of Problem 20 is shifted so that its vertex is at the point $(-2, 5)$.

** 22. Find the equation of the parabola with vertex at $(1, 2)$ and directrix the line $x = y$.

23. The parabola in Problem 20 is translated so that its vertex is now at $(3, -1)$. Find its new focus and directrix.

24. The **latus rectum** of a parabola is the chord passing through the focus that is perpendicular to the axis. Compute the length of the latus rectum of the parabola $x^2 = 6y$.

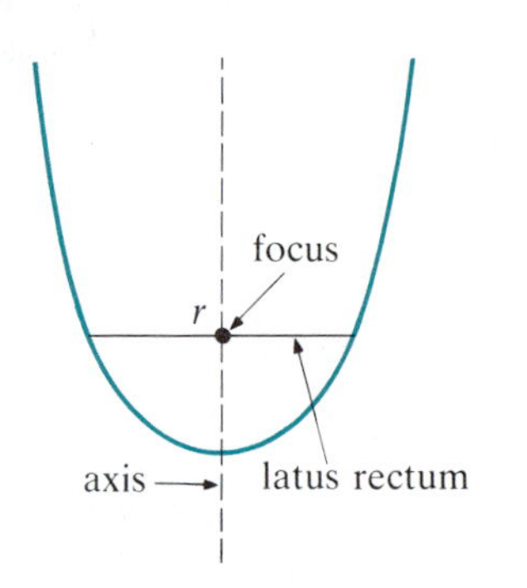

* 25. Show that if a parabola has the equation $x^2 = 4cy$ or $y^2 = 4cx$, then the length of its latus rectum is $|4c|$.

26. The tops of two towers of a suspension bridge (like the one in Figure 14) are 100 feet above water level and 375 feet apart. The lowest point of the parabolic cable connecting the two towers is 40 feet above the water. How high is a point on the cable that is 60 feet (horizontally) from one of the towers?

27. The receiver of a parabolic signal receptor is at the focus, which is 2 feet from the vertex. If the receptor is placed as in the figure, find an equation for the cross-sectional parabola that lies in the xy-plane.

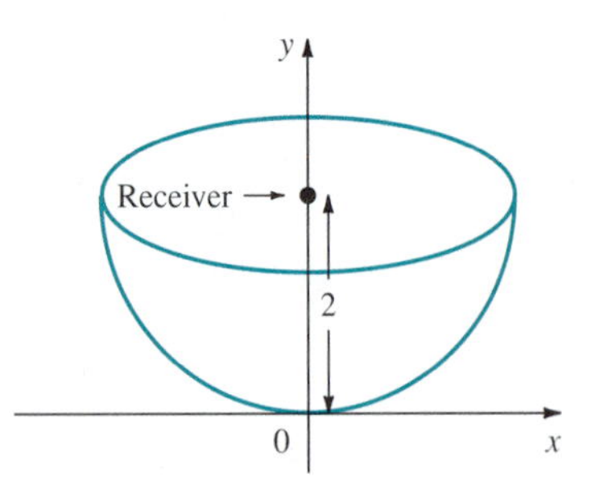

28. A missile is fired from ground level. Its path is a parabola opening downward. The missile reaches a height of 1200 meters and travels 15,000 meters (15 kilometers) horizontally. When the missile first reaches a height of 500 meters, how far is it (horizontally) from the firing site?

29. If an object is dropped from rest from an initial height of h_0 feet, then, after t seconds, its height, $h(t)$, is given by

$$h(t) = h_0 - 16t^2 \text{ feet}$$

(a) Graph this equation for $h_0 = 1000$ feet.
(b) When does the object hit the ground?

* 30. A bomber, flying at 350 miles per hour, releases a bomb from an altitude of 28,000 feet. How far does the bomb travel horizontally before it hits the ground?

The graphs of eight parabolas are given. Match the graphs with the equations given in Problems 31–38.

31. $x^2 + 10x + 3y + 13 = 0$
32. $x^2 - 4x - 5y - 11 = 0$
33. $4x - y^2 = 0$
34. $x^2 - 3y = 0$
35. $2x + y^2 = 0$
36. $y^2 + 4y + x = 0$
37. $y^2 - 2x - 4y + 6 = 0$
38. $x^2 + 4y = 0$

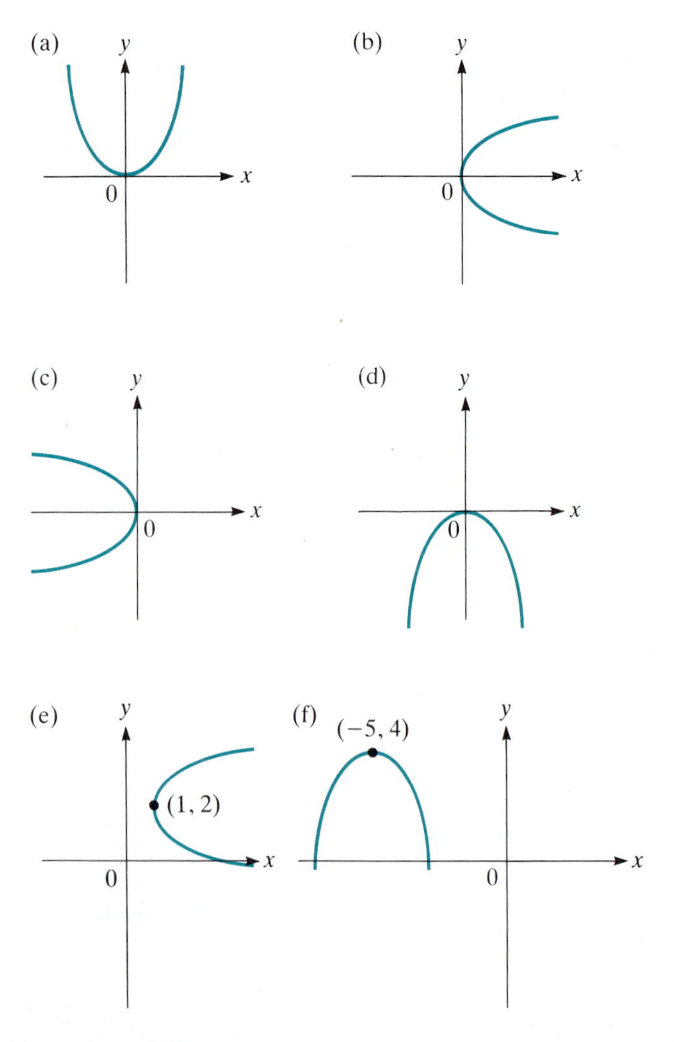

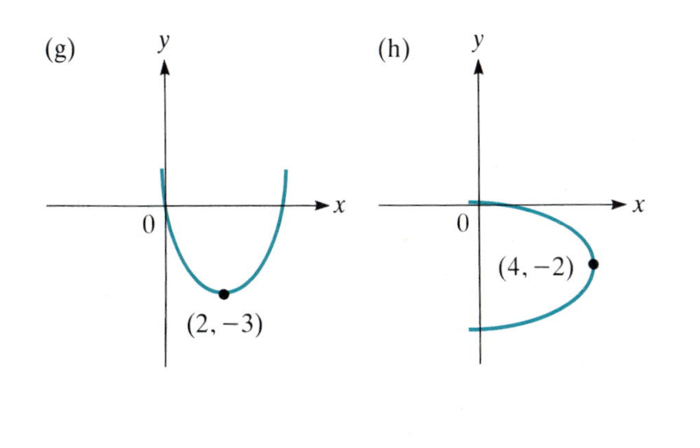

Graphing Calculator Problems

In Problems 39–48 obtain the graph of each parabola on a graphing calculator. Note that in order to graph the parabola $y^2 = 4cx$, it is necessary to graph *two* functions. Read Examples 9, 10, and 11 in Appendix A for more details.

39. $x^2 = 20y$
40. $y^2 = 20x$
41. $y^2 = -3x$
42. $x^2 = -4.7y$
43. $3x^2 = -17y$
44. $5y^2 = 23y$
45. $(x + 2)^2 = -5(y + 1)$
46. $(y - 3)^2 = 4(x + \frac{1}{2})$
47. $x^2 + 3x + 2y = 0$
48. $y^2 + 6y - 3x = 6$

5.4 The Hyperbola

Definition of a Hyperbola

A **hyperbola** is a set of points (x, y) with the property that the positive difference between the distances from (x, y), and each of two given (distinct) points is a constant. Each of the two given points is called a **focus** of the hyperbola.

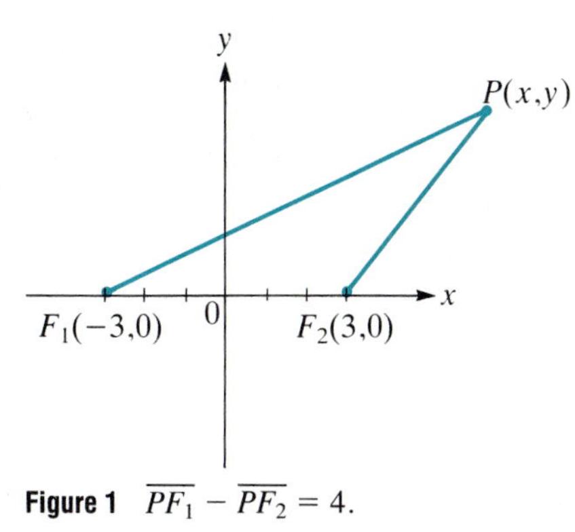

Figure 1 $\overline{PF_1} - \overline{PF_2} = 4.$

EXAMPLE 1 ***Finding the Equation of a Hyperbola with Given Foci***

Find an equation of the hyperbola whose foci are the points $(-3, 0)$ and $(3, 0)$ and in which the difference of the distances from a point (x, y) on the hyperbola to the foci is equal to 4.

SOLUTION If P is a point on the hyperbola and F_1 and F_2 denote the foci as in Figure 1, then the difference of the distances from P to the foci is

$$\overline{PF_1} - \overline{PF_2} = 4$$

Using the distance formula, we obtain

$$\sqrt{(x + 3)^2 + y^2} - \sqrt{(x - 3)^2 + y^2} = 4$$

$$\sqrt{(x + 3)^2 + y^2} = \sqrt{(x - 3)^2 + y^2} + 4$$

$$(x + 3)^2 + y^2 = (x - 3)^2 + y^2 + 8\sqrt{(x - 3)^2 + y^2} + 16$$

We squared both sides

$$x^2 + 6x + 9 + y^2 = x^2 - 6x + 9 + y^2 + 8\sqrt{(x - 3)^2 + y^2} + 16$$

We multiplied through

$$6x = -6x + 8\sqrt{(x - 3)^2 + y^2} + 16$$

We subtracted $x^2 + 9 + y^2$ from both sides

$$12x - 16 = 8\sqrt{(x - 3)^2 + y^2}$$

We rearranged terms

$$\frac{3}{2}x - 2 = \sqrt{(x-3)^2 + y^2} \qquad \text{We divided by 8}$$

$$\frac{9}{4}x^2 - 6x + 4 = (x-3)^2 + y^2 \qquad \text{We squared again}$$

$$\frac{9}{4}x^2 - 6x + 4 = x^2 - 6x + 9 + y^2$$

$$\left(\frac{9}{4} - 1\right)x^2 - y^2 = 5 \qquad \text{We combined terms}$$

$$\frac{5}{4}x^2 - y^2 = 5$$

$$\frac{x^2}{4} - \frac{y^2}{5} = 1 \qquad \text{We divided by 5} \qquad (1)$$

Equation (1) is the **standard equation** of the hyperbola. Here we assumed that $\overline{PF_1} > \overline{PF_2}$. If $\overline{PF_2} > \overline{PF_1}$, we obtain the same equation. You should verify this.

In order to help graph the hyperbola, we make several observations. First, from equation (1) we have

$$\frac{x^2}{4} = 1 + \frac{y^2}{5} \quad \text{or} \quad x^2 = 4 + \frac{4}{5}y^2$$

Since $y^2 \geq 0$ for all real numbers y, we have $x^2 \geq 4$ so that

$$x \geq 2 \quad \text{or} \quad x \leq -2$$

That is, no point on the hyperbola has an x-coordinate in the interval $(-2, 2)$. Second, if we replace x by $-x$ in (1), we obtain the same equation. This means that the graph is symmetric about the y-axis so that the hyperbola has two symmetric branches: one for $x > 0$ and one for $x < 0$. These correspond to the two cases $\overline{PF_1} - \overline{PF_2} = 4$ and $\overline{PF_2} - \overline{PF_1} = 4$.

We now solve (1) for y:

$$\frac{y^2}{5} = \frac{x^2}{4} - 1$$

$$y^2 = \frac{5}{4}x^2 - 5$$

$$y = \pm\sqrt{\frac{5}{4}x^2 - 5}$$

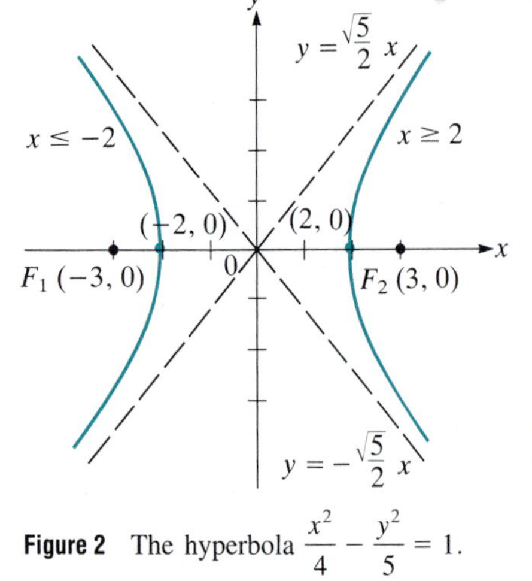

Figure 2 The hyperbola $\frac{x^2}{4} - \frac{y^2}{5} = 1$.

Suppose $|x|$ is large, then $\frac{5}{4}x^2 - 5 \approx \frac{5}{4}x^2$. To see this clearly, insert some numbers. For example, if $x = \pm 1000$, then $\frac{5}{4}x^2 - 5 = 1{,}249{,}995$ while $\frac{5}{4}x^2 = 1{,}250{,}000$. Thus, for x large

$$y \approx \pm\sqrt{\frac{5}{4}x^2} = \pm\frac{\sqrt{5}}{2}x$$

The lines $y = \frac{\sqrt{5}}{2}x$ and $y = -\frac{\sqrt{5}}{2}x$ are called **oblique asymptotes** for the hyperbola. Putting this all together, we obtain the graph in Figure 2.

To calculate the equation of a more general hyperbola, we place the axes so that the foci are the points $(c, 0)$ and $(-c, 0)$, and the difference of the distances from a point (x, y) on the hyperbola to the foci is equal to $2a > 0$. In Figure 3, we assume that $\overline{PF_2} > \overline{PF_1}$. Then

$$\text{difference of distances} = \overline{PF_2} - \overline{PF_1} = 2a$$

so

$$\overline{PF_2} = 2a + \overline{PF_1}$$

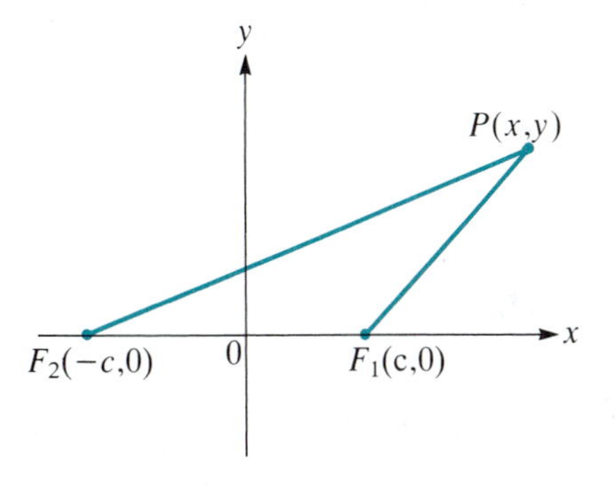

Figure 3 $\overline{PF_2} - \overline{PF_1} = 2a$.

But, as the shortest distance between two points is a straight line, we have

$$2a + \overline{PF_1} = \overline{PF_2} < \overline{PF_1} + \overline{F_1F_2} = \overline{PF_1} + 2c$$

Therefore,

$$2a + \overline{PF_1} < 2c + \overline{PF_1}$$

or

$$2a < 2c \quad \text{and} \quad c > a$$

If (x, y) is a point on the hyperbola,

$$\overline{PF_2} - \overline{PF_1} = 2a$$

or

$$\sqrt{(x + c)^2 + y^2} - \sqrt{(x - c)^2 + y^2} = 2a$$

(Again, we assumed that $\overline{PF_2} > \overline{PF_1}$ so that $\overline{PF_2} - \overline{PF_1}$ gives us a positive distance.) Then we obtain, successively,

$$\sqrt{(x + c)^2 + y^2} = \sqrt{(x - c)^2 + y^2} + 2a$$

$$(x + c)^2 + y^2 = (x - c)^2 + y^2 + 4a\sqrt{(x - c)^2 + y^2} + 4a^2 \quad \text{We squared}$$

$$x^2 + 2cx + c^2 + y^2 = x^2 - 2cx + c^2 + y^2 + 4a\sqrt{(x - c)^2 + y^2} + 4a^2$$

$$4cx - 4a^2 = 4a\sqrt{(x - c)^2 + y^2} \quad \text{We simplified}$$

$$\frac{c}{a}x - a = \sqrt{(x - c)^2 + y^2} \quad \text{We divided by } 4a$$

$$\frac{c^2}{a^2}x^2 - 2cx + a^2 = (x - c)^2 + y^2 \quad \text{We squared again}$$

$$\frac{c^2}{a^2}x^2 - 2cx + a^2 = x^2 - 2cx + c^2 + y^2$$

$$\left(\frac{c^2}{a^2} - 1\right)x^2 - y^2 = c^2 - a^2 \tag{2}$$

Finally, since $c > a > 0$, $c^2 - a^2 > 0$, and we can define the positive number b by

$$b^2 = c^2 - a^2$$

Then dividing both sides of (2) by $c^2 - a^2$, we obtain, since $\dfrac{c^2}{a^2} - 1 = \dfrac{c^2 - a^2}{a^2}$:

Standard Equation of a Hyperbola Centered at the Origin

$$\frac{x^2}{a^2} - \frac{y^2}{b^2} = 1 \tag{3}$$

A similar derivation shows that if we assume that $\overline{PF}_1 > \overline{PF}_2$, we also obtain equation (3).

These two cases correspond to the right and left branches of the hyperbola sketched in Figure 4. Note that the hyperbola given by (3) is symmetric about

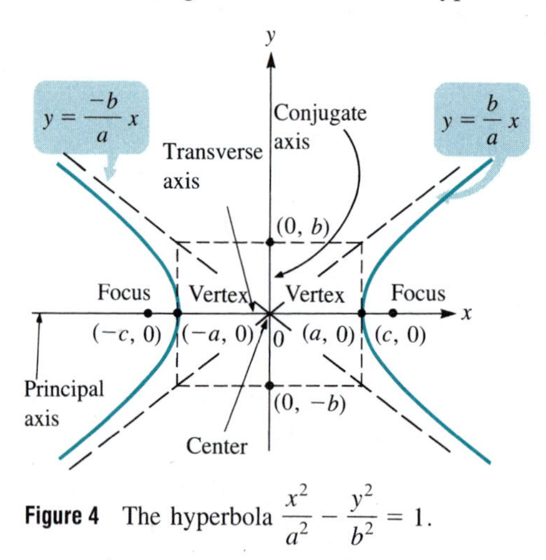

Figure 4 The hyperbola $\dfrac{x^2}{a^2} - \dfrac{y^2}{b^2} = 1$.

both the x-axis and the y-axis. The **principal axis** is the line containing the foci. The **vertices** of the parabola are the points of intersection of the hyperbola and its principal axis. The midpoint of the line segment joining the foci is called the **center** of the hyperbola. The **transverse axis** of the hyperbola is the line segment joining the vertices. The **conjugate axis** is the line joining the points $(0, -b)$ and $(0, b)$.

As in Example 1, we can write (3) as

$$y = \pm\sqrt{\frac{b^2}{a^2}x^2 - b^2} \approx \pm\sqrt{\frac{b^2}{a^2}x^2} = \pm\frac{b}{a}x \text{ for } |x| \text{ large}$$

Thus, the lines $y = \pm\dfrac{b}{a}x$ are oblique asymptotes to the hyperbola. To make it easier to sketch the hyperbola in Figure 4, we draw the rectangle with sides $x = \pm a$, $y = \pm b$. The lines that pass through the diagonals of this rectangle are the asymptotes of the hyperbola.

Finally, we note that the hyperbola given by (3) is called a hyperbola with **horizontal transverse axis.**

EXAMPLE 2 *Sketching a Hyperbola with a Horizontal Transverse Axis*

Discuss the curve given by $x^2 - 4y^2 = 9$.

SOLUTION Dividing by 9, we obtain $(x^2/9) - (4y^2/9) = 1$, or

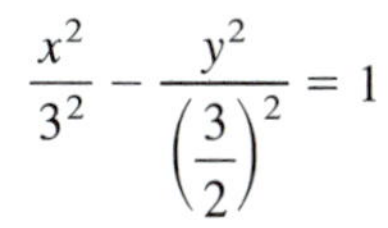

$$\frac{x^2}{3^2} - \frac{y^2}{\left(\frac{3}{2}\right)^2} = 1$$

This is the equation of a hyperbola with $a = 3$ and $b = \frac{3}{2}$. Then $c^2 = a^2 + b^2 = 9 + 9/4 = 45/4$ so that the foci are $(\sqrt{45}/2, 0)$ and $(-\sqrt{45}/2, 0)$. The vertices are $(3, 0)$ and $(-3, 0)$, and the asymptotes are $y = \pm\frac{1}{2}x$. The curve is sketched in Figure 5.

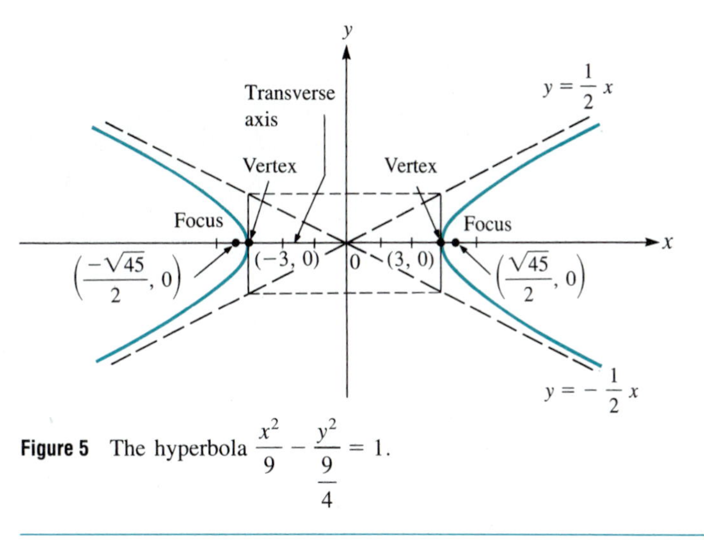

Figure 5 The hyperbola $\dfrac{x^2}{9} - \dfrac{y^2}{\frac{9}{4}} = 1$.

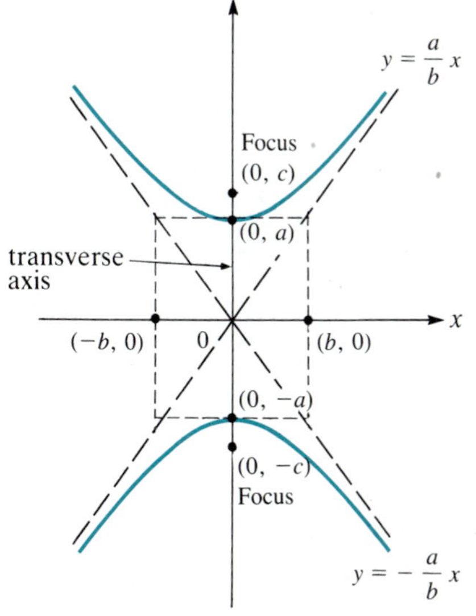

Figure 6 The hyperbola $\dfrac{y^2}{a^2} - \dfrac{x^2}{b^2} = 1$.

As with the ellipse and parabola, the roles of x and y can be reversed. The graph of the equation

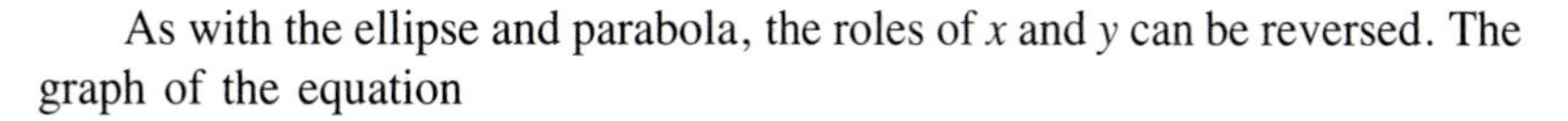

$$\frac{y^2}{a^2} - \frac{x^2}{b^2} = 1 \tag{4}$$

is sketched in Figure 6. In (4) the transverse axis is on the y-axis. From (2) we have $c^2 = a^2 + b^2$, and the foci are at $(0, c)$ and $(0, -c)$. The vertices of the hyperbola given by (4) are the points $(0, a)$ and $(0, -a)$. The asymptotes are the lines $y = \pm(a/b)x$. This is a hyperbola with a **vertical transverse axis.**

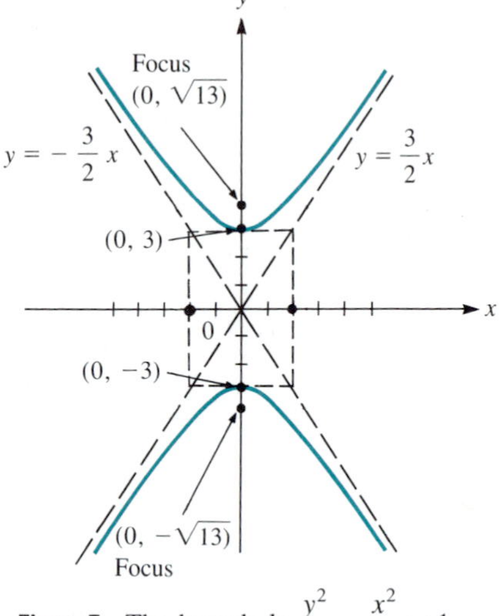

Figure 7 The hyperbola $\frac{y^2}{9} - \frac{x^2}{4} = 1$.

EXAMPLE 3 *A Hyperbola with a Vertical Transverse Axis*

Discuss the curve given by $4y^2 - 9x^2 = 36$.

SOLUTION Dividing by 36, we obtain

$$\frac{y^2}{9} - \frac{x^2}{4} = 1$$

Hence $a = 3$, $b = 2$, and $c^2 = 9 + 4 = 13$ so that $c = \sqrt{13}$. The foci are $(0, \sqrt{13})$ and $(0, -\sqrt{13})$, and the vertices are $(0, 3)$ and $(0, -3)$. The asymptotes are the lines $y = \frac{3}{2}x$ and $y = -\frac{3}{2}x$. This curve is sketched in Figure 7.

The hyperbolas we have sketched to this point are **standard hyperbolas.** Our results are summarized in Table 1.

Table 1
Standard Hyperbolas Centered at the Origin

Equations	Description	Picture
$\frac{x^2}{a^2} - \frac{y^2}{b^2} = 1$	Hyperbola with foci at $(c, 0)$ and $(-c, 0)$, where $c^2 = a^2 + b^2$; transverse axis is on the x-axis; center at origin; asymptotes $y = \pm(b/a)x$; curve opens to right and left; horizontal transverse axis	See Figure 4
$\frac{y^2}{a^2} - \frac{x^2}{b^2} = 1$	Hyperbola with foci at $(0, c)$ and $(0, -c)$, where $c^2 = a^2 + b^2$; transverse axis is on the y-axis; center at origin; asymptotes $y = \pm(a/b)x$; curve opens at top and bottom; vertical transverse axis	See Figure 6
$\frac{x^2}{a^2} - \frac{y^2}{b^2} = 0$	Degenerate hyperbola; graph consists of two lines: $y = \pm(b/a)x$	y; $y = \frac{b}{a}x$; x; 0; $y = -\frac{b}{a}x$

The hyperbolas we have so far discussed have had their centers at the origin. Other hyperbolas can be obtained by a translation of the axes. The hyperbola

$$\frac{(x - x_0)^2}{a^2} - \frac{(y - y_0)^2}{b^2} = 1 \qquad (5)$$

has its center at (x_0, y_0) and has a horizontal transverse axis.

The hyperbola

$$\frac{(y - y_0)^2}{a^2} - \frac{(x - x_0)^2}{b^2} = 1 \qquad (6)$$

has its center at (x_0, y_0) and has a vertical transverse axis.

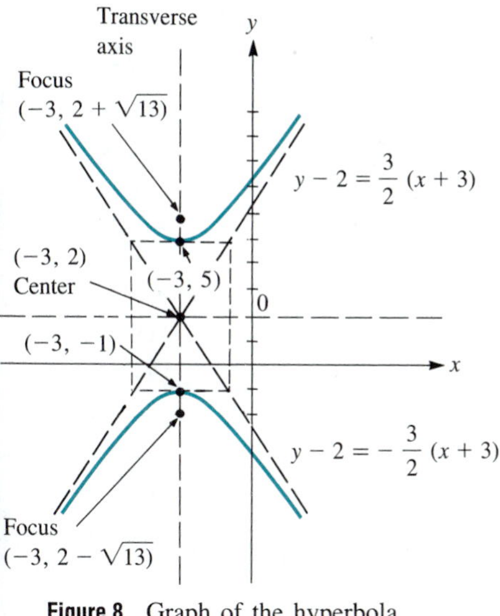

Figure 8 Graph of the hyperbola

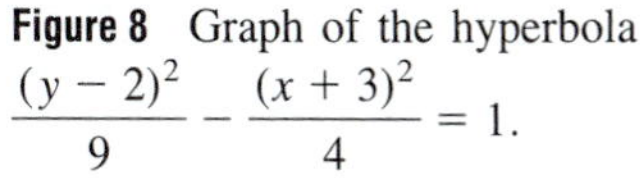

$$\frac{(y-2)^2}{9} - \frac{(x+3)^2}{4} = 1.$$

EXAMPLE 4 *Translating a Hyperbola*

Describe the hyperbola $\dfrac{(y-2)^2}{9} - \dfrac{(x+3)^2}{4} = 1$.

SOLUTION This curve is the hyperbola of Example 3 shifted three units to the left and two units up so that its center is at $(-3, 2)$. It is sketched in Figure 8. ■

EXAMPLE 5 *Determining the Nature of a Translated Conic by First Completing Two Squares*

Describe the curve $x^2 - 4y^2 - 4x - 8y - 9 = 0$.

SOLUTION We have
$$(x^2 - 4x) - 4(y^2 + 2y) - 9 = 0$$

or completing the squares,

$$(x-2)^2 - 4 - 4[(y+1)^2 - 1] - 9 = 0$$
$$(x-2)^2 - 4(y+1)^2 = 9$$
$$\frac{(x-2)^2}{9} - \frac{4}{9}(y+1)^2 = 1$$

and

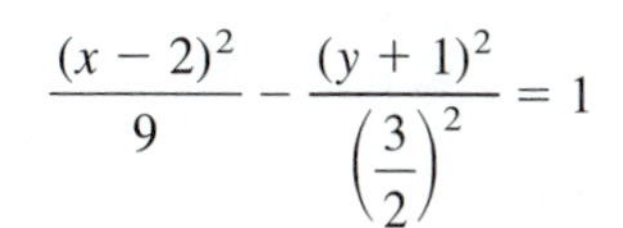

$$\frac{(x-2)^2}{9} - \frac{(y+1)^2}{\left(\frac{3}{2}\right)^2} = 1$$

This is the equation of a hyperbola with center at $(2, -1)$ and horizontal transverse axis. Except for the translation, it is the hyperbola of Example 2 and is sketched in Figure 9.

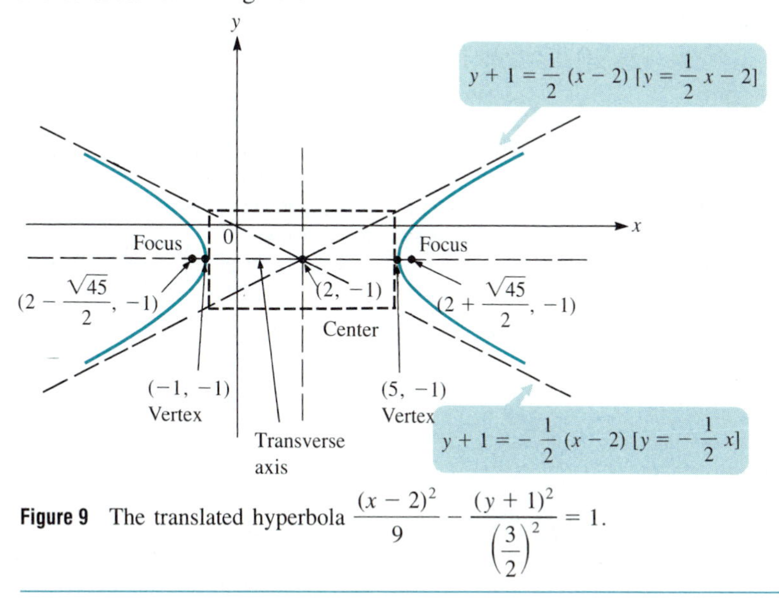

Figure 9 The translated hyperbola $\dfrac{(x-2)^2}{9} - \dfrac{(y+1)^2}{\left(\frac{3}{2}\right)^2} = 1$.

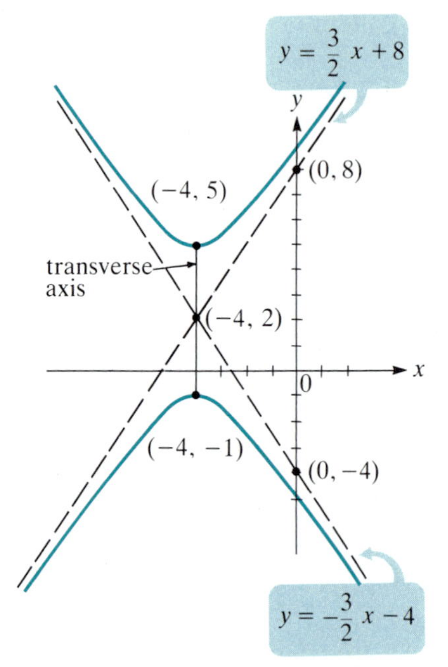

Figure 10 The translated hyperbola $\frac{(y-2)^2}{9} - \frac{(x+4)^2}{4} = 1$.

EXAMPLE 6 ***Finding the Standard Equation of a Translated Hyperbola***

Find the standard equation of the hyperbola with vertices at $(-4, -1)$ and $(-4, 5)$ and asymptotes the lines $y = \frac{3}{2}x + 8$ and $y = -\frac{3}{2}x - 4$.

SOLUTION As in Figure 10, the vertices lie on the vertical line $x = -4$, so the transverse axis is vertical. There are two ways to find the center. The simpler way is to observe that the center is the midpoint of the vertices. The x-coordinate is -4, and the y-coordinate (the average of -1 and 5) is $\frac{-1+5}{2} = 2$. Thus the center is at $(-4, 2)$. The other way is to find the point of intersection of the asymptotes. Now

$$2a = \text{distance between vertices} = 5 - (-1) = 6 \text{ so } a = 3$$

To find b, we observe that the asymptotes have the equations

$$y - 2 = \pm\frac{a}{b}(x + 4)$$

In our case, we must have $\frac{a}{b} = \frac{3}{2}$. Since $a = 3$, we see that $b = 2$. Therefore, the standard equation of the hyperbola is

$$\frac{(y-2)^2}{3^2} - \frac{(x+4)^2}{2^2} = 1 \quad \text{or} \quad \frac{(y-2)^2}{9} - \frac{(x+4)^2}{4} = 1$$

FOCUS ON

The Hyperbola in the Real World

Hyperbolas do not appear in physical constructions as often as ellipses and parabolas, but nevertheless they are useful. Hyperbolas do frequently appear as the graphs of important equations in physics, chemistry, biology, business, and economics. Examples include Ohm's law and supply and demand curves. In Einstein's theory of special relativity, an observer in an inertial reference frame sees a particle in a parallel force field follow a hyperbolic path in space time. The British physicist Ernest Rutherford (1871–1937) developed his now-accepted model of the atom by measuring the hyperbolic orbits of scattered positive-charged particles.

Capillary action is the elevation of the surface of liquids in fine tubes, and so on, due to surface tension and other forces. Suppose, as in Figure 11, that two pieces of glass are joined along one pair of edges and are slightly separated along the other pair. If the glass configuration is placed vertically in a dish of colored water, then capillary action will force the water to rise in such a way as to form a hyperbola. Try it.

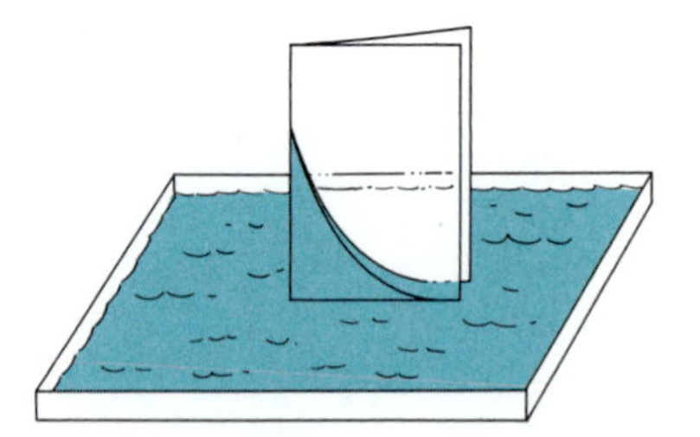

Figure 11 Colored water is drawn up by capillary action.

Hyperbolas are very useful in certain types of navigation. In order to explain why, we first do an example.

EXAMPLE 7

An explosion was heard on two ships 1 kilometer apart. Sailors on Ship B heard the explosion $1\frac{1}{2}$ seconds before those on Ship A. Relative to the two ships, where did the explosion occur?

SOLUTION The speed of sound in air (at 20°C) is approximately 340 meters/sec. In $1\frac{1}{2}$ seconds the sound traveled $1\frac{1}{2} \times 340 = 510$ meters. Therefore, the explosion took place at a point 510 meters closer to Ship B than to Ship A. In Figure 12, we draw a coordinate system and place A and B on the x-axis, equidistant from the origin. Since $\overline{AB} = 1$ km $= 1000$ m, the coordinates of A and B are $(-500, 0)$ and $(500, 0)$. The explosion took place at a

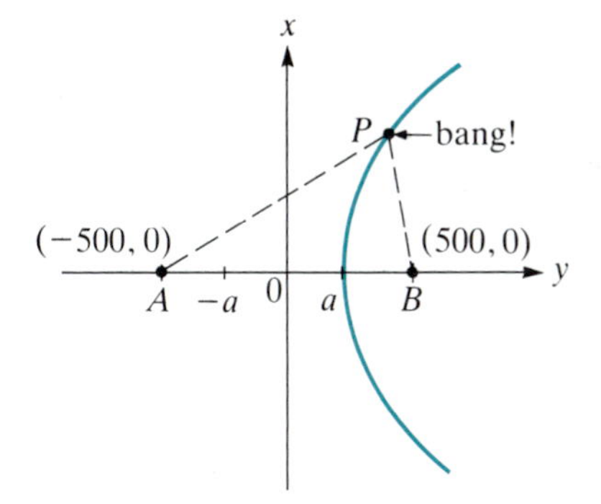

Figure 12 One branch of the hyperbola $\dfrac{x^2}{65{,}025} - \dfrac{y^2}{184{,}975} = 1$.

point P such that $\overline{PA} - \overline{PB} = 510$. Thus, P is on one branch of the hyperbola with foci $(-500, 0)$ and $(500, 0)$ such that the difference of the distances to the foci is $2a = 510$. Thus, $a = \frac{510}{2} = 255$, $c = 500$, $b^2 = c^2 - a^2 = 250{,}000 - 65{,}025 = 184{,}975$. The equation of this hyperbola is

$$\frac{x^2}{65{,}025} - \frac{y^2}{184{,}975} = 1$$

and the point P is on the branch of the hyperbola containing points that are closer to B than to A.

Of course, we have not located the point P precisely. However, if we have a third ship, Ship C, that hears the explosion, then we can obtain two more hyperbolas (one for Ships A and C and one for Ships B and C), and the point of the explosion is the single point at which the three hyperbolas intersect.

Because of the technique illustrated in the last example, the hyperbola is very useful in navigation — particularly in the LORAN (LOng RAnge Navigation) system. During World War II, LORAN served as a navigational aid for the strategic night bombing of Germany and for the long-range bombing of Japan from islands in the South Pacific. LORAN was used to draw highly accurate navigational maps, and one of these may have been used by the crew of the Enola Gay when it dropped the first atomic bomb over Hiroshima.†

†LORAN is now available in small, two-seat, and four-seat aircraft.

The LORAN map in Figure 13 shows two sets of confocal hyperbolas (that is, hyperbolas with the same foci) with the foci

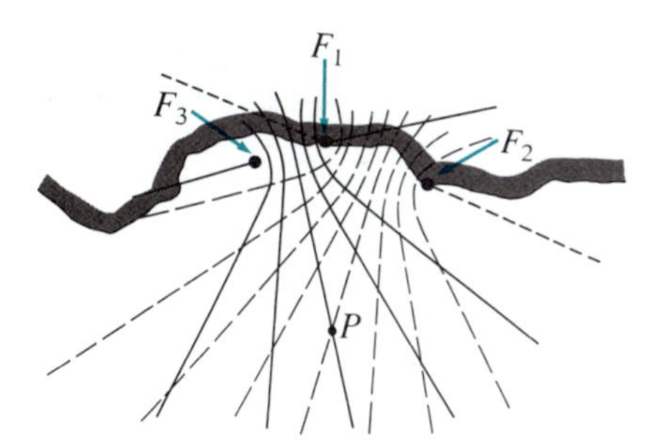

Figure 13 A LORAN map with hyperbolas of constant time difference. Two hyperbolas are needed to get a cross-fix at point P.

located at three radio broadcasting stations. As in Example 7, LORAN is based on the time difference between the reception of signals sent simultaneously from the stations in each broadcasting pair. A ship records the time difference as the signals from one pair (F_1 and F_2, say) arrive and determines its own position on one branch of a hyperbola. It uses the signals from the other pair (F_1 and F_3) to determine its position on one branch of a second hyperbola. The point of intersection of the two hyperbolas is the location of the ship.

LORAN has several advantages over other navigational systems. Radio waves are not affected by clouds or fog that do hamper celestial and visual navigation. Sun-spot activity and atmospheric storms can bend radio signals so that direction finders may be inaccurate. However, these storms do not seriously affect the *velocity* of the radio signals, so measurements based on time differences remain accurate.

Like the ellipse and parabola, the hyperbola has a useful **reflection property.** A light ray approaching (or leaving) one focus will be reflected toward (or away from) the other focus. This is illustrated in Figure 14, in which the direction of the arrows may be reversed.

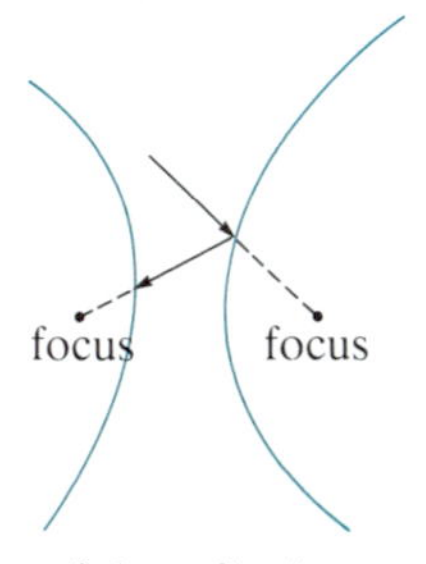

Figure 14 Illustration of the reflection property of the hyperbola.

The reflection property is exploited in the design of telescopes. In 1672, the French sculptor and astronomer Guillaume Cassegrain designed a reflecting telescope with a large parabolic mirror and a smaller hyperbolic mirror both sharing a common focus. This is illustrated in Figure 15. The same principle is used in the 200-inch Hale telescope on Mount Palomar in California.

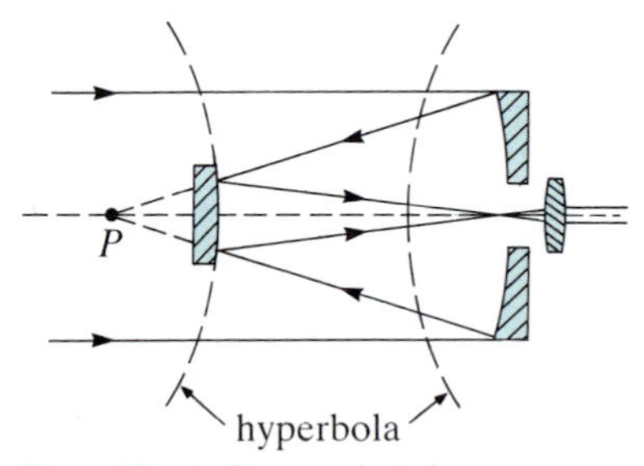

Figure 15 A Cassegrain telescope.

Hyperbolas appear in three-dimensional guises as well. In Figure 16, we provide a computer-drawn sketch of a **hyperboloid of one sheet.** This is a solid with the following properties: cross-sections (slices) parallel to the *xy*-plane are ellipses, while cross-sections perpendicular to the *xz*-plane are hyperbolas. The hyperboloid of one sheet is the design standard for all nuclear cooling towers.

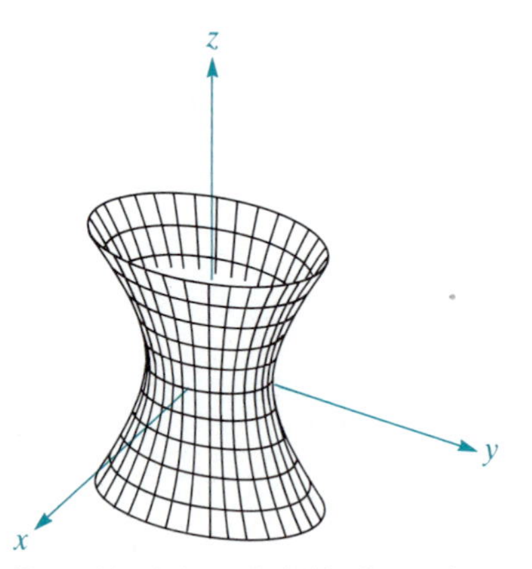

Figure 16 A hyperboloid of one sheet.

For more details of these and other interesting applications of the hyperbola, see Lee Whitt's delightful article, "The Standup Conic Presents: The Hyperbola and Applications," in *The UMAP Journal,* Vol. 5, No. 1, 1984, pp. 9–21.

Problems 5.4

Readiness Check

I. The graph of $\frac{y^2}{9} - \frac{x^2}{16} = 1$ is a hyperbola that opens _____.

a. to the left
b. to the left and right
c. upward
d. upward and downward

II. The transverse axis of the hyperbola satisfying $x^2 - y^2 = 1$ is _____.

a. the line segment between $(-\sqrt{2}, 0)$ and $(\sqrt{2}, 0)$
b. the line segment between $(-1, 0)$ and $(1, 0)$
c. the line $x = 0$
d. the line $y = 0$

III. The vertices of the hyperbola satisfying $\frac{x^2}{16} - \frac{y^2}{9} = 1$ are _____.

a. $(0, 0)$
b. $y = \frac{3x}{4}$ and $y = -\frac{3x}{4}$
c. $(-5, 0)$ and $(5, 0)$
d. $(-4, 0)$ and $(4, 0)$

IV. The asymptotes of the hyperbola satisfying $\frac{x^2}{16} - \frac{y^2}{9} = 1$ are _____.

a. $y = \frac{3x}{4}$ and $y = -\frac{3x}{4}$
b. $y = \frac{4x}{3}$ and $y = -\frac{4x}{3}$
c. $y = x$ and $y = -x$
d. the x-axis and the y-axis

V. Which of the following are the vertices of $9y^2 - x^2 + 36 = 0$?

a. $(0, \pm 2)$
b. $(0, \pm 3)$
c. $(\pm 6, 0)$
d. $(\pm 1, 0)$

Answers to Readiness Check

I. d II. b III. d IV. a V. c

In Problems 1–20 the equation of a hyperbola is given. Find its foci, transverse axis, conjugate axis, center, vertices, and asymptotes. Then sketch it.

1. $\frac{x^2}{16} - \frac{y^2}{25} = 1$
2. $\frac{y^2}{16} - \frac{x^2}{25} = 1$
3. $\frac{y^2}{25} - \frac{x^2}{16} = 1$
4. $\frac{x^2}{25} - \frac{y^2}{16} = 1$
5. $y^2 - x^2 = 1$
6. $x^2 - y^2 = 1$
7. $x^2 - 4y^2 = 9$
8. $4y^2 - x^2 = 9$
9. $y^2 - 4x^2 = 9$
10. $4x^2 - y^2 = 9$
11. $2x^2 - 3y^2 = 4$
12. $3x^2 - 2y^2 = 4$
13. $2y^2 - 3x^2 = 4$
14. $3y^2 - 2x^2 = 4$
15. $(x - 1)^2 - 4(y + 2)^2 = 4$
16. $\frac{(y + 3)^2}{4} - \frac{(x - 2)^2}{9} = 1$
17. $4x^2 + 8x - y^2 - 6y = 21$
18. $-4x^2 - 8x + y^2 - 6y = 20$
19. $2x^2 - 16x - 3y^2 + 12y = 45$
20. $2y^2 - 16y - 3x^2 + 12x = 45$

In Problems 21–30 find the standard equation of the indicated hyperbola.

21. foci: $(-4, 0)$, $(4, 0)$
    vertices: $(-3, 0)$, $(3, 0)$
22. foci: $(-1, 0)$, $(1, 0)$
    vertices: $(-\frac{1}{2}, 0)$, $(\frac{1}{2}, 0)$
23. foci: $(0, -3)$, $(0, 3)$
    vertices: $(0, -2)$, $(0, 2)$
24. foci: $(0, -6)$, $(0, 6)$
    vertices: $(0, -4)$, $(0, 4)$
25. foci: $(-1, 1)$, $(5, 1)$
    vertices: $(0, 1)$, $(4, 1)$
26. foci: $(-3, -1)$, $(-3, 5)$
    vertices: $(-3, 1)$, $(-3, 3)$
27. vertices: $(-2, 0)$, $(2, 0)$
    asymptotes: $y = \pm x$
28. vertices: $(0, -3)$, $(0, 3)$
    asymptotes: $y = \pm 2x$
29. vertices: $(1, 1)$, $(5, 1)$
    asymptotes: $y = 2x - 5$, $y = -2x + 7$
30. vertices: $(-2, -1)$, $(-2, 9)$
    asymptotes: $y = \frac{5}{2}x + 9$, $y = -\frac{5}{2}x - 1$
31. Find the standard equation of the hyperbola obtained by translating the hyperbola of Problem 21 so that its center is at $(4, -3)$.
32. Find the standard equation of the hyperbola obtained by translating the hyperbola of Problem 24 so that its center is at $(-3, 2)$.
33. The **eccentricity** e of a hyperbola is defined by

$$e = \frac{\text{distance between foci}}{\text{distance between vertices}}$$

Show that the eccentricity of any hyperbola is greater than 1.

34. For the hyperbola $x^2/a^2 - y^2/b^2 = 1$, show that $e = \sqrt{a^2 + b^2}/a = \sqrt{1 + (b/a)^2}$.

In Problems 35–44 find the eccentricity of the given hyperbola.

35. the hyperbola of Problem 1
36. the hyperbola of Problem 4
37. the hyperbola of Problem 5
38. the hyperbola of Problem 8
39. the hyperbola of Problem 9
40. the hyperbola of Problem 12
41. the hyperbola of Problem 13
42. the hyperbola of Problem 16
43. the hyperbola of Problem 19
44. the hyperbola of Problem 20

* 45. Find the equation of the hyperbola with center at $(0, 0)$ and axis parallel to the x-axis that passes through the points $(1, 2)$ and $(5, 12)$.

46. Find the equation of the curve having the property that the difference of the distances from any point on the curve to the points $(1, -2)$ and $(4, 3)$ is 5.

47. Show that the graph of the curve $x^2 + 4x - 3y^2 + 6y = c$ is (a) a hyperbola if $c \neq -1$, (b) a pair of straight lines if $c = -1$. (c) If $c = -1$, find the equations of the lines.

* 48. Find conditions relating the numbers a, b, and c such that the graph of the equation $2x^2 + ax - 3y^2 + by = c$ is (a) a hyperbola, (b) a pair of straight lines.

* 49. The speed of sound in sea water (at 25°C) is 1533 meters/second ($\approx$ 5030 ft/sec). Submarine A heard the sound of an exploding depth charge 2 seconds before Submarine B heard the sound. The submarines are 4 kilometers apart, and the depth charge was dropped by an enemy destroyer. Find an equation for the hyperbola that contains the point at which the destroyer dropped the charge. [Hint: Draw a coordinate system and put A and B on the x-axis, equidistant from the origin.]

** 50. In the coordinate system of Problem 49, assume that the positive x-axis points east. Submarine C, located 2 km due north of Submarine A, heard the sound of the depth charge 1 second after Submarine A heard it. Find the exact location of the destroyer at the moment it dropped the charge.

Graphing Calculator Problems

In Problems 51–60 obtain the graph of the given hyperbola on a graphing calculator. Before you begin, read Example 12 in Appendix A.

51. $\frac{x^2}{10} - \frac{y^2}{47} = 1$
52. $\frac{y^2}{43} - \frac{x^2}{19} = 137$
53. $3x^2 - 4y^2 = 7$
54. $12y^2 - 11x^2 = 34$
55. $-43x^2 + 19y^2 = 73$
56. $4(y + 2)^2 - 3(x - 7)^2 = 8$
57. $5(x + 1.9)^2 - 16(y + 1.1)^2 = 42$
58. $3x^2 - 4x - y^2 + 7y = 10$
59. $6x - 11x^2 + 4y^2 + 6y = 12$
60. $3.7x^2 - 4.9x + 12.9y - 6.3y^2 = 38.5$

5.5 Second-Degree Equations and Rotation of Axes

The curves we have considered so far in this chapter had their axes parallel to the two coordinate axes. This is not always the case. The curves sketched in Figure 1 are, respectively, an ellipse, a parabola, and a hyperbola. To obtain

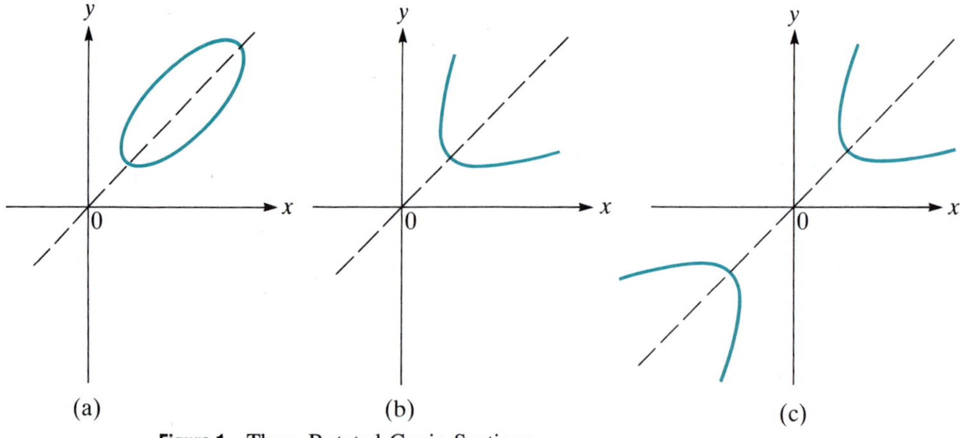

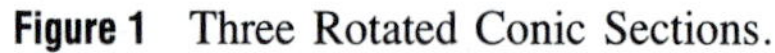

Figure 1 Three Rotated Conic Sections.

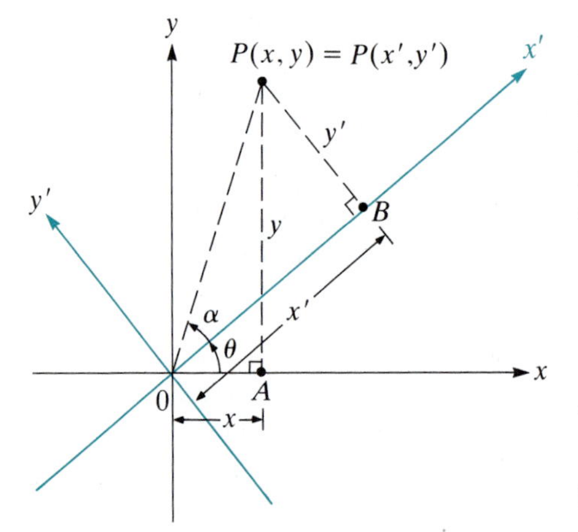

Figure 2 The x'- and y'-axes are obtained by rotating the x- and y-axes through an angle of θ.

curves like those pictured, we must rotate the coordinate axes through an appropriate angle. How do we do so? Suppose that the x- and y-axes are rotated through an angle of θ with respect to the origin (see Figure 2). Let $P(x, y)$ represent a typical point in the coordinates x and y. We now seek a representation of that point in the "new" coordinates x' and y'. From Figure 2 we see that

$$\cos(\theta + \alpha) = \frac{\overline{OA}}{\overline{OP}}$$

$$x = \overline{OA} = \overline{OP}\cos(\theta + \alpha)$$

and, similarly,

$$y = \overline{AP} = \overline{OP}\sin(\theta + \alpha)$$

But (see pp. 158 and 159)

$$\cos(\theta + \alpha) = \cos\theta\cos\alpha - \sin\theta\sin\alpha$$

and

$$\sin(\theta + \alpha) = \sin\alpha\cos\theta + \cos\alpha\sin\theta$$

Thus

$$x = \overline{OP}(\cos\theta\cos\alpha - \sin\theta\sin\alpha) \tag{1}$$

and

$$y = \overline{OP}(\sin\alpha\cos\theta + \cos\alpha\sin\theta) \tag{2}$$

Now from the right triangle OBP we find that

$$\sin\alpha = \frac{y'}{\overline{OP}} \quad \text{or} \quad \overline{OP}\sin\alpha = y'$$

Similarly,

$$\cos \alpha = \frac{x'}{\overline{OP}} \quad \text{or} \quad \overline{OP} \cos \alpha = x'$$

Substituting these last expressions into (1) and (2) yields

Converting from $x'y'$-Coordinates to xy-Coordinates

$$x = x' \cos \theta - y' \sin \theta \tag{3}$$

$$y = x' \sin \theta + y' \cos \theta \tag{4}$$

Equations (3) and (4) can be solved simultaneously to express the "new" coordinates x' and y' in terms of the "old" coordinates x and y. Switching from (x', y') to (x, y) is the same as rotating through an angle of $-\theta$. Since $\cos(-\theta) = \cos \theta$ and $\sin(-\theta) = -\sin \theta$, we obtain, from (3) and (4),

Converting from xy-Coordinates to New (Rotated) $x'y'$-Coordinates

$$x' = x \cos \theta + y \sin \theta \tag{5}$$

$$y' = -x \sin \theta + y \cos \theta \tag{6}$$

EXAMPLE 1 *Obtaining the Equation of a Rotated Ellipse*

Find the equation of the curve obtained from the graph of $x^2 - xy + y^2 = 10$ by rotating the axes through an angle of $\pi/4$.

SOLUTION Since $\cos \pi/4 = \sin \pi/4 = 1/\sqrt{2}$, we obtain, from equations (3) and (4),

$$x = \frac{x' - y'}{\sqrt{2}} \quad \text{and} \quad y = \frac{x' + y'}{\sqrt{2}}$$

Substitution of these into the equation $x^2 - xy + y^2 = 10$ yields

$$\left(\frac{x' - y'}{\sqrt{2}}\right)^2 - \left(\frac{x' - y'}{\sqrt{2}}\right)\left(\frac{x' + y'}{\sqrt{2}}\right) + \left(\frac{x' + y'}{\sqrt{2}}\right)^2 = 10$$

or

$$\frac{(x')^2 - 2x'y' + (y')^2}{2} - \left[\frac{(x')^2 - (y')^2}{2}\right] + \frac{(x')^2 + 2x'y' + (y')^2}{2} = 10$$

which after simplification yields

$$\frac{(x')^2 + 3(y')^2}{2} = 10 \quad \text{or} \quad (x')^2 + 3(y')^2 = 20$$

and, finally,

$$\frac{(x')^2}{20} + \frac{(y')^2}{\frac{20}{3}} = 1$$

In the new coordinates x' and y' this is the equation of an ellipse with $a = \sqrt{20}$, $b = \sqrt{\frac{20}{3}}$, $c = \sqrt{20 - \frac{20}{3}} = \sqrt{\frac{40}{3}}$, and foci at $(\sqrt{\frac{40}{3}}, 0)$ and $(-\sqrt{\frac{40}{3}}, 0)$. [This ellipse can be obtained by rotating the ellipse $(x^2/20) + (y^2/\frac{20}{3}) = 1$ through an angle of $\pi/4$.] It is sketched in Figure 3. We can obtain the foci in the coordinates (x, y) by using equations (3) and (4). Since

$$x = \frac{x' - y'}{\sqrt{2}} \quad \text{and} \quad y = \frac{x' + y'}{\sqrt{2}}$$

we find that $(\sqrt{\frac{40}{3}}, 0)$ in the coordinates (x', y') comes from $(\sqrt{\frac{20}{3}}, \sqrt{\frac{20}{3}})$ in the coordinates (x, y) and $(-\sqrt{\frac{40}{3}}, 0)$ comes from $(-\sqrt{\frac{20}{3}}, -\sqrt{\frac{20}{3}})$. The major axis is on the line $y = x$.

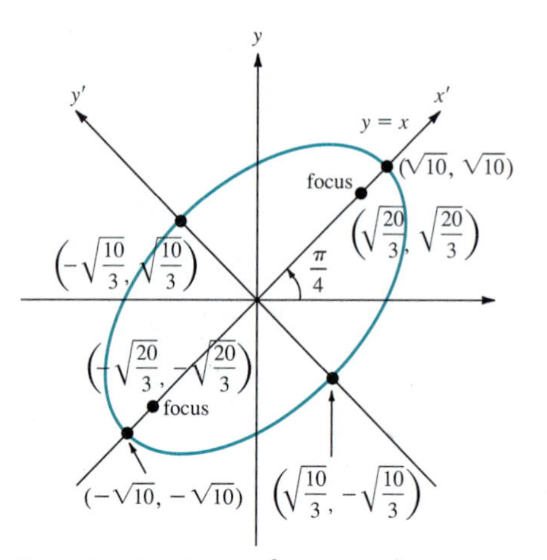

Figure 3 The ellipse $x^2 - xy + y^2 = 10$ or $(x')^2/20 + (y')^2/(20/3) = 1$. ■

EXAMPLE 2 *Obtaining the Equation of a Rotated Hyperbola*

Find the equation of the curve obtained from the graph of $xy = 1$ by rotating the axes through an angle of $\pi/4$.

SOLUTION As in Example 1, we have $x = (x' - y')/\sqrt{2}$ and $y = (x' + y')/\sqrt{2}$, so that

$$1 = xy = \left(\frac{x' - y'}{\sqrt{2}}\right)\left(\frac{x' + y'}{\sqrt{2}}\right) = \frac{(x')^2}{2} - \frac{(y')^2}{2}$$

Thus in the new (rotated) coordinate system we obtain the hyperbola $[(x')^2/2] - [(y')^2/2] = 1$. This curve is sketched in Figure 4. We may say that the equation $xy = 1$ is really the equation of a ''standard'' hyperbola rotated through an angle of $\pi/4$.

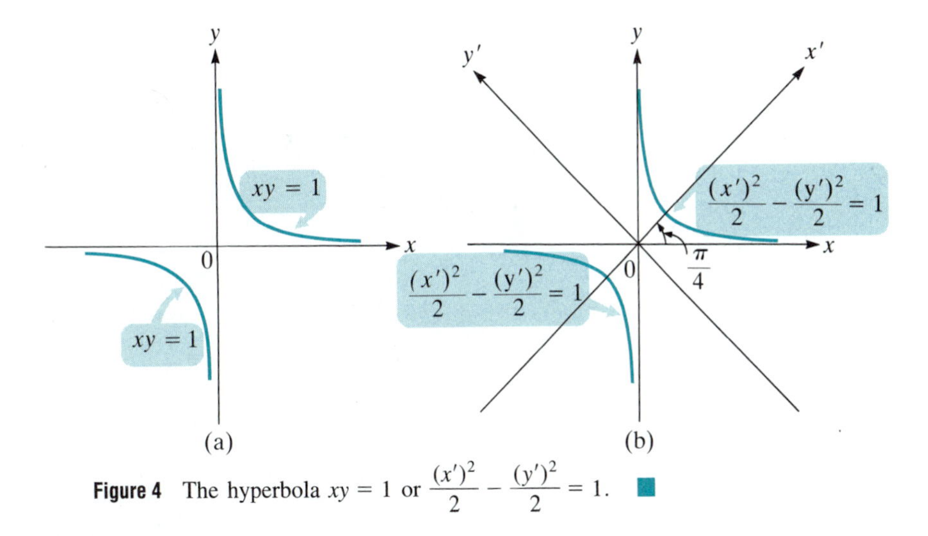

Figure 4 The hyperbola $xy = 1$ or $\frac{(x')^2}{2} - \frac{(y')^2}{2} = 1$. ■

EXAMPLE 3 *Finding the Equation of a Rotated Parabola*

Find the equation of the parabola $y = x^2$ if the coordinate axes are rotated through an angle of 30°.

SOLUTION From (3) and (4) we obtain

$$x = x' \cos 30° - y' \sin 30° = \frac{\sqrt{3}}{2}x' - \frac{1}{2}y'$$

$$y = x' \sin 30° + y' \cos 30° = \frac{1}{2}x' + \frac{\sqrt{3}}{2}y'$$

Substituting these values into the equation $y = x^2$ yields

$$\frac{1}{2}x' + \frac{\sqrt{3}}{2}y' = \left(\frac{\sqrt{3}}{2}x' - \frac{1}{2}y'\right)^2 = \frac{3}{4}x'^2 - \frac{\sqrt{3}}{2}x'y' + \frac{1}{4}y'^2$$

or

$$\frac{3}{4}x'^2 - \frac{\sqrt{3}}{2}x'y' + \frac{1}{4}y'^2 - \frac{1}{2}x' - \frac{\sqrt{3}}{2}y' = 0$$

In this example, we took an equation in standard form and rotated the axes so that the conic was hard to recognize. This is usually not a very useful thing to do.

At the end of this section, we will show you how to choose a rotation that will simplify the equation of the conic; that is, we will show you how to eliminate the xy-term. Without the xy-term the conic can be written in a standard form in which the directrix or axes of the conic are parallel to the new coordinate axes.

Examples 1 and 2 illustrate the fact that the equations of our three basic curves can take forms other than the standard forms given in Sections 5.2,

5.3, and 5.4. It turns out that any second-degree equation $Ax^2 + Bxy + Cy^2 + Dx + Ey + F = 0$ can be written in a standard form whose graph is a circle, an ellipse, a parabola, a hyperbola, or a degenerate form such as a line, a pair of lines, a point, or an empty set of points. We will not discuss this fact further except to state the following theorem. (You are asked to prove the theorem in Problem 24.)

Theorem: Determining the Type of Conic from the Discriminant

Consider the second-degree equation

$$Ax^2 + Bxy + Cy^2 + Dx + Ey + F = 0 \tag{7}$$

(i) If $B^2 - 4AC = 0$, then (7) is the equation of a parabola, a line, or two parallel lines or is imaginary.†
(ii) If $B^2 - 4AC < 0$, then (7) is the equation of a circle, an ellipse, or a single point or is imaginary.
(iii) If $B^2 - 4AC > 0$, then (7) is the equation of a hyperbola or two intersecting lines.

The number $B^2 - 4AC$ is called the **discriminant** of the conic section.

EXAMPLE 4 ***Determining the Type of Conic Section and Writing It in Standard Form***

Determine the type of curve represented by the equation

$$16x^2 - 24xy + 9y^2 + 100x - 200y + 100 = 0 \tag{8}$$

Then write the equation in a standard form by finding an appropriate translation and rotation of axes.

SOLUTION Here $A = 16$, $B = -24$, and $C = 9$, so that $B^2 - 4AC = (-24)^2 - 4(16)(9) = 576 - 576 = 0$. Thus the equation represents a parabola (or a degenerate form of a parabola). To write it in a standard form, we first rotate the axes through an appropriate angle θ to eliminate the xy term in (8). To determine θ, we substitute

$$x = x' \cos\theta - y' \sin\theta \qquad \text{and} \qquad y = x' \sin\theta + y' \cos\theta$$

in equation (8). We then obtain

$$\begin{aligned} 16(x' \cos\theta - y' \sin\theta)^2 - 24(x' \cos\theta - y' \sin\theta)(x' \sin\theta + y' \cos\theta) \\ + 9(x' \sin\theta + y' \cos\theta)^2 + 100(x' \cos\theta - y' \sin\theta) \\ - 200(x' \sin\theta + y' \cos\theta) + 100 = 0 \end{aligned}$$

† By "imaginary" we mean that the graph contains no real points; for example, $x^2 + 2xy + y^2 + 1 = 0$ satisfies $B^2 - 4AC = 0$; but there are no real values of x and y that satisfy $0 = x^2 + 2xy + y^2 + 1 = (x + y)^2 + 1$. (Explain why.)

$$
\begin{aligned}
16[x'^2 \cos^2\theta - 2x'y' \cos\theta \sin\theta + y'^2 \sin^2\theta] \\
- 24[x'^2 \cos\theta \sin\theta - x'y' \sin^2\theta + x'y' \cos^2\theta - y'^2 \sin\theta \cos\theta] \\
+ 9[x'^2 \sin^2\theta + 2x'y' \sin\theta \cos\theta + y'^2 \cos^2\theta] + (100 \cos\theta)x' \\
- (100 \sin\theta)y' - (200 \sin\theta)x' - (200 \cos\theta)y' + 100 = 0
\end{aligned}
$$

The idea now is to choose θ so that the coefficient of the term $x'y'$ is zero. This term is

$$
\begin{aligned}
-32 \cos\theta \sin\theta + 24 \sin^2\theta - 24 \cos^2\theta + 18 \sin\theta \cos\theta \\
= -24(\cos^2\theta - \sin^2\theta) - 14 \sin\theta \cos\theta
\end{aligned}
$$

Eq. (2) on p. 165
↓

Setting this expression equal to zero and using the fact that $\cos 2\theta = \cos^2\theta - \sin^2\theta$ and $\sin 2\theta = 2 \sin\theta \cos\theta$, we obtain

$$24 \cos 2\theta + 7 \sin 2\theta = 0$$

or dividing by $\cos 2\theta$ and rearranging terms,

$$
\begin{aligned}
24 \frac{\cos 2\theta}{\cos 2\theta} + \frac{7 \sin 2\theta}{\cos 2\theta} &= 0 \\
24 + 7 \tan 2\theta &= 0 \\
\tan 2\theta &= -\frac{24}{7}
\end{aligned}
$$

If $\tan 2\theta = -\frac{24}{7}$, then $\cos 2\theta = -\frac{7}{25}$,† and also

Eq. (12) on p. 168
↓

$$\cos\theta = \sqrt{\frac{1 + \cos 2\theta}{2}} = \sqrt{\frac{9}{25}} = \frac{3}{5}$$

Furthermore, $\sin\theta = \frac{4}{5}$. At this point we do not need to know θ but only the values of $\cos\theta$ and $\sin\theta$. Next,

$$x = x' \cos\theta - y' \sin\theta = \tfrac{1}{5}(3x' - 4y')$$

and

$$y = x' \sin\theta + y' \cos\theta = \tfrac{1}{5}(4x' + 3y')$$

We substitute these expressions into (7) to obtain (after simplification)

$$25(y')^2 - 200y' - 100x' + 100 = 0$$

To further simplify, we divide by 25 and complete the square:

$$
\begin{aligned}
(y')^2 - 8y' - 4x' + 4 &= 0 \\
(y' - 4)^2 - 16 - 4x' + 4 &= 0 \\
(y' - 4)^2 &= 4x' + 12 = 4(x' + 3)
\end{aligned}
$$

† Alternatively, we could choose $\cos 2\theta = \frac{7}{25}$, yielding $\cos\theta = \frac{4}{5}$ and $\sin\theta = -\frac{3}{5}$. This, however, is merely a 90° rotation of the coordinate axes obtained with the choice $\cos 2\theta = -\frac{7}{25}$.

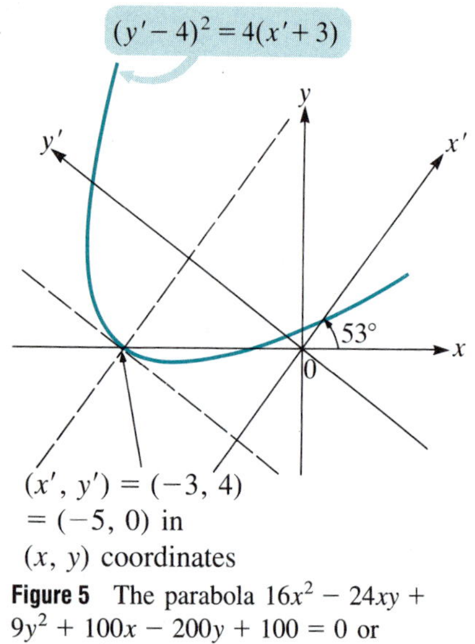

Figure 5 The parabola $16x^2 - 24xy + 9y^2 + 100x - 200y + 100 = 0$ or $(y' - 4)^2 = 4(x' + 3)$

This is the equation of a parabola with vertex at $(-3, 4)$ in the new (rotated) coordinate system. To sketch the parabola, we need to calculate the angle of rotation given by $\theta = \cos^{-1}\frac{3}{5} \approx 0.927 \approx 53°$. The curve is sketched in Figure 5.

We note here that the xy term in (7) can always be eliminated by a rotation of axes through an angle θ, where θ is given by the equation (9).

Rotating the Axes Through θ Will Eliminate the xy-term When

$$\cot 2\theta = \frac{A - C}{B} \tag{9}$$

The proof of this fact follows exactly as in the derivation of the angle of rotation in Example 4 and is left as an exercise. (See Problem 21.)

We conclude by pointing out that instead of rotating axes, we can rotate curves, keeping the axes fixed. Note the following important fact:

Rotating a curve through an angle θ has the effect of rotating the axes through an angle $-\theta$.

Problems 5.5

Readiness Check

I. Suppose the x- and y-axes are rotated through an angle of $\pi/6$. Which of the following gives the new coordinates x' and y' in terms of x and y?

a. $x' = \frac{1}{2}x + \frac{\sqrt{3}}{2}y$, $y' = -\frac{\sqrt{3}}{2}x + \frac{1}{2}y$

b. $x' = \frac{\sqrt{3}}{2}x + \frac{1}{2}y$, $y' = \frac{1}{2}x + \frac{\sqrt{3}}{2}y$

c. $x' = \frac{\sqrt{3}}{2}x + \frac{1}{2}y$, $y' = -\frac{1}{2}x + \frac{\sqrt{3}}{2}y$

d. $x' = -\frac{\sqrt{3}}{2}x + \frac{1}{2}y$, $y' = \frac{1}{2}x + \frac{\sqrt{3}}{2}y$

e. $x' = \frac{\sqrt{3}}{2}x - \frac{1}{2}y$, $y' = \frac{1}{2}x + \frac{\sqrt{3}}{2}y$

II. Suppose the x- and y-axes are rotated through an angle of $2\pi/3$. Which of the following gives x and y in terms of the new coordinates x' and y'?

a. $x = \frac{1}{2}x' + \frac{\sqrt{3}}{2}y'$ $y = -\frac{\sqrt{3}}{2}x' + \frac{1}{2}y'$

b. $x = \frac{\sqrt{3}}{2}x' - \frac{1}{2}y'$ $y = \frac{1}{2}x' + \frac{\sqrt{3}}{2}y'$

c. $x = -\frac{1}{2}x' + \frac{\sqrt{3}}{2}y'$ $y = -\frac{\sqrt{3}}{2}x' - \frac{1}{2}y'$

d. $x = -\frac{1}{2}x' - \frac{\sqrt{3}}{2}y'$ $y = \frac{\sqrt{3}}{2}x' - \frac{1}{2}y'$

e. $x = \frac{1}{2}x' - \frac{\sqrt{3}}{2}y'$ $y = -\frac{\sqrt{3}}{2}x' - \frac{1}{2}y'$

III. The equation $x^2 + xy + y^2 + x + y + 1 = 0$ is the equation of ______.
a. an ellipse b. a parabola c. a hyperbola

IV. The equation $2x^2 + 5xy + y^2 + 3x - 2y + 10 = 0$ is the equation of ______.
a. an ellipse b. a parabola c. a hyperbola

V. The equation $x^2 + 2xy + y^2 - 2x + 3y = 0$ is the equation of ______.
a. an ellipse b. a parabola c. a hyperbola

Answers to Readiness Check

I. c II. d III. a IV. c V. b

1. Describe the curve obtained from the graph of $4x^2 - 2xy + 4y^2 = 45$ by rotating the axes through an angle of $\pi/4$.
2. Describe the curve obtained from the graph of $x^2 + 2\sqrt{3}xy - y^2 = 4$ by rotating the axes through an angle of $\pi/6$.
3. What is the equation of the line obtained from the line $2x - 3y = 6$ by rotating the axes through an angle of $\pi/6$?
4. Find the equation of the line obtained from the line $ax + by + c = 0$ by rotating the axes through an angle θ.
5. Find the equation of a parabola obtained from the parabola $y^2 = -12x$ if the axes are rotated until the axis of the parabola coincides with the line $y = \sqrt{3}x$.
6. * Find the equation of the ellipse whose major axis is on the line $y = -x$, which is obtained by rotating the ellipse $(x^2/25) + (y^2/16) = 1$. [Hint: Find the angle through which the axes must be rotated to accomplish this.]

In Problems 7–20, find a rotation of coordinate axes in which the given equation written in the new coordinates has no xy term. Describe each curve and then sketch it.

7. $4x^2 + 4xy + y^2 = 9$
8. $4x^2 + 4xy - y^2 = 9$
9. $3x^2 - 2xy - 5 = 0$
10. $xy = 2$
11. $xy = a,\ a > 0$
12. $xy = a,\ a < 0$
13. $4x^2 + 4xy + y^2 + 20x - 10y = 0$
14. $x^2 + 4xy + 4y^2 - 6 = 0$
15. $2x^2 + xy + y^2 = 4$
16. $9x^2 + 6xy + y^2 + 10x - 30y = 0$
17. $3x^2 - 6xy + 5y^2 = 36$
18. $x^2 - 3xy + 4y^2 = 1$
19. $3y^2 - 4xy + 30y - 20x + 40 = 0$
20. $6x^2 + 5xy - 6y^2 + 7 = 0$
21. Show that if $A \neq C$, the xy term in the second-degree equation (7) will be eliminated by rotation through an angle θ if θ is given by

$$\cot 2\theta = \frac{A - C}{B}$$

22. Show that if $A = C$ in Problem 21, then the xy term will be eliminated by a rotation through an angle of either $\pi/4$ or $-\pi/4$.
23. * Suppose that a rotation converts $Ax^2 + Bxy + Cy^2$ into $A'(x')^2 + B'(x'y') + C'(y')^2$.
    (a) Show that $A + C = A' + C'$
    (b) Show that $B^2 - 4AC = (B')^2 - 4A'C'$.
24. * Use the result of Problem 23 to prove the theorem on p. 312. [Hint: Rotate the axes so that $B' = 0$.]
25. Show that, for every constant $k \neq 0$, $xy = k$ is the equation of a hyperbola.

5.6 Polar Equations of the Conic Sections

In Section 4.7, we introduced the polar coordinate representation of a point in the plane. For a number of applications in space mechanics and other areas, it is useful to write the equations of the conic sections in polar coordinates. For example, as we saw in Section 5.2, the orbit of the earth around the sun is an ellipse with one of its foci at the sun. If we set up a polar axis with the pole at the sun, then the equation of the ellipse takes a particularly simple form.

In order to write the equations of conic sections in polar coordinates, it is first necessary to define the conic sections in an alternative way. Recall from p. 289 that a parabola is defined as the set of points equidistant from a fixed point (focus) and a fixed line that does not contain the focus (directrix). We now generalize this definition.

Theorem: Determining the Type of Conic Section from Its Eccentricity

Let L be a fixed line and let F be a fixed point not on L. Let e denote a fixed positive number. Finally, let C be the set of all points $P(x, y)$ such that

$$\frac{\text{distance from } P \text{ to } F}{\text{distance from } P \text{ to } L} = e \tag{1}$$

Then

$$\text{if } e = 1,\ C \text{ is a parabola}$$
$$\text{if } 0 < e < 1,\ C \text{ is an ellipse}$$
$$\text{if } e > 1,\ C \text{ is a hyperbola}$$

The point F is called the **focus** of the conic, the line L is called the **directrix,** and the number e is called the **eccentricity** of the conic.

Proof

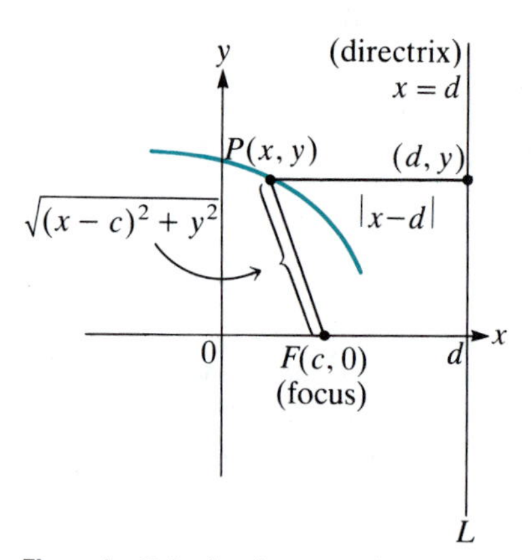

Figure 1 F is the focus, and the line $x = d$ is the directrix of the curve.

We place the focus and directrix so that the focus is the point $(c, 0)$ and the directrix is the vertical line $x = d$, as in Figure 1. Then, using the distance formula and Figure 1, we obtain

$$\frac{\text{distance from } P \text{ to } F}{\text{distance from } P \text{ to } L} = \frac{\sqrt{(x-c)^2 + y^2}}{|x-d|} = e$$

$$\sqrt{(x-c)^2 + y^2} = e|x-d|$$

$$(x-c)^2 + y^2 = e^2(x-d)^2 \quad \text{We squared both sides}$$

$$x^2 - 2cx + c^2 + y^2 = e^2(x^2 - 2dx + d^2) = e^2x^2 - 2de^2x + e^2d^2$$

$$x^2(1-e^2) + y^2 + (2de^2 - 2c)x + c^2 - e^2d^2 = 0 \qquad (2)$$

This is the equation of a conic. This can be written in the form

$$Ax^2 + Bxy + Cy^2 + Dx + Ey + F = 0$$

where $A = 1 - e^2$, $B = 0$, $C = 1$, $D = 2de^2 - 2c$, $E = 0$, and $F = c^2 - e^2d^2$. The discriminant (see p. 312) of equation (2) is

$$B^2 - 4AC = 0^2 - 4(1-e^2) = -4(1-e^2)$$

Then, from the theorem on p. 312,

(i) if $e = 1$, $-4(1-e^2) = -4(1-1) = 0$ and equation (2) is the equation of a parabola.
(ii) if $0 < e < 1$, $e^2 < 1$, so $1 - e^2 > 0$ and $-4(1-e^2) < 0$ so equation (2) is the equation of an ellipse.
(iii) if $e > 1$, $e^2 > 1$, so $1 - e^2 < 0$ and $-4(1-e^2) > 0$, so equation (2) is the equation of a hyperbola. ■

NOTE (1) If $e = 1$, then equation (1) says that

$$\text{distance from } P \text{ to focus} = \text{distance from } P \text{ to directrix}$$

This is our original definition of the parabola so nothing new was proved in this case.

(2) In Problem 21, you are asked to show that if we pick two numbers a and b with $a > b > 0$ and set $c = \sqrt{a^2 - b^2}$, $e = \dfrac{c}{a}$, and $d = \dfrac{a}{e}$, then equation (1) becomes $\dfrac{x^2}{a^2} + \dfrac{y^2}{b^2} = 1$. This is the standard

equation of an ellipse with eccentricity defined as on p. 282 and $0 < e < 1$.

(3) In Problem 22, you are asked to show that if we pick two positive numbers a and b and set $c = \sqrt{a^2 + b^2}$, $e = \dfrac{c}{a}$, and $d = \dfrac{a}{e}$, then equation (1) becomes $\dfrac{x^2}{a^2} - \dfrac{y^2}{b^2} = 1$. This is the standard equation of a hyperbola with the eccentricity given as in Problem 33 on p. 307 and $e > 1$.

We now show how the conic sections can be written in polar coordinates. We choose a polar axis with the focus F at the pole and consider four cases:

Case 1 The directrix is perpendicular to the polar axis and lies p units to the left of the pole (focus). Then, as in Figure 2,

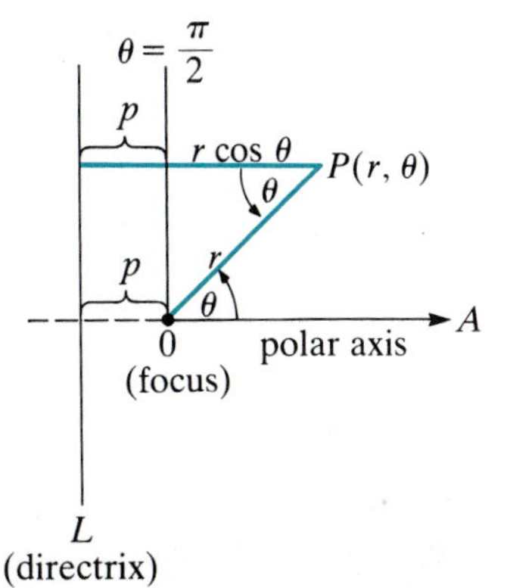

Figure 2 The directrix L is perpendicular to the polar axis and lies p units to the left of the pole.

$$e = \frac{\text{distance from } P(r, \theta) \text{ to } 0}{\text{distance from } P(r, \theta) \text{ to } L} = \frac{r}{p + r\cos\theta}$$

$$e(p + r\cos\theta) = r$$

$$ep + r(e\cos\theta) = r$$

$$r(1 - e\cos\theta) = ep$$

or

$$r = \frac{ep}{1 - e\cos\theta}$$

There are three other cases that can be evaluated as we just evaluated Case 1. The results are given in the table below.

Polar Equations for Conic Sections with the Focus at the Pole

Directrix	Standard Equation	Axis (parabola), Major Axis (ellipse), or Transverse Axis (hyperbola)	
Perpendicular to the polar axis and p units to the left of the pole	$r = \dfrac{ep}{1 - e\cos\theta}$	Horizontal (on the polar axis)	(3)
Perpendicular to the polar axis and p units to the right of the pole	$r = \dfrac{ep}{1 + e\cos\theta}$	Horizontal (on the polar axis)	(4)
Parallel to the polar axis and p units below it	$r = \dfrac{ep}{1 - e\sin\theta}$	Vertical $\left(\text{on the line } \theta = \dfrac{\pi}{2}\right)$	(5)
Parallel to the polar axis and p units above it	$r = \dfrac{ep}{1 + e\sin\theta}$	Vertical $\left(\text{on the line } \theta = \dfrac{\pi}{2}\right)$	(6)

EXAMPLE 1 ***Identifying and Sketching a Conic from Its Polar Equation***

Identify and sketch the conic whose graph is

$$r = \frac{6}{1 - 2\cos\theta} \tag{7}$$

SOLUTION Equation (7) is of the form (3) with $e = 2$, so it is the equation of a hyperbola with horizontal transverse axis. Since $ep = 2p = 6$, $p = 3$. Thus, the focus is at the pole, and the directrix is three units to the left of the horizontal ray $\theta = \frac{\pi}{2}$. The vertices lie on the transverse axis which lies on the polar axis. Points on the polar axis correspond to $\theta = 0$ and $\theta = \pi$. Setting $\theta = 0$ and π in (7) yields

$$r = \frac{6}{1 - 2\cos 0} = \frac{6}{1 - 2} = -6$$

and

$$r = \frac{6}{1 - 2\cos\pi} = \frac{6}{1 - 2(-1)} = \frac{6}{3} = 2$$

Thus, in polar coordinates, the vertices are $(-6, 0) \overset{\text{Eq. (1) on p. 264}}{=} (6, \pi)$ and $(2, \pi)$.

In equation (7), we see that r becomes very large as the denominator $1 - 2\cos\theta$ gets close to 0. But

$$1 - 2\cos\theta = 0$$

means that

$$2\cos\theta = 1$$

and

$$\cos\theta = \frac{1}{2}$$

or

$$\theta = \frac{\pi}{3} \quad \text{and} \quad \theta = \frac{5\pi}{3}.$$

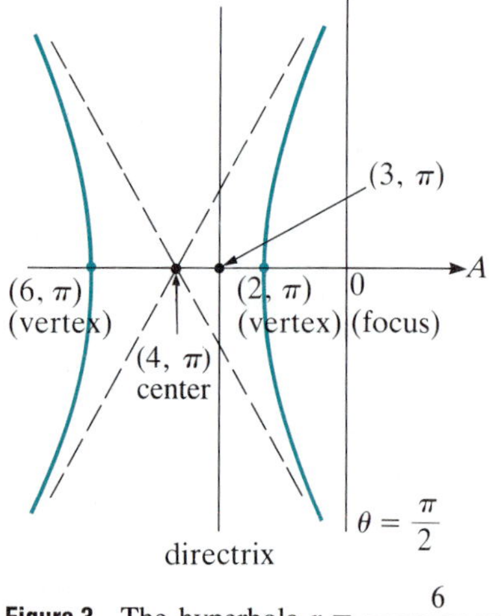

Figure 3 The hyperbola $r = \frac{6}{1 - 2\cos\theta}$.

The asymptotes for the hyperbola are therefore the two lines that make angles of $\frac{\pi}{3}$ and $\frac{5\pi}{3}$ with the polar axis and pass through the point $(4, \pi)$ (the center). With this information, we can sketch the curve as in Figure 3. ■

EXAMPLE 2 ***Identifying and Sketching a Conic from Its Polar Equation***

Identify and sketch the conic whose graph is

$$r = \frac{3}{2 + \sin\theta} \tag{8}$$

SOLUTION To get the equation into one of the standard forms, we divide numerator and denominator by 2 to obtain

$$r = \frac{\frac{3}{2}}{1 + \frac{1}{2}\sin\theta} \tag{9}$$

This is equation (6) with $e = \frac{1}{2}$ so we have the equation of an ellipse with eccentricity $\frac{1}{2}$ and a vertical major axis. Since $ep = \frac{1}{2}p = \frac{3}{2}$, we have $p = 3$, and the directrix is parallel to the polar axis and 3 units above it. The major axis is vertical and contains the foci, and so lies on the line $\theta = \frac{\pi}{2}$. Vertical lines correspond to the values $\theta = \frac{\pi}{2}$ and $\theta = \frac{3\pi}{2}$. From equation (8)

$$r\left(\frac{\pi}{2}\right) = \frac{3}{2 + \sin\frac{\pi}{2}} = \frac{3}{2+1} = 1$$

$$r\left(\frac{3\pi}{2}\right) = \frac{3}{2 + \sin\frac{3\pi}{2}} = \frac{3}{2-1} = 3$$

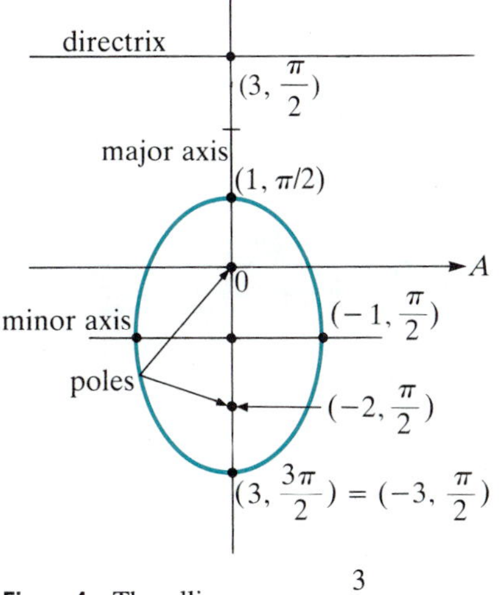

Figure 4 The ellipse $r = \dfrac{3}{2 + \sin\theta}$.

Thus, the vertices are the points $\left(1, \frac{\pi}{2}\right)$ and $\left(3, \frac{3\pi}{2}\right) = \left(-3, \frac{\pi}{2}\right)$. The center is the midpoint of the vertices:

$$\text{center} = \left(\frac{1 + (-3)}{2}, \frac{\pi}{2}\right) = \left(-1, \frac{\pi}{2}\right) = \left(1, \frac{3\pi}{2}\right)$$

The minor axis is horizontal and contains the center. The ellipse is sketched in Figure 4. ■

EXAMPLE 3 *Finding the Polar Equation of a Parabola*

Find the polar equation of the parabola with focus at the pole and directrix a line parallel to the polar axis and 4 units below it. Sketch the curve.

SOLUTION Since the conic is a parabola, we have $e = 1$, and the other information leads to equation (5) with $p = 4$:

$$r = \frac{4}{1 - \sin\theta}$$

The axis contains the focus and is vertical so it must be the line $\theta = \frac{\pi}{2}$. The vertex is a point on the axis that lies halfway between the focus $\left[\text{the point } \left(0, \frac{\pi}{2}\right)\right]$ and the directrix $\left[\text{which contains the point } \left(-4, \frac{\pi}{2}\right)\right]$. Thus, the

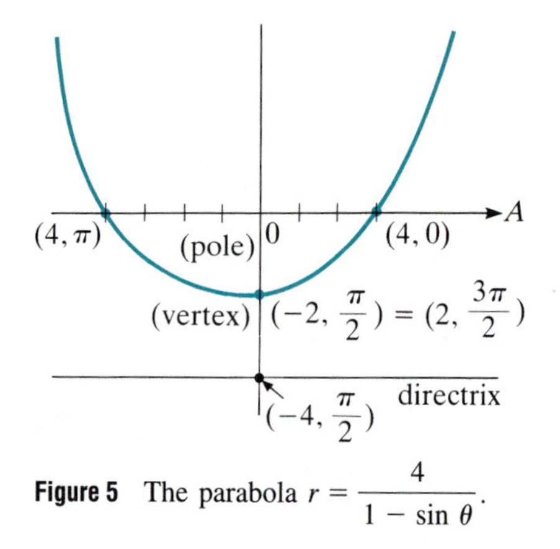

Figure 5 The parabola $r = \dfrac{4}{1 - \sin\theta}$.

vertex is $\left(-2, \dfrac{\pi}{2}\right) = \left(2, \dfrac{3\pi}{2}\right)$. The parabola is sketched in Figure 5. Note that the parabola intersects the polar axis when $\theta = 0$ or π; $r(0) = 4$, and $r(\pi) = 4$.

The Polar Equations of Planetary Motion

Suppose a planet is orbiting the sun in an elliptical orbit of eccentricity e. In order to find a polar equation for the orbit, we draw a polar axis with the sun at the pole. For consistency, we orient the axis so the directrix is parallel to and to the left of the line $\theta = \dfrac{\pi}{2}$. Then, from (3), the polar equation of the orbit is

$$r = \frac{ep}{1 - e\cos\theta}, \quad 0 < e < 1 \tag{10}$$

We wish to write (10) in terms of a, where $2a$ is the length of the major axis of the ellipse. The major axis is on the polar axis so the vertices on the major axis correspond to $\theta = 0$ and $\theta = \pi$ in (10):

$$r(0) = \frac{ep}{1 - e\cos 0} = \frac{ep}{1 - e} \quad \text{and} \quad r(\pi) = \frac{ep}{1 - e\cos\pi} = \frac{ep}{1 + e}$$

From Figure 6 we find that

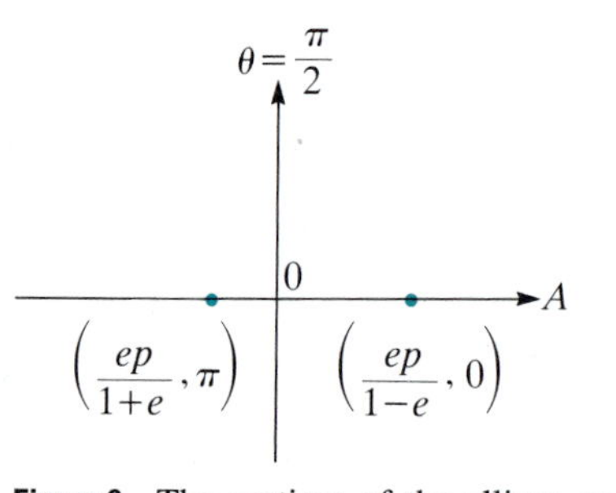

Figure 6 The vertices of the ellipse are the points $\left(\dfrac{ep}{1+e}, \pi\right)$ and $\left(\dfrac{ep}{1-e}, 0\right)$.

$$\begin{aligned} 2a = \text{distance between vertices} &= \frac{ep}{1-e} + \frac{ep}{1+e} \\ &= ep\left(\frac{1}{1-e} + \frac{1}{1+e}\right) = ep\left(\frac{1+e}{(1-e)(1+e)} + \frac{1-e}{(1+e)(1-e)}\right) \\ &= ep\left(\frac{1+e+1-e}{1-e^2}\right) = \frac{2ep}{1-e^2} \end{aligned}$$

Then

$$\frac{2ep}{1-e^2} = 2a$$

$$ep = 2a\left(\frac{1-e^2}{2}\right) = a(1-e^2)$$

and (10) becomes

Polar Equation of an Ellipse with Eccentricity *e* and Major Axis of Length 2*a*

$$r = \frac{a(1-e^2)}{1 - e\cos\theta} \tag{11}$$

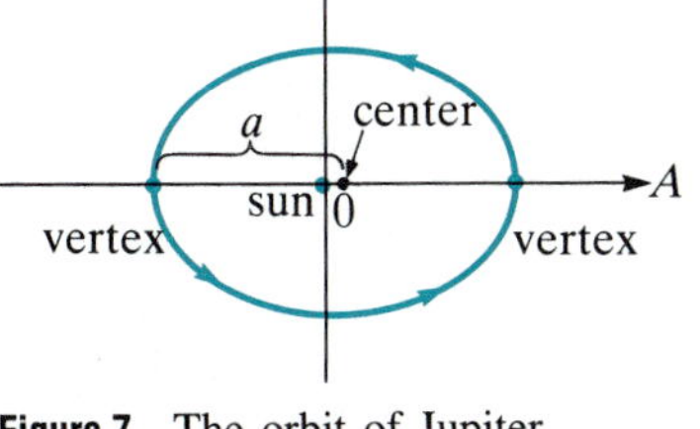

Figure 7 The orbit of Jupiter

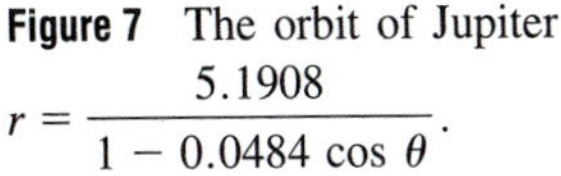

$$r = \frac{5.1908}{1 - 0.0484\cos\theta}.$$

EXAMPLE 4 *Finding a Polar Equation for the Orbit of Jupiter*

Find a polar equation for the orbit of Jupiter given that $e = 0.0484$ and $2a = 10.406$ AU. [1 AU = 1 astronomical unit = the distance between the earth and the sun ≈ 93 million miles ≈ 1.496×10^{11} meters.] How close does Jupiter get to the sun? That is, what is its perihelion?

SOLUTION Inserting $e = 0.0484$ and $a = 5.203$ into equation (11) yields

$$r = \frac{5.203(1 - 0.0484^2)}{1 - 0.0484\cos\theta} \approx \frac{5.1908}{1 - 0.0484\cos\theta}\ \text{AU} \qquad \textbf{(12)}$$

The orbit is sketched in Figure 7.

In equation (12) we see that r is smallest when $1 - 0.0484\cos\theta$ is largest; this occurs when $-0.0484\cos\theta$ is largest or when $\cos\theta = -1$ and $\theta = \pi$. Inserting $\theta = \pi$ into (12) yields

$$\begin{array}{l}\text{closest distance}\\ \text{between Jupiter}\\ \text{and the sun}\end{array} \approx \frac{5.1908}{1 + 0.0484} \approx 4.95116\ \text{AU} \approx 460{,}458{,}000\ \text{miles}$$

Problems 5.6

Readiness Check

I. The equation $r = \dfrac{8}{1 - 2\cos\theta}$ is the equation of _____.

a. a parabola b. an ellipse c. a hyperbola

II. The equation $r = \dfrac{8}{2 - \cos\theta}$ is the equation of _____.

a. a parabola b. an ellipse c. a hyperbola

III. The equation $r = \dfrac{8}{2 - 2\cos\theta}$ is the equation of _____.

a. a parabola b. an ellipse c. a hyperbola

IV. Which of the following is true about the directrix of the hyperbola $r = \dfrac{12}{1 - 3\sin\theta}$?

a. It is perpendicular to the polar axis and 4 units to the left of the pole.
b. It is perpendicular to the polar axis and 4 units to the right of the pole.
c. It is parallel to the polar axis and 4 units below it.
d. It is parallel to the polar axis and 4 units above it.

V. The directrix of the ellipse $r = \dfrac{18}{6 + 2\cos\theta}$ is perpendicular to the polar axis and _____ units to the right of the pole.

a. 2 b. 3 c. 6 d. 9 e. 18

In Problems 1–12, identify and sketch each conic with focus on the pole. On your sketch indicate the directrix, center or vertex, vertices (for an ellipse or hyperbola) and axis, major axis, or transverse axis.

1. $r = \dfrac{8}{1 - \cos\theta}$

2. $r = \dfrac{8}{1 - 3\cos\theta}$

3. $r = \dfrac{8}{3 - \cos\theta}$

4. $r = \dfrac{4}{5 + \sin\theta}$

5. $r = \dfrac{4}{1 + 5\sin\theta}$

6. $r = \dfrac{4}{1 + \sin\theta}$

7. $r = \dfrac{3}{1 - \sin\theta}$

8. $r = \dfrac{3}{4 - 4\sin\theta}$

Answers to Readiness Check

I. c II. b III. a IV. c V. d

9. $r = \dfrac{3}{4 - \sin\theta}$
10. $r = \dfrac{5}{3 + 2\cos\theta}$
11. $r = \dfrac{5}{2 + 3\cos\theta}$
12. $r = \dfrac{5}{3 + 3\cos\theta}$

In Problems 13–20 find a polar equation for the conic with focus at the pole and satisfying the given conditions.

13. $e = 2$; directrix parallel to the polar axis and 3 units above it.
14. parabola; directrix perpendicular to the polar axis and 1 unit to the right of the pole.
15. parabola; directrix parallel to the polar axis and 2 units above it.
16. $e = \frac{1}{3}$; directrix parallel to the polar axis and 1 unit above it.
17. $e = \frac{2}{3}$; directrix perpendicular to the polar axis and 5 units to the left of the pole.
18. $e = 4$; directrix perpendicular to the polar axis and 3 units to the right of the pole.
19. $e = 1.5$; directrix parallel to the polar axis and 2 units below it.
20. $e = 0.9$; directrix perpendicular to the polar axis and 6 units to the right of the pole.
21. If $a > b > 0$, $c = \sqrt{a^2 - b^2}$, $e = \dfrac{c}{a}$ and $d = \dfrac{a}{e}$, show that equation (1) becomes $\dfrac{x^2}{a^2} + \dfrac{y^2}{b^2} = 1$.
22. If $a > 0$ and $b > 0$, $c = \sqrt{a^2 + b^2}$, $e = \dfrac{c}{a}$ and $d = \dfrac{a}{e}$, show that equation (1) becomes $\dfrac{x^2}{a^2} - \dfrac{y^2}{b^2} = 1$.

* 23. Let e, p, and δ be constants. Write the equation

$$r = \frac{ep}{1 - e\cos(\theta - \delta)}$$

in Cartesian coordinates. [Hint: Use the results of Section 5.5.]

24. The equation

$$r = \frac{6}{1 - 4\cos\theta}$$

is the equation of a hyperbola. Show that the asymptotes of the hyperbola make angles of $\cos^{-1}\frac{1}{4}$ and $\cos^{-1}(-\frac{1}{4})$ with the polar axis.

25. Show that the asymptotes of the hyperbola $r = \dfrac{ep}{1 - e\cos\theta}$, $e > 1$ make angles of $\cos^{-1}\dfrac{1}{e}$ and $\cos^{-1}\left(-\dfrac{1}{e}\right)$ with the polar axis.
26. For the orbit of Saturn, $e = 0.0543$ and $2a = 19.078$ AU. Write a polar equation for this orbit and find Saturn's closest (perihelion) and farthest (aphelion) distances from the sun.
27. Answer the questions in Problem 26 for the planet Mercury: $e = 0.2056$ and $2a = 0.7742$ AU.
28. Do the same for the planet Pluto: $e = 0.2481$ and $2a = 15.27$ AU.
29. Show that for a planet travelling around the sun, the perihelion is $a(1 - e)$ and the aphelion is $a(1 + e)$.
30. A satellite orbits the earth in an elliptical orbit with the earth at one focus. The satellite gets as close as 200 miles from the earth's surface and as far as 94,000 miles from the earth's surface. Find a polar equation for the orbit.

In Problems 31–36, find polar equations for each conic.

31. $\dfrac{x^2}{4} - y^2 = 1$
32. $\dfrac{x^2}{4} + \dfrac{y^2}{9} = 1$
33. $\dfrac{x^2}{16} - \dfrac{y^2}{9} = 1$
34. $\dfrac{x^2}{16} - \dfrac{y^2}{25} = 1$
35. $\dfrac{x^2}{9} + \dfrac{y^2}{25} = 1$
36. $x^2 + \dfrac{y^2}{4} = 1$

5.7 Plane Curves and Parametric Equations

In Section 4.4, we discussed vectors in $\mathbb{R}^2$ that could be written as

$$\mathbf{v} = (a, b) = a\mathbf{i} + b\mathbf{j} \tag{1}$$

In this section, we see what happens when the numbers a and b are replaced by functions $f_1(t)$ and $f_2(t)$.

Vector Function in $\mathbb{R}^2$

Let f_1 and f_2 be functions of the real variable t. Then for all values of t for which $f_1(t)$ and $f_2(t)$ are defined, we define the **vector-valued function f** by

$$\mathbf{f}(t) = (f_1(t), f_2(t)) = f_1(t)\mathbf{i} + f_2(t)\mathbf{j}. \tag{2}$$

The **domain** of **f** is the intersection of the domains of f_1 and f_2. It is a set of real numbers. The **range** of **f** is a set of vectors in $\mathbb{R}^2$.

NOTE For simplicity, we will refer to vector-valued functions as **vector functions.**

EXAMPLE 1 *Finding the Domain of a Vector-Valued Function*

Let $\mathbf{f}(t) = f_1(t)\mathbf{i} + f_2(t)\mathbf{j} = (1/t)\mathbf{i} + \sqrt{t+1}\mathbf{j}$. Find the domain of **f.**

SOLUTION The domain of **f** is the set of all t for which f_1 and f_2 are defined. Since $f_1(t)$ is defined for $t \neq 0$ and $f_2(t)$ is defined for $t \geq -1$, we see that the domain of **f** is the set $\{t\colon t \geq -1 \text{ and } t \neq 0\}$.

Let **f** be a vector function. Then for each t in the domain of **f,** the endpoint of the vector $f_1(t)\mathbf{i} + f_2(t)\mathbf{j}$ is a point (x, y) in the xy-plane, where

$$x = f_1(t) \qquad \text{and} \qquad y = f_2(t). \tag{3}$$

Plane Curves and Parametric Equations

Suppose that the interval $[a, b]$ is in the domain of the function **f** and that both f_1 and f_2 are continuous† in $[a, b]$. Then the set of points $(f_1(t), f_2(t))$ for $a \leq t \leq b$ is called a **plane curve** C. Equation (2) is called the **vector equation** of C, while equations (3) are called the **parametric equations** or **parametric representation** of C. In this context, the variable t is called a **parameter.**

NOTE Any plane curve can be thought of as the range of a vector function whose domain is restricted to the interval $[a, b]$.

EXAMPLE 2 *Parametric Equations of the Unit Circle*

Describe the curve given by the vector equation

$$\mathbf{f}(t) = (\cos t)\mathbf{i} + (\sin t)\mathbf{j}, \qquad 0 \leq t \leq 2\pi. \tag{4}$$

† A function f is **continuous** in $[a,b]$ if its graph is unbroken as it moves from the point $(a,f(a))$ to the point $(b,f(b))$.

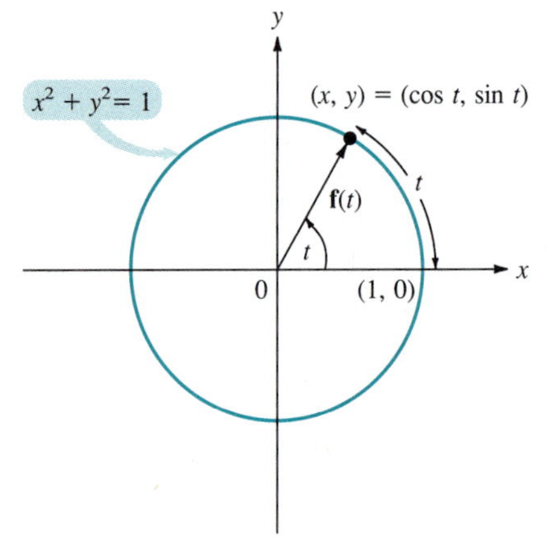

Figure 1 If $x = \cos t$ and $y = \sin t$, then $x^2 + y^2 = 1$.

SOLUTION Recall from p. 237 that if $\mathbf{v} = a\mathbf{i} + b\mathbf{j}$, then the length or magnitude of $\mathbf{v}$, denoted by $|\mathbf{v}|$, is given by $|\mathbf{v}| = \sqrt{a^2 + b^2}$. Thus

$$|\mathbf{f}(t)| = \sqrt{\cos^2 t + \sin^2 t} = \sqrt{1} = 1$$

Moreover, if we write the curve in its parametric representation, we find that

$$x = \cos t, \qquad y = \sin t, \tag{5}$$

and since $\cos^2 t + \sin^2 t = 1$, we have

$$x^2 + y^2 = 1,$$

which is the equation of the unit circle. This curve is sketched in Figure 1. Note that in the sketch the parameter t represents both the length of the arc from (1, 0) to the endpoint of the vector and the angle (measured in radians) the vector makes with the positive x-axis. The representation $x^2 + y^2 = 1$ is called the *Cartesian equation* of the curve given by (5).

NOTE As t increases from 0 to 2π, we move around the unit circle in the counterclockwise direction. If t were instead restricted to the range $0 \le t \le \pi$, then we would not get the entire circle. Rather, we would stop at the point $(\cos \pi, \sin \pi) = (-1, 0)$, which would give us the upper semicircle only.

On the other hand, suppose that $0 \le t \le 4\pi$. Starting at (1, 0) when $t = 0$, we get back to (1, 0) when $t = 2\pi$. Then, because $\cos t$ and $\sin t$ are periodic of period 2π, we simply go around the circle again. That is, the curve given by $\mathbf{f}(t) = (\cos t)\mathbf{i} + (\sin t)\mathbf{j}$, $0 \le t \le 4\pi$ is the unit circle. It appears as in Figure 1, but, in this case, it is traversed *twice*. Finally, the curve given by $\mathbf{f}(t) = (\cos t)\mathbf{i} + (\sin t)\mathbf{j}$, $0 \le t \le 2n\pi$ is the unit circle traversed n times.

Cartesian Equation of a Plane Curve

A **Cartesian equation** of the curve $f(t) = x(t)\mathbf{i} + y(t)\mathbf{j}$ is an equation relating the variables x and y only.

EXAMPLE 3 *Obtaining a Cartesian Equation from the Parametric Equations of a Curve*

Describe and sketch the curve given parametrically by $x = t + 3$, $y = t^2 - t + 2$.

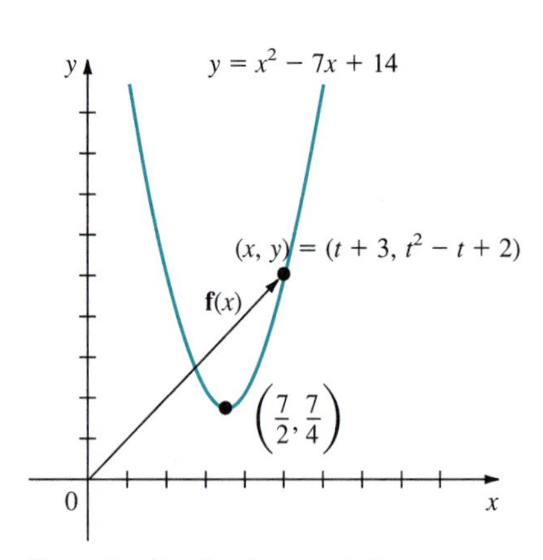

Figure 2 Graph of $x = t + 3$, $y = t^2 - t + 2$.

SOLUTION With problems of this type, the easiest thing to do is to write t as a function of x or y, if possible. Since $x = t + 3$, we immediately see that $t = x - 3$ and $y = t^2 - t + 2 = (x - 3)^2 - (x - 3) + 2 = x^2 - 7x + 14$. This is the Cartesian equation of the curve and is the equation of a parabola. It is sketched in Figure 2. Note that in the Cartesian equation of the parabola, the parameter t does not appear. ■

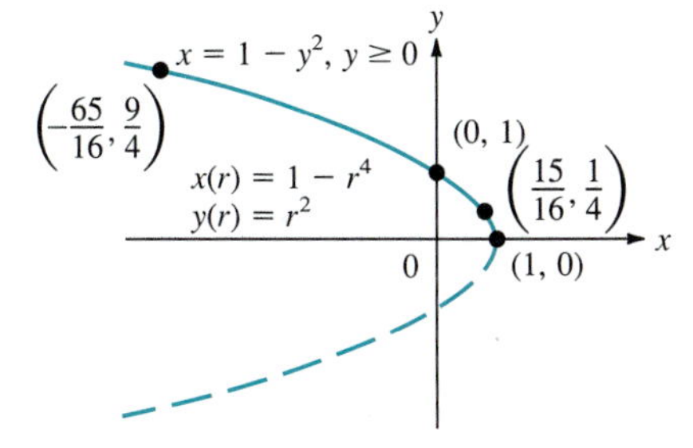

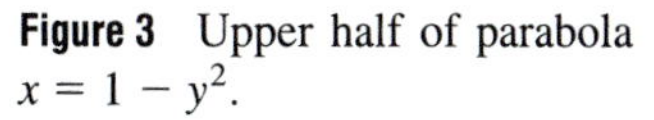
Figure 3 Upper half of parabola $x = 1 - y^2$.

EXAMPLE 4 *Obtaining a Cartesian Equation from the Parametric Equations of a Curve*

Describe and sketch the curve given by the vector equation

$$\mathbf{f}(r) = (1 - r^4)\mathbf{i} + r^2\mathbf{j}. \qquad (6)$$

SOLUTION First, we note that the parameter in this problem is r instead of t. This makes absolutely no difference. Now to get a feeling for the shape of this curve, we display in Table 1 values of x and y for various values of r. Plotting some of these points leads to the sketch in Figure 3. To write the Cartesian equation for this curve, we square both sides of the equation $y = r^2$ to obtain $y^2 = r^4$ and $x = 1 - r^4 = 1 - y^2$, which is the equation of the parabola sketched in Figure 3. Note that this curve is *not* the graph of the parabola $x = 1 - y^2$ since the parametric representation $y = r^2$ requires that y be nonnegative. Thus, the curve described by (6) is only the *upper half* of the parabola described by the equation $x = 1 - y^2$.

Table 1

r	0	$\pm\frac{1}{2}$	± 1	$\pm\frac{3}{2}$	± 2
$x = 1 - r^4$	1	$\frac{15}{16}$	0	$-\frac{65}{16}$	-15
$y = r^2$	0	$\frac{1}{4}$	1	$\frac{9}{4}$	4

Problems 5.7

Readiness Check

I. The graph of $x = 1 + t$, $y = 3 - 2t$ is the same as the graph of ______.
 a. $x + y = 4$
 b. $2x + y = 5$
 c. $y = -2x + 1$ and $x \geq 0$
 d. $|2x| + |y| = 5$

II. The graph of $x = 1 + t^2$, $y = 3 + t^2$ is the same as the graph of ______.
 a. $y = x + 2$
 b. $x = y + 2$
 c. $y = x + 2$ and $x \geq 0$
 d. $y = x + 2$ and $x \geq 1$

III. The graph of $x = \cos t$, $y = -\cos t$ is the same as the graph of ______.
 a. $x = -y$
 b. $x = -y$ and $x \geq 0$
 c. $y = -x$ and $x \leq 1$
 d. $y = -x$ and $-1 \leq x \leq 1$

IV. The unit circle $x^2 + y^2 = 1$ can be parameterized by ______.
 a. $x = \sin t$, $y = \cos t$
 b. $x = \cos(t - \pi/2)$, $y = \sin(t - \pi/2)$
 c. $x = -\cos t$, $y = \sin t$
 d. $x = \cos 2t$, $y = \sin 2t$

Answers to Readiness Check

I. b II. d III. d IV. a, b, c, d

In Problems 1–4, find the domain of the given vector-valued function.

1. $\mathbf{f}(t) = \frac{1}{t}\mathbf{i} + \frac{1}{t-1}\mathbf{j}$
2. $\mathbf{f}(t) = \sqrt{t}\mathbf{i} + \frac{1}{t}\mathbf{j}$
3. $\mathbf{f}(s) = \frac{1}{s^2-1}\mathbf{i} + (s^2-1)\mathbf{j}$
4. $\mathbf{f}(s) = e^{1/s}\mathbf{i} + e^{-1/(s+1)}\mathbf{j}$ [Hint: $e \approx 2.718$; see Section 7.2]

In Problems 5–18, find the Cartesian equation for each curve; then sketch the curve. [Skip problems 13, 14 and 16 if you have not covered the material in Chapter 7.]

5. $\mathbf{f}(t) = t^2\mathbf{i} + 2t\mathbf{j}$
6. $\mathbf{f}(t) = (2t-3)\mathbf{i} + t^2\mathbf{j}$
7. $\mathbf{f}(t) = t^2\mathbf{i} + t^3\mathbf{j}$
8. $\mathbf{f}(t) = 3(\sin t)\mathbf{i} + 3(\cos t)\mathbf{j}$
9. $\mathbf{f}(t) = (2t-1)\mathbf{i} + (4t+3)\mathbf{j}$
10. $\mathbf{f}(t) = (t^4+t^2+1)\mathbf{i} + t^2\mathbf{j}$
11. $\mathbf{f}(t) = t^2\mathbf{i} + t^8\mathbf{j}$
12. $\mathbf{f}(t) = t^3\mathbf{i} + (t^9-1)\mathbf{j}$
13. $\mathbf{f}(t) = t\mathbf{i} + e^t\mathbf{j}$
14. $\mathbf{f}(t) = e^t\mathbf{i} + e^{2t}\mathbf{j}$
15. $\mathbf{f}(t) = (t^2+t-3)\mathbf{i} + \sqrt{t}\mathbf{j}$

* 16. $\mathbf{f}(t) = e^t(\sin t)\mathbf{i} + e^t(\cos t)\mathbf{j}$
[Hint: Show that $|\mathbf{f}(t)| = e^t$.]

17. $\mathbf{f}(t) = \mathbf{i} + (\tan t)\mathbf{j}$

* 18. $\mathbf{f}(t) = \frac{6t}{1+t^3}\mathbf{i} + \frac{6t^2}{1+t^3}\mathbf{j}$

19. Verify that a vector equation for the ellipse $(x^2/a^2) + (y^2/b^2) = 1$ is

$$\mathbf{f}(t) = a(\cos t)\mathbf{i} + b(\sin t)\mathbf{j}.$$

20. Verify that a parametric representation for the hyperbola $(x^2/a^2) - (y^2/b^2) = 1$ is

$$x(\theta) = a \sec \theta \quad \text{and} \quad y(\theta) = b \tan \theta.$$

* 21. Suppose that a wheel of radius r is rolling in a straight line without slipping. Let P be a fixed point on the wheel which is distance s from the center. As the wheel rotates, the point P traces out a curve known as a **trochoid.**† Look at the figure in the next column. Suppose that P moves through an angle of α radians to reach its position shown in part (b) of the figure.

a. Show that the new position vector is $\overrightarrow{0P} = \overrightarrow{0R} + \overrightarrow{RC} + \overrightarrow{CP}$.
b. Show that $\overrightarrow{0R} = \alpha r\mathbf{i}$.
c. Show that $\overrightarrow{CP} = s[(\cos\theta)\mathbf{i} + (\sin\theta)\mathbf{j}]$.
d. Show that $\overrightarrow{0P} = [\alpha r + s\cos\theta]\mathbf{i} + [r + s\sin\theta]\mathbf{j}$.
e. Show that $\theta = (3\pi/2) - \alpha$.

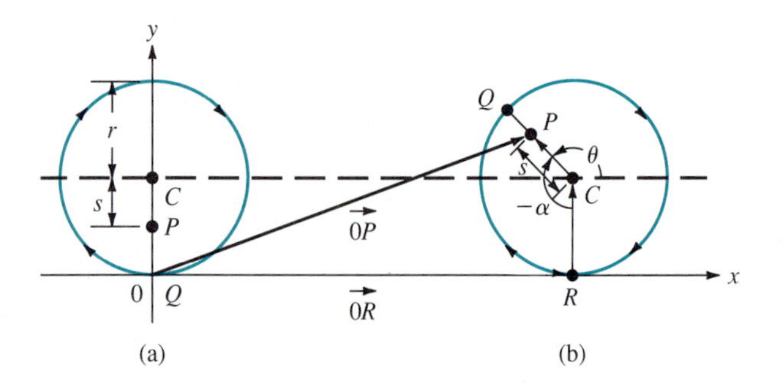

f. Show that $\cos\theta = -\sin\alpha$ and $\sin\theta = -\cos\alpha$.
g. Show that the trochoid satisfies the parametric equations

$$x = r\alpha - s\sin\alpha \quad \text{and} \quad y = r - s\cos\alpha.$$

22. A point is located 25 cm from the center of a wheel 1 m in diameter. Find a parametric representation of the trochoid traced by that point as the wheel rolls.

* 23. A trochoid generated by a point P on the circumference of the wheel (i.e., $s = r$) has the special name **cycloid.**†
a. Find parametric equations for the cycloid.
b. Verify that the portion of the cycloid for $0 \le x \le 2r$ has the Cartesian equation

$$x = r\cos^{-1}\left(\frac{r-y}{y}\right) - \sqrt{2ry - y^2}.$$

24. A point is located on the circumference of a wheel 1 m in diameter. Find a parametric representation for the cycloid traced by that point as the wheel rolls.

* 25. A **hypocycloid**‡ is a curve generated by the motion of a point P on the circumference of a circle that rolls internally (without slipping) on a larger circle (see below). Assume

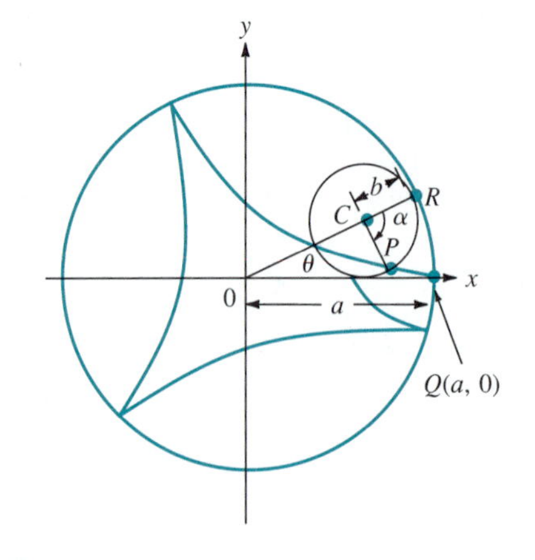

†From the Greek word *trochos,* meaning *wheel.*

†From the Greek word *kyklos,* meaning *circle.*
‡The *hypo* comes from the Greek *hupo,* meaning *under.*

that the radius of the larger circle is a while that of the smaller circle is b. Let θ be the angle indicated in the figure. Verify that the hypocycloid has parametric representation

$$x = (a - b)\cos\theta + b\cos\left[\left(\frac{a-b}{b}\right)\theta\right],$$

$$y = (a - b)\sin\theta - b\sin\left[\left(\frac{a-b}{b}\right)\theta\right].$$

[Hint: First show that $a\theta = b\alpha$.]

26. If $a = 4b$ in the previous figure, then the curve generated is called a **hypocycloid of four cusps.** Sketch this curve.
27. Verify that the hypocycloid of four cusps has the Cartesian equation

$$x^{2/3} + y^{2/3} = a^{2/3}.$$

[Hint: Use the identities $\cos 3\theta = 4\cos^3\theta - 3\cos\theta$ and $\sin 3\theta = 3\sin\theta - 4\sin^3\theta$ together with the facts that $a - b = 3b$ and $(a - b)/b = 3$.]

28. The figure in the next column shows three points P, Q, R on a straight line.
    a. Show that $\overrightarrow{PR} = t\overrightarrow{PQ}$ for some real number t.
    b. Show that $\overrightarrow{0R} = \overrightarrow{0P} + \overrightarrow{PR} = \overrightarrow{0P} + t\overrightarrow{PQ}$.
    c. Use the results of parts (a) and (b) to show that the line passing through the points (x_1, y_1) and (x_2, y_2) satisfies the parametric equations

$$x = x_1 + t(x_2 - x_1) \quad \text{and}$$
$$y = y_1 + t(y_2 - y_1).$$

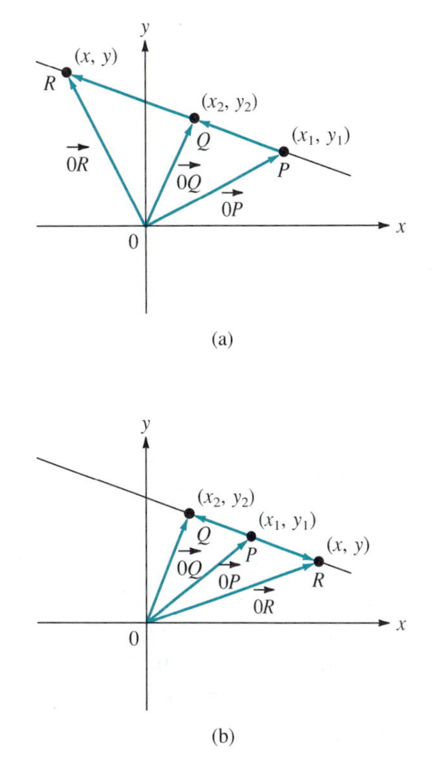

In Problems 29–34, use the result of Problem 27 to find a parametric representation of the straight line that passes through the two given points.

29. (2, 4); (1, 6) 30. (−3, 2); (0, 4)
31. (3, 5); (−1, −7) 32. (4, 6); (7, 9)
33. (−2, 3); (4, 7) 34. (−4, 0); (3, −2)
35. Use the substitution $\tan\theta = \sin t$ to show that the lemniscate $r^2 = \cos 2\theta$ has the Cartesian parametric representation

$$x = \frac{\cos t}{1 + \sin^2 t} \quad \text{and} \quad y = \frac{\sin t \cos t}{1 + \sin^2 t}.$$

Summary Outline of Chapter 5

- An **ellipse** is the set of points (x, y) such that the sum of the distances from (x, y) to two given points is fixed. Each of the two points is called a **focus** of the ellipse. p. 278
- **The standard equation of the ellipse** is $\frac{x^2}{a^2} + \frac{y^2}{b^2} = 1$. p. 280
- If $a > b$, the line segment joining $(-a, 0)$ to $(a, 0)$ is the **major axis,** the line segment joining $(0, -b)$ to $(0, b)$ is the **minor axis,** and the points $(-a, 0)$ and $(a, 0)$ are **vertices** of the ellipse. If $b > a$, the major and minor axes are reversed. If $a = b$, the ellipse is a **circle.** The intersection of the axes is the **center** of the ellipse. pp. 280, 281

- The **eccentricity,** e, of an ellipse is given by $e = \dfrac{c}{a}$ if $a \geq b$ and $e = \dfrac{c}{b}$ if $b \geq a$, where $2c$ is the distance between the foci. p. 282

- A **translated ellipse** has the standard equation

$$\frac{(x - x_0)^2}{a^2} + \frac{(y - y_0)^2}{b^2} = 1$$ p. 283

- A **parabola** is the set of points (x, y) equidistant from a fixed point called the **focus** and a fixed line (that does not contain the focus) called the **directrix.** p. 289

- The **standard equations** of a parabola are $x^2 = 4cy$ (which opens upward if $c > 0$ and downward if $c < 0$) and $y^2 = 4cx$ (which opens to the right if $c > 0$ and to the left if $c < 0$). pp. 290, 291

- The line about which a parabola is symmetric is the **axis** of the parabola. p. 290

- The point at which the axis and parabola intersect is the **vertex** of the parabola. p. 290

- A **translated parabola** takes the standard form $(x - x_0)^2 = 4c(y - y_0)$ or $(y - y_0)^2 = 4c(x - x_0)$. p. 292

- A **hyperbola** is a set of points (x, y) with the property that the positive difference between the distances from (x, y) and each of the two distinct points, called **foci,** is a constant. p. 297

- The **principal axis** of a hyperbola is the line containing the foci. The points of intersection of the principal axis and the hyperbola are the **vertices,** and the line segment joining the vertices is the **transverse axis.** The midpoint of the line segment joining the foci is the **center** of the hyperbola. p. 300

- The **standard equations** of a hyperbola are

$$\frac{x^2}{a^2} - \frac{y^2}{b^2} = 1 \quad \text{and} \quad \frac{y^2}{a^2} - \frac{x^2}{b^2} = 1$$ pp. 300–302

- A **translated hyperbola** takes the standard form

$$\frac{(x - x_0)^2}{a^2} - \frac{(y - y_0)^2}{b^2} = 1 \quad \text{or} \quad \frac{(y - y_0)^2}{a^2} - \frac{(x - x_0)^2}{b^2} = 1$$ p. 302

- When the coordinate axes (x-axis and y-axis) are **rotated** through an angle of θ to obtain new x'- and y'-axes, we have

$$x = x' \cos\theta - y' \sin\theta$$
$$y = x' \sin\theta + y' \cos\theta$$ p. 309

and

$$x' = x \cos\theta + y \sin\theta$$
$$y' = -x \sin\theta + y \cos\theta$$ p. 309

- The second-degree equation $Ax^2 + Bxy + Cy^2 + Dx + Ey + F = 0$ is the equation of

(i) a parabola, line, two parallel lines, or is imaginary if $B^2 - 4AC = 0$.
(ii) a circle, ellipse, single point, or is imaginary if $B^2 - 4AC < 0$.
(iii) a hyperbola or two intersecting lines if $B^2 - 4AC > 0$. p. 312

- The polar equation $r = \dfrac{ep}{1 \pm e \cos\theta}$ is the equation of

  a parabola with horizontal axis if $e = 1$
  an ellipse with horizontal major axis if $0 < e < 1$
  a hyperbola with horizontal transverse axis if $e > 1$.

  In each case, the focus is at the pole, and the directrix is perpendicular to the polar axis. pp. 316, 317

- The polar equation $r = \dfrac{ep}{1 \pm e \sin\theta}$ is the equation of:

  a parabola with vertical axis if $e = 1$
  an ellipse with vertical major axis if $0 < e < 1$
  a hyperbola with vertical transverse axis if $e > 1$.

  In each case, the focus is at the pole, and the directrix is parallel to the polar axis. pp. 316, 317

- **Vector Function in $\mathbb{R}^2$** Let f_1 and f_2 be functions of the real variable t. Then for all values of t for which $f_1(t)$ and $f_2(t)$ are defined, we define the **vector-valued function f** by

  $$\mathbf{f}(t) = (f_1(t), f_2(t)) = f_1(t)\mathbf{i} + f_2(t)\mathbf{j}. \qquad (*)$$

  The **domain** of **f** is the intersection of the domains of f_1 and f_2. It is a set of real numbers. The **range** of **f** is a set of vectors in $\mathbb{R}^2$. p. 323

- **Plane Curves and Parametric Equations** Let **f** be a vector function. Then for each t in the domain of **f,** the endpoint of the vector $f_1(t)\mathbf{i} + f_2(t)\mathbf{j}$ is a point (x, y) in the xy-plane, where

  $$x = f_1(t) \quad \text{and} \quad y = f_2(t). \qquad (**)$$

  Suppose that the interval $[a, b]$ is in the domain of the function **f** and that both f_1 and f_2 are continuous in $[a, b]$. Then the set of points $(f_1(t), f_2(t))$ for $a \le t \le b$ is called a **plane curve** C. Equation $(*)$ is called the **vector equation** of C, while equations $(**)$ are called the **parametric equations** or **parametric representation** of C. In this context, the variable t is called a **parameter.** p. 323

- **Cartesian Equation of a Plane Curve** A **Cartesian equation** of the curve $f(t) = x(t)\mathbf{i} + y(t)\mathbf{j}$ is an equation relating the variables x and y only. p. 324

Review Exercises for Chapter 5

In Exercises 1–14, identify the type of conic. If it is an ellipse (or circle), give its foci, center, vertices, major and minor axes, and eccentricity. If it is a parabola, give its focus, directrix, axis, and vertex. If it is a hyperbola, give its foci, conjugate axis, transverse axis, center, vertices, asymptotes, and eccentricity. Finally, sketch the curve (if it is not degenerate).

1. $\dfrac{x^2}{9} + \dfrac{y^2}{16} = 1$
2. $\dfrac{x^2}{9} - \dfrac{y^2}{16} = 1$
3. $\dfrac{x^2}{9} - \dfrac{y}{16} = 0$
4. $\dfrac{y^2}{16} - \dfrac{x}{9} = 0$
5. $\dfrac{y^2}{9} - \dfrac{x^2}{16} = 1$
6. $\dfrac{x^2}{16} + \dfrac{y^2}{9} = 1$
7. $\dfrac{(x-1)^2}{4} + \dfrac{(y+1)^2}{9} = 1$
8. $\dfrac{(x+2)^2}{25} - \dfrac{(y+3)^2}{4} = 1$
9. $\dfrac{(x+2)^2}{25} + \dfrac{(y-5)^2}{25} = 0$
10. $x^2 + 2x + y^2 + 2y = 0$
11. $x^2 + 2x - y^2 + 2y = 0$
12. $x^2 + 2x - 2y = 0$
13. $4x^2 + 4x + 3y^2 + 24y = 5$
14. $-3x^2 + 6x + 2y^2 + 4y = 6$
15. Find the equation of an ellipse with foci at $(3, 0)$ and $(-3, 0)$ and eccentricity 0.6.

16. Find the equation of the parabola with focus $(3, 0)$ and directrix the line $x = -4$.
17. Find the equation of the hyperbola with foci $(0, 3)$ and $(0, -3)$ and vertices $(0, 2)$ and $(0, -2)$.
18. Find the equation of the hyperbola centered at $(0, 0)$ with vertices at $(0, 3)$ and $(0, -3)$ that is asymptotic to the lines $y = \pm 5x$.
19. Describe the curve obtained from the graph of $2x^2 + xy + 2y^2 = 10$. [Hint: Rotate through an angle of $\pi/4$.]
20. Find the equation of the line obtained by rotating the line $2x - 3y = 7$ through an angle of $\pi/3$.

In Exercises 21–26, find a rotation of coordinate axes in which the given equation written in the new coordinates has no xy term. Describe each curve and then sketch it.

21. $x^2 + 4xy + 4y^2 = 9$
22. $x^2 + 4xy - 4y^2 = 9$
23. $xy = -3$
24. $4x^2 + 3xy + y^2 = 1$
25. $-2x^2 + 3xy + 4y^2 = 5$
26. $x^2 + 2xy + y^2 = 6$

In Exercises 27–31, identify and sketch each conic with focus at the pole. On your sketch indicate the directrix, center or vertex, vertices (for an ellipse or hyperbola) and axis, and major axis or transverse axis.

27. $r = \dfrac{6}{1 + 2\cos\theta}$
28. $r = \dfrac{5}{3 - 2\sin\theta}$
29. $r = \dfrac{3}{2 - 2\cos\theta}$
30. $r = \dfrac{2}{4 + 2\sin\theta}$
31. $r = \dfrac{3}{2 + 5\sin\theta}$

In Exercises 32–34, find a polar equation for the conic with focus at the pole and satisfying the given conditions.

32. parabola; directrix parallel to the polar axis and 2 units above it.
33. $e = 2$; directrix perpendicular to the polar axis and 1 unit to the left of the pole.
34. $e = \frac{1}{2}$; directrix perpendicular to the polar axis and 3 units to the right of the pole.

In Problems 35–42, find a Cartesian equation for the curve and then sketch the curve in the xy-plane.

35. $\mathbf{f}(t) = t\mathbf{i} + 2t\mathbf{j}$
36. $\mathbf{f}(t) = (2t - 6)\mathbf{i} + t^2\mathbf{j}$
37. $\mathbf{f}(t) = t^2\mathbf{i} + (2t - 6)\mathbf{j}$
38. $\mathbf{f}(t) = t^2\mathbf{i} + t^4\mathbf{j}$
39. $\mathbf{f}(t) = (\cos 4t)\mathbf{i} + (\sin 4t)\mathbf{j}$
40. $\mathbf{f}(t) = 4(\sin t)\mathbf{i} + 9(\sin t)\mathbf{j}$
41. $\mathbf{f}(t) = t^6\mathbf{i} + t^2\mathbf{j}$
42. $\mathbf{f}(t) = e^t(\cos t)\mathbf{i} + e^t(\sin t)\mathbf{j}$ [If you have seen the material in Section 7.2]

Chapter 6

Complex Numbers

6.1 Introduction to Complex Numbers

Consider the quadratic equation

$$ax^2 + bx + c = 0 \tag{1}$$

According to the quadratic formula, the solutions to (1) are

$$x = \frac{-b \pm \sqrt{b^2 - 4ac}}{2a}$$

The number $b^2 - 4ac$ is called the **discriminant** of the quadratic equation.

EXAMPLE 1 *Using the Quadratic Formula*

The solutions to $2x^2 + 3x - 6 = 0$ are

$$x = \frac{-3 \pm \sqrt{3^2 - 4(2)(-6)}}{2 \cdot 2} = \frac{-3 \pm \sqrt{57}}{4}$$

or

$$x = \frac{-3 + \sqrt{57}}{4} \approx 1.1375 \quad \text{and} \quad x = \frac{-3 - \sqrt{57}}{4} \approx -2.6375$$

We observe that if the discriminant, $b^2 - 4ac$, is negative, then the equation has no real roots. The problem is that no negative number has a real square root. For example,

$$x^2 + 16 = 0$$

leads to

$$x^2 = -16$$

or

$$x = \pm\sqrt{-16}$$

But there is no real number whose square is -16. However, as we will soon see, the equation $x^2 + 16 = 0$ *does* have two roots.

In order to describe these roots, we must extend our number system. We do this by introducing a new kind of number.

Definition of the Imaginary Unit

The **imaginary unit** i is defined by

$$i^2 = -1 \tag{2}$$

With this definition we write

$$i = \sqrt{-1} \tag{2'}$$

We interpret this to mean that i is the square root of -1. Using this definition, we can define the square root of a negative number.

Principal Square Root of a Negative Number

If $a > 0$ (so $-a < 0$), then the **principal square root** of $-a$ is given by

$$\sqrt{-a} = \sqrt{a}i, \qquad a > 0 \tag{3}$$

WARNING Do *not* put the i under the radical (the square root sign) in (3). It is $\sqrt{a}i$, *not* $\sqrt{ai}$. ■

EXAMPLE 2 ***The Principal Square Roots of Four Negative Numbers***

(a) $\sqrt{-4} = \sqrt{4}i = 2i$

(b) $\sqrt{-16} = \sqrt{16}i = 4i$

(c) $\sqrt{-\frac{1}{9}} = \sqrt{\frac{1}{9}}i = \frac{1}{3}i$

(d) $\sqrt{-57} = \sqrt{57}i \approx 7.549834435i$

FOCUS ON
The Word *Imaginary*

You should not be troubled by the term *imaginary.* It's just a name. The British mathematician Alfred North Whitehead, in the chapter on imaginary numbers in his *Introduction to Mathematics*, wrote:

> *At this point it may be useful to observe that a certain type of intellect is always worrying itself and others by discussion as to the applicability of technical terms. Are the incommensurable numbers properly called numbers? Are the positive and negative numbers really numbers? Are the imaginary numbers imaginary, and are they numbers?—are types of such futile questions. Now, it cannot be too clearly understood that, in science, technical terms are names arbitrarily assigned, like Christian names to children. There can be no question of the names being right or wrong. They may be judicious or injudicious; for they can sometimes be so arranged as to be easy to remember, or so as to suggest relevant and important ideas. But the essential principle involved was quite clearly enunciated in Wonderland to Alice by Humpty Dumpty, when he told her, apropos of his use of words, 'I pay them extra and make them mean what I like'. So we will not bother as to whether imaginary numbers are imaginary, or as to whether they are numbers, but will take the phrase as the arbitrary name of a certain mathematical idea, which we will now endeavour to make plain.*

Definition of a Complex Number

A **complex number** is an expression that can be written in the form

$$a + bi \tag{4}$$

where a and b are real numbers.

a is called the **real part** of the complex number.

b is called the **imaginary part** of the complex number.

EXAMPLE 3 *Four Complex Numbers*

The following are complex numbers:

(a) $1 + i$ real part $= 1$; imaginary part $= 1$
(b) $2 - 7i$ real part $= 2$; imaginary part $= -7$
(c) 6 real part $= 6$; imaginary part $= 0$ A real number
(d) $3i$ real part $= 0$; imaginary part $= 3$ A pure imaginary number

Two complex numbers are **equal** if their real and imaginary parts are equal. That is,

$$a + bi = c + di \quad \text{if and only if} \quad a = c \quad \text{and} \quad b = d$$

We make two observations:

1. In part (c) of Example 3 we wrote the number 6 as a complex number. There is nothing special about the number 6, and we see that *every real number is a complex number*. Thus the definition of a complex number extends our number system.
2. The number $3i$ in part (d) of Example 3 is a complex number with real part zero. Such a number is called **pure imaginary.**

The Algebra of Complex Numbers

When we add, subtract, multiply, and divide complex numbers, we use the same rules we use for algebraic expressions.

Rules for Addition and Subtraction

$$(a + bi) + (c + di) = (a + c) + (b + d)i \qquad (5)$$

$$(a + bi) - (c + di) = (a - c) + (b - d)i \qquad (6)$$

EXAMPLE 4 ***Adding Two Complex Numbers***

Add $(2 + 5i) + (7 - 2i)$.

SOLUTION $(2 + 5i) + (7 - 2i) = (2 + 7) + [5 + (-2)]i = 9 + 3i.$

Use equation (5) ■

EXAMPLE 5 ***Subtracting Two Complex Numbers***

Compute $(-2 + 3i) - (5 + 8i)$.

SOLUTION $(-2 + 3i) - (5 + 8i) = (-2 - 5) + (3 - 8)i = -7 - 5i.$

Use equation (6)

Multiplication

The distributive properties used to multiply algebraic expressions hold for complex numbers as well. We use them for complex number multiplication.

EXAMPLE 6 ***Multiplying Two Complex Numbers***

Compute $(3 + 4i)(5 + 2i)$.

SOLUTION

$$\begin{aligned}(3 + 4i)(5 + 2i) &= 3(5 + 2i) + (4i)(5 + 2i)\\ &= (3)(5) + (3)(2i) + (4i)(5) + (4i)(2i)\\ &= 15 + 6i + 20i + 8i^2\end{aligned}$$

But $i^2 = -1$, so $8i^2 = -8$, and we may continue:

$$= 15 + 26i - 8 = 7 + 26i$$ ■

EXAMPLE 7 ***Obtaining Powers of i***

We can multiply i by itself repeatedly to obtain powers of i:

$$\begin{array}{ll} i^1 = i & i^5 = i \\ i^2 = -1 & i^6 = -1 \\ i^3 = -i & i^7 = -i \\ i^4 = 1 & i^8 = 1 \end{array}$$

etc.

For example,

$$i^{103} = i^{100}i^3 = (i^4)^{25}i^3 = 1^{25}i^3 = i^3 = -i \quad \blacksquare$$

EXAMPLE 8 ***A Product of Two Complex Numbers That Is a Real Number***

Compute $(2 + 3i)(2 - 3i)$.

SOLUTION
$$\begin{aligned}(2 + 3i)(2 - 3i) &= 2(2 - 3i) + 3i(2 - 3i)\\ &= 4 - 6i + 6i - 9i^2 = 4 - 9(-1)\\ &= 4 + 9 = 13\end{aligned}$$

Definition of the Complex Conjugate

The **complex conjugate** of the number $a + bi$ is the number $a - bi$. It is denoted by $\overline{a + bi}$. That is,

$$\overline{a + bi} = a - bi$$

We compute

$$\begin{aligned}(a + bi)(a - bi) &= a^2 - abi + abi - b^2i^2 = a^2 - b^2(-1)\\ &= a^2 + b^2\end{aligned}$$

Thus, *when we multiply a nonzero complex number by its conjugate, we get a positive real number:*

The Product of a Complex Number and Its Conjugate Is a Real Number

$$(a + bi)(\overline{a + bi}) = (a + bi)(a - bi) = a^2 + b^2 \tag{7}$$

Equation (7) enables us to factor expressions that we couldn't factor before. This is true because the product of complex conjugates is the sum of two squares.

EXAMPLE 9 ***Factoring a Sum of Squares***

Factor $x^2 + 16$.

SOLUTION In (7) set $x^2 = a^2$ and $16 = b^2$. Then

$$x = a, \qquad 4 = b$$

and

$$x^2 + 16 = (x + 4i)(x - 4i)$$

WARNING It is true that

$$\sqrt{a}\sqrt{b} = \sqrt{ab} \qquad \text{if } a > 0,\ b > 0$$

This equation is *false* if $a < 0$ and $b < 0$. For example,

$$\sqrt{-1}\sqrt{-1} = i \cdot i = i^2 = -1$$

but

$$\sqrt{(-1)(-1)} = \sqrt{1} = 1$$

That is,

$$\sqrt{-1}\sqrt{-1} \neq \sqrt{(-1)(-1)} \quad \blacksquare$$

However, the following is true: If $a > 0$ and $b < 0$ or $a < 0$ and $b > 0$, then

$$\sqrt{ab} = \sqrt{a}\sqrt{b}$$

EXAMPLE 10 ***Simplifying a Radical***

$$\sqrt{-80} = \sqrt{5(-16)} = \sqrt{5}\sqrt{-16} = \sqrt{5}(4i) = 4\sqrt{5}i$$

Division of Complex Numbers

To divide two complex numbers, multiply the numerator and denominator by the conjugate of the denominator. This results in a positive real number in the denominator.

EXAMPLE 11 ***Dividing Two Complex Numbers***

Compute $\dfrac{4-5i}{2+3i}$.

SOLUTION

$$\frac{4-5i}{2+3i} = \frac{(4-5i)(2-3i)}{(2+3i)(2-3i)} \overset{\text{See Example 8}}{=} \frac{(8-15)+(-12-10)i}{13}$$

$$= \frac{-7-22i}{13} = -\frac{7}{13} - \frac{22}{13}i \quad \blacksquare$$

EXAMPLE 12 ***Dividing Two Complex Numbers***

Compute $\dfrac{3-7i}{2i}$.

SOLUTION The conjugate of $2i$ is $-2i$. We obtain

$$\frac{3-7i}{2i} = \frac{(3-7i)(-2i)}{(2i)(-2i)} = \frac{14i^2-6i}{-4i^2}$$

$$= \frac{-14-6i}{4} = -\frac{7}{2} - \frac{3}{2}i$$

Let z be a nonzero complex number with conjugate $\bar{z}$. Then

$$1 = z \cdot \frac{1}{z} \underset{\text{Multiply and divide by } \bar{z}}{=} z \cdot \frac{\bar{z}}{z\bar{z}} \tag{8}$$

The number $\frac{1}{z}$ is called the **inverse** of z. From (8), we see that the inverse of z is given by

$$\frac{1}{z} = \frac{\bar{z}}{z\bar{z}} \tag{9}$$

EXAMPLE 13 *Computing the Inverse of a Complex Number*

Compute the inverse of $2 - 3i$.

SOLUTION $\overline{2 - 3i} = 2 + 3i$ and so, from (9), we obtain

$$\frac{1}{2-3i} = \frac{\overline{2-3i}}{(2-3i)(\overline{2-3i})} \underset{\text{From (7)}}{=} \frac{2+3i}{4+9} = \frac{2+3i}{13} = \frac{2}{13} + \frac{3}{13}i$$

Quadratic Equations with Complex Roots

We have seen that the quadratic equation $ax^2 + bx + c = 0$ has two distinct real roots if the discriminant $b^2 - 4ac > 0$ and one double root if $b^2 - 4ac = 0$. We now discuss the case $b^2 - 4ac < 0$.

The Solutions to a Quadratic Equation with a Negative Discriminant

If the discriminant, $b^2 - 4ac$, is negative, then $4ac - b^2 > 0$, and two solutions to the quadratic equation (1) are

$$x = \frac{-b + \sqrt{4ac - b^2}\,i}{2a} \quad \text{and} \quad x = \frac{-b - \sqrt{4ac - b^2}\,i}{2a} \tag{10}$$

EXAMPLE 14 *A Quadratic Equation with Complex Conjugate Roots*

Solve the quadratic equation $x^2 + 2x + 2 = 0$.

SOLUTION The discriminant is $4 - 4 \cdot 2 = -4 < 0$. Then the roots are

$$x = \frac{-2 \pm \sqrt{-4}}{2} = \frac{-2 \pm \sqrt{4}\,i}{2} = \frac{-2 \pm 2i}{2} = -1 \pm i$$

We check one of them:

$$(-1 + i)^2 + 2(-1 + i) + 2 = 1 - 2i + i^2 - 2 + 2i + 2$$
$$= 1 - 2i - 1 - 2 + 2i + 2 = 0 \quad ■$$

EXAMPLE 15 ***Quadratic Equation With Complex Conjugate Roots***

Solve the quadratic equation $3x^2 + 5x + 4 = 0$.

SOLUTION $b^2 - 4ac = 5^2 - 4 \cdot 3 \cdot 4 = 25 - 48 = -23$. Thus solutions are

$$x = \frac{-5 + \sqrt{23}i}{6} \quad \text{and} \quad x = \frac{-5 - \sqrt{23}i}{6}$$

We may write these two solutions together as $x = \dfrac{-5 \pm \sqrt{23}i}{6}$

FOCUS ON
A Short History of Complex Numbers

Complex numbers came into use in the same way we have used them—as solutions to quadratic equations. One of the first to study quadratic equations was the Greek mathematician Diophantus of Alexandria, who lived about A.D. 250. Diophantus wrote a treatise called *Arithmetica,* in which he solved 189 problems involving quadratic equations, but did not describe a general method for solving such equations. In his solutions, he gave answers only when the roots were positive rational numbers. When roots were negative, irrational, or complex, he ignored them. In fact, when an equation had two positive rational roots, he gave only the larger of them.

The first real use of complex numbers as solutions to quadratic equations was given by the Italian mathematician Gerolamo Cardano (1501–1576). Cardano was born in Pavia in 1501 as the illegitimate son of a jurist and developed into a man of passionate contrasts. In his autobiography *De Vita Propria (Book of My Life),* he described himself as high-tempered, devoted to erotic pleasures, vindictive, humorless, and cruel. He gambled and played chess as an escape from "poverty, chronic illness and injustice." He commenced his turbulent professional life as a doctor, studying, teaching, and writing mathematics while practicing his profession. He once traveled as far as Scotland, and, upon his return to Italy, he successively held important chairs at the Universities of Pavia and Bologna. In 1570 he was imprisoned for heresy for casting the horoscope of Jesus. Shortly thereafter, he was hired as the personal astrologer of the Pope. (Not all mathematicians lead dull lives!) He died in Rome in 1576, by his own hand, one story says, so as to fulfill his earlier astrological prediction of the date of his death.

GEROLAMO CARDANO
(New York Public Library Collection)

LEONHARD EULER
(Library of Congress)

Cardano introduced complex numbers as a way to solve the following problem: Find two numbers whose sum is 10 and whose product is 40. If x and y denote the two numbers, we have

$$x + y = 10 \quad \text{or} \quad y = 10 - x$$
$$xy = 40$$
$$x(10 - x) = 40$$
$$-x^2 + 10x = 40$$
$$0 = x^2 - 10x + 40$$
$$x = \frac{10 \pm \sqrt{100 - 160}}{2} = \frac{10 \pm \sqrt{-60}}{2}$$
$$= \frac{10 \pm \sqrt{4(-15)}}{2} = \frac{10 \pm 2\sqrt{-15}}{2}$$
$$= 5 \pm \sqrt{-15}$$

Cardano obtained these two roots, but didn't really know what to do with them. He wrote, "putting aside the mental tortures involved," multiply $5 + \sqrt{-15}$ and $5 - \sqrt{-15}$ to obtain the product $25 - (-15) = 40$. He added, "So progresses arithmetic subtlety the end of which, as is said, is as refined as it is useless."

Complex numbers continued to plague mathematicians for the next three centuries. The German mathematician Gottfried Wilhelm Leibniz (1646–1716), codiscoverer (with Newton) of calculus, called the number i an "amphibean between being and nonbeing." In 1768 the great Swiss mathematician Leonhard Euler (1707–1783) wrote

> *Because all conceivable numbers are either greater than 0 or less than 0 or equal to 0, then it is clear that the square roots of negative numbers cannot be included among the possible numbers. Consequently we must say that these are impossible numbers. And this circumstance leads us to the concept of these numbers, which by their nature are impossible, and ordinarily are called imaginary or fancied numbers, because they exist only in the imagination.*

Actually, Euler made some use of complex numbers even though he called them "impossible." For example, in solving Cardano's problem he also obtained the "solution" $5 \pm \sqrt{-15}$. He then used this result to prove that the problem cannot be solved. Euler was also the first to use the letter i to denote $\sqrt{-1}$.

Complex numbers were not put on a concrete footing until the 19th century. By that time mathematicians and physicists had found a link between complex numbers and trigonometric functions. Today the study of complex numbers and complex functions is an important tool in solving a wide variety of problems involving cyclical or periodic activity. Such applications arise in all branches of science as well as in business and economics.

Problems 6.1

Readiness Check

I. Which of the following is true about $\sqrt{-16}$?
a. Its conjugate is $-4i$.
b. Its complex form is $4 + i$.
c. When it is written in $a + bi$ form, $b = -4$.
d. Its conjugate is -16.

II. Which of the following equations has solutions $\pm 7i$?
a. $x^2 = 49$ b. $x^2 - 7ix = 0$
c. $x^2 + 49 = 0$ d. $x^2 = -7ix$

III. Which of the following equations has the solutions $2 + i$ and $2 - i$?
a. $x^2 = 5 - 4x$ b. $x^2 - 4x + 5 = 0$
c. $x^2 - 2x + 1 = 0$ d. $x^2 + 4x - 2 = 0$

IV. $(1 + i)(1 - i) =$ _____.
a. 2 b. -2 c. $2i$ d. $-2i$

V. $\dfrac{1 + i}{1 - i} =$ _____.
a. 2 b. -2 c. 1 d. -1 e. i f. $-i$

In Problems 1–4 write each number in the form bi.

1. $\sqrt{-25}$ 2. $\sqrt{-100}$ 3. $\sqrt{-5}$ 4. $\sqrt{-80}$

In Problems 5–14 write the conjugate of each number.

5. $3 - 7i$ 6. $-2i$ 7. 5 8. $5 + 6i$ 9. $-2 - \frac{1}{3}i$ 10. $10 + \sqrt{53}i$

11. $\sqrt{73}i$ 12. $\sqrt{73}$ 13. $\dfrac{4 + \sqrt{2}i}{7}$

14. $\dfrac{2 - \sqrt{7}i}{12}$

Answers to Readiness Check

I. a II. c III. b IV. a V. e

In Problems 15–18 write each number in the form $a + bi$.

15. $3 - \sqrt{-64}$
16. $-5 + \sqrt{-4}$
17. $\dfrac{-4 + \sqrt{-64}}{8}$
18. $\dfrac{4 - \sqrt{-49}}{2}$

In Problems 19–52 carry out the indicated operations and write each answer in the form $a + bi$.

19. $(2 + 3i) + (4 + 5i)$
20. $(7 - 2i) + (8 + 5i)$
21. $\left(\frac{1}{3} + \frac{1}{5}i\right) + \left(\frac{1}{6} - \frac{1}{4}i\right)$
22. $(4 - i) - (2 + i)$
23. $(6 + \sqrt{3}i) - (6 - \sqrt{3}i)$
24. $(5 + \sqrt{2}i) + (5 - \sqrt{2}i)$
25. $(1 + i)(1 + 2i)$
26. $(1 + i)^2$
27. $(1 - i)^2$
28. $(3 - 4i)^2$
29. $(4 + 2i)^2$
30. $(3 + 4i)(3 - 4i)$
31. $(7 + 2i)(3 - 5i)$
32. $(4 - 3i)(7 + 5i)$
33. $\left(\frac{1}{2} + i\right)\left(-3 + \frac{1}{4}i\right)$
34. $(i + 5)(-3i + 7)$
35. $(2i - 3)(5 + 2i)$
36. i^5
37. i^{11}
38. i^{1002}
39. $(-i)^{15}$
40. $(1 + i)^3$
41. $(1 - i)^4$
42. $\dfrac{1}{i}$
43. $\dfrac{1}{1 + i}$
44. $\dfrac{1}{3 + 2i}$
45. $\dfrac{8 - 3i}{2i}$
46. $\dfrac{5 + i}{4i}$
47. $\dfrac{1 + i}{1 - i}$
48. $\dfrac{1 - i}{1 + i}$
49. $\dfrac{3 + 2i}{2 - 3i}$
50. $\dfrac{5 + i}{7 - 3i}$
51. $\dfrac{5 + 7i}{2 + 4i}$
52. $\dfrac{1}{3 - i} - \dfrac{i}{4 + 2i}$

In Problems 53–62 compute the inverse of each complex number.

53. $2i$
54. $-3i$
55. $1 - i$
56. $1 + 4i$
57. $2 - 5i$
58. $3 + 7i$
59. $4 - 6i$
60. $10 + 5i$
61. $a - bi,\ b \neq 0$
62. $a + bi,\ b \neq 0$

In Problems 63–66 solve for u and v, where u and v represent real numbers. [Hint: Equate real and imaginary parts.]

63. $u + v - 2i = 5 + 2vi$
64. $ui - 3v + 5 = 11 - 3i$
65. $2u - 3vi = 7ui + v - 2i$
66. $u + 2iv = 2u + iv + 3$

In Problems 67–78 find all solutions to the given quadratic equation.

67. $x^2 + 4 = 0$
68. $y^2 + 25 = 0$
69. $z^2 + z + 2 = 0$
70. $2u^2 + 2u + 1 = 0$
71. $p^2 - 6p + 10 = 0$
72. $3x^2 + 2x + 5 = 0$
73. $2y^2 - 3y + 5 = 0$
74. $8v^2 + 4v + 3 = 0$
75. $\frac{1}{2}x^2 + \frac{1}{3}x + \frac{1}{4} = 0$
76. $0.3y^2 + 0.4y + 0.26 = 0$
77. $12.72z^2 - 8.06z + 16.58 = 0$
78. $10^{-4}x^2 - 10^{-6}x + 10^{-5} = 0$
79. Show that if a complex number is equal to its conjugate, then the number is real.
80. Show that if a complex number is equal to the negative of its conjugate, then the number is pure imaginary.
* 81. Find the flaw in this "proof" that $i = 0$:

$$i = i$$
$$\sqrt{-1} = \sqrt{-1}$$
$$\sqrt{\frac{-1}{1}} = \sqrt{\frac{1}{-1}}$$
$$\frac{\sqrt{-1}}{\sqrt{1}} = \frac{\sqrt{1}}{\sqrt{-1}}$$
$$\frac{i}{1} = \frac{1}{i} = \frac{-(-1)}{i} = \frac{-i^2}{i}$$
$$i = -i$$
$$2i = 0$$
$$i = 0$$

6.2 The Polar Form of a Complex Number

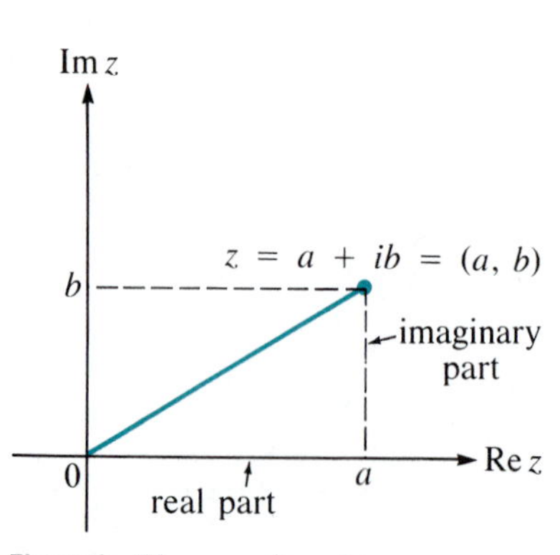

Figure 1 The complex plane.

In Section 6.1, we discussed complex numbers. These are numbers that can be written in the form

$$z = a + bi \tag{1}$$

where a and b are real numbers and $i = \sqrt{-1}$. This form is called the **Cartesian form** of the complex number. The number a is called the **real part** of z, denoted by Re z, and b is called the **imaginary part** of z, denoted by Im z.

We can plot a complex number z in the xy-plane by plotting Re z along the x-axis and Im z along the y-axis. Thus, each complex number can be thought of as a point or vector, (a, b), in the xy-plane. With this representation, the xy-plane is called the **complex plane** (Figure 1). Some representative points are plotted in Figure 2.

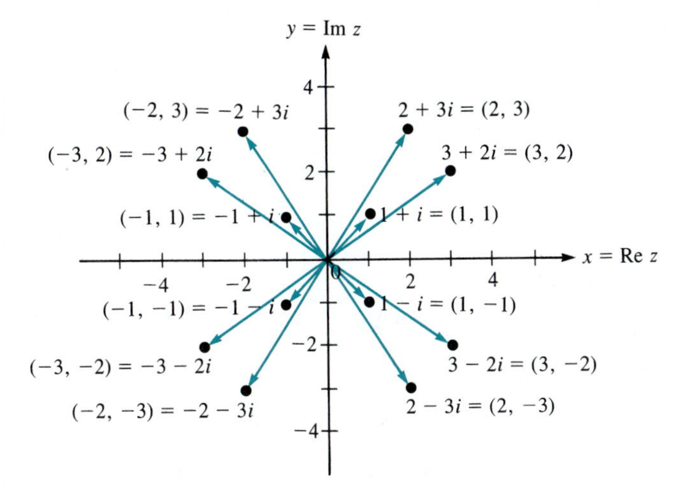

Figure 2 Twelve points in the complex plane.

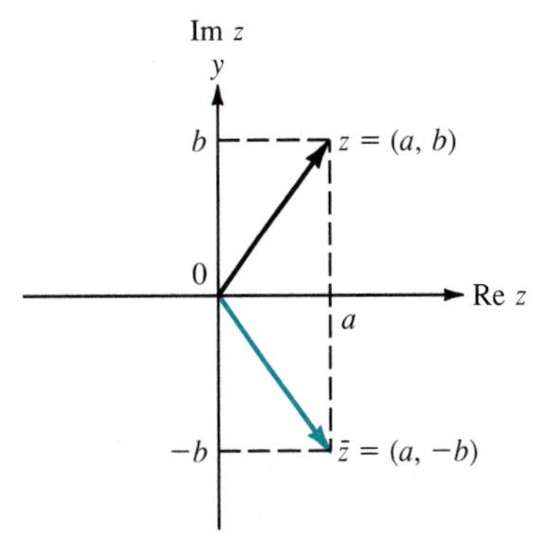

Figure 3 $\bar{z}$ is the reflection of z about the x-axis (Re z-axis).

Recall from p. 335 that the **complex conjugate** of $z = a + bi$, denoted by $\bar{z}$, is given by

$$\bar{z} = a - bi = (a, -b)$$

In Figure 3, we show graphically the relationship between z and $\bar{z}$. $\bar{z}$ is the reflection of z about the real axis.

Absolute Value and Argument

Since each complex number can be thought of as a vector, it has a magnitude and direction. We define the **absolute value** or **magnitude** of z, denoted by $|z|$, to be the distance between (a, b) and the origin. That is,

Absolute Value or Magnitude of z

$$\text{absolute value of } z = |z| = \sqrt{a^2 + b^2} \tag{2}$$

EXAMPLE 1 ***Finding the Absolute Values of Six Numbers***

Find the absolute value of (a) 2, (b) -2, (c) i, (d) $1 + i$, (e) $-1 - \sqrt{3}i$, and (f) $-2 + 7i$.

(a) $|2| = |2 + 0i| = \sqrt{2^2 + 0^2} = \sqrt{4} = 2$

(b) $|-2| = |-2 + 0i| = \sqrt{(-2)^2 + 0^2} = \sqrt{4} = 2$

(c) $|i| = |0 + 1 \cdot i| = \sqrt{0^2 + 1^2} = \sqrt{1} = 1$

(d) $|1 + i| = \sqrt{1^2 + 1^2} = \sqrt{2}$

(e) $|-1 - \sqrt{3}i| = \sqrt{(-1)^2 + (-\sqrt{3})^2} = \sqrt{1 + 3} = \sqrt{4} = 2$

(f) $|-2 + 7i| = \sqrt{(-2)^2 + 7^2} = \sqrt{4 + 49} = \sqrt{53}$

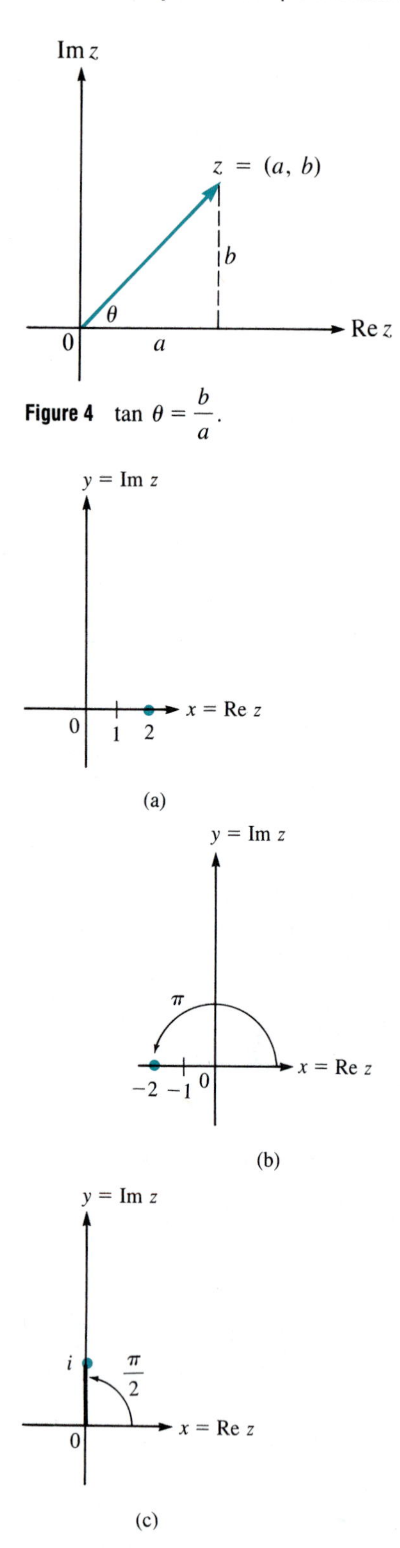

Figure 4 $\tan\theta = \dfrac{b}{a}$.

The **argument** of $z \neq 0$, denoted by **arg z,** is defined as the angle θ between the line $0z$ and the positive real axis (see Figure 4). By convention, we choose θ such that $-\pi < \theta \leq \pi$. From Figure 4, we see that $\tan\theta = b/a$.

Argument of z

$$\text{argument of } z = \arg(a + bi) = \theta$$

where

$$\tan\theta = \frac{b}{a} \qquad \text{if } a \neq 0 \text{ and } -\pi < \theta \leq \pi \tag{3}$$

$$\arg 0 \text{ is not defined}$$

EXAMPLE 2 ***Finding the Arguments of Six Complex Numbers***

Find the argument of (a) 2, (b) −2, (c) i, (d) $1 + i$, (e) $-1 - \sqrt{3}i$, and (f) $-2 + 7i$.

SOLUTION We plot the six numbers in Figure 5. We see immediately that

(a) $\arg 2 = 0$ (b) $\arg(-2) = \pi$ (c) $\arg i = \dfrac{\pi}{2}$

(d) $\arg(1 + i) = \tan^{-1}\left(\dfrac{1}{1}\right) = \tan^{-1} 1 = \dfrac{\pi}{4}$

(e) $\tan\theta = \dfrac{b}{a} = \dfrac{-\sqrt{3}}{-1} = \sqrt{3}$. Since $\theta = \arg(-1 - \sqrt{3}i)$ is in the third quadrant, we know from equation (27) on p. 131 that

$$\theta = \tan^{-1}\sqrt{3} + \pi = \frac{\pi}{3} + \pi = \frac{4\pi}{3}$$

But $\arg\theta$ must be between $-\pi$ and π. In Figure 5(e) we see that $\dfrac{4\pi}{3}$ has the same terminal side as $-\dfrac{2\pi}{3}$. Thus $\arg(-1 - \sqrt{3}i) = -\dfrac{2\pi}{3}$ $(= -120°)$.

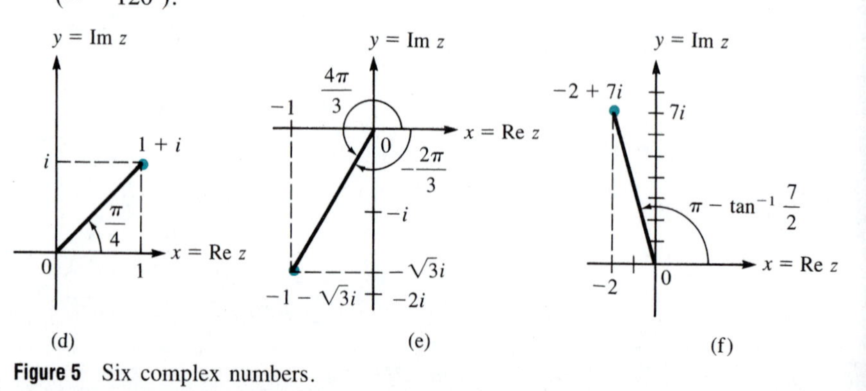

Figure 5 Six complex numbers.

(f) $\tan\theta = \frac{b}{a} = \frac{7}{-2} = -3.5$ and $\theta = \arg(-2+7i)$ is in the second quadrant so, from equation (26) on p. 131 (taking the positive value 3.5)

$$\begin{aligned}\arg(-2+7i) &= \pi - \tan^{-1}(3.5) \approx \pi - 1.2925\\ &= 1.8491\ (\approx 106°)\end{aligned}$$

Let us look again at $\arg z = \arg(a+bi)$. Remember that by the definition of $\tan^{-1} x$, $-\frac{\pi}{2} < \tan^{-1} x < \frac{\pi}{2}$.

Case 1 $a, b > 0$ Then z is in the first quadrant, and $\arg z = \tan^{-1}\frac{b}{a}$.

Case 2 $a > 0, b < 0$ Then z is in the fourth quadrant, $-\frac{\pi}{2} < \arg z < 0$, and, therefore, $\arg z = \tan^{-1}\frac{b}{a}$.

Case 3 $a < 0, b > 0$ Then z is in the second quadrant, so $\frac{\pi}{2} < \arg z < \pi$. But $\frac{b}{a} < 0$, so $-\frac{\pi}{2} < \tan^{-1}\frac{b}{a} < 0$. Then we see that $\arg z = \pi + \tan^{-1}\frac{b}{a}$.

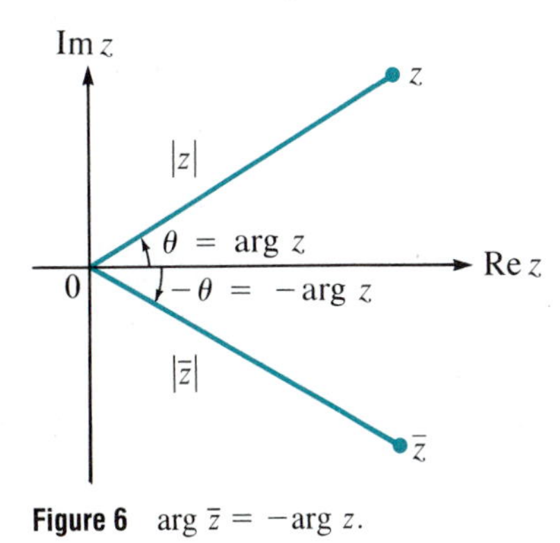

Figure 6 $\arg \bar{z} = -\arg z$.

Case 4 $a < 0, b < 0$ Then z is in the third quadrant and $\frac{b}{a} > 0$. Since $\tan^{-1}\frac{b}{a}$ is in $\left(0, \frac{\pi}{2}\right)$, we subtract π to obtain a value in $\left(-\pi, \frac{-\pi}{2}\right)$. That is, $\arg z = \tan^{-1}\frac{b}{a} - \pi$.

Finally, we repeat that

$$\arg 0 \text{ is undefined}$$

The following facts are useful and follow from Figure 6.

Theorem

Let $z = a + bi$ be a complex number. Then,

$$|\bar{z}| = |z| \tag{4}$$

and

$$\arg \bar{z} = -\arg z \tag{5}$$

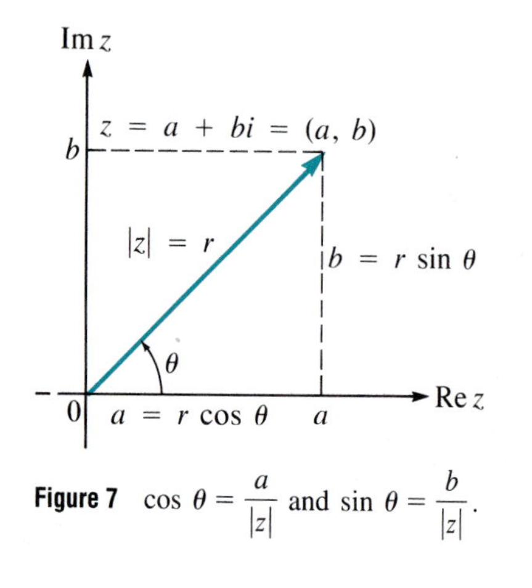

Figure 7 $\cos\theta = \dfrac{a}{|z|}$ and $\sin\theta = \dfrac{b}{|z|}$.

We now denote $|z|$ by r and $\arg z$ by θ. Then, as in Figure 7, we see that

$$\cos\theta = \frac{a}{r} \quad \text{and} \quad \sin\theta = \frac{b}{r}$$

so

$$a = r\cos\theta \quad \text{and} \quad b = r\sin\theta$$

But $z = a + bi$, so we may write $z = r\cos\theta + ir\sin\theta = r(\cos\theta + i\sin\theta)$. This representation of z is called the **polar form** of z.

Polar Form of a Complex Number

If $z = a + bi$, then z may be written

$$z = r(\cos\theta + i\sin\theta) \tag{6}$$

where $r = |z| = \sqrt{a^2 + b^2}$ and $\theta = \arg z$

EXAMPLE 3 *Writing Six Complex Numbers in Polar Form*

Write each number in polar form: (a) 2, (b) -2, (c) i, (d) $1 + i$, (e) $-1 - \sqrt{3}i$, (f) $-2 + 7i$.

SOLUTION We use the results of Examples 1 and 2 and use Figure 5 on p. 342.

(a) $|2| = 2$ and $\theta = 0$, so $2 = 2(\cos 0 + i\sin 0)$

(b) $|-2| = 2$ and $\theta = \pi$, so $-2 = 2(\cos\pi + i\sin\pi)$

(c) $|i| = 1$ and $\theta = \dfrac{\pi}{2}$, so $i = 1\left(\cos\dfrac{\pi}{2} + i\sin\dfrac{\pi}{2}\right) = \cos\dfrac{\pi}{2} + i\sin\dfrac{\pi}{2}$

(d) $|1 + i| = \sqrt{2}$ and $\theta = \dfrac{\pi}{4}$, so $1 + i = \sqrt{2}\left(\cos\dfrac{\pi}{4} + i\sin\dfrac{\pi}{4}\right)$

(e) $|-1 - \sqrt{3}i| = 2$ and $\theta = -\dfrac{2\pi}{3}$, so

$$-1 - \sqrt{3}i = 2\left[\cos\left(\frac{-2\pi}{3}\right) + i\sin\left(\frac{-2\pi}{3}\right)\right] = 2\cos\frac{2\pi}{3} - 2i\sin\frac{2\pi}{3}$$

(f) $|-2 + 7i| = \sqrt{53}$ and $\theta \approx 1.8491$, so
$-2 + 7i \approx \sqrt{53}(\cos 1.8491 + i\sin 1.8491)$ ■

EXAMPLE 4 *Converting from Polar to Cartesian Form*

Convert each number from polar form to Cartesian form.

(a) $\sqrt{3}\left(\cos\dfrac{\pi}{6} + i\sin\dfrac{\pi}{6}\right)$ (b) $\sqrt{2}\left[\cos\left(-\dfrac{3\pi}{4}\right) + i\sin\left(-\dfrac{3\pi}{4}\right)\right]$

(c) $5\left[\cos\left(\dfrac{-\pi}{2}\right) + i\sin\left(\dfrac{-\pi}{2}\right)\right]$

(d) $4(\cos 2.5 + i\sin 2.5)$

SOLUTION

(a) $\sqrt{3}\left(\cos\frac{\pi}{6} + i\sin\frac{\pi}{6}\right) = \sqrt{3}\left(\frac{\sqrt{3}}{2} + i\cdot\frac{1}{2}\right) = \frac{3}{2} + \frac{\sqrt{3}}{2}i$

(b) $\sqrt{2}\left[\cos\left(-\frac{3\pi}{4}\right) + i\sin\left(-\frac{3\pi}{4}\right)\right] = \sqrt{2}\left(-\frac{1}{\sqrt{2}} - \frac{i}{\sqrt{2}}\right) = -1 - i$

(c) $5\left[\cos\left(\frac{-\pi}{2}\right) + i\sin\left(\frac{-\pi}{2}\right)\right] = 5[0 + i(-1)] = -5i$

(d) $4(\cos 2.5 + i\sin 2.5) \approx -3.2046 + 2.3939i$

Before continuing, we prove two useful facts.

Theorem

Let z be a complex number.

(i) $z\bar{z} = |z|^2$ **(7)**

(ii) If $z = r(\cos\theta + i\sin\theta)$, then

$$\bar{z} = r(\cos\theta - i\sin\theta) \tag{8}$$

Proof

(i) Let $z = a + bi$. Then $\bar{z} = a - bi$, and

$$z\bar{z} = (a + bi)(a - bi) = a^2 + abi - abi - i^2b^2 \underset{i^2=-1}{=} a^2 + b^2 = |z|^2$$

(ii) From (5), if $\arg z = \theta$, then $\arg\bar{z} = -\theta$. Also,

$$\begin{aligned} |\bar{z}| &= |z| = r, \text{ so} \\ \bar{z} &= r[\cos(-\theta) + i\sin(-\theta)] \\ &= r(\cos\theta - i\sin\theta) \end{aligned}$$

■

There are many reasons for writing complex numbers in polar form. We will see here and in the next section that the polar form enables us to take products, quotients, powers, and roots of complex numbers relatively easily.

Theorem: Products and Quotients of Complex Numbers in Polar Form

Let $z_1 = r_1(\cos\theta_1 + i\sin\theta_1)$ and $z_2 = r_2(\cos\theta_2 + i\sin\theta_2)$. Then

(i) $z_1z_2 = r_1r_2[\cos(\theta_1 + \theta_2) + i\sin(\theta_1 + \theta_2)]$ **(9)**

If $z_2 \neq 0$, then

(ii) $\frac{z_1}{z_2} = \frac{r_1}{r_2}[\cos(\theta_1 - \theta_2) + i\sin(\theta_1 - \theta_2)]$ **(10)**

In words,

(i) The absolute value of the product is the *product* of the absolute values, and the argument of the product is the *sum* of the arguments.
(ii) The absolute value of the quotient is the *quotient* of the absolute values, and the argument of the quotient is the *difference* of the arguments.

Proof

(i) $z_1 z_2 = r_1(\cos\theta_1 + i\sin\theta_1)r_2(\cos\theta_2 + i\sin\theta_2)$

$i^2 = -1$ ↓

$$= r_1r_2[\cos\theta_1\cos\theta_2 + i\sin\theta_1\cos\theta_2 + i\sin\theta_2\cos\theta_1 + i^2\sin\theta_1\sin\theta_2]$$
$$= r_1r_2[(\cos\theta_1\cos\theta_2 - \sin\theta_1\sin\theta_2) + i(\sin\theta_1\cos\theta_2 + \sin\theta_2\cos\theta_1)]$$
$$= r_1r_2[\cos(\theta_1+\theta_2) + i\sin(\theta_1+\theta_2)]$$

We use the sum identities for sine and cosine (pp. 159 and 158)

From (7) ↓

(ii) $$\frac{z_1}{z_2} = \frac{z_1}{z_2}\cdot\frac{\bar{z}_2}{\bar{z}_2} = \frac{z_1\bar{z}_2}{|z_2|^2}$$
$$= \frac{r_1(\cos\theta_1 + i\sin\theta_1)r_2(\cos\theta_2 - i\sin\theta_2)}{r_2^{\ 2}}$$
$$= \frac{r_1}{r_2}[(\cos\theta_1 + i\sin\theta_1)(\cos\theta_2 - i\sin\theta_2)]$$
$$= \frac{r_1}{r_2}(\cos\theta_1\cos\theta_2 + i\sin\theta_1\cos\theta_2 - i\sin\theta_2\cos\theta_1 - i^2\sin\theta_1\sin\theta_2)$$

$i^2 = -1$ ↓

$$= \frac{r_1}{r_2}[(\cos\theta_1\cos\theta_2 + \sin\theta_1\sin\theta_2) + i(\sin\theta_1\cos\theta_2 - \sin\theta_2\cos\theta_1)]$$
$$= \frac{r_1}{r_2}[\cos(\theta_1-\theta_2) + i\sin(\theta_1-\theta_2)]$$

We use the difference identities for sine and cosine (pp. 159 and 156)

■

EXAMPLE 5 *Computing the Product of Two Complex Numbers*

Compute the product of

$$z_1 = 5(\cos 1 + i\sin 1) \text{ and } z_2 = 3(\cos 2.6 + i\sin 2.6)$$

SOLUTION From (9), we have

$$z_1z_2 = 5\cdot 3[\cos(1+2.6) + i\sin(1+2.6)]$$
$$= 15(\cos 3.6 + i\sin 3.6)$$ ■

EXAMPLE 6 *Computing the Quotient of Two Complex Numbers*

Compute the quotient $\frac{z_1}{z_2}$, where

$$z_1 = 12(\cos 55° + i \sin 55°)$$

and

$$z_2 = 4[\cos(-48°) + i \sin(-48°)]$$

SOLUTION From (10), we obtain

$$\begin{aligned}\frac{z_1}{z_2} &= \frac{12}{4}\{\cos[55° - (-48°)] + i \sin[55° - (-48)]\}\\ &= 3(\cos 103° + i \sin 103°)\end{aligned}$$

Problems 6.2

Readiness Check

I. $|7 - 2i| =$ ______.
a. $7 + 2i$ b. 53 c. 5 d. $\sqrt{53}$

II. Which of the following is true of the complex number $z = 8 - 2i$?
a. $\bar{z} = -8 + 2i$ b. $|z| = 68$
c. $z \cdot \bar{z} = 68$ d. $z \cdot \bar{z} = 2\sqrt{17}$

III. Which of the following is the polar form for -3?
a. $-3(\cos \pi + i \sin \pi)$
b. $3\left[\cos\left(-\frac{\pi}{2}\right) + i \sin\left(-\frac{\pi}{2}\right)\right]$
c. $3[\cos \pi + i \sin \pi]$
d. $-3\left[\cos\left(\frac{\pi}{2}\right) + i \sin\left(\frac{\pi}{2}\right)\right]$

IV. Which of the following is true about $z = [\cos(-\pi/2) + i \sin(-\pi/2)]$?
a. Its magnitude is 0.
b. It is located on the negative x-axis.
c. $\arg z = \frac{\pi}{2}$.
d. Its Cartesian form is $-i$.

V. Which of the following is the approximate polar form of $z = -1 - 3i$?
a. $\sqrt{10}[\cos(-1.9) + i \sin(-1.9)]$
b. $2(\cos 1.2 + i \sin 1.2)$
c. $\sqrt{10}(\cos 72° + i \sin 72°)$
d. $2(\cos 108° + i \sin 108°)$

In Problems 1–22 find the absolute value, argument, and polar form of the given complex number. Use a calculator where indicated.

1. $5i$
2. $-5i$
3. 8
4. -6
5. $1 - i$
6. $-1 + i$
7. $-1 - i$
8. $1 + \sqrt{3}i$
9. $-1 + \sqrt{3}i$
10. $\sqrt{3} - i$
11. $-\sqrt{3} - i$
12. $-3 + 4i$
13. $4 + 3i$
14. $5 - 12i$
15. $3 + 5i$
16. $6 - i$
17. $5 + 2i$
18. $12 + 5i$
19. $-5 + 3i$
20. $50 + 50i$
21. $-100 + 100i$
22. $30 - 30\sqrt{3}i$

In Problems 23–40 write each complex number in Cartesian form.

23. $4\left(\cos\frac{\pi}{2} + i \sin\frac{\pi}{2}\right)$
24. $\sqrt{5}\left(\cos\frac{\pi}{2} + i \sin\frac{\pi}{2}\right)$
25. $12(\cos \pi + i \sin \pi)$
26. $\sqrt{10}(\cos 0 + i \sin 0)$
27. $8\sqrt{2}\left(\cos\frac{\pi}{3} + i \sin\frac{\pi}{3}\right)$

Answers to Readiness Check

I. d II. c III. c IV. d V. a

28. $6\left(\cos\frac{\pi}{6} + i\sin\frac{\pi}{6}\right)$

29. $4\left(\cos\frac{5\pi}{6} + i\sin\frac{5\pi}{6}\right)$

30. $4\left[\cos\left(-\frac{5\pi}{6}\right) + i\sin\left(\frac{-5\pi}{6}\right)\right]$

31. $4\left(\cos\frac{2\pi}{3} + i\sin\frac{2\pi}{3}\right)$

32. $4\left[\cos\left(-\frac{2\pi}{3}\right) + i\sin\left(-\frac{2\pi}{3}\right)\right]$

33. $\frac{1}{4}\left(\cos\frac{\pi}{4} + i\sin\frac{\pi}{4}\right)$

34. $\frac{1}{4}\left[\cos\left(-\frac{3\pi}{4}\right) + i\sin\left(-\frac{3\pi}{4}\right)\right]$

35. $\frac{1}{4}\left(\cos\frac{3\pi}{4} + i\sin\frac{3\pi}{4}\right)$

36. $\frac{1}{4}\left[\cos\left(-\frac{\pi}{4}\right) + i\sin\left(-\frac{\pi}{4}\right)\right]$

37. $5(\cos 2 + i\sin 2)$

38. $\sqrt{7}(\cos(-2.7) + i\sin(-2.7))$

39. $6\left[\cos\left(-\frac{\pi}{15}\right) + i\sin\left(-\frac{\pi}{15}\right)\right]$

40. $3\left(\cos\frac{19\pi}{20} + i\sin\frac{19\pi}{20}\right)$

In Problems 41–45 compute z_1z_2 and $\frac{z_1}{z_2}$. Leave the answer in polar form.

41. $z_1 = 4\left(\cos\frac{\pi}{3} + i\sin\frac{\pi}{3}\right)$; $z_2 = 3\left(\cos\frac{\pi}{6} + i\sin\frac{\pi}{6}\right)$

42. $z_1 = 5\left(\cos\frac{3\pi}{4} + i\sin\frac{3\pi}{4}\right)$;
$z_2 = 10\left[\cos\left(-\frac{\pi}{4}\right) + i\sin\left(-\frac{\pi}{4}\right)\right]$

43. $z_1 = \cos 2 + i\sin 2$; $z_2 = \frac{2}{3}[\cos(-1) + i\sin(-1)]$

44. $z_1 = \frac{1}{3}[\cos(-37°) + i\sin(-37°)]$;
$z_2 = 3(\cos 112° + i\sin 112°)$

45. $z_1 = 5[\cos(-95°) + i\sin(-95°)]$;
$z_2 = \frac{1}{7}(\cos 9° + i\sin 9°)$

46. Prove the **triangle inequality:** $|z_1 + z_2| \le |z_1| + |z_2|$. [Hint: Draw z_1, z_2, and $z_1 + z_2$ in the complex plane using the parallelogram rule discussed on p. 241. Use the fact that a straight line is the shortest distance between two points.]

* 47. Show that the triangle inequality is an equality for nonzero numbers z_1 and z_2 if and only if $\arg z_1 = \arg z_2$.

48. Show that the circle of radius 1 centered at the origin (the unit circle) is the set of points in the complex plane that satisfy $|z| = 1$.

49. For any complex number z_0 and real number a, describe $\{z: |z - z_0| = a\}$.

50. Describe $\{z: |z - z_0| \le a\}$, where z_0 and a are as in Problem 49.

6.3 Powers and Roots of a Complex Number; De Moivre's Theorem

In this section, we use the polar form of a complex number to compute powers and roots of the number. To do so, we need an important result. Let $z = r(\cos\theta + i\sin\theta)$. Then, from the multiplication rule (9) on p. 345

$$\begin{aligned} z^2 = z \cdot z &= [r(\cos\theta + i\sin\theta)][r(\cos\theta + i\sin\theta)] \\ &= r^2[\cos(\theta+\theta) + i\sin(\theta+\theta)] \\ &= r^2(\cos 2\theta + i\sin 2\theta) \\ z^3 = z^2 \cdot z &= r^2(\cos 2\theta + i\sin 2\theta)r(\cos\theta + i\sin\theta) \\ &= r^3(\cos 3\theta + i\sin 3\theta) \end{aligned}$$

$$z^4 = z^3 \cdot z = r^3(\cos 3\theta + i \sin 3\theta)r(\cos \theta + i \sin \theta)$$
$$= r^4(\cos 4\theta + i \sin 4\theta)$$

Continuing this process, we obtain De Moivre's theorem.†

De Moivre's Theorem

$$[r(\cos \theta + i \sin \theta)]^n = r^n(\cos n\theta + i \sin n\theta), \quad (1)$$

where n is a positive integer

EXAMPLE 1 *Computing Powers of Complex Numbers*

Compute (a) $(1 + i)^5$ and (b) $(1 + i)^{100}$.

SOLUTION In Example 3(d) of Section 6.2, we saw that

$$1 + i = \sqrt{2}\left(\cos \frac{\pi}{4} + i \sin \frac{\pi}{4}\right)$$

Then, from (1),

(a) $(1 + i)^5 = (\sqrt{2})^5\left(\cos \frac{5\pi}{4} + i \sin \frac{5\pi}{4}\right) = 4\sqrt{2}\left(-\frac{1}{\sqrt{2}} - \frac{1}{\sqrt{2}}i\right)$

$= -4 - 4i$

(b) $(1 + i)^{100} = (\sqrt{2})^{100}\left(\cos \frac{100\pi}{4} + i \sin \frac{100\pi}{4}\right)$

$= 2^{50}(\cos 25\pi + i \sin 25\pi) = 2^{50}(\cos \pi + i \sin \pi)$

$= 2^{50}(-1 + 0i) = -2^{50}$ ■

EXAMPLE 2 *Computing a Power of a Complex Number*

Find an approximate value for $(-2 + 1.3i)^{10}$.

SOLUTION

$$|-2 + 1.3i| = \sqrt{(-2)^2 + (1.3)^2} = \sqrt{5.69}$$

† Abraham De Moivre (1667–1754) was a French mathematician well known for his work in probability theory, infinite series, and trigonometry. He was so highly regarded that Newton often told those who came to him with questions on mathematics. "Go to M. De Moivre; he knows these things better than I do."

Since $-2 + 1.3i$ is in the second quadrant, it follows from Case 3 on p. 343 that

$$\arg(-2 + 1.3i) = \tan^{-1}\left(-\frac{1.3}{2}\right) + \pi$$
$$= \tan^{-1}(-0.65) + \pi \approx 2.5652$$

Then

$$(-2 + 1.3i)^{10} \approx [\sqrt{5.69}(\cos 2.5652 + i \sin 2.5652)]^{10}$$
$$\approx 5.69^5(\cos 25.652 + i \sin 25.652)$$
$$\approx 5964.3(0.86819 + 0.49624i)$$
$$\approx 5178.1 + 2959.7i$$

Finding *n*th Roots

The complex number w is an ***n*th root** of z if $w^n = z$. For example, 3 is a fourth root of 81 because $3^4 = 81$. Suppose $w = s(\cos \phi + i \sin \phi)$ is an nth root of $z = r(\cos \theta + i \sin \theta)$. Then, by De Moivre's theorem,

$$w^n = s^n(\cos n\phi + i \sin n\phi)$$
$$= r(\cos \theta + i \sin \theta) = z$$

It then follows that $s^n = r$ and

$$s = r^{1/n}$$

The number $r^{1/n}$ is called the **principal *n*th root of *r*.** We must also have

$$\cos n\phi = \cos \theta \qquad \text{and} \qquad \sin n\phi = \sin \theta$$

These two equations hold if and only if $n\phi$ and θ differ by a multiple of 2π. That is,

$$n\phi - \theta = 2k\pi \quad \text{For some integer } k$$
$$n\phi = \theta + 2k\pi$$
$$\phi = \frac{\theta}{n} + \frac{2k\pi}{n} \tag{2}$$

Equation (2) holds for every integer k. However if $k = n$, for example,

$$\cos \phi = \cos\left(\frac{\theta}{n} + \frac{2n\pi}{n}\right) = \cos\left(\frac{\theta}{n} + 2\pi\right) = \cos\frac{\theta}{n}$$

and

$$\sin \phi = \sin\left(\frac{\theta}{n} + \frac{2n\pi}{n}\right) = \sin\frac{\theta}{n}$$

Then

$$w_1 = r^{1/n}\left(\cos\frac{\theta}{n} + i \sin\frac{\theta}{n}\right)$$

is the same number as

$$w_2 = r^{1/n}\left[\cos\left(\frac{\theta}{n} + \frac{2n\pi}{n}\right) + i \sin\left(\frac{\theta}{n} + \frac{2n\pi}{n}\right)\right]$$

Similarly, if $k = n + 1$, then

$$\cos \phi = \cos\left(\frac{\theta}{n} + \frac{2(n+1)\pi}{n}\right) = \cos\left(\frac{\theta}{n} + \frac{2\pi}{n} + \frac{2n\pi}{n}\right)$$
$$= \cos\left(\frac{\theta + 2\pi}{n}\right)$$

which is the same value obtained when $k = 1$. We get different values of $\cos \phi$ and $\sin \phi$ for $k = 0, 1, 2, \ldots, n - 1$. The values repeat for $k = n$, $n + 1, \ldots$. In sum,

The nth Roots of a Complex Number

The complex number $z = r(\cos \theta + i \sin \theta)$ has exactly n nth roots w_1, $w_2, \ldots, w_n$ given by

$$w_k = r^{1/n}\left[\cos\left(\frac{\theta + 2k\pi}{n}\right) + i \sin\left(\frac{\theta + 2k\pi}{n}\right)\right] \tag{3}$$

for $k = 0, 1, 2, \ldots, n - 1$.

EXAMPLE 3 *Finding All the Sixth Roots of 1*

Find the six sixth roots of 1.

SOLUTION Since $\arg 1 = 0$, we have $\theta = 0$ and

$$\frac{\theta + 2k\pi}{n} = 0 + \frac{2k\pi}{6}, \qquad k = 0, 1, 2, 3, 4, 5$$
$$= 0, \frac{2\pi}{6}, \frac{4\pi}{6}, \frac{6\pi}{6}, \frac{8\pi}{6}, \frac{10\pi}{6}$$
$$= 0, \frac{\pi}{3}, \frac{2\pi}{3}, \pi, \frac{4\pi}{3}, \frac{5\pi}{3}$$

Now $1^{1/6} = 1$, so the six sixth roots of 1 are

$$w_1 = \cos 0 + i \sin 0 = 1$$
$$w_2 = \cos\frac{\pi}{3} + i \sin\frac{\pi}{3} = \frac{1}{2} + \frac{\sqrt{3}}{2}i$$
$$w_3 = \cos\frac{2\pi}{3} + i \sin\frac{2\pi}{3} = -\frac{1}{2} + \frac{\sqrt{3}}{2}i$$
$$w_4 = \cos \pi + i \sin \pi = -1$$
$$w_5 = \cos\frac{4\pi}{3} + i \sin\frac{4\pi}{3} = -\frac{1}{2} - \frac{\sqrt{3}}{2}i$$
$$w_6 = \cos\frac{5\pi}{3} + i \sin\frac{5\pi}{3} = \frac{1}{2} - \frac{\sqrt{3}}{2}i$$

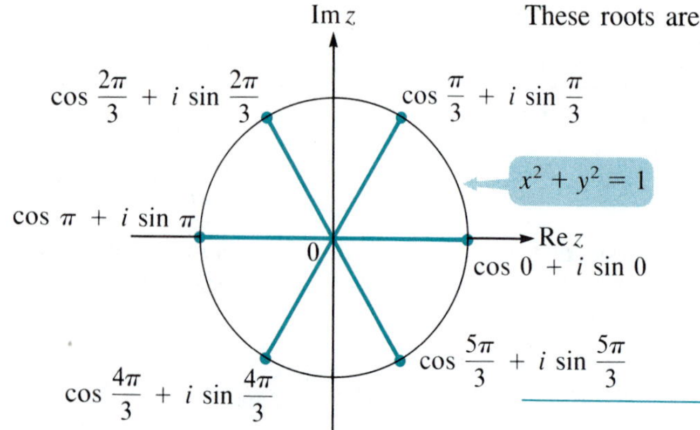

Figure 1 The six sixth roots of 1 lie on the unit circle with their arguments $\frac{\pi}{3}$ units apart.

These roots are illustrated in Figure 1. We check one answer:

$$
\begin{aligned}
(w_2)^6 = (w_2{}^2)^3 &= \left[\left(\frac{1}{2} + \frac{\sqrt{3}}{2}i\right)^2\right]^3 \\
&= \left(-\frac{1}{2} + \frac{\sqrt{3}}{2}i\right)\left(-\frac{1}{2} + \frac{\sqrt{3}}{2}i\right)\left(-\frac{1}{2} + \frac{\sqrt{3}}{2}i\right) \\
&= \left(-\frac{1}{2} - \frac{\sqrt{3}}{2}i\right)\left(-\frac{1}{2} + \frac{\sqrt{3}}{2}i\right) \\
&= \frac{1}{4} + \frac{3}{4} = 1
\end{aligned}
$$

The sixth roots of 1 are called the **sixth roots of unity.** In general,

The n nth Roots of Unity

The n nth roots of unity are given by

$$w_k = \cos\frac{2k\pi}{n} + i\sin\frac{2k\pi}{n}, \qquad k = 0, 1, 2, \ldots, n-1 \qquad \textbf{(4)}$$

EXAMPLE 4 *Finding the Two Square Roots of i*

Find the two square roots of i.

SOLUTION $\arg i = \frac{\pi}{2}$ and $|i| = 1$. If $\theta = \frac{\pi}{2}$, then

$$\frac{\theta}{2} = \frac{\pi}{4} \qquad \text{and} \qquad \frac{\theta + 2\pi}{2} = \frac{5\pi}{4}$$

so

$$w_1 = \cos\frac{\pi}{4} + i\sin\frac{\pi}{4} = \frac{1}{\sqrt{2}}(1 + i)$$

$$w_2 = \cos\frac{5\pi}{4} + i\sin\frac{5\pi}{4} = -\frac{1}{\sqrt{2}}(1 + i)$$

The two roots are sketched in Figure 2. Note, as a check, that

$$w_1{}^2 = \left[\frac{1}{\sqrt{2}}(1 + i)\right]^2 = \frac{1}{2}(1 + 2i + i^2) = \frac{1}{2}(2i) = i \quad ■$$

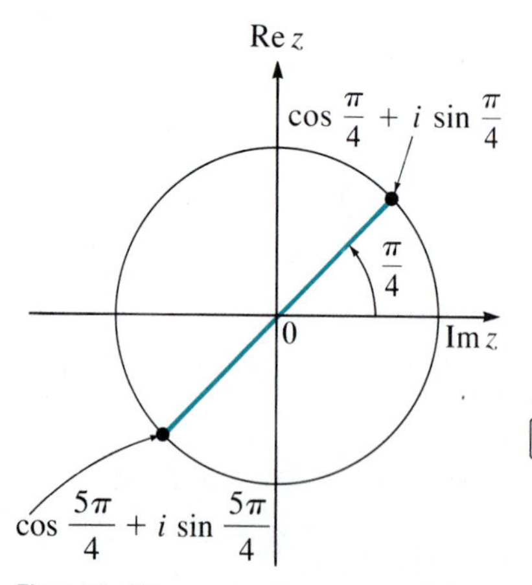

Figure 2 The two square roots of i.

EXAMPLE 5 *Calculating the Five Fifth Roots of a Complex Number*

Find, approximately, the five fifth roots of $-1 - i$.

NOTE This is equivalent to asking: Find all roots of $z^5 = -1 - i$.

SOLUTION $|-1 - i| = \sqrt{2}$ and, from Case 4 on p. 343

$$\theta = \arg(-1 - i) = \tan^{-1}\left(\frac{-1}{-1}\right) - \pi = \frac{\pi}{4} - \pi = -\frac{3\pi}{4}$$

Then

$$r^{1/5} = (2^{1/2})^{1/5} = 2^{1/10}$$

and

$$\theta = \frac{-\dfrac{3\pi}{4} + 2k\pi}{5} = \pi\left(\frac{8k - 3}{20}\right), \qquad k = 0, 1, 2, 3, 4$$

We obtain, using a calculator,

$$w_1 = 2^{1/10}\left[\cos\left(\frac{-3\pi}{20}\right) + i \sin\left(\frac{-3\pi}{20}\right)\right] \approx 0.9550 - 0.4866i$$

$$w_2 = 2^{1/10}\left[\cos\frac{\pi}{4} + i \sin\frac{\pi}{4}\right] = 2^{1/10}\left(\frac{1}{\sqrt{2}} + \frac{i}{\sqrt{2}}\right) \approx 0.7579 + 0.7579i$$

$$w_3 = 2^{1/10}\left(\cos\frac{13\pi}{20} + i \sin\frac{13\pi}{20}\right) \approx -0.4866 + 0.9550i$$

$$w_4 = 2^{1/10}\left(\cos\frac{21\pi}{20} + i \sin\frac{21\pi}{20}\right) \approx -1.0586 - 0.1677i$$

$$w_5 = 2^{1/10}\left(\cos\frac{29\pi}{20} + i \sin\frac{29\pi}{20}\right) \approx -0.1677 - 1.0586i$$

The roots are sketched in Figure 3.

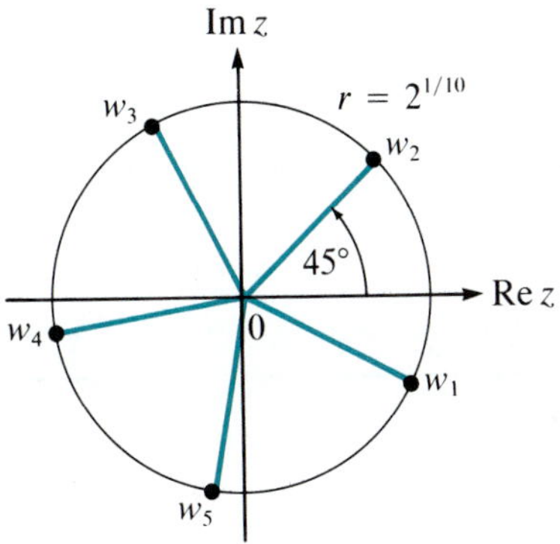

Figure 3 The five fifth roots of $-1 - i$ lie on a circle of radius $2^{1/10}$.

Problems 6.3

Readiness Check

I. Which of the following is the argument for the result of $(-2 + 2i)^3$?

a. $\frac{27\pi}{64}$ b. $\frac{\pi}{4}$ c. $-\frac{3\pi}{4}$ d. $6\sqrt{2}$

II. Which of the following is another fifth root of a number if $2(\cos 16° + i \sin 16°)$ is one of its fifth roots?

a. $2(\cos 160° + i \sin 160°)$
b. $2(\cos 98° + i \sin 98°)$
c. $2[\cos(-150°) + i \sin(-150°)]$
d. $2(\cos 52° + i \sin 52°)$

III. Which of the following is one of the sixth roots of i?

a. $-\frac{\sqrt{2}}{2} - \frac{\sqrt{2}}{2}i$ b. $-\frac{\sqrt{2}}{2} + \frac{\sqrt{2}}{2}i$ c. $\frac{\sqrt{3}}{2} + \frac{1}{2}i$ d. $\frac{\sqrt{3}}{2} - \frac{1}{2}i$

IV. Which of the following is true about $z = -i$?

a. $z^2 = i$
b. i is a cube root of z.
c. $\bar{z} = -1$
d. It is located on the negative real axis.

V. If $\sqrt{7}(\cos \pi/4 + i \sin \pi/4)$ is one square root of $7i$, then which of the following angles should be added to its argument to get the other square root of $7i$?

a. π b. $\frac{3\pi}{2}$ c. $\frac{\pi}{2}$ d. 0

Answers to Readiness Check

I. b II. a III. b IV. b V. a

In Problems 1–10 use De Moivre's theorem to express each number in the form $a + bi$, where a and b are real.

1. $(1+i)^{29}$
2. $(-1-i)^{18}$
3. $(1-i)^{15}$
4. $(-3+3i)^5$
5. $(\sqrt{3}+i)^{15}$
6. $(-\sqrt{3}+i)^{13}$
7. $(-1-\sqrt{3}i)^{10}$
8. $(1+2i)^{12}$
9. $(-4+3i)^5$
10. $\left(\frac{1}{2}+\frac{1}{3}i\right)^6$

Find all solutions of the equations in Problems 11–34. If a calculator is used, round all answers to 4 significant figures.

11. $z^2 = 1$
12. $z^4 = 1$
13. $z^3 = 1$
14. $z^5 = 1$
15. $z^2 = -i$
16. $z^2 = 1 + i$
17. $z^2 = 2 - i$
18. $z^2 = \sqrt{3} + i$
19. $z^3 = 2 + i$
20. $z^3 = 1 + \sqrt{3}i$
21. $z^4 = 1$
22. $z^4 = -i$
23. $z^8 = i$
24. $z^4 = -1 + i$
25. $z^4 = 1 - i$
26. $z^5 = 32$
27. $z^5 = 32i$
28. $z^6 = -64$
29. $z^6 = -64i$
30. $z^4 = 16 + 16i$
31. $z^4 = -16 + 16i$
32. $z^4 = -16 - 16i$
33. $z^4 = 16 - 16i$
34. * $z^5 = 3 + 2i$

35. ** Let z_k be any nth root of unity. Prove that
$$1 + z_k + z_k^2 + \cdots + z_k^{n-1} = 0, \qquad z_k \neq 1$$

36. ** If $1, z_1, z_2, \ldots, z_{n-1}$ are the nth roots of unity, show that
$$(z - z_1)(z - z_2) \cdot \cdots \cdot (z - z_{n-1}) = 1 + z + z^2 + \cdots + z^{n-1}$$

37. * Let $z = a + bi$. For $b \neq 0$, we define $\operatorname{sgn} b = \frac{b}{|b|}$; that is,
$$\operatorname{sgn} b = \begin{cases} 1, & b > 0 \\ -1, & b < 0 \end{cases}$$
Show that the two values for $\sqrt{z}$ are given by
$$\sqrt{z} = \pm\left[\sqrt{\frac{(|z| + a)}{2}} + \left(\operatorname{sgn} b\sqrt{\frac{(|z| - a)}{2}}\right)i\right]$$

In Problems 38–42 use the result of Problem 37 to compute the two square roots of the given number.

38. $\sqrt{3i}$
39. $\sqrt{2+3i}$
40. $\sqrt{-1+i}$
41. $\sqrt{-4-5i}$
42. $\sqrt{2-i}$

Summary Outline of Chapter 6

- **Imaginary Unit i:** Defined by $i = \sqrt{-1}$ — p. 332
  If $a > 0$, then $\sqrt{-a} = \sqrt{a}\,i$
- **Complex Number:** Number of the form $a + bi$ where a and b are real, $i = \sqrt{-1}$ — p. 333
- **Algebra of Complex Numbers:**
  $(a + bi) + (c + di) = (a + c) + (b + d)i$ — p. 334
  $(a + bi) - (c + di) = (a - c) + (b - d)i$ — p. 334
  $(a + bi)(c + di) = (ac - bd) + (ad + bc)i$ — p. 334
  $(a + bi)(a - bi) = a^2 + b^2$ — p. 335
  $\frac{a+bi}{c+di} = \frac{(a+bi)(c-di)}{(c+di)(c-di)} = \frac{ac+bd}{c^2+d^2} + \frac{bc-ad}{c^2+d^2}i$ — p. 336
- **Complex Conjugate:** $\bar{z} = a - bi$ is the complex conjugate of $z = a + bi$ — p. 335
- If $z = a + bi$ is a complex number, then the **absolute value** of $z = |z| = \sqrt{a^2 + b^2}$. — p. 341
- The **argument** of $z = a + bi$, denoted by **arg z,** is the angle θ between 0z and the positive real axis.
  $\tan\theta = \frac{b}{a}$ and $-\pi < \arg z \leq \pi$ — p. 342
- **Polar Form of a Complex Number:**
  $z = r(\cos\theta + i\sin\theta)$ where $r = |z|$ and $\theta = \arg z$ — p. 344

- **More Complex Algebra:** p. 345

  $z\bar{z} = |z|^2$ and $\bar{z} = r(\cos\theta - i\sin\theta)$

  $z_1z_2 = |z_1||z_2|[\cos(\arg z_1 + \arg z_2) + i\sin(\arg z_1 + \arg z_2)]$

  $\dfrac{z_1}{z_2} = \dfrac{|z_1|}{|z_2|}[\cos(\arg z_1 - \arg z_2) + i\sin(\arg z_1 - \arg z_2)]$

- **De Moivre's Theorem:** p. 349

  $[r(\cos\theta + i\sin\theta)]^n = r^n(\cos n\theta + i\sin n\theta)$

- **n^{th} Roots of a Complex Number z:** p. 351

  $w_1, w_2, \ldots, w_n$ where

  $w_k = r^{1/n}\left[\cos\left(\dfrac{\theta + 2k\pi}{n}\right) + i\sin\left(\dfrac{\theta + 2k\pi}{n}\right)\right]$, $k = 0, 1, 2, \ldots, n-1$;

  $r = |z|$ and $\theta = \arg z$

Review Exercises for Chapter 6

1. Write each number in the form bi.
   (a) $\sqrt{-64}$
   (b) $\sqrt{-75}$

In Exercises 2–10, write each number in the form $a + bi$.

2. $\dfrac{3 + \sqrt{-49}}{4}$
3. $(3 - 2i) + (4 - 6i)$
4. $(7 + 2i) - (-1 + 2i)$
5. $(3 - i)(4 + 5i)$
6. $(8 + 5i)\left(\dfrac{1}{2} - \dfrac{1}{3}i\right)$
7. i^7
8. $\dfrac{1}{3 - 2i}$
9. $\dfrac{1 - i}{1 + i}$
10. $\dfrac{1}{2 + i} - \dfrac{3i}{2 - 3i}$

In Exercises 11–21, find the absolute value, argument, and polar representation of the given complex number.

11. -5
12. $6i$
13. $-\sqrt{2}\,i$
14. $2 + 2i$
15. $-2 + 2i$
16. $2 - 2i$
17. $-2 - 2i$
18. $3 + 8i$
19. $4 - 7i$
20. $7 - 4i$
21. $-3 - 5i$

In Exercises 22–25, write each complex number in Cartesian form.

22. $7\left(\cos\dfrac{\pi}{2} + i\sin\dfrac{\pi}{2}\right)$
23. $12\left[\cos\left(-\dfrac{\pi}{6}\right) + i\sin\left(-\dfrac{\pi}{6}\right)\right]$
24. $\cos\dfrac{5\pi}{6} + i\sin\dfrac{5\pi}{6}$
25. $6\left[\cos\left(-\dfrac{3\pi}{4}\right) + i\sin\left(-\dfrac{3\pi}{4}\right)\right]$

In Exercises 26 and 27, compute z_1z_2 and z_1/z_2. Leave the answer in polar form.

26. $z_1 = 2\left(\cos\dfrac{\pi}{4} + i\sin\dfrac{\pi}{4}\right)$; $z_2 = 6\left(\cos\dfrac{2\pi}{3} + i\sin\dfrac{2\pi}{3}\right)$
27. $z_1 = 8(\cos 2 + i\sin 2)$; $z_2 = 4(\cos 0.85 + i\sin 0.85)$

In Exercises 28–31, use De Moivre's theorem to express each number in the form $a + bi$, where a and b are real.

28. $(1 - i)^{10}$
29. $(\sqrt{3} - i)^{12}$
30. $\left(\dfrac{1}{2} + \dfrac{\sqrt{3}}{2}i\right)^{50}$
31. $(-2 + 7i)^6$

In Exercises 32–37, find all solutions of each equation.

32. $z^4 = 1$
33. $z^3 = -1 + i$
34. $z^5 = 32 + 32\sqrt{3}i$
35. $z^2 = -9i$
36. $z^2 = 1 + 2i$
37. $z^5 = -i$

Chapter 7

Exponential and Logarithmic Functions

7.1 Exponential Functions

In this chapter, we introduce some of the most important functions in mathematics. First, we review some facts from algebra.

Let a be a positive real number.

(a) If $x = n$, *a positive integer,* then

$$a^x = a^n = \underbrace{a \cdot a \cdot a \cdot \cdots \cdot a}_{n \text{ factors}}$$

(b) If $x = 0$, then

$$a^x = a^0 = 1$$

(c) If $x = -n$, *where* n *is a positive integer,* then

$$a^x = a^{-n} = \frac{1}{a^n}$$

(d) If $x = 1/n$, *where* n *is a positive integer,* then

$$a^x = a^{1/n} = \text{the } n\text{th root of } a$$

(e) If $x = m/n$ (m *and* n *are positive integers*), then

$$a^x = a^{m/n} = (a^{1/n})^m$$

(f) If $x = -m/n$, $n \neq 0$, *a negative rational number,* then

$$a^x = a^{-m/n} = \frac{1}{a^{m/n}}$$

Thus, a^x $(a > 0)$ is defined if x is a rational number. If x is not a rational number, then we have not as yet defined a^x. However, we can define an approximation to a^x by first approximating x as a decimal and then computing a to the power of this decimal. With the aid of a calculator, this is quite easily done.

EXAMPLE 1 *Approximating a^x Where x Is Irrational*

Use the procedure outlined above to approximate $4^{\sqrt{2}}$.

SOLUTION We find that $\sqrt{2} = 1.414213562 \ldots$. Thus, $\sqrt{2}$ can be approximated, successively, by 1, 1.4, 1.41, 1.414, . . . , and, since each of these numbers is a rational number, we can compute 4^1, $4^{1.4}$, and so on. Some results are given in Table 1.

Table 1

r	1	1.4	1.41	1.414	1.4142	1.414213562
4^r	4	6.964404506	7.06162397	7.100890698	7.102859756	7.102993298

We can obtain this approximation on a calculator by the following key sequence:

[4] [y^x] [2] [$\sqrt{x}$] [=]

On a calculator carrying 10 digits, this results in the value 7.102993301.†

NOTE Some calculators require parentheses to obtain this number

[4] [y^x] [(] [2] [$\sqrt{x}$] [)] [=]

The procedure described above really provides us with the definition of a^x when $a > 0$ and x is irrational. We simply define a^x as the "limit" of a^r as r approximates x to more and more decimal places.

We can now define an exponential function.

Definition of an Exponential Function

Let $a \neq 1$ be a positive real number. Then the function f defined by $f(x) = a^x$ is called an **exponential function with base *a*.** Since $y = a^x$ is defined for every real number x, we see that domain of $f(x) = a^x$ is $\mathbb{R}$.

NOTE In an exponential function, the exponent is the variable. In the power function $y = x^n$, the base is the variable, and the exponent is constant.

EXAMPLE 2 *The Graph of 2^x*

Sketch the graph of the function $y = 2^x$.

† Some calculators that do not carry as many internal (that is, undisplayed) digits might give an answer that differs from this one in the last digit. The answer given here is correct.

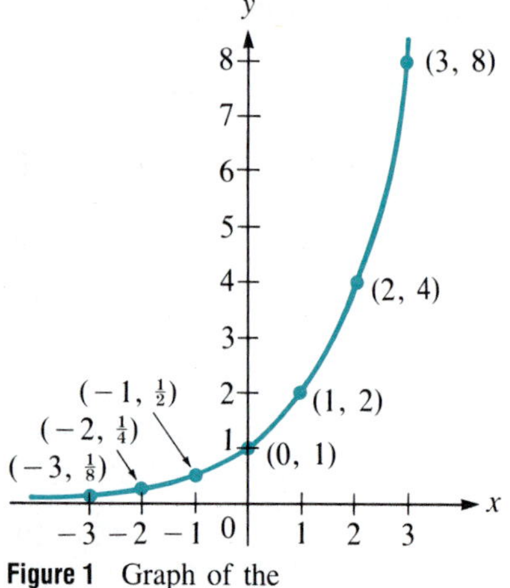

Figure 1 Graph of the exponential function $f(x) = 2^x$.

SOLUTION We provide some values of 2^x in Table 2. We plot these values and then draw a curve joining the points to obtain the sketch in Figure 1. ■

Table 2

x	-10	-5	-2	-1	0	$\frac{1}{2}$	1	$\frac{3}{2}$	2	3	5	10
2^x	0.001	0.03	0.25	0.5	1	1.4142	2	2.8284	4	8	32	1024

EXAMPLE 3 *The Graph of* $(\frac{1}{2})^x$

Sketch the graph of the function $y = (\frac{1}{2})^x$.

SOLUTION We see that $(\frac{1}{2})^x = 1/2^x = 2^{-x}$. Thus, if $f(x) = 2^x$, then $2^{-x} = f(-x)$, and the graph of $(\frac{1}{2})^x$ is the graph of 2^x reflected about the y-axis (see p. 34). The graph is given in Figure 2.

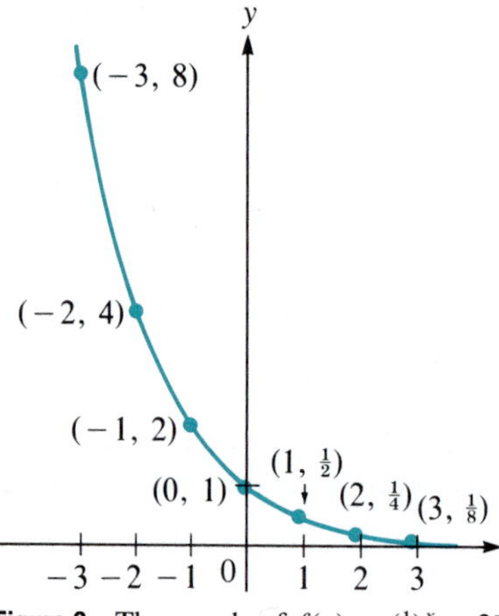

Figure 2 The graph of $f(x) = (\frac{1}{2})^x = 2^{-x}$ is the reflection about the y-axis of the graph of 2^x.

NOTE Let $f(x) = n^x$ and $g(x) = \left(\frac{1}{n}\right)^x$, where n is a positive integer. Then $f(-x) = n^{-x} = \frac{1}{n^x} = g(x)$ so, as in Figures 1 and 2, the graph of $\left(\frac{1}{n}\right)^x$ is the graph of n^x reflected about the y-axis.

EXAMPLE 4 *The Graph of* 10^x

Sketch the graph of $y = 10^x$.

SOLUTION We give some values of 10^x in Table 3 and draw the graph in Figure 3.

Table 3

x	-3	-2	-1	0	0.25	0.5	0.75	1	1.5	2	3
10^x	0.001	0.01	0.1	1	1.778	3.162	5.623	10	31.62	100	1000

■

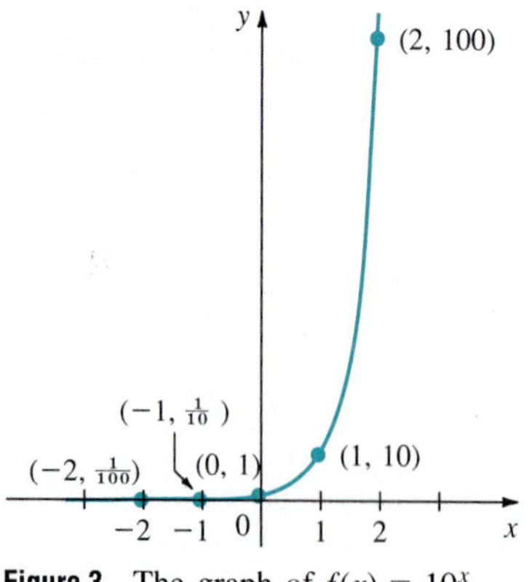

Figure 3 The graph of $f(x) = 10^x$.

EXAMPLE 5 *The Graph of* $(\frac{1}{10})^x$

Sketch the graph of $y = (\frac{1}{10})^x$.

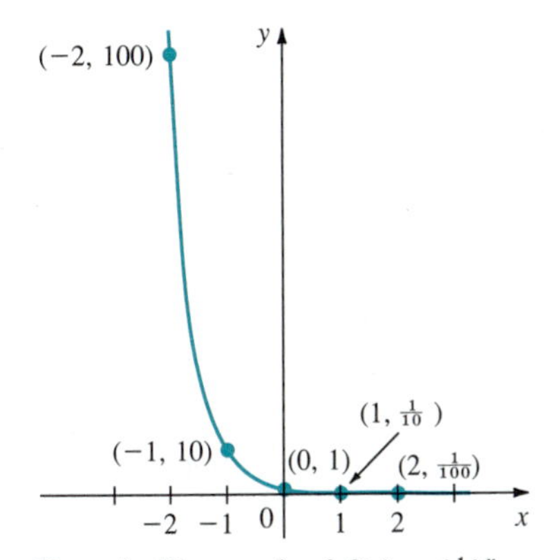

Figure 4 The graph of $f(x) = (\frac{1}{10})^x = 10^{-x}$ is the reflection about the y-axis of the graph of 10^x.

SOLUTION As in Example 3, we can obtain the graph by reflecting the graph of 10^x about the y-axis. We do this in Figure 4.

If you look at Figures 1 and 3, you may observe that the graphs of 2^x and 10^x are very similar. The only difference is that 10^x increases faster than 2^x as x increases and that 10^x decreases faster than 2^x when x is negative and x decreases. The functions $(\frac{1}{2})^x$ and $(\frac{1}{10})^x$ behave very similarly. These facts are not surprising after we observe that all exponential functions share some interesting properties. We cite some of these properties below. These all follow from properties of rational exponents.

Properties of Exponential Functions

Let $a > 0$ and let x and y be real numbers. Then

	Illustration
1. $a^x > 0$ and the range of $f(x) = a^x$ is $\{y: y > 0\}$.	
2. $a^{-x} = \dfrac{1}{a^x}$.	$2^{-3} = \dfrac{1}{2^3} = \dfrac{1}{8}$
3. $a^{x+y} = a^x a^y$.	$2^{3+4} = 2^3 2^4 = 8 \cdot 16 = 128$
4. $a^{x-y} = \dfrac{a^x}{a^y}$.	$2^{5-2} = \dfrac{2^5}{2^2} = \dfrac{32}{4} = 8$
5. $a^0 = 1$.	$2^0 = 1$
6. $a^1 = a$.	$2^1 = 2$
7. $a^{xy} = (a^x)^y = (a^y)^x$.	$2^{3 \cdot 2} = (2^3)^2 = 8^2 = 64$

8. If $a > 1$, a^x is an increasing function.†
9. If $0 < a < 1$, a^x is a decreasing function.
10. The graph of $y = a^x$ passes through the point (0, 1). That is, $f(0) = 1$.
11. **Exponentiation property:** If $x = y$, then $a^x = a^y$.
12. If $a > 1$, $a^x > 1$ if $x > 0$ and $0 < a^x < 1$ if $x < 0$.

Compound Interest Formulas

Suppose that $\$P$ is invested for t years at an annual interest rate of r (usually given as a number between 0 and 1). According to the **simple interest formula,**

$$I = Prt$$

where I is the interest earned, P is the **principal,** r is the rate of interest, and t is the time the investment is held (usually measured in years).

For example, if \$2000 is invested for 3 years at an annual interest rate of 12%, then $P = \$2000$, $t = 3$ and $r = 0.12$ so the simple interest earned is given by

$$I = Prt = (2000)(0.12)(3) = \$720$$

† We discussed increasing and decreasing functions in Section 1.6 (p. 47).

Compound interest is interest paid on interest previously earned as well as on the original investment. Suppose that interest is paid annually. Then if P dollars are invested, the interest after one year ($t = 1$) is rP dollars, and the amount in the account is $P + rP$ dollars. After two years the interest is paid on $P + rP$ dollars (the P dollars originally invested plus the rP dollars earned after the first year). That is

$$\text{interest paid at end of second year} = r(P + rP) = rP(1 + r)$$

This means that

$$\begin{aligned}\begin{matrix}\text{total amount of investment}\\ \text{after 2 years}\end{matrix} &= P + \begin{matrix}\text{interest after}\\ \text{first year}\end{matrix} + \begin{matrix}\text{interest after}\\ \text{second year}\end{matrix}\\ &= P + rP + rP(1 + r)\\ &= P(1 + r) + rP(1 + r) = (P + rP)(1 + r)\\ &= P(1 + r)(1 + r) = P(1 + r)^2\end{aligned}$$

If $A(t)$ denotes the value (amount) of our investment after t years, then we can continue the computation above to show that

Compound Interest Formula

$$A(t) = P(1 + r)^t \qquad (1)$$

where P is the original principal, r is the rate of interest, t is the time the investment is held, and $A(t)$ is the total value of the investment after t years.

EXAMPLE 6 *Computing the Value in Three Years with Annual Compounding*

What is the value of a \$2000 investment after 3 years if it is invested at 12% interest compounded annually?

SOLUTION Here $P = 2000$, $r = 0.12$, and $t = 3$ in formula (1). Thus

$$A(3) = 2000(1 + 0.12)^3 = 2000(1.12)^3$$

$$\overset{\text{Calculator}}{\downarrow}$$

$$= 2000(1.404928) \approx 2809.86$$

Thus the investment is worth \$2809.86 after 3 years.

In practice, interest is compounded more frequently than annually. If it is paid m times a year, then in each interest period the rate of interest is r/m and in t years there are tm pay periods. Then, similar to formula (1), we have the

Compound Interest Formula: Compounding m Times a Year

$$A(t) = P\left(1 + \frac{r}{m}\right)^{mt} \tag{2}$$

where

P is the original principal

r is the annual interest rate

t is the number of years the investment is held

m is the number of times interest is compounded each year

$A(t)$ is the amount (in dollars) after t years.

EXAMPLE 7 *Computing the Value in Five Years with Annual, Quarterly, Monthly, and Daily Compounding*

$2000 is invested for 5 years at 6% interest. How much is the investment worth if interest is compounded (a) annually? (b) quarterly? (c) monthly? (d) daily?

SOLUTION

(a) Interest is compounded just once a year. Thus we can use either formula (1) or formula (2) with $m = 1$. Here $P = 2000$, $r = 0.06$, and $t = 5$. From (2), we have

$$\$A(5) = \$2000(1 + 0.06)^5 = \$2000(1.06)^5 = \$2000(1.338225578)$$
$$= \$2676.45$$

Note that the total interest earned over the 5-year period is $676.45.

(b) Interest is compounded 4 times a year. Thus $m = 4$ in (2), and we have

$$\$A(5) = \$2000\left(1 + \frac{0.06}{4}\right)^{(4)(5)}$$

(with arrows labelled m pointing to the 4 in the denominator and the 4 in the exponent)

$$= \$2000(1.015)^{20} = \$2000(1.346855007) = \$2693.71$$

The total interest is now $693.71.

(c) Here $m = 12$ (12 months in a year), so

$$\$A(5) = \$2000\left(1 + \frac{0.06}{12}\right)^{(12)(5)}$$

(with arrows labelled m pointing to the 12 in the denominator and the 12 in the exponent)

$$= \$2000(1.005)^{60} = \$2000(1.348850153) = \$2697.70$$

The total interest is $697.70.

(d) Here $m = 365$, so

$$\$A(5) = \$2000\left(1 + \frac{0.06}{365}\right)^{(365)(5)} = \$2000(1.000164384)^{1825}$$

$$= \$2000(1.349825523) = \$2699.65$$

In this case, the total interest paid is \$699.65.

As the preceding example indicates, the more frequently interest is compounded, the more the investment increases in value. In Table 4, we show the value after 10 years and the interest earned on a \$1000 investment at 8% annual interest for different numbers of payment periods each year.

Table 4
Value of a \$1000 Investment Compounded m Times a Year for 10 Years at an Annual Rate of 8%

m (number of times interest is compounded each year)	Value of \$1000 After 10 Years at 8% Interest (\$)	Total Interest Earned (\$)
1 (annually)	2158.92	1158.92
2 (semiannually)	2191.12	1191.12
4 (quarterly)	2208.04	1208.04
12 (monthly)	2219.64	1219.64
52 (weekly)	2224.17	1224.17
365 (daily)	2225.35	1225.35
8,760 (hourly)	2225.53	1225.53
525,600 (each minute)	2225.54	1225.54

Table 4 is revealing. It suggests that though there is a considerable difference when we change from annual to semiannual compounding (a difference in this example of \$32.20), the difference becomes negligible as we increase the number of interest periods. For example, the difference between monthly compounding and hourly compounding is only \$5.89. The numbers in Table 4 suggest that, after a point, little is gained by increasing the number of interest periods per year.

Many bank advertisements contain statements like "our 8% savings plan carries an effective interest rate or yield of $8\frac{1}{3}$%." The **effective interest rate,** or **yield,** is the rate of simple interest received over a one-year period. For example, \$100 would be worth \$108 if that sum is invested for one year at 8%

interest compounded annually. But if it is compounded quarterly, for instance, then it is worth

$$\$100(1.02)^4 = \$108.24$$

after one year. Thus the interest paid is \$8.24, and the effective interest rate is $8.24\% \approx 8\frac{1}{4}\%$.

We see that for most problems there are two rates of interest: the *quoted* rate and the *effective* rate. The first of these is often called the **nominal** rate of interest. Thus, as we have seen, a nominal rate of 8% provides an effective rate of 8.24% when interest is compounded quarterly.

EXAMPLE 8 *Determining the Effective Rate of Interest*

If money is invested at a nominal rate of 15% compounded monthly, what is the effective rate of interest?

SOLUTION Starting with P dollars, there will be $P(1 + 0.15/12)^{12} \approx 1.161P$ dollars after 1 year. The increase is $0.161P = 16.1\%$ of P. Thus, P dollars will have grown by approximately 16.1% after 1 year. This is the effective rate of interest.

Table 5 gives the effective interest rates if a sum is invested at a nominal rate of 8% compounded m times a year.

Table 5

m (number of times interest is paid per year)	Effective Interest Rate (based on 8%) (%)
1	8.000
2	8.160
4	8.24322
8	8.28567
12	8.29995
24	8.31430
52	8.32205
365	8.32776
1,000	8.32836
10,000	8.32867
1,000,000	8.32871

Problems 7.1

Readiness Check

I. Which of the following is an exponential function?
   a. $y = x^{\sqrt{3}}$
   b. $y = \sqrt{2 + x^4}$
   c. $y = \sqrt{2}^x$
   d. $y = x^{-3}$

II. Which of the following is true about the graph of $y = 2^{-x}$?
   a. It is an increasing function.
   b. It has an x-intercept but no y-intercept.
   c. It has the same shape but does not decrease as fast as $y = (\frac{1}{10})^x$.
   d. As $x \to \infty$, $y \to -\infty$.

III. Which of the following functions would result if the graph of $f(x) = 2^x$ was shifted to the right 2 units and down 1 unit?
   a. $f(x) = 2^{x-1} + 2$
   b. $f(x) = 2^{x-2} - 1$
   c. $f(x) = 2^{x+1}$
   d. $f(x) = 2^{x-3}$

IV. Which of the following is true about the graph below if $a > 0$, $a \neq 1$, and $y = a^x$?
   a. It passes through (1, 0).
   b. Its range is all real numbers.
   c. $a > 1$
   d. $0 < a < 1$

y
$y = a^x$
0
x

V. Which of the following is graphed below?
   a. $f(x) = 1 + 2^{x-2}$
   b. $f(x) = 1 - 2^{x+2}$
   c. $f(x) = 2^{x+2} + 1$
   d. $f(x) = 2^{x-2} - 1$

y
$y = 1$
0
x
$x = 2$

In Problems 1–14 draw a sketch of the given exponential function.

1. $y = 3^x$
2. $y = (\frac{1}{3})^x$
3. $y = (\frac{1}{5})^x$
4. $y = 5^x$
5. $f(x) = (7.2)^x$
6. $f(x) = (0.623)^x$
7. $f(x) = 3 \cdot 2^x$
8. $f(x) = 4 \cdot 10^x$
9. $y = -2 \cdot 10^x$
10. $y = 10 \cdot 2^x$
11. $y = 2^{x-1}$ [Hint: Shift a graph as in Section 1.4.]
12. $y = 3^{x-2}$
13. $y = 3 \cdot 10^{x+1} + 5$
14. $y = 4 \cdot 2^{1-x} - 1$

In Problems 15–22 use a calculator to estimate the given number to as many decimal places of accuracy as are carried on the machine.

15. $10^{2.2}$
16. $(3.8)^{4.7}$
17. $4^{-1.6}$
18. $(\frac{1}{2})^{5.1}$
19. $(0.35)^{0.42}$
20. $(53.21)^{-0.152}$
21. $3^{\sqrt{2}}$
22. $3^{\sqrt{2}}$

Answers to Readiness Check

I. c II. c III. b IV. d V. a

The remaining problems all require the use of a calculator.

In Problems 23–32 compute the value of an investment after t years and the total interest paid if P dollars is invested at a nominal interest rate of $r\%$ compounded m times a year.

23. $P = \$5000,\ r = 6\%,\ t = 4,\ m = 1$
24. $P = \$5000,\ r = 6\%,\ t = 4,\ m = 4$
25. $P = \$5000,\ r = 6\%,\ t = 4,\ m = 12$
26. $P = \$8000,\ r = 11\%,\ t = 4,\ m = 1$
27. $P = \$8000,\ r = 11\%,\ t = 4,\ m = 4$
28. $P = \$8000,\ r = 11\%,\ t = 4,\ m = 12$
29. $P = \$8000,\ r = 11\%,\ t = 4,\ m = 100$
30. $P = \$10{,}000,\ r = 8\frac{1}{2}\%,\ t = 6,\ m = 1$
31. $P = \$10{,}000,\ r = 8\frac{1}{2}\%,\ t = 6,\ m = 4$
32. $P = \$10{,}000,\ r = 8\frac{1}{2}\%,\ t = 6,\ m = 12$
33. Calculate the percentage difference in return on investment if P dollars is invested for 10 years at 6% compounded annually and quarterly.
34. As a gimmick to lure depositors, a bank offers 5% interest compounded daily in comparison with its competitor, who offers $5\frac{1}{8}\%$ compounded annually. Which bank would you choose?
35. Suppose the competitor in Problem 34 now compounds $5\frac{1}{8}\%$ semiannually. Which bank would you choose?
36. If \$20,000 is invested in bonds yielding 8% compounded quarterly, what will the bonds be worth in 9 years?
** 37. A certain government bond sells for \$750 and can be redeemed for \$1000 in 8 years. Assuming quarterly compounding, what is the nominal rate of interest paid?
38. A Roman deposited 1¢ in a bank at the beginning of the year A.D. 1. If the bank paid a meager 2% interest compounded quarterly, what would the investment be worth at the beginning of 1992?
39. Mrs. Jones has just invested \$400 in a 5-year term deposit (account A) paying 12% per year compounded twice per year, and she has invested another \$400 in a 5-year deposit (account B) at 11% per year compounded monthly.
    (a) Calculate the effective interest rate (as a percentage) for each account. Give answers correct to two decimal places.
    (b) Which investment is worth more after 5 years? by how much (to the nearest cent)?
40. (a) On November 1, 1975, Mr. Smith invested \$10,000 in a 10-year certificate that paid 11% interest per year compounded quarterly. When this matured on November 1, 1985, he reinvested the entire accumulated amount in Canada Savings Bonds with an interest rate of 7% compounded annually. To the nearest dollar, what was Mr. Smith's accumulated amount on November 1, 1990?
    (b) If Mr. Smith had made a single investment of \$10,000 in 1975 that matured in 1990 and had an effective rate of interest of 9%, would his accumulated amount be more or less than that in part (a)? by how much (to the nearest dollar)?

Four exponential functions are graphed below. Each is the graph of one of the functions given in Problems 41–44. Match each function with its corresponding graph.

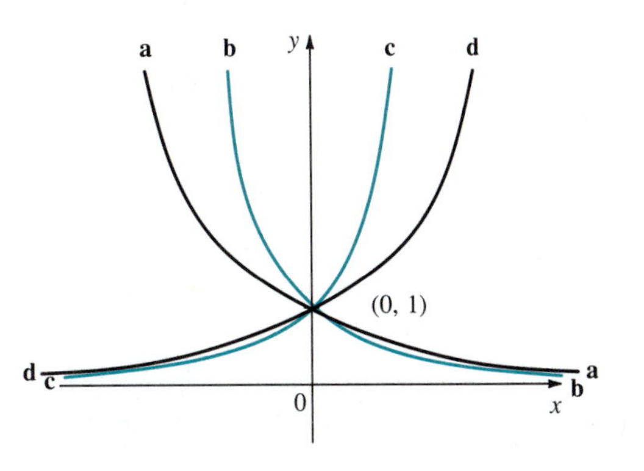

41. $f(x) = 3^x$
42. $f(x) = 5^x$
43. $f(x) = (\frac{1}{3})^x$
44. $f(x) = (\frac{1}{5})^x$

Graphing Calculator Problems

Before beginning, read Appendix A.

45. Use your calculator to sketch 2^x, 3^x, and 10^x on the same screen.
46. Sketch 2^{-x}, 3^{-x}, and 10^{-x} on the same screen.
47. Sketch 5^x and $(\frac{1}{5})^x$ on the same screen.
48. Sketch 2^x, 2^{x-1}, and $2^x - 1$ on the same screen.

In Problems 49–62 sketch each exponential function.

49. 3.7^x
50. 3.7^{x-2}
51. 3.7^{x+3}
52. $3.7^x + 5$
53. -3.7^x
54. 3.7^{-x}
55. 3.7^{2-x}
56. $4 - 3.7^{1-x}$
57. $4.2^{x/2}$
58. $3^{x/5}$
59. $-1.5^{1.5x}$
60. 2^{-2x}
61. $10^{x/100}$
62. $3^{-0.002x}$

7.2 The Natural Exponential Function

In this section, we introduce one of the most important functions in mathematics: the function e^x. To motivate the definition of the number e, we look again at the compound interest formula (equation (2) in Section 7.1)

$$A(t) = P\left(1 + \frac{r}{m}\right)^{mt} \tag{1}$$

In (1), we set $P = 1$, $r = 1$ (corresponding to 100% interest), and $t = 1$. Then (1) becomes

$$A(1) = \left(1 + \frac{1}{m}\right)^{m} \tag{2}$$

which is the value after one year of \$1 invested at 100% interest compounded m times a year. For example, if $m = 1$, then interest is paid once a year (annually) and

$$A(1) = \left(1 + \frac{1}{1}\right)^{1} = 2^1 = \$2$$

This is no surprise since the investor has the original \$1 plus \$1 interest (100% of \$1) paid at the end of the year. If $m = 2$, then

$$A(1) = \left(1 + \frac{1}{2}\right)^{2} = \left(\frac{3}{2}\right)^{2} = \frac{9}{4} = \$2.25$$

Again, this is reasonable. Now interest of 50% is paid twice a year. After six months the investor has $\$1 + \frac{1}{2}(\$1) = \$1.50$, and after one year she has \$1.50 plus 50% of \$1.50 or $\$1.50 + \$0.75 = \$2.25$.

In Table 1, we compute values of $\left(1 + \frac{1}{m}\right)^{m}$ for a number of values for m.

Table 1

m	$1/m$	$(1 + 1/m)^m$
1	1	2
2	0.5	2.25
5	0.2	2.48832
10	0.1	2.59374246
100	0.01	2.704813829
1,000	0.001	2.716923932
10,000	0.0001	2.718145927
100,000	0.00001	2.718268237
1,000,000	0.000001	2.718280469
1,000,000,000	0.000000001	2.718281827

It seems that the expression $(1 + 1/m)^m$ gets closer and closer to a fixed number. This number is denoted by e.

The Number e

Definition of e

The number $\boldsymbol{e}$ is defined to be the number approached by the expression $(1 + 1/m)^m$ as m increases without bound. To 10 decimal places

$$e \approx 2.7182818285$$

The number e was first discovered by the great Swiss mathematician Leonhard Euler (1707–1783), who described the number in 1728.†

The Function e^x

Once we know the number e, we can write the function $y = e^x$. This function arises in an astonishingly wide variety of applications. We will see how later in this section.

Values of e^x can be found on a scientific calculator in one of two ways:

To Find e^x on a Calculator

(a) If there is an [eˣ] key, use it directly.
(b) If not, then there is a [ln] or [ln x] key. Press [INV] [ln x] or [2nd F] [ln x] to obtain e^x.

NOTE Sometimes the [eˣ] key is called [exp x].

As we will see in Section 7.4, e^x is the inverse of a function called the natural logarithm function $\ln x$, so [INV] [ln x] gives us e^x.

EXAMPLE 1 *Computing e^x for Three Values of x*

Compute (a) e^2, (b) $e^{0.46}$, and (c) $e^{-3.14}$ on a calculator.

SOLUTION

(a) By pressing [2] [INV] [ln x], we obtain 7.389056099. We achieve the same result on a calculator with an [eˣ] key by pressing [2] [eˣ].
(b) [.] [4] [6] [INV] [ln x] yields 1.584073985.
(c) [3] [.] [1] [4] [+/−] displays −3.14, and then pressing [INV] [ln x] yields 0.043282797.

Alternatively, we could first compute $e^{3.14}$ and then use the reciprocal key [1/x] to compute $e^{-3.14} = 1/e^{3.14}$.

NOTE On some calculators [INV] [ln x] or [2nd F] [ln x] must be pressed *before* entering the number 2, 0.46 or −3.14.

---

† See the biographical sketch of Euler on p. 373.

In Table 1 at the back of the book, we provide values for e^x and e^{-x} $(= 1/e^x)$ for a wide range of numbers. However, each of these values can be obtained with more accuracy on a calculator.

The graph of $y = e^x$ is given in Figure 1(a). Since $2 < e < 3$, the graph of e^x is sandwiched between the graphs of 2^x and 3^x. We have

$$2^x < e^x < 3^x \qquad \text{if } x > 0$$
$$3^x < e^x < 2^x \qquad \text{if } x < 0$$

This is indicated in Figure 1(b).

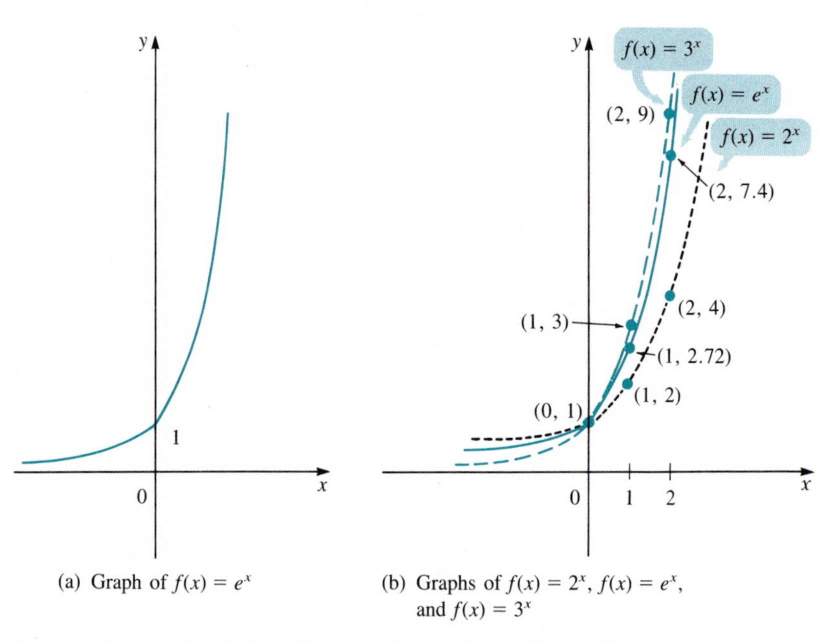

(a) Graph of $f(x) = e^x$

(b) Graphs of $f(x) = 2^x$, $f(x) = e^x$, and $f(x) = 3^x$

Figure 1 The graph of e^x lies between the graphs of 2^x and 3^x.

EXAMPLE 2 *Graphing a Shifted Exponential Function*

Sketch the graph of $f(x) = e^{x-1} - 2$.

SOLUTION We do this in two steps.

Step 1 The graph of e^{x-1} is the graph of e^x shifted 1 unit to the right (Figure 2(a)).

Step 2 The graph of $e^{x-1} - 2$ is the graph of e^{x-1} shifted 2 units down (Figure 2(b)).

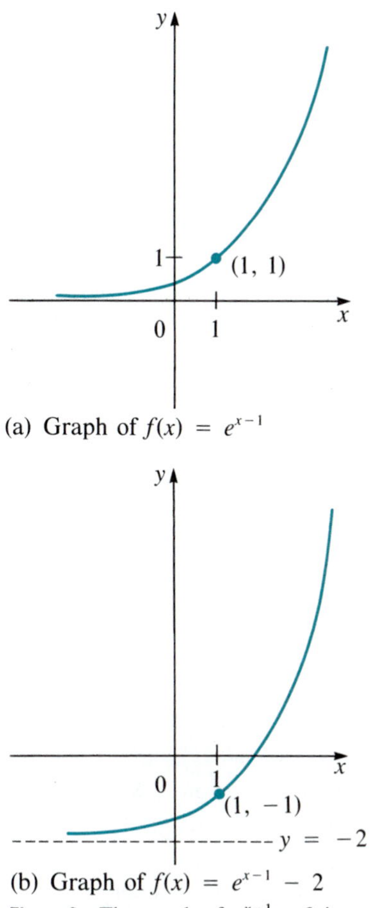

(a) Graph of $f(x) = e^{x-1}$

(b) Graph of $f(x) = e^{x-1} - 2$

Figure 2 The graph of $e^{x-1} - 2$ is obtained by shifting the graph of e^x 1 unit to the right and 2 units down.

Growth Rates of Power and Exponential Functions

Which grows faster: a power function or an exponential function? To get some idea, we give, in Table 2, values for x^2, x^5, x^{10}, and e^x.

Table 2
Some Approximate Values for x^2, x^5, x^{10}, and e^x

x	x^2	x^5	x^{10}	e^x
1	1	1	1	2.71828
10	100	100,000	10^{10}	22,026
50	2500	312,500,000	9.7656×10^{16}	5.1847×10^{21}
100	10,000	10^{10}	10^{20}	2.688×10^{43}
250	62,500	9.7656×10^{11}	9.5367×10^{23}	3.746×10^{108}
1000	1,000,000	10^{15}	10^{30}	1.9701×10^{434}

It seems that e^x grows much faster than the function x^{10}. In fact the following is true:

$$\frac{e^x}{x^n} \to \infty \text{ as } x \to \infty$$

It is difficult to prove this fact without using calculus. However, Table 2 certainly suggests that it is plausible.

Continuous Compounding of Interest

We return to equation (1):

$$A(t) = P\left(1 + \frac{r}{m}\right)^{mt}$$

What happens as the number of interest payment periods (m) increases without bound? That is, what happens if interest is compounded *continuously?* Let $k = m/r$. Then $m = kr$, $r/m = 1/k$, and

$$A(t) = P\left(1 + \frac{r}{m}\right)^{mt} = P\left(1 + \frac{1}{k}\right)^{krt} = P\left[\left(1 + \frac{1}{k}\right)^{k}\right]^{rt}$$

But, as k gets large, $(1 + 1/k)^k$ approaches e. Thus we have the following:

Formula for Continuous Compounding

If P dollars is invested at a rate of interest r **compounded continuously,** then, after t years, the investment is worth

$$A(t) = Pe^{rt} \quad (3)$$

EXAMPLE 3 ***The Value After Five Years with Continuous Compounding***

If \$2000 is invested at 6% interest compounded continuously, what is the investment worth after 5 years?

SOLUTION $A(5) = 2000e^{(0.06)5} = 2000e^{0.3}$
$= 2000(1.349858808) \approx \$2699.72$

NOTE In Example 7 in Section 7.1, we found that if interest were compounded daily, then the investment would be worth \$2699.65. The difference is 7¢. Continuous compounding sounds better but, as we see here, there really isn't much difference between frequent compounding and continuous compounding. ■

EXAMPLE 4 ***The Value After Ten Years with Continuous Compounding***

Suppose \$5000 is invested in a bond yielding $8\frac{1}{2}\%$ annually. What will the bond be worth in 10 years if interest is compounded continuously?

SOLUTION $A(t) = Pe^{rt} = 5000e^{(0.085)(10)} = 5000e^{0.85} = \$11{,}698.23$

In Table 5 on p. 363, we gave the effective interest rate of 8% interest compounded m times a year. We can extend the table to include continuous compounding by adding the result of the next example.

EXAMPLE 5 ***Computing the Effective Interest Rate with Continuous Compounding***

If money is invested at 8% compounded continuously, what is the effective interest rate?

SOLUTION One dollar invested will be worth $e^{r \cdot 1} = e^{0.08} = \$1.083287068$ after 1 year. The effective interest rate is 8.32871%.

Exponential Growth and Decay

We now give one of the reasons the exponential function is so important. Let $y = f(x)$ represent some quantity that is growing or declining, such as the volume of a substance, the population of a certain species, the value of an investment, or the mass of a decaying radioactive substance.

We define the **relative rate of growth** of y as follows:

$$\text{Relative rate of growth} = \frac{\text{actual rate of growth}}{\text{size of } f(x)}$$

When we say that a population is growing at 8% a year or an investment pays 8% interest, compounded continuously, or an ice cube is evaporating at a rate of 8% a minute, we are talking about the *relative* rate of growth.

The following remarkable fact is proved in a calculus course:

Constant Relative Rate of Growth = Exponential Growth

If the relative rate of growth of $y(t)$ is a constant, k, and if $y(t)$ is changing continuously, then

$$y(t) = y(0)e^{kt} \tag{4}$$

The value $y(0)$ of y at $t = 0$ is called the **initial value** of the quantity. If $k > 0$, then y is said to be **growing exponentially.** If $k < 0$, then y is said to be **decaying exponentially.**

These two ideas are illustrated in Figure 3.

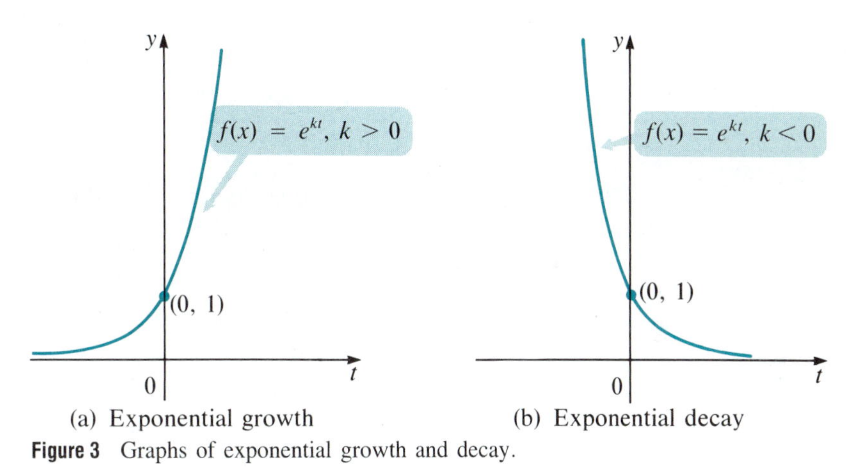

Figure 3 Graphs of exponential growth and decay.

EXAMPLE 6 *Population Growth*

A bacteria population is growing at a constant relative rate equal to 10% of its population each day. Its initial size is 10,000 organisms. How many bacteria are present after 10 days? after 30 days?

SOLUTION The relative rate of growth here is $10\% = 0.1$. Thus, if $P(t)$ denotes the population at time t, we have, from (4) with $k = 0.1$,

$$P(t) = P(0)e^{0.1t} \tag{5}$$

But the initial population is $P(0) = 10{,}000$, so (5) becomes

$$P(t) = 10{,}000e^{0.1t}$$

Now we can answer the questions:

$$\text{After 10 days,} \quad P(10) = 10{,}000e^{(0.1)(10)} = 10{,}000e^{1} \approx 10{,}000(2.7183) = 27{,}183$$

$$\text{After 30 days,} \quad P(30) = 10{,}000e^{(0.1)(30)} = 10{,}000e^{3} \approx 10{,}000(20.0855) = 200{,}855 \quad ■$$

EXAMPLE 7 *Computing the Volume of a Melting Block of Ice*

A block of ice is melting at a constant relative rate and loses 5% of its volume each minute. If its initial volume is 20 cubic inches (in^3), what is its volume after (a) 5 minutes? (b) 20 minutes? (c) 1 hour?

SOLUTION The block is *losing* 5% of its volume each minute, so $k = -0.05$ in formula (4). If $V(t)$ denotes the volume after t minutes, then $V(0) = 20$ (this is given) and

$$V(t) = 20e^{-0.05t}$$

(a) $V(5) = 20e^{-0.05(5)} = 20e^{-0.25} \approx 15.576 \text{ in}^3$.
(b) $V(20) = 20e^{-0.05(20)} = 20e^{-1} \approx 7.358 \text{ in}^3$.

1 hour = 60 minutes ↓

(c) $V(60) = 20e^{-0.05(60)} = 20e^{-3} \approx 0.9957 \text{ in}^3$.

We will see many more examples of exponential growth and decay in Section 7.6, after we have discussed the natural logarithm function.

NOTE There may be some confusion between the relative growth rate and the percentage growth per unit time. To illustrate this difference, let us compute the growth in a population of 100,000 over one year when (a) the relative growth rate is 50% and (b) there is a 50% increase in the population each year:

(a) By (4) with $y(0) = 100{,}000$ and $k = 0.5$,

$$\text{population after 1 year} = y(1) = 100{,}000e^{0.5} \approx 164{,}872$$

which is an increase of 64,872 individuals.

(b) 50% of 100,000 = 50,000 so

$$\text{population after 1 year} = 100{,}000 + 50{,}000 = 150{,}000$$

Can you explain the difference between the increase of 64,872 in (a) and 50,000 in (b)? The answer is similar to the difference between simple and compound interest. A relative rate of growth implies that populations are changing continuously. For example, after 1 month (1/12 year) the population is

$$y\left(\frac{1}{12}\right) = 100{,}000e^{0.5(1/12)} \approx 104{,}255$$

The second month begins with a population of 104,255, which grows at a 50% rate. This is like "paying interest on interest." On the other hand, a growth rate of 50% per year means that each year the population increases by 50%. This is analogous to paying 50% simple interest at the end of each year. The two concepts are different.

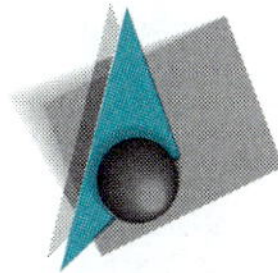

FOCUS ON

Leonhard Euler (1707–1783)

The great Swiss mathematician Leonhard Euler (pronounced "Oiler") was born in Basel, Switzerland, in 1707. Euler's father was a clergyman who hoped that his son would follow him into the ministry. The father was adept at mathematics, however, and together with Johann Bernoulli, instructed young Leonhard in that subject. Euler also studied theology, astronomy, physics, medicine, and several Eastern languages.

In 1727, Euler applied to and was accepted for a chair of medicine and physiology at the St. Petersburg Academy. The day Euler arrived in Russia, however, Catherine I—founder of the Academy—died, and the Academy was plunged into turmoil. By 1730, Euler was pursuing his mathematical career from the chair of natural philosophy. Accepting an invitation from Frederick the Great, Euler went to Berlin in 1741 to head the Prussian Academy. Twenty-five years later, he returned to St. Petersburg, where he died in 1783 at the age of 76.

The most prolific writer in the history of mathematics, Euler found new results in virtually every branch of pure and applied mathematics. Although German was his native language, he wrote mostly in Latin and occasionally in French. His amazing productivity did not decline even when he became totally blind in 1766. During his lifetime, Euler published 530 books and papers. When he died, he left so many unpublished manuscripts that the St. Petersburg academy was still publishing his work in its *Proceedings* almost half a century later. Euler's work enriched such diverse areas as hydraulics, celestial mechanics, lunar theory, and the theory of music, as well as mathematics.

Euler had a phenomenal memory. As a young man, he memorized the entire *Aeneid* by Virgil (in Latin), and many years later could still recite the entire work. He was able to solve astonishingly complex mathematical problems in his head and is said to have solved, again in his head, problems in astronomy that stymied Newton. The French academician François Arago once commented that Euler could calculate without effort "just as men breathe, as eagles sustain themselves in the air."

Euler wrote in a mathematical language that is largely in use today. Among symbols first used by him are

- $f(x)$ for functional notation
- e for the base of the natural exponential and logarithm functions
- Σ for the summation sign
- i to denote $\sqrt{-1}$

Euler's textbooks were models of clarity. His texts included the *Introductio in analysin infinitorum* (1748), his *Institutiones calculi differentialis* (1755), and the three-volume *Institutiones calculi integralis* (1768–1774). This and others of his works served as models for many of today's mathematics textbooks.

It is said that Euler did for mathematical analysis what Euclid did for geometry. It is no wonder that so many later mathematicians expressed their debt to him.

Problems 7.2

Readiness Check

I. Which of the following is the definition of e?
   a. The number that $(1 + x)^{1/x}$ approaches as x approaches ∞.
   b. The number that $\left(1 - \dfrac{1}{x}\right)^x$ approaches as x approaches ∞.
   c. The number that $\left(1 + \dfrac{1}{x}\right)^x$ approaches as x approaches ∞.
   d. The number that $(1 - x)^{1/x}$ approaches as x approaches ∞.

II. Which of the following is true about $f(x) = e^{-x}$?
   a. Its graph is decreasing.
   b. Its graph is the reflection of $f(x) = e^x$ about the y-axis.
   c. If $x > 0$, it takes negative values.
   d. It is equal to $f(x) = \dfrac{-1}{e^x}$.

III. Which of the following is an increasing function?
   a. $f(x) = e^{x-3}$ b. $f(x) = e^{-\pi x}$
   c. $f(x) = 1 - e^{3x}$ d. $f(x) = -e^{3x}$

IV. Which of the following is true?
   a. $2 < e^{-1} < 3$ b. $e > \pi$
   c. $\sqrt{2} < e < \sqrt{3}$ d. $\dfrac{1}{3} < \dfrac{1}{e} < \dfrac{1}{2}$

V. Which function's graph is given in the figure below?
   a. e^{-x} b. $e^x - 2$
   c. $-e^x$ d. $-e^{-x}$
   e. $e^{-x} - 2$

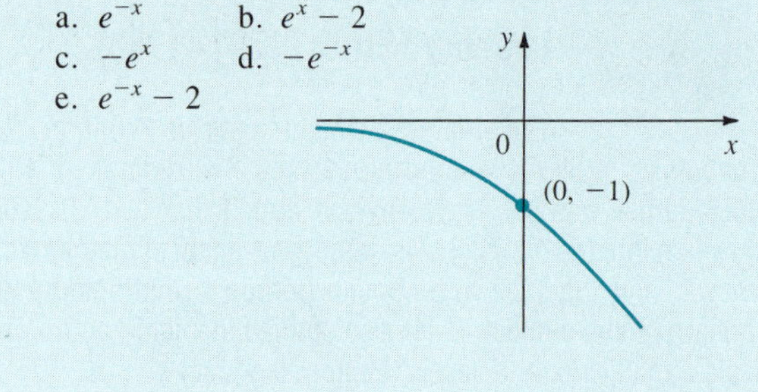

In Problems 1–10 use a calculator to compute e^x to six significant digits.

1. $e^{3.15}$ 2. $e^{-0.6}$ 3. $e^{\sqrt{3}}$ 4. $e^{12.02}$
5. $e^{29.4}$ 6. $e^{-4.17}$ 7. $e^{-15.9}$ * 8. e^e
* 9. e^{π} * 10. π^e

Match each function in Problems 11–20 with one of the ten graphs given below.

11. $-e^x$ 12. $-e^{-x}$ 13. $1 - e^x$ 14. e^{x+1}
15. e^{x-1} 16. $2 - e^{x-1}$ 17. $e^x - 2$ 18. $2e^x$
19. $-2e^x$ 20. $2 + e^{-x}$

(c) (1, 1)

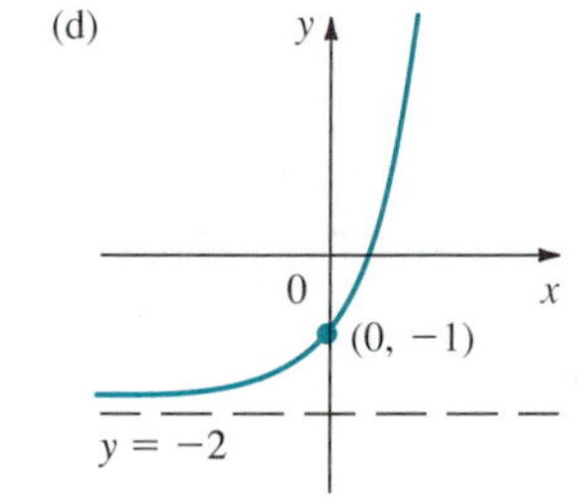

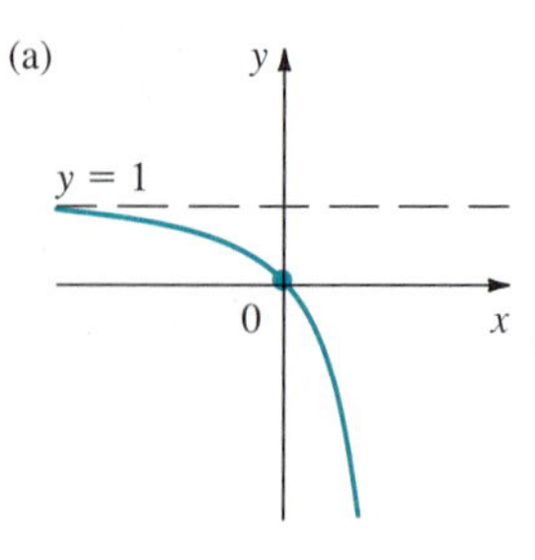

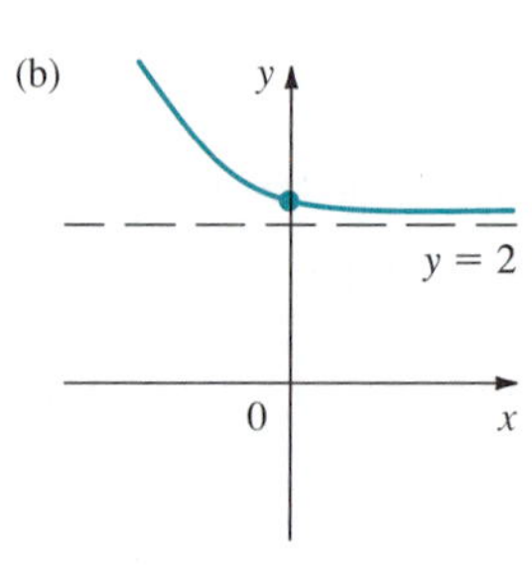

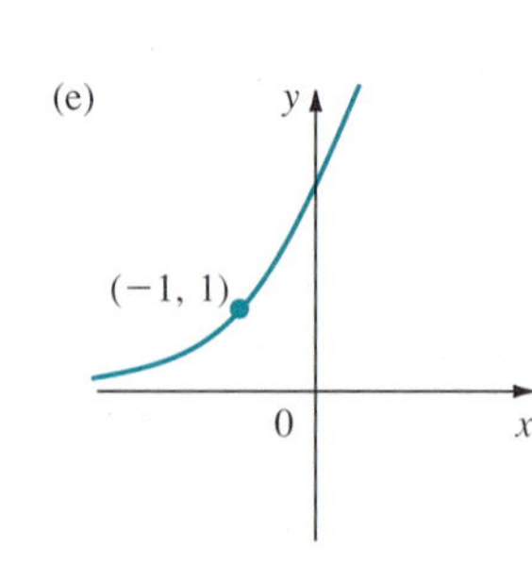

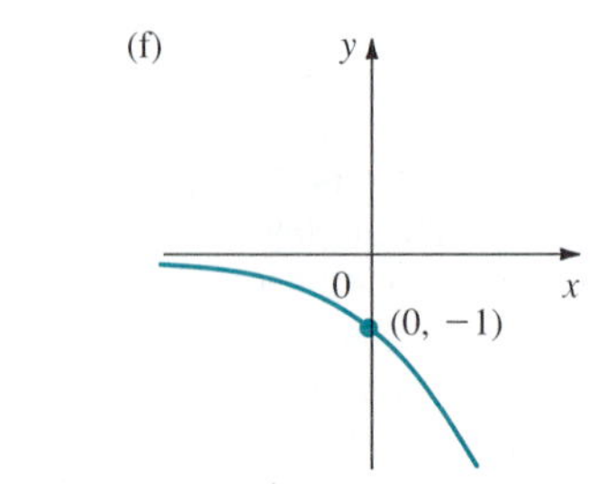

Answers to Readiness Check

I. c II. a, b III. a IV. d V. c

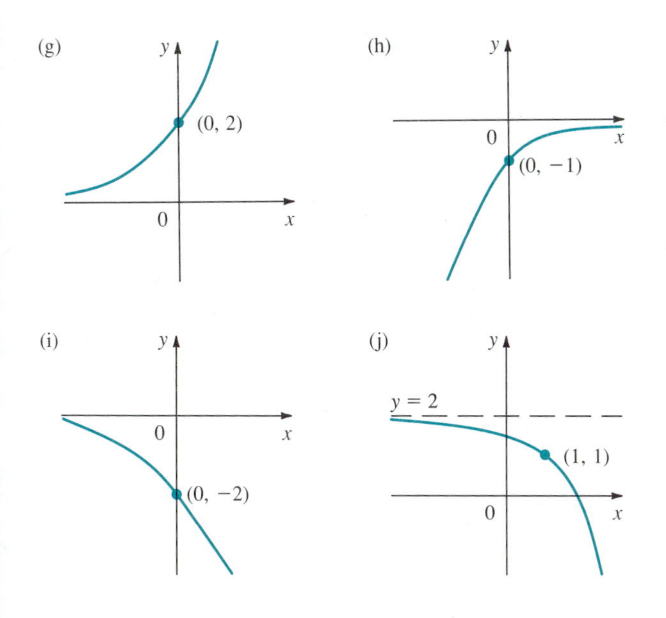

In Problems 21–28 sketch the graph of the exponential function.

21. e^{2x} 22. $e^{-x/2}$ 23. $3e^{1.5x}$ 24. $-e^x$
25. e^{x-2} 26. e^{x+3} 27. $e^x + 5$ * 28. $3 - e^{2-x}$

29. The exponential e^x can be estimated for x in $[-\frac{1}{2}, \frac{1}{2}]$ by the formula

$$e^x \approx \left(\left\{\left[\left(\frac{x}{5}+1\right)\frac{x}{4}+1\right]\frac{x}{3}+1\right\}\frac{x}{2}+1\right)x+1$$

(a) Calculate an approximate value for $e^{0.13}$.
(b) Calculate an approximate value for $e^{-0.37}$.

In Problems 30–33 compute the value of an investment after t years and the total interest paid if P dollars is invested at a nominal interest rate of $r\%$, compounded continuously.

30. $P = \$5000$, $r = 6\%$, $t = 4$
31. $P = \$9000$, $r = 10\%$, $t = 5$
32. $P = \$10{,}000$, $r = 6.5\%$, $t = 8$
33. $P = \$10{,}000$, $r = 13\%$, $t = 8$

Use a calculator to solve Problems 34–49.

34. A sum of \$5000 is invested at a return of 7% per year, compounded continuously. What is the investment worth after 8 years?
35. If \$10,000 is invested in bonds yielding 9% compounded continuously, what will the bonds be worth in 8 years?
36. An investor buys a bond that pays 12% annual interest compounded continuously. If she invests \$10,000 now, what will her investment be worth in (a) 1 year? (b) 4 years? (c) 10 years?
37. As a gimmick to lure depositors, a bank offers 5% interest compounded continuously in comparison with its competitor, which offers $5\frac{1}{8}\%$ compounded annually. Which bank would you choose?
38. If money is invested at 10% compounded continuously, what is the effective interest rate? [The effective interest rate is defined on p. 362.]
* 39. After how many years will the bond in Problem 36 be worth \$20,000? [Hint: We will see an easy way to solve this problem in Section 7.4. For now, use trial and error and give an answer to the nearest tenth of a year.]
40. A Roman deposited 1¢ in a bank at the beginning of the year A.D. 1. If the bank paid a meager 1% interest, compounded continuously, what would the investment be worth at the beginning of 1992?
41. Mrs. Jones has just invested \$400 in a 5-year term deposit (account A) paying 12% per year compounded twice per year, and she has invested another \$400 in a 5-year deposit (account B) at 11% per year compounded continuously.
(a) Calculate the effective interest rate (as a percentage) for each account. Give answers correct to two decimal places.
(b) Which investment is worth more after 5 years? by how much (to the nearest cent)?
42. The population of a certain city was 800,000 in 1985. If the population grows at a constant rate of 2% a year, what was the population in 1990? What will be the population in the year 2010? [Hint: Treat 1985 as year 0.]
43. The estimated world population in 1986 was 4,845,000,000. Assume that the population grows at a constant rate of 1% per year. Estimate the world population in (a) 1990, (b) 2000, and (c) 2010.
44. The mass of a radioactive substance is declining at a rate of 25% a week. The mass is initially 20 grams. What is the mass after (a) 2 weeks? (b) 10 weeks? (c) 1 year?
* 45. The temperature difference between a hot coal and the surrounding air declines 12% each minute. At noon, the temperature of the coal is 180°F, and the temperature of the air is 50°F. Assuming that the air temperature does not change, estimate the temperature of the coal at (a) 12:05 P.M., (b) 12:15 P.M., (c) 12:30 P.M.
46. The **hyperbolic sine** function, denoted by $\sinh x$, is defined by

$$\sinh x = \frac{e^x - e^{-x}}{2}$$

Compute (a) $\sinh 0$, (b) $\sinh 1$, (c) $\sinh(-\frac{1}{2})$, (d) $\sinh(-4)$, (e) $\sinh 2.37$.
47. (a) Show that $\sinh x$ is an odd function.
(b) Plot values of $\sinh x$ on graph paper for x between 0 and 5 in increments of 0.2.
(c) Sketch the graph of $y = \sinh x$.

48. The **hyperbolic cosine** function, denoted by $\cosh x$, is defined by

$$\cosh x = \frac{e^x + e^{-x}}{2}$$

Compute (a) $\cosh 0$, (b) $\cosh 1$, (c) $\cosh \frac{1}{2}$, (d) $\cosh 2.37$, (e) $\cosh 5$.

49. (a) Show that $\cosh x$ is an even function.
   (b) Show that $\cosh 0 = 1$ and $\cosh x > 1$ if $x \neq 0$.
   (c) Plot values of $\cosh x$ on graph paper for x between 0 and 5 in increments of 0.2.
   (d) Sketch the graph of $y = \cosh x$.

* 50. The **hyperbolic tangent** function, denoted by $\tanh x$, is defined by

$$\tanh x = \frac{\sinh x}{\cosh x} = \frac{e^x - e^{-x}}{e^x + e^{-x}}$$

   (a) Show that $\tanh x$ is defined for every real number.
   (b) Show that $-1 < \tanh x < 1$ for every real number x.

Graphing Calculator Problems

Before beginning, read Example 4 in Appendix A.

51. Sketch $e^{x/3}$ and $e^{-x/3}$ on the same screen.
52. Sketch e^{2x} and $-e^{2x}$ on the same screen.
53. Sketch e^x, e^{2x}, and e^{3x} on the same screen.
54. Sketch e^{-x}, e^{-2x}, and e^{-3x} on the same screen.
55. Sketch $e^{x/2}$, $e^{(x-1)/2}$, and $e^{(x+1)/2}$ on the same screen.
56. Sketch $e^{1.5x}$, $e^{1.5x} + 3$, and $e^{1.5x} - 3$ on the same screen.

In Problems 57–73 obtain a sketch of each exponential function.

57. $e^{1.6x}$
58. $e^{1.6(x-2)}$
59. $-e^{1.6x}$
60. $e^{-1.6x}$
61. $2 - e^{1.6(1-x)}$
62. $e^{-0.23x}$
63. $e^{-3.5x}$
64. $e^{0.57x}$
65. $-e^{0.9(x-0.6)}$
66. $1 - 2e^{-\frac{1}{2}x}$
67. $-3 + 4e^{-0.8(x+0.4)}$
68. e^{x^2}
69. e^{-x^2}
70. $e^{1/x}$
71. $-e^{-1/x}$
72. $e^{\sqrt{x}}$
73. $e^{-\sqrt{x}}$

7.3 Logarithmic Functions

In Section 1.6, we discussed inverse functions. We proved that if a function f is increasing or decreasing, then f has an inverse. In Section 7.1, we saw that the function defined by $f(x) = a^x$ is increasing if $a > 1$ and decreasing if $0 < a < 1$. Putting these two facts together, we conclude that

The function $f(x) = a^x$ has an inverse for $a > 0$, $a \neq 1$

What does such an inverse look like? We know that $a^x > 0$ for every real x. Thus the domain of the inverse of a^x is the set of positive real numbers. Put another way, if $y > 0$ is given, we can find a unique x such that $a^x = y$.

EXAMPLE 1 ***Solving for the Exponent***

Solve for x: (a) $2^x = 8$, (b) $3^x = \frac{1}{9}$, (c) $(\frac{1}{2})^x = 8$.

SOLUTION (a) We see that $2^x = 8$ if and only if $x = 3$. Similarly, in (b), $3^x = \frac{1}{9}$ if and only if $x = -2$, and in (c), $(\frac{1}{2})^x = 8$ if and only if $x = -3$ $\left[(\frac{1}{2})^{-3} = \frac{1}{(\frac{1}{2})^3} = \frac{1}{(\frac{1}{8})} = 8\right]$.

We now reverse the roles of x and y† in $y = a^x$ and define an important new function, the inverse of $f(x) = a^x$.

† We reverse the roles of x and y so that we can write the inverse function with x as the independent variable and y as the dependent variable. This is the process we used to find inverse functions in Section 1.6.

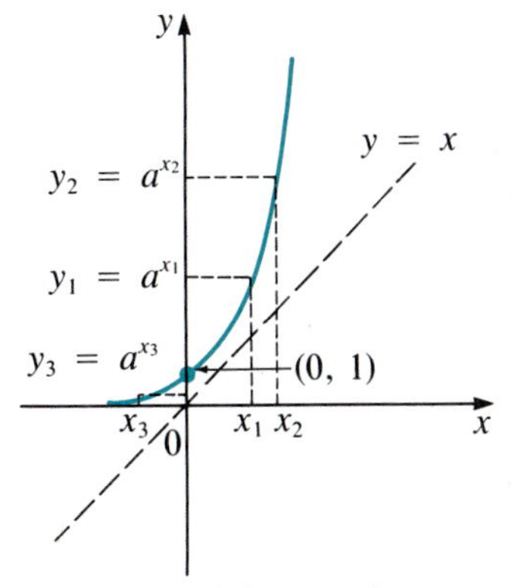

(a) Graph of $f(x) = a^x$

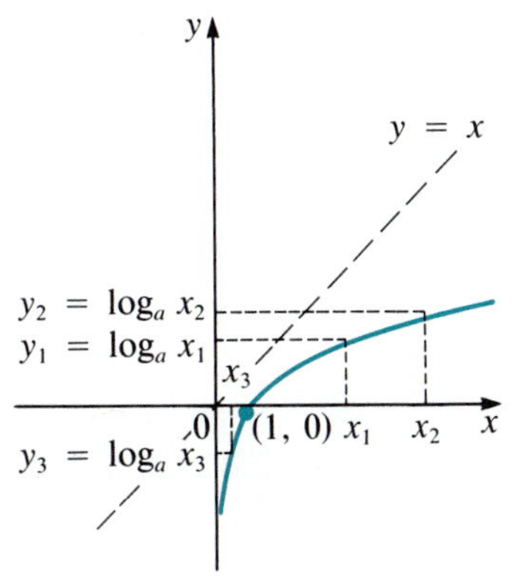

(b) Graph of $f(x) = \log_a x$

Figure 1 The graph of $\log_a x$ is the reflection about the line $y = x$ of the graph of a^x.

Definition of the Logarithm to the Base *a*

If $x = a^y$ with $a > 0$ and $a \neq 1$, then the **logarithm to the base *a* of** x is y. This is written

$$y = \log_a x \tag{1}$$

where $\log_a x$ is defined for $x > 0$.

This means that if

$$x > 0, \; a > 0, \quad \text{and} \quad a \neq 1$$

then

$$y = \log_a x \quad \text{if and only if} \quad x = a^y$$

By the reflection property discussed in Section 1.6 (p. 49), the graph of $y = \log_a x$ is the reflection of the graph of $y = a^x$ about the line $y = x$. In Figure 1, typical graphs for $y = a^x$ and $y = \log_a x$ are given for $a > 1$.

We stress the following facts about the logarithmic functions.

$$y = \log_a x \quad \text{is equivalent to} \quad x = a^y \tag{2}$$

$$y = a^x \quad \text{is equivalent to} \quad x = \log_a y \tag{3}$$

Think of $y = \log_a x$ as an answer to the question: To what power must a be raised to obtain the number x? This immediately implies that

$$a^{\log_a x} = x \qquad \text{for every positive real number } x \tag{4}$$

$$\log_a a^x = x \qquad \text{for every real number } x \tag{5}$$

EXAMPLE 2 ***Illustration of Properties of Logarithms***

(a) $2 = \log_3 9$ so $9 = 3^2$ From (2)

(b) $\dfrac{1}{16} = 2^{-4}$ so $-4 = \log_2 \dfrac{1}{16}$ From (3)

(c) $4^{\log_4 7} = 7$ From (4)

(d) $\log_{10} 10^{12} = 12$ From (5)

We also stress that $\log_a x$ is only defined for $x > 0$ because the equation $a^y = x$ has no solution when $x \leq 0$.† For example, $\log_2 (-1)$ is not defined because there is no real number y such that $2^y = -1$. We have

$$\text{domain of } \log_a x = \{x\text{: } x > 0\}$$

EXAMPLE 3 ***Writing Equations Using Logarithmic Notation***

We rewrite the exponential equations solved in Example 1 using logarithmic notation: (a) $\log_2 8 = 3$, (b) $\log_3 \frac{1}{9} = -2$, and (c) $\log_{1/2} 8 = -3$. ■

† On p. 22, we discussed three reasons for restricting the domain of a function. The logarithmic function provides a fourth reason.

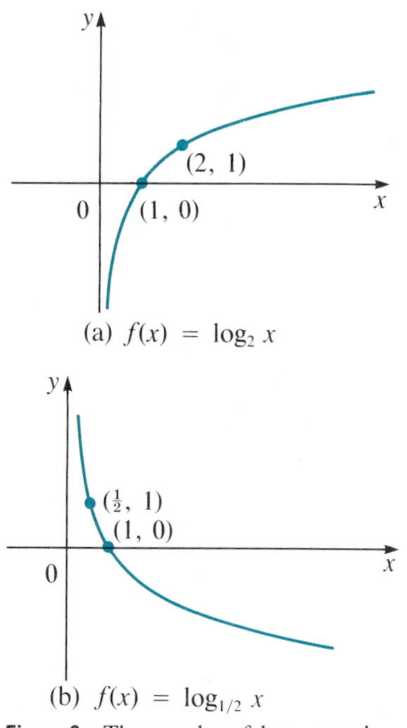

Figure 2 The graphs of $\log_2 x$ and $\log_{1/2} x$.

EXAMPLE 4 ***The Graphs of Two Logarithmic Functions***

The graphs of $y = \log_2 x$ and $y = \log_{1/2} x$ are given in Figure 2.

It follows from facts about the function a^x and the graph of $\log_a x$ that

$\log_a x$ is an increasing function if $a > 1$, and a decreasing function if $0 < a < 1$.

EXAMPLE 5 ***Writing a Logarithmic Equation in Exponential Form***

Change to exponential form: (a) $\log_9 3 = \frac{1}{2}$ and (b) $\log_4 64 = 3$.

SOLUTION

(a) $\log_9 3 = \frac{1}{2}$ is equivalent to $9^{1/2} = 3$.
(b) $\log_4 64 = 3$ is equivalent to $4^3 = 64$. ■

EXAMPLE 6 ***Writing an Exponential Equation in Logarithmic Form***

Change to logarithmic form: (a) $5^4 = 625$ and (b) $(\frac{1}{2})^{-4} = 16$.

SOLUTION

(a) $5^4 = 625$ is equivalent to $\log_5 625 = 4$.
(b) $(\frac{1}{2})^{-4} = 16$ is equivalent to $\log_{1/2} 16 = -4$.

We note the following fact: $y = \log_a x$ defines a one-to-one function. That is,

$$\text{If} \quad \log_a x_1 = \log_a x_2, \quad \text{then } x_1 = x_2. \tag{6}$$

Properties of Logarithms

Logarithms have many useful properties. Four properties follow immediately from the definition of the logarithm. We have seen that $a^{\log_a x} = x$ and $\log_a a^x = x$ (equations (4) and (5)). Since $a^1 = a$, $1 = \log_a a$ from (2). Also, $a^0 = 1$, so $\log_a 1 = 0$.

Four Basic Properties of $\log_a x$

	Property	*Illustration*
1.	$a^{\log_a x} = x$	$2^{\log_2 27.3} = 27.3$
2.	$\log_a a^x = x$	$\log_5 5^{\sqrt{7}} = \sqrt{7}$
3.	$\log_a a = 1$	$\log_{12} 12 = 1$
4.	$\log_a 1 = 0$	$\log_{1/2} 1 = 0$

The next four properties are especially useful in computations.

Computational Properties of $\log_a x$

For $x > 0$, $w > 0$, and r a real number,

5. $\log_a xw = \log_a x + \log_a w$
6. $\log_a \frac{x}{w} = \log_a x - \log_a w$
7. $\log_a \frac{1}{x} = -\log_a x$
8. $\log_a x^r = r \log_a x$

NOTE Property 7 is a special case of both Property 6 and Property 8.

Proof of Property 5

Let $u = \log_a x$ and $v = \log_a w$. Then, from (2),

$$x = a^u \qquad \text{and} \qquad w = a^v$$

so

Property 3 on p. 359
↓

$$xw = a^u a^v = a^{u+v}$$

Thus,

Property 2
↓

$$\log_a xw = \log_a a^{u+v} = u + v = \log_a x + \log_a w \quad \blacksquare$$

Proof of Property 6

With u and v as in Property 5 above,

Property 4 on p. 359
↓

$$\frac{x}{w} = \frac{a^u}{a^v} = a^{u-v}$$

so, from (2),

$$\log_a \frac{x}{w} = \log_a \frac{a^u}{a^v} = \log_a a^{u-v} = u - v = \log_a x - \log_a w \quad \blacksquare$$

Proof of Property 7

From Property 6 (already proven)

Property 4
↓

$$\log_a \frac{1}{x} = \log_a 1 - \log_a x = 0 - \log_a x = -\log_a x \quad \blacksquare$$

Proof of Property 8

Let $u = \log_a x$. Then $x = a^u$, and

$$x^r = (a^u)^r \overset{\text{Property 7 on p. 359}}{=} a^{ur}.$$

Thus,

$$\log_a x^r = \log_a a^{ur} = ur = ru = r \log_a x \quad \blacksquare$$

EXAMPLE 7 *Illustrating That $\log_a xw = \log_a x + \log_a w$*

Since $2^5 = 32$, we have $\log_2 32 = 5$. Also, $32 = 8 \cdot 4$ so, from Property 5,

$$\log_2 32 = \log_2 8 \cdot 4 = \log_2 8 + \log_2 4 = 3 + 2 = 5 \quad \blacksquare$$

EXAMPLE 8 *Using the Fact That $\log_a x^r = r \log_a x$*

Compute $\log_5 \sqrt[3]{25}$.

SOLUTION $\sqrt[3]{25} = 25^{1/3}$ so, from Property 8,

$$\log_5 25^{1/3} = \frac{1}{3} \log_5 25 = \frac{1}{3} \log_5 5^2 = \frac{1}{3} \cdot 2 = \frac{2}{3} \quad \blacksquare$$

EXAMPLE 9 *Using Properties of Logarithms to Compute New Logarithmic Values from Given Ones*

$\log_{10} 2 \approx 0.3010$ and $\log_{10} 3 \approx 0.4771$. Using these values approximate (a) $\log_{10} 6$, (b) $\log_{10} \frac{2}{3}$, (c) $\log_{10} 5$, (d) $\log_{10} 8$, (e) $\log_{10} 108$.

SOLUTION

(a) $\log_{10} 6 = \log_{10} 3 \cdot 2 \overset{\text{Property 5}}{=} \log_{10} 3 + \log_{10} 2$
$\approx 0.4771 + 0.3010 = 0.7781$

(b) $\log_{10} \frac{2}{3} \overset{\text{Property 6}}{=} \log_{10} 2 - \log_{10} 3 \approx 0.3010 - 0.4771 = -0.1761$

(c) $\log_{10} 5 = \log_{10} \frac{10}{2} = \log_{10} 10 - \log_{10} 2 \overset{\text{Property 3}}{=} 1 - \log_{10} 2$
$\approx 1 - 0.3010 = 0.6990$

(d) $\log_{10} 8 = \log_{10} 2^3 \overset{\text{Property 8}}{=} 3 \log_{10} 2 = 3(0.3010) = 0.9030$

(e) $\log_{10} 108 = \log_{10} 4 \cdot 27 = \log_{10} 2^2 \cdot 3^3 = \log_{10} 2^2 + \log_{10} 3^3$
$= 2 \log_{10} 2 + 3 \log_{10} 3 \approx 2(0.3010) + 3(0.4771)$
$= 2.0333$

NOTE Approximate values for $\log_{10} 2$ and $\log_{10} 3$ can be obtained from a calculator or from Table 3 at the back of the book.

EXAMPLE 10 ***Solving an Equation Involving a Logarithm and an Exponent***

Find x if $x = 6^{-(1/2)\log_6 25}$.

SOLUTION

$$-\frac{1}{2}\log_6 25 \overset{\text{Property 8}}{=} \log_6 25^{-1/2} = \log_6 \frac{1}{25^{1/2}} = \log_6 \frac{1}{5}$$

Then

$$x = 6^{-(1/2)\log_6 25} = 6^{\log_6 1/5} \overset{\text{Property 1}}{=} \frac{1}{5} \quad ■$$

EXAMPLE 11 ***Simplifying a Logarithmic Expression***

Write $\log_a \dfrac{x^4y^6}{\sqrt{x^{2/3}y^8}}$ in terms of $\log_a x$ and $\log_a y$.

SOLUTION

$$\begin{aligned}
\log_a \frac{x^4y^6}{\sqrt{x^{2/3}y^8}} &= \log_a x^4y^6 - \log_a \sqrt{x^{2/3}y^8} && \text{Property 6}\\
&= \log_a x^4 + \log_a y^6 - \log_a (x^{2/3}y^8)^{1/2} && \text{Property 5}\\
&= 4\log_a x + 6\log_a y - \tfrac{1}{2}\log_a x^{2/3}y^8 && \text{Property 8}\\
&= 4\log_a x + 6\log_a y - \tfrac{1}{2}(\log_a x^{2/3} + \log_a y^8) && \text{Property 5}\\
&= 4\log_a x + 6\log_a y - \tfrac{1}{2}(\tfrac{2}{3}\log_a x + 8\log_a y) && \text{Property 8}\\
&= 4\log_a x + 6\log_a y - \tfrac{1}{3}\log_a x - 4\log_a y\\
&= \tfrac{11}{3}\log_a x + 2\log_a y \quad ■
\end{aligned}$$

EXAMPLE 12 ***Combining Logarithmic Terms***

Write the following expression as a single logarithm:

$$\tfrac{1}{2}\log_a x + 4\log_a y - 3\log_a z$$

SOLUTION

$$\begin{aligned}
\tfrac{1}{2}\log_a x + 4\log_a y - 3\log_a z &= \log_a x^{1/2} + \log_a y^4 - \log_a z^3 && \text{Property 8}\\
&= \log_a x^{1/2}y^4 - \log_a z^3 && \text{Property 5}\\
&= \log_a \frac{x^{1/2}y^4}{z^3} && \text{Property 6}
\end{aligned}$$

WARNING Two common errors are made by students who first study logarithms. Do not make them.

COMMON ERROR 1

$$\log_a (x + y) \neq \log_a x + \log_a y$$

Incorrect	Correct
$\log_{10} 20 = \log_{10} (10 + 10)$	$\log_{10} 20 = \log_{10} 2 \cdot 10$
	$= \log_{10} 2 + \log_{10} 10$
This is the incorrect step ↓	See Example 9 ↓
$= \log_{10} 10 + \log_{10} 10$	$\approx 0.3010 + 1 = 1.3010$
$= 1 + 1 = 2$	

In general, *the logarithm of the sum of two numbers cannot be simplified*. If you obtain the expression $\log_a (x + y)$, leave it alone. That's as far as you can go. No rule of logarithms will make the answer simpler.

COMMON ERROR 2

$$\frac{\log_a x}{\log_a y} \neq \log_a \frac{x}{y}$$

Incorrect	Correct
$\dfrac{\log_{10} 20}{\log_{10} 10} = \log_{10} \dfrac{20}{10}$	$\dfrac{\log_{10} 20}{\log_{10} 10} \approx \dfrac{1.3010}{1} = 1.3010$
$= \log_{10} 2 \approx 0.3010$	$\log_{10} 20 = \log_{10} 10 + \log_{10} 2 \approx 1.3010$
This is the incorrect step	

Problems 7.3

Readiness Check

I. $2^{\log_2 16} =$ _____.

a. 2 b. 4 c. 16 d. 8 e. 32

II. Which of the following is true about $f(x) = \log_b x$?

a. $f(x)$ is always positive.
b. It is the inverse function of $f(x) = x^b$.
c. If $b > 1$, its graph is decreasing.
d. Its graph has an x-intercept but no y-intercept.

III. $\log_5 5^{125} =$ _____.

a. 5 b. 125 c. $\dfrac{125}{5} = 25$ d. $125 \cdot 5 = 625$ e. 3

IV. Which of the following is true of the graph of $f(x) = \log_2 (x - 1)$?

a. It decreases from left to right.
b. It passes through (2, 0).
c. It is symmetric with respect to the x-axis.
d. It is asymptotic to the negative y-axis.

Answers to Readiness Check

I. c II. d III. b IV. b

In Problems 1–10, change each equation to an exponential form. For example, $\log_9 3 = \frac{1}{2}$ is equivalent to $9^{1/2} = 3$.

1. $\log_{16} 4 = \frac{1}{2}$
2. $\log_2 32 = 5$
3. $\log_{1/2} 16 = -4$
4. $\log_3 \frac{1}{3} = -1$
5. $\log_{12} 1 = 0$
6. $\log_{10} 100 = 2$
7. $\log_{10} 0.001 = -3$
8. $\log_{1/10} 100 = -2$
9. $\log_e e^2 = 2$
10. $\log_{e^2} e = \frac{1}{2}$

In Problems 11–16, change each equation to a logarithmic form.

11. $2^5 = 32$
12. $(\frac{1}{2})^3 = \frac{1}{8}$
13. $(\frac{1}{3})^{-2} = 9$
14. $10^3 = 1000$
15. $10^{0.301029995} = 2$
16. $e^{0.69314718} = 2$

In Problems 17–27, solve for the unknown variable. (Do not use tables or a calculator.)

17. $y = \log_3 27$
18. $x = \log_4 16$
19. $u = \log_{1/2} 4$
20. $x = \dfrac{1}{3} \log_7 \dfrac{1}{7}$
21. $s = \log_{1/4} 2$
22. $v = \log_{81} 3$
23. $y = \log_e e^5$
24. $u = \log_{10} 0.01$
25. $y = 3 \log_5 \dfrac{1}{125}$
26. $y = \pi \log_\pi \dfrac{1}{\pi^4}$
27. $u = \log_3 \dfrac{1}{3^{4.5}}$

In Problems 28–31 a function is given. Match each function with one of the graphs given below.

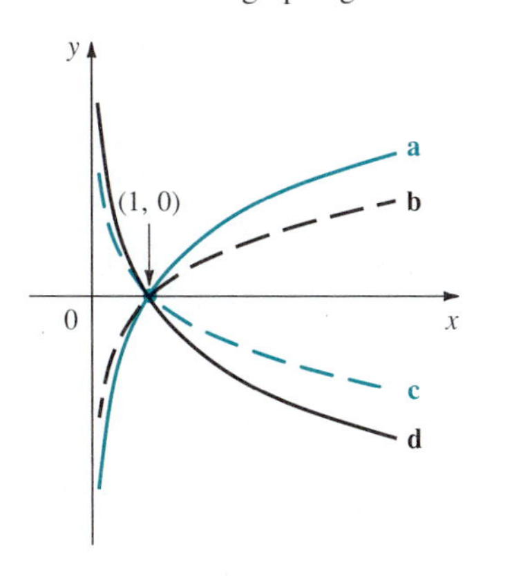

28. $\log_2 x$
29. $\log_3 x$
30. $\log_{1/2} x$
31. $\log_{1/3} x$

In Problems 32–36, use the fact that $\log_e 2 \approx 0.6931$ and $\log_e 6 \approx 1.7918$ to approximate each logarithm.

32. $\log_e 3$
33. $\log_e 8$
34. $\log_e \frac{1}{2}$
35. $\log_e 36$
36. $\log_e \frac{27}{16}$

In Problems 37–46, use the fact that $\log_{10} 3.45 \approx 0.5378$ to approximate each logarithm.

37. $\log_{10} 345$ [Hint: $345 = 3.45 \times 10^2$]
38. $\log_{10} 3{,}450{,}000$
39. $\log_{10} 0.345$ [Hint: $0.345 = 3.45 \times 10^{-1}$]
40. $\log_{10} 0.00000345$
41. $\log_{10} \sqrt{3.45}$
42. $\log_{10} \sqrt{34.5}$
43. $\log_{10} \dfrac{1}{\sqrt{3450}}$
44. $\log_{10} (345)^2$
45. $\log_{10} \left(\dfrac{1}{34.5}\right)^3$
46. $\log_{10} [(345)^{2/3}(3.45)^{3/4}]$

In Problems 47–54, write each expression in terms of $\log_a x$, $\log_a y$, and $\log_a z$.

47. $\log_a x^4y^3$
48. $\log_a \dfrac{x}{y^2}$
49. $\log_a \sqrt{x^4y^{3/2}}$
50. $\log_a \dfrac{xy^5}{z^2}$
51. $\log_a \dfrac{\sqrt{x}\sqrt[3]{y}}{\sqrt[5]{z}}$
52. $\log_a \dfrac{x^4}{y^7z^3}$
53. $\log_a \left(\dfrac{xy}{z}\right)^{4/5}$
54. $\log_a x^{20}y^{30}z^{-2}$

In Problems 55–68, write each expression as a single logarithm.

55. $\log_a x + \log_a \dfrac{1}{x}$
56. $2 \log_a u + 3 \log_a v^2 - 6 \log_a u^{1/2}$
57. $\log_a x + 2 \log_a y$
58. $3 \log_a x - \frac{1}{3} \log_a y$
59. $\log_a 4x + \log_a 5w$
60. $2 \log_a \dfrac{1}{x} - 3 \log_a \dfrac{1}{z}$
61. $\log_a x + 2 \log_a y + 3 \log_a z$
62. $3 \log_a u - 4 \log_a v + \dfrac{3}{5} \log_a w$
63. $\log_a (x^2 - 4) - \log_a (x + 2)$
64. $2 \log_a (w^2 - 1) - 3 \log_a (w + 1) - \log_a (w - 1)$
65. $\log_a (x^2 - 2y) + \log_a (y + 2)$
66. $\log_a (x - 1) + \dfrac{1}{2} \log_a (y - 2) + \dfrac{1}{3} \log_a (z - 3)$
67. $\log_a (x^2 - y^2) - 4 \log_a (x - y) + 3 \log_a (x + y)$
68. $2 \log_a (x^3 - y^3) - \log_a (x - y) - \log_a (x + y)$

7.4 Common and Natural Logarithms

Although any positive number ($\neq 1$) can be used as a base for a logarithm, two bases are used almost exclusively. The first of these is base 10. Logarithms to the base 10 are called **common logarithms.** The second, and more important, base for logarithms is the base e. This base is used in calculus. Logarithms to the base e are called **natural logarithms.**

Common Logarithms

Common logarithms are denoted by $\log x$. That is,

Definition of the Common Logarithm Function

$$\log x = \log_{10} x, \qquad x > 0 \tag{1}$$

Logarithms were discovered by the Scottish mathematician John Napier, Baron of Merchiston (1550–1617). In one of Napier's earliest notes, the following table appears:

	I	II	III	IIII	V	VI	VII	⋯
1	2	4	8	16	32	64	128	

You should note that the number represented by each Roman numeral above is the logarithm to the base 2 of the number beneath it. Alternatively, 2 to the power of the Roman numeral above is equal to the number below. In 1615, Henry Briggs (1561–1631) suggested to Napier that 10 be used as the base for the logarithm. Soon Napier discovered the importance of logarithms to the base 10 for computations since our numbering system is founded on the base 10. Because he foresaw the practical usefulness of logarithms in trigonometry and astronomy, he abandoned other mathematical pursuits and set himself the difficult task of producing a table of common logarithms—a task that took him 25 years to complete. When the tables were completed, they created great excitement on the European continent and were immediately used by two of the great astronomers of the day: the Dane Tycho Brahe and the German Johannes Kepler.

Until recently, common logarithms were frequently used for arithmetic computations. However, logarithms are now rarely used in this way because arithmetic computations can be carried out more accurately and much faster on a hand-held calculator. They are still used in certain scientific formulas. Most scientific calculators have a [log] or [log x] key for computing common logarithms.

FOCUS ON

Why Common Logarithms Were Once More Important Than They Are Today

Common logarithms were once very important in computations because base 10 is the basis for our number system. If one knows the common logarithms of all the numbers from 1 to 10, one can determine the common logarithm of every positive real number.

All algebra students 30 years ago had access to tables showing the common logarithms (usually to four decimal places) of all numbers from 1 to 10, in increments of 0.01. We give this in Table 3 at the back of the book. Thus one could look up, for example, log 2.73, log 1.02, and log 9.65. These values could then be used to perform some very difficult computations. One example is given in Example 2 below.

EXAMPLE 1 ***Obtaining Common Logarithms on a Calculator***

Use a calculator to compute (a) log 4.7 (b) log 0.02456 (c) log 10,584 and (d) $\log e$.

SOLUTION Using the [log] key, we obtain

(a) $\log 4.7 = 0.672097857$
(b) $\log 0.02456 = -1.609771638$
(c) $\log 10{,}584 = 4.024649831$
(d) Using $e = 2.718281828$,† we have $\log e = 0.434294481$

EXAMPLE 2 ***Calculation Using Common Logarithms (An Example from the Past)***

Approximate

$$x = \frac{(395)^{1/3}(1280)^{3/4}}{(89)^{1/2}}$$

SOLUTION Taking common logarithms and using the rules of the last section, we have

$$\log x = \frac{1}{3}\log 395 + \frac{3}{4}\log 1280 - \frac{1}{2}\log 89$$

Now we use a table:

From Table 3 ↓

$$\log 395 = \log (3.95 \times 10^2) = \log 3.95 + \log 10^2 = 0.5966 + 2 = 2.5966$$
$$\log 1280 = \log (1.28 \times 10^3) = \log 1.28 + \log 10^3 = 0.1072 + 3 = 3.1072$$
$$\log 89 = \log (8.9 \times 10^1) = \log 8.9 + \log 10 = 0.9494 + 1 = 1.9494$$

†To get e on a calculator, press [1] [INV] [ln x] or [1] [2nd F] [ln x] (or [2nd F] [ln x] [1] or something similar to one of these).

Then $$\log x = \frac{1}{3}(2.5966) + \frac{3}{4}(3.1072) - \frac{1}{2}(1.9494)$$

$$= 0.8655 + 2.3304 - 0.9747 = 2.2212$$

Thus

Property 1
↓
$$x = 10^{\log x} = 10^{2.2212} = 10^2 10^{0.2212} = (100)10^{0.2212}$$

What is $10^{0.2212}$? If $y = 10^{0.2212}$, then $\log y = 0.2212$. We use the table again. We would find

$$\log 1.66 = 0.2201 \qquad \text{and} \qquad \log 1.67 = 0.2227$$

Now, 0.2212 is closer to 0.2201 than to 0.2227. Thus, $10^{0.2212} \approx 1.66$ and

$$x \approx (100)(1.66) = 166$$

Our answer is accurate as far as it goes. Using a calculator, the answer, correct to 10 digits, is

$$x = 166.4350685$$

Natural Logarithms

The second, and more important, base for logarithms is the base e. This is the base used in calculus. The number e was discussed in Section 7.2. Logarithms to the base e are called **natural logarithms** and are denoted by $\ln x$. That is,

Definition of the Natural Logarithm Function

$$\ln x = \log_e x, \qquad x > 0 \qquad \textbf{(2)}$$

We stress that the function $y = \ln x$ is the inverse of the function $y = e^x$. It is the logarithmic function encountered in the overwhelming majority of applications. One reason for this is that $y = e^{kx}$ is the function that describes constant relative growth, as we saw in Section 7.2.

We can rewrite the four statements on p. 378 in terms of the natural logarithmic function:

Basic Properties of the Natural Logarithm Function

Property	*Illustration*
(a) $y = \ln x$ is equivalent to $x = e^y$	If $y = \ln 8$, then $8 = e^y$
(b) $y = e^x$ is equivalent to $x = \ln y$	If $y = e^{-3}$, then $-3 = \ln y$
(c) $e^{\ln x} = x$ for every positive real number x	$e^{\ln 17.4} = 17.4$
(d) $\ln e^x = x$ for every real number x	$\ln e^{-8.2} = -8.2$

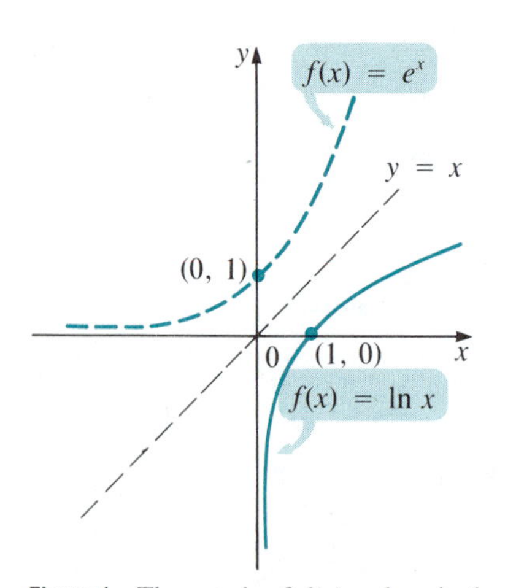

Figure 1 The graph of $f(x) = \ln x$ is the reflection about the line $y = x$ of the graph of e^x.

A graph of $\ln x$ is given in Figure 1. It is the reflection about the line $y = x$ of the graph of e^x.

EXAMPLE 3 *Finding the Domain of a Logarithmic Function*

Find the domain of each function.
(a) $f(x) = \ln(x - 2)$ (b) $f(x) = \ln(3 - x)$
(c) $f(x) = \ln(x^2 - 4)$

SOLUTION $\ln x$ is defined for $x > 0$.

(a) $\ln(x - 2)$ is defined if $x - 2 > 0$ or $x > 2$ so domain of $\ln(x - 2) = \{x: x > 2\}$
(b) We need $3 - x > 0$ or $3 > x$ so domain of $\ln(3 - x) = \{x: x < 3\}$
(c) $x^2 - 4 > 0$ if $x^2 > 4$ or $|x| > 2$. This is true for $x > 2$ or $x < -2$. Thus domain of $\ln(x^2 - 4) = \{x: x < -2 \text{ or } x > 2\}$ ■

EXAMPLE 4 *Sketching a Logarithmic Graph*

Sketch the graph of $f(x) = 2 - \ln(1 - x)$.

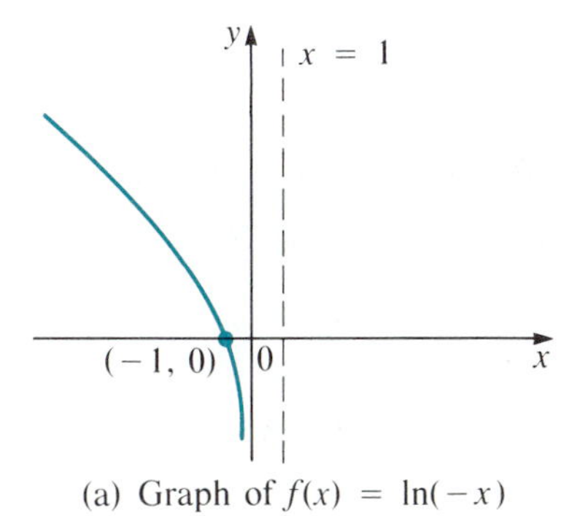

(a) Graph of $f(x) = \ln(-x)$

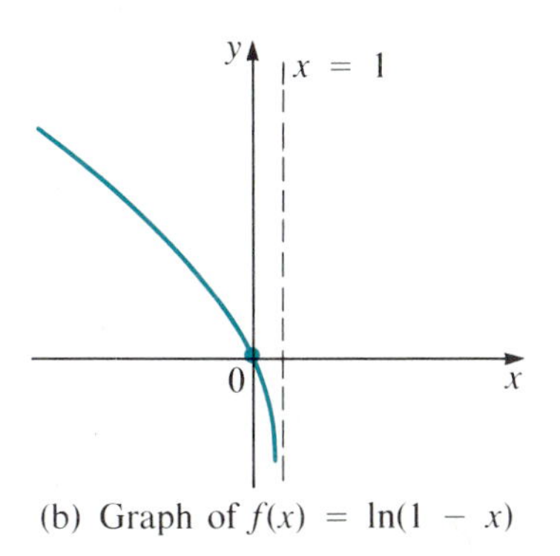

(b) Graph of $f(x) = \ln(1 - x)$

SOLUTION We do this in four steps.

Step 1 The graph of $y = \ln(-x)$ is given in Figure 2(a). It is the graph of $\ln(x)$ reflected about the y-axis (see p. 34).

Step 2 The graph of $y = \ln(1 - x) = \ln(-(x - 1))$ is the graph $y = \ln(-x)$ shifted 1 unit to the right (see p. 33). It is given in Figure 2(b).

Step 3 The graph of $y = -\ln(1 - x)$ is the graph of $y = \ln(1 - x)$ reflected about the x-axis (see p. 34) and is given in Figure 2(c).

Step 4 The graph of $y = 2 - \ln(1 - x)$ is the graph of $y = -\ln(1 - x)$ shifted up 2 units. It is given in Figure 2(d).

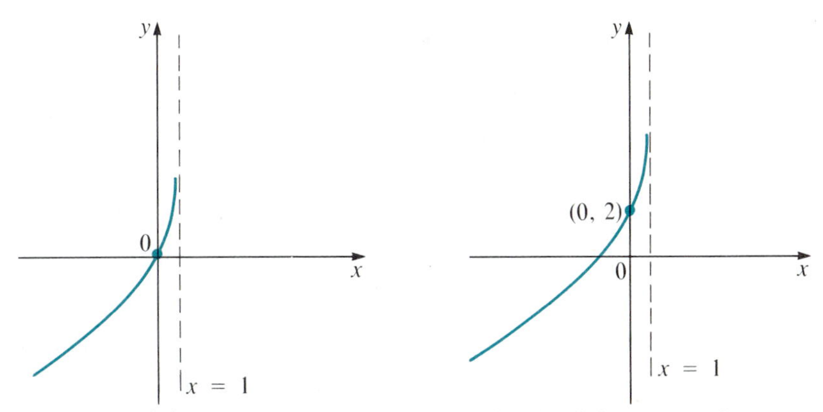

(c) Graph of $f(x) = -\ln(1 - x)$ (d) Graph of $f(x) = 2 - \ln(1 - x)$

Figure 2 The graph of $2 - \ln(1 - x)$ is obtained from the graph of $\ln x$ by reflecting about the y-axis, shifting 1 unit to the right, reflecting about the x-axis, and shifting up 2 units (in that precise order).

Because $\ln x$ is so important, we repeat the rules of logarithms written in terms of $\ln x$.

Properties of Logarithms for $\ln x = \log_e x$

Property	*Illustration*
1. $e^{\ln x} = x$	$e^{\ln 2} = 2$
2. $\ln e^x = x$	$\ln e^{1.7} = 1.7$
3. $\ln e = 1$	
4. $\ln 1 = 0$	
5. $\ln xw = \ln x + \ln w$	$\ln 6 = \ln 2 \cdot 3 = \ln 2 + \ln 3$
6. $\ln \dfrac{x}{w} = \ln x - \ln w$	$\ln 2 = \ln \dfrac{6}{3} = \ln 6 - \ln 3$
7. $\ln \dfrac{1}{x} = -\ln x$	$\ln \dfrac{1}{5} = -\ln 5$
8. $\ln x^r = r \ln x$	$\ln 81 = \ln 3^4 = 4 \ln 3$

EXAMPLE 5 *Using Properties of Logarithms to Compute New Logarithms from Given Ones*

Given that $\ln 2 \approx 0.6931$ and $\ln 5 \approx 1.6094$, compute (a) $\ln 10$, (b) $\ln \frac{2}{5}$, (c) $\ln 8$, and (d) $\ln \frac{16}{25}$.

SOLUTION

(a) $\ln 10 = \ln (2 \cdot 5) \overset{\text{Property 5}}{=} \ln 2 + \ln 5 \approx 0.6931 + 1.6094 = 2.3025$

(b) $\ln \frac{2}{5} \overset{\text{Property 6}}{=} \ln 2 - \ln 5 \approx 0.6931 - 1.6094 = -0.9163$

(c) $\ln 8 = \ln 2^3 \overset{\text{Property 8}}{=} 3 \ln 2 \approx 3(0.6931) = 2.0793$

(d) $\ln \frac{16}{25} = \ln 16 - \ln 25 = \ln 2^4 - \ln 5^2$
$= 4 \ln 2 - 2 \ln 5 \approx 4(0.6931) - 2(1.6094)$
$= -0.4464$ ■

EXAMPLE 6 *Combining Logarithmic Terms*

Write these as a single logarithm:

(a) $\ln (x - 1) - 2 \ln (x + 5)$
(b) $\ln x + 2 \ln (x + 1) + 3 \ln (x + 2)$

SOLUTION

(a) $\ln (x - 1) - 2 \ln (x + 5) \overset{\text{Property 8}}{=} \ln (x - 1) - \ln (x + 5)^2 \overset{\text{Property 6}}{=} \ln \dfrac{x - 1}{(x + 5)^2}$

$$\text{(b)}\ \ln x + 2 \ln (x+1) + 3 \ln (x+2) \overset{\text{Property 8}}{=} \ln x + \ln (x+1)^2 + \ln (x+2)^3 \overset{\text{Property 5}}{=} \ln [x(x+1)^2(x+2)^3]$$

All scientific calculators have [ln] or [ln x] buttons.

EXAMPLE 7 *Finding Values of ln x on a Calculator*

Use a calculator to compute (a) ln 4.7, (b) ln 0.02486, (c) ln 2486, (d) ln 10,584, and (e) ln 10.

SOLUTION

(a) $\ln 4.7 = 1.547562509$
(b) $\ln 0.02486 = -3.694495193$
(c) $\ln 2486 = 7.818430272$
(d) $\ln 10{,}584 = 9.267098706$
(e) $\ln 10 = 2.302585093$

In Table 1, we provide some values of $\ln x$. Of course, these and other values can be obtained with greater accuracy on a calculator. A more complete table appears in Table 2 at the back of the book.

Table 1
Sample Values Taken by the Natural Logarithm Function ln x

x	0.0	0.1	0.5	1	2	5	10	50	100	1000	100,000	10^{50}
$\ln x$	undefined	−2.3026	−0.6931	0	0.6931	1.6094	2.3026	3.9120	4.6052	6.9078	11.5129	115.1293

EXAMPLE 8 *The Growth Rate of ln x*

Suppose $y = 4 \ln x$.

(a) What happens to y if x doubles?
(b) What happens to y if x is halved?

SOLUTION

(a) If x doubles, then x becomes $2x$ and

$$y = 4 \ln 2x = 4(\ln 2 + \ln x) = 4 \ln 2 + 4 \ln x$$

Thus, y increases by $4 \ln 2 \approx 2.77$.

(b) Now x is replaced by $\frac{x}{2}$ and

$$y = 4 \ln \frac{x}{2} = 4(\ln x - \ln 2) = 4 \ln x - 4 \ln 2$$

Thus, y is reduced by $4 \ln 2 \approx 2.77$.

Computing Logarithms to Other Bases

Sometimes we need to compute something like $\log_2 35$. The following three rules are very useful.

Change of Base Properties

Suppose $a > 0$, $b > 0$, and $x > 0$.

9. $\log_a b = \dfrac{1}{\log_b a}$ 10. $\log_a x = \dfrac{\ln x}{\ln a}$ 11. $\log_a x = \dfrac{\log x}{\log a}$

Proof of Property 9

Let $y = \log_a b$. Then

$$\begin{aligned} a^y &= b && \text{Definition of the logarithm} \\ \log_b a^y &= \log_b b && \text{Take the logarithm to the base } b \text{ of each side} \\ y \log_b a &= 1 && \text{Properties 8 and 3 on pp. 379 and 378} \\ y &= \frac{1}{\log_b a} \\ \log_a b &= \frac{1}{\log_b a} && \text{Since } y = \log_a b \quad \blacksquare \end{aligned}$$

Proof of Property 10

Let $u = \log_a x$. Then

$$\begin{aligned} a^u &= x \\ \ln a^u &= \ln x && \text{Take the natural logarithm of each side} \\ u \ln a &= \ln x && \text{Property 8} \\ u &= \frac{\ln x}{\ln a} \\ \log_a x &= \frac{\ln x}{\ln a} && \text{Since } u = \log_a x \quad \blacksquare \end{aligned}$$

Proof of Property 11

Prove as for Property 10, but take common logarithms of each side instead. ■

EXAMPLE 9

Compute $\log_2 35$.

SOLUTION

Method 1 Using Property 10, we have

From a calculator

$$\log_2 35 = \frac{\ln 35}{\ln 2} = \frac{3.555348061}{0.69314718} = 5.129283017$$

Method 2 Using Property 11, we obtain

From a calculator

$$\log_2 35 = \frac{\log 35}{\log 2} = \frac{1.544068044}{0.301029995} = 5.129283017$$

Check If $\log_2 35 = 5.129283017$, then $2^{5.129283017} = 35$. We can verify on our calculator that this is true.

Problems 7.4

Readiness Check

I. Which of the following is a property of natural logarithms?
a. $\ln x + \ln w = \ln (x + w)$ b. $\ln 0 = 1$
c. $\ln (1/x) = 1/\ln x$ d. $\ln x^r = r \ln x$

II. Which of the following is equal to x for *every* real number x?
a. $e^{\ln x}$ b. $\ln e^x$ c. $10^{\log x}$ d. $\ln x^e$

III. If $x = e^{4t}$, then $t =$ _____.
a. $e^{(1/4)x}$ b. $\dfrac{\ln x}{\ln 4}$ c. $\dfrac{\ln x}{4}$ d. $\ln \dfrac{x}{4}$
e. None of these.

IV. If $x = 5^{4u}$, then $u =$ _____.
a. $\dfrac{\ln x}{4 \ln 5}$ b. $\dfrac{\ln x}{\ln 4 \ln 5}$
c. $\ln x - \ln 4 - \ln 5 = \ln x - \ln 20$ d. $\dfrac{\ln \frac{x}{4}}{\ln 5}$
e. $5 \ln \dfrac{x}{4}$

V. If $y = \ln (x + 5)$, then $x =$ _____.
a. $e^y + 5$ b. $5e^y$ c. $e^y - \ln 5$
d. $e^y - 5$ e. $\dfrac{e^y}{\ln 5}$

In Problems 1–12, solve for the unknown variable. Do not use a calculator.

1. $x = \log 1000$
2. $y = \log 0.001$
3. $z = \log 10^{20}$
4. $u = \log 10^{-8}$
5. $v = 10^{\log 3.4}$
6. $w = 10^{\log 10^{10}}$
7. $x = e^{\ln 0.235}$
8. $x = \ln e^{6.4}$
9. $x = \ln e^{\pi}$
10. $x = e^{2 \ln 3}$
11. $x = \ln e^{6.4}$
12. $x = \ln e^{-3.7}$

In Problems 13–22, write as a single logarithm.

13. $\ln (x + 3) + \ln x$
14. $\ln (x + 1) - \ln (x - 1) + \ln (x^2 - 1)$
15. $\ln x + \ln y + \ln z$
16. $3 \ln x - 4 \ln y$
17. $\dfrac{1}{2} \ln z + \dfrac{3}{4} \ln w - 2 \ln (x^2 + 1)$
18. $\ln 5 + \ln 2 + \ln 3$

Answers to Readiness Check

I. d II. b ($\ln x$ and $\log x$ are not defined for $x \le 0$) III. c IV. a V. d

19. $\ln 20 - \ln 2 - \ln 5$
20. $\ln \dfrac{1}{2} - 2 \ln \dfrac{1}{x}$
21. $3 \ln \dfrac{2}{w} - 4 \ln \dfrac{z}{3}$
22. $\dfrac{1}{2} \ln (x^2 - 1) - \dfrac{2}{3} \ln (x + 1) + \dfrac{3}{4} \ln (z^3 - 2)$

In Problems 23–35 find the domain of the given logarithmic function.

23. $f(x) = \ln (x + 1)$
24. $f(x) = \ln (x - 7)$
25. $f(x) = \ln (1 - x)$
26. $f(x) = \ln (x^2 + 1)$
27. $f(x) = \ln |x|$
28. $f(x) = \ln |x^2 - 1|$
29. $f(x) = \ln (x^2 - 1)$
30. $f(x) = \ln (x^2 - 2x + 1)$
31. $f(x) = \ln (x^2 - x - 12)$
32. $f(x) = \ln (x^2 + 4x - 21)$
33. $f(x) = \ln (x^2 + 4x + 10)$
34. $f(x) = \ln (-x^2 - 2x - 2)$
35. $f(x) = \ln \left(\dfrac{1}{x}\right)$

In Problems 36–44 nine functions are listed. Match each function with one of the graphs given below.

36. $\ln (-x)$
37. $\ln (x - 1)$
38. $\ln (x + 1)$
39. $\ln x + 1$
40. $\ln x - 1$
41. $\ln x$
42. $-\ln x$
43. $1 - \ln x$
44. $\ln x^2$

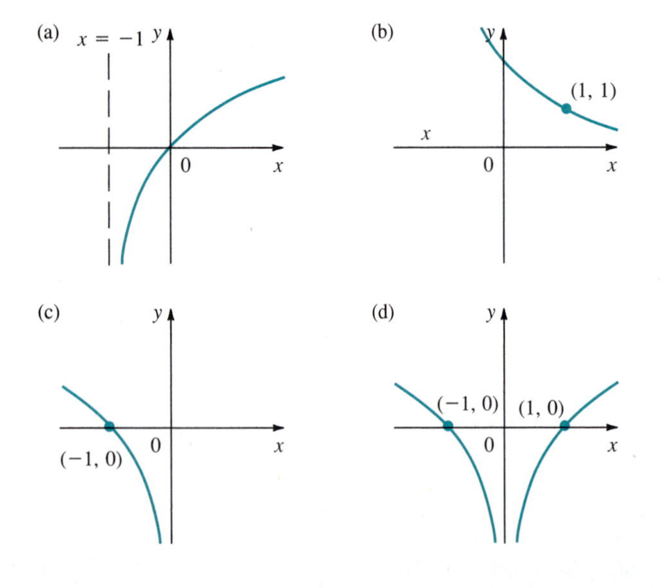

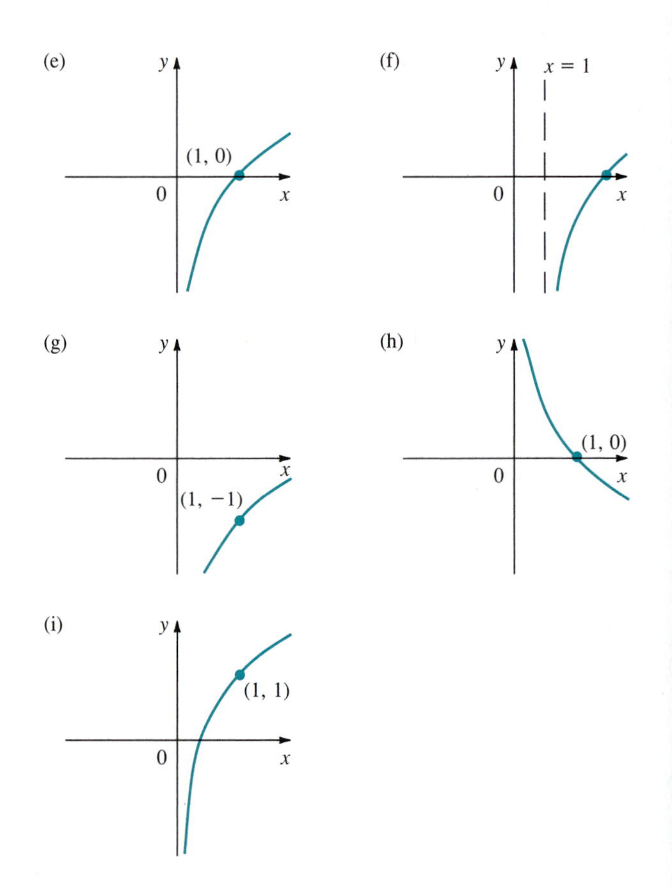

In Problems 45–54, solve for x. Do not use any table or calculator.

45. $\log x = 4 \log 2 + 2 \log 3$
46. $\log x = 4 \log \frac{1}{2} - 3 \log \frac{1}{3}$
47. $\log x = a \log b + b \log a$
48. $\log x = \log 1 + \log 2 + \log 3 + \log 4$
49. $\log x = \log 2 - \log 3 + \log 5 - \log 7$
50. $\ln x^3 = 2 \ln 5 - 3 \ln 2$
51. $\ln \sqrt[3]{x} = 2 \ln 3 + \ln 6$
52. $\ln x^2 - \ln x = \ln 18 - \ln 6$
53. $\ln x^{3/2} - \ln x = \ln 9 - \ln 3$
54. $\ln x^3 - \ln x - \ln 2x = \ln 32 - 2 \ln 4$

In Problems 55–57 do not use a calculator.

55. Given $\log 6.84 \approx 0.8351$. Approximate (a) $\log 684$ (b) $\log 6{,}840{,}000$ (c) $\log 0.0684$ (d) $\log 0.000000684$.
56. Given $\log 2.97 \approx 0.4728$. Approximate (a) $\log 2970$ (b) $\log (297 \times 10^3)$ (c) $\log 0.297$ (d) $\log (0.0297 \times 10^{-5})$.
57. Given $\log 9.52 \approx 0.9786$. Approximate (a) $\log 95.2$ (b) $\log 95{,}200$ (c) $\log 0.0952$ (d) $\log (95.2 \times 10^{-6})$.

In Problems 58–68, use the logarithms given in Problems 55–57 to approximate the common logarithm of the given expression.

58. $\frac{1}{684}$
59. $\frac{1}{0.00952}$
60. $\frac{100}{2970}$
61. $\sqrt{29,700}$
62. $\sqrt[3]{0.952}$
63. $\frac{1}{\sqrt{68.4}}$
64. $(952)^{3/4}$
65. $(68.4)(29.7)(95.2)$
66. $\frac{0.684}{0.0952}$
* 67. $\frac{\sqrt{6840}\sqrt[3]{95.2}}{(0.297)^5}$
* 68. $\frac{(29.7)^{3/8}(95,200)^{3/10}}{(68,400)^2}$

69. Let $y = 3 \ln x$. What happens to y if (a) x triples? (b) x is replaced by $\frac{x}{10}$?

70. Let $y = 6 \ln \frac{1}{\sqrt{x}}$. What happens to y if (a) x is multiplied by 4? (b) x is replaced by $0.2x$?

In Problems 71–86, find the given value on a calculator.

71. $\log 57.24$
72. $\log 0.02315$
73. $\log 50002$
74. $\log 0.00371496$
75. $\ln 1000$
76. $\ln 37.85$
77. $\ln 0.0325$
78. $\ln 18.18$
79. $\log \pi$
80. $\ln \pi$
81. $\ln e^{\pi}$
* 82. $\ln \pi^{e}$
* 83. $\log \pi^{\pi}$
* 84. $\log e^{\pi}$
* 85. $\log \pi^{e}$
86. $\ln \ln 10$

In Problems 87–94, use Property 10 or Property 11 to compute the given logarithm.

87. $\log_3 4$
88. $\log_4 3$
89. $\log_6 25$
90. $\log_{1/2} 5$
91. $\log_{1/4} \frac{1}{3}$
92. $\log_{0.3} 57.4$
93. $\log_{28.2} 39.5$
94. $\log_{6.4} 1753$

95. Given that $\log_2 5 \approx 2.3219$, approximate (a) $\log_5 2$ (b) $\log_5 4$ (c) $\log_5 \frac{1}{8}$.

96. Given that $\log_3 8 \approx 1.8928$, approximate (a) $\log_8 3$ (b) $\log_8 27$ (c) $\log_8 \frac{1}{9}$.

* 97. Natural logarithms can be approximated on a hand calculator even if the calculator does not have an ln key. If $\frac{1}{2} \le x \le \frac{3}{2}$ and if $A = \frac{x-1}{x+1}$, then a good approximation to $\ln x$ is

$$\ln x \approx \left[\left(\frac{3A^2}{5} + 1\right) \cdot \frac{A^2}{3} + 1\right]2A, \qquad \frac{1}{2} \le x \le \frac{3}{2}$$

(a) Use this formula to calculate $\ln 0.8$ and $\ln 1.2$.
(b) Using the facts about logarithms, calculate (approximately)

$$\ln 2 = \ln\left(\frac{3}{2} \cdot \frac{4}{3}\right)$$

(c) Using the result of (b), calculate $\ln 3$ and $\ln 8$.

* 98. Show, using logarithms, that $a^b = b^a$, where $a = [1 + 1/n]^n$ and $b = [1 + 1/n]^{n+1}$. What does this result prove when $n = 1$?

* 99. The quantity $n! = n(n-1)(n-2) \cdot \cdots \cdot 3 \cdot 2 \cdot 1$ grows very rapidly as n increases. According to **Stirling's formula,** when n is large,

$$n! \approx \sqrt{2\pi n}\left(\frac{n}{e}\right)^n$$

Use Stirling's formula to estimate 100! and 200!. [Hint: Use common logarithms.]

100. Watch carefully. Suppose that $0 < A < B$. Because the logarithm is an increasing function, we have
(a) $\log A < \log B$; then
(b) $10A \cdot \log A < 10B \cdot \log B$,
(c) $\log A^{10A} < \log B^{10B}$,
(d) $A^{10A} < B^{10B}$.
On the other hand, we run into trouble with particular choices of A and B. For instance, choose $A = \frac{1}{10}$ and $B = \frac{1}{2}$. Clearly, $0 < A < B$, but $A^{10A} = \left(\frac{1}{10}\right)^1 = \frac{1}{10}$ is greater than $B^{10B} = \left(\frac{1}{2}\right)^5 = \frac{1}{32}$. Where was the first false step made?

101. In Problem 46 of Section 7.2 (p. 375), we defined the function

$$\sinh x = \frac{e^x - e^{-x}}{2}$$

(a) If $\sinh x = 2$, show that $e^x - e^{-x} = 4$.
(b) Show that $e^{2x} - 4e^x - 1 = 0$.
(c) Let $w = e^x$ to obtain the quadratic equation

$$w^2 - 4w - 1 = 0$$

and find two values of w that satisfy this equation.
(d) Show that only one of the values obtained in (c) is useful.
(e) Take a natural logarithm to find the unique value of x for which $\sinh x = 2$.

In Problems 102–104, use the technique of Problem 101 to solve for x.

102. $\sinh x = \frac{1}{2}$
103. $\sinh x = 4$
104. $\sinh x = -10$

In Problem 48 of Section 7.2 (p. 376) we defined the function

$$\cosh x = \frac{e^x + e^{-x}}{2}$$

In Problems 105 and 106, modify the technique of Problem 101 to obtain two solutions for x.

105. $\cosh x = 2$ 106. $\cosh x = 5$

Graphing Calculator Problems

107. Sketch $\ln\left(\frac{x}{2}\right)$ and $-\ln\left(\frac{x}{2}\right)$ on the same screen.
108. Sketch $\ln 3x$ and $\ln(-3x)$ on the same screen.
109. Sketch $\ln x$, $\ln(x - 1.5)$, and $\ln(x + 1.5)$ on the same screen.
110. Sketch $\ln(1.7x)$, $\ln(1.7x) + 1$, and $\ln(1.7x) - 1$ on the same screen.

In Problems 111–125 obtain a sketch of the given function.

111. $\log 4x$
112. $4 \log x$
113. $\log\left(\frac{x-1}{2}\right)$
114. $\log 50(x - 1)$
115. $-3 \log(-x)$
116. $\ln(2 - x)$
117. $-2 + 3 \ln(x + 1)$
118. $2.3 + 12 \ln\left(\frac{x}{6} - 1\right)$
119. $\ln x^{10}$
120. $\ln(x^2 + 1)$
121. $\ln e^x$
122. $\ln \ln x$
123. $\log_4 x$ $\left[\text{Hint: } \log_4 x = \frac{\ln x}{\ln 4}.\right]$
124. $\log_{1/2} x$
125. $\log_2(x^2 + 1)$

7.5 Equations Involving Exponential and Logarithmic Functions

Exponential and logarithmic equations arise in a great variety of applications. We saw some of these already. In this section we see how to solve some types of these equations.

Exponential Equations

Definition

An **exponential equation** is an equation involving one or more exponential functions.

EXAMPLE 1 ***Solving an Exponential Equation***

Solve for x if

$$2^x = 8$$

SOLUTION Since $2^3 = 8$ (and 2^x is a one-to-one function), we conclude that $x = 3$.

It is rare that we can solve an exponential equation by inspection, as in Example 1. When we cannot, we can often solve it by taking logarithms, as the next three examples illustrate.

EXAMPLE 2 *Solving an Exponential Equation by Taking Logarithms*

Find x if

$$5^x = 30$$

SOLUTION We take natural logarithms of both sides:

$$\begin{aligned} 5^x &= 30 \\ \ln 5^x &= \ln 30 \\ x \ln 5 &= \ln 30 && \text{Since } \ln x^r = r \ln x \\ x &= \frac{\ln 30}{\ln 5} && \text{Divide by } \ln 5 \\ x &= \frac{3.401197382}{1.609437912} \approx 2.113 && \text{Values obtained on a calculator} \end{aligned}$$

We can also obtain this answer by taking common logarithms:

$$\begin{aligned} 5^x &= 30 \\ \log 5^x &= \log 30 \\ x \log 5 &= \log 30 \\ x &= \frac{\log 30}{\log 5} = \frac{1.447121255}{0.698970004} \approx 2.113 \end{aligned}$$

The reason we get the same answer in both cases is that by Properties 10 and 11 on p. 390,

$$\frac{\log 30}{\log 5} = \frac{\log_{10} 30}{\log_{10} 5} = \frac{\dfrac{\ln 30}{\ln 10}}{\dfrac{\ln 5}{\ln 10}} = \frac{\ln 30}{\ln 5}$$

In the following examples we will take natural logarithms when logarithms are called for because, as we saw in Example 2, either logarithm leads to the same answer. [Sometimes, however, it is more convenient to use one form rather than the other, as the next example illustrates.]

EXAMPLE 3 *Solving an Exponential Function*

Solve the following equation for r:

$$e^{2(r-4)} = 65$$

SOLUTION

$$e^{2(r-4)} = 65$$

$$\ln e^{2(r-4)} = \ln 65 \quad \text{Take natural logarithms of both sides}$$

$$2(r-4) = \ln 65 \quad \ln e^x = x \quad [\text{but } \log e^x \neq x\text{—that's why it is better to use } \ln x \text{ here}]$$

$$2r - 8 = \ln 65$$

$$2r = \ln 65 + 8$$

$$r = \frac{\ln 65 + 8}{2}$$

$$\approx \frac{4.17439 + 8}{2} \approx 6.087 \quad \ln 65 \text{ obtained on a calculator}$$

NOTE In solving exponential equations involving the function e^x, we will always take natural logarithms because $\ln e^u = u$ for any real number [but $\log e^u = u \log e$—which is a more complicated expression].

EXAMPLE 4 *Solving an Exponential Equation*

Solve the following equation for u:

$$3^u = 5^{u-4}$$

SOLUTION

$$3^u = 5^{u-4}$$

$$\ln 3^u = \ln 5^{u-4} \quad \text{Take natural logarithms on both sides}$$

$$u \ln 3 = (u-4)\ln 5 \quad \ln x^r = r \ln x$$

$$u \ln 3 = u \ln 5 - 4 \ln 5 \quad \text{Multiply through}$$

$$u \ln 5 - u \ln 3 = 4 \ln 5 \quad \text{Rearrange terms}$$

$$u(\ln 5 - \ln 3) = 4 \ln 5 \quad \text{Factor out the } u$$

$$u \ln \frac{5}{3} = 4 \ln 5 \quad \ln x - \ln y = \ln \frac{x}{y}$$

$$u = \frac{4 \ln 5}{\ln \frac{5}{3}} \approx 12.6026 \quad \text{Values obtained on a calculator}$$

Check You can check this answer on a calculator by using the $\boxed{y^x}$ key.

$$3^u \approx 3^{12.6026} = 1{,}030{,}310.972$$

$$5^{u-4} \approx 5^{8.6026} = 1{,}030{,}289.703$$

The slight difference in answers occurs because we rounded. If you store the value $u = \dfrac{4 \ln 5}{\ln \frac{5}{3}} \approx 12.60264041$, then on most calculators you will obtain the same value for 3^u and 5^{u-4}. ■

EXAMPLE 5 ***An Exponential Equation That Leads to a Quadratic Equation***

Solve the following equation for x:

$$e^x - 10e^{-x} = 3$$

SOLUTION

$$\begin{aligned} e^x - 10e^{-x} &= 3 \\ e^x e^x - 10e^{-x}e^x &= 3e^x && \text{Multiply both sides by } e^x \\ e^{2x} - 10e^{-x+x} &= 3e^x && e^u e^v = e^{u+v} \\ e^{2x} - 10e^0 &= 3e^x \\ e^{2x} - 10 &= 3e^x && e^0 = 1 \\ e^{2x} - 3e^x - 10 &= 0 \end{aligned}$$

If we set $u = e^x$, then $u^2 = e^{2x} = (e^x)^2$, and we see that this is an equation of quadratic type:

$$\begin{aligned} u^2 - 3u - 10 &= 0 && \text{Set } u = e^x \\ (u - 5)(u + 2) &= 0 && \text{Factor} \\ u = 5 \quad &\text{or} \quad u = -2 \\ e^x = 5 \quad &\text{or} \quad e^x = -2 && u = e^x \end{aligned}$$

But $e^x > 0$, so the equation $e^x = -2$ is impossible, and we are left with

$$\begin{aligned} e^x &= 5 \\ \ln e^x &= \ln 5 && \text{Take logarithms of both sides} \\ x &= \ln 5 \end{aligned}$$

Check

$$\begin{aligned} e^x - 10e^{-x} &= e^{\ln 5} - 10e^{-\ln 5} \\ &= 5 - 10e^{\ln 1/5} && e^{\ln x} = x \text{ and } -\ln x = \ln \frac{1}{x} \\ &= 5 - 10 \cdot \frac{1}{5} && e^{\ln x} = x \\ &= 5 - 2 = 3 \end{aligned}$$

NOTE There is no way to solve the equation by first taking logarithms. We stress that $\ln(e^x - 10e^{-x}) \neq \ln e^x - \ln 10e^{-x}$.

EXAMPLE 6 *An Exponential Equation with Two Solutions*

Solve the following equation for x: $\quad e^{2x} + 10e^{-2x} = 7$

SOLUTION We proceed as in Example 5:

$$\begin{aligned} e^{2x} + 10e^{-2x} &= 7 \\ e^{4x} + 10 &= 7e^{2x} && \text{Multiply both sides by } e^{2x} \\ e^{4x} - 7e^{2x} + 10 &= 0 \\ u^2 - 7u + 10 &= 0 && \text{Set } u = e^{2x} \\ (u-2)(u-5) &= 0 && \text{Factor} \\ (e^{2x}-2)(e^{2x}-5) &= 0 && u = e^{2x} \\ e^{2x} &= 2 \\ 2x &= \ln 2 \qquad 2x = \ln 5 \\ x &= \frac{1}{2}\ln 2 \qquad x = \frac{1}{2}\ln 5 \end{aligned}$$

The two solutions are $x = \frac{1}{2}\ln 2$ and $x = \frac{1}{2}\ln 5$. ■

EXAMPLE 7 *An Exponential Equation with No Solution*

Solve the following equation for x: $\quad e^x + e^{-x} + 2 = 0$

SOLUTION Since $e^x > 0$ and $e^{-x} > 0$ for every real number, there is no real x for which $e^x + e^{-x} = -2$, and the equation has no solution. We can see this in another way by proceeding as in Examples 5 and 6:

$$\begin{aligned} e^x + e^{-x} + 2 &= 0 \\ e^{2x} + 1 + 2e^x &= 0 && \text{Multiply by } e^x \\ u^2 + 2u + 1 &= 0 && \text{Set } u = e^x \\ (u+1)^2 &= 0 && \text{Factor} \\ u + 1 &= 0 \\ u &= -1 \\ e^x &= -1 && u = e^x \end{aligned}$$

which is impossible, since $e^x > 0$ for every real number x.

Logarithmic Equations

Definition

A **logarithmic equation** is an equation involving one or more logarithmic functions.

EXAMPLE 8 ***Solving a Logarithmic Equation***

Find x if

$$\ln x = 2$$

SOLUTION The expression $\ln x = \log_e x = 2$ means that

$$e^2 = x$$

That is, $x = e^2 \approx 7.389$.

In solving logarithmic equations, the following facts are useful. They are, essentially, the definition of $\log_a x$. The first property is equation (6) on p. 379.

$$\text{If } \log_a u = \log_a v \quad \text{then} \quad u = v \tag{1}$$

$$\text{If } u = \log_a v \quad \text{then} \quad v = a^u \tag{2}$$

In particular,

$$\text{if } u = \log v \quad \text{then} \quad v = 10^u \tag{3}$$

and

$$\text{if } u = \ln v \quad \text{then} \quad v = e^u \tag{4}$$

EXAMPLE 9 ***Solving a Logarithmic Equation***

Find x if

SOLUTION

$$3 \ln (x - 2) = 1.5$$

$$\ln (x - 2) = \frac{1.5}{3} = 0.5$$

$$x - 2 = e^{0.5} \qquad \text{From (4)}$$

$$x = 2 + e^{0.5} \approx 3.6487 \quad ■$$

EXAMPLE 10 ***Solving a Logarithmic Equation***

Solve for y:

$$2 = \log \frac{1}{y}$$

SOLUTION From Property 7 on p. 379, we have

$$2 = \log \frac{1}{y} = -\log y$$

$$\log y = -2$$

$$y = 10^{-2} = \frac{1}{100} \qquad \text{From (3)}$$

EXAMPLE 11 *Solving a Logarithmic Equation*

Solve the following equation for x:

$$\log(4x + 10) - \log(x - 2) = 1$$

SOLUTION

$$\log(4x + 10) - \log(x - 2) = 1$$

$$\log \frac{4x + 10}{x - 2} = 1 \qquad \text{Property 6 on p. 379}$$

$$\frac{4x + 10}{x - 2} = 10^1 = 10 \qquad \text{From 3}$$

$$4x + 10 = 10(x - 2)$$

$$4x + 10 = 10x - 20$$

$$6x = 30$$

$$x = 5$$

Check $4x + 10 = 30 > 0$ and $x - 2 = 3 > 0$ so that both $\log(4x + 10)$ and $\log(x - 2)$ are defined. This check must always be carried out. Otherwise, the answer might not make sense. Then, for $x = 5$,

$$\log(4x + 10) - \log(x - 2) = \log 30 - \log 3 = \log \frac{30}{3} = \log 10 = 1$$ ■

EXAMPLE 12 *A Logarithmic Equation with No Solution*

Solve the following equation for x:

$$\ln(2x + 3) - \ln(x + 7) = 3$$

SOLUTION

$$\ln(2x + 3) - \ln(x + 7) = 3$$

$$\ln \frac{2x + 3}{x + 7} = 3 \qquad \text{Property 6 on p. 388}$$

$$\frac{2x + 3}{x + 7} = e^3 \qquad \text{From 4}$$

$$2x + 3 = e^3(x + 7) \qquad \text{Remember that } e^3 \text{ is a real number}$$

$$2x + 3 = e^3x + 7e^3$$

$$2x - e^3x = 7e^3 - 3 \qquad \text{Rearrange terms}$$

$$x(2 - e^3) = 7e^3 - 3$$

$$x = \frac{7e^3 - 3}{2 - e^3} \approx -7.608$$

Check If $x = -7.608$, then $2x + 3 \approx -12.2 < 0$ and $x + 7 = -0.6 < 0$ so neither $\ln(2x + 3)$ nor $\ln(x + 7)$ is defined, and $x \approx -7.6$ cannot be a solution to the original equation. We conclude that the equation has no solution.

One final note: there are many exponential and logarithmic equations that cannot be solved by the methods of this section. For example, consider the equation

$$e^x + 10^x = 8 \tag{5}$$

We see that $e^0 + 10^0 = 1 + 1 = 2 < 8$ and $e^1 + 10^1 = e + 10 \approx 12.718 > 8$. Thus there is one solution to (5) in the interval (0, 1). But how do we find it? None of the techniques discussed in this section will work. For example, we can try the following:

$$e^x = 8 - 10^x$$
$$\ln e^x = \ln (8 - 10^x)$$
$$x = \ln (8 - 10^x) \tag{6}$$

But equation (6) is no easier to solve than equation (5). In fact, it is worse because $\ln (8 - 10^x)$ is not defined if $8 - 10^x < 0$ or $10^x > 8$ or $x > \log 8 \approx 0.903$.

The best we can do is to approximate the solution by trial and error or by an appropriate numerical technique—unless we have access to a graphing calculator. To four decimal places, the solution is $x \approx 0.7669$.

Problems 7.5

Readiness Check

I. What can be concluded from the fact that $\log x = 2.5$?
a. $x = e^{2.5}$ b. $x = 2.5^{10}$
c. $x = 10^{2.5}$ d. $x = \log 2.5$

II. What can be concluded from the fact that $\ln x = -0.27$?
a. $x = e^{-0.27}$ b. $x = -0.27^{10}$
c. $x = 10^{-0.27}$ d. $x = -\ln 0.27$

III. Which are the solutions to the equation $\ln x^2 = \ln x$?
a. $x = -1, 1$ b. $x = 0, 1$ c. $x = 1$
d. $x = 0$ e. There are no real solutions.

IV. Which are the solutions to $\log x^2 - \log (x^2 + 1) = 1$?
a. $x = 0$ b. $x = 1$ c. $x = 0, 1$
d. $x = \pm 1$ e. There are no real solutions.

In Problems 1–73 solve for the unknown variable. Use a calculator where necessary. Give each answer to 3-decimal-place accuracy.

1. $10^x = 100$
2. $e^x = e^3$
3. $4^{-x} = 4$
4. $e^x = \dfrac{1}{e}$
5. $\log_2 x = 8$
6. $\ln x = 2$
7. $\ln x = -1$
8. $\ln x = e$
9. $\log x = 1$
10. $\log x = -2$
11. $\log x = 0$
12. $\log x = 3$
13. $e^x = 10$
14. $e^x = 0.2$
15. $e^x = -1$
16. $4^x = 12$
17. $7^x = 14$
18. $2.7^x = 8.92$
19. $231^x = 8$
20. $0.019^x = 6$
21. $\ln x = 1.6$
22. $\log x = 0.23$
23. $\log x = -1.57$
24. $\ln x = -2.95$
25. $e^{x-1} = 2$
26. $e^{2x} = 5$
27. $e^{t/2} = 4$
28. $e^{t^2} = 16$
29. $10e^{-h/5} = 50$
30. $50e^{0.21t} = 100$
31. $\ln x + \ln 3 = 1$
32. $\ln x + \ln (x - 1) = \ln 6$
33. $\ln v + \ln \dfrac{1}{v} = \ln (v - 3)$
34. $\ln (2x - 3) = 4$
35. $2 \ln (0.172w) = 0.856$
36. $\dfrac{\ln v}{\ln 2} = \dfrac{\ln 3}{\ln 5}$

Answers to Readiness Check

I. c II. a III. c IV. e

37. $\ln(z-3) - \ln(z+4) = \ln 2$
38. $\ln(z-3) - \ln(z+4) = 3$
39. $e^x + 8e^{-x} = 6$
40. $e^{2x} + 3e^{-2x} = 4$
41. $e^{(1/2)x} - 6e^{-(1/2)x} + 1 = 0$
42. $e^{3x} - 14e^{-3x} + 5 = 0$
43. $e^x + 2e^{-x} + 3 = 0$
44. $e^{x/5} + 12e^{-x/5} + 7 = 0$
45. $z = 1.3^{\log_{1.3} 48}$
46. $p = e^{\log_e 37.4}$
47. $q = 10^{\log_{10} 0.0023}$
48. $r = 6^{\log_6 \sqrt{2}}$
49. $s = 3^{(1/2)\log_3 16}$
50. $t = 10^{-\log_{10} 4}$
51. $u = e^{-(1/2)\ln 100}$
52. $2 \log_3 v = 4$
53. $3 \log_5 2x = 1$
54. $y = e^{\log_e e}$
55. $\log_2 u^4 = 4$
56. $\log_y 64 = 3$
57. $\log_w 125 = -3$
58. $\log_q 32 = -5$
59. $5 \cdot 2^{x-1} = 20$
60. $e^x e^{x+1} = e^2$
61. $2 \log_4 x + 3 = 0$
62. $\log_3 (x-2) = 2$
63. $\log_{10}(x+5) = -3$
64. $3^{2y} 3^{\log_3 1/3} = 9$
65. $2 + \log_5 5^{1/x} = 5$
66. $10^{t^2+2t-8} = 1,\ t > 0$
67. $10^{t^2+2t-8} = 1,\ t < 0$
68. $\log_2 x + \log_2 (x-1) = 3$
69. $\log_4 x - \log_4 (x+1) = 1$
70. $\log 5x = \log (x^2+6)$
71. $\ln(x+3) = \ln(2x-5)$
72. $\log(\log(v+1)) = 0$
73. $10^{10^x} = 10$

Graphing Calculator Problems

In Problems 74–84 find all solutions (if any) to each equation by sketching the appropriate functions on your calculator. Give each answer to 2-decimal-place accuracy. Before beginning, read (or reread) Example 7 in Appendix A.

74. $e^x = x + 2$
75. $\ln x = \frac{1}{2}x - 1$
76. $x = e^{1/x}$
77. $e^x + \ln x = 5$
78. $e^x - \ln x = 2$
79. $e^{-x} + x = 1.5$
80. $xe^{-x} = 0.1$
81. $e^x = 2x^2 + 3x + 5$
82. $e^{x/5} = 10x^4 + 8x^3 + 20$
83. $e^{0.1x} = x^5 + 1000$
84. $5^x + 3^x = 100$

7.6 Applications of Exponential and Logarithmic Functions

In this section, we illustrate why logarithmic functions, especially the function $\ln x$, are so important by showing some applications. We begin by modifying problems we have seen before.

Interest Rate Problems

Recall the compound interest formulas discussed in Sections 7.1 and 7.2.

Compounding m times a year $$A(t) = P\left(1 + \frac{r}{m}\right)^{mt} \quad (1)$$

Compounding continuously $$A(t) = Pe^{rt} \quad (2)$$

where P is the initial amount invested, r is the rate of interest, t is the amount of time (in years) the investment is held, m is the number of times that interest is compounded each year, and $A(t)$ is the value of the investment after t years.

EXAMPLE 1 *Computing Doubling Time*

Money is invested at 8% compounded quarterly. How long does it take to double?

SOLUTION If P dollars is invested, then the investment has doubled when $A(t) = 2P$. From (1), with $r = 0.08$ and $m = 4$,

$$A(t) = P\left(1 + \frac{0.08}{4}\right)^{4t}$$

$$2P = P(1.02)^{4t}$$

$$2 = (1.02)^{4t} \qquad \text{Divide by } P$$

We can now continue by taking logarithms. Logarithms to any base will work. We solve the problems using natural logarithms.

$$2 = (1.02)^{4t} \tag{3}$$

$$\ln 2 = \ln (1.02)^{4t} \qquad \text{Take natural logarithms of both sides}$$

$$\ln 2 = 4t \ln (1.02) \qquad \text{Use Property 8 on p. 388}$$

$$t = \frac{\ln 2}{4 \ln (1.02)}$$

$$t \approx \frac{0.69314718}{4(0.019802627)}$$

$$\approx 8.750697195 \qquad \text{From a calculator}$$

That is, money doubles in about 8.75 years if it is invested at 8% interest compounded quarterly. ■

EXAMPLE 2 ***Computing the Time It Takes for an Investment to Triple***

A bond pays 12% interest compounded continuously. If \$10,000 is initially invested, when will the bond be worth \$30,000?

SOLUTION From (2), we have

$$A(t) = Pe^{rt} = 10{,}000e^{0.12t} \qquad (0.12 = 12\%) \tag{4}$$

We seek a number t such that $A(t) = 30{,}000$. That is,

$$30{,}000 = 10{,}000e^{0.12t}$$

$$e^{0.12t} = 3 \qquad \text{Divide both sides by 10,000}$$

$$\ln e^{0.12t} = \ln 3 \qquad \text{Take natural logarithms of both sides}$$

$$0.12t = \ln 3 \qquad \text{Property 2 on p. 388}$$

$$t = \frac{\ln 3}{0.12} = \frac{1.098612289}{0.12} \approx 9.155 \text{ years}$$

We have shown that an investment triples in approximately 9.155 years if it is invested at 12% interest compounded continuously.

In general, we can easily compute how long it takes for a sum to increase by a factor of k if it is invested at $r\%$ compounded continuously. Written as a decimal, $r\% = r/100$. Then, for P dollars to increase to kP dollars, we must find a t^* such that

$$kP = Pe^{(r/100)t^*}$$
$$k = e^{(r/100)t^*}$$
$$\ln k = rt^*/100$$

Multiplying an Investment by a Factor of *k*

$$t^* = \frac{\ln k}{r/100} = \frac{100 \ln k}{r} \text{ years} \tag{5}$$

is the amount of time over which money invested at $r\%$ compounded continuously will increase by a factor of k.

EXAMPLE 3 ***The Time for an Investment to Increase by a Factor of 4***

How long does it take for a sum of money to increase by a factor of 4 if it is invested at 13% compounded continuously?

SOLUTION Here $k = 4$ and $r = 13$, so from (5),

$$t^* = \frac{100 \ln 4}{13} \approx \frac{138.63}{13} \approx 10.66 \text{ years}$$

In particular,

Doubling Time

If money is invested at $r\%$ interest compounded continuously, it will double in $\dfrac{100 \ln 2}{r}$ years.

For example, money invested at 10% will double in $100(\ln 2)/10 = 10 \ln 2$ years. In Table 1, see the doubling time at various rates of interest.

Table 1

Interest Rate (%)	Doubling Time (years)	Interest Rate (%)	Doubling Time (years)
1	69.3	8	8.7
2	34.7	10	6.9
3	23.1	12	5.8
4	17.3	15	4.6
5	13.9	18	3.9
6	11.6	25	2.8
7	9.9	50	1.4

Exponential Growth and Decay

In Section 7.2, we used the formula of exponential or constant relative growth:

$$y(t) = y(0)e^{kt} \tag{6}$$

We use this formula in the examples below.

Population Growth

EXAMPLE 4 ***Population Growth***

A bacteria population grows exponentially. A population that is initially 10,000 grows to 25,000 after 2 hours. What is the population (a) after 5 hours? (b) after 24 hours?

SOLUTION From (6), we have

$$P(t) = P(0)e^{kt} \tag{7}$$

where $P(t)$ denotes the population after t hours. We are told that $P(0) = 10{,}000$, so (7) becomes

$$P(t) = 10{,}000e^{kt} \tag{8}$$

This problem is different from the problems in Section 7.2 (and more realistic) because now the constant k is not given to us. However, we do have one more piece of information. We are told that $P(2) = 25{,}000$. Inserting $t = 2$ into (8), we obtain

$$\begin{aligned} P(2) = 25{,}000 &= 10{,}000e^{2k} \\ e^{2k} &= \frac{25{,}000}{10{,}000} = 2.5 \\ \ln e^{2k} &= \ln 2.5 \\ 2k &= \ln 2.5 \\ k &= \frac{\ln 2.5}{2} \end{aligned}$$

Then

$$P(t) = 10{,}000e^{kt} = 10{,}000e^{(\ln 2.5/2)t} = 10{,}000e^{(t/2)\ln 2.5}$$

Property 8 on p. 388 ↓ Property 1 ↓

$$= 10{,}000e^{\ln (2.5)^{t/2}} = 10{,}000(2.5)^{t/2}$$

$\boxed{y^x}$ key on a calculator ↓

(a) $P(5) = 10{,}000(2.5)^{5/2} \approx 98{,}821$

(b) $P(24) = 10{,}000(2.5)^{24/2} = 10{,}000(2.5)^{12} \approx 596{,}046{,}448$

NOTE We can compute $k = \dfrac{\ln 2.5}{2} = 0.458145365$, but it is not necessary in this problem. The value of k is interesting because it tells us that the population is growing at the very high rate of approximately 45.8% an hour. ■

EXAMPLE 5 *Estimating the Future Population of India*

The population of India was estimated to be 574,220,000 in 1974 and 746,388,000 in 1984. Assume that the relative growth rate is constant.

(a) Estimate the population in 1994.
(b) When will the population reach 1.5 billion?

SOLUTION

(a) From (7), we have

$$P(t) = P(0)e^{kt}$$

Treat the year 1974 as year zero. Then 1984 = year 10. We are told that

$$P(0) = 574{,}220{,}000 \qquad \text{and} \qquad P(10) = 746{,}388{,}000$$

Thus

$$P(t) = 574{,}220{,}000e^{kt}$$

$$P(10) = 746{,}388{,}000 = 574{,}220{,}000e^{10k}$$

$$e^{10k} = \frac{746{,}388{,}000}{574{,}220{,}000} = 1.299829334$$

$$\ln e^{10k} = \ln 1.299829334$$

$$10k = \ln 1.299829334$$

$$k = \frac{\ln 1.299829334}{10} \approx 0.026$$ so the population grew by about 2.6% a year between 1974 and 1984

The year 1994 is year 20. Thus

$$P(20) = 574{,}220{,}000e^{20k} = 574{,}220{,}000e^{(20 \ln 1.299829334)/10}$$

$$= 574{,}220{,}000e^{2 \ln 1.299829334} = 574{,}220{,}000e^{\ln (1.299829334)^2}$$

Property 1 → $= (574{,}220{,}000)(1.299829334)^2$

$$= (574{,}220{,}000)(1.689556297) = 970{,}177{,}017$$

(b) We seek a number t such that $P(t) = 1{,}500{,}000{,}000$. That is,

$$P(t) = 1{,}500{,}000{,}000 = 574{,}220{,}000e^{kt}$$

$$e^{kt} = \frac{1{,}500{,}000{,}000}{574{,}220{,}000} = 2.612239211$$

$$\ln e^{kt} = \ln 2.612239211$$

$$kt = \ln 2.612239211$$

$$t = \frac{\ln 2.612239211}{k}$$

$$= \frac{\ln 2.612239211}{\ln 1.299829334/10} = \frac{10 \ln 2.612239211}{\ln 1.299829334}$$

$$= 36.6 \text{ years}$$

Then $\qquad$ Year $36.6 = 1974 + 36.6 = 2010.6$

We conclude that the population of India would reach 1.5 billion sometime in the year 2010 if its population growth rate continued at the same rate as it was between 1974 and 1984.

NOTE It is tempting to round 1.299829334 to 1.3. If we do this, we obtain

$$P(20) \approx (574{,}220{,}000)(1.3)^2 = (574{,}220{,}000)(1.69) = 970{,}431{,}800.$$

In doing this, we gained 314,783 people (970,431,800 − 970,177,017). This isn't a very large percentage of the total

$$\left(\frac{314{,}783}{970{,}177{,}017} \approx 0.0003 = 0.03\%\right),$$

but it's still a lot of people. This illustrates that you have to be careful when you round. In this example, since you are using a calculator anyway, there is no good reason to round until the end.

Newton's Law of Cooling

Newton's law of cooling states that the rate of change of the temperature of an object is proportional to the temperature difference between the body and its surrounding medium. That is, the temperature difference changes at a constant relative rate. An object that is hotter than its surroundings will cool off; one that is cooler will warm up. Thus the temperature difference will decrease exponentially as a function of time. Let $T(t)$ denote the temperature of the object at time t, and let T_S denote the temperature of the surroundings (T_S is assumed to be constant throughout). If the initial temperature is $T(0)$, then the initial temperature difference is $T(0) - T_S$. Thus, in equation (6) we may set $y(t) = T(t) - T_S$ to obtain

$$T(t) - T_S = (T(0) - T_S)e^{-kt}$$

or

$$T(t) = T_S + (T(0) - T_S)e^{-kt} \qquad \textbf{(9)}$$

The minus sign in the exponent indicates that the temperature difference decreases as t increases.

EXAMPLE 6 *Using Newton's Law of Cooling to Calculate the Temperature of Milk*

When a bottle of milk was taken out of the refrigerator, its temperature was 40°F. An hour later, its temperature was 50°F. The temperature of the air is 70°F, and this temperature does not change.

(a) What was the temperature of the milk after 2 hours?
(b) After how many hours was its temperature 65°F?

SOLUTION Here, $T(0) = 40$, $T_S = 70$, and $T(0) - T_S = -30$. Thus equation (9) becomes

$$T(t) = 70 - 30e^{-kt} \tag{10}$$

But $T(1) = 50$, so

$$50 = 70 - 30e^{-k}$$

$$30e^{-k} = 20$$

$$e^{-k} = \frac{20}{30} = \frac{2}{3}$$

$$\ln e^{-k} = \ln \frac{2}{3}$$

$$-k = \ln \frac{2}{3}$$

Then (10) becomes

$$T(t) = 70 - 30e^{(-k)t}$$

Property 8

$$= 70 - 30e^{t(\ln (2/3))} = 70 - 30e^{\ln (2/3)^t}$$

Property 1

$$= 70 - 30\left(\tfrac{2}{3}\right)^t$$

(a) $T(2) = 70 - 30(\frac{2}{3})^2 = 70 - 30(\frac{4}{9}) = 70 - \frac{40}{3} \approx 56.7$°F
(b) We seek t such that $T(t) = 65$. That is,

$$T(t) = 65 = 70 - 30\left(\frac{2}{3}\right)^t$$

$$30\left(\frac{2}{3}\right)^t = 5$$

$$\left(\frac{2}{3}\right)^t = \frac{5}{30} = \frac{1}{6}$$

$$t \ln \frac{2}{3} = \ln \frac{1}{6} \qquad \text{Take natural logarithms of both sides}$$

$$t = \frac{\ln \frac{1}{6}}{\ln \frac{2}{3}} = \frac{-1.791759469}{-0.405465108} \approx 4.42 \text{ hours}$$

$$\approx 4 \text{ hours } 25 \text{ minutes}$$

Carbon Dating

Carbon dating is a technique used by archaeologists, geologists, and others who want to estimate the ages of certain artifacts and fossils they uncover. The technique is based on certain properties of the carbon atom. In its natural state, the nucleus of the carbon atom ^{12}C has six protons and six neutrons. The **isotope** carbon-14, ^{14}C, is produced through cosmic-ray bombardment of nitrogen in the atmosphere. Carbon-14 has 6 protons and 8 neutrons and is **radioactive.** It decays by beta emission. That is, when an atom of ^{14}C decays, it gives up an electron to form a stable nitrogen atom ^{14}N. We make the assumption that the ratio of ^{14}C to ^{12}C in the atmosphere is constant. This assumption has been shown experimentally to be approximately valid, for although ^{14}C is being constantly lost through **radioactive decay** (as this process is often termed), new ^{14}C is constantly being produced. Living plants and animals do not distinguish between ^{12}C and ^{14}C, so at the time of death the ratio of ^{12}C to ^{14}C in an organism is the same as the ratio in the atmosphere. However, this ratio changes after death since ^{14}C is converted to ^{12}C but no further ^{14}C is taken in.

It has been observed that ^{14}C decays at a rate proportional to its mass and that its **half-life** is approximately 5580 years.† That is, if a substance starts with 1 gram of ^{14}C, then 5580 years later it would have $\frac{1}{2}$ gram of ^{14}C, the other $\frac{1}{2}$ gram having been converted to ^{14}N. Moreover, the relative rate of decay is constant.

EXAMPLE 7 *Estimating the Age of a Fossil*

A fossil is unearthed, and it is determined that the amount of ^{14}C present is 40% of what it would be for a similarly sized living organism. What is the approximate age of the fossil?

SOLUTION Let $M(t)$ denote the mass of ^{14}C present in the fossil. Since the ^{14}C decays at a constant relative rate, we have, from (6),

$$M(t) = M_0 e^{-kt} \tag{11}$$

The minus sign in the exponent indicates that the amount of ^{14}C present is decreasing as t increases.

Our first task is to determine k. We know that ^{14}C has a half-life of 5580 years. This means that M_0 grams initially ($t = 0$) decay to $\frac{1}{2}M_0$ grams 5580 years later ($t = 5580$), so

$$M(5580) = \frac{1}{2}M_0 = M_0 e^{-5580k}$$

$$e^{-5580k} = \frac{1}{2} \qquad \text{Divide by } M_0$$

† This number was first determined in 1941 by the American chemist W. S. Libby, who based his calculations on the wood from sequoia trees, whose ages were determined by rings marking years of growth.

$$\ln e^{-5580k} = \ln \frac{1}{2} \qquad \text{Take natural logarithms of both sides}$$

$$-5580k = \ln \frac{1}{2}$$

and

$$-k = \frac{\ln \frac{1}{2}}{5580}$$

Then

$$e^{-kt} = e^{(t \ln \frac{1}{2})/5580} \overset{\text{Property 8}}{=} e^{\ln (1/2)^{t/5580}} \overset{\text{Property 1}}{=} \left(\frac{1}{2}\right)^{t/5580}$$

and (11) becomes

$$M(t) = M_0\left(\frac{1}{2}\right)^{t/5580} \qquad \textbf{(12)}$$

Now we are told that after t years (from the death of the fossilized organism to the present), $M(t) = 0.4M_0$, and we are asked to determine t. Then

$$0.4M_0 = M_0\left(\frac{1}{2}\right)^{t/5580}$$

$$0.4 = \left(\frac{1}{2}\right)^{t/5580} \qquad \text{Divide by } M_0$$

$$\ln 0.4 = \ln \left(\frac{1}{2}\right)^{t/5580} \qquad \text{Take natural logarithms of both sides}$$

$$\ln 0.4 = \frac{t}{5580} \ln \frac{1}{2}$$

$$5580 \ln 0.4 = t \ln \frac{1}{2} = t \ln 0.5$$

$$t = \frac{5580 \ln 0.4}{\ln 0.5} = \frac{5580(-0.916290731)}{-0.69314718} \approx 7376 \text{ years}$$

The carbon-dating method has been used successfully on numerous occasions. It was this technique that established that the Dead Sea scrolls were prepared and buried about 2000 years ago.

An Application of Common Logarithms: The pH of a Substance

The acidity of a liquid substance is measured by the concentration of the hydrogen ion $[H^+]$ in the substance. This concentration is usually measured in terms of moles per liter (mol/L). A standard way to describe this acidity is to define the pH of a substance by $\text{pH} = -\log [H^+]$.

Distilled water has an approximate H^+ concentration of 10^{-7} mol/L, so its pH is $-\log 10^{-7} = -(-7) = 7$.

A substance with a pH under 7 is termed an **acid,** and one with a pH above 7 is called a **base.**

It is standard to write the pH of a substance with two-decimal-place accuracy.

EXAMPLE 8 *Finding the pH of a Substance*

Find the pH of a substance whose hydrogen ion concentration is 0.05 mol/L.

SOLUTION We have

$$\begin{aligned} \text{pH} &= -\log (0.05) \\ &= -(-1.301029996) = 1.30 \quad \text{Two-decimal-place accuracy} \end{aligned}$$

EXAMPLE 9 *Finding the Hydrogen Ion Concentration Given the pH*

The pH of the juice of a certain type of lemon is 2.30. What is the hydrogen ion concentration for this kind of lemon juice?

SOLUTION We have

$$\begin{aligned} \text{pH} = -\log [H^+] &= 2.30 \\ \log [H^+] &= -2.30 \\ [H^+] &= 10^{-2.30} \quad \text{Definition of } \log x \\ [H^+] &= 0.005 \text{ mol/L} \quad \text{From a calculator} \end{aligned}$$

FOCUS ON

The Mathematical Model of Constant Relative Rate of Growth

In many of the examples given in this section, we assumed that the relative rate of growth was constant so that

$$y(x) = y(0)e^{kx}$$

Suppose that $k > 0$. Let's give it a value, say $k = 0.1$. This means that something—a population or investment, for example, is growing continuously at a rate of 10% per time period. Did you stop to consider the fact that this rate of growth *cannot possibly continue* for an indefinite period of time?

To illustrate this fact, suppose that a certain insect population grows at 10% a month and starts with 1000 individuals. Then the population after t months is

$$P(t) = 1000e^{0.1t}$$

The population after t months is given in Table 2. We see that after 240 months (20 years) there would be over 26 trillion insects.

Table 2

t	$P(t) = 1000e^{0.1t}$
10	2718
20	7389
30	20,086
60	403,429
120	162,754,791
240	2.649×10^{13}
480	7.017×10^{23}

After 40 years there would be over 7×10^{23} insects. That is an unimaginably large number. To give you some idea, let us do some calculations.

(1) The radius of the earth is approximately 4000 miles.
(2) The surface area of a sphere is given by $S = 4\pi r^2$ so surface area of the earth $\approx 4\pi(4000)^2 \approx 200{,}000{,}000$ square miles.
(3) There are 5280^2 square feet in a square mile so the surface area of the earth in square feet $\approx (200{,}000{,}000)(5280^2) \approx 5.6 \times 10^{15}$ square feet.
(4) We divide:

$$\frac{7 \times 10^{23} \text{ insects}}{5.6 \times 10^{15} \text{ sq ft}} \approx 125{,}000{,}000 \text{ insects per square foot.}$$

That is, 125 million insects occupying every square foot on earth! Where would they fit? What would they eat?

The situation is ridiculous. This should come as no surprise because as $t \to \infty$, $e^{kt} \to \infty$ if $k > 0$, so any population (or anything else) grows without bound if it is growing exponentially.

Since nothing on earth can possibly grow without bound, it seems that the model of this section is useless. This is not true. Many quantities grow exponentially—like continuously compounded interest, for example. But nothing can grow exponentially indefinitely. There are always limits to growth. Any population growing exponentially will eventually run out of space or food and suffer other effects from overcrowding. Then nature will force the growth rate to change.

But unlimited growth is not the only problem with this model. Is the assumption that growth continues at a constant relative rate a reasonable one? If a bond pays 7% compounded continuously for 20 years, then you are certain that, at least for 20 years, your money will grow exponentially at the constant rate of 7%.

However, for another kind of problem, the answer is likely to be no. In Example 5, we found that the population of India grew at an average rate of 2.6% a year from 1974 to 1984. In order to answer the questions in that example, we assumed that the growth rate would remain the same at least until the year 2010. This is very unlikely. There are many factors that influence population growth rates: climatic conditions such as drought might lead to insufficient food and deaths by famine; education in birth control might lead to smaller families; other factors may lead to greater population growth rates. You can undoubtedly think of a few.

So the assumption of indefinite growth at a constant rate is unrealistic for many different types of problems. Models based on this assumption may be very useful. But they are often useful only for limited periods of time. As with all mathematical models, you must question the validity of the assumptions inherent in the model before you make predictions based on the model.

Problems 7.6

Readiness Check

I. Which of the following is the concentration of hydrogen ions present in a certain substance if pH $= -\log\,[H^+]$ and its pH is 7.5?
a. $10^{7.5}$ b. $10^{-7.5}$ c. log 7.5 d. −log 7.5

II. Which of the following is true about $P(t) = P(0)e^{kt}$ if $k < 0$ and $t > 0$?
a. $P(0) < P(t)$
b. $P(t) < P(0)$
c. $P(t) = P(0)$
d. No determination can be made until t and $P(0)$ are known.

III. Which of the following is $A(t) = Pe^{rt}$ if $P = \$3000$, $r = 6.7\%$, and $t = 3$ months $= \frac{1}{4}$ year?
a. \$3050.67 b. \$3667.87 c. \$3801.97
d. \$3025.17

IV. Which of the following is true about $M(t) = M_0e^{kt}$ if $k > 0$?
a. $M(t)$ increases as t increases.
b. $M(t)$ decreases as t increases.
c. $M(t)$ remains constant as t increases.
d. The change in $M(t)$ cannot be determined until values of M_0, k, and t are given.

1. How long would it take an investment to increase by half if it is invested at 4% compounded monthly?
2. What must be the nominal interest rate in order that an investment triple in 15 years if interest is compounded semiannually?
3. How long will it take an investment to triple (assuming quarterly compounding) for each interest rate?
(a) 2% (b) 5% (c) 8% (d) 10% (e) 15%
4. Answer the questions of Problem 3 if interest is compounded monthly.
5. Answer the questions of Problem 3 if interest is compounded continuously.
6. A bank account pays 8.5% annual interest compounded monthly. How large a deposit must be made now in order that the account contain exactly \$10,000 at the end of 1 year?

Answers to Readiness Check

I. b II. b III. a IV. a

7. Answer the question in Problem 6 if interest is compounded continuously.
8. A doting father wants his newly born daughter to have what $10,000 would buy on the date of her birth as a gift for her 21st birthday. He decides to accomplish this by making a single initial payment into a trust fund set up specially for this purpose.
   (a) Assuming that the rate of inflation is to be 6% (effective annual rate), what sum will have the same buying power in 20 years as $10,000 does at the date of her birth?
   (b) What is the amount of the single initial payment into the trust if its interest is compounded continuously at a rate of 10%?
9. What is the most a banker should pay for a $10,000 note due in 5 years if he can invest a like amount of money at 9% compounded annually?
10. A sum of $20,000 is invested at a steady rate of return with interest compounded continuously. If the investment is worth $35,000 in 2 years, what is the annual interest rate?
11. A certain government bond sells for $750 and can be redeemed for $1000 in 8 years. Assuming continuous compounding, what is the rate of interest paid?
12. How long would it take an investment to increase by half if it is invested at 5% compounded continuously?
13. What must be the interest rate in order that an investment triple in 15 years if interest is continuously compounded?
14. A bacterial population grows at a constant relative rate. It is 7000 at noon and 10,000 at 3 P.M.
   (a) What is the population at 6 P.M.?
   (b) When will the population reach 20,000?
15. The estimated world population in 1986 was 4,845,000,000. Assume that the population grows at a constant rate of 1.9%. When will the world population reach 8 billion?
16. In Problem 15, at what constant rate would the population grow if it reached 6 billion in the year 2000?
17. According to the official U.S. census, the population of the United States was 226,549,448 in 1980 and 248,709,873 in 1990. Assume that population grows at a constant relative rate.
   (a) Estimate the population in the year 2000.
   (b) When will the population reach one-half billion?
18. The population of the United States was 76,212,168 in 1900 and 92,228,496 in 1910. If population had grown at a constant percentage until 1990, what would the population have been in 1980 and 1990? Compare this answer with the data given in Problem 17. To explain this discrepancy is a problem in history, not algebra.
19. The population of Australia was approximately 13,400,000 in 1974 and 16,643,000 in 1990. Assume constant relative growth.
   (a) Predict the population in the year 2000.
   (b) When will the population be 20 million?
20. The population of New York State was 17,558,165 in 1980 and 17,990,455 in 1990. Assume a constant relative growth in population.
   (a) Predict the population in 1995.
   (b) When will the population reach 18.5 million?
21. * The population of Florida was 9,747,197 in 1980 and 12,937,926 in 1990. Assuming that Florida continues its high growth rate, when will the population of Florida exceed the population of New York (see Problem 20)?
22. In 1920, the consumer price index (CPI) for perishable goods was 213.4, with 1913 assigned the base rate of 100.† Assuming that inflation was constant (that is, assume continuous compounding), find the rate of inflation of the price of perishable goods between 1913 and 1920.
23. In 1920, the CPI for construction materials was 262.0. Again, with CPI = 100 in 1913, find the rate of inflation of the price of construction materials between 1913 and 1920.
24. The average annual earnings of full-time employees in the United States was $4743 in 1960 and $7564 in 1970. Assuming a continuous increase in earnings at a constant rate during the period 1960–1970, what was the rate of increase in wages?
25. The CPI was 88.7 in 1960 and 116.3 in 1970 (1967 = 100). Assuming a continuous increase in prices at a constant rate between 1960 and 1970, find the rate of inflation.
26. The *real* increase in earnings is defined as the percentage increase in wages minus the rate of inflation. Using the data in Problems 24 and 25, determine the real percentage increase in the average U.S. worker's earnings between 1960 and 1970.
27. Assume that the figures in Problems 24 and 25 hold, except that the CPI in 1970 is unknown. What would the CPI be if it were known that workers between 1960 and 1970 experienced no gain or loss in real income?
28. When the air temperature is 60°F, an object cools from 170°F to 130°F in half an hour.
   (a) What will be the temperature after 1 hour?
   (b) When will the temperature be 80°F?
29. A hot coal (temperature 150°C) is immersed in a liquid (temperature −10°C). After 30 seconds the temperature of the coal is 60°C. Assume that the liquid is kept at −10°C.
   (a) What is the temperature of the coal after 2 minutes?
   (b) When will the temperature of the coal be 0°C?

---

†This means that an average item costing $1.00 in 1913 would cost 2.13\frac{4}{10}$ in 1920.

30. A fossilized leaf contains 65% of a "normal" amount of ^{14}C. How old is the fossil?
31. Forty percent of a radioactive substance disappears in 100 years.
    (a) What is its half-life?
    (b) After how many years will 90% be gone?
32. Salt decomposes in water into sodium $[Na^+]$ and chloride $[Cl^-]$ ions at a rate proportional to its mass. Suppose there were 35 kg of salt initially and 21 kg after 12 hours.
    (a) How much salt would be left after 1 day?
    (b) After how many hours would there be less than $\frac{1}{4}$ kg of salt left?
33. Radioactive beryllium is sometimes used to date fossils found in deep-sea sediment. The mass of radioactive beryllium satisfies equation (11) with $k = 1.5 \times 10^{-7}$. What is the half-life of beryllium?
34. In a certain medical treatment, a tracer dye is injected into the pancreas to measure its function rate. A normally active pancreas will secrete 4% of the dye each minute. A physician injects 0.4 gram of the dye, and 30 minutes later 0.15 gram remains. How much dye would remain if the pancreas were functioning normally?
35. Atmospheric pressure is a function of altitude above sea level and satisfies an equation of the form (6), where $P(a)$ denotes pressure at altitude a. The pressure is measured in millibars (mbar). At sea level ($a = 0$), $P(0)$ is 1013.25 mbar, which means that the atmosphere at sea level will support a column of mercury 1013.25 millimeters (mm) high at a standard temperature of 15°C. At an altitude of $a = 1500$ m, the pressure is 845.6 mbar.
    (a) What is the pressure at $a = 4000$ m?
    (b) What is the pressure at 10 km?
    (c) In California, the highest and lowest points are Mount Whitney (4418 m) and Death Valley (86 m below sea level). What is the difference in their atmospheric pressures?
    (d) What is the atmospheric pressure at Mount Everest (elevation 8848 m)?
    (e) At what elevation is the atmospheric pressure equal to 1 mbar?

* 36. A bacteria population is known to grow exponentially. The following data were collected:

Number of Days	Number of Bacteria
5	1054
10	2018
20	7405

(a) What was the initial population?
(b) If the present growth rate were to continue, what would be the population after 60 days?

* 37. A bacteria population is declining exponentially. The following data were collected:

Number of Hours	Number of Bacteria
12	5969
24	3563
48	1269

(a) What was the initial population?
(b) How many bacteria are left after 1 week?
(c) When will there be no bacteria left (that is, when is $P(t) < 1$)?

38. What is the pH of a substance with a hydrogen ion concentration of 3.6×10^{-6}?
39. What is the pH of a substance with a hydrogen ion concentration of 0.6×10^{-7}?
40. Milk has a pH of 6.5. What is its hydrogen ion concentration?
41. Beer has a pH of 4.5. What is its hydrogen ion concentration?
42. Milk of magnesia has a pH of 10.5. What is its hydrogen ion concentration?
43. A general psychophysical relation was established in 1834 by the German physiologist Ernest Weber and given a more precise phrasing later by the German physicist Gustav Fechner. By the **Weber-Fechner law,** $S = c \log (R + d)$, where S is the intensity of a sensation, R is the strength of the stimulus producing it, and c and d are constants. The Greek astronomer Ptolemy catalogued stars according to their visual brightness in six categories or **magnitudes.** A star of the first magnitude was about $2\frac{1}{2}$ times as bright as a star of the second magnitude, which in turn was about $2\frac{1}{2}$ times as bright as a star of the third magnitude, and so on. Let b_n and b_m denote the apparent brightness of two stars having magnitudes n and m, respectively. Modern astronomers have established the Weber-Fechner law relating the relative brightness to the difference in magnitudes as

$$(m - n) = 2.5 \log \left(\frac{b_n}{b_m}\right)$$

(a) Using this formula, calculate the ratio of brightness for two stars of the second and fifth magnitudes, respectively.
(b) If star A is five times as bright to the naked eye as star B, what is the difference in their magnitudes?
(c) How much brighter is Sirius (magnitude 1.4) than a star of magnitude 2.15?
(d) The Nova Aquilae in a 2–3-day period in June 1918 increased in brightness about 45,000 times. How many magnitudes did it rise?

* (e) The bright star Castor appears to the naked eye as a single star but can be seen with the aid of a telescope to be really two stars whose magnitudes have been calculated to be 1.97 and 2.95. What is the magnitude of the two combined? [Hint: Brightnesses, but not magnitudes, can be added.]

* 44. The subjective impression of loudness can be described by a Weber-Fechner law. Let I denote the intensity of a sound. The least intense sound that can be heard is $I_0 = 10^{-12}$ watt/m^2 at a frequency of 1000 cycles/sec. (This value is called the **threshold of audibility.**) If L denotes the loudness of a sound, measured in decibels,† then $L = 10 \log (I/I_0)$.

(a) If one sound has twice the intensity of another, what is the ratio of the perceived loudness of the two sounds?

(b) If one sound appears to be twice as loud as another, what is the ratio of their intensities?

(c) Ordinary conversation sounds 6 times as loud as a low whisper. What is the actual ratio of intensity of their sounds?

Summary Outline of Chapter 7

- An **exponential function** is a function of the form $f(x) = a^x$, where $a > 0$, $a \neq 1$. p. 357

- **The number *e*** is the number approached by the expression $\left(1 + \frac{1}{m}\right)^m$ as $m \to \infty$. To 10 decimal places $e = 2.7182818285$. The function $y = e^x$ is the most important exponential function in applications. p. 367

- The function $y(t) = y(0)e^{kt}$ is the equation of **exponential growth** (if $k > 0$) or **exponential decay** (if $k < 0$). p. 371

- The **logarithm function** $y = \log_a x$ is the inverse of $y = a^x$. $y = \log_a x$ if and only if $x = a^y$. p. 377

- **Properties of Logarithmic Functions**

  $a^{\log_a x} = x \qquad \log_a a^x = x \qquad \log_a a = 1 \qquad \log_a 1 = 0$ pp. 378, 379

  $\log_a xw = \log_a x + \log_a w \qquad \log_a \frac{x}{w} = \log_a x - \log_a w$ p. 379

  $\log_a \frac{1}{x} = -\log_a x \qquad \log_a x^r = r \log_a x \qquad \log_a b = \frac{1}{\log_b a}$ p. 379

  $\log_a x = \frac{\log x}{\log a} = \frac{\ln x}{\ln a}$ p. 390

- **Common Logarithms** are logarithms to the base 10. p. 384

- **Natural Logarithms** are logarithms to the base e. p. 386

- **Properties of The Natural Logarithm** p. 388

  $e^{\ln x} = x \qquad \ln e^x = x \qquad \ln e = 1 \qquad \ln 1 = 0$

  $\ln xw = \ln x + \ln w \qquad \ln \frac{x}{w} = \ln x - \ln w \qquad \ln \frac{1}{x} = -\ln x \qquad \ln x^r = r \ln x$

Review Exercises for Chapter 7

In Exercises 1–8, draw a sketch of the given exponential or logarithmic function.

1. $y = 4^x$
2. $y = (\frac{1}{4})^x$
3. $y = 3 \cdot 5^x$
4. $y = -2 \cdot 2^{-x}$
5. $y = \log_2 (x + 3)$
6. $y = \log (1 - x)$
7. $y = 2 - \ln x$
8. $y = \ln (3 - x) - 1$

In Exercises 9–16, use a calculator to estimate the given number to as many decimal places of accuracy as your calculator carries.

9. $e^{1.7}$
10. $10^{3.45}$
11. $(\frac{1}{3})^{2.3}$
12. $2.4^{\sqrt{5}}$
13. $\log 28.4$
14. $\log 0.0032$
15. $\ln 1,000,000$
16. $\ln 0.00235$

†decibel (dB) = $\frac{1}{10}$ bel, named after Alexander Graham Bell (1847–1922), inventor of the telephone.

In Exercises 17–28, solve for the given variable.

17. $y = \log_2 16$
18. $y = \log_{1/3} 9$
19. $\frac{1}{2} = \log_x 5$
20. $y = e^{\ln 17.2}$
21. $\log x = 10^{-9}$
22. $\log_x 32 = -5$
23. $2 \ln (x + 3) = 4$
24. $e^{-2(x+1)} = 2$
25. $\ln x - \ln (x + 1) = \ln 2$
26. $3 \ln x = \ln 4 + \ln 2$
27. $\ln (y + 2) + \ln y = \ln 3$
28. $\ln z^3 - 2 \ln z = 1$
29. If $y = 3 \ln x$, what happens to y if x doubles?
30. If $y = -4 \ln x$, what happens to y if x is cut in half?
31. Convert each equation to an exponential form.
    (a) $\log_9 27 = \frac{3}{2}$
    (b) $\log_{1/2} 8 = -3$
32. Change each equation to a logarithmic form.
    (a) $5^3 = 125$
    (b) $3^{-2} = \frac{1}{9}$

In Exercises 33–38, write as a single logarithm.

33. $\log_3 x + \log_3 (x - 2)$
34. $\log x + \log y$
35. $\log (x + 1) - \log (x + 3) + \log z^2$
36. $3 \ln z - 4 \ln (x + 1) + 5 \ln (x^2 - 1)$
37. $3 \ln (x + 1) - \frac{1}{2} \ln (y + 4) + \frac{1}{3} \ln (z + 12)$
38. $\ln a - \ln b + \ln c - \ln d$

In Exercises 39–42 compute each logarithm.

39. $\log_2 7$
40. $\log_{1/2} \frac{1}{3}$
41. $\log_4 37$
42. $\log_{90} 2$

The remaining exercises all require the use of a calculator.

43. What is the simple interest paid on \$5000 invested at 7% for 6 years?
44. What is the simple interest paid on \$8000 invested at $7\frac{1}{2}$% for 12 years?

In Exercises 45–48, compute the value of an investment after t years and the total interest paid if P dollars is invested at a nominal rate of r% compounded m times a year.

45. $P = \$8000$, $r = 5\frac{1}{2}\%$, $t = 4$, $m = 1$
46. $P = \$10{,}000$, $r = 6\frac{1}{2}\%$, $t = 4$, $m = 2$
47. $P = \$6000$, $r = 10\%$, $t = 5$, continuously
48. $P = \$25{,}000$, $r = 7\frac{3}{4}\%$, $t = 15$, continuously
49. If money is invested at 8% compounded continuously, what is the effective interest rate?
50. What is the effective interest rate of 8% compounded monthly?
51. How long will it take an investment to double (assuming quarterly compounding) if the interest rate is (a) 3%? (b) $6\frac{1}{2}$%? (c) 8%? (d) 12%?
52. Answer the questions of Exercise 51 if money is compounded continuously.
53. The relative rate of growth of a population is 15%. If the initial population is 10,000, what is the population after 5 years? after 10 years?
54. In Exercise 53, how long will it take for the population to double?
55. When a cake is taken out of the oven, its temperature is 125°C. Room temperature is 23°C. The temperature of the cake is 80°C after 10 minutes.
    (a) What will be its temperature after 20 minutes?
    (b) How long will the cake take to cool to 25°C?
56. A fossil contains 35% of the normal amount of ^{14}C. What is its approximate age?
57. What is the half-life of an exponentially decaying substance that loses 20% of its mass in one week?
58. How long will it take the substance in Exercise 57 to lose 75% of its mass? 95% of its mass?

Appendix A

Graphing Using a Calculator

It has long been possible to generate graphs of a wide variety of functions on a computer. Recently, hand-held calculators with graphing capabilities have become available. In this appendix, we shall discuss some techniques that will help you use the graphing calculator more effectively. We shall also discuss some ways that graphing calculators can be used to solve algebraic problems like approximating the zeros of functions, finding where two curves intersect, and solving inequalities.

The graphing calculators currently available are produced by (in alphabetical order) Casio, Hewlett-Packard, Sharp, and Texas Instruments. This appendix is generic; that is, it is intended for use with any graphing calculator. All applications cited in this appendix can be carried out with any of the calculators currently available. Therefore, our discussion will focus on graphing techniques rather than on specific keystrokes; that is, we will not tell you which buttons to push. For that reason:

> It is essential that you read the instruction manual that accompanies your graphing calculator before you read any further.

Problems that will require you to use a graphing calculator can be found in selected sections of this text or in the graphing calculator supplement that is available from the publisher.

I. Obtaining a Graph on a Calculator

In order to obtain a graph on a calculator, two things must be done (not necessarily in the order given here).

A. Enter the Function to Be Graphed

Read your manual to learn the procedure that must be used to enter functions. There will be a special way to enter the function variable, which is most often denoted by X.

B. Determine the Range and Scale

You must tell the calculator the range of values over which you wish the function to be graphed and specify the scale on the x- and y-axes.

The range is given by entering the smallest and largest values to be taken for x and y. The scale is the length represented by each tick mark on an axis. If the x-scale is 2, for example, then the distance between two successive tick marks represents a length of 2 units. In this appendix, we use the following notation:

$x_{\min}$ = minimum value of x
$x_{\max}$ = maximum value of x
x_{scl} = the scale on the x-axis

y_{min} = minimum value of y
y_{max} = maximum value of y
y_{scl} = the scale on the y-axis

Most calculators have a key (sometimes labeled RANGE) that must be pressed in order to enter range and scale values.

NOTE If you graph a function that you have entered but do not specify range and scale values, then one of three things will happen. First, the calculator may use the range and scale values that were entered for the previous graph that was sketched. Second, the calculator may use some "standard" built-in range and scale values. These are called "standard defaults." On one TI calculator, for example, the standard defaults for both the x- and y-axes are the intervals $[-10,10]$ with a scale of 1. Third, if the calculator has built-in graphs, then preset ranges and scales will be used whenever one of these graphs is sketched. For example, on one Casio calculator the graph of $y = \ln x$ is built in. When this function is sketched, the calculator uses the following range and scale values:

$$x_{min} = -1 \qquad y_{min} = -1.6$$
$$x_{max} = 8.4 \qquad y_{max} = 2.368$$
$$x_{scl} = 2 \qquad y_{scl} = 1$$

Before pressing a graphing key, check the range and scale values. A number of things can happen if you do not enter these values yourself.

The hardest part about graphing on a calculator is choosing appropriate range values. Our first example illustrates why care in choosing these values is essential.

EXAMPLE 1 *Finding an Appropriate Range in Order to Generate a Graph*

Sketch the graph of $y = -x^3 + 2x^2 + 5x - 6 = -(x + 2)(x - 1)(x - 3)$.

SOLUTION Suppose that we do not notice that the cubic can be factored and we arbitrarily choose the following range and scale values:

Ranges: $-2 \le x \le 2$; $-3 \le y \le 3$

Scales: Each x-axis tick represents 1 unit, and each y-axis tick represents 2 units. That is,

$$x_{min}: -2 \qquad y_{min}: -3$$
$$x_{max}: 2 \qquad y_{max}: 3$$
$$x_{scl}: 1 \qquad y_{scl}: 2$$

After the function is entered, the graph in Figure 1 is obtained.

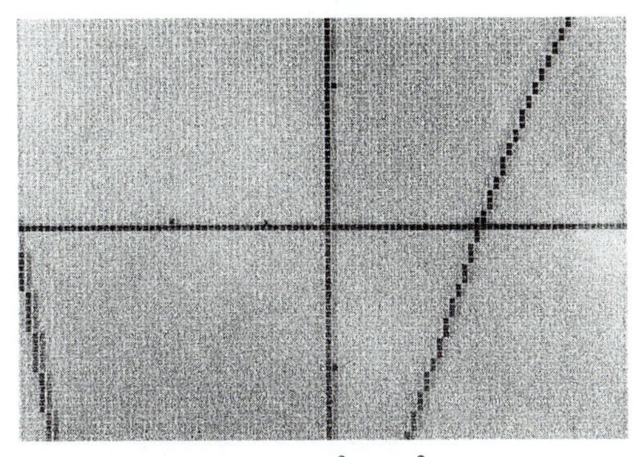

Figure 1 Graph of $y = -x^3 + 2x^2 + 5x - 6$ for $-2 \leq x \leq 2,\ -3 \leq y \leq 3$

This graph is accurate but not very useful. We need to see what happens outside of our rather limited range. Let us greatly expand the ranges:

$$-20 \leq x \leq 20;\ -200 \leq y \leq 200;\ \text{with } x_{\text{scl}} = 1 \text{ and } y_{\text{scl}} = 20.$$

The graph now appears as in Figure 2:

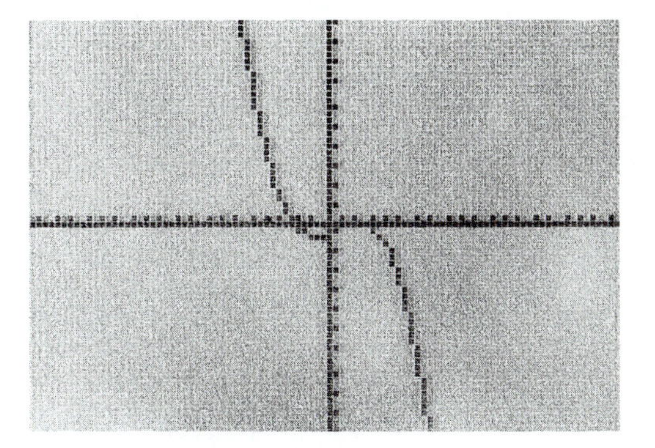

Figure 2 Graph of $y = -x^3 + 2x^2 + 5x - 6$ for $-20 \leq x \leq 20,\ -200 \leq y \leq 200$

This graph looks different but is still not what we want. Its appearance should not be surprising. For $|x|$ large, $-x^3 + 2x^2 + 5x - 6 \approx -x^3$, so since we used a large range of x-values, we have obtained a graph that looks like the graph of $-x^3$.

We can get a much more revealing graph by thinking a bit before entering range values. If $f(x) = -x^3 + 2x^2 + 5x - 6$, then, for example, $f(-4) = 70$ and $f(4) = -18$. It is not hard to see that if $x < -4$, then $f(x) > 70$, and if $x > 4$, then $f(x) < -18$. Thus, most of the interesting behavior of this function occurs for $-4 \leq x \leq 4$ and $-18 \leq y \leq 70$. Setting these range values and letting $x_{\text{scl}} = 1$ and $y_{\text{scl}} = 2$, we obtain the graph in Figure 3.

This is the type of graph we want. We can clearly see the zeros at -2, 1, and 3. Other interesting behavior, like intervals over which the function is increasing or decreasing, is plainly shown.

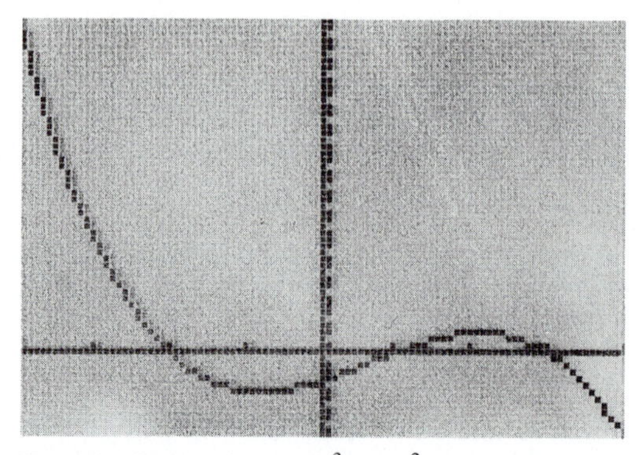

Figure 3 Graph of $y = -x^3 + 2x^2 + 5x - 6$ for $-4 \le x \le 4,\ -18 \le y \le 70$

NOTE On some calculators, there is a "zoom" or "factor" feature. This feature allows you to zoom in or zoom out by a factor you set. If you zoom out in Figure 1, for example, then you get a more global picture of the graph. If you zoom out too much, you might get a picture like the one in Figure 2. Then you could zoom in to obtain the more accurate graph in Figure 3. You can experiment with the zoom feature, zooming in and out, until you get a graph that looks accurate. You can save a lot of confusion, however, if you first think about what reasonable range values should be. ■

EXAMPLE 2 *Sketching the Graph of a Rational Function Having a Vertical Asymptote*

Sketch the graph of $y = f(x) = \dfrac{x^2 - 2x + 5}{x + 2}$.

SOLUTION We first note that the function is not defined at $x = -2$. As x gets close to -2 from either side, $|y|$ gets large. Also, $|y|$ gets large as $|x|$ gets large. To see this, we divide to obtain

$$\frac{x^2 - 2x + 5}{x + 2} = x - 4 + \frac{13}{x + 2}.$$

Since $\dfrac{13}{x+2} \to 0$ as $x \to \pm\infty$, we see that, for x large,

$$\frac{x^2 - 2x + 5}{x + 2} \approx x - 4.$$

(The symbol $x \to \infty$ is explained on page 113.)

You must be very careful when you enter functions in your calculator. Without parentheses around the numerator and denominator of this function, you would not be graphing the function you want to graph. Specifically, if you enter

$$y = x^2 - 2x + 5 \div x + 2,$$

you will obtain the graph of

$$y = x^2 - 2x + \frac{5}{x} + 2.$$

This is very different from the correctly entered function

$$y = (x^2 - 2x + 5) \div (x + 2).$$

There are many ranges of values that will give us a suitable graph. Each one must include $x = -2$ and allow for reasonably large values for $|y|$. Here is one set of range values:

$$-12 \leq x \leq 10, \ -25 \leq y \leq 15$$

We use scales of 1 on each axis.

The graph is given in Figure 4.

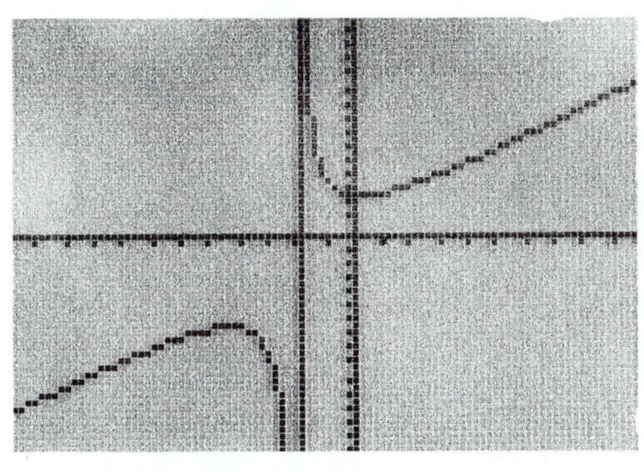

(a) Graph showing the vertical asymptote

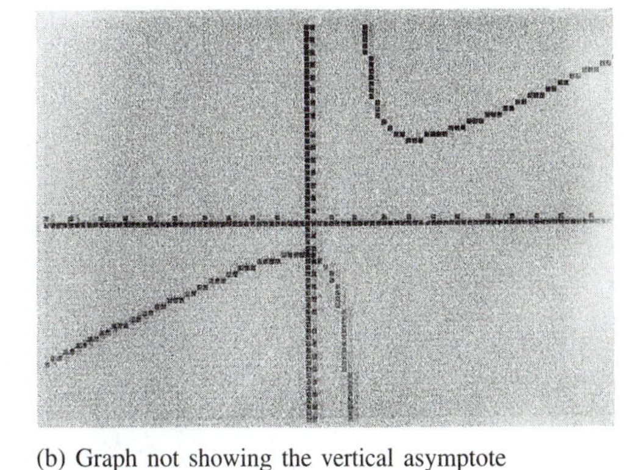

(b) Graph not showing the vertical asymptote

Figure 4 Graph of $y = \dfrac{x^2 - 2x + 5}{x + 2}$ for $-12 \leq x \leq 10, \ -25 \leq y \leq 15$

If your calculator shows a vertical line at $x = -2$, it is because the calculator is attempting to connect the points on the graph. Thus, some calculators may show the vertical asymptote (Figure 4(a)) at $x = -2$ and some may not (Figure 4(b)).† ■

EXAMPLE 3 *Sketching the Graph of a Rational Function with Horizontal and Vertical Asymptotes*

Sketch the graph of $y = \dfrac{2x^2 - 3x + 5}{x^2 - 1}$.

† One TI calculator draws the asymptote if the range values for x are set to $[-12, 10]$ but does not draw it if the range values are set to $[-10, 10]$. Thus the appearance or nonappearance of asymptotes depends both on the calculator used and the range settings.

SOLUTION We first note that $x^2 - 1 = 0$ when $x = 1$ and $x = -1$. Since the numerator is not zero at either of these values, the lines $x = 1$ and $x = -1$ are vertical asymptotes. If we divide numerator and denominator by x^2, we obtain

$$\frac{2x^2 - 3x + 5}{x^2 - 1} = \frac{2 - \dfrac{3}{x} + \dfrac{5}{x^2}}{1 - \dfrac{1}{x^2}}.$$

The terms $\dfrac{3}{x}$, $\dfrac{5}{x^2}$, and $\dfrac{1}{x^2} \to 0$ as $x \to \pm\infty$, so $\dfrac{2x^2 - 3x + 5}{x^2 - 1} \to 2$ as $x \to \pm\infty$. Thus $y = 2$ is a horizontal asymptote to the graph.

This means that for x large in either direction, y is close to 2. On the other hand, $|y| \to \infty$ as x gets near 1 or -1 so we must allow for large values of y.

In Figure 5, we provide the graph for the values $-3 \le x \le 3$, $-20 \le y \le 20$.

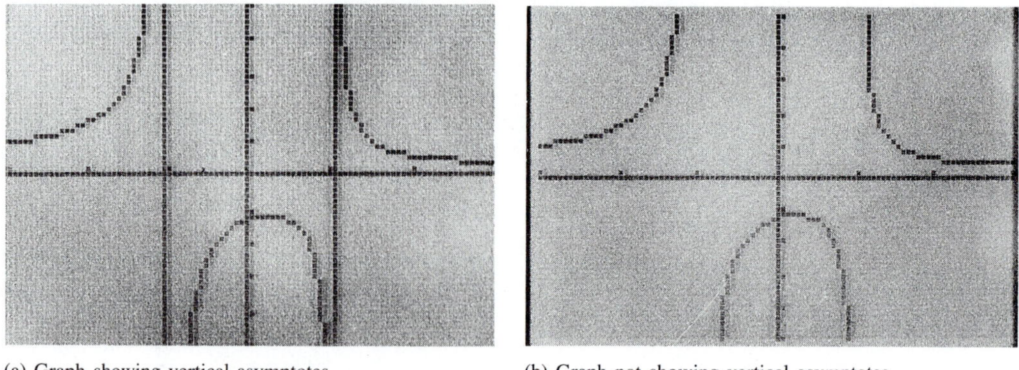

(a) Graph showing vertical asymptotes (b) Graph not showing vertical asymptotes

Figure 5 Graph of $y = \dfrac{2x^2 - 3x + 5}{x^2 - 1}$ for $-3 \le x \le 3$, $-20 \le y \le 20$, $x_{scl} = 1$, $y_{scl} = 5$

This graph gives us an accurate picture. However, if we want to see what happens as $|x|$ gets larger (to see the approach to the asymptote $y = 2$), then we can increase the x-values and decrease the y-values.

In Figure 6, we give the graph for $-12 \le x \le 12$ and $-10 \le y \le 10$.

Now we see clearly that the graph approaches a horizontal line as $|x|$ gets large. However, we lose an accurate picture in the interval $-1 < x < 1$.

Which of the graphs in Figures 5 and 6 is better? The answer here is not clear. If you want to see what happens to the function as $|x|$ becomes large, then Figure 6 is better. However, Figure 5 provides a clearer picture of the shape of the curve and is closer to the one that would appear in a textbook. A

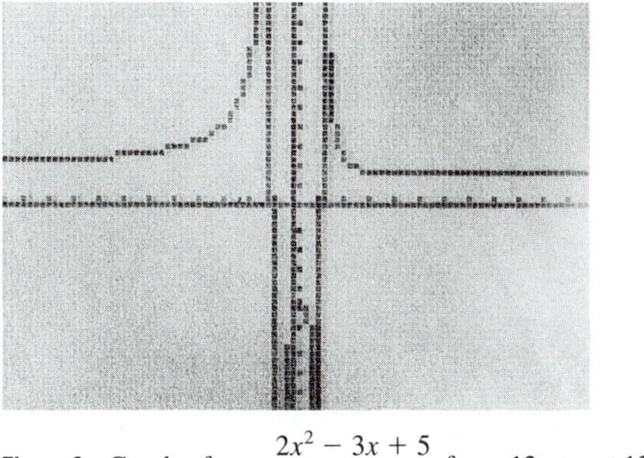

Figure 6 Graph of $y = \dfrac{2x^2 - 3x + 5}{x^2 - 1}$ for $-12 \le x \le 12,\ -10 \le y \le 10$

nice feature of a graphing calculator, especially one with a "zoom" feature, is that it allows you to experiment with different range values in order to obtain a graph that best suits your needs. ■

EXAMPLE 4 *Sketching the Graph of an Exponential Function*

Sketch the graph of $y = f(x) = e^{x^2+1}$.

SOLUTION $x^2 + 1 \ge 1$ so $e^{x^2+1} \ge e^1 = e \approx 2.718$. Thus the minimum value for y is 2.718. Also $f(-2) = f(2) = e^5 \approx 148$. One suitable set of ranges is

$$-2 \le x \le 2;\ 0 \le y \le 20.$$

We allow the value $y = 0$ so that the graph will clearly exhibit the minimum value at $x = 0$. We use the scale values $x_{\text{scl}} = 0.25$ and $y_{\text{scl}} = 1$.

Consult your calculator manual to see how to enter the function e^{x^2+1}. (A discussion of how to obtain exponential values on a calculator appears on page 367.) When e^{x^2+1} is entered, the graph in Figure 7 appears.

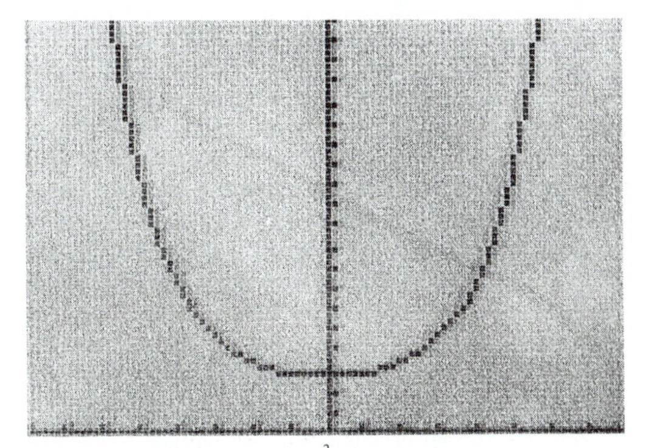

Figure 7 Graph of $y = e^{x^2+1}$ for $-2 \le x \le 2,\ 0 \le y \le 20$ ■

EXAMPLE 5 *Sketching the Graph of a Trigonometric Function*

Sketch the graph of $y = f(x) = 4 \sin x + 2 \cos 2x$.

SOLUTION We drew this graph in Figure 4 on page 199. Since $\sin x$ is periodic of period 2π and $\cos 2x$ is periodic of period π, $f(x)$ is periodic of period 2π (explain why) and we can obtain two complete cycles by letting the range of x include an interval having length of at least $4\pi \approx 12.6$. A suitable range is $-7 \le x \le 7$. Also, $|4 \sin x| \le 4$ and $|2 \cos 2x| \le 2$, so $|4 \sin x + 2 \cos 2x| \le 6$, and a suitable range for y is $-6 \le y \le 6$. Since we are dealing with a function of period 2π, a suitable x-scale is $\pi/4 \approx 0.785$. A suitable y-scale is 1. With these values, we obtain the sketch in Figure 8.

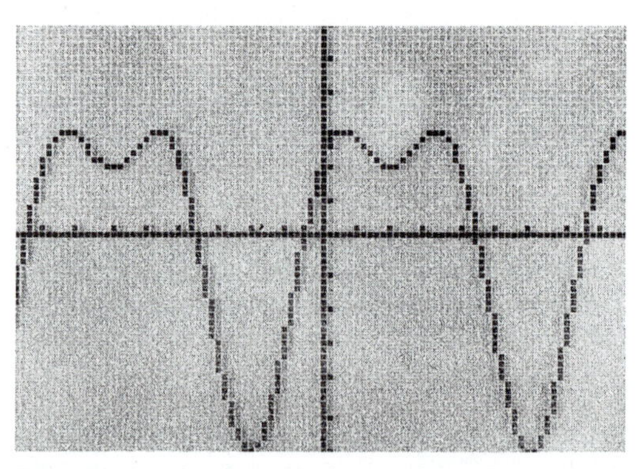

Figure 8 Graph of $y = 4 \sin x + 2 \cos 2x$, $-7 \le x \le 7$, $-6 \le y \le 6$

II. Other Uses of Calculator Graphing

There are many types of problems that can be solved on a calculator by using suitable graphs. Among these are (1) finding the zeros of a function and (2) finding points of intersections of graphs. The two problems are really the same because finding where $f(x) = g(x)$ is equivalent to finding the zeros of $f(x) - g(x)$. We will, however, treat these as distinct problems and give examples of each.

EXAMPLE 6 *Using Graphs to Find the Zeros of a Polynomial*

Find all zeros of $f(x) = x^3 - 3x^2 + 7x - 8$.

SOLUTION $f(0) = -8, f(x) < 0$ if $x < 0, f(1) = -3, f(2) = 2$, and $f(3) = 13$. Reasonable ranges are $-2 \le x \le 3$ and $-10 \le y \le 15$. Using a scale of 1 on each axis, we obtain the graph in Figure 9. (The numbers on the x-axis were added to make things clearer.)

The graph suggests that there is one and only one zero between 1 and 2. (We already knew that there was at least one zero because $f(1) < 0$ and $f(2) > 0$.)

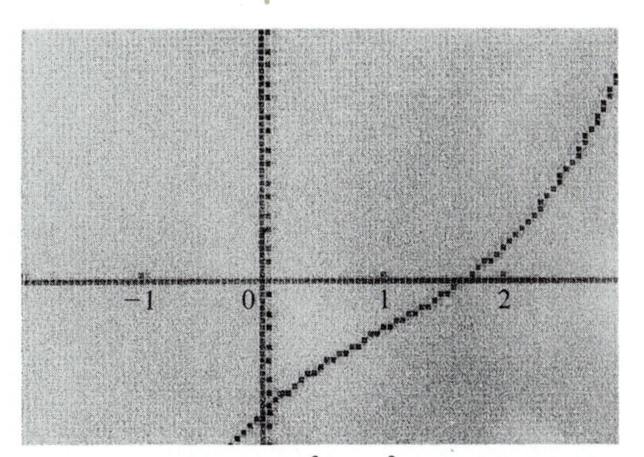

Figure 9 Graph of $y = x^3 - 3x^2 + 7x - 8$ for $-2 \le x \le 3$, $-10 \le y \le 15$

There are several ways to approximate this zero more closely. If your calculator has a zoom feature, you can zoom in on the part of the graph where the curve crosses the x-axis and find the zero to at least 4 or 5 decimal place accuracy with little difficulty. Whether or not you have this feature, or any similar one, you can improve your estimate by changing the range values along the x-axis. This is what we do now.

Here are some new range and scale values:

$$1 \le x \le 2;\ x_{scl} = 0.05;\ -3 \le y \le 2;\ y_{scl} = 0.1.$$

From the graph in Figure 10, we see that the graph crosses the x-axis in the interval $1.6 < x < 1.7$.

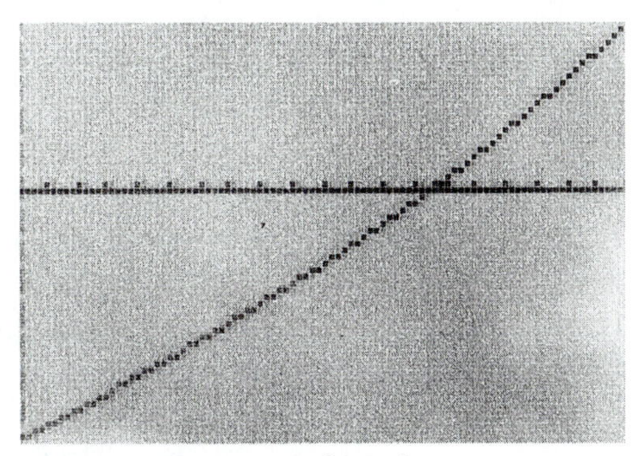

Figure 10 Graph of $y = x^3 - 3x^2 + 7x - 8$ for $1 \le x \le 2$, $-3 \le y \le 2$, $x_{scl} = 0.05$, $y_{scl} = 0.1$

To get more precision, we can change range and scale again:

$$1.6 \le x \le 1.7;\ x_{scl} = 0.01$$
$$-0.5 \le y \le 0.5;\ y_{scl} = 0.1$$

Now we see, in Figure 11, that the zero falls between 1.67 and 1.68. We stop here because there are faster ways to get more precision if more precision is needed.

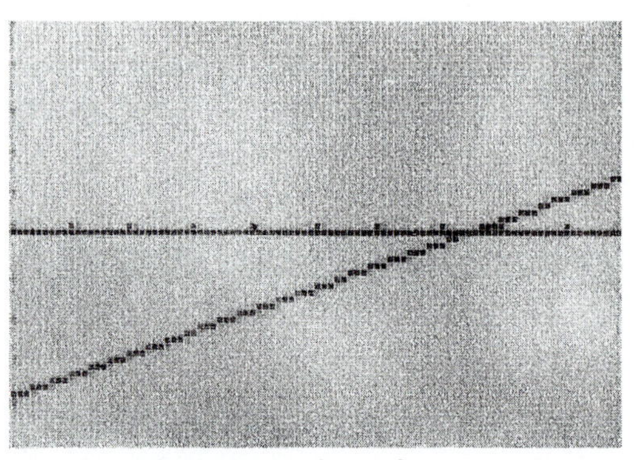

Figure 11 Graph of $y = x^3 - 3x^2 + 7x - 8$ for $1.6 \leq x \leq 1.7$, $-0.5 \leq y \leq 0.5$, $x_{\text{scl}} = 0.01$, $y_{\text{scl}} = 0.1$

A NOTE ON ACCURACY We say that a number x_A approximates a number x with k *decimal places of accuracy* if

$$|x - x_A| < \frac{1}{2} \times 10^{-k}.$$

For example, suppose we wish to approximate $\pi \approx 3.14159265359$.

A First Approximation

$$\pi \approx 3.1$$

Then

$$|\pi - 3.1| \approx 0.04159265359.$$

Now $10^{-1} = 0.1$ and $\frac{1}{2} \times 10^{-1} = 0.05$.

So

$$|\pi - 3.1| < 0.05 = \frac{1}{2} \times 10^{-1}.$$

So 3.1 approximates π with one decimal place accuracy.

A Second Approximation

$$\pi \approx 3.1415$$

Then

$$|\pi - 3.1415| \approx 0.00009265359.$$

Now

$$10^{-4} = 0.0001 \quad \text{and} \quad \frac{1}{2} \times 10^{-4} = 0.00005.$$

$$10^{-3} = 0.001 \quad \text{and} \quad \frac{1}{2} \times 10^{-3} = 0.0005.$$

So since $0.00005 < 0.00009265359 < 0.0005$, we see that 3.1415 approximates π with three (but not four) decimal place accuracy.

ANOTHER NOTE ON THE ZOOM FEATURE On a calculator with a zoom feature, the zero in Example 5 can be approximated quicker than by rescaling manually. If we zoom in on the graph in Figure 9 by a scale of 10 in each axis, with the zooming centered at the place where the curve crosses the x-axis, then we obtain the graph in Figure 12.

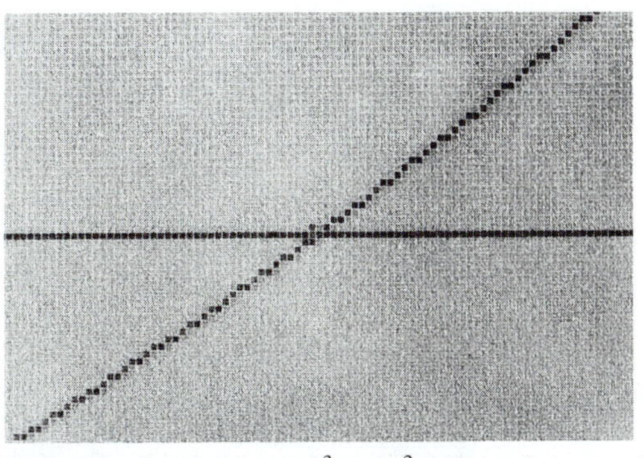

Figure 12 Graph of $y = x^3 - 3x^2 + 7x - 8$ (zoomed in from Figure 9)

Using the trace feature on the calculator, we can move along the curve to locate two points near where the curve crosses the x-axis. One point should be to the left of the zero, and the other should be to the right. We can find, in Figure 12, that $1.65 <$ the zero < 1.69. If we zoom in again by a factor of 10, we get the graph in Figure 13.

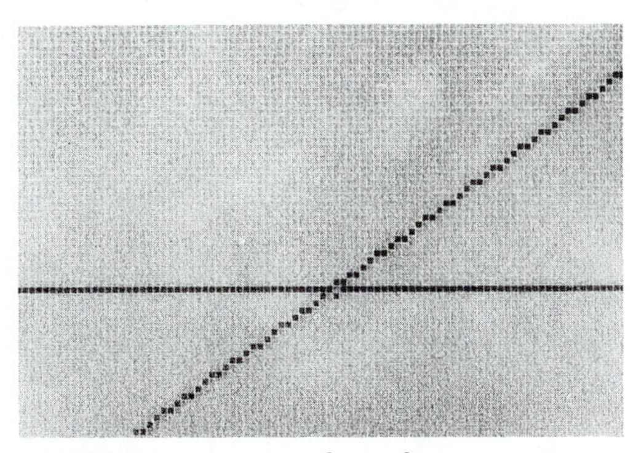

Figure 13 Graph of $y = x^3 - 3x^2 + 7x - 8$ (zoomed in from Figure 12)

Then we find that the zero is between $x = 1.671$ and $x = 1.674$. Thus we know that, to two decimal places, $x \approx 1.67$.

The reason that there are no tick marks in Figures 11 and 12 is that, in zooming in, the calculator we used did not change the scale values. Therefore, the scales stayed at 1 unit on each axis. Since the range of x-values is far less than 1 unit, no tick marks appear.

Alternatively, if your calculator has a zoom feature, it may also have a "box" feature. This allows you to draw a rectangle on the screen and then zoom in on the rectangle within. If you blow up a small rectangle around the zero, you can see more precisely where the curve crosses the x-axis. By doing this repeatedly, you can approximate the zero quickly and accurately.

We will say no more about zoom and box features in this appendix, but you should use them if they are available on your calculator. ■

EXAMPLE 7 *Finding the Zeros of a Polynomial*

Find all real zeros of $p(x) = 2x^4 + 1.5x^3 - 9x^2 - x + 5$.

SOLUTION We first observe that the $2x^4$ term dominates the other terms for x large. For example, if $x = 5$, then $2x^4 = 1250$ while $1.5x^3 - 9x^2 - x + 5 = -37.5$. The range values $-5 \leq x \leq 5$ include all the "interesting" behavior in the graph. If we set the y values at $-10 \leq y \leq 10$ with $x_{\text{scl}} = 1$ and $y_{\text{scl}} = 1$, we obtain the graph in Figure 14.

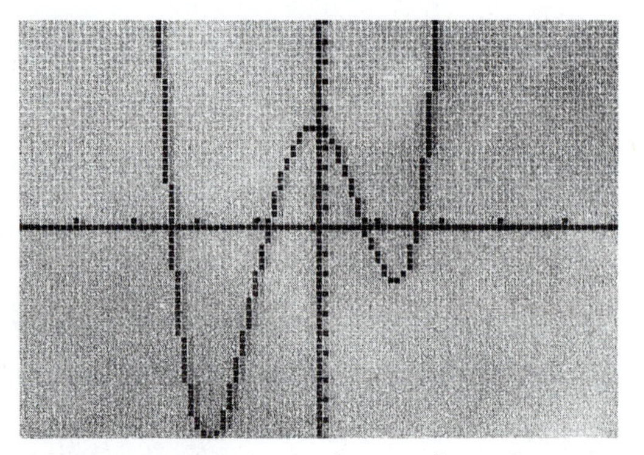

Figure 14 Graph of $y = 2x^4 + 1.5x^3 - 9x^2 - x + 5$ for $-5 \leq x \leq 5$, $-10 \leq y \leq 10$

Evidently $p(x)$ has four zeros. There is one between -3 and -2, one between -1 and $-\frac{1}{2}$, one between $\frac{1}{2}$ and 1, and one between 1 and 2. We can obtain each zero by choosing x-values near the zero.

FIRST ZERO We set $-3 \leq x \leq -2$, $-1 \leq y \leq 1$ and $x_{\text{scl}} = y_{\text{scl}} = 0.1$ to obtain the graph in Figure 15.

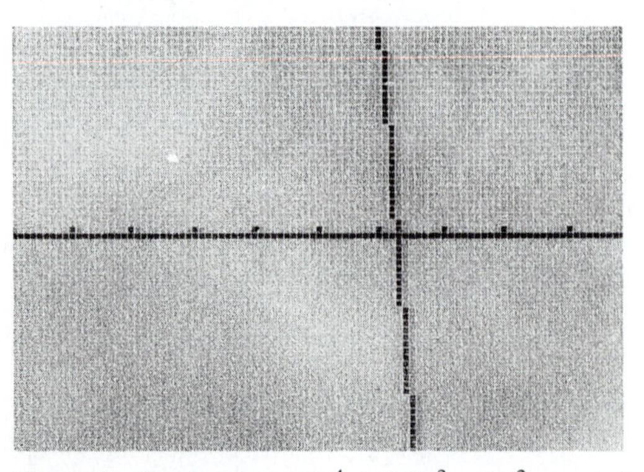

Figure 15 Graph of $y = 2x^4 + 1.5x^3 - 9x^2 - x + 5$, $-3 \leq x \leq -2$, $-1 \leq y \leq 1$, $x_{\text{scl}} = 0.1$

The zero is between -2.4 and -2.3. Rescaling again for $-2.4 \leq x \leq -2.3$, $-1 \leq y \leq 1$, $x_{scl} = y_{scl} = 0.01$, we obtain the graph in Figure 16.

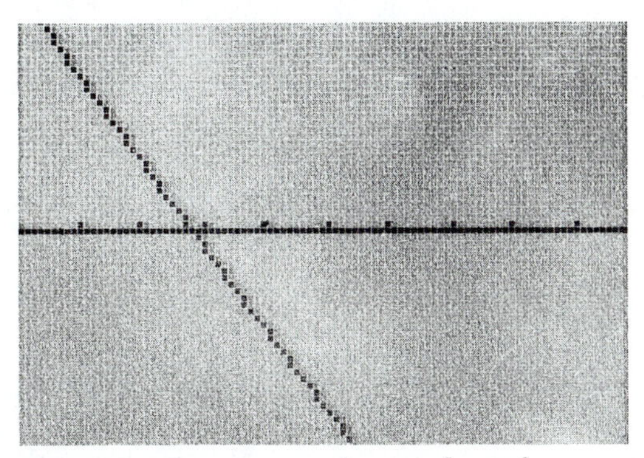

Figure 16 Graph of $y = 2x^4 + 1.5x^3 - 9x^2 - x + 5$, $-2.4 \leq x \leq -2.3$, $x_{scl} = 0.01$

Now we see that the zero is between -2.38 and -2.37. Continuing in this manner, we find, to three decimal place accuracy, $x \approx -2.371$.

SECOND ZERO Since the zero is between -1 and -0.5, we rescale with $-1.0 \leq x \leq -0.5$, $-1 \leq y \leq 1$, $x_{scl} = y_{scl} = 0.1$. The result is sketched in Figure 17.

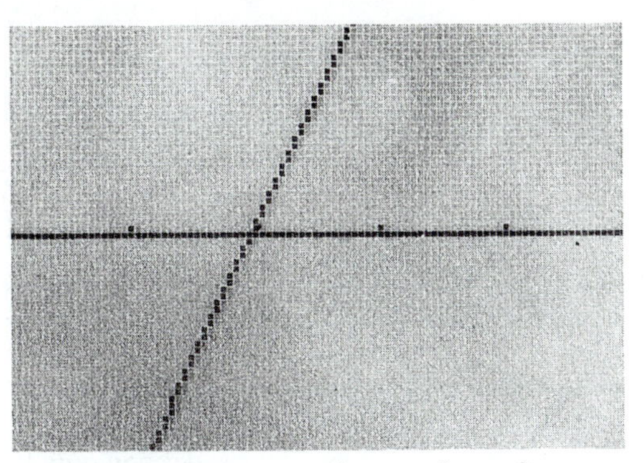

Figure 17 Graph of $y = 2x^4 + 1.5x^3 - 9x^2 - x + 5$, $-1.0 \leq x \leq -0.5$

The zero is very close to -0.8. We can continue to find that, to three decimal places, $x \approx -0.807$.

THIRD ZERO This zero is between $\frac{1}{2}$ and 1, so in Figure 18, we sketch the curve for $0.5 \leq x \leq 1.0$ and $x_{scl} = 0.1$.
The curve seems to cross the x-axis near $x = 0.8$. To three decimal places, $x = 0.803$.

FOURTH ZERO Since this zero is between 1 and 2, we set $1 \leq x \leq 2$ with $x_{scl} = 0.1$ to obtain the graph in Figure 19.
The last zero is near 1.6. We find that, to three decimal places, $x \approx 1.626$.

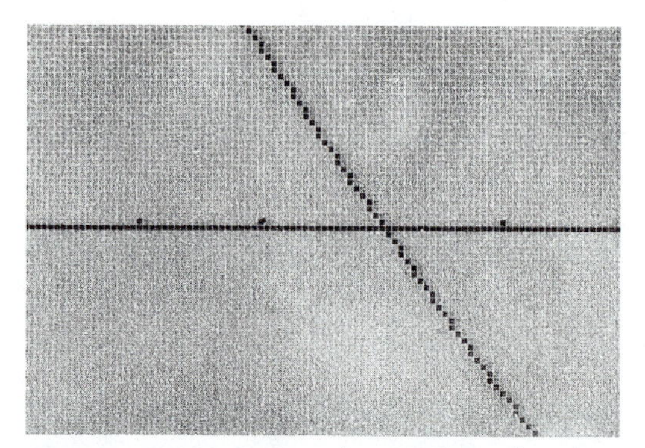

Figure 18 Graph of $y = 2x^4 + 1.5x^3 - 9x^2 - x + 5$, $0.5 \le x \le 1.0$, $x_{scl} = 0.1$

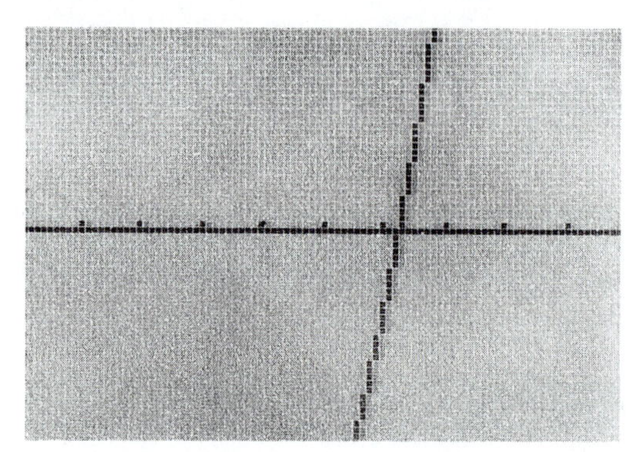

Figure 19 Graph of $y = 2x^4 + 1.5x^3 - 9x^2 - x + 5$, $1 \le x \le 2$, $x_{scl} = 0.1$

Thus, to three decimal places, the four zeros of $p(x) = 2x^4 + 1.5x^3 - 9x^2 - x + 5$ are -2.371, -0.807, 0.803, and 1.626. ■

EXAMPLE 8 *Finding Points of Intersection of Two Graphs*

Find, to two decimal place accuracy, all points of intersection of the graphs of $y = 2 - x$ and $y = \ln x$.

SOLUTION $\ln x$ is defined only when $x > 0$ (see page 377). Also, if $x > 2$, $2 - x < 0$ and $\ln x > \ln 2 \approx 0.6931 > 0$, so there are no points of intersection for $x > 2$. We use the following ranges and scales on our initial graphs:

x_{min}: 0	y_{min}: -2
x_{max}: 2	y_{max}: 2
x_{scl}: 0.5	y_{scl}: 1

Depending on the calculator, we can sketch both curves on the same screen in one of two ways.

ON SOME CALCULATORS You can enter several functions. When the "graph" key is pressed, the functions will be sketched, one after the other. (It is most common that four functions can be entered.)

ON OTHER CALCULATORS Simply sketch one function and do not clear it. Then, sketch the second function. As long as the range and scale values are not changed, both functions will appear on the same graph.
The graphs of both functions are given in Figure 20.

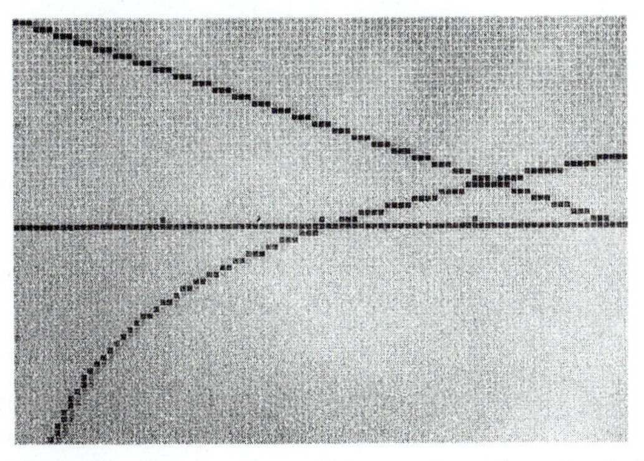

Figure 20 Graphs of $y = 2 - x$ and $y = \ln x$ for $0 < x \leq 2$, $-2 \leq y \leq 2$, $x_{scl} = 0.5$, $y_{scl} = 1$

It appears that the graphs intersect between 1.4 and 1.6 and probably closer to 1.6.

In Figure 21, we draw the graphs for x in the interval [1.5, 1.6]:

x_{min}: 1.5 y_{min}: 0.4 ($\ln 1.5 \approx 0.405$)
x_{max}: 1.6 y_{max}: 0.5 ($\ln 1.6 \approx 0.47$)
x_{scl}: 0.01 y_{scl}: 0.01

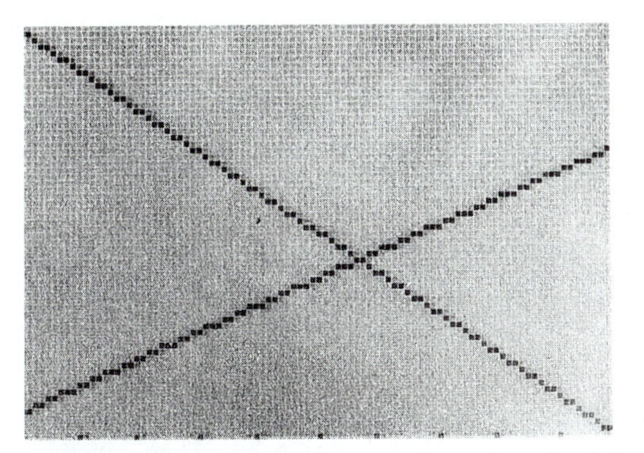

Figure 21 Graphs of $y = 2 - x$ and $y = \ln x$ for $1.5 \leq x \leq 1.6$, $0.4 \leq y \leq 0.5$, $x_{scl} = y_{scl} = 0.01$

The graphs seem to cross between 1.55 and 1.56. To be more precise, we sketch the graphs for x in the interval [1.55, 1.56] (see Figure 22).

x_{min}: 1.55 y_{min}: 0.43 ($\ln 1.55 \approx 0.438$)
x_{max}: 1.56 y_{max}: 0.45 ($\ln 1.56 \approx 0.445$)
x_{scl}: 0.001 y_{scl}: 0.002

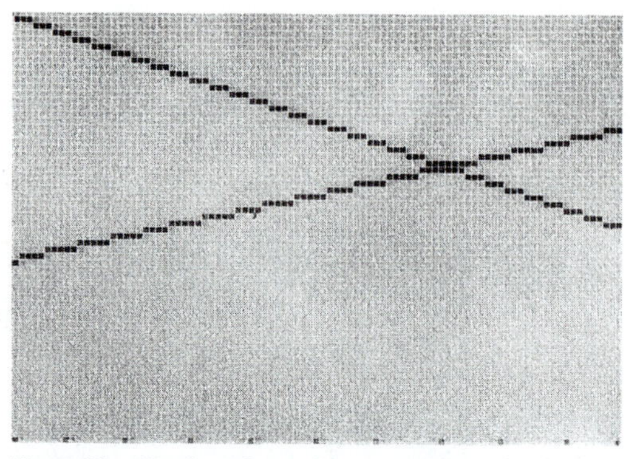

Figure 22 Graphs of $y = 2 - x$ and $y = \ln x$ for $1.55 \le x \le 1.56$, $0.43 \le y \le 0.45$, $x_{scl} = 0.001$, $y_{scl} = 0.002$

We see that the graphs intersect near the value 1.557. Therefore, to two decimal places, the graphs intersect at $x = 1.56$.

NOTE 1 Using a numerical method, we can show that, to 9 decimal places, the curves intersect at $x = 1.557145599$.

NOTE 2 This problem seemed more difficult than problems involving finding zeros because it is hard to see where the graphs intersect in relation to tick marks along the x-axis. This difficulty can be avoided in one of two ways: First, if your calculator has a trace function, and most do, then you can use it to find, approximately, the x- and y-coordinates of the point of intersection. If you do this, you don't have to worry about the tick marks.

Second, you can change the problem. If $2 - x = \ln x$, then

$$f(x) = 2 - x - \ln x = 0.$$

That is, you can change the problem into one of finding a zero of a function. This, as we have seen, is easier. ■

EXAMPLE 9 *Determining the Number of Points of Intersection of Two Graphs*

At how many points in the interval $[-2\pi, 2\pi]$ do the graphs of $y = \tan 2x$ and $y = \frac{1}{2}x$ intersect?

SOLUTION The range of x-values is given as $-2\pi \le x \le 2\pi \approx -6.28 \le x \le 6.28$. Then since $y = \frac{1}{2}x$ for one of the graphs, we have $-\pi \le y \le \pi$ or $-3.14 \le y \le 3.14$. We use scales of $\frac{\pi}{4} \approx 0.785$ on the x-axis and 1 on the y-axis.

We can plot both curves on the same screen as in Example 8. When we do so, we obtain the graphs in Figure 23. In Figure 23(a), the calculator has sketched the vertical asymptotes at $x = -\frac{7\pi}{4}, -\frac{5\pi}{4}, -\frac{3\pi}{4}, -\frac{\pi}{4}, \frac{\pi}{4}, \frac{3\pi}{4}$

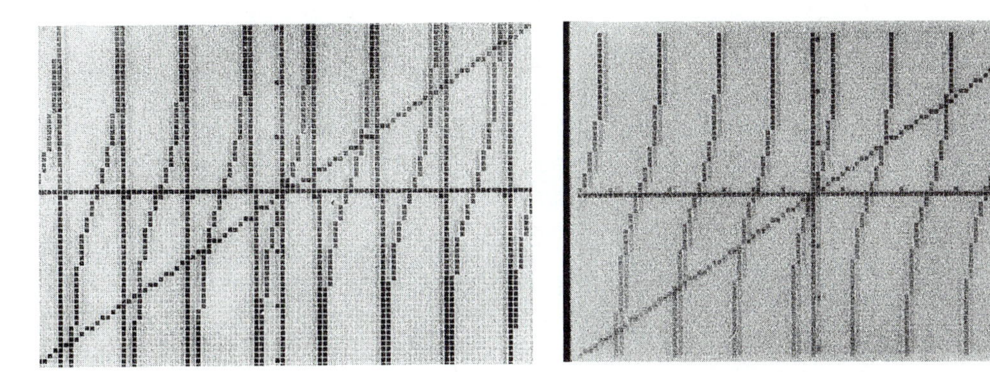

(a) Asymptotes sketched

(b) Asymptotes not sketched

Figure 23 Graphs of $y = \tan 2x$ and $y = \frac{1}{2}x$ showing seven points of intersection in the interval $[-2\pi, 2\pi]$

and $\frac{5\pi}{4}$. In Figure 23(b), the calculator omitted the asymptotes. In either case, we see that there are seven points of intersection in the interval $[-2\pi, 2\pi]$. ■

EXAMPLE 10 *Using Graphs to Solve a Quadratic Inequality*

Solve the quadratic inequality

$$4 + 4x - x^2 > 1 - 2x.$$

SOLUTION We observe that

$$f(x) > g(x) \text{ is equivalent to } f(x) - g(x) > 0.$$

It is simpler to solve the equivalent inequality

$$(4 + 4x - x^2) - (1 - 2x) > 0$$

or

$$4 + 4x - x^2 - 1 + 2x > 0$$

or

$$3 + 6x - x^2 > 0.$$

Before continuing, we observe that $y = f(x) > 0$ for precisely those values of x at which the graph of f lies above the x-axis.

We therefore graph $f(x) = 3 + 6x - x^2$ and see where it lies above the x-axis.

We use the range and scale values:

x_{min}: -2 $\quad$ y_{min}: -4

x_{max}: 10 $\quad$ y_{max}: 12

x_{scl}: 1 $\quad$ y_{scl}: 1

The graph is given in Figure 24.

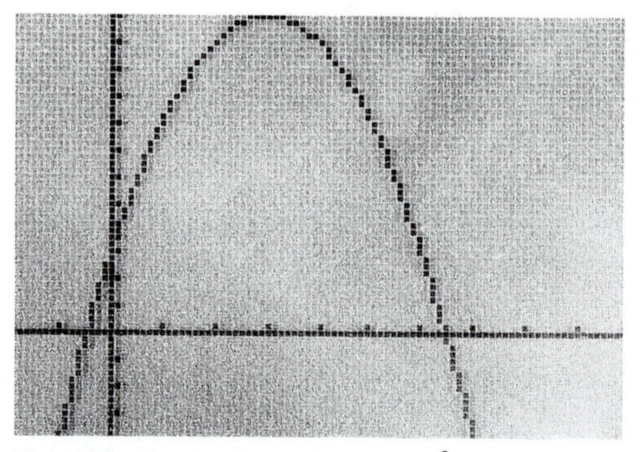

Figure 24 Graph of $y = 3 + 6x - x^2$ for $-2 \leq x \leq 10$, $-4 \leq y \leq 12$, $x_{\text{scl}} = y_{\text{scl}} = 1$

We see that $3 + 6x - x^2$ has two zeros, one between -1 and 0 and the other between 6 and 7. Moreover, $3 + 6x - x^2$ is above the x-axis between these two zeros.

We can find these zeros more precisely by using the method of Examples 6 and 7. To two decimal places, the zeros are -0.46 and 6.46. Thus the solution set of the inequality is, approximately,

$$-0.46 < x < 6.46.$$

We can check this answer by obtaining the points of intersection algebraically:

$$3 + 6x - x^2 = 0$$

$$x^2 - 6x - 3 = 0 \qquad \text{Multiply both sides by } -1$$

$$x = \frac{6 \pm \sqrt{36 - 4(1)(-3)}}{2} \qquad \text{Quadratic formula}$$

$$x = \frac{6 \pm \sqrt{48}}{2} = \frac{6 \pm 4\sqrt{3}}{2} = 3 \pm 2\sqrt{3}$$

The points of intersection are

$$x = 3 - 2\sqrt{3} \approx -0.464101615 \quad \text{and} \quad x \approx 3 + 2\sqrt{3} = 6.464101615.$$

This illustrates that the graphical technique is useful for obtaining an approximate solution to a problem. To obtain very accurate answers, other techniques may be more efficient.

III. Graphing Conic Sections

As long as your calculator can plot two functions on the same screen, you can use it to obtain the graphs of conic sections. We illustrate this first with a simple example.

EXAMPLE 11 *Graphing a Circle*

Graph the circle $x^2 + y^2 = 4$.

SOLUTION This is the circle of radius 2 centered at the origin. We cannot graph this directly since we must enter a function in the form $y = f(x)$.†

$$y^2 = 4 - x^2$$
$$y = \pm\sqrt{4 - x^2} \qquad (1)$$

Equation (1) is not the equation of a function. In fact, it is the equation of *two* functions:

$$y_1 = \sqrt{4 - x^2} \qquad \text{The upper half of the circle}$$
$$y_2 = -\sqrt{4 - x^2} \qquad \text{The lower half of the circle}$$

If we plot these two functions together, we obtain the entire circle. Using the range values $-2 \le x \le 2$ and $-2 \le y \le 2$ with $x_{scl} = y_{scl} = 1$, we obtain the graph in Figure 25.

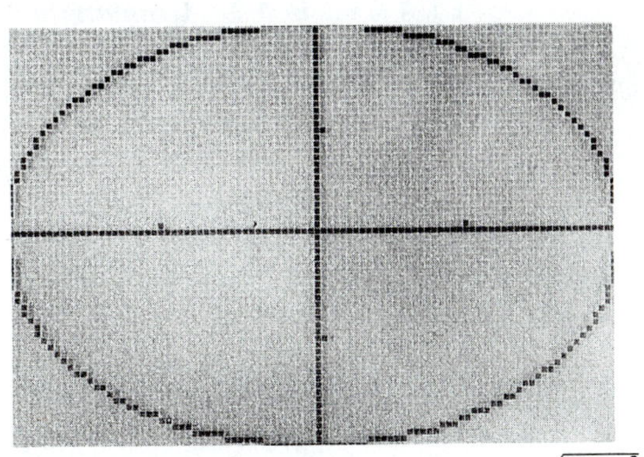

Figure 25 Graphs of the two functions $y_1 = \sqrt{4 - x^2}$ and $y_2 = -\sqrt{4 - x^2}$, $-2 \le x \le 2$, $-2 \le y \le 2$

This figure appears as an ellipse since the built-in scales on the x- and y-axes are not the same. (The rectangular screen is longer along the x-axis than along the y-axis.) There are two ways to fix this. The easier way can be used if your calculator has a function key labeled SQUARE. By using this function, the calculator will automatically rescale along the x-axis or the y-axis to make units along both axes have approximately the same length. (That is, a tick

† Some calculators can sketch graphs given in parametric form. On such a calculator, we could graph the circle in another way. We will not discuss this method here, but we discuss how to obtain polar graphs using a parametric form beginning on page A.25.

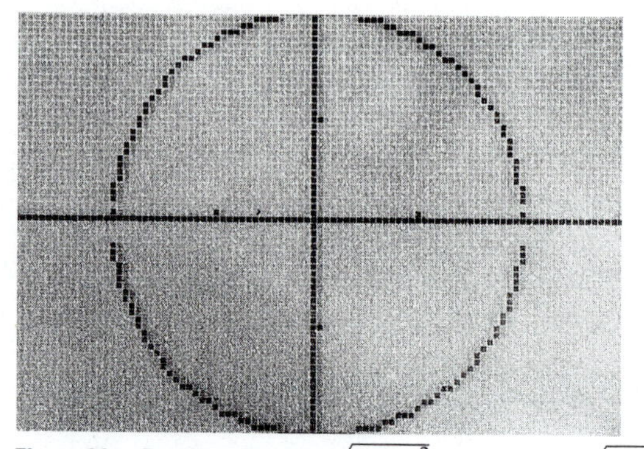

Figure 26 Graphs of $y_1 = \sqrt{4 - x^2}$ and $y_2 = -\sqrt{4 - x^2}$, rescaled

mark corresponding to a scale value of 1 will have approximately the same length on each axis.) When this is done, the graph appears as in Figure 26.

Alternatively, rescale the screen yourself. In this case, the circle in Figure 26 is obtained if the x-range is expanded to $-3 \le x \le 3$. ■

EXAMPLE 12 *Graphing a Hyperbola*

Graph the hyperbola

$$\frac{y^2}{9} - \frac{x^2}{16} = 1 \tag{2}$$

SOLUTION We discuss hyperbolas in Section 5.4. To graph this hyperbola, we first solve equation (2) for y:

$$\frac{y^2}{9} = 1 + \frac{x^2}{16}$$

$$y^2 = 9 + \frac{9}{16}x^2$$

$$y = \pm\sqrt{9 + \frac{9}{16}x^2}$$

Since $x^2 \ge 0$, $\sqrt{9 + \frac{9}{16}x^2} \ge \sqrt{9} = 3$. Thus either $y \ge 3$ or $y \le -3$. x can take on any real value. As in Example 11, we graph the two functions

$$y_1 = \sqrt{9 + \frac{9}{16}x^2} \quad \text{and} \quad y_2 = -\sqrt{9 + \frac{9}{16}x^2}$$

with range values, $-20 \le x \le 20$, $-6 \le y \le 6$, and $x_{\text{scl}} = y_{\text{scl}} = 1$. The graph is given in Figure 27.

If we "square" the graph as in Example 11, we obtain the more accurate sketch in Figure 28.

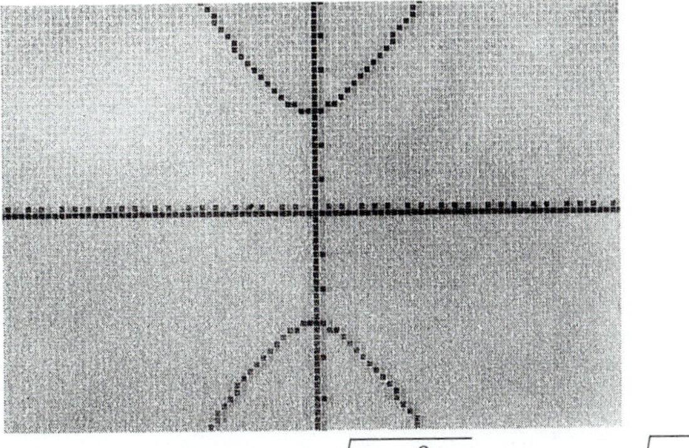

Figure 27 Graphs of $y_1 = \sqrt{9 + \frac{9}{16}x^2}$ and $y_2 = -\sqrt{9 + \frac{9}{16}x^2}$ for $-20 \le x \le 20,\ -6 \le y \le 6$

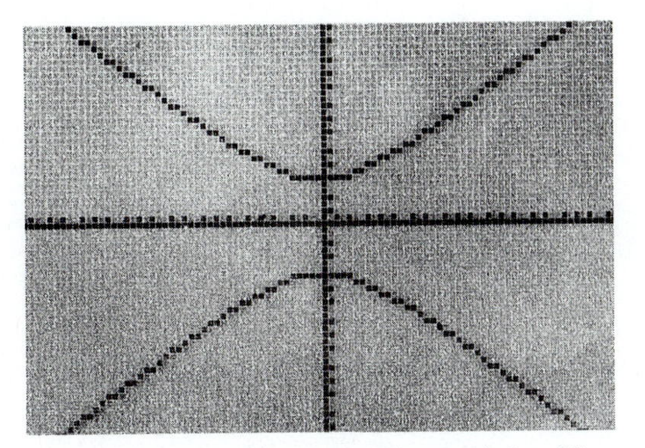

Figure 28 Graph of the hyperbola rescaled: $-20 \le x \le 20,\ -13.33 \le y \le 13.33$

EXAMPLE 13 *Graphing an Ellipse*

Sketch the graph of the second-degree equation

$$9x^2 + 36x + 4y^2 - 8y + 4 = 0. \tag{3}$$

SOLUTION We graphed this curve in Example 6 in Section 5.2 (see p. 285). To write equation (3) in a form that can be entered on a calculator, we first complete the squares.

$$\begin{aligned}
9x^2 + 36x + 4y^2 - 8y + 4 &= 0 \\
9(x^2 + 4x) + 4(y^2 - 2y) &= -4 \\
9(x^2 + 4x + 4) - 9 \cdot 4 + 4(y^2 - 2y + 1) - 4 \cdot 1 &= -4 \\
9(x + 2)^2 - 36 + 4(y - 1)^2 - 4 &= -4 \\
9(x + 2)^2 + 4(y - 1)^2 &= 36 \\
4(y - 1)^2 &= 36 - 9(x + 2)^2
\end{aligned}$$

$$(y-1)^2 = 9 - \frac{9}{4}(x+2)^2$$

$$y - 1 = \pm\sqrt{9 - \frac{9}{4}(x+2)^2}$$

$$y = 1 \pm \sqrt{9 - \frac{9}{4}(x+2)^2}$$

Before sketching these two curves, we observe that we must have

$$9 - \frac{9}{4}(x+2)^2 \geq 0$$

$$\frac{9}{4}(x+2)^2 \leq 9$$

$$(x+2)^2 \leq 4$$

$$|x+2| \leq 2$$

$$-2 \leq x + 2 \leq 2$$

$$-4 \leq x \leq 0$$

This is our range of values for x.

Also,

$$0 \leq (x+2)^2 \leq 4$$

If $(x+2)^2 = 4$, then $\sqrt{9 - \frac{9}{4}(x+2)^2} = 0$.
If $(x+2)^2 = 0$, then $\sqrt{9 - \frac{9}{4}(x+2)^2} = \sqrt{9} = 3$.
Thus

$$0 \leq \sqrt{9 + \frac{9}{4}(x+2)^2} \leq 3$$

so

$$1 \leq 1 + \sqrt{9 + \frac{9}{4}(x+2)^2} \leq 4$$

and

$$-2 \leq 1 - \sqrt{9 + \frac{9}{4}(x+2)^2} \leq 1.$$

Thus a suitable range of y-values is $-2 \leq y \leq 4$. Using these values, we obtain the ellipse in Figure 29.

Finally, if we square to equalize the spacing along the axes, we obtain the curve in Figure 30.

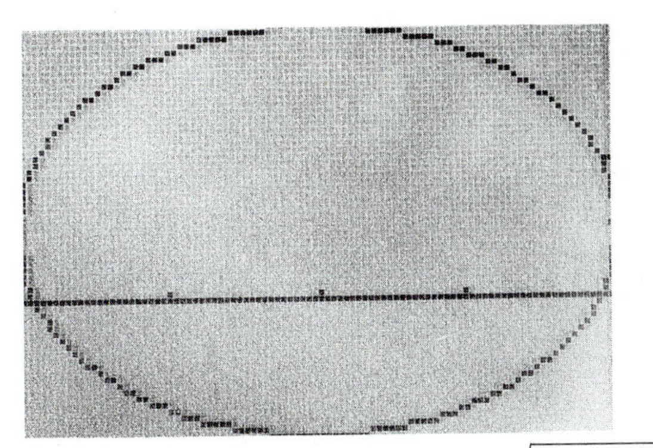

Figure 29 Graphs of the curves $y_1 = 1 + \sqrt{9 - \frac{9}{4}(x+2)^2}$ and $y_2 = 1 - \sqrt{9 - \frac{9}{4}(x+2)^2}$, $-4 \le x \le 0$, $-2 \le y \le 4$, $x_{scl} = y_{scl} = 1$

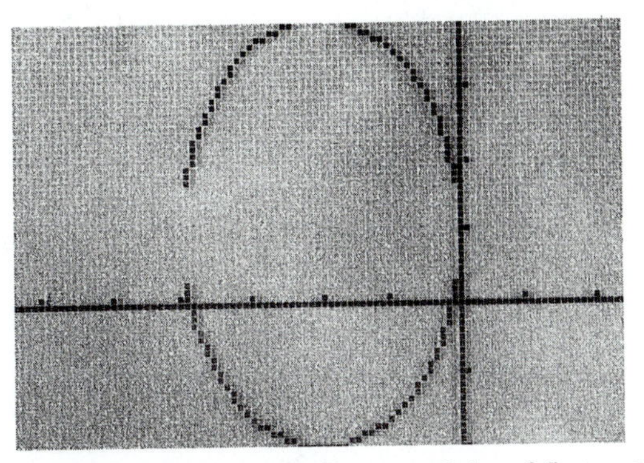

Figure 30 Graph of the ellipse rescaled: $-6.5 \le x \le 2.5$, $-2 \le y \le 4$

IV. Graphing Calculators Cannot Graph All Functions Accurately

We have seen that a remarkable amount of information can be obtained by graphing a function on a calculator. However, as our next example illustrates, graphing calculators do have their limitations.

EXAMPLE 14 *A Polynomial Whose Graph Cannot Be Obtained Accurately on a Graphing Calculator*

Graph the polynomial $p(x) = -\frac{x^5}{40} + x^4 + 3x^3 - 5x^2 - 10x + 4$, and find each of its zeros to one decimal place accuracy.

SOLUTION We set our initial range and scale values as follows:

$$x_{min} = -5 \qquad y_{min} = -15$$
$$x_{max} = 5 \qquad y_{max} = 15$$
$$x_{scl} = 1 \qquad y_{scl} = 5$$

The graph in Figure 31 is obtained.

We can see four zeros quite clearly. To one decimal place, they are -3.2, -1.7, 0.4, and 1.9.

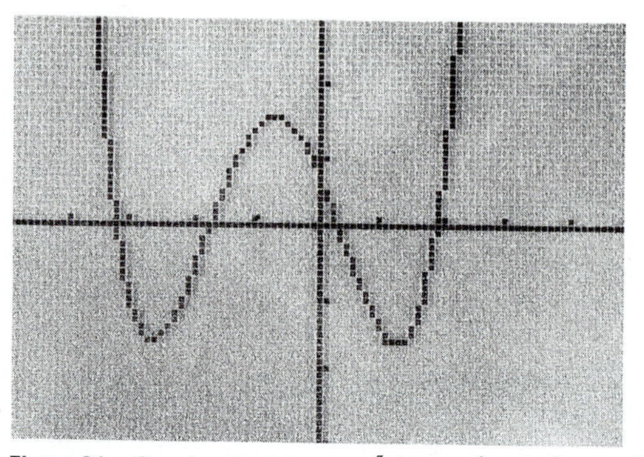

Figure 31 Graph of $p(x) = -x^5/40 + x^4 + 3x^3 - 5x^2 - 10x + 4$ for $-5 \leq x \leq 5$, $-15 \leq y \leq 15$

But something is clearly wrong. If x is positive and large, then the term $-x^5/40$ will dominate the other terms. For example, if $x = 100$, then $-x^5/40 = -250{,}000{,}000$ while $x^4 + 3x^3 - 5x^2 - 10x + 4 = 102{,}949{,}004$ so $p(100) = -147{,}050{,}996$. We see that the graph of $p(x)$ must eventually turn downward and become (and stay) negative as x increases. Thus $p(x)$ has a fifth zero. To find it, we change our range and scale values as follows (after some trial and error):

$$x_{\min} = -5 \qquad y_{\min} = -100{,}000$$
$$x_{\max} = 45 \qquad y_{\max} = 350{,}000$$
$$x_{\text{scl}} = 5 \qquad y_{\text{scl}} = 100{,}000$$

We obtain the sketch in Figure 32.

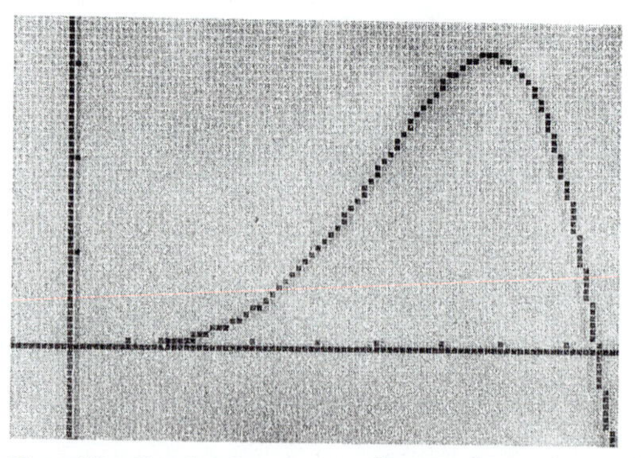

Figure 32 Graph of $p(x) = -x^5/40 + x^4 + 3x^3 - 5x^2 - 10x + 4$ for $-5 \leq x \leq 45$, $-100{,}000 \leq y \leq 350{,}000$, $x_{\text{scl}} = 5$, $y_{\text{scl}} = 100{,}000$

From this figure, we can determine that $p(x)$ reaches a maximum height of approximately 312,000 when x is near 34. It then decreases rapidly and is zero at $x = 42.7$ (rounded to 1 decimal place).

However, in obtaining a graph that showed the behavior of $p(x)$ for x greater than 30, we completely lost the picture obtained in Figure 31. The reason for this should be clear. In the graph in Figure 31, $p(x)$ reaches an approximate minimum value of -8 at $x \approx -3$ and $x \approx 1.3$ and a maximum value of approximately 8 at $x \approx -0.7$. But the number 8 is negligible compared to the maximum height of 312,000 at $x \approx 34$. It is simply impossible to scale the x- and y-axes to show the behavior for $-4 \leq x \leq 4$ and $30 \leq x \leq 50$ simultaneously on your calculator screen. If we show one, we lose the other. In fact, an accurate sketch drawn *to scale* that portrayed the graph of $p(x)$ accurately would have to be drawn on a very large piece of paper indeed. For example, if we used a scale of 1 mm = 1 unit (this is very small—there are 25 mm in an inch), then we would need a positive y-axis that extended 312,000 mm = 312 m $\approx$ 1024 ft, which is almost a fifth of a mile! And if we use a scale much smaller than 1 mm per unit, then, as in Figure 32, we simply would not see how $p(x)$ behaved for $-4 \leq x \leq 4$. The graph in this interval would appear as part of the x-axis.

This example illustrates the fact that a bit of thought is needed when using even the best available graphing device. You should, among other things, think about what happens to the graph of a function as $|x|$ gets large. It may be that no range setting will give you an accurate picture. ■

V. Graphing Polar Equations

In Section 4.8, we discussed the graphs of polar equations. It is possible to graph polar equations on a calculator that can draw parametric graphs.

An equation is given *parametrically* if it is written in the form

$$\begin{aligned} x &= f(t) \\ f &= g(t) \end{aligned} \qquad \text{for } a \leq t \leq b.$$

The number t is called a *parameter*. We will not discuss parametric equations in detail here. Rather, we will show how they can be used to obtain polar graphs.

Recall from equation (2) on page 265 that if the point (r, θ) is given in polar coordinates and (x, y) denotes the point in Cartesian coordinates, then

$$x = r \cos \theta \quad \text{and} \quad y = r \sin \theta. \tag{4}$$

Suppose we have the polar equation $r = f(\theta)$. Then, from (4), we obtain

$$x = r \cos \theta = f(\theta) \cos \theta$$

and

$$y = r \sin \theta = f(\theta) \sin \theta.$$

On calculators with parametric graphing capability, the parameter is usually denoted by T. Then, after setting the mode to parametric graphing, we can graph $r = f(\theta)$ by graphing the functions

$$\begin{aligned} x &= f(T) \cos T \\ y &= f(T) \sin T. \end{aligned} \tag{5}$$

In addition to x- and y-range and scale values, we also need to set T_{min}, T_{max}, and T_{step} values. The last setting tells the calculator how many points to plot. For example, if $T_{min} = 0$, $T_{max} = 1$, and $T_{step} = 0.1$, then the calculator will plot x- and y-values for $T = 0, 0.1, 0.2, \ldots, 0.9$, and 1.0. If $T_{step} = 0.2$, then it will plot values for $T = 0, 0.2, 0.4, 0.6, 0.8$, and 1.0.

If we are plotting a polar graph, then because $\cos\theta$ and $\sin\theta$ are periodic of period $2\pi \approx 6.28$, we will set $T_{min} = 0$ and $T_{max} = 6.28$ in our graphs. In addition, we will set $T_{step} = 0.1$. Finally, we set x_{min}, x_{max}, x_{scl}, y_{min}, y_{max}, and y_{scl} in the usual way.

In the three examples that follow, we will not discuss how to obtain range and scale values. These are obtained by the kind of thoughtful analysis that we have described for other graphing problems in this appendix.

EXAMPLE 15 *Graphing a Circle in Polar Coordinates*

Graph the circle $r = \cos\theta$.

SOLUTION We can show that $r = \cos\theta$ is the equation of a circle because, from (4),

$$x = r\cos\theta = \cos\theta\,(\cos\theta) = \cos^2\theta,\quad y = r\sin\theta = (\cos\theta)\sin\theta$$

and

$$x^2 + y^2 = \cos^4\theta + \cos^2\theta\sin^2\theta = \cos^2\theta\,(\cos^2\theta + \sin^2\theta)$$

$$\cos^2\theta + \sin^2\theta = 1$$
$$\downarrow$$
$$= \cos^2\theta = x$$

Then, completing the square,

$$x^2 - x - y^2 = 0$$

$$\left(x - \frac{1}{2}\right)^2 - \frac{1}{4} + y^2 = 0$$

$$\left(x - \frac{1}{2}\right)^2 + y^2 = \frac{1}{4}$$

This is the equation of a circle centered at $(\frac{1}{2}, 0)$ with radius $\frac{1}{2}$. To graph it, we set

$$T_{min} = 0 \qquad x_{min} = -1 \qquad y_{min} = -1$$
$$T_{max} = 6.28 \qquad x_{max} = 1 \qquad y_{max} = 1$$
$$T_{step} = 0.1 \qquad x_{scl} = 0.5 \qquad y_{scl} = 0.5$$
$$x(T) = \cos^2 T$$
$$y(T) = \cos T\sin T$$

The graph in Figure 33 is obtained.

As in Example 11, our circle doesn't look very circular. But if we use the squaring function in the calculator, we obtain the graph in Figure 34.

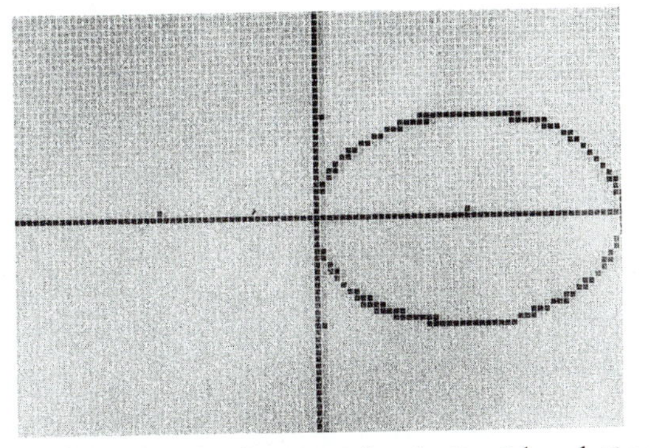

Figure 33 Graph of $r = \cos\theta$, $-1 \le x \le 1$, $-1 \le y \le 1$

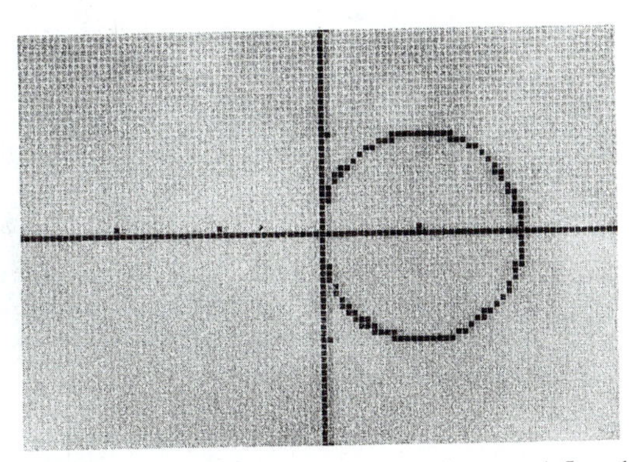

Figure 34 Graph of $r = \cos\theta$, $-1.5 \le x \le 1.5$, $-1 \le y \le 1$ ■

EXAMPLE 16 *Graphing a Limaçon with Loop*

Graph the polar equation $r = 1 + 2\sin\theta$.

SOLUTION We set the range and scale values as follows:

$$T_{\text{min}} = 0 \qquad x_{\text{min}} = -2 \qquad y_{\text{min}} = -1$$
$$T_{\text{max}} = 6.28 \qquad x_{\text{max}} = 2 \qquad y_{\text{max}} = 3$$
$$T_{\text{step}} = 0.1 \qquad x_{\text{scl}} = 1 \qquad y_{\text{scl}} = 1$$

Then we enter

$$x(T) = (1 + 2\sin T)\cos T$$
$$y(T) = (1 + 2\sin T)\sin T$$

The graph in Figure 35 is generated.

We recognize this as the graph of a limaçon with loop (see page 270). We get the more accurate graph in Figure 36 by using the squaring function.

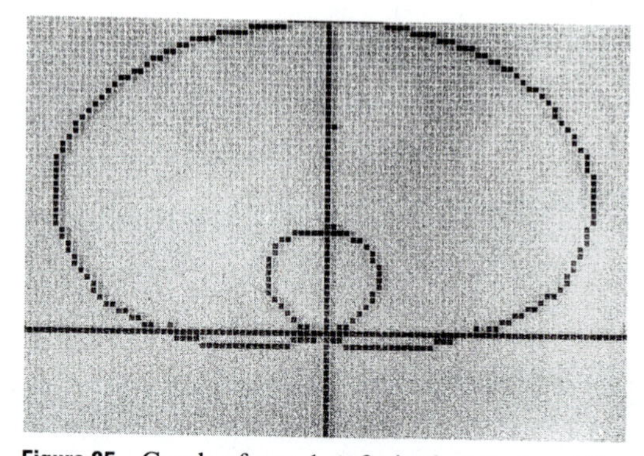

Figure 35 Graph of $r = 1 + 2 \sin \theta$, $-2 \le x \le 2$, $-1 \le y \le 3$

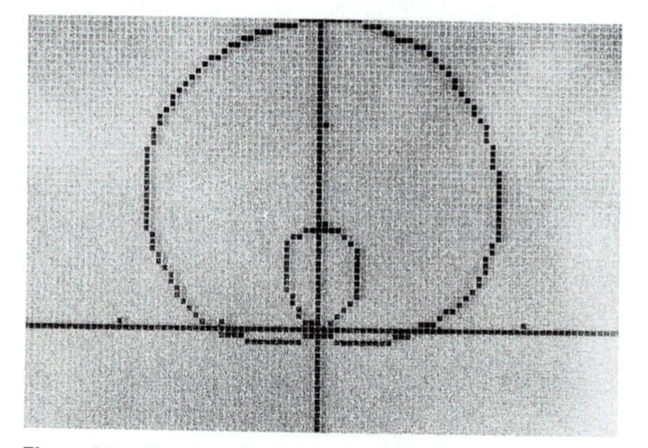

Figure 36 Graph of $r = 1 + 2 \cos \theta$, $-3 \le x \le 3$, $-1 \le y \le 3$ ■

EXAMPLE 17 *Graphing an Eight-Leafed Rose*

Graph the polar equation $r = 3 \cos 4\theta$.

SOLUTION We set the following range and scale values:

$$T_{\text{min}} = 0 \qquad x_{\text{min}} = -3 \qquad y_{\text{min}} = -3$$
$$T_{\text{max}} = 6.28 \qquad x_{\text{max}} = 3 \qquad y_{\text{max}} = 3$$
$$T_{\text{step}} = 0.1 \qquad x_{\text{scl}} = 1 \qquad y_{\text{scl}} = 1$$

We then enter

$$x(T) = 3 \cos 4T \cos T$$
$$y(T) = 3 \cos 4T \sin T$$

to obtain the graph in Figure 37.

The curve in Figure 37 is called an **eight-leafed rose.** By using the squaring function, we obtain the more accurate graph in Figure 38.

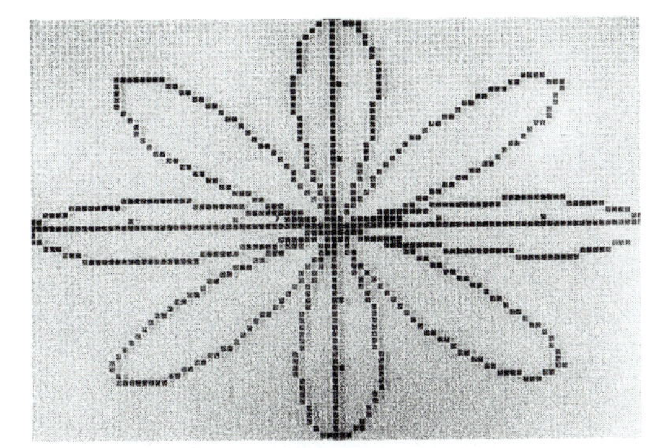

Figure 37 Graph of $r = 3 \cos 4\theta$, $-3 \le x \le 3$, $-3 \le y \le 3$

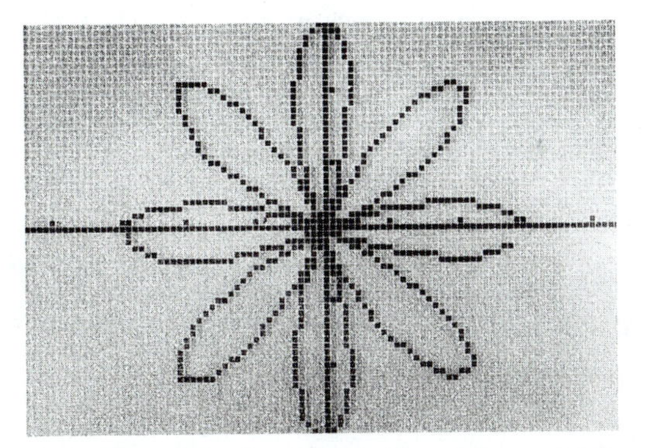

Figure 38 Graph of $r = 3 \cos 4\theta$, $-4.5 \le x \le 4.5$, $-3 \le y \le 3$

Appendix A Problems

In Problems 1–27 obtain sketches of each graph by first setting the appropriate calculator range and scale.

1. $y = x^2 + 2$
2. $y = -x^2 + 4x + 2$
3. $y = x^3 - x$
4. $y = 2x^3 - 3x^2 + 4x - 5$
5. $y = -x^4 + 2x^2 - x + 3$
6. $y = x^5 - x^4 - x^3 + x^2 + x - 2$
7. $y = \dfrac{x}{x+1}$
8. $y = \dfrac{x^2 - 3}{x}$
9. $y = \dfrac{x+2}{x-4}$
10. $y = \dfrac{x^3 - x^2 + 1}{x - 3}$
11. $y = \dfrac{x-2}{x^2 - 1}$
12. $y = e^{3x-2}$
13. $y = \ln(x^2 - 4)$
14. $3 - 2\ln(1 - x) + 4e^x$
15. $\dfrac{e^{-1/x}}{x+2}$
16. $\dfrac{1 - \ln x}{x^2 + 4}$
17. $e^{x - \ln x}$
18. $y = 1 - \sin x$
19. $y = 2 \sin x - 5 \cos 3x$
20. $y = \sin x + \sin 3x + \sin 5x$
21. $y = x \cos x$
22. $y = e^{-x} \cos x$
23. $y = e^{2x} \sin 3x$
24. $y = \tan^2 \dfrac{x}{3}$
25. $y = 3 \sec 5x$ $\left[\textit{Hint: } \sec x = \dfrac{1}{\cos x}\right]$
26. $y = \cot \dfrac{x}{2}$
27. $y = \csc(x - 3)$

In Problems 28–37 approximate, to two decimal place accuracy, all zeros of each polynomial by sketching appropriate graphs. [*Note:* To be sure that you have two decimal place accuracy, you should approximate each zero to three decimal places.]

28. $x^2 - 3x - 10$
29. $6x^2 + 5x - 56$

30. $x^3 + x^2 + 5$
31. $x^3 - x^2 - 1$
32. $3x^5 + 2x^4 + x^2 + 3$
33. $x^7 - x^2 - 12$
34. $6x^4 - 2x^3 + 3x^2 - 4x - 2$
35. $x^3 - 6x^2 - 15x + 4$
36. $x^3 + 14x^2 + 60x + 78$
37. $4x^4 - 4x^3 - 23x^2 + x + 10$

In Problems 38–43 find, by graphing, the number of solutions to each equation, and approximate each solution to two decimal place accuracy.

38. $\ln x = x - 4$
39. $e^{-x} = \ln x$
40. $\frac{x^2 - 4}{x^2 - 9} = \frac{x}{4}$
41. $x^2 - 3x + 5 = \ln(x^4 + x + 5)$
42. $\sin 2x = 1 - x^2$
43. $1 + 3\cos x = \sqrt{x + 5}$

In Problems 44–52 find an approximate solution to each inequality.

44. $2x + 3 > 5$
45. $|3x - 5| < 4$ [*Hint:* Use the Abs key or its equivalent on your calculator.]
46. $\left|\frac{x}{2} - 5\right| < 8$
47. $x^2 - 2 > x + 5$
48. $\frac{1}{x + 2} > \frac{x}{x + 1}$
49. $\frac{x}{x^2 - 3} < \frac{1}{2 - x}$
50. $2x^2 + 3x - 5 < 3 - 7x$
51. $x^3 - 2x^2 + 7x - 5 > 2x^2 - x + 4$
52. $3x^3 + 7 < 1 - 2x + x^4$

In Problems 53–67 sketch the graph of each conic section.

53. $x^2 + y^2 = 25$
54. $(x + 2)^2 + (y - 3)^2 = 12$
55. $x^2 + 2x + y^2 - 4y = 20$
56. $2x^2 + 3y^2 = 12$
57. $7x^2 - 11y^2 = 47$
58. $19y^2 - 43x^2 = 6$
59. $\frac{x^2}{37} + \frac{y^2}{121} = 1$
60. $7x^2 - 3y = 4$
61. $3y^2 + 12x = 6$
62. $\frac{x}{2} - \frac{y^2}{9} = 3$
63. $2x^2 + 7x + 5y^2 - 2y = 8$
64. $3x^2 + 6x + 9y - 7 = 0$
65. $2y^2 - 8y - 12x^2 + 12x = 16$
66. $17x^2 - 32x - 43y - 11y^2 = 33$
67. $\frac{1}{2}x^2 + 4x + \frac{1}{3}y^2 - 3y = 5$

In Problems 68–81 sketch the graph of each polar equation.

68. $r = 4$
69. $r = 3\sin\theta$
70. $r = 2\cos\theta - 2\sin\theta$
71. $r = 1 - 4\sin\theta$
72. $r = -2 - 5\cos\theta$
73. $r = 4 + 3\sin\theta$
74. $r = 4\cos 2\theta$
75. $r = 2\sin 3\theta$
76. $r = \theta$
77. $r = \frac{\theta}{3}$
78. $r^2 = 4\cos 2\theta$ [*Hint:* Graph $r = \sqrt{4\cos 2\theta}$ and $r = -\sqrt{4\cos 2\theta}$ separately.]
79. $r^2 = -9\sin 2\theta$
80. $r^2 = \theta$
81. $r = 1 + 2\sec\theta$

Table 1

Exponential Functions

x	e^x	e^{-x}	x	e^x	e^{-x}
0.00	1.0000	1.0000	3.0	20.086	0.0498
0.05	1.0513	0.9512	3.1	22.198	0.0450
0.10	1.1052	0.9048	3.2	24.533	0.0408
0.15	1.1618	0.8607	3.3	27.113	0.0369
0.20	1.2214	0.8187	3.4	29.964	0.0334
0.25	1.2840	0.7788	3.5	33.115	0.0302
0.30	1.3499	0.7408	3.6	36.598	0.0273
0.35	1.4191	0.7047	3.7	40.447	0.0247
0.40	1.4918	0.6703	3.8	44.701	0.0224
0.45	1.5683	0.6376	3.9	49.402	0.0202
0.50	1.6487	0.6065	4.0	54.598	0.0183
0.55	1.7333	0.5769	4.1	60.340	0.0166
0.60	1.8221	0.5488	4.2	66.686	0.0150
0.65	1.9155	0.5220	4.3	73.700	0.0136
0.70	2.0138	0.4966	4.4	81.451	0.0123
0.75	2.1170	0.4724	4.5	90.017	0.0111
0.80	2.2255	0.4493	4.6	99.484	0.0101
0.85	2.3396	0.4274	4.7	109.95	0.0091
0.90	2.4596	0.4066	4.8	121.51	0.0082
0.95	2.5857	0.3867	4.9	134.29	0.0074
1.0	2.7183	0.3679	5.0	148.41	0.0067
1.1	3.0042	0.3329	5.1	164.02	0.0061
1.2	3.3201	0.3012	5.2	181.27	0.0055
1.3	3.6693	0.2725	5.3	200.34	0.0050
1.4	4.0552	0.2466	5.4	221.41	0.0045
1.5	4.4817	0.2231	5.5	244.69	0.0041
1.6	4.9530	0.2019	5.6	270.43	0.0037
1.7	5.4739	0.1827	5.7	298.87	0.0033
1.8	6.0496	0.1653	5.8	330.30	0.0030
1.9	6.6859	0.1496	5.9	365.04	0.0027
2.0	7.3891	0.1353	6.0	403.43	0.0025
2.1	8.1662	0.1225	6.5	665.14	0.0015
2.2	9.0250	0.1108	7.0	1096.6	0.0009
2.3	9.9742	0.1003	7.5	1808.0	0.0006
2.4	11.023	0.0907	8.0	2981.0	0.0003
2.5	12.182	0.0821	8.5	4914.8	0.0002
2.6	13.464	0.0743	9.0	8103.1	0.0001
2.7	14.880	0.0672	9.5	13,360	0.00007
2.8	16.445	0.0608	10.0	22,026	0.00005
2.9	18.174	0.0550			

Table 2

Natural Logarithms

n	$\log_e n$	n	$\log_e n$	n	$\log_e n$
0.0	—	4.5	1.5041	9.0	2.1972
0.1	−2.3026	4.6	1.5261	9.1	2.2083
0.2	−1.6094	4.7	1.5476	9.2	2.2192
0.3	−1.2040	4.8	1.5686	9.3	2.2300
0.4	−0.9163	4.9	1.5892	9.4	2.2407
0.5	−0.6931	5.0	1.6094	9.5	2.2513
0.6	−0.5108	5.1	1.6292	9.6	2.2618
0.7	−0.3567	5.2	1.6487	9.7	2.2721
0.8	−0.2231	5.3	1.6677	9.8	2.2824
0.9	−0.1054	5.4	1.6864	9.9	2.2925
1.0	0.0000	5.5	1.7047	10	2.3026
1.1	0.0953	5.6	1.7228	11	2.3979
1.2	0.1823	5.7	1.7405	12	2.4849
1.3	0.2624	5.8	1.7579	13	2.5649
1.4	0.3365	5.9	1.7750	14	2.6391
1.5	0.4055	6.0	1.7918	15	2.7081
1.6	0.4700	6.1	1.8083	16	2.7726
1.7	0.5306	6.2	1.8245	17	2.8332
1.8	0.5878	6.3	1.8405	18	2.8904
1.9	0.6419	6.4	1.8563	19	2.9444
2.0	0.6931	6.5	1.8718	20	2.9957
2.1	0.7419	6.6	1.8871	25	3.2189
2.2	0.7885	6.7	1.9021	30	3.4012
2.3	0.8329	6.8	1.9169	35	3.5553
2.4	0.8755	6.9	1.9315	40	3.6889
2.5	0.9163	7.0	1.9459	45	3.8067
2.6	0.9555	7.1	1.9601	50	3.9120
2.7	0.9933	7.2	1.9741	55	4.0073
2.8	1.0296	7.3	1.9879	60	4.0943
2.9	1.0647	7.4	2.0015	65	4.1744
3.0	1.0986	7.5	2.0149	70	4.2485
3.1	1.1314	7.6	2.0281	75	4.3175
3.2	1.1632	7.7	2.0412	80	4.3820
3.3	1.1939	7.8	2.0541	85	4.4427
3.4	1.2238	7.9	2.0669	90	4.4998
3.5	1.2528	8.0	2.0794	95	4.5539
3.6	1.2809	8.1	2.0919	100	4.6052
3.7	1.3083	8.2	2.1041	200	5.2983
3.8	1.3350	8.3	2.1163	300	5.7038
3.9	1.3610	8.4	2.1282	400	5.9915
4.0	1.3863	8.5	2.1401	500	6.2146
4.1	1.4110	8.6	2.1518	600	6.3969
4.2	1.4351	8.7	2.1633	700	6.5511
4.3	1.4586	8.8	2.1748	800	6.6846
4.4	1.4816	8.9	2.1861	900	6.8024

Using the Common Logarithm Tables (Table 3)

Table 3 contains four-decimal-place approximations for common logarithms of numbers between 1.00 and 9.99 in intervals of 0.01. The use of these tables is illustrated in the following example.

Example

Approximate each of the following:

(a) log 33.4 (b) log 33,400 (c) log 0.0334

Solution

(a) Express the number in scientific notation

$$33.4 = 3.34 \times 10^1$$

Therefore, $\log 33.4 = \log (3.34 \times 10^1)$
$= \log 3.34 + \log 10^1$

To find the log of 3.34 in Table 3, look down the left-hand column to $n = 3.3$ and over to the right to $n = 4$. The log of 3.34 is shown as .5237. The log of 10^1 is, of course, 1.

Therefore, $\log 33.4 = \log 3.34 + \log 10^1$
$= .5237 \quad + 1$
$= 1.5237$

(b) $\log 33{,}400 = \log (3.34 \times 10^4)$
$= \log 3.34 \quad + \log 10^4$
$= .5237 \quad + 4$
$= 4.5237$

(c) $\log 0.0334 = \log 3.334 + \log 10^{-2}$
$= 0.5237 + (-2) = -1.4763$

Table 3

Common Logarithms

n	0	1	2	3	4	5	6	7	8	9
1.0	.0000	.0043	.0086	.0128	.0170	.0212	.0253	.0294	.0334	.0374
1.1	.0414	.0453	.0492	.0531	.0569	.0607	.0645	.0682	.0719	.0755
1.2	.0792	.0828	.0864	.0899	.0934	.0969	.1004	.1038	.1072	.1106
1.3	.1139	.1173	.1206	.1239	.1271	.1303	.1335	.1367	.1399	.1430
1.4	.1461	.1492	.1523	.1553	.1584	.1614	.1644	.1673	.1703	.1732
1.5	.1761	.1790	.1818	.1847	.1875	.1903	.1931	.1959	.1987	.2014
1.6	.2041	.2068	.2095	.2122	.2148	.2175	.2201	.2227	.2253	.2279
1.7	.2304	.2330	.2355	.2380	.2405	.2430	.2455	.2480	.2504	.2529
1.8	.2553	.2577	.2601	.2625	.2648	.2672	.2695	.2718	.2742	.2765
1.9	.2788	.2810	.2833	.2856	.2878	.2900	.2923	.2945	.2967	.2989
2.0	.3010	.3032	.3054	.3075	.3096	.3118	.3139	.3160	.3181	.3201
2.1	.3222	.3243	.3263	.3284	.3304	.3324	.3345	.3365	.3385	.3404
2.2	.3424	.3444	.3464	.3483	.3502	.3522	.3541	.3560	.3579	.3598
2.3	.3617	.3636	.3655	.3674	.3692	.3711	.3729	.3747	.3766	.3784
2.4	.3802	.3820	.3838	.3856	.3874	.3892	.3909	.3927	.3945	.3962
2.5	.3979	.3997	.4014	.4031	.4048	.4065	.4082	.4099	.4116	.4133
2.6	.4150	.4166	.4183	.4200	.4216	.4232	.4249	.4265	.4281	.4298
2.7	.4314	.4330	.4346	.4362	.4378	.4393	.4409	.4425	.4440	.4456
2.8	.4472	.4487	.4502	.4518	.4533	.4548	.4564	.4579	.4594	.4609
2.9	.4624	.4639	.4654	.4669	.4683	.4698	.4713	.4728	.4742	.4757
3.0	.4771	.4786	.4800	.4814	.4829	.4843	.4857	.4871	.4886	.4900
3.1	.4914	.4928	.4942	.4955	.4969	.4983	.4997	.5011	.5024	.5038
3.2	.5051	.5065	.5079	.5092	.5105	.5119	.5132	.5145	.5159	.5172
3.3	.5185	.5198	.5211	.5224	.5237	.5250	.5263	.5276	.5289	.5302
3.4	.5315	.5328	.5340	.5353	.5366	.5378	.5391	.5403	.5416	.5428
3.5	.5441	.5453	.5465	.5478	.5490	.5502	.5514	.5527	.5539	.5551
3.6	.5563	.5575	.5587	.5599	.5611	.5623	.5635	.5647	.5658	.5670
3.7	.5682	.5694	.5705	.5717	.5729	.5740	.5752	.5763	.5775	.5786
3.8	.5798	.5809	.5821	.5832	.5843	.5855	.5866	.5877	.5888	.5899
3.9	.5911	.5922	.5933	.5944	.5955	.5966	.5977	.5988	.5999	.6010
4.0	.6021	.6031	.6042	.6053	.6064	.6075	.6085	.6096	.6107	.6117
4.1	.6128	.6138	.6149	.6160	.6170	.6180	.6191	.6201	.6212	.6222
4.2	.6232	.6243	.6253	.6263	.6274	.6284	.6294	.6304	.6314	.6325
4.3	.6335	.6345	.6355	.6365	.6375	.6385	.6395	.6405	.6415	.6425
4.4	.6435	.6444	.6454	.6464	.6474	.6484	.6493	.6503	.6513	.6522
4.5	.6532	.6542	.6551	.6561	.6571	.6580	.6590	.6599	.6609	.6618
4.6	.6628	.6637	.6646	.6656	.6665	.6675	.6684	.6693	.6702	.6712
4.7	.6721	.6730	.6739	.6749	.6758	.6767	.6776	.6785	.6794	.6803
4.8	.6812	.6821	.6830	.6839	.6848	.6857	.6866	.6875	.6884	.6893
4.9	.6902	.6911	.6920	.6928	.6937	.6946	.6955	.6964	.6972	.6981
5.0	.6990	.6998	.7007	.7016	.7024	.7033	.7042	.7050	.7059	.7067
5.1	.7076	.7084	.7093	.7101	.7110	.7118	.7126	.7135	.7143	.7152
5.2	.7160	.7168	.7177	.7185	.7193	.7202	.7210	.7218	.7226	.7235
5.3	.7243	.7251	.7259	.7267	.7275	.7284	.7292	.7300	.7308	.7316
5.4	.7324	.7332	.7340	.7348	.7356	.7364	.7372	.7380	.7388	.7396

Common Logarithms

n	0	1	2	3	4	5	6	7	8	9
5.5	.7404	.7412	.7419	.7427	.7435	.7443	.7451	.7459	.7466	.7474
5.6	.7482	.7490	.7497	.7505	.7513	.7520	.7528	.7536	.7543	.7551
5.7	.7559	.7566	.7574	.7582	.7589	.7597	.7604	.7612	.7619	.7627
5.8	.7634	.7642	.7649	.7657	.7664	.7672	.7679	.7686	.7694	.7701
5.9	.7709	.7716	.7723	.7731	.7738	.7745	.7752	.7760	.7767	.7774
6.0	.7782	.7789	.7796	.7803	.7810	.7818	.7825	.7832	.7839	.7846
6.1	.7853	.7860	.7868	.7875	.7882	.7889	.7896	.7903	.7910	.7917
6.2	.7924	.7931	.7938	.7945	.7952	.7959	.7966	.7973	.7980	.7987
6.3	.7993	.8000	.8007	.8014	.8021	.8028	.8035	.8041	.8048	.8055
6.4	.8062	.8069	.8075	.8082	.8089	.8096	.8102	.8109	.8116	.8122
6.5	.8129	.8136	.8142	.8149	.8156	.8162	.8169	.8176	.8182	.8189
6.6	.8195	.8202	.8209	.8215	.8222	.8228	.8235	.8241	.8248	.8254
6.7	.8261	.8267	.8274	.8280	.8287	.8293	.8299	.8306	.8312	.8319
6.8	.8325	.8331	.8338	.8344	.8351	.8357	.8363	.8370	.8376	.8382
6.9	.8388	.8395	.8401	.8407	.8414	.8420	.8426	.8432	.8439	.8445
7.0	.8451	.8457	.8463	.8470	.8476	.8482	.8488	.8494	.8500	.8506
7.1	.8513	.8519	.8525	.8531	.8537	.8543	.8549	.8555	.8561	.8567
7.2	.8573	.8579	.8585	.8591	.8597	.8603	.8609	.8615	.8621	.8627
7.3	.8633	.8639	.8645	.8651	.8657	.8663	.8669	.8675	.8681	.8686
7.4	.8692	.8698	.8704	.8710	.8716	.8722	.8727	.8733	.8739	.8745
7.5	.8751	.8756	.8762	.8768	.8774	.8779	.8785	.8791	.8797	.8802
7.6	.8808	.8814	.8820	.8825	.8831	.8837	.8842	.8848	.8854	.8859
7.7	.8865	.8871	.8876	.8882	.8887	.8893	.8899	.8904	.8910	.8915
7.8	.8921	.8927	.8932	.8938	.8943	.8949	.8954	.8960	.8965	.8971
7.9	.8976	.8982	.8987	.8993	.8998	.9004	.9009	.9015	.9020	.9025
8.0	.9031	.9036	.9042	.9047	.9053	.9058	.9063	.9069	.9074	.9079
8.1	.9085	.9090	.9096	.9101	.9106	.9112	.9117	.9122	.9128	.9133
8.2	.9138	.9143	.9149	.9154	.9159	.9165	.9170	.9175	.9180	.9186
8.3	.9191	.9196	.9201	.9206	.9212	.9217	.9222	.9227	.9232	.9238
8.4	.9243	.9248	.9253	.9258	.9263	.9269	.9274	.9279	.9284	.9289
8.5	.9294	.9299	.9304	.9309	.9315	.9320	.9325	.9330	.9335	.9340
8.6	.9345	.9350	.9355	.9360	.9365	.9370	.9375	.9380	.9385	.9390
8.7	.9395	.9400	.9405	.9410	.9415	.9420	.9425	.9430	.9435	.9440
8.8	.9445	.9450	.9455	.9460	.9465	.9469	.9474	.9479	.9484	.9489
8.9	.9494	.9499	.9504	.9509	.9513	.9518	.9523	.9528	.9533	.9538
9.0	.9542	.9547	.9552	.9557	.9562	.9566	.9571	.9576	.9581	.9586
9.1	.9590	.9595	.9600	.9605	.9609	.9614	.9619	.9624	.9628	.9633
9.2	.9638	.9643	.9647	.9652	.9657	.9661	.9666	.9671	.9675	.9680
9.3	.9685	.9689	.9694	.9699	.9703	.9708	.9713	.9717	.9722	.9727
9.4	.9731	.9736	.9741	.9745	.9750	.9754	.9759	.9763	.9768	.9773
9.5	.9777	.9782	.9786	.9791	.9795	.9800	.9805	.9809	.9814	.9818
9.6	.9823	.9827	.9832	.9836	.9841	.9845	.9850	.9854	.9859	.9863
9.7	.9868	.9872	.9877	.9881	.9886	.9890	.9894	.9899	.9903	.9908
9.8	.9912	.9917	.9921	.9926	.9930	.9934	.9939	.9943	.9948	.9952
9.9	.9956	.9961	.9965	.9969	.9974	.9978	.9983	.9987	.9991	.9996

Table 4

Trigonometric Functions

Degrees	Radians	Sin	Tan	Cot	Cos		
0	0	0	0	—	1.0000	1.5708	90
1	0.0175	0.0175	0.0175	57.290	0.9998	1.5533	89
2	0.0349	0.0349	0.0349	28.636	0.9994	1.5359	88
3	0.0524	0.0523	0.0524	19.081	0.9986	1.5184	87
4	0.0698	0.0698	0.0699	14.301	0.9976	1.5010	86
5	0.0873	0.0872	0.0875	11.430	0.9962	1.4835	85
6	0.1047	0.1045	0.1051	9.5144	0.9945	1.4661	84
7	0.1222	0.1219	0.1228	8.1443	0.9925	1.4486	83
8	0.1396	0.1392	0.1405	7.1154	0.9903	1.4312	82
9	0.1571	0.1564	0.1584	6.3138	0.9877	1.4137	81
10	0.1745	0.1736	0.1763	5.6713	0.9848	1.3963	80
11	0.1920	0.1908	0.1944	5.1446	0.9816	1.3788	79
12	0.2094	0.2079	0.2126	4.7046	0.9781	1.3614	78
13	0.2269	0.2250	0.2309	4.3315	0.9744	1.3439	77
14	0.2443	0.2419	0.2493	4.0108	0.9703	1.3265	76
15	0.2618	0.2588	0.2679	3.7321	0.9659	1.3090	75
16	0.2793	0.2756	0.2867	3.4874	0.9613	1.2915	74
17	0.2967	0.2924	0.3057	3.2709	0.9563	1.2741	73
18	0.3142	0.3090	0.3249	3.0777	0.9511	1.2566	72
19	0.3316	0.3256	0.3443	2.9042	0.9455	1.2392	71
20	0.3491	0.3420	0.3640	2.7475	0.9397	1.2217	70
21	0.3665	0.3584	0.3839	2.6051	0.9336	1.2043	69
22	0.3840	0.3746	0.4040	2.4751	0.9272	1.1868	68
23	0.4014	0.3907	0.4245	2.3559	0.9205	1.1694	67
24	0.4189	0.4067	0.4452	2.2460	0.9135	1.1519	66
25	0.4363	0.4226	0.4663	2.1445	0.9063	1.1345	65
26	0.4538	0.4384	0.4877	2.0503	0.8988	1.1170	64
27	0.4712	0.4540	0.5095	1.9626	0.8910	1.0996	63
28	0.4887	0.4695	0.5317	1.8807	0.8829	1.0821	62
29	0.5061	0.4848	0.5543	1.8040	0.8746	1.0647	61
30	0.5236	0.5000	0.5774	1.7321	0.8660	1.0472	60
31	0.5411	0.5150	0.6009	1.6643	0.8572	1.0297	59
32	0.5585	0.5299	0.6249	1.6003	0.8480	1.0123	58
33	0.5760	0.5446	0.6494	1.5399	0.8387	0.9948	57
34	0.5934	0.5592	0.6745	1.4826	0.8290	0.9774	56
35	0.6109	0.5736	0.7002	1.4281	0.8192	0.9599	55
36	0.6283	0.5878	0.7265	1.3764	0.8090	0.9425	54
37	0.6458	0.6018	0.7536	1.3270	0.7986	0.9250	53
38	0.6632	0.6157	0.7813	1.2799	0.7880	0.9076	52
39	0.6807	0.6293	0.8098	1.2349	0.7771	0.8901	51
40	0.6981	0.6428	0.8391	1.1918	0.7660	0.8727	50
41	0.7156	0.6561	0.8693	1.1504	0.7547	0.8552	49
42	0.7330	0.6691	0.9004	1.1106	0.7431	0.8378	48
43	0.7505	0.6820	0.9325	1.0724	0.7314	0.8203	47
44	0.7679	0.6947	0.9657	1.0355	0.7193	0.8029	46
45	0.7854	0.7071	1.0000	1.0000	0.7071	0.7854	45
		Cos	Cot	Tan	Sin	Radians	Degrees

Answers to Selected Odd-Numbered Problems

Chapter 1

Problems 1.1, page 8

1. (1, 2) **3.** (0, 8) **5.** [−2, 0] **7.** (0, 5]
9. [−1.32, 4.16) **11.** $(-\infty, 0)$ **13.** $(-\infty, 3]$
15. $(-\infty, -5)$ **17.** (2, 17) **19.** [0, 4)
21. $(0, \infty)$ **23.** $(-\infty, -5]$ **25.** (−2.3, −0.2]
27. 4 **29.** −5 **31.** $\pi - 2$

Problems 1.2, page 14

1. IV

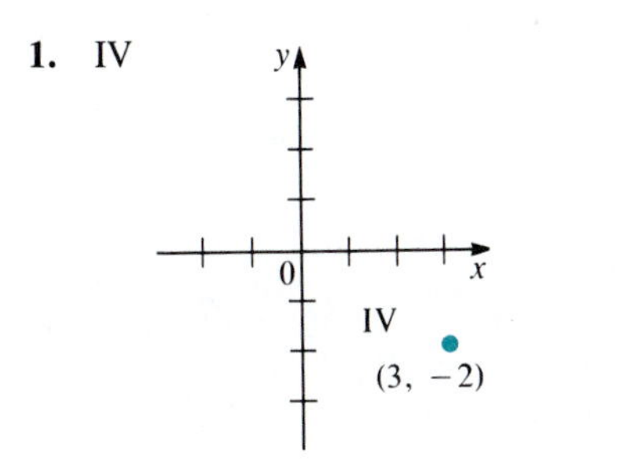

3. on x-axis

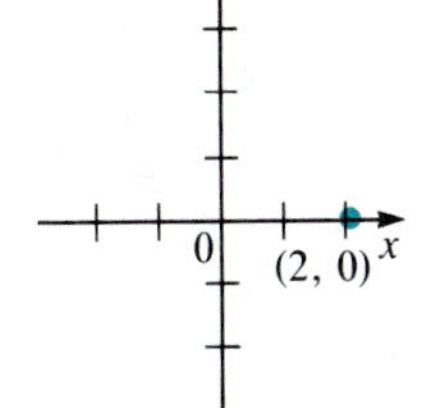

5. III

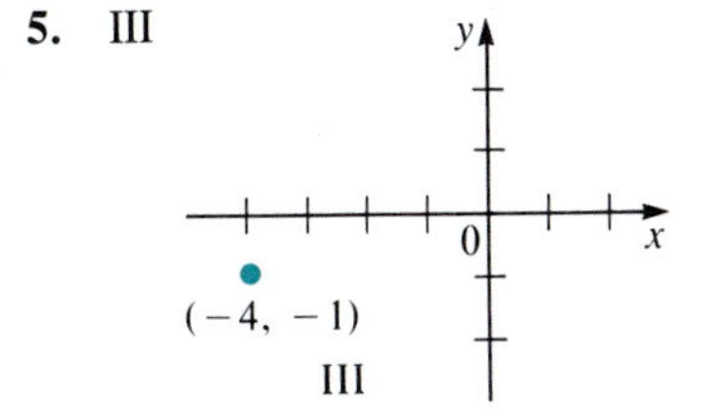

7. I

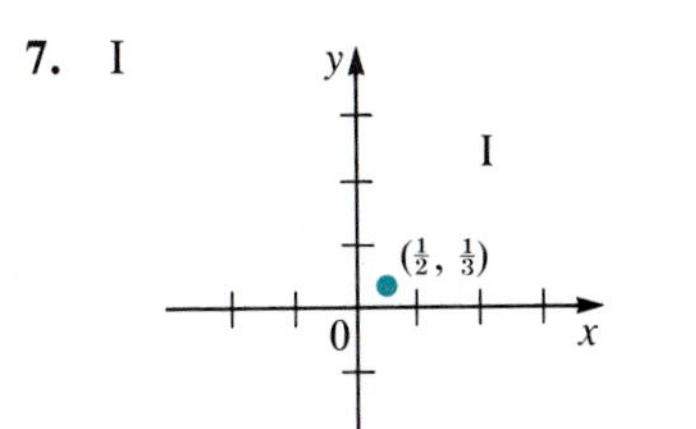

9. on y-axis

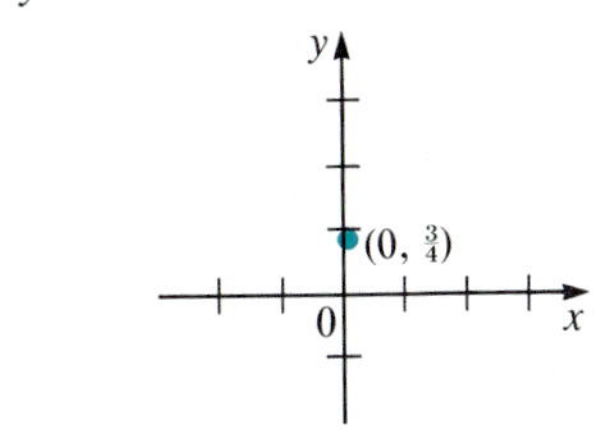

11. 5 **13.** $\sqrt{202}$ **15.** $\sqrt{29}$ **17.** $\sqrt{c^2 + d^2}$
19. $\sqrt{70.5805} \approx 8.40$
21. $x^2 + (y - 2)^2 = 1$

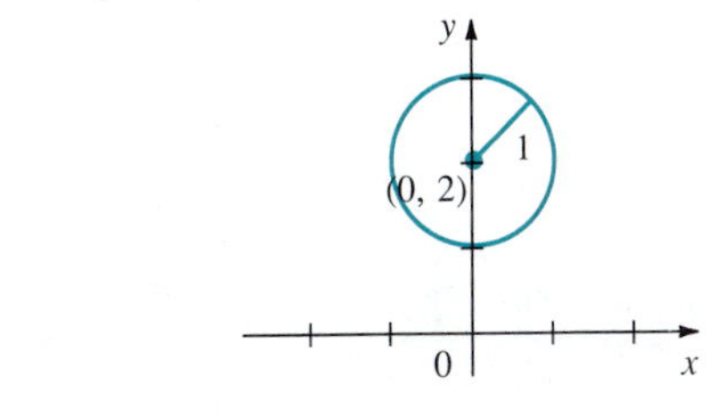

23. $(x - 1)^2 + (y - 1)^2 = 2$

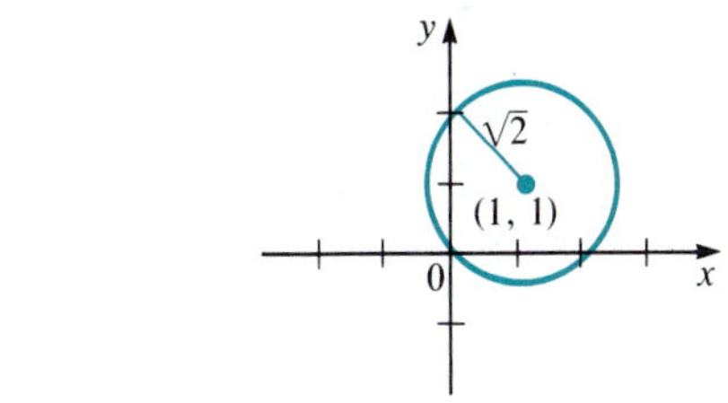

25. $(x + 1)^2 + (y - 4)^2 = 25$

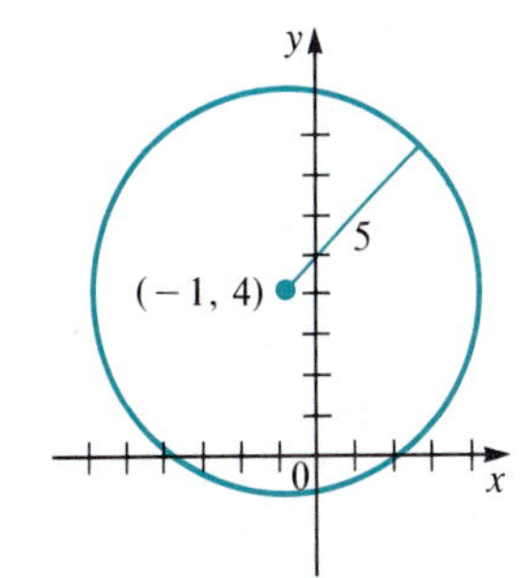

27. $(x - \pi)^2 + (y - 2\pi)^2 = \pi$

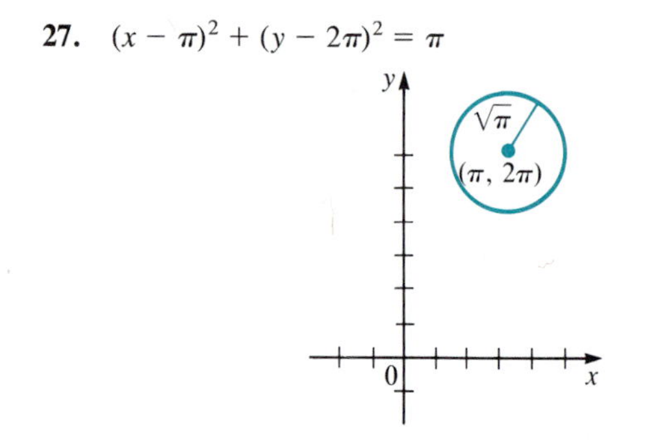

29. $(x - 3)^2 + (y + 2)^2 = 16$

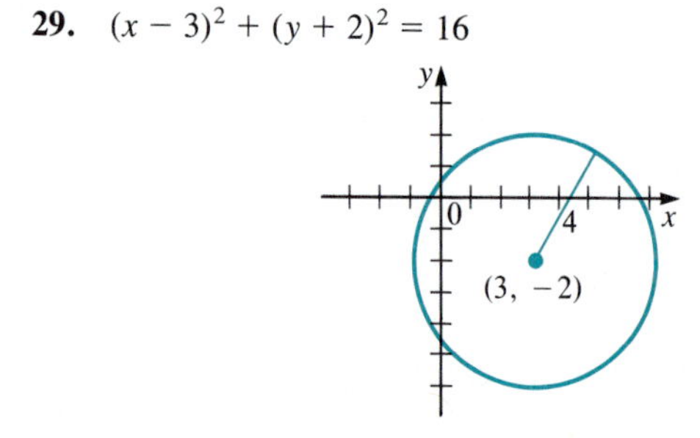

31. center (0, 3), radius $\sqrt{6}$
33. center $(-\frac{1}{2}, \frac{1}{4})$, radius = $\sqrt{66}/4$
35. $(x - 3)^2 + (y + 7)^2 = 34$

Problems 1.3, page 24

1. $1, \dfrac{1}{2}, -1, -\dfrac{1}{4}, \dfrac{1}{1+x^2}, \dfrac{1}{1+\sqrt{x}} = \dfrac{1-\sqrt{x}}{1-x}$
3. $-1, 7, 17, -\frac{1}{2}, 2w - 1, 2w^{10} - 1$
5. $0, 16, 16, 25, s^{4/5}, s^4 - 4s^3 + 6s^2 - 4s + 1$
7. $1, 0, 2, 2\sqrt{2}, n^{3/2}, \sqrt{\dfrac{1+w}{w}}$
9. $1, 3, 91, 31, n^4 - n^2 + 1, \dfrac{1}{n^6} - \dfrac{1}{n^3} + 1$
11. function
13. not a function because $g(a)$ is not unique
15. function
17. not a function with domain D because it is not defined at w
19. function **21.** yes **23.** yes **25.** no
27. yes **29.** yes **31.** yes
33. $x^n = a$ has two solutions if n is even **35.** yes
37. yes **39.** yes **41.** no
43. no (the y-axis passes through two points on the graph)
45. domain: all real numbers; range: all real numbers
47. domain: all real numbers except 0; range: $(0, \infty)$
49. domain: all real numbers except -1; range: all real numbers except 0
51. domain: $[1, \infty)$; range: $[0, \infty)$
53. domain: all real numbers except 0; range: $(0, \infty)$
55. domain: all real numbers; range: $[0, \infty)$
57. domain: all real numbers; range: $[0, \infty)$
59. domain: all real numbers; range: all integers
61. 1.419936, 54.372225, −1676.384304
63. −0.980796045, 0.227833604
65. $x^3 + 3x^2\Delta x + 3x(\Delta x)^2 + (\Delta x)^3$, $3x^2 + 3x\Delta x + (\Delta x)^2$
67. domain: all real numbers except 0; range: $\{1, -1\}$
69. $A(W) = 25W - W^2$; domain: $(0, 25)$; range: $\left(0, \dfrac{625}{4}\right)$
71.
$$d(t) = \begin{cases} 30\sqrt{t^2 - 6t + 18}, & 0 \le t < 3 \\ 180 - 30t, & 3 \le t \le 6 \\ 30t - 180, & 6 < t \le 9 \\ 30\sqrt{t^2 - 18t + 90}, & 9 < t \le 12 \end{cases}$$
73. (a) 1183 (b) 1187 (c) 1203 (d) 1248 (e) 1204
75. even **77.** even **79.** neither **81.** even
83. even **85.** even **87.** odd **89.** odd
91. neither

Problems 1.4, page 36

1.

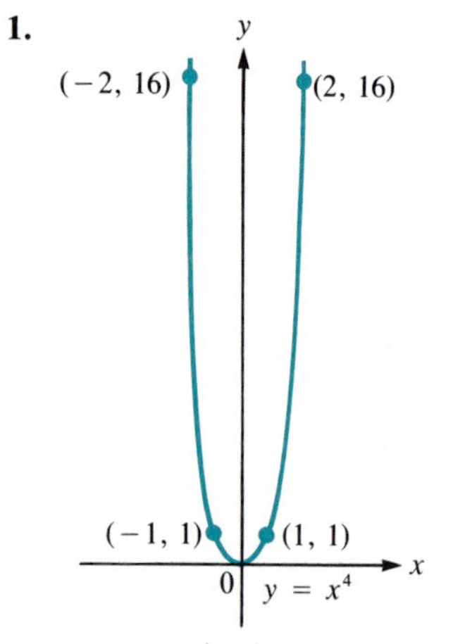

symmetric about y-axis

3.

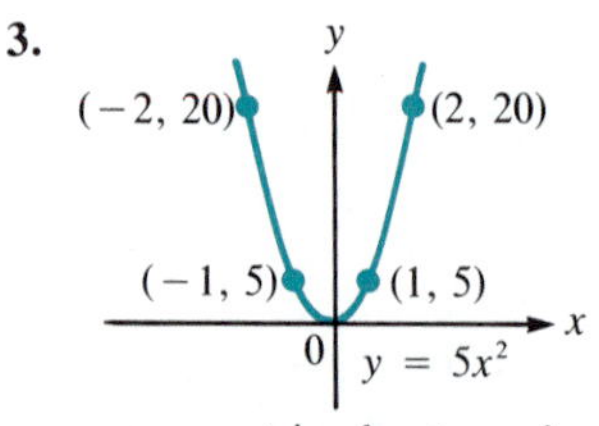

symmetric about y-axis

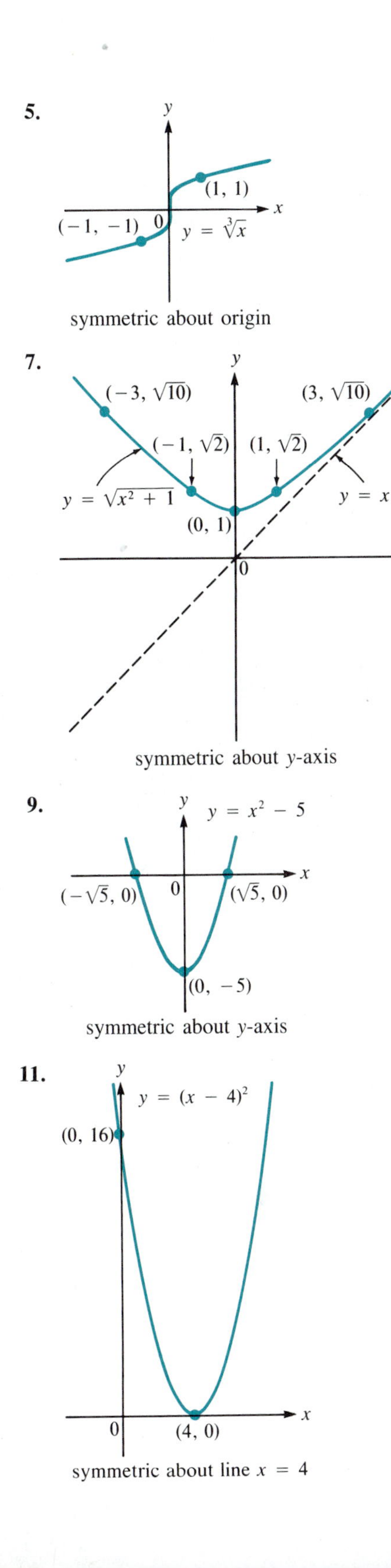
5.
y
(1, 1)
x
(−1, −1)
0
$y = \sqrt[3]{x}$
symmetric about origin
7.
y
$(-3, \sqrt{10})$
$(3, \sqrt{10})$
$(-1, \sqrt{2})$
$(1, \sqrt{2})$
$y = \sqrt{x^2 + 1}$
$y = x$
(0, 1)
0
x
symmetric about y-axis
9.
y
$y = x^2 - 5$
x
$(-\sqrt{5}, 0)$
0
$(\sqrt{5}, 0)$
(0, −5)
symmetric about y-axis
11.
y
$y = (x - 4)^2$
(0, 16)
x
0
(4, 0)
symmetric about line x = 4

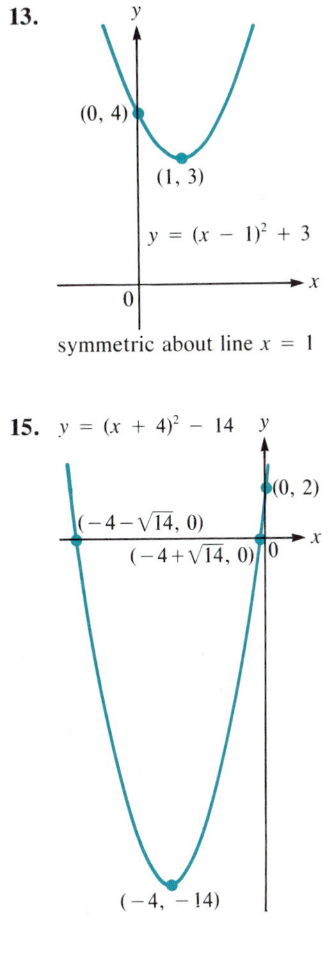
13.
y
(0, 4)
(1, 3)
$y = (x - 1)^2 + 3$
x
0
symmetric about line x = 1
15.
$y = (x + 4)^2 - 14$
y
(0, 2)
$(-4-\sqrt{14}, 0)$
x
$(-4+\sqrt{14}, 0)$
0
(−4, −14)

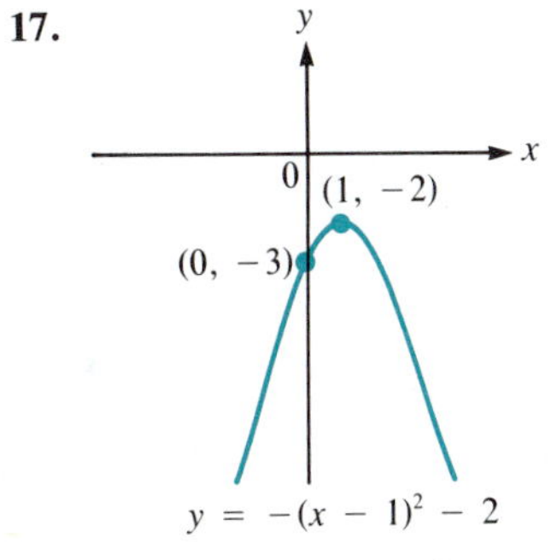
17.
y
x
0
(1, −2)
(0, −3)
$y = -(x - 1)^2 - 2$

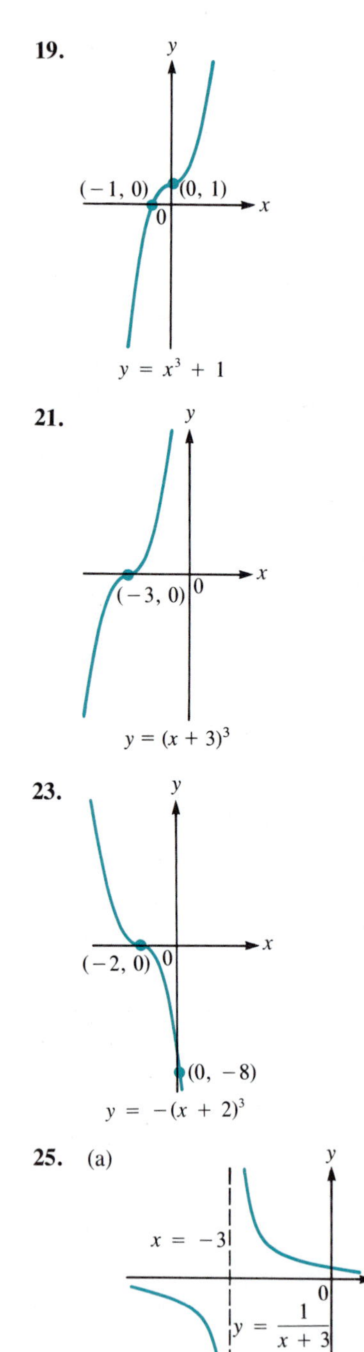

19.
y
x
(−1, 0)
(0, 1)
0
y = x³ + 1
21.
y
x
(−3, 0)
0
y = (x + 3)³
23.
y
x
(−2, 0)
0
(0, −8)
y = −(x + 2)³
25. (a)
y
x
x = −3
0
y = 1/(x + 3)

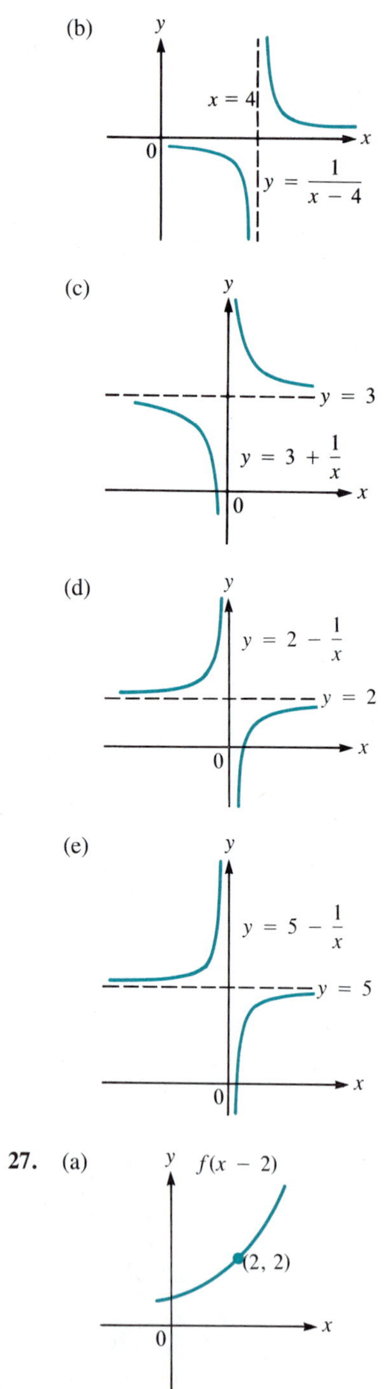

(b)
y
x
x = 4
0
y = 1/(x − 4)
(c)
y
y = 3
y = 3 + 1/x
x
0
(d)
y
y = 2 − 1/x
y = 2
x
0
(e)
y
y = 5 − 1/x
y = 5
x
0
27. (a)
y
f(x − 2)
(2, 2)
x
0

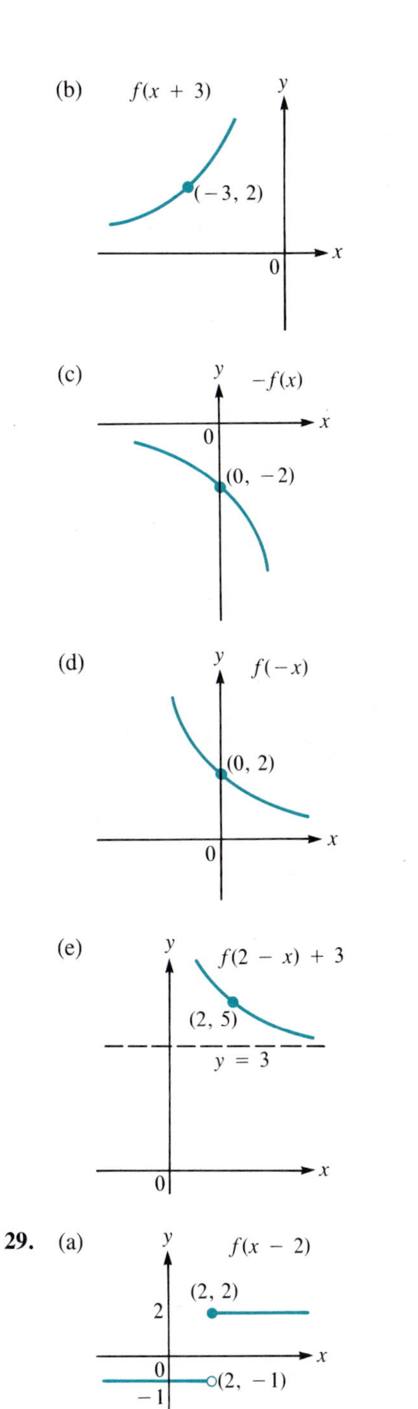
(b)
f(x + 3)
(−3, 2)
(c)
−f(x)
(0, −2)
(d)
f(−x)
(0, 2)
(e)
f(2 − x) + 3
(2, 5)
y = 3
29. (a)
f(x − 2)
(2, 2)
2
−1
(2, −1)

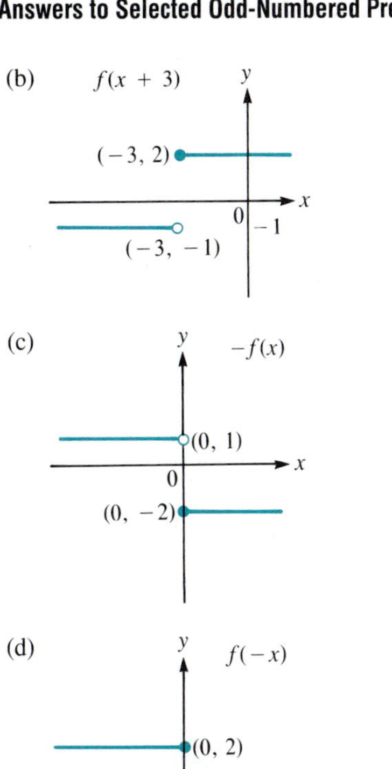
(b)
f(x + 3)
(−3, 2)
(−3, −1)
−1
(c)
−f(x)
(0, 1)
(0, −2)

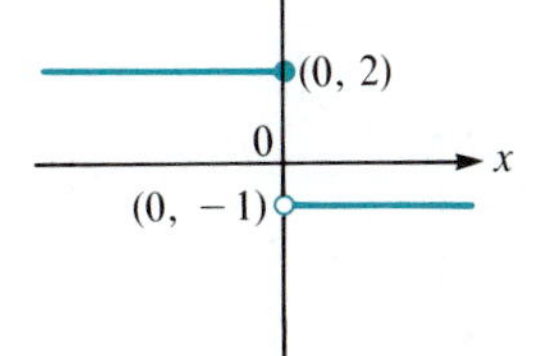
(d)
f(−x)
(0, 2)
(0, −1)

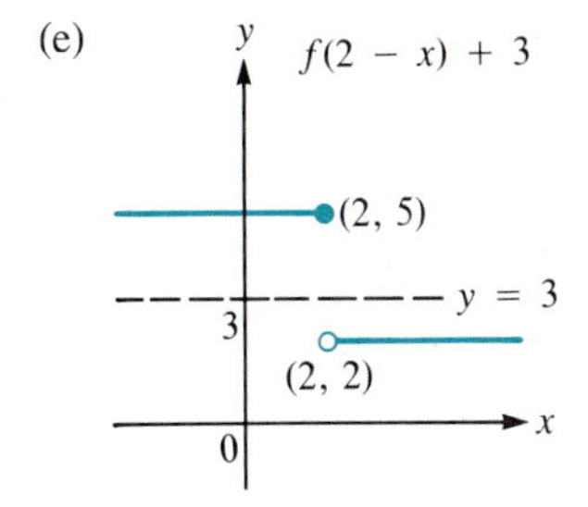
(e)
f(2 − x) + 3
(2, 5)
y = 3
3
(2, 2)

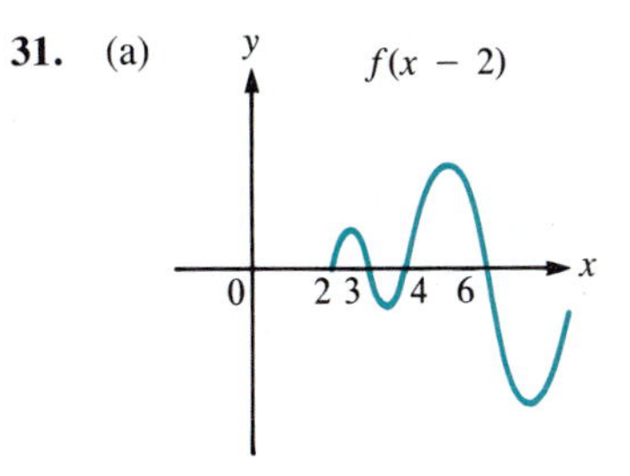
31. (a)
f(x − 2)
2 3 4 6

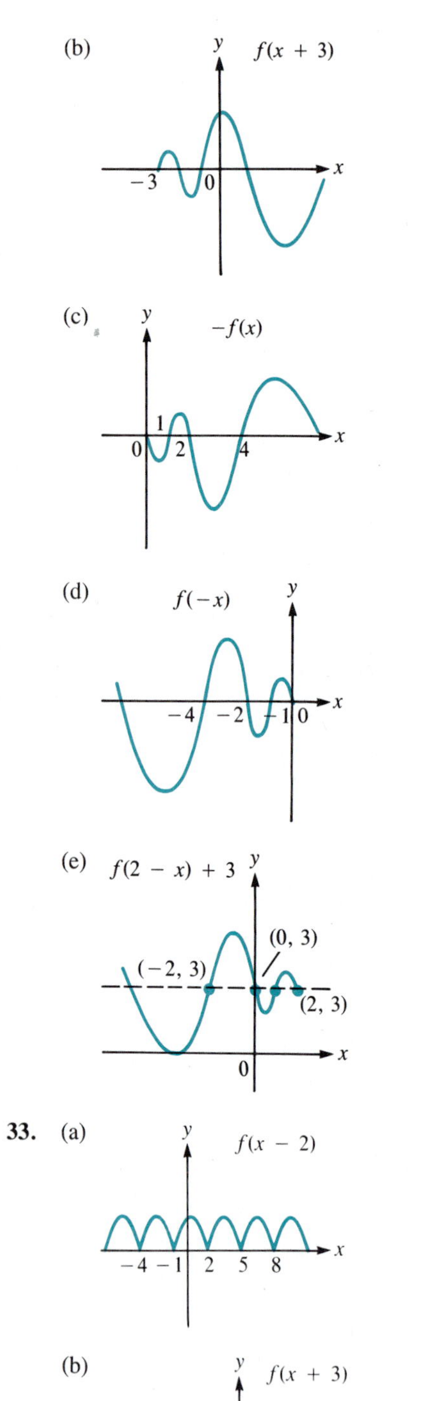
(b)
$f(x + 3)$
-3
0
(c)
$-f(x)$
0
1
2
4
(d)
$f(-x)$
-4
-2
-1
0
(e)
$f(2 - x) + 3$
$(0, 3)$
$(-2, 3)$
$(2, 3)$
0
33. (a)
$f(x - 2)$
-4 -1 2 5 8

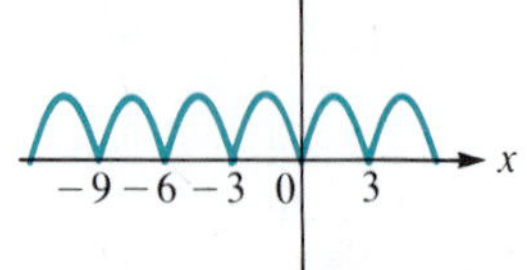
(b)
$f(x + 3)$
-9 -6 -3 0 3

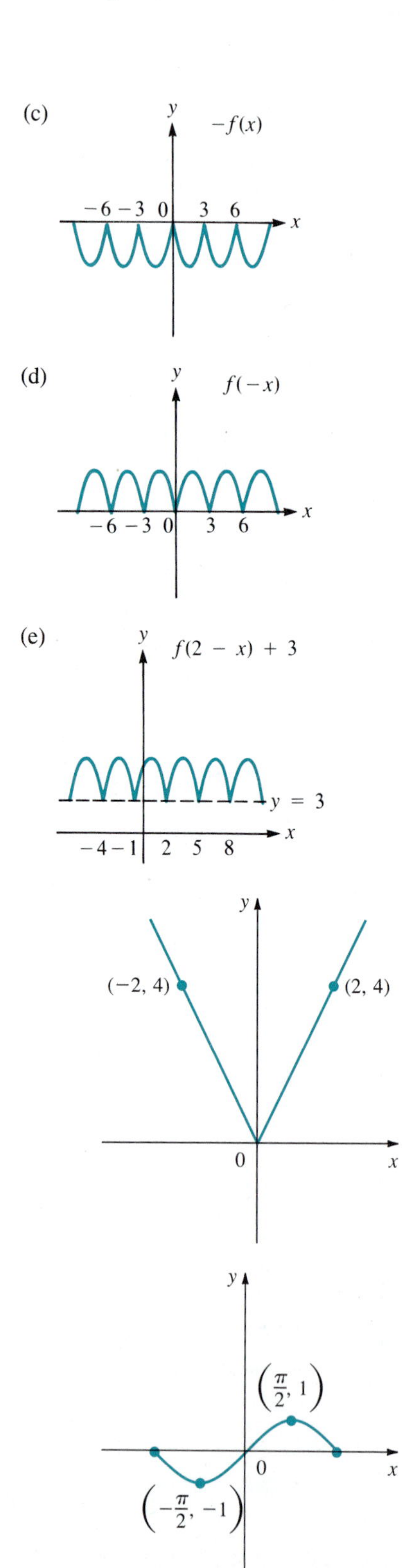
(c)
$-f(x)$
-6 -3 0 3 6
(d)
$f(-x)$
-6 -3 0 3 6
(e)
$f(2 - x) + 3$
$y = 3$
-4 -1 2 5 8
$(-2, 4)$
$(2, 4)$
0
$\left(\frac{\pi}{2}, 1\right)$
$\left(-\frac{\pi}{2}, -1\right)$
0

35.

37.

39.

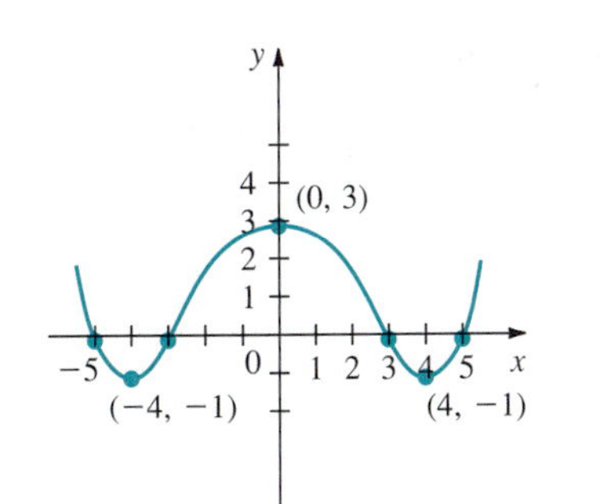

41.

43. $y = \sqrt[3]{x} \quad -9 \le x \le 9,\ -2.5 \le y \le 2.5$

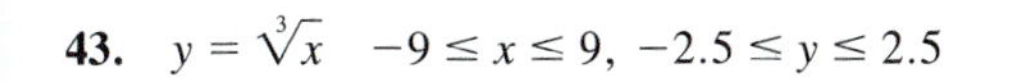

45.

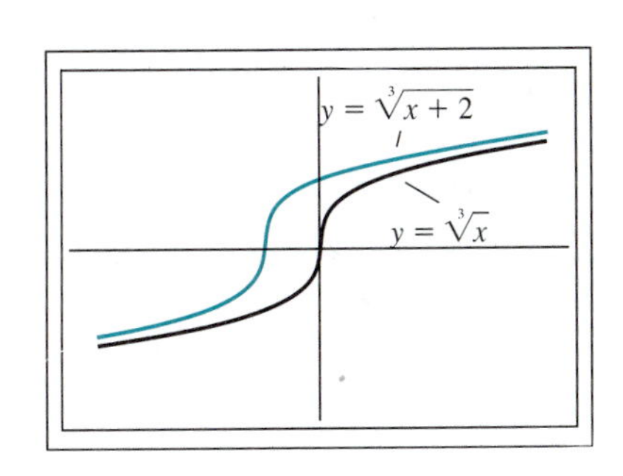

47.

49.

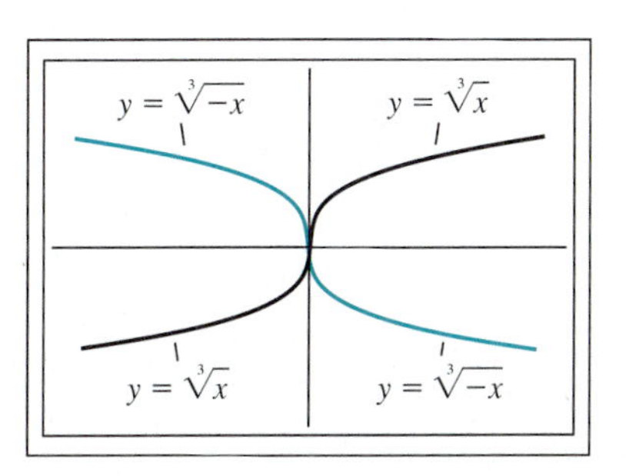

51.

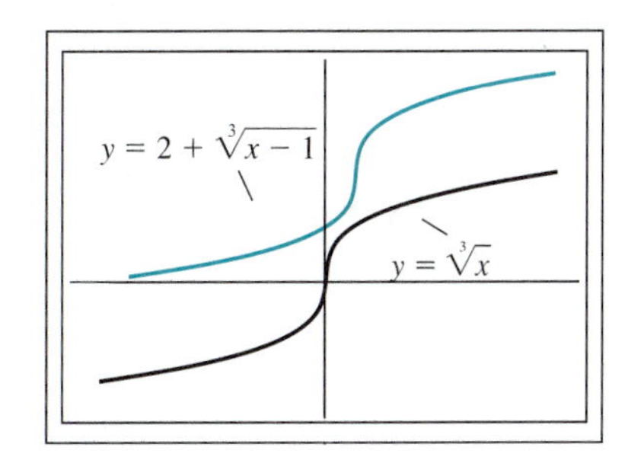

53. $y = -(x-2)^3 + 2(x-2)^2 + 5(x-2) - 6$
$-2 \le x \le 6,\ -18 \le y \le 70$

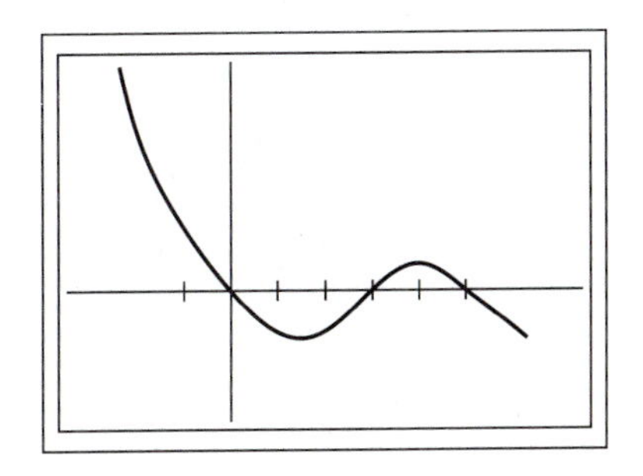

55. $y = -x^3 + 2x^2 + 5x - 5$ $\quad -4 \le x \le 4,\ -17 \le y \le 71$

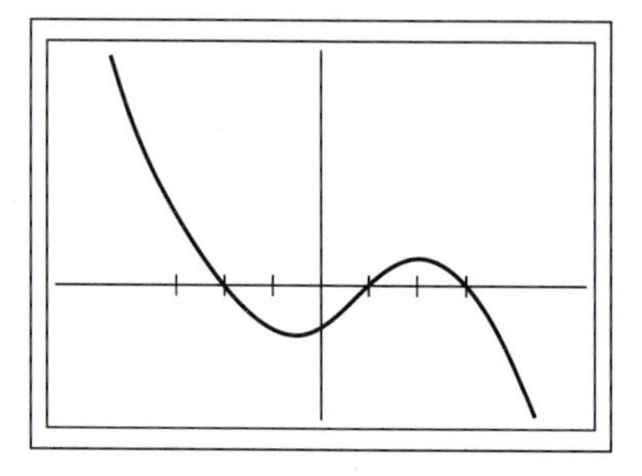

57. $y = x^3 - 2x^2 - 5x + 6$ $\quad -4 \le x \le 4,\ -70 \le y \le 18$

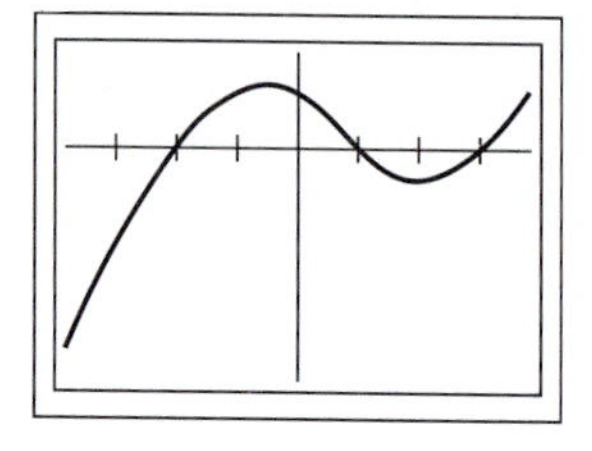

59. $y = -(2 - x)^3 + 2(2 - x)^2 + 5(2 - x) - 6$
$-2 \le x \le 6,\ -18 \le y \le 70$

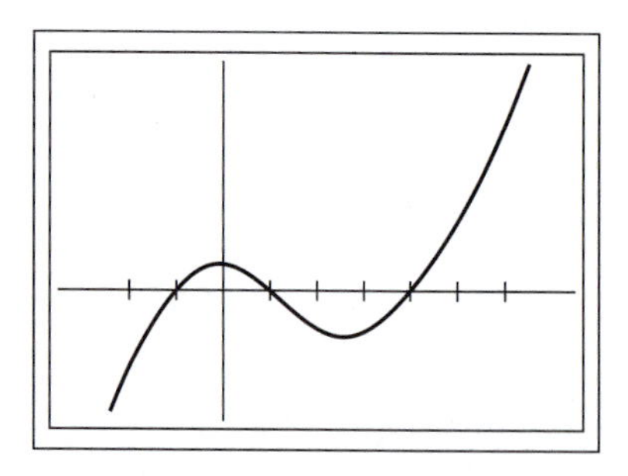

61. $y = -(1 + x)^3 + 2(1 + x)^2 + 5(1 + x) - 10$
$-5 \le x \le 3,\ -22 \le y \le 66$

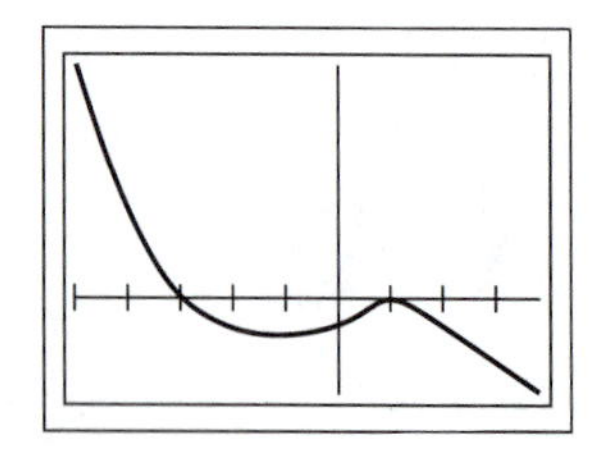

Problems 1.5, page 42

1. $f + g$: $-x + 3$; domain: all real numbers
$f - g$: $5x + 3$; domain: all real numbers
$f \cdot g$: $-6x^2 - 9x$; domain: all real numbers
$\dfrac{f}{g}$: $-\dfrac{2x + 3}{3x}$; domain: all real numbers except 0

3. $f + g$: 14; domain: all real numbers
$f - g$: -6; domain: all real numbers
fg: 40; domain: all real numbers
f/g: 0.4; domain: all real numbers

5. $f + g$: $x + 1/x$; domain: all real numbers except 0
$f - g$: $x - 1/x$; domain: all real numbers except 0
fg: 1; domain: all real numbers except 0
f/g: x^2; domain: all real numbers except 0

7. $f + g$: $\sqrt{x + 1} + \sqrt{1 - x}$; domain: $[-1, 1]$
$f - g$: $\sqrt{x + 1} - \sqrt{1 - x}$; domain: $[-1, 1]$
$f \cdot g$: $\sqrt{1 - x^2}$; domain: $[-1, 1]$
$\dfrac{f}{g}$: $\dfrac{\sqrt{1 - x^2}}{1 - x}$; domain $[-1, 1)$

9. $f + g$: $x^5 + 2 - |x|$; domain: all real numbers
$f - g$: $x^5 + |x|$; domain: all real numbers
fg: $x^5 - x^5|x| - |x| + 1$; domain: all real numbers
f/g: $(x^5 + 1)/(1 - |x|)$; domain: all real numbers except 1 and -1

11. $f + g$: $\sqrt[5]{x + 2} + \sqrt[4]{x - 3}$; domain: $[3, \infty)$
$f - g$: $\sqrt[5]{x + 2} - \sqrt[4]{x - 3}$; domain: $[3, \infty)$
fg: $\sqrt[5]{x + 2}\sqrt[4]{x - 3}$; domain: $[3, \infty)$
f/g: $\sqrt[5]{x + 2}/\sqrt[4]{x - 3}$; domain: $(3, \infty)$

13. $f \circ g$: $3x - 6$; domain: all real numbers
$g \circ f$: $3x - 2$; domain: all real numbers

15. $f \circ g$: 5; domain: all real numbers
$g \circ f$: 8; domain: all real numbers

17. $f \circ g$: $1/2x$; domain: all real numbers except 0
$g \circ f$: $1/2x$; domain: all real numbers except 0

19. $f \circ g$: $2x + 2$; domain: all real numbers
$g \circ f$: $2x - 1$; domain: all real numbers

21. $f \circ g$: $\dfrac{x + 1}{x - 1}$; domain: all real numbers except 0, 1
$g \circ f$: $\dfrac{2}{x}$; domain: all real numbers except 0, 2

23. $f \circ g$: $\sqrt{1 - \sqrt{x - 1}}$; domain: $[1, 2]$
$g \circ f$: $\sqrt{\sqrt{1 - x} - 1}$; domain: $(-\infty, 0]$

27. $g_1(x) = x - 5$; $g_2(x) = 5 - x$

29. Domain of k is $[0, \infty)$; $h(x) = \sqrt{x}$, $g(x) = x + 1$, $f(x) = x^{5/7}$

31. $ad + b = bc + d$

33. (a) $C(p) = 168{,}000 - 3200p$
(b) $R(p) = 20{,}000p - 400p^2$
(c) $P(p) = -400p^2 + 23{,}200p - 168{,}000$
(d) \$29.00; \$168,400

Problems 1.6, page 49

1. $\dfrac{x+1}{2} = \dfrac{1}{2}x + \dfrac{1}{2}$ **3.** $\dfrac{12x+3}{8} = \dfrac{3}{2}x + \dfrac{3}{8}$

5. $-\frac{11}{2}x + \frac{3}{2}$ **7.** $3/x$ **9.** $\dfrac{1-x}{x} = \dfrac{1}{x} - 1$

11. $4 - 3/x$ **13.** $\sqrt[3]{4/x - 2}$ **15.** $x^2 - 2;\ x \geq 0$

17. $(1 - x^2)/2,\ x \geq 0$ **19.** $1/x^3 + 7$

21. $x/(1 - x)$ **23.** $\sqrt[3]{x} - 3$

25. $-\sqrt{x-1}$ on $(-\infty, 0]$; $\sqrt{x-1}$ on $[0, \infty)$

27. $4 - \sqrt{x}$ on $(-\infty, 4]$; $4 + \sqrt{x}$ on $[4, \infty)$

29. $-\sqrt[4]{x}$ on $(-\infty, 0]$; $\sqrt[4]{x}$ on $[0, \infty)$

31. $-\sqrt[4]{x+5}$ on $(-\infty, 0]$; $\sqrt[4]{x+5}$ on $[0, \infty)$

33. $\frac{7}{2} - \sqrt{x + \frac{25}{4}}$ on $(-\infty, \frac{7}{2}]$; $\frac{7}{2} + \sqrt{x + \frac{25}{4}}$ on $[\frac{7}{2}, \infty)$

35. $-\sqrt{-2 + \sqrt{x}}$ on $(-\infty, 0]$; $\sqrt{-2 + \sqrt{x}}$ on $[0, \infty]$

37. $-x$ on $(-\infty, 0]$; x on $[0, \infty)$ **43.** yes **45.** no

47. no **49.** yes **51.** yes

53.

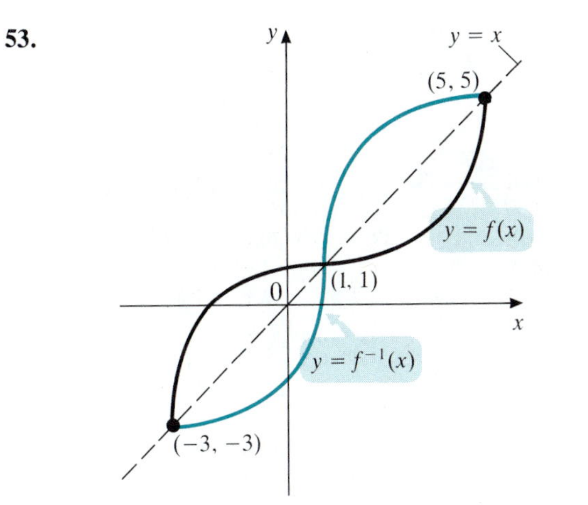

55.

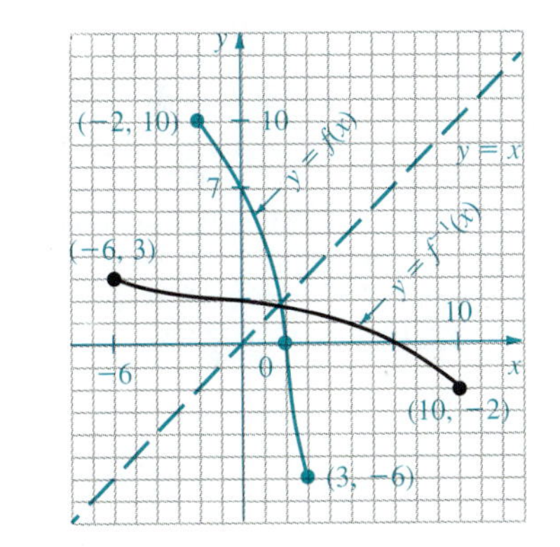

Review Exercises for Chapter 1, page 53

1. $\sqrt{10}$ **3.** $2\sqrt{2}$ **5.** 11

7. center $(3, -5)$, radius $\sqrt{2}$

9. yes; domain: all real numbers; range: $[-8, \infty)$

11. yes; domain: $[-2, \infty)$; range: $[0, \infty)$

13. yes; domain: all real numbers; range: $[-\frac{1}{2}, \frac{1}{2}]$

15. domain: $\{x : x \leq -\sqrt{6} \text{ or } x \geq \sqrt{6}\}$; range: $[0, \infty)$

19. odd

21. even

23.

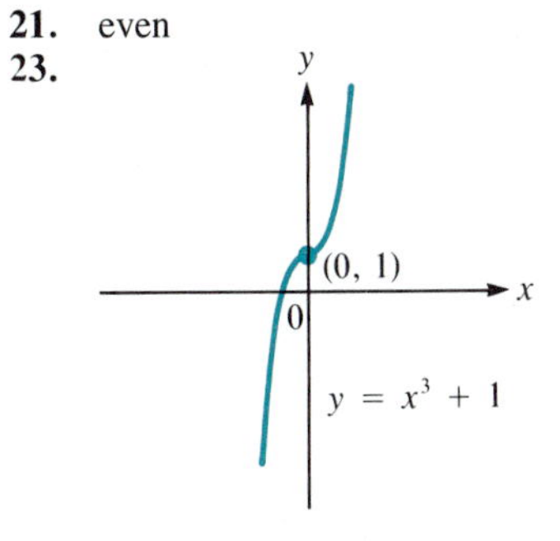

25.

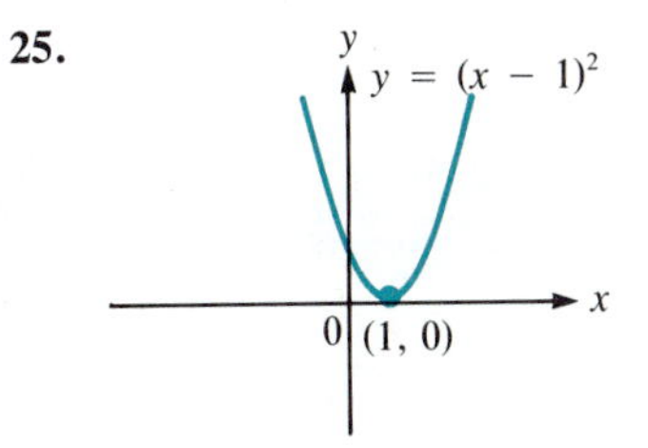

27.

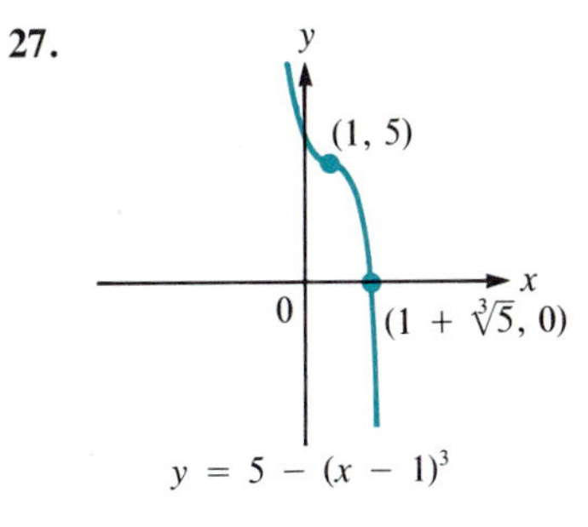

29.

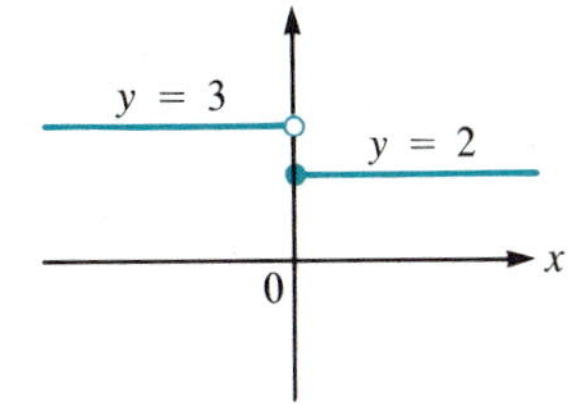

31. (a)

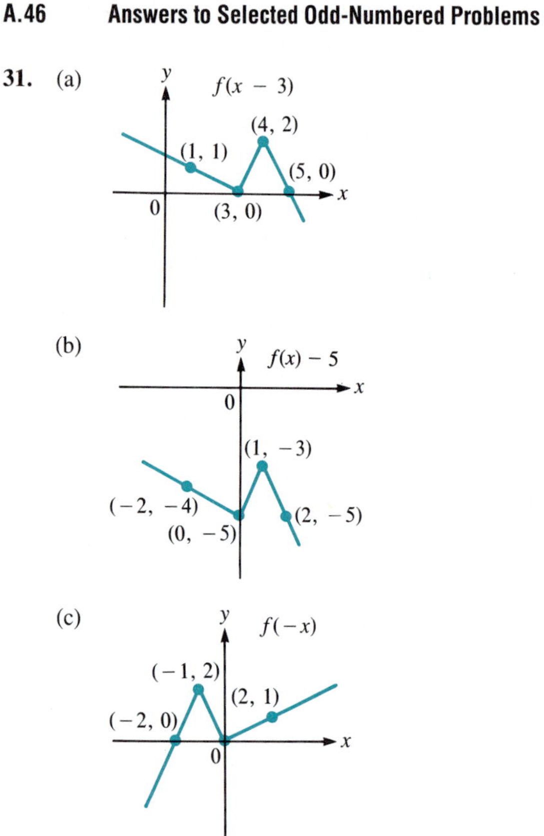

(d)

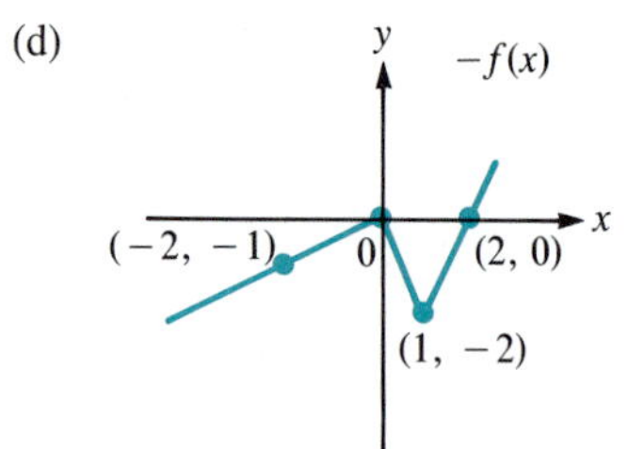

(e)

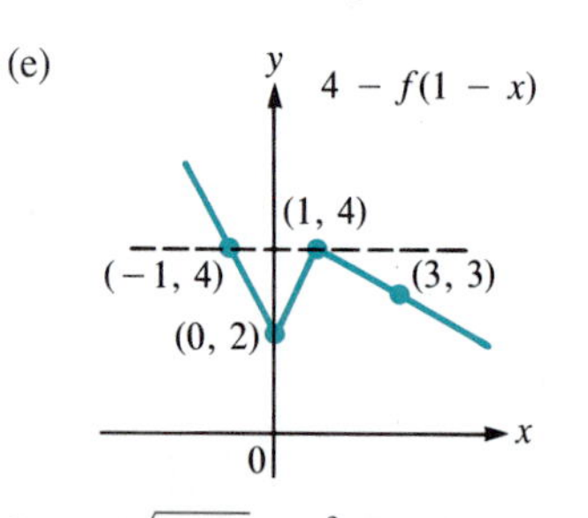

33. $f + g$: $\sqrt{x+1} + x^3$ domain: $[-1, \infty)$
$f - g$: $\sqrt{x+1} - x^3$ domain: $[-1, \infty)$
fg: $x^3\sqrt{x+1}$ domain: $[-1, \infty)$
g/f: $x^3/\sqrt{x+1}$ domain: $(-1, \infty)$
$f \circ g$: $\sqrt{x^3+1}$ domain: $[-1, \infty)$
$g \circ f$: $(x+1)^{3/2}$ domain: $[-1, \infty)$

35. $(x+1)/4$ **37.** $2/x$ **39.** $x^2 + 2,\ x \geq 0$
41. $f^{-1}(x) = -\sqrt{(x-3)/2}$ on $(-\infty, 0]$
$f^{-1}(x) = \sqrt{(x-3)/2}$ on $[0, \infty)$
43.

y
y = x
$y = f^{-1}(x)$
(5, 6)
(6, 5)
$y = f(x)$
7, 5, 3, 0, 2, 4, 6, x

Chapter 2

Problems 2.1, page 66

1. I **3.** III **5.** I **7.** II **9.** IV
11. II **13.** II **15.** 400°, −320°
17. 335°, −385° **19.** 260°, −460°
21. $7\pi/3,\ -5\pi/3$ **23.** $10\pi/3,\ -2\pi/3$
25. $25\pi/8,\ -7\pi/8$ **27.** 43° **29.** 23.4°
31. 43°45′ **33.** 159° **35.** 70° **37.** 131°38′
39. 42°21′ **41.** 121°25′30″ **43.** 24.5042°
45. −20.1806° **47.** −48.2942° **49.** $\pi/18$
51. $5\pi/6$ **53.** 3π **55.** 0.4712 **57.** −1.2392
59. 22.5° **61.** −67.5° **63.** 300° **65.** 70°
67. −13.1780° **69.** 572.9578°
71. $\pi/3$ cm ≈ 1.0472 cm **73.** 2.8 m
75. $\frac{28\pi}{3}$ in ≈ 29.3215 in **77.** $5\pi/6$ ft ≈ 2.6180 ft
79. 24.4346 m **81.** 1.6667 radians ≈ 95.4930°
83. 6 radians ≈ 343.7747°
85. (a) 90π radians/min (b) 4.5π radians = 810°
(c) 315π in/min ≈ 989.6 in/min
(d) 15.75π in ≈ 49.48 in
87. (a) 80π radians/min (b) 60π m/min ≈ 188.5 m/min
(c) 8π m ≈ 25.1 m
89. 342.36 rpm **91.** 16π in ≈ 50.2655 in
93. 72π m/min ≈ 226.1947 m/min ≈ 13.5717 km/hr
95. (a) $10\pi/16.69 \approx 107.8° = 1.8823$ radians
(b) 440,395.2 miles/Earth day ≈ 18,349.8 miles/Earth h
(c) 1.906×10^9 miles
97. 3403.9 miles **99.** 2466.25 miles
101. 500/3960 ≈ 7°14′4″ ≈ 0.1263 rad

Problems 2.2, page 76

1. $\sin\theta = \frac{15}{17}$; $\cos\theta = \frac{8}{17}$; $\tan\theta = \frac{15}{8}$; $\sec\theta = \frac{17}{8}$;
$\csc\theta = \frac{17}{15}$; $\cot\theta = \frac{8}{15}$
3. $\sin\theta = 4/\sqrt{17}$; $\cos\theta = 1/\sqrt{17}$; $\tan\theta = 4$;
$\sec\theta = \sqrt{17}$; $\csc\theta = \sqrt{17}/4$; $\cot\theta = \frac{1}{4}$

5. $\sin\theta = 2\sqrt{10}/7$; $\cos\theta = \frac{3}{7}$; $\tan\theta = 2\sqrt{10}/3$; $\sec\theta = \frac{7}{3}$; $\csc\theta = 7/2\sqrt{10}$; $\cot\theta = 3/2\sqrt{10}$

7. $\sin\theta = \sqrt{5}/3$; $\cos\theta = \frac{2}{3}$; $\tan\theta = \sqrt{5}/2$; $\sec\theta = \frac{3}{2}$; $\csc\theta = 3/\sqrt{5}$; $\cot\theta = 2/\sqrt{5}$

9. $\sin\theta = \sqrt{5}/3$; $\cos\theta = \frac{2}{3}$; $\tan\theta = \sqrt{5}/2$; $\sec\theta = \frac{3}{2}$; $\csc\theta = 3/\sqrt{5}$; $\cot\theta = 2/\sqrt{5}$

11.

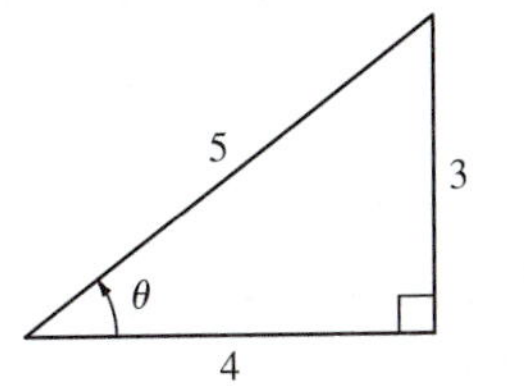

$\cos\theta = \frac{4}{5}$; $\tan\theta = \frac{3}{4}$; $\sec\theta = \frac{5}{4}$; $\csc\theta = \frac{5}{3}$; $\cot\theta = \frac{4}{3}$

13.

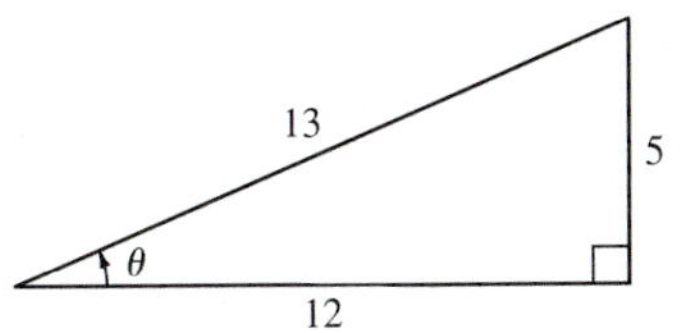

$\sin\theta = \frac{5}{13}$; $\cos\theta = \frac{12}{13}$; $\tan\theta = \frac{5}{12}$; $\sec\theta = \frac{13}{12}$; $\csc\theta = \frac{13}{5}$

15.

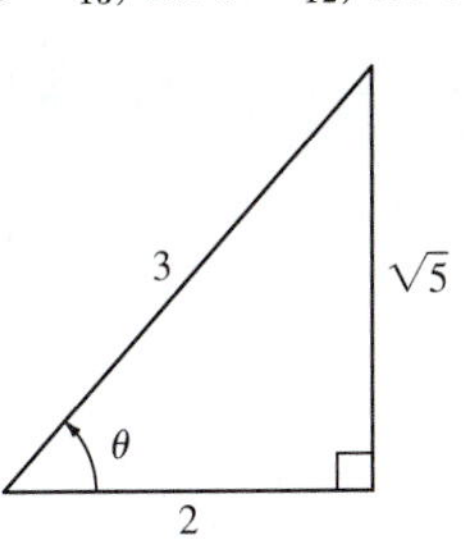

$\sin\theta = \sqrt{5}/3$; $\tan\theta = \sqrt{5}/2$; $\sec\theta = \frac{3}{2}$; $\csc\theta = 3/\sqrt{5} = 3\sqrt{5}/5$; $\cot\theta = 2/\sqrt{5} = 2\sqrt{5}/5$

17.

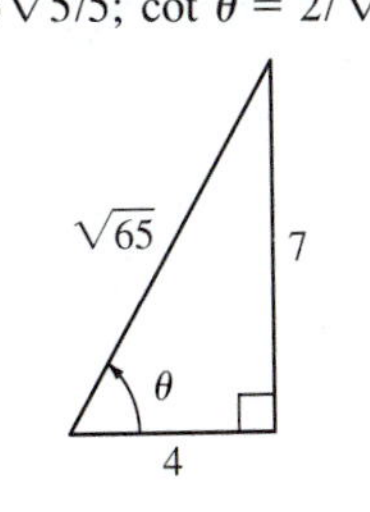

$\sin\theta = 7/\sqrt{65}$; $\cos\theta = 4/\sqrt{65}$; $\tan\theta = \frac{7}{4}$; $\sec\theta = \sqrt{65}/4$; $\csc\theta = \sqrt{65}/7$

19.

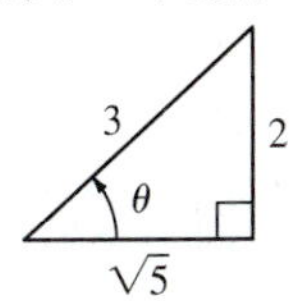

$\sin\theta = \frac{2}{3}$; $\cos\theta = \sqrt{5}/3$; $\tan\theta = 2/\sqrt{5}$; $\sec\theta = 3/\sqrt{5}$; $\cot\theta = \sqrt{5}/2$

21.

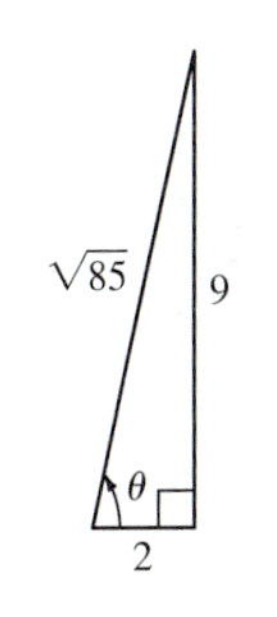

$\sin\theta = 9/\sqrt{85}$; $\cos\theta = 2/\sqrt{85}$; $\sec\theta = \sqrt{85}/2$; $\csc\theta = \sqrt{85}/9$; $\cot\theta = \frac{2}{9}$

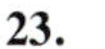

23.

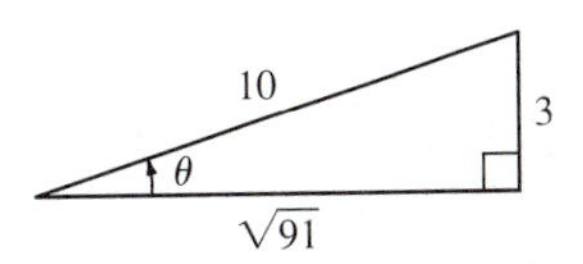

$\cos\theta = \sqrt{91}/10$; $\tan\theta = 3/\sqrt{91}$; $\sec\theta = 10/\sqrt{91}$; $\csc\theta = \frac{10}{3}$; $\cot\theta = \sqrt{91}/3$

25.

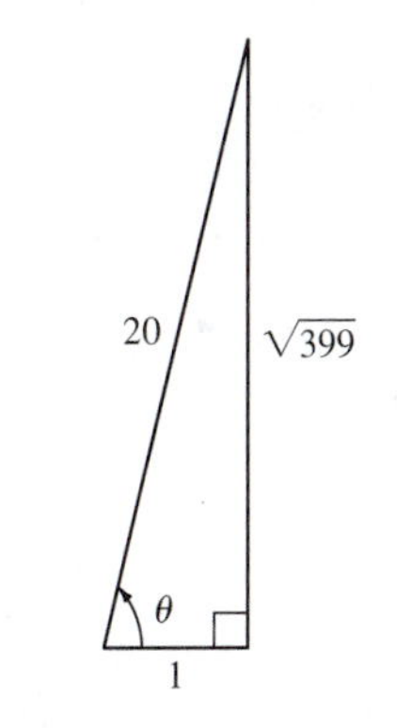

$\sin\theta = \sqrt{399}/20$; $\cos\theta = \frac{1}{20}$; $\tan\theta = \sqrt{399}$; $\csc\theta = 20/\sqrt{399}$; $\cot\theta = 1/\sqrt{399}$

27.

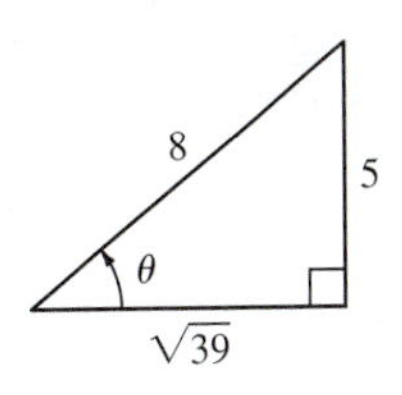

$\sin\theta = \frac{5}{8}$; $\cos\theta = \sqrt{39}/8$; $\tan\theta = 5/\sqrt{39}$; $\sec\theta = 8/\sqrt{39}$; $\cot\theta = \sqrt{39}/5$

29. $\tan\theta = \dfrac{\sqrt{4 - x^2}}{x}$ **31.** $\sec\theta = \dfrac{4}{y}$

33. $\cos\theta = \sqrt{\dfrac{3 - u^2}{3}} = \frac{1}{3}\sqrt{9 - 3u^2}$

35. no solution since $7/\sqrt{2} > 1$

Problems 2.3, page 87

1. $\cos\theta = 1/\sqrt{5}$; $\sin\theta = 2/\sqrt{5}$

3. $\cos\theta = -2/\sqrt{5}$; $\sin\theta = 1/\sqrt{5}$

5. $\cos\theta = -3/\sqrt{13}$; $\sin\theta = -2/\sqrt{13}$

7. $\cos\theta = 3/\sqrt{10}$; $\sin\theta = -1/\sqrt{10}$

9. $\cos\theta = -\frac{1}{2}$; $\sin\theta = \sqrt{3}/2$

11. $(-1, 0)$; $\cos\theta = -1$; $\sin\theta = 0$

13. $(-1, 0)$; $\cos\theta = -1$; $\sin\theta = 0$

15. $(0, 1)$; $\cos\theta = 0$; $\sin\theta = 1$

17. $(0, -1)$; $\cos\theta = 0$; $\sin\theta = -1$

19. $\left(\frac{\sqrt{2}}{2}, \frac{\sqrt{2}}{2}\right) = \left(\frac{1}{\sqrt{2}}, \frac{1}{\sqrt{2}}\right)$; $\cos\theta = \frac{\sqrt{2}}{2} = \frac{1}{\sqrt{2}}$; $\sin\theta = \frac{\sqrt{2}}{2} = \frac{1}{\sqrt{2}}$

21. $\left(\frac{\sqrt{3}}{2}, \frac{1}{2}\right)$; $\cos\theta = \frac{\sqrt{3}}{2}$; $\sin\theta = \frac{1}{2}$

In Problems 23–39 the sign of $\cos\theta$ is given first.

23. positive; positive **25.** negative; negative
27. negative; positive **29.** negative; positive
31. positive; positive **33.** negative; positive
35. negative; positive **37.** negative; positive
39. negative; negative **41.** 0.8
43. $-\sqrt{0.4375} \approx -0.6614$ **45.** $\sqrt{0.8556} \approx 0.9250$

47. not possible since $\sin\theta = \frac{5}{4} > 1$ **49.** IV

51. I

53. $\sin 37° = 0.601815023$; $\cos 37° = 0.79863551$

55. $\sin 263° = -0.992546151$; $\cos 263° = -0.121869343$

57. $\cos 1.5 \approx 0.070117$; $\sin 1.5 \approx 0.997391$. To six decimal places, $\cos 1.5 = 0.070737$ (error $= 0.00062 \approx 0.88\%$) and $\sin 1.5 = 0.997495$ (error $= 0.000104 \approx 0.01\%$).

59. $50° = 0.872664626$ rad; $\cos 50° \approx 0.642779$ and $\sin 50° \approx 0.766044$. To six decimal places, $\cos 50° = 0.642788$ (error $= 0.000009 \approx 0.0014\%$) and $\sin 50° = 0.766044$ (error $= 0$).

61. $\cos(-0.9) \approx 0.621599$ and $\sin(-0.9) \approx -0.783326$. To six decimal places, $\cos(-0.9) = 0.621610$ (error $= 0.000011 = 0.0018\%$) and $\sin(-0.9) = -0.783327$ (error $= -0.000001 = 0.00013\%$).

65. $\sqrt{3} \approx 1.732$ **67.** 1.285575219

69. 1.931851653 **71.** 1.845379677

73. (a) $(-1, 0)$ (b) $(0, -1)$
(c) $(\cos 2, \sin 2) = (-0.416146836, 0.909297426)$
(d) $(\cos 20, \sin 20) = (0.408082061, 0.91294525)$

Problems 2.4, page 99

1. $\pi/3$ **3.** $\pi/3$ **5.** $\pi/10$ **7.** 20° **9.** 17°

In Problems 11–37 the values are given in the order $\sin\theta$, $\cos\theta$.

11. $\frac{1}{2}$; $-\sqrt{3}/2$ **13.** 1; 0 **15.** $-\frac{1}{2}$; $-\sqrt{3}/2$
17. -1; 0 **19.** $\sqrt{2}/2$; $-\sqrt{2}/2$ **21.** $-\frac{1}{2}$; $\sqrt{3}/2$
23. 0; -1 **25.** $-\sqrt{3}/2$; $-\frac{1}{2}$ **27.** $-\sqrt{2}/2$; $\sqrt{2}/2$
29. 0; -1 **31.** 1; 0 **33.** -1; 0 **35.** $\sqrt{3}/2$; $\frac{1}{2}$
37. $-\frac{1}{2}$; $\sqrt{3}/2$

39. $\cos\theta$ is decreasing, $\sin\theta$ is decreasing

41. Both are decreasing

43. $\cos\theta$ is increasing, $\sin\theta$ is decreasing

45. Both are increasing **47.** 0.6208, 5.6624

49. 2.5208, 3.7624 **51.** $\pi/3$, $5\pi/3$, $7\pi/3$, $11\pi/3$

53. $\pi/4$, $3\pi/4$, $9\pi/4$, $11\pi/4$, $17\pi/4$

59. $[0, \pi/6)$ and $(5\pi/6, \pi]$

61.

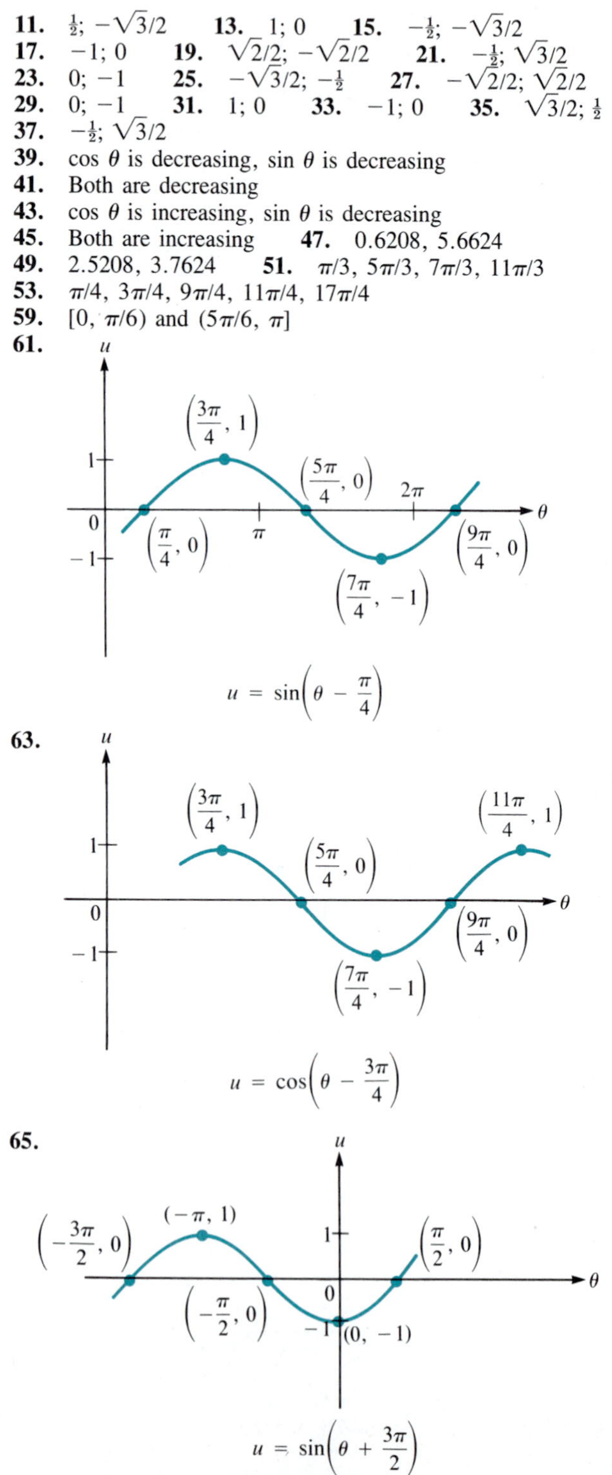

67.

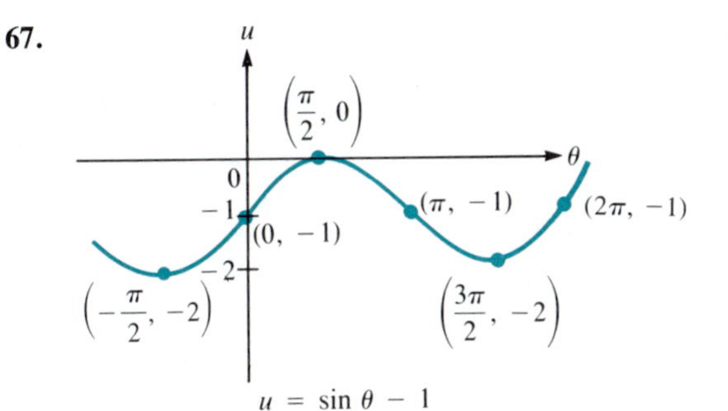

69.

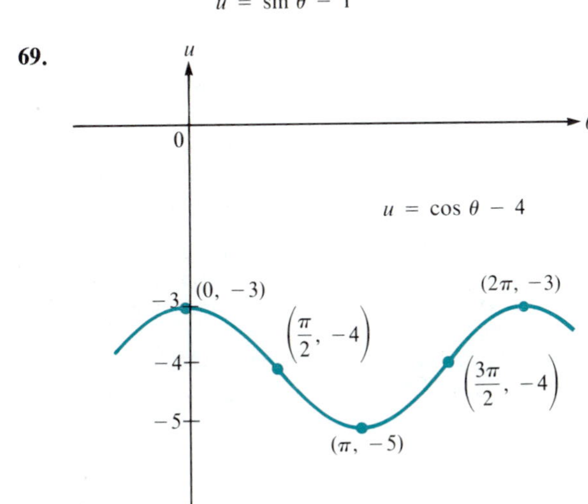

71. $\theta = \pi/2 + 2k\pi$, $k = 0, \pm 1, \pm 2, \ldots$

73. $\pi/6 + 2k\pi$, $11\pi/6 + 2k\pi$, $k = 0, \pm 1, \pm 2, \ldots$

Problems 2.5, page 110

In Problems 1–9 the answers are given in the order tan θ, cot θ, sec θ, csc θ.

1. $1, 1, \sqrt{2}, \sqrt{2}$ **3.** $1/\sqrt{3}, \sqrt{3}, -2/\sqrt{3}, -2$

5. undefined, 0, undefined, -1

7. $-1, -1, \sqrt{2}, -\sqrt{2}$

9. $-\sqrt{3}, -1/\sqrt{3}, 2, -2/\sqrt{3}$

In Problems 11–33 the answers are given in the order sin θ, cos θ, tan θ, cot θ, sec θ, csc θ.

11. $\frac{1}{3}, 2\sqrt{2}/3, 1/2\sqrt{2}, 2\sqrt{2}, 3/2\sqrt{2}, 3$

13. $4/\sqrt{17}, 1/\sqrt{17}, 4, \frac{1}{4}, \sqrt{17}, \sqrt{17}/4$

15. $\sqrt{5}/3, -2/3, -\sqrt{5}/2, -2/\sqrt{5}, -3/2, 3/\sqrt{5}$

17. $5/\sqrt{29}, -2/\sqrt{29}, -5/2, -2/5, -\sqrt{29}/2, \sqrt{29}/5$

19. $\frac{3}{5}, \frac{4}{5}, \frac{3}{4}, \frac{4}{3}, \frac{5}{4}, \frac{5}{3}$ **21.** $\frac{3}{5}, \frac{4}{5}, \frac{3}{4}, \frac{4}{3}, \frac{5}{4}, \frac{5}{3}$; I

23. $-\frac{3}{5}, \frac{4}{5}, -\frac{3}{4}, -\frac{4}{3}, \frac{5}{4}, -\frac{5}{3}$; IV

25. $-\frac{8}{17}, \frac{15}{17}, -\frac{8}{15}, -\frac{15}{8}, \frac{17}{15}, -\frac{17}{8}$; IV

27. $-\frac{8}{17}, -\frac{15}{17}, \frac{8}{15}, \frac{15}{8}, -\frac{17}{15}, -\frac{17}{8}$; III

29. $\frac{24}{25}, -\frac{7}{25}, -\frac{24}{7}, -\frac{7}{24}, -\frac{25}{7}, \frac{25}{24}$; II

31. $-0.9897, 0.1432, -6.9113, -0.1447, 6.9832, -1.0104$; IV

33. $-0.2397, 0.9709, -0.2469, -4.0507, 1.0300, -4.1722$; IV

35. -1.010108666 **37.** 0.793551147

39. 1.305407289 **41.** 5.67128182

43. 3.34120995 **57.** $0 \le \theta \le \pi/3$

59. $0 < \theta < \pi/6$

Problems 2.6, page 118

1. $-\sqrt{2}$ **3.** -1 **5.** does not exist **7.** $\sqrt{3}$

9. $1/\sqrt{3}$ **11.** $\pi/4 + k\pi$, $k = 0, \pm 1, \pm 2, \ldots$

13. $\pi/2 + 2k\pi$, $k = 0, \pm 1, \pm 2, \ldots$

15.

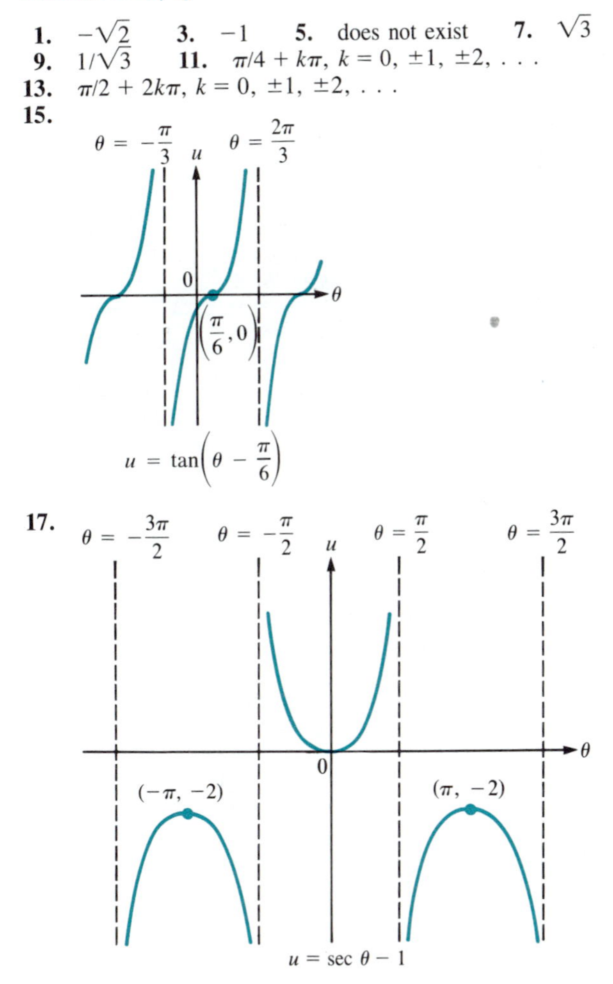

19.

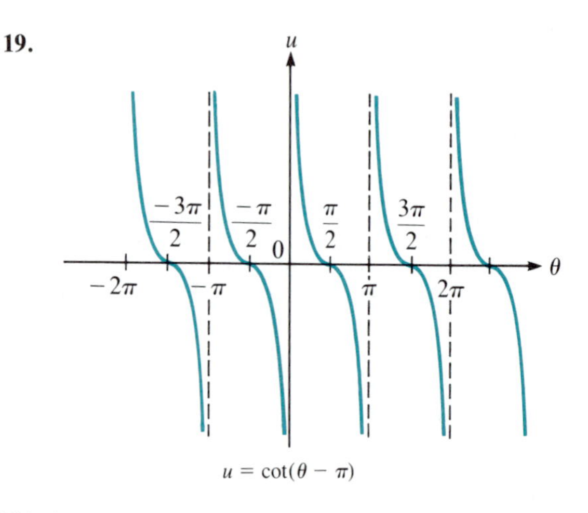

$u = \cot(\theta - \pi)$

This is identical to the graph of $\cot\theta$.

Problems 2.7, page 131

1. $\pi/6$ **3.** $2\pi/3$ **5.** $\pi/6$ **7.** $-\pi/4$
9. 0 **11.** $\frac{4}{5}$ **13.** $3/\sqrt{34}$
15. no solution; $\sin^{-1} 5$ is undefined
17. $\sqrt{1-x^2}$ (assuming $-1 \le x \le 1$) **19.** $\frac{5}{4}$
21. no solution; $\cos^{-1}\frac{9}{7}$ is undefined **23.** 10^{-55}
25. $5\pi/6$ **27.** 0 **29.** $\sqrt{101}$ **31.** $7/\sqrt{33}$
33. $-\frac{1}{2}$ **35.** $\sqrt{11}/6$ **37.** $\sqrt{a^2-x^2}/a$ if $a \ge |x|$
39. $(x+1)/\sqrt{2x+1}$ **41.** 71.9408°; 1.2556
43. −24.8346°; −0.4334 **45.** 69.6769°; 1.2161
47. 7.1250°; 0.1244 **49.** 51.7575°; 0.9033
51. 168.4630°, 2.9402 **53.** −11.5370°; −0.2014
55. 228.1897°; 3.9827 **57.** −26.5651°; −0.4636
59. 256.3910°, 4.4749 **61.** $13\pi/3$, $17\pi/3$
63. $\tan^{-1} 7 + 10\pi \approx 32.8448$, $\tan^{-1} 7 + 11\pi \approx 35.9864$

Problems 2.8, page 140

1. $\alpha = 56°$, $x = 2.9651$, $y = 3.5766$
3. $\alpha = 61.72°$, $x = 2.6900$, $y = 5.6777$
5. $\beta = 34.17$, $x = 19.8824$, $y = 29.2891$
7. $\theta = 30.96°$, $\alpha = 59.04°$, $x = 58.3095$
9. $\alpha = 70.90°$, $\beta = 19.10°$, $z = 519.7115$
11. 70.2 ft **13.** 148.8 m **15.** 83.4 ft
17. 80.5 m **19.** 3354.1 ft **21.** 66.4 ft
23. 300.0 m **25.** 11.6 m **27.** 11.1 cm
29. 32.0° **31.** 272.1 ft
33. 30.4 ft (the distance from the edge of the moat to the bottom of the window)
35. 228.8 miles (using the value 6.6 miles in the computation)
37. 35,870.5 miles **39.** 2.6 m **41.** 26.4 km
43. N56°41′E **45.** S84°37′W **47.** 40°36′
49. $\frac{1}{2}\tan 28.67° \approx 0.27$ **51.** 4 ft **53.** 19.5 ft

Review Exercises for Chapter 2, page 146

1. II **3.** IV **5.** IV **7.** 136°18′
9. 127.4786°
11. (a) $11\pi/12$ (b) $-3\pi/8$ (c) $28\pi/45 \approx 1.9548$
13. (a) 70π rad/min ≈ 219.9 rad/min
(b) 210π m/min ≈ 659.7 m/min
(c) 42π m ≈ 131.9 m
15. 3407.4 miles

In Exercises 17–43 the answers are given in the order $\sin\theta$, $\cos\theta$, $\tan\theta$, $\cot\theta$, $\sec\theta$, $\csc\theta$.

17. $\frac{1}{2}$, $\sqrt{3}/2$, $1/\sqrt{3}$, $\sqrt{3}$, $2/\sqrt{3}$, 2
19. 0, 1, 0, undefined, 1, undefined
21. $-1/2$, $\sqrt{3}/2$, $-1/\sqrt{3}$, $-\sqrt{3}$, $2/\sqrt{3}$, -2
23. 0, −1, 0, undefined, −1, undefined
25. −1, 0, undefined, 0, undefined, −1
27. $-\sqrt{2}/2$, $\sqrt{2}/2$, −1, −1, $\sqrt{2}$, $-\sqrt{2}$
29. −0.6506, −0.7594, 0.8568, 1.1672, −1.3168, −1.5370
31. −0.6018, −0.7986, 0.7536, 1.3270, −1.2521, −1.6616
33. −0.9822, −0.1881, 5.2221, 0.1915, −5.3170, −1.0182
35. 0.9482, 0.3178, 2.9833, 0.3352, 3.1465, 1.0547
37. $4/\sqrt{17}$, $1/\sqrt{17}$, 4, 1/4, $\sqrt{17}$, $\sqrt{17}/4$
39. $-\sqrt{3}/2$, $-\frac{1}{2}$, $\sqrt{3}$, $1/\sqrt{3}$, −2, $-2/\sqrt{3}$
41. $-2\sqrt{2}/3$, $\frac{1}{3}$, $-2\sqrt{2}$, $-1/2\sqrt{2}$, 3, $-3/2\sqrt{2}$
43. $1/\sqrt{37}$, $-6/\sqrt{37}$, −1/6, −6, $-\sqrt{37}/6$, $\sqrt{37}$
45. 0.2208, 2.9208 **47.** $\pi/6$, $11\pi/6$, $13\pi/6$, $23\pi/6$
49. $\tan\theta = -\frac{3}{4}$, $\cot\theta = -\frac{4}{3}$, $\sec\theta = -\frac{5}{4}$, $\csc\theta = \frac{5}{3}$
51.

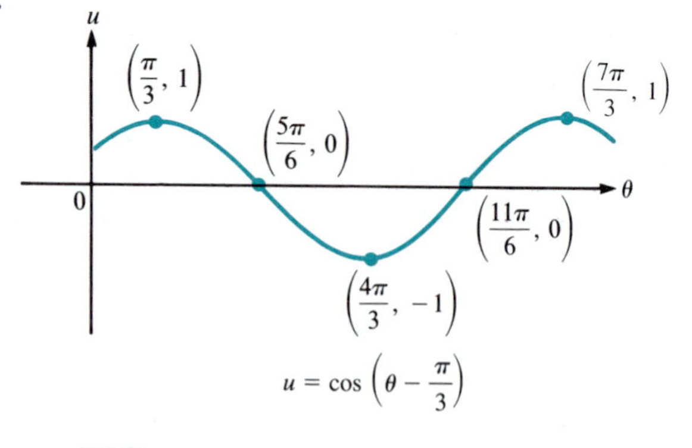

$u = \cos\left(\theta - \frac{\pi}{3}\right)$

53. $x/\sqrt{9+x^2}$ **55.** $\pi/6$ **57.** $\pi/4$
59. $\sqrt{39}/8$ **61.** $1/\sqrt{10}$ **63.** $-\sqrt{55}/3$
65. $\sqrt{16-x^2}/4$, $|x| < 4$
67. $y = 4.6672$, $x = 3.5753$, $\theta = 50°$
69. $y = 2.9521$, $x = 9.5543$, $\theta = 72.8°$ **71.** 42.85 m
73. 57.8 ft **75.** 212.3 miles **77.** S11°11′W

Chapter 3

Problems 3.1, page 154

1. $\sin x$ **3.** $(\cos^2\theta)/(\sin\theta)$
5. $(\cos t + 1)/(\cos t - \sin t)$
7. $(\sin u + 1)/(\sin u - 1)$
9. $(\cos s + \sin s)/(\cos s - \sin s)$ **11.** $1 - \cos^2 x$
13. $(\cos x + 1)/(\cos x - 1)$ **15.** $2 - \cos^2 x$
17. $(1 - \sin^2 x)/\sin x$ **19.** $\sin^2 x/(1 - \sin^2 x)$
81. Choose any number in $(\pi, 2\pi)$; for example, $\sin(3\pi/2) = -1$, but $\sqrt{\sin^2(3\pi/2)} = 1$.
83. Choose any number in $(\pi, 2\pi)$; for example, $\sin(3\pi/2) = -1$, but $\sqrt{1 - \cos^2(3\pi/2)} = 1$.
85. Choose any number in $(-\pi/2, 0)$ or $(\pi/2, \pi)$; for example, $\cot(-\pi/4) = -1$, but $\sqrt{\csc^2(-\pi/4) - 1} = 1$.
87. Choose any number in $(0, 2\pi)$ except $\pi/2$. For example, $\sqrt{\sin^2\pi + \cos^2\pi} = 1$, but $\sin\pi + \cos\pi = -1$.
89. $\ln(1/(\cos x)) = -\ln\cos x$, which has the opposite sign as $1/(\ln\cos x)$ unless $\ln\cos x = 0$, in which case $1/(\ln\cos x)$ is not defined.
91. $\sin(\sec 0) = \sin 1 \approx 0.8415 \neq 1$
93. Whenever $\sin x \geq 0$; if $x \in [0, 2\pi)$, then $\sin x \geq 0$ for $x \in [0, \pi]$.
95. $a\cos\theta$ **97.** $2\sec\theta$ **99.** $\frac{2}{3}\sec\theta$
101. $a\tan\theta$ **103.** $1/|8\tan^3\theta| = |\cot^3\theta|/8,\ \theta \neq 0$
105. $(\sqrt{3}\cos^2\theta)/(\sin\theta) = \sqrt{3}\cos\theta\cot\theta$
107. $\tan^n\theta$
109. $(\tan\theta)/(a^{n-1}\sec^n\theta) = a^{1-n}\tan\theta\cos^n\theta$

Problems 3.2, page 162

1. $(1 - \sqrt{3})/2\sqrt{2} = (\sqrt{2} - \sqrt{6})/4$
3. $(1 + \sqrt{3})/(1 - \sqrt{3})$ **5.** $(\sqrt{3} - 1)/(\sqrt{3} + 1)$
7. $(1 - \sqrt{3})/2\sqrt{2} = (\sqrt{2} - \sqrt{6})/4$
9. $(1 - \sqrt{3})/2\sqrt{2} = (\sqrt{2} - \sqrt{6})/4$ **11.** $\cos 69°$
13. $\sin(-0.8) = -\sin 0.8$ **15.** $\cos 0.8$
17. $\tan\frac{3}{2}$ **19.** $\sin 3x$ **21.** $\tan(-3\alpha) = -\tan 3\alpha$
23. $\sin(\alpha + \beta) = \frac{63}{65}$, $\sin(\alpha - \beta) = -\frac{33}{65}$, $\cos(\alpha + \beta) = -\frac{16}{65}$, $\cos(\alpha - \beta) = \frac{56}{65}$, $\tan(\alpha + \beta) = -\frac{63}{16}$, $\tan(\alpha - \beta) = -\frac{33}{56}$; quadrant of $\alpha + \beta$ is II, quadrant of $\alpha - \beta$ is IV
25. $\sin(\alpha + \beta) = \frac{56}{65}$, $\sin(\alpha - \beta) = \frac{16}{65}$, $\cos(\alpha + \beta) = -\frac{33}{65}$, $\cos(\alpha - \beta) = -\frac{63}{65}$, $\tan(\alpha + \beta) = -\frac{56}{33}$, $\tan(\alpha - \beta) = -\frac{16}{63}$; quadrant of $\alpha + \beta$ is II, quadrant of $\alpha - \beta$ is II
27. $\sin(\alpha + \beta) = -\frac{63}{65}$, $\sin(\alpha - \beta) = \frac{33}{65}$, $\cos(\alpha + \beta) = -\frac{16}{65}$, $\cos(\alpha - \beta) = -\frac{56}{65}$, $\tan(\alpha + \beta) = \frac{63}{16}$, $\tan(\alpha - \beta) = -\frac{33}{56}$, quadrant of $\alpha + \beta$ is III, quadrant of $\alpha - \beta$ is II
55. $\sqrt{3}/2$ **57.** $(\sqrt{2} - \sqrt{6})/4$
59. $\tan(\pi/2)$ is undefined
61. $(3\sqrt{7} + 12)/(9 - 4\sqrt{7})$ **63.** $\frac{4}{5}$
65. $(x^2 + 2x + 1)/\sqrt{x^2 + 1}\sqrt{(x + 2)^2 + 1}$
67. $2x^2 + \sqrt{1 - 4x^2}\sqrt{1 - x^2}$
71. $\cos 4x = \cos^4 x - 6\sin^2 x\cos^2 x + \sin^4 x$ (this can also be written in terms of $\sin x$ only or $\cos x$ only by using the identity $\sin^2 x + \cos^2 x = 1$)

Problems 3.3, page 170

1. $\sqrt{2 - \sqrt{3}}/2$
3. $\sqrt{(2 - \sqrt{3})/(2 + \sqrt{3})} = 2 - \sqrt{3}$
5. $\sqrt{2 + \sqrt{2 + \sqrt{3}}}/2$ **7.** $\sqrt{2 + \sqrt{2}}/2$
9. $-\sqrt{(2 - \sqrt{3})/(2 + \sqrt{3})} = -2 + \sqrt{3}$
11. $2/\sqrt{2 + \sqrt{2}} = \sqrt{4 - 2\sqrt{2}}$
13. $\sin 2\theta = \frac{24}{25}$
$\cos 2\theta = \frac{7}{25}$
$\tan 2\theta = \frac{24}{7}$
$\sin\theta/2 = 1/\sqrt{10} = \sqrt{10}/10$
$\cos\theta/2 = 3/\sqrt{10} = 3\sqrt{10}/10$
$\tan\theta/2 = \frac{1}{3}$
15. $\sin 2\theta = \frac{120}{169}$
$\cos 2\theta = \frac{119}{169}$
$\tan 2\theta = \frac{120}{119}$
$\sin\theta/2 = 5/\sqrt{26} = 5\sqrt{26}/26$
$\cos\theta/2 = -1/\sqrt{26} = -\sqrt{26}/26$
$\tan\theta/2 = -5$
17. $\sin 2\theta = 12\sqrt{10}/49$
$\cos 2\theta = \frac{31}{49}$
$\tan 2\theta = 12\sqrt{10}/31$
$\sin\theta/2 = \sqrt{(7 + 2\sqrt{10})/14}$
$\cos\theta/2 = -\sqrt{(7 - 2\sqrt{10})/14}$
$\tan\theta/2 = (-7 - 2\sqrt{10})/3$
19. $\sin 2\theta = -4\sqrt{5}/9$; $\sin\theta/2 = \sqrt{30}/6$
$\cos 2\theta = -\frac{1}{9}$; $\cos\theta/2 = \sqrt{6}/6$
$\tan 2\theta = 4\sqrt{5}$; $\tan\theta/2 = \sqrt{5}$
21. $\sin 2\theta = \sqrt{15}/8$; $\sin\theta/2 = \sqrt{6}/4$
$\cos 2\theta = -\frac{7}{8}$; $\cos\theta/2 = \sqrt{10}/4$
$\tan 2\theta = -\sqrt{15}/7$; $\tan\theta/2 = \sqrt{15}/5$
23. $4\cos^3 x - 3\cos x$
25. $(4\sin x - 8\sin^3 x)\cos x = 4\sin x\cos x(\cos^2 x - \sin^2 x)$
27. $\frac{1}{8}\sin\theta\cos\theta - \frac{1}{8}\sin\theta\cos\theta\cos 4\theta$ **53.** 8.0349

Problems 3.4, page 178

1. $\frac{1}{2}(\sin 80° - \sin 20°)$ **3.** $\frac{1}{2}(\sin 100° + \sin 8°)$
5. $\frac{1}{2}(\sin 11\pi/28 - \sin 3\pi/28)$ **7.** $\frac{1}{2}(\cos\frac{13}{6} + \cos\frac{5}{6})$
9. $\frac{1}{2}(\sin 7\alpha + \sin\alpha)$ **11.** $\frac{3}{2}(\sin 2x + \sin 2)$
13. $\frac{1}{4}(\cos t + \cos 3t + \cos 5t + \cos 7t)$
15. $2\cos 25°\cos 10°$ **17.** $2\cos 4\cos 1$
19. $2\cos 5\pi/48\sin\pi/48$ **21.** $2\cos 7x/2\sin 3x/2$
23. $2\cos(7\theta/2)\sin(3\theta/2)$
25. $\sqrt{13}\cos(t - 0.9828)$; $\sqrt{13}\sin(t + 0.5880)$
27. $\sqrt{13}\cos(t - 4.1244)$; $\sqrt{13}\sin(t + 3.7296)$
29. $5\cos(2\theta - 0.6435)$; $5\sin(2\theta + 0.9273)$
31. $13\cos(\alpha/2 - 1.9656)$; $13\sin(\alpha/2 + 5.8884)$
33. $(\sqrt{5}/4)\cos(4\beta - 5.8195)$; $(\sqrt{5}/4)\sin(4\beta + 2.0344)$

35. $\sqrt{58}\cos(\pi x/2 - 3.5465)$; $\sqrt{58}\sin(\pi x/2 + 4.3075)$
59. $y(\text{sum}) = 4\sin 396\pi t \cos 132\pi t$

Problems 3.5, page 184

In Problems 1–17, the answers are correct for $k = 0, \pm 1, \pm 2, \ldots$.

1. $\pi/2 + k\pi$ **3.** $2\pi/3 + 2k\pi$, $4\pi/3 + 2k\pi$
5. $5\pi/24 + k\pi/4 = -\pi/24 + k\pi/4$ **7.** $\pi/4 + k\pi$
9. $7\pi/6 + 2k\pi$, $-\pi/6 + 2k\pi$ **11.** $\pi/6 + k\pi/3$
13. $0.7297 + 2k\pi$, $2.4119 + 2k\pi$
15. $\pm 0.9884 + 2k\pi$ **17.** $\pi/4 + k\pi$
21. $\pi/6$, $5\pi/6$, $7\pi/6$, $11\pi/6$ **23.** $0, \pi, 2\pi$
25. $0, \pi, 2\pi$
27. $\pi/9$, $5\pi/9$, $7\pi/9$, $11\pi/9$, $13\pi/9$, $17\pi/9$
29. $\pi/3$, π, $5\pi/3$ **31.** $2\pi/3$, $4\pi/3$, π
33. $\pi/3$, $\pi/2$, $3\pi/2$, $5\pi/3$ **35.** $\pi/3$, $5\pi/3$
37. no solutions
39. 1.2661, 5.0171, 2.4981, 3.7851 [$\cos^{-1} 0.3$, $2\pi - \cos^{-1} 0.3$, $\cos^{-1}(-0.8)$, $2\pi - \cos^{-1}(-0.8)$]
41. no solutions
43. 0.9785, 5.3047, 1.9372, 4.3460 [$\cos^{-1} 2/(\sqrt{21} - 1)$, $2\pi - \cos^{-1}(2/(\sqrt{21} - 1))$, $\cos^{-1}(-2/(\sqrt{21} + 1))$, $2\pi - \cos^{-1}(-2/(\sqrt{21} + 1))$]
45. $\tan^{-1}((5 + \sqrt{33})/4) \approx 1.2144$,
$\pi + \tan^{-1}((5 - \sqrt{33})/4) \approx 2.9576$,
$\pi + \tan^{-1}((5 + \sqrt{33})/4) \approx 4.3560$,
$2\pi + \tan^{-1}((5 - \sqrt{33})/4) \approx 6.0992$
47. $0, 2\pi, 2\pi/3$ **49.** $\pi/4$, $3\pi/4$, $5\pi/4$, $7\pi/4$
51. $\pi/2$, $3\pi/2$ **53.** $\frac{1}{2}$ **55.** 21°33′
57. $\cos^{-1}\frac{1}{16} \approx 86.417°$ **59.** 25.32%
61. (a) P(0) = 12,000 (b) 43.66% **63.** 3.50, 5.92
65. 0.35, 2.11, 2.90, 4.51 **67.** 0.36, 2.80

Problems 3.6, page 194

1. $y = \cos 2t$; period $= \pi$
3. $y = 2\sin \pi t$; period $= 2$
5. $y = -3\sin 4t$; period $= \pi/2$
7. $y = 10\sin 5t$; period $= 2\pi/5$
9. $y = -\cos(t/4)$; period $= 8\pi$
11. $y = 2\sin(t/\pi)$; period $= 2\pi^2$

In Problems 13–25 the answers are given in the order amplitude, period, and phase shift $|\delta|/\omega$.

13. $2, 2\pi, 0$

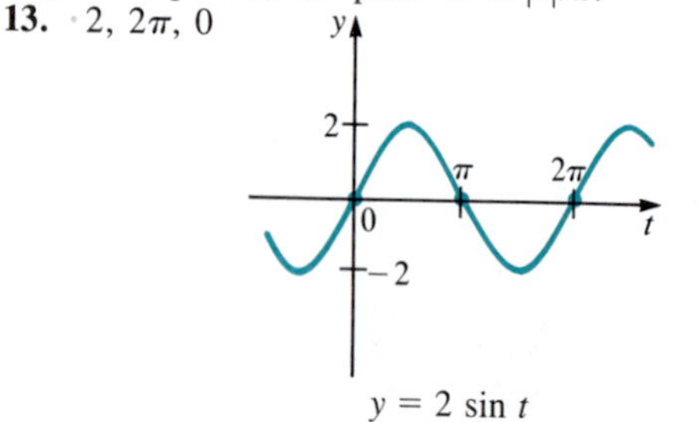

$y = 2\sin t$

15. $2, 2\pi, 0$

$y = -2\sin t$

17. $2, \pi, 0$

$y = 2\sin 2t$

19. $\frac{1}{2}, 4\pi, 0$

$y = \frac{1}{2}\sin\frac{1}{2}t$

21. $1, 2, 0$

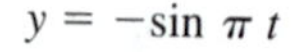

$y = -\sin \pi t$

23. $1, 2\pi, \pi$ or $1, 2\pi, 0$ [$\cos(t - \pi) = -\cos t$]

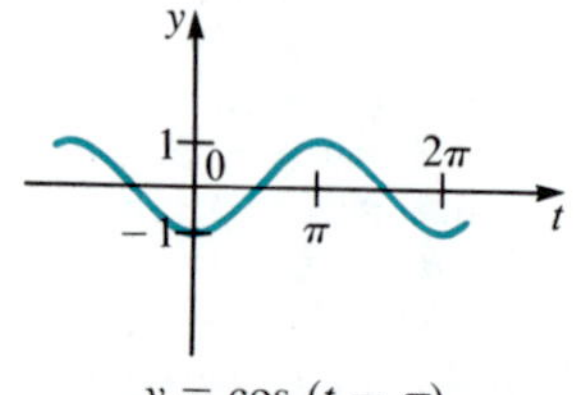

$y = \cos(t - \pi)$

25. 1, 4π, $\pi/2$

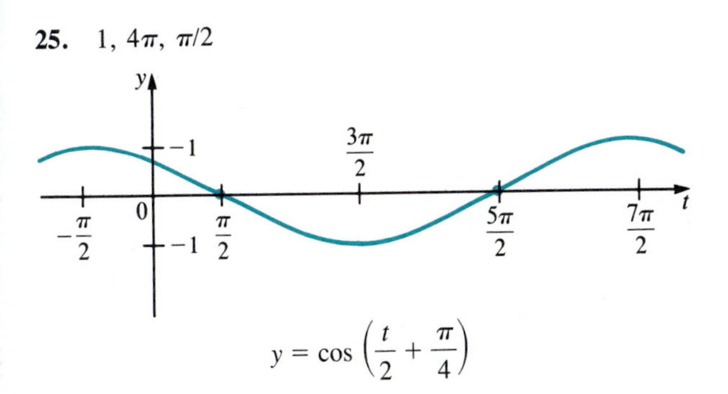

$$y = \cos\left(\frac{t}{2} + \frac{\pi}{4}\right)$$

In Problems 27, 29, and 31 we give solutions with $A > 0$ and $0 < \delta < 2\pi$

27. $A = 2$, $\omega = 2$, $\delta = \pi/2$
29. $A = 3$, $\omega = \frac{1}{2}$, $\delta = 7\pi/6$
31. $A = \frac{1}{3}$, $\omega = 1/\pi$, $\delta = 1$
33. $y = 3 \cos 2.6x$

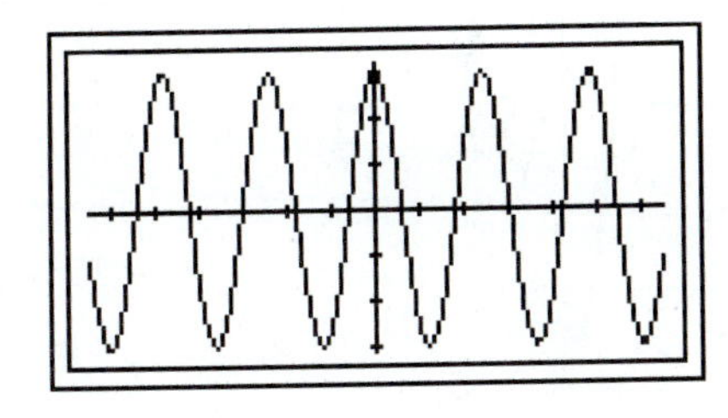

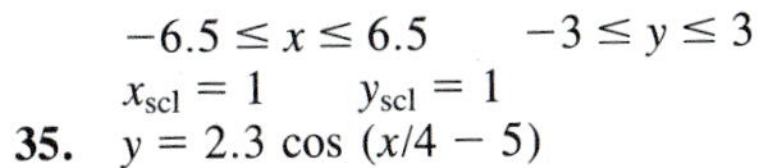

$-6.5 \le x \le 6.5$ $\quad -3 \le y \le 3$
$x_{scl} = 1$ $\quad y_{scl} = 1$

35. $y = 2.3 \cos (x/4 - 5)$

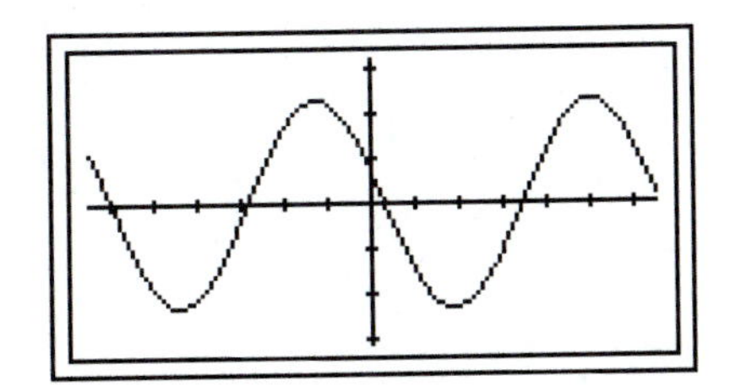

$-26 \le x \le 26$ $\quad -3 \le y \le 3$
$x_{scl} = 4$ $\quad y_{scl} = 1$

37. $y = 0.02 \sin (x/137 + 0.15\pi)$

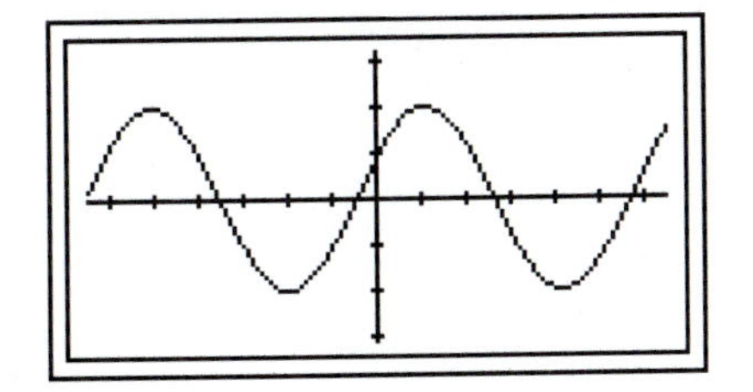

$-900 \le x \le 900$ $\quad -.03 \le y \le .03$
$x_{scl} = 140$ $\quad y_{scl} = .01$

39. $y = (\frac{1}{50}) \cos \pi^3 x$

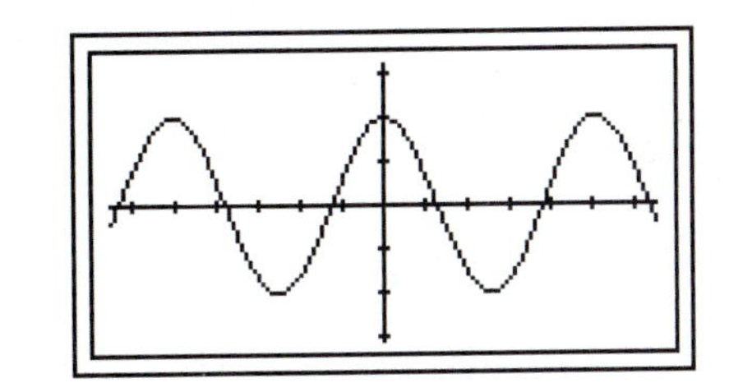

$-.26 \le x \le .26$ $\quad -.03 \le y \le .03$
$x_{scl} = .04$ $\quad y_{scl} = .01$

41. $y = 4 \sin (2x + 0.78)$

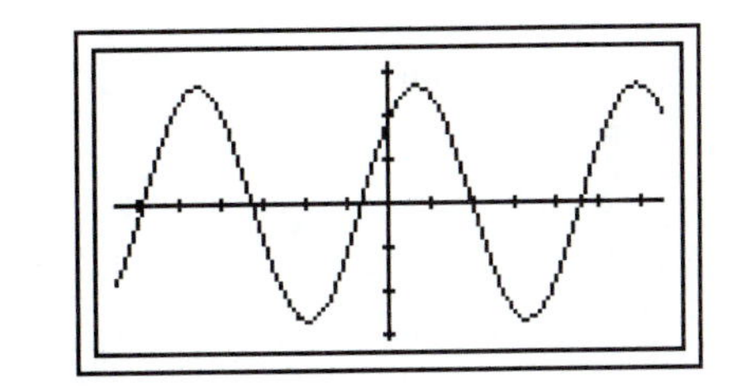

$-4 \le x \le 4$ $\quad -4.7 \le y \le 4.7$
$x_{scl} = .6$ $\quad y_{scl} = 1.5$

Problems 3.7, page 201

1.

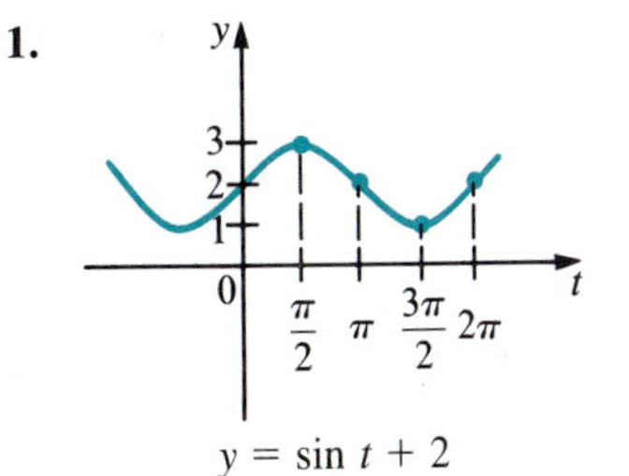

$y = \sin t + 2$

3.

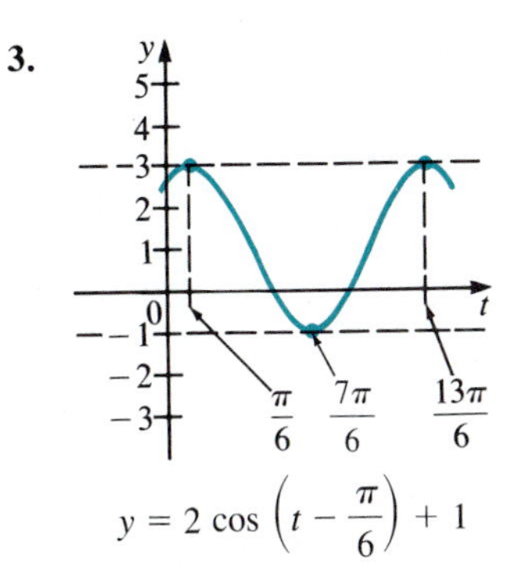

$$y = 2 \cos\left(t - \frac{\pi}{6}\right) + 1$$

5.

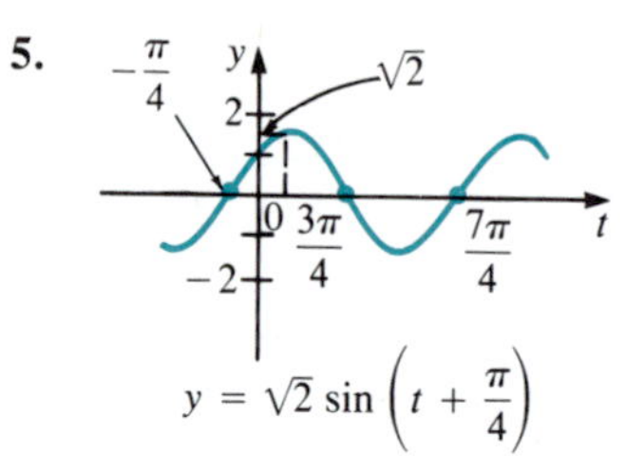

$y = \sqrt{2}\sin\left(t + \frac{\pi}{4}\right)$

7.

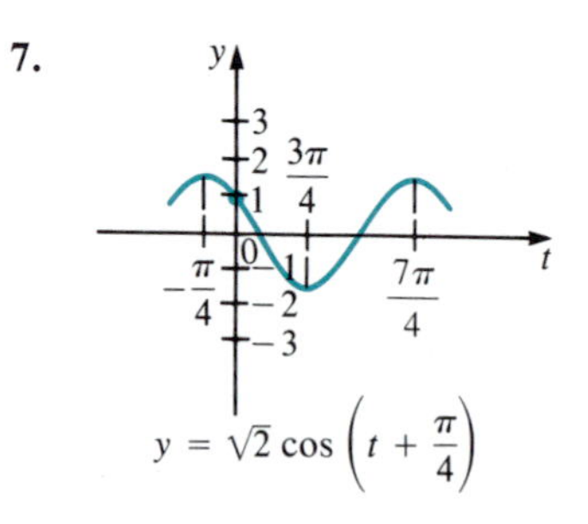

$y = \sqrt{2}\cos\left(t + \frac{\pi}{4}\right)$

9.

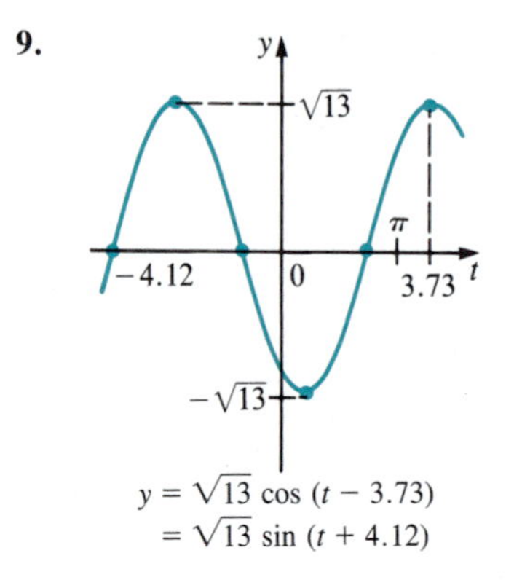

$y = \sqrt{13}\cos(t - 3.73)$
$= \sqrt{13}\sin(t + 4.12)$

11.

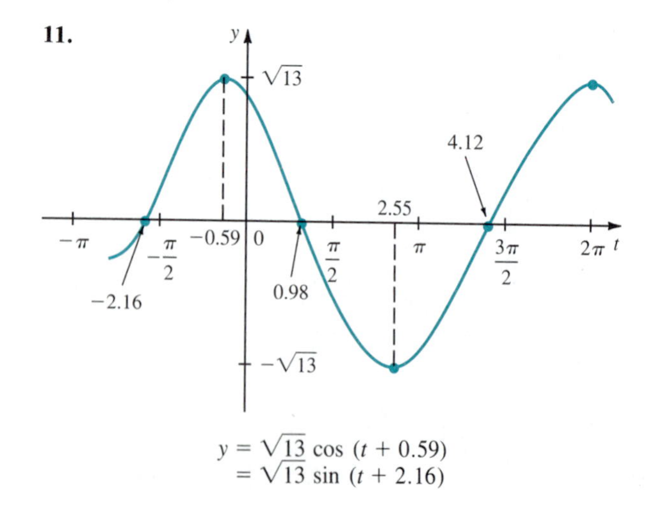

$y = \sqrt{13}\cos(t + 0.59)$
$= \sqrt{13}\sin(t + 2.16)$

13.

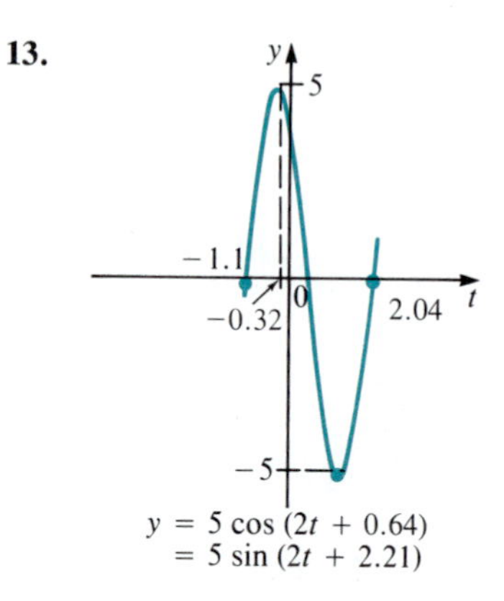

$y = 5\cos(2t + 0.64)$
$= 5\sin(2t + 2.21)$

15.

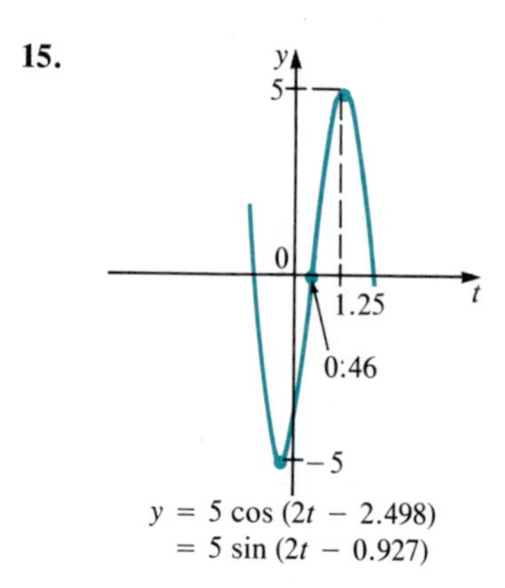

$y = 5\cos(2t - 2.498)$
$= 5\sin(2t - 0.927)$

17.

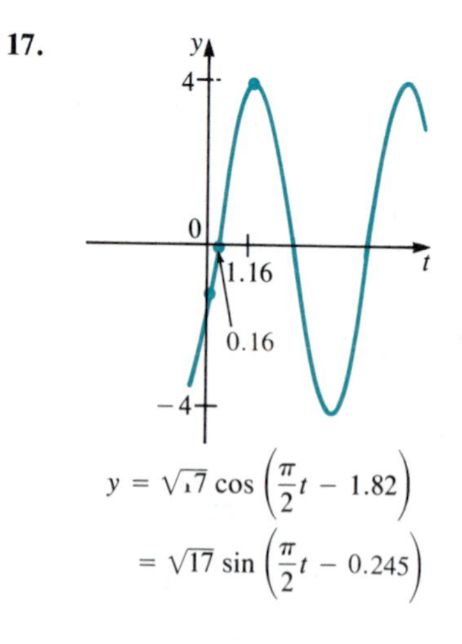

$y = \sqrt{17}\cos\left(\frac{\pi}{2}t - 1.82\right)$
$= \sqrt{17}\sin\left(\frac{\pi}{2}t - 0.245\right)$

19.

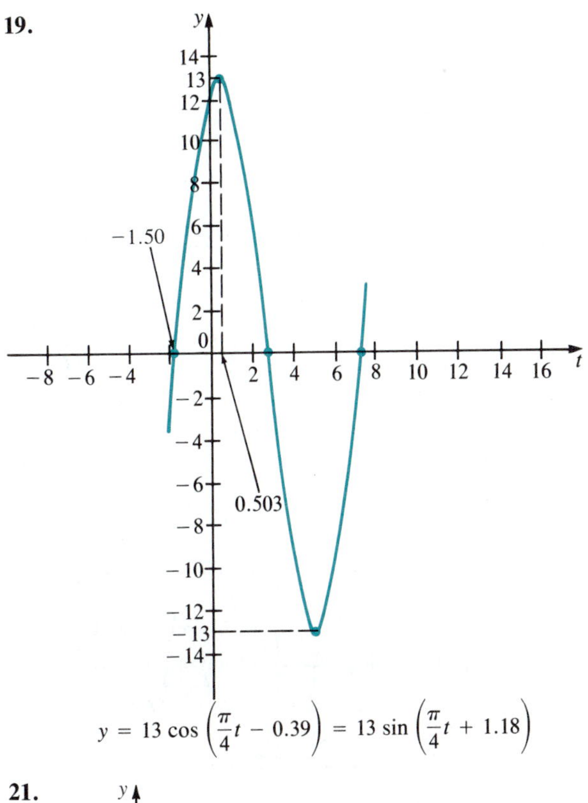

$$y = 13\cos\left(\frac{\pi}{4}t - 0.39\right) = 13\sin\left(\frac{\pi}{4}t + 1.18\right)$$

21.

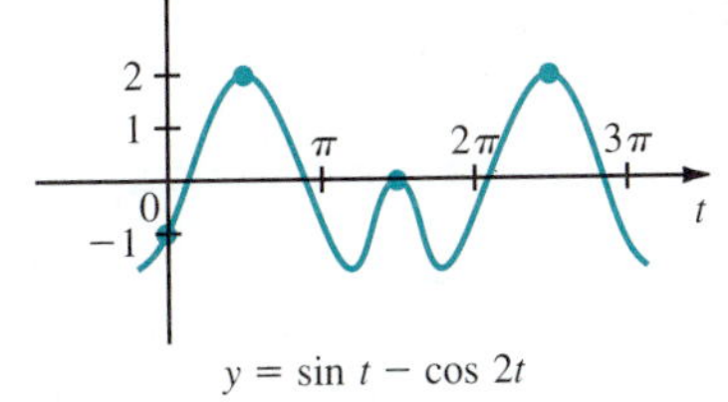

$y = \sin t - \cos 2t$

23.

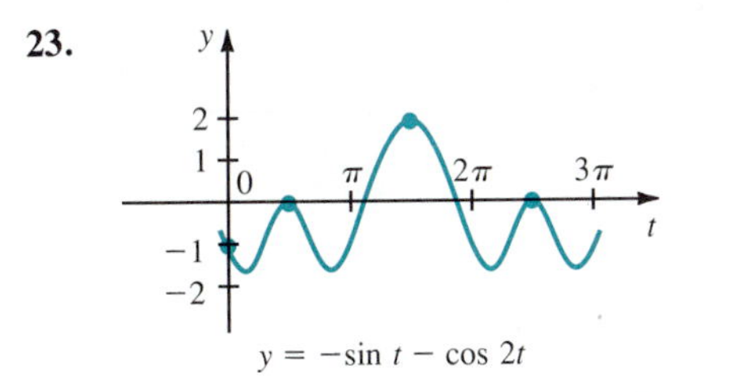

$y = -\sin t - \cos 2t$

25.

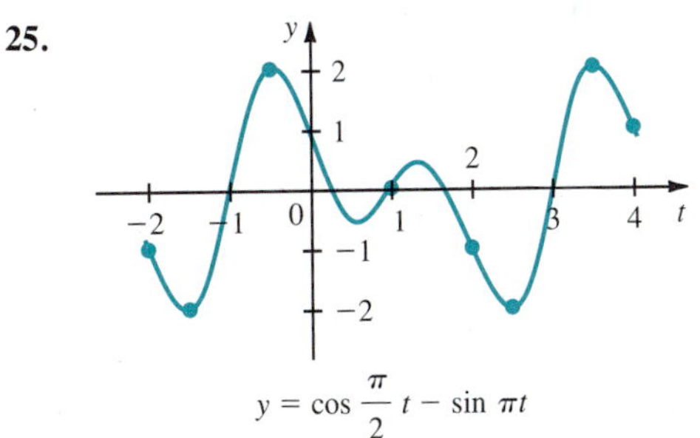

$$y = \cos\frac{\pi}{2}t - \sin \pi t$$

27. $3\sin(2x) - 5\cos(3x)$

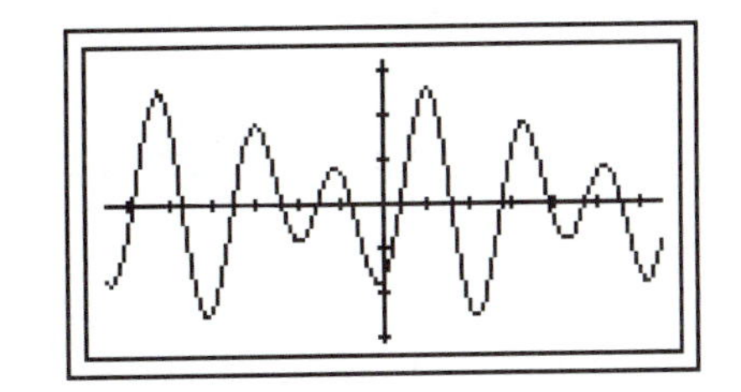

$-6.5 \le x \le 6.5 \qquad -9.5 \le y \le 9.5$
$x_{scl} = 1 \qquad y_{scl} = 3$

29. $12\cos(x/2) + 8\cos(2x/3)$

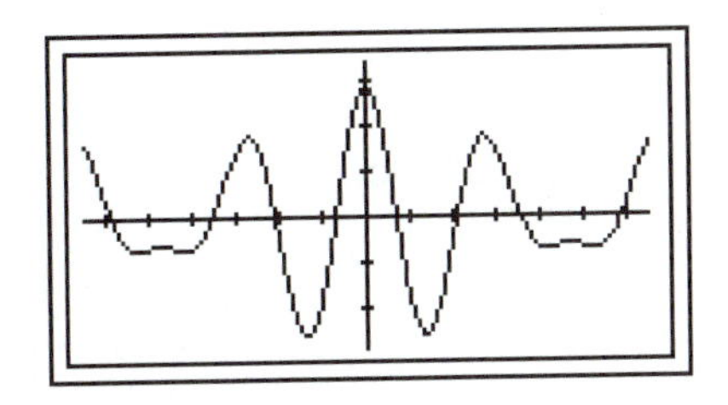

$-26 \le x \le 26 \qquad -22 \le y \le 22$
$x_{scl} = 4 \qquad y_{scl} = 7$

31. $3\sin(2x) - \cos x + 5\sin(4x)$

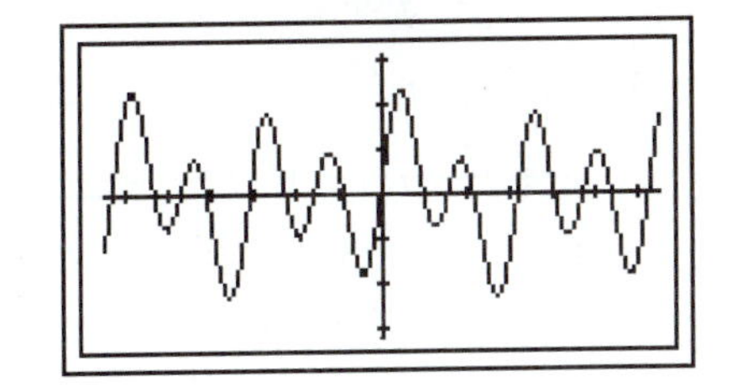

$-6.5 \le x \le 6.5 \qquad -11 \le y \le 11$
$x_{scl} = 1 \qquad y_{scl} = 3.5$

33. $\sin x + (\sin)\, 2x(/2) + (\sin)\, 3x(/3)$

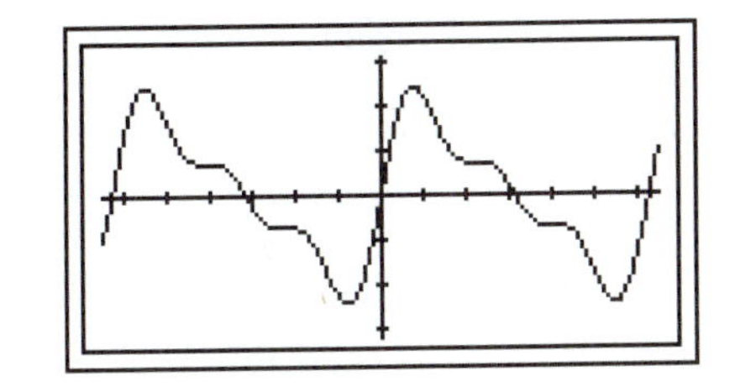

$-6.5 \le x \le 6.5 \qquad -2 \le y \le 2$
$x_{scl} = 1 \qquad y_{scl} \doteq .6$

35. $f(x) = 2(\sin x - \frac{1}{2}\sin 2x + \frac{1}{3}\sin 3x - \frac{1}{4}\sin 4x + \cdots)$

$-4\pi \le x \le 4\pi(-12.566 \le x \le 12.566) \qquad -3 \le y \le 3$
$x_{scl} = 1.93 \qquad y_{scl} = 1$

(Note: The range and scale values are the same for all parts of this problem.)

(a) first term

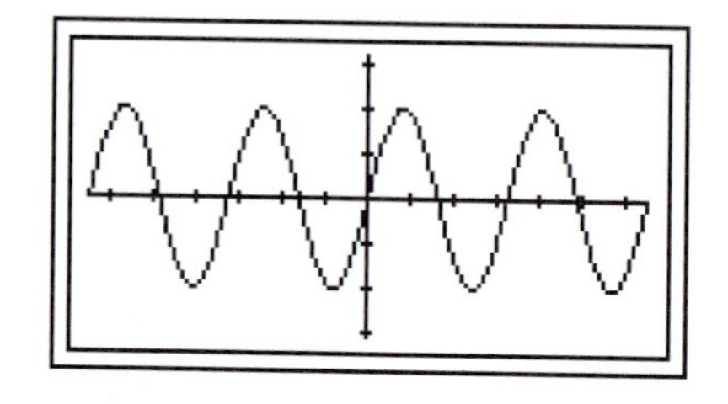

$y = 2 \sin x$

(b) S_2

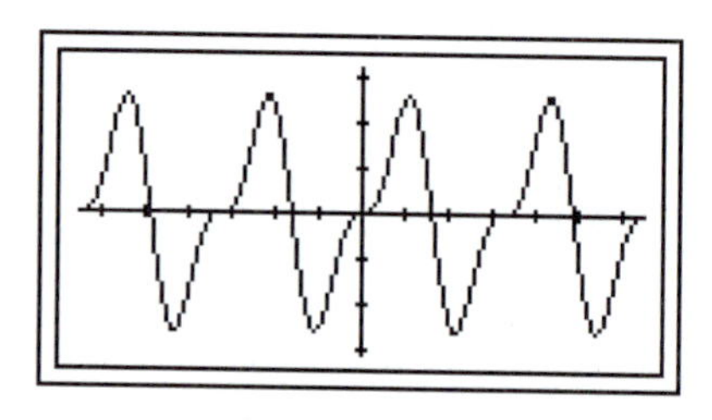

$y = 2(\sin x - .5 \sin (2x))$

(c) S_3

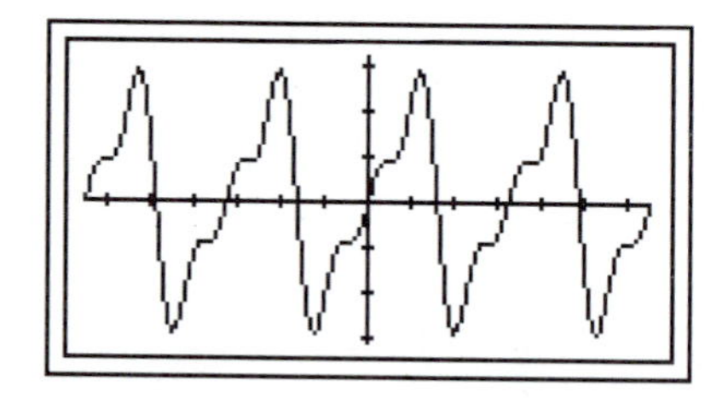

$y = 2(\sin x - .5 \sin (2x) + (\frac{1}{3}) \sin (3x))$

(d) S_4

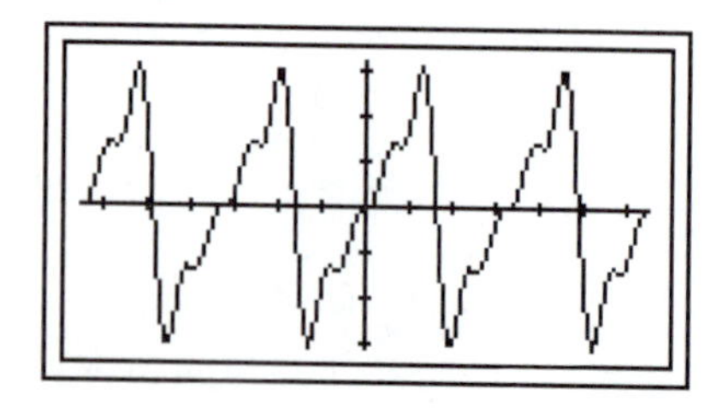

$y = 2(\sin x - .5 \sin (2x) + (\frac{1}{3}) \sin (3x) - .25 \sin (4x))$

37. $f(x) = \dfrac{4}{\pi}(\sin \pi x + \frac{1}{3} \sin 3\pi x + \frac{1}{5} \sin 5\pi x + \cdots)$

$-4 \leq x \leq 5 \qquad -2 \leq y \leq 2$

$x_{\text{scl}} = 0.71 \qquad y_{\text{scl}} = 0.67$

(Note: The range and scale values are the same for all parts of this problem.)

(a) first term

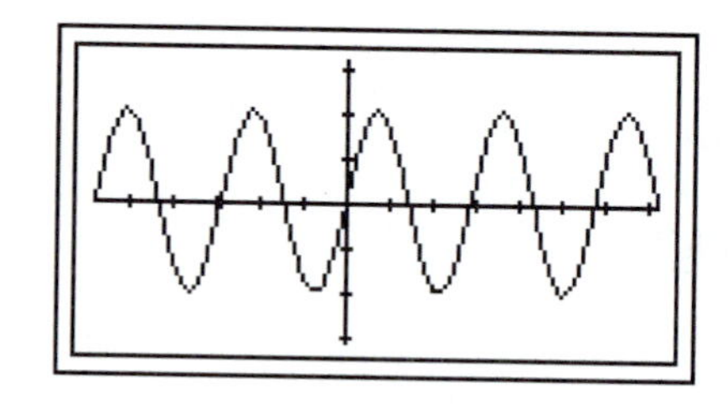

$y = (4/\pi) \sin (\pi x)$

(b) S_2

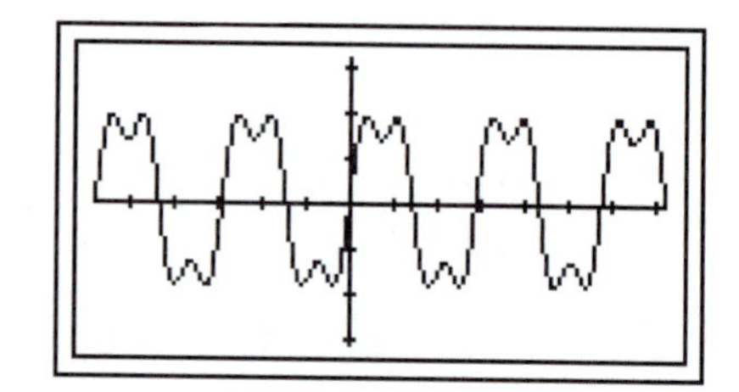

$y = (4/\pi)(\sin (\pi x) + (\frac{1}{3}) \sin (3\pi x))$

(c) S_3

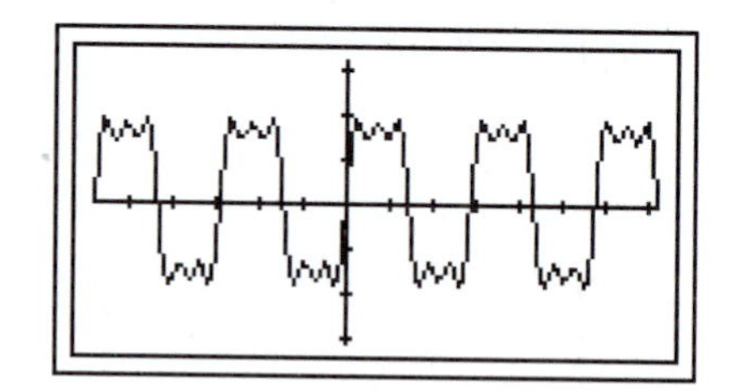

$y = (4/\pi)(\sin (\pi x) + (\frac{1}{3}) \sin (3\pi x) + .2 \sin (5\pi x))$

(d) S_4

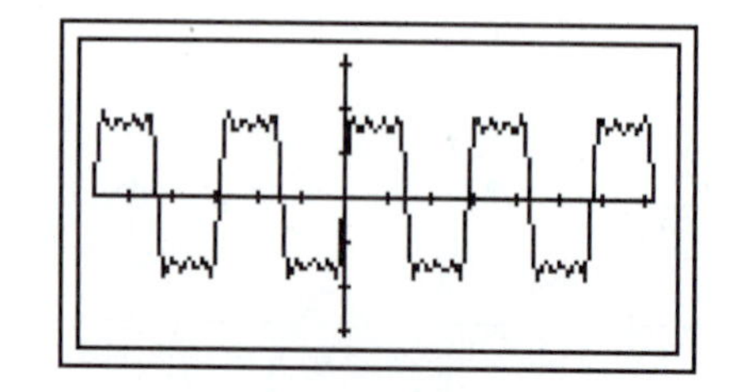

$y = (4/\pi)(\sin (\pi x) + (\frac{1}{3}) \sin (3\pi x) + .2 \sin (5\pi x) + (\frac{1}{7}) \sin (7\pi x))$

Problems 3.8, page 206

1. (a) 1 cycle/sec (b) 2 amp
3. (a) $\frac{1}{4}$ cycle/sec (b) 50 amp
5. (a) after $\frac{1}{2}\tan^{-1} 10 \approx 0.74$ years (b) 622
7. (a) after $(2/\pi)\tan^{-1}\frac{2}{3} \approx 0.3743$ years (b) 1728
9. (a) 100 cycles/sec (b) 47.553 volts
11. $y = 5 \sin 2112\,\pi t$ **13.** $y = \frac{1}{5}\sin(\pi t/100)$

Review Exercises for Chapter 3, page 209

1. $(\sin x)/(\cos^2 x)$ **3.** $-1/(\cos x + 1)^2$
9. whenever $\cos x > 0$ (quadrants I and IV)
11. $4 \sec\theta$ **13.** $(\sqrt{3}+1)/(\sqrt{3}-1) = 2 + \sqrt{3}$
15. $\sqrt{2} + 1$ **17.** $-\sin 4°$
21. $\sin(\alpha+\beta) = \frac{2}{3} - 2/3\sqrt{5}$; $\sin(\alpha-\beta) = \frac{2}{3} + 2/3\sqrt{5}$; $\cos(\alpha+\beta) = 4/3\sqrt{5} + \frac{1}{3}$; $\cos(\alpha-\beta) = 4/3\sqrt{5} - \frac{1}{3}$; $\tan(\alpha+\beta) = (2\sqrt{5}-2)/(4+\sqrt{5})$; $\tan(\alpha-\beta) = (2\sqrt{5}+2)/(4-\sqrt{5})$; $\alpha+\beta$ in quadrant I; $\alpha-\beta$ in quadrant I
23. $(-6 - 10\sqrt{2})/(3\sqrt{5} - 4\sqrt{10})$
25. $\sin 2\theta = \frac{24}{25}$, $\cos 2\theta = \frac{7}{25}$, $\tan 2\theta = \frac{24}{7}$, $\sin(\theta/2) = 1/\sqrt{10}$, $\cos(\theta/2) = 3/\sqrt{10}$, $\tan(\theta/2) = \frac{1}{3}$
27. $\sin 2\theta = 3\sqrt{7}/8$; $\cos 2\theta = \frac{1}{8}$; $\tan 2\theta = 3\sqrt{7}$; $\sin(\theta/2) = \sqrt{14}/4$; $\cos(\theta/2) = -\sqrt{2}/4$; $\tan(\theta/2) = -\sqrt{7}$
31. $\frac{1}{2}(\sin 70° + \sin 40°)$ **33.** $\frac{1}{2}(\cos 3x - \cos 7x)$
35. $2 \sin 35° \cos 15°$
37. $\sqrt{34}\cos(t + 0.5404195) = \sqrt{34}\sin(t + 2.111215827)$
39. $10\cos(3\theta - 4.068887872) = 10\sin(3\theta + 3.785093762)$
43. $4\pi/3$, $5\pi/3$
45. $\pi/4$, $7\pi/12$, $11\pi/12$, $5\pi/4$, $19\pi/12$, $23\pi/12$
47. 0.8481, 2.2935 **49.** 0, π, 2π, $2\pi/3$, $4\pi/3$
51. π **53.** 0, $4\pi/3$
55.

$y = 2 \sin 3t$

57.

$y = 3\cos\left(2t - \frac{\pi}{6}\right)$

59.

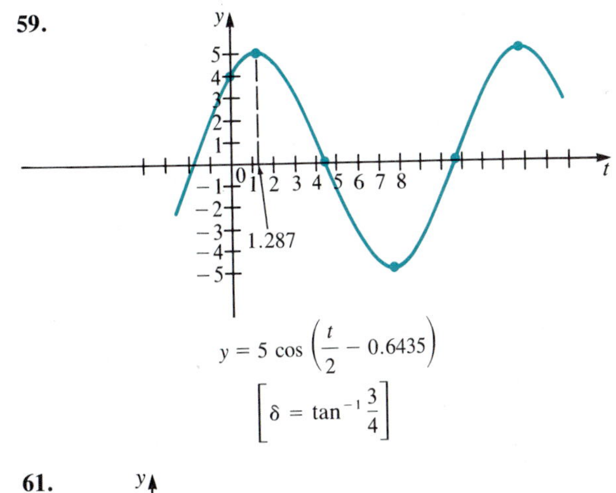

$y = 5\cos\left(\frac{t}{2} - 0.6435\right)$

$\left[\delta = \tan^{-1}\frac{3}{4}\right]$

61.

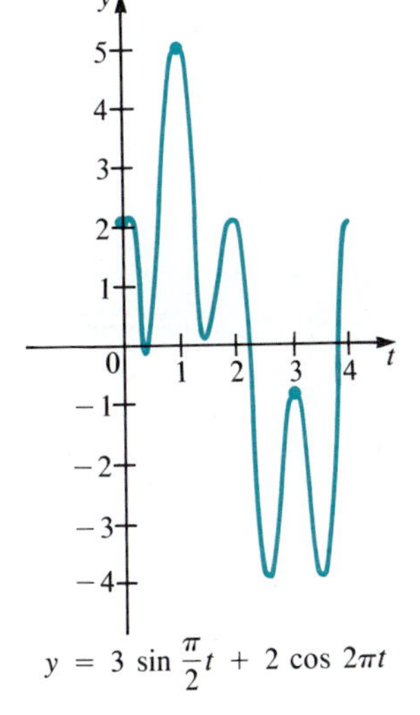

$y = 3\sin\frac{\pi}{2}t + 2\cos 2\pi t$

Chapter 4

Problems 4.1, page 217

1. $C = 65°$, $a = 3.7972$, $b = 6.5197$
3. $a = 646.85$, $c = 595.29$, $C = 65°$
5. $C = 17.3°$, $B = 122.7°$, $b = 17.011$
7. $C = 34.3°$, $B = 35.7°$, $b = 6.2069$
9. $A = 60°$, $b = 14.188$, $c = 15.698$
11. $a = 26.966$, $b = 140.74$, $B = 65°$
13. $a = 159{,}210$, $c = 222{,}280$, $B = 82.1°$
15. $C = 152°$, $b = 7.7903$, $c = 10.693$
17. $C = 51.5°$, $B = 18.5°$, $b = 2.0214$
19. no such triangle
21. $c = 64.340$, $A = 19.9°$, $C = 133.1°$
23. two triangles:
$A = 42.9°$, $C = 110.1°$, $c = 41.376$
$A = 137.1°$, $C = 15.9°$, $c = 12.084$

25. $a = 1.8586$, $A = 37.3°$, $B = 40.7°$
27. no such triangle
29. (a) $x > 10/\sin 62° \approx 11.326$ (b) $x = 11.326$ or $x \le 10$ (c) $10 < x < 11.326$
31. area $\approx$ 1.3620 square units
33. area $\approx$ 8.8274 square units
35. area $\approx$ 13.575 square units

Problems 4.2, page 225

1. 3.9101, angle opposite $4 \approx 51.6°$, angle opposite $5 \approx 78.4°$
3. 35.708, angle opposite $160 \approx 51.1°$, angle opposite $180 \approx 118.9°$
5. angle opposite $10 \approx 128.7°$, angle opposite $4 \approx 18.2°$, angle opposite $7 \approx 33.1°$
7. not possible: $147 + 185 = 332 < 353$
9. $a = 3.3577$, $B = 85.1°$, $C = 52.9°$
11. $c = 105.59$, $B = 97.6°$, $A = 44.4°$
13. $c = 5374.8$, $A = 21.7°$, $B = 6.3°$
15. $A = 32.0°$, $B = 42.4°$, $C = 105.6°$
17. $A = 139.3°$, $B = 20.4°$, $C = 20.4°$
19. 30 square units
21. no such triangle: $8 + 13 = 21$
23. 0.33223 square units **25.** 4,379,250 square units
31. 7.6 cm

Problems 4.3, page 232

1. 55.4 ft
3. 1098 ft (the length of the hill, not the elevation at the top)
5. 2.23 Å **7.** 1.439 to 1
9. (a) 86.1 m (b) 62.2 m **11.** 2273 km
13. 48.6 km **15.** 83.3° **17.** N87°14′E
19. S22°0′W **21.** (a) 117.4 ft (b) 2598.0 ft^2
23. 40.4°
25. The ship is 1087.9 km from the first lighthouse (and 1097.4 km from the second).
27. 19.7 km **29.** S37°15′W **31.** N10°43′E
33. 6080.6 m; S59°48′W

Problems 4.4, page 245

1. $\sqrt{10}$, $\tan^{-1} 3 \approx 1.249$, $\mathbf{i} + 3\mathbf{j}$
3. $\sqrt{10}$; $2\pi + \tan^{-1}(-3) \approx 5.034$; $\mathbf{i} - 3\mathbf{j}$
5. $\sqrt{34}$, $\tan^{-1}(\frac{5}{3}) \approx 1.030$; $3\mathbf{i} + 5\mathbf{j}$
7. $\sqrt{41}$, $\pi + \tan^{-1}(-0.8) \approx 2.467$; $-5\mathbf{i} + 4\mathbf{j}$
9. 4, $\pi/2$, $4\mathbf{j}$

11. (3, 3)

13. (4, 4)

15. (−2, −5)

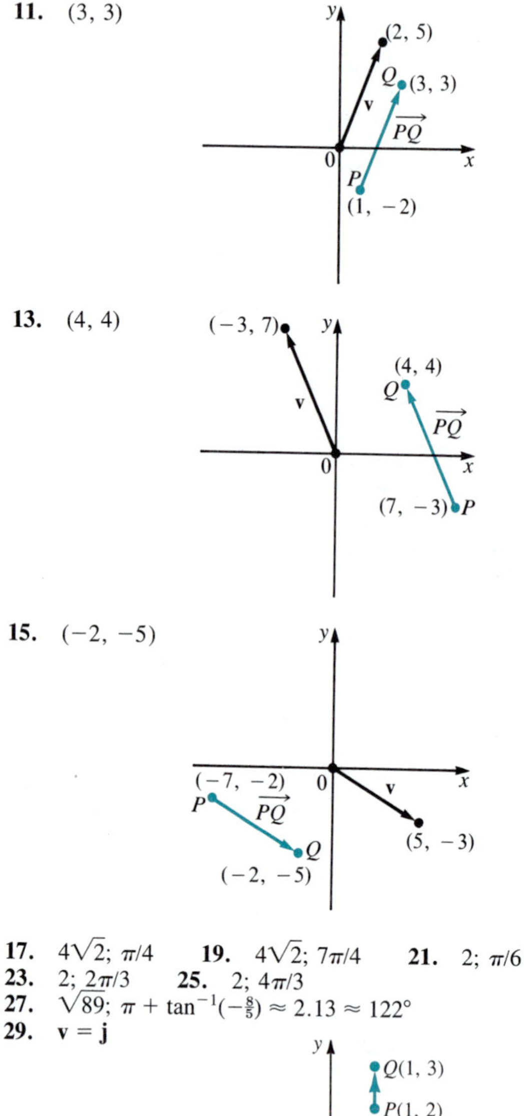

17. $4\sqrt{2}$; $\pi/4$ **19.** $4\sqrt{2}$; $7\pi/4$ **21.** 2; $\pi/6$
23. 2; $2\pi/3$ **25.** 2; $4\pi/3$
27. $\sqrt{89}$; $\pi + \tan^{-1}(-\frac{8}{5}) \approx 2.13 \approx 122°$
29. $\mathbf{v} = \mathbf{j}$

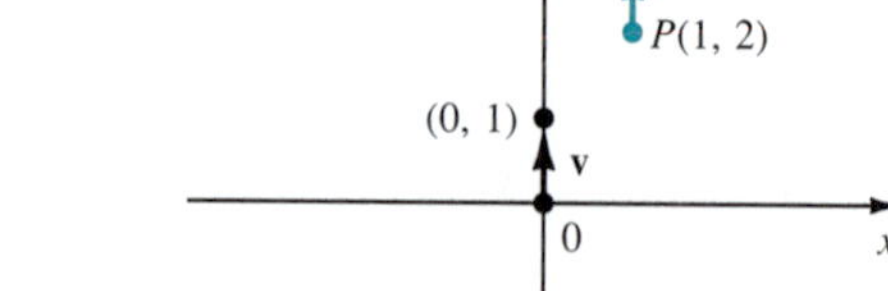

31. $\mathbf{v} = -6\mathbf{i} + \mathbf{j}$

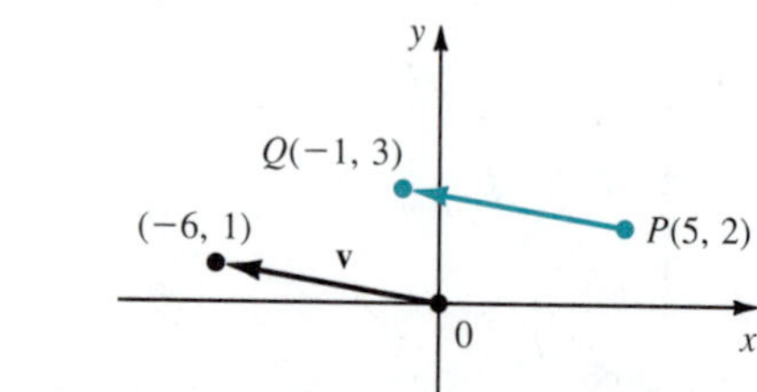

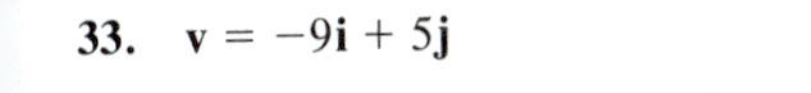

33. $\mathbf{v} = -9\mathbf{i} + 5\mathbf{j}$

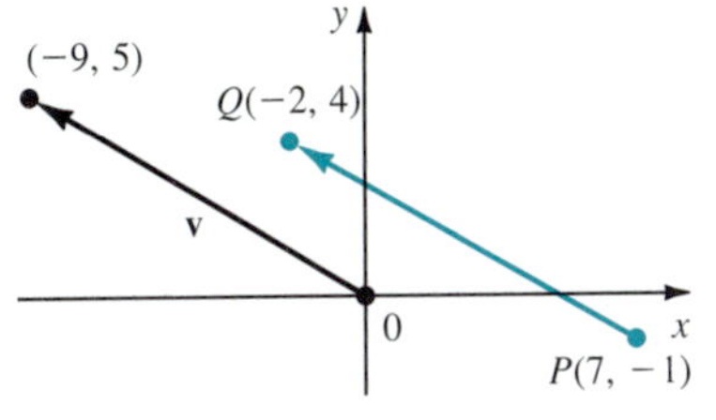

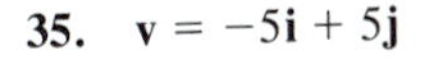

35. $\mathbf{v} = -5\mathbf{i} + 5\mathbf{j}$

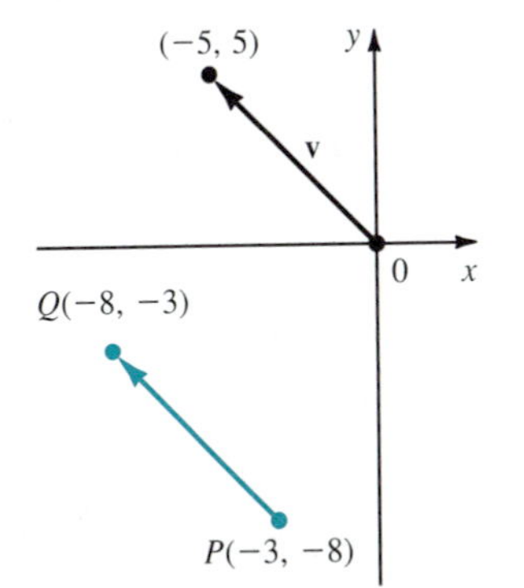

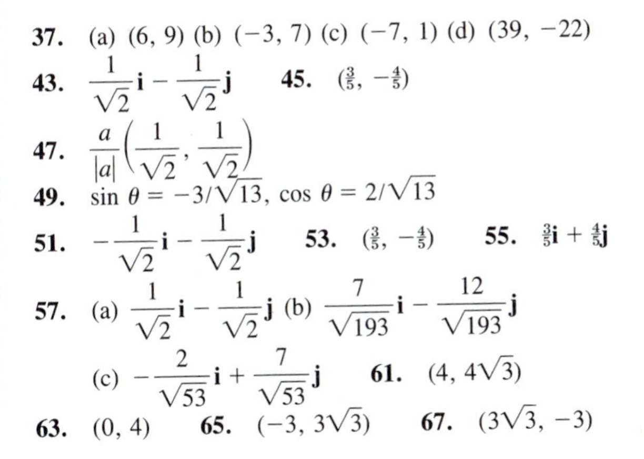

37. (a) (6, 9) (b) (−3, 7) (c) (−7, 1) (d) (39, −22)

43. $\frac{1}{\sqrt{2}}\mathbf{i} - \frac{1}{\sqrt{2}}\mathbf{j}$ **45.** $(\frac{3}{5}, -\frac{4}{5})$

47. $\frac{a}{|a|}\left(\frac{1}{\sqrt{2}}, \frac{1}{\sqrt{2}}\right)$

49. $\sin\theta = -3/\sqrt{13}$, $\cos\theta = 2/\sqrt{13}$

51. $-\frac{1}{\sqrt{2}}\mathbf{i} - \frac{1}{\sqrt{2}}\mathbf{j}$ **53.** $(\frac{3}{5}, -\frac{4}{5})$ **55.** $\frac{3}{5}\mathbf{i} + \frac{4}{5}\mathbf{j}$

57. (a) $\frac{1}{\sqrt{2}}\mathbf{i} - \frac{1}{\sqrt{2}}\mathbf{j}$ (b) $\frac{7}{\sqrt{193}}\mathbf{i} - \frac{12}{\sqrt{193}}\mathbf{j}$
(c) $-\frac{2}{\sqrt{53}}\mathbf{i} + \frac{7}{\sqrt{53}}\mathbf{j}$ **61.** $(4, 4\sqrt{3})$

63. (0, 4) **65.** $(-3, 3\sqrt{3})$ **67.** $(3\sqrt{3}, -3)$

Problems 4.5, page 252

1. (a) 62.835 lb (b) 104.574 lb **3.** $45.1° = 45°6'$
5. 135.7 lb **7.** $54.8° = 54°49'$
9. vertical 17.397 lb, horizontal 19.322 lb
11. (a) 24.6 lb (b) $37.6° = 37°34'$
13. 9092.7 kg (using $g = 9.8$ m/sec^2)
15. 413.4 mph; N48°16′W **17.** 486.4 mph; S37°34′E
19. 377.2 km/hr; N20°46′W
21. 388.4 ft/min; N48°58′E or N48°58′W **23.** 45.1°

Problems 4.6, page 261

1. 0; 0 **3.** 0; 0 **5.** 20; $\frac{20}{29}$
7. −22; $-22/5\sqrt{53}$ **9.** 100; $20/\sqrt{481}$
11. $\mathbf{u} \cdot \mathbf{v} = \alpha\beta - \beta\alpha = 0$
13. parallel

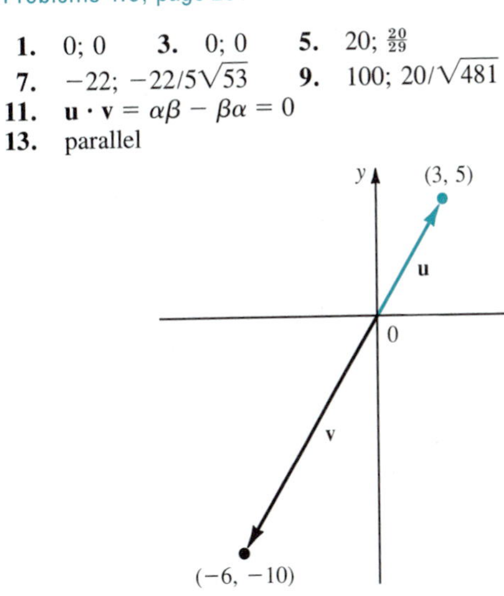

15. neither

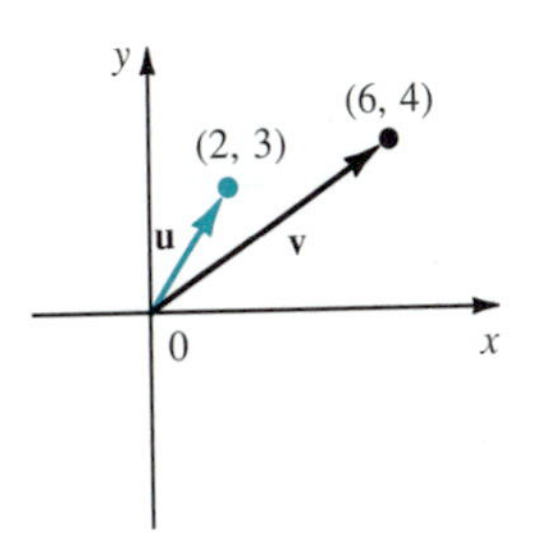

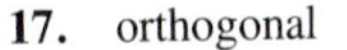

17. orthogonal

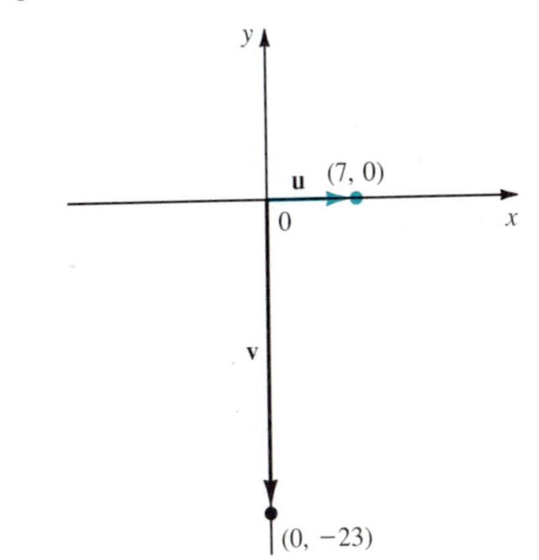

19. parallel if $\alpha \neq 0$

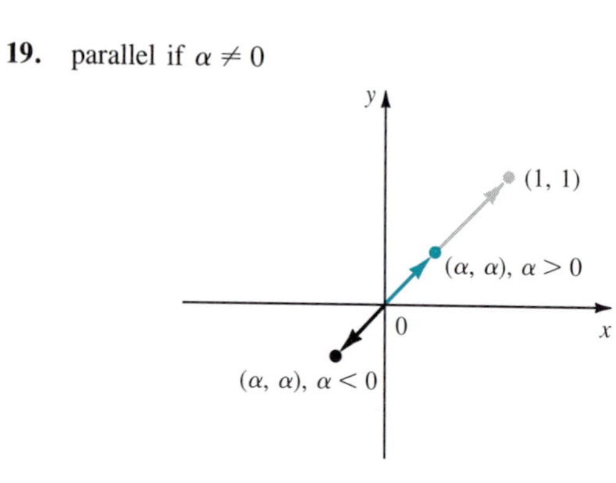

21. (a) $-\frac{3}{4}$ (b) $\frac{4}{3}$ (c) $\frac{1}{7}$ (d) $(-96 + \sqrt{7500})/78 \approx -0.12$
25. $\frac{3}{2}\mathbf{i} + \frac{3}{2}\mathbf{j}$ **27.** $\mathbf{0}$ **29.** $-\frac{2}{13}\mathbf{i} + \frac{3}{13}\mathbf{j}$
31. $\frac{14}{5}\mathbf{i} + \frac{28}{5}\mathbf{j}$ **33.** $-\frac{14}{5}\mathbf{i} + \frac{28}{5}\mathbf{j}$
35. $[(\alpha + \beta)/2]\mathbf{i} + [(\alpha + \beta)/2]\mathbf{j}$
37. $[(\alpha - \beta)/2]\mathbf{i} + [(\alpha - \beta)/2]\mathbf{j}$ **39.** $a_1a_2 + b_1b_2 > 0$
41. $\text{Proj}_{\overrightarrow{PQ}}\,\overrightarrow{RS} = \frac{51}{25}\mathbf{i} + \frac{68}{25}\mathbf{j}$; $\text{Proj}_{\overrightarrow{RS}}\,\overrightarrow{PQ} = -\frac{17}{26}\mathbf{i} + \frac{85}{26}\mathbf{j}$
47. -12 J **49.** $(8\sqrt{3} + 4)$ J **51.** $3\sqrt{2}$ J
53. $12/\sqrt{13}$ J **55.** $500(\sin 20°/\sin 30°) \approx 342$ N
57. tugboat 1: $(500)(\cos 20°)(750) \approx 352{,}385$ J;
tugboat 2: $(342)(\cos 30°)(750) \approx 222{,}136$ J
59. (a) 62.8 lb (b) -1570.9 ft-lb **61.** 1545.7 ft-lb
63. (a) 126.9 lb (b) 38,058 ft-lb

Problems 4.7, page 267

1. (2, 0) **3.** (−5, 0) **5.** $(-3\sqrt{3}, -3)$
7. (−5, 0) **9.** (5, 0) **11.** $(-3/\sqrt{2}, 3/\sqrt{2})$
13. $(-3/\sqrt{2}, 3/\sqrt{2})$ **15.** (0, −1) **17.** (0, −1)
19. (2, 0) **21.** $(3, \pi)$ **23.** $(2\sqrt{2}, 3\pi/4)$
25. $(2\sqrt{2}, 5\pi/4)$ **27.** $(1, 3\pi/2)$ **29.** $(4, 2\pi/3)$
31. $(4, 4\pi/3)$ **33.** $(4, 11\pi/6)$ **35.** $(4, 7\pi/6)$

Problems 4.8, page 271

1. circle centered at origin with radius 5
3. circle centered at origin with radius 4
5. straight line through origin with slope $-1/\sqrt{3}$

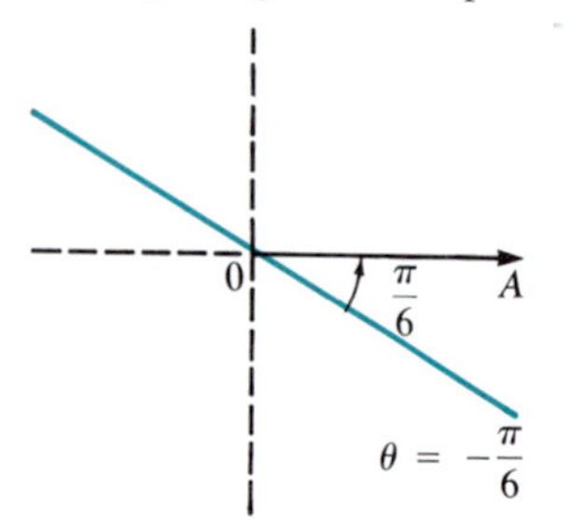

7. circle centered $(0, \frac{5}{2})$ with radius $\frac{5}{2}$
9. circle centered at $(0, -\frac{5}{2})$ with radius $\frac{5}{2}$
11. circle centered at $(\frac{5}{2}, \frac{5}{2})$ with radius $5/\sqrt{2}$

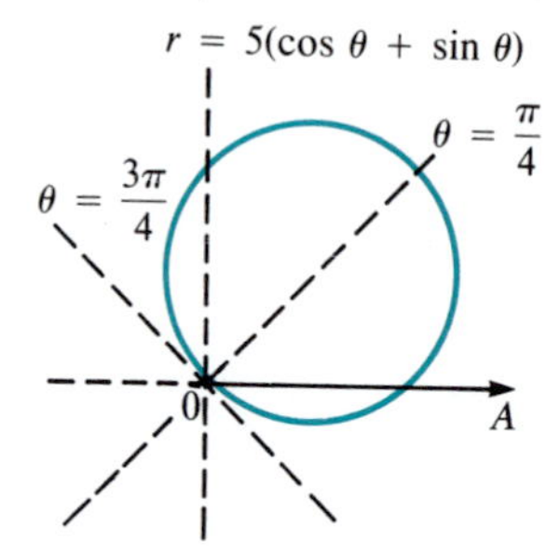

13. circle centered at $(\frac{5}{2}, -\frac{5}{2})$ with radius $5/\sqrt{2}$
15. cardioid, symmetric about $\theta = \pi/2$

17. cardioid obtained by reflecting curve in problem 15 about polar axis
19. same graph as obtained in Problem 15
21. same graph as obtained in Problem 17
23. limaçon with loop, symmetric about $\theta = \dfrac{\pi}{2}$

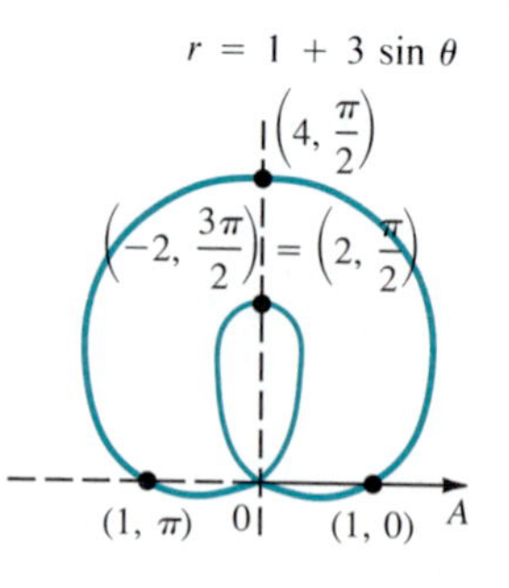

25. limaçon with loop, symmetric about polar axis

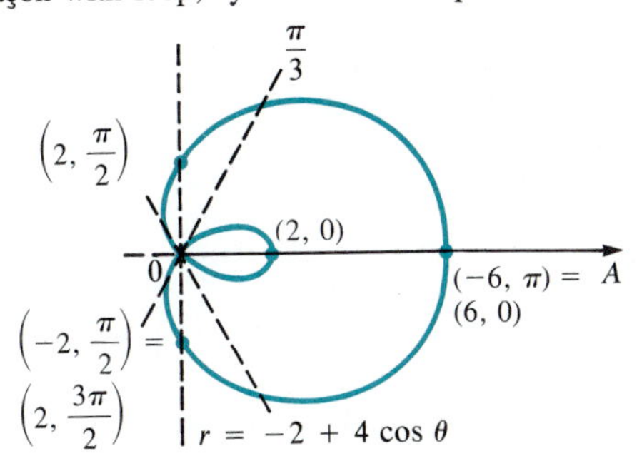

27. limaçon with loop, symmetric about polar axis

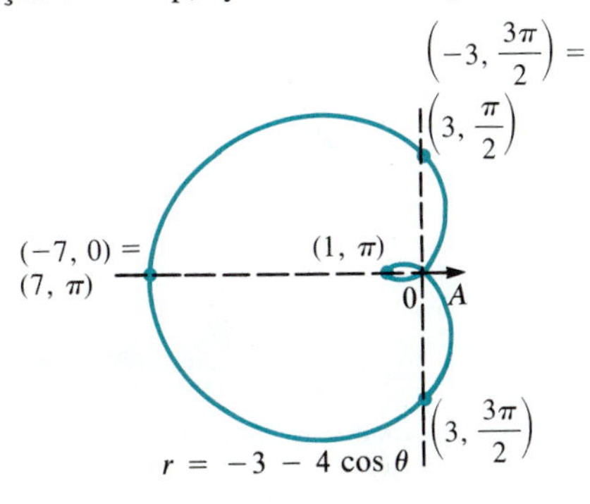

29. limaçon with loop, symmetric about $\theta = \frac{\pi}{2}$

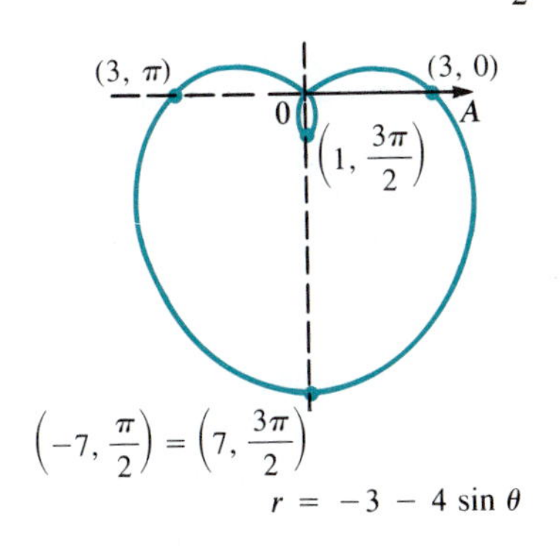

31. limaçon without loop, symmetric about polar axis

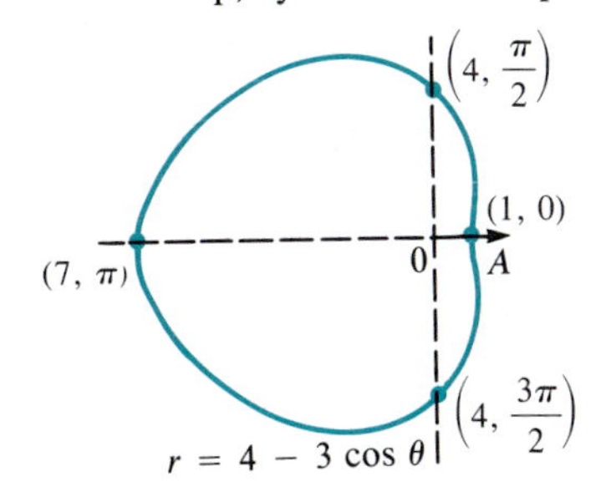

33. limaçon without loop, symmetric about $\theta = \frac{\pi}{2}$

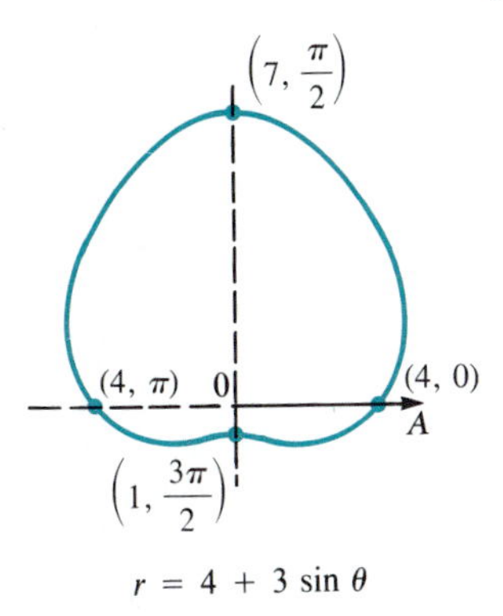

35. four-leafed rose, symmetric about polar axis, pole, and $\theta = \frac{\pi}{2}$

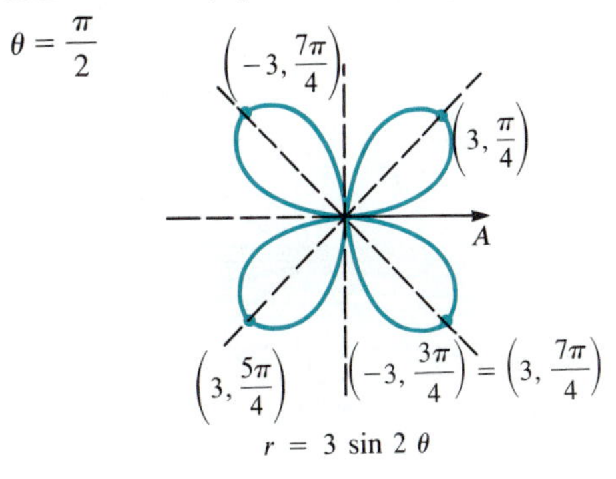

37. four-leafed rose, symmetric about polar axis, pole, and $\theta = \frac{\pi}{2}$

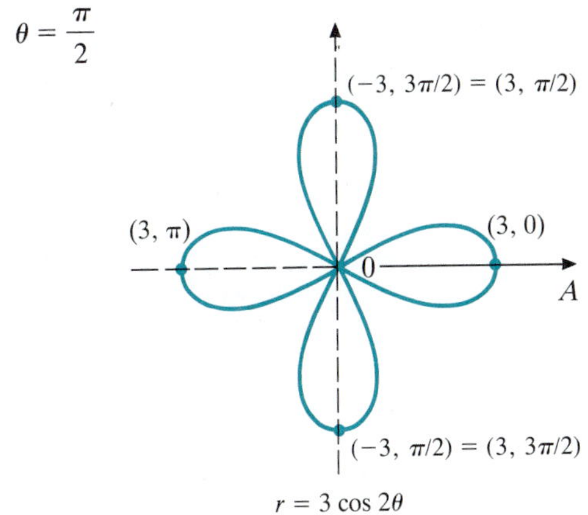

39. symmetric about $\theta = \dfrac{\pi}{2}$

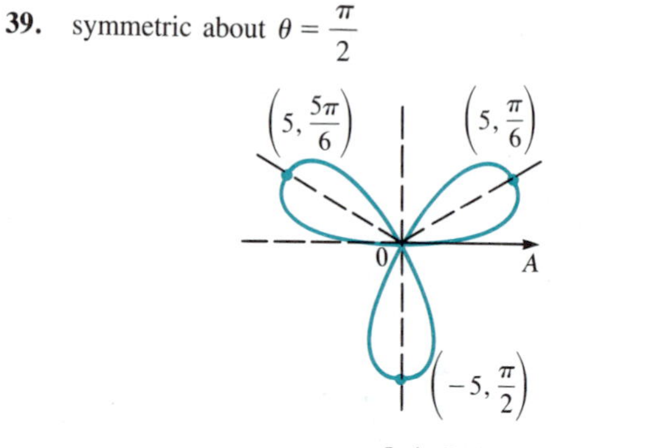

$r = 5 \sin 3\,\theta$

41. the graph of Problem 39 reflected about the polar axis

43. eight-leafed rose, symmetric about polar axis, pole, and $\theta = \dfrac{\pi}{2}$

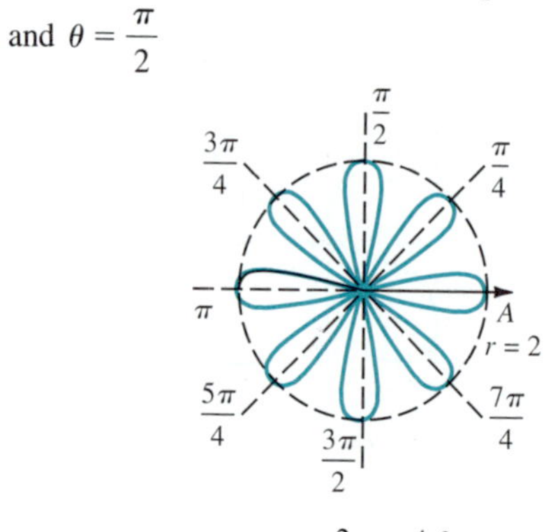

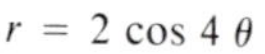

$r = 2 \cos 4\,\theta$

45. same graph as in Problem 43

47.

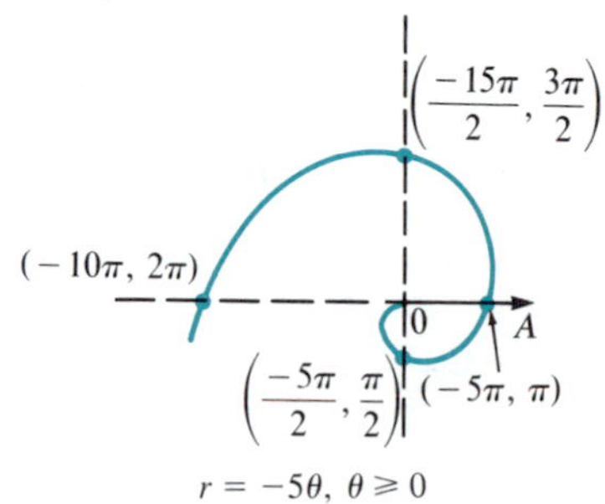

$r = -5\theta,\ \theta \geq 0$

49.

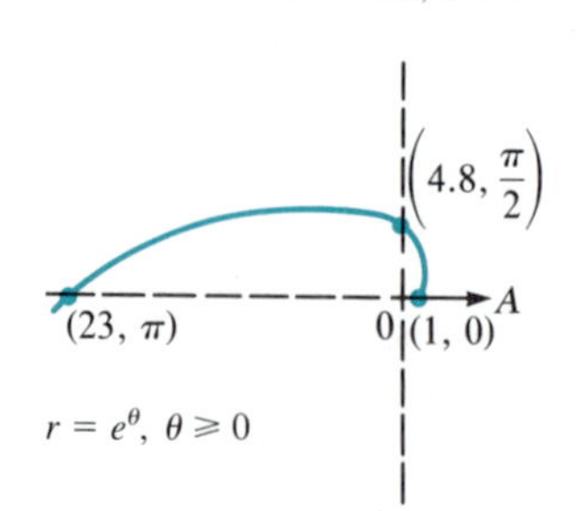

51.

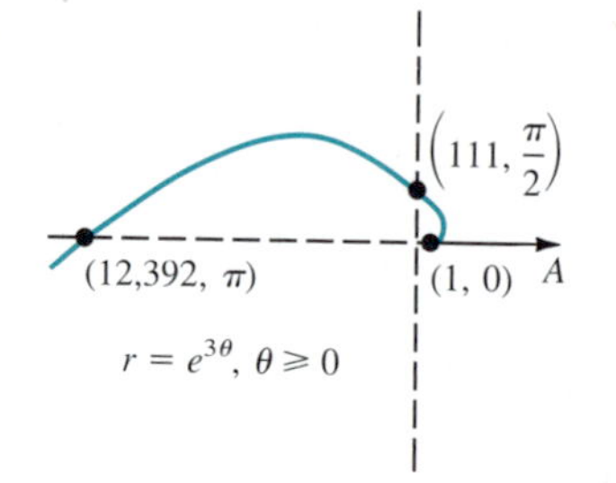

53. lemniscate, symmetric about pole

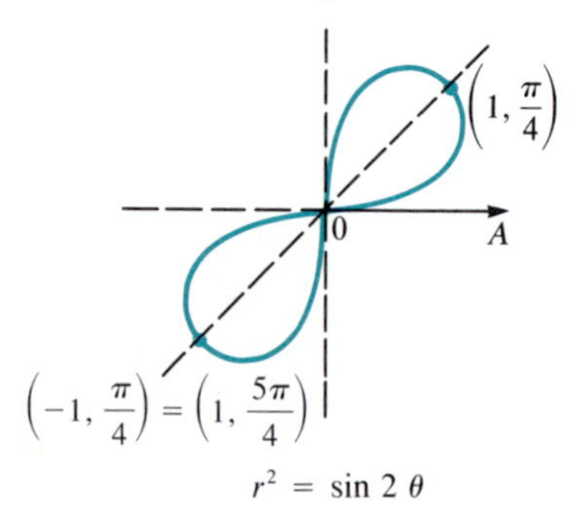

$r^2 = \sin 2\,\theta$

55. lemniscate, symmetric about pole

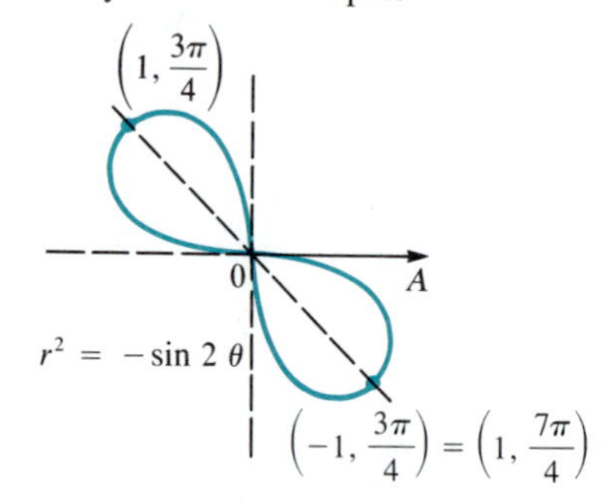

57. lemniscate, symmetric about polar axis, pole, and $\theta = \dfrac{\pi}{2}$

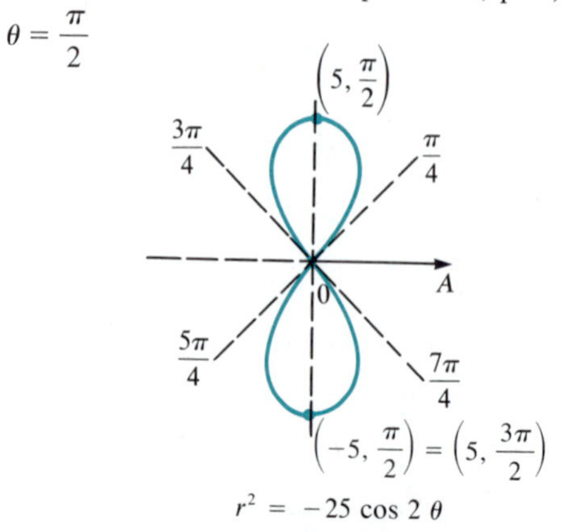

$r^2 = -25 \cos 2\,\theta$

59. cissoid, symmetric about polar axis

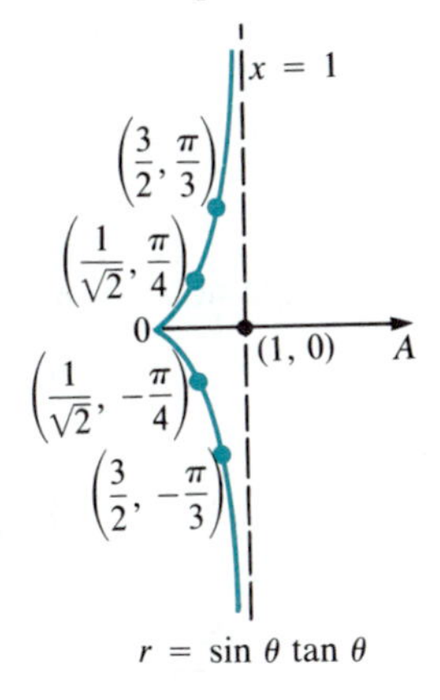

61. conchoid, symmetric about $\theta = \dfrac{\pi}{2}$

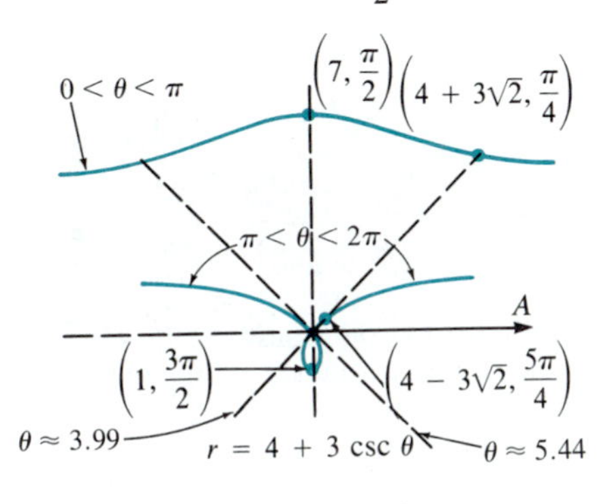

63.

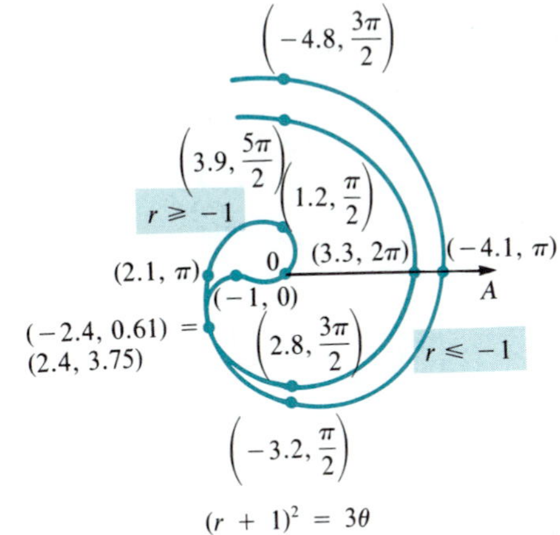

65. two circles (called *osculating circles*) with centers at $(0, \pm\frac{1}{2})$, radii $= \frac{1}{2}$, symmetric about polar axis, pole, and $\theta = \dfrac{\pi}{2}$

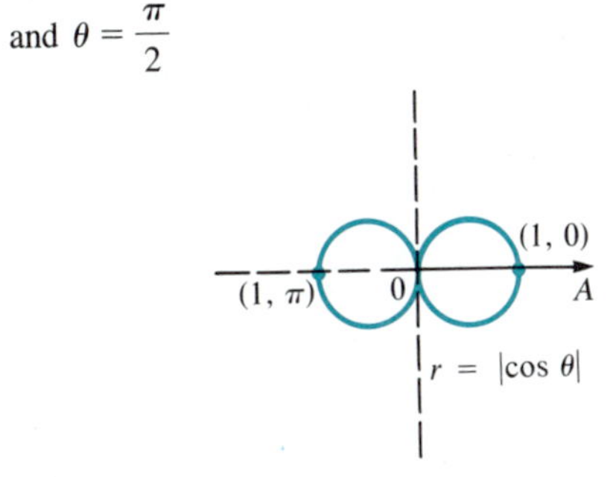

69. $r = -4\cos\theta + 3\sin\theta$

73. (a) the line $y = 3$ (b) the line $x = -2$ (c) the line $x = \frac{8}{3}$ (d) the line $2y - 3x = 4$ (e) the line $-y + 2x = 4$

Review Exercises for Chapter 4, page 274

1. $B = 70°$, $a = 4.2727$, $b = 7.1954$
3. 6.7710, 33.7°, 76.3° (opposite 7)
5. 1.8794, 2.5321 (opposite 120°), 40°
7. 24.1° (opposite 4), 30.8° (opposite 5), 125.1°
9. $b = 10.909$, $c = 16.906$, $A = 45°$
11. $c = 3.7041$, $A = 80.8°$, $B = 52.2°$
13. $A = 53.6°$, $B = 13.3°$, $C = 113.1°$
15. $c = 32.564$, $B = 64.5°$, $C = 78.5°$ and $c = 15.355$, $B = 115.5°$, $C = 27.5°$
17. $c = 39.833$, $B = 29.2°$, $C = 103.8°$
19. no such triangle **21.** $\sqrt{216} \approx 14.697$
23. 46.46 ft **25.** 4656.9 m **27.** 153 km
29. 179.4 miles; N13°30′E **31.** $3\sqrt{2}$, $3\pi/4$
33. 2, $\pi/6$ **35.** (6, 14)
37. (a) (9, 12) (b) $(-7, -3)$ (c) (6, 27)
39. $(2/\sqrt{29})\mathbf{i} + (5/\sqrt{29})\mathbf{j}$ **41.** $\frac{3}{5}\mathbf{i} + \frac{4}{5}\mathbf{j}$
43. $(\sqrt{2}/2)\mathbf{i} - (\sqrt{2}/2)\mathbf{j}$ if $a > 0$; $(-\sqrt{2}/2)\mathbf{i} + (\sqrt{2}/2)\mathbf{j}$ if $a < 0$
45. $-(5/\sqrt{29})\mathbf{i} - (2/\sqrt{29})\mathbf{j}$
47. $-(10/\sqrt{149})\mathbf{i} + (7/\sqrt{149})\mathbf{j}$ **49.** $\mathbf{j}$
51. $-(7\sqrt{3}/2)\mathbf{i} + \frac{7}{2}\mathbf{j}$ **53.** 0; 0
55. -14; $-14/\sqrt{205}$ **57.** neither
59. parallel (opposite directions)
61. parallel (same direction) **63.** $7\mathbf{i} + 7\mathbf{j}$
65. $\frac{15}{13}\mathbf{i} + \frac{10}{13}\mathbf{j}$ **67.** $-\frac{3}{2}\mathbf{i} - \frac{7}{2}\mathbf{j}$ **69.** $-\sqrt{2}$ J
71. $-33\sqrt{3} - 11 \approx -68.16$ J **73.** 300 lb
75. 116.6 ft **77.** N30°30′E
79. N22°53′E; 534.6 mph **81.** $(3\sqrt{3}/2, 3/2)$
83. $(-2, 2\sqrt{3})$ **85.** (4, 0) **87.** $(2, 11\pi/6)$
89. $(6\sqrt{2}, 5\pi/4)$
91. circle of radius 8 centered at the pole
93. circle of radius 1 centered at (1, 0)

95. limaçon, without loop, symmetric about $\theta = \pi/2$

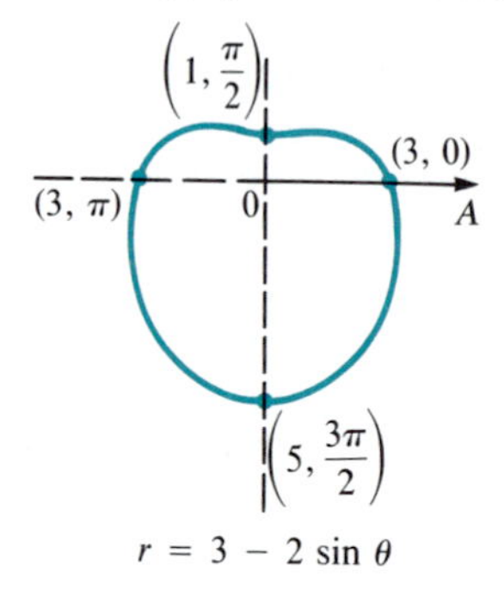

$r = 3 - 2 \sin \theta$

97. four-leafed rose, symmetric about polar axis, pole, and $\theta = \pi/2$

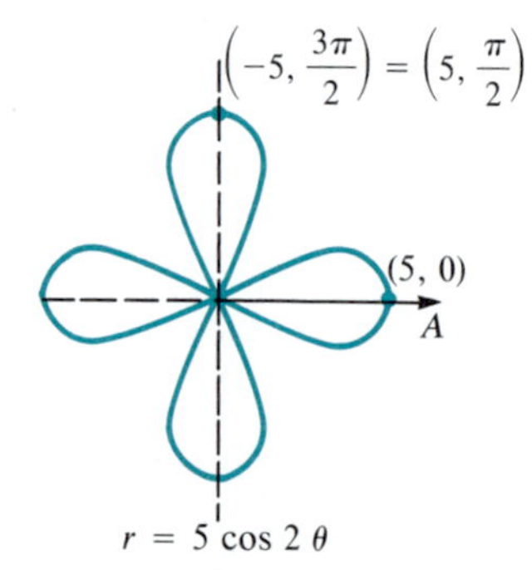

$r = 5 \cos 2\theta$

99. eight-leafed rose, symmetric about polar axis, pole, and $\theta = \pi/2$

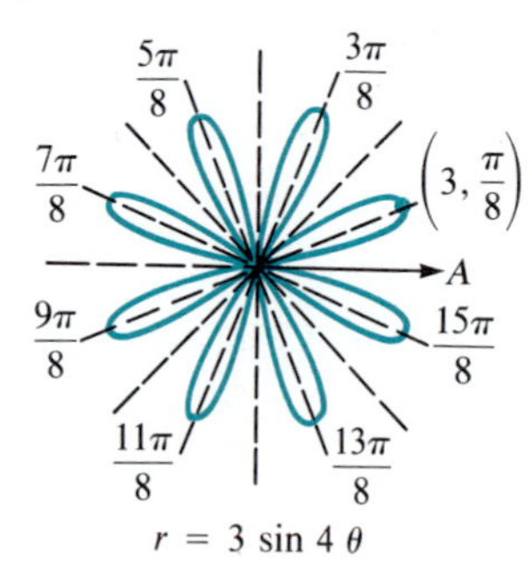

$r = 3 \sin 4\theta$

101. $x^2 + y^2 + 6x - 4y = 0$ or $(x + 3)^2 + (y - 2)^2 = 13$

Chapter 5

Problems 5.2, page 287

1. center: (0, 0)
foci: (0, ±3)
vertices: (0, ±5)
major axis: line segment between (0, −5) and (0, 5)
minor axis: line segment between (−4, 0) and (4, 0)

$e = \frac{3}{5} = 0.6$

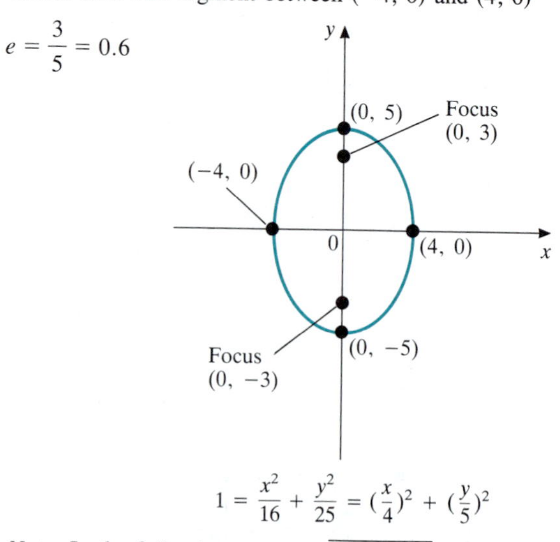

$1 = \frac{x^2}{16} + \frac{y^2}{25} = (\frac{x}{4})^2 + (\frac{y}{5})^2$

Note: In the following answers, $\overline{(a, b)(c, d)}$ denotes the line segment between (a, b) and (c, d).

3. center: (0, 0)
foci: $(0, \pm 2\sqrt{2})$
vertices: (0, ±3)
major axis: $\overline{(0, -3)(0, 3)}$
minor axis: $\overline{(-1, 0)(1, 0)}$

$e = \frac{2\sqrt{2}}{3} \approx 0.94$

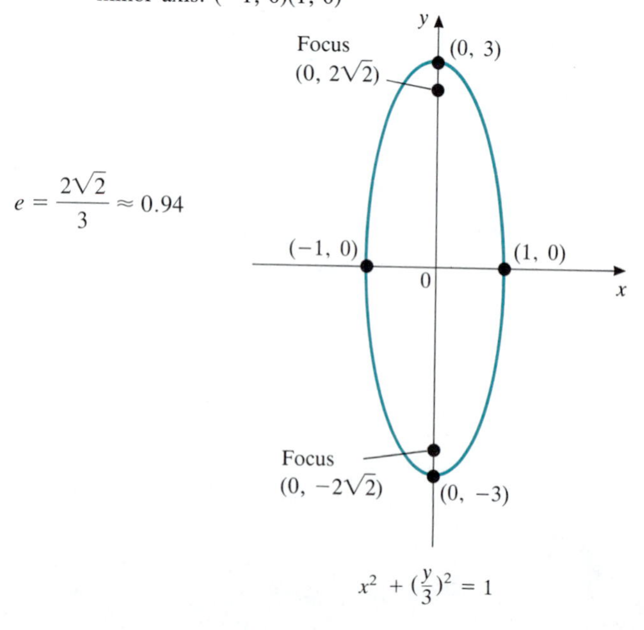

$x^2 + (\frac{y}{3})^2 = 1$

5. center: $(0, 0)$
foci: $(\pm 2\sqrt{3}, 0)$
vertices: $(\pm 4, 0)$
major axis: $\overline{(-4, 0)(4, 0)}$
minor axis: $\overline{(0, -2)(0, 2)}$

$e = \dfrac{\sqrt{3}}{2} \approx 0.87$

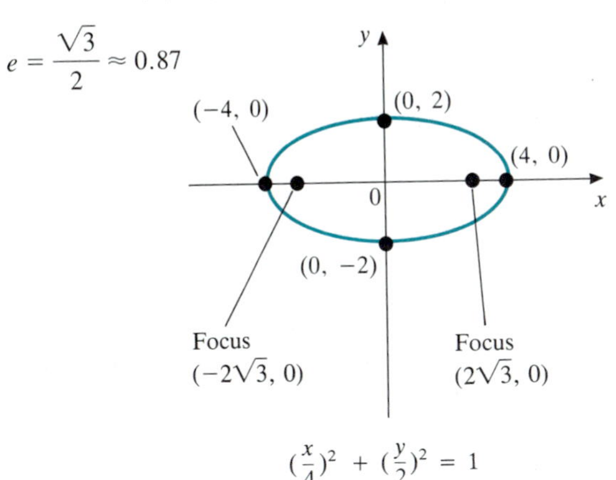

$(\frac{x}{4})^2 + (\frac{y}{2})^2 = 1$

7. center: $(1, -3)$
foci: $(1, -6)$, $(1, 0)$
vertices: $(1, -8)$, $(1, 2)$
major axis: $\overline{(1, -8)(1, 2)}$
minor axis: $\overline{(-3, -3)(5, -3)}$

$e = \dfrac{3}{5} = 0.6$

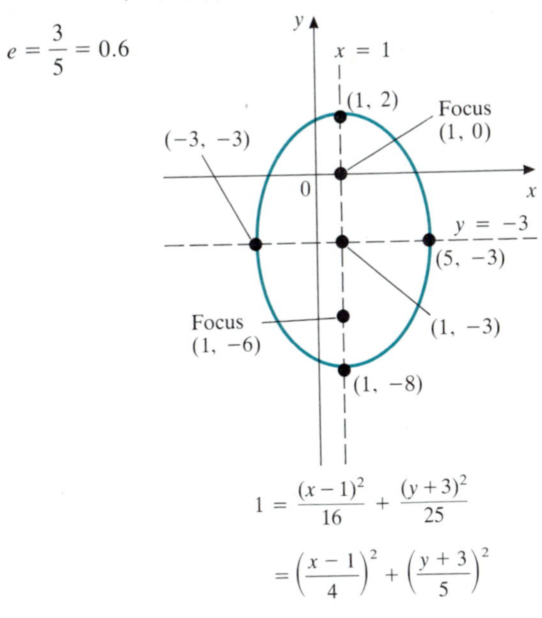

$$1 = \frac{(x-1)^2}{16} + \frac{(y+3)^2}{25}$$
$$= \left(\frac{x-1}{4}\right)^2 + \left(\frac{y+3}{5}\right)^2$$

9. The graph is a circle, centered at $(0, 0)$ with radius 1 (the unit circle).

11. center: $(0, 0)$
foci: $(\pm 3\sqrt{3}/2, 0)$
vertices: $(\pm 3, 0)$
major axis: $\overline{(-3, 0)(3, 0)}$
minor axis: $\overline{(0, -\frac{3}{2})(0, \frac{3}{2})}$

$e = \dfrac{\sqrt{3}}{2} \approx 0.87$

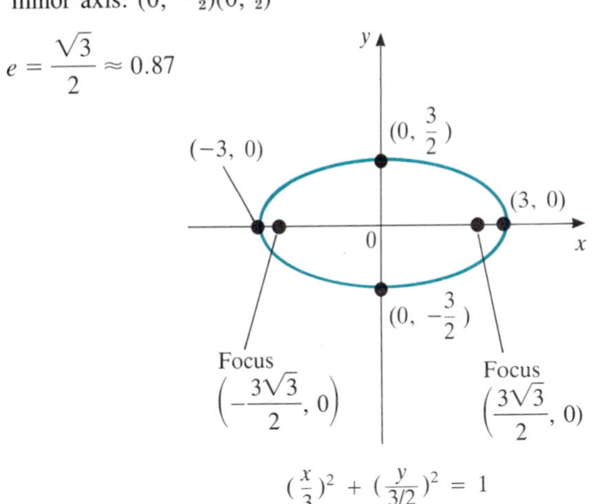

$(\frac{x}{3})^2 + (\frac{y}{3/2})^2 = 1$

13. center: $(-1, -3)$
foci: $(-1, -3 \pm 2\sqrt{3})$
vertices: $(-1, -7)$, $(-1, 1)$
major axis: $\overline{(-1, -7)(-1, 1)}$
minor axis: $\overline{(-3, -3)(1, -3)}$

$e = \dfrac{\sqrt{3}}{2} \approx 0.87$

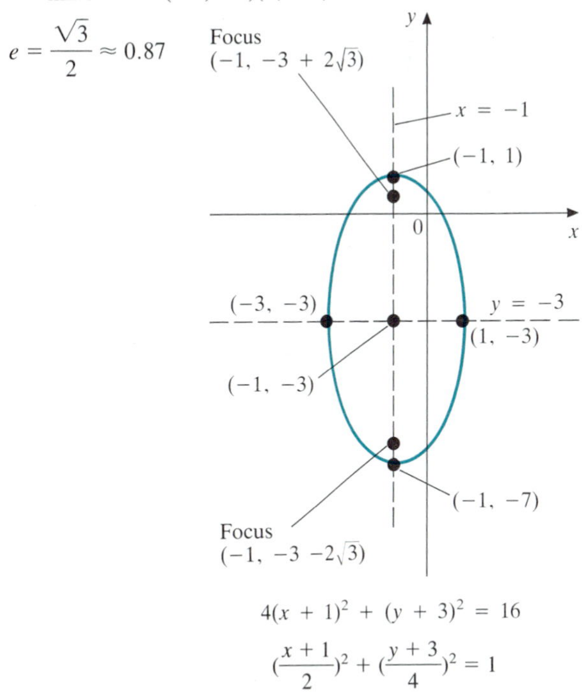

$$4(x+1)^2 + (y+3)^2 = 16$$
$$(\frac{x+1}{2})^2 + (\frac{y+3}{4})^2 = 1$$

15. center: $(-1, 3)$
foci: $(-1, 3 \pm 2\sqrt{3})$
vertices: $(-1, -1), (-1, 7)$
major axis: $\overline{(-1, -1)(-1, 7)}$
minor axis: $\overline{(-3, 3)(1, 3)}$

$e = \dfrac{\sqrt{3}}{2} \approx 0.87$

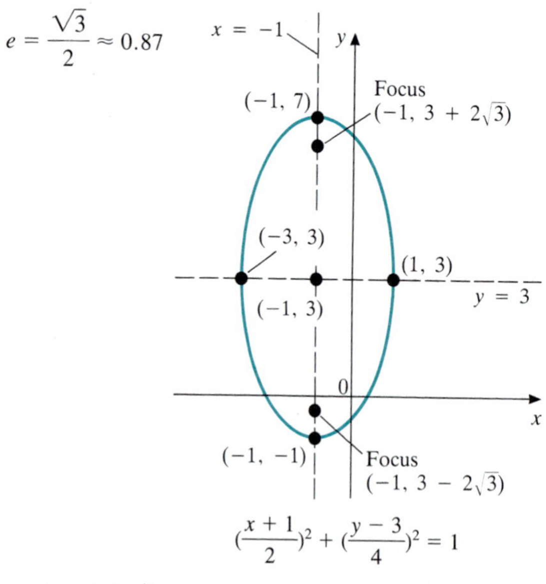

$(\frac{x+1}{2})^2 + (\frac{y-3}{4})^2 = 1$

17. center: $(-2, \frac{1}{4})$
foci: $(-2 \pm \sqrt{\frac{325}{48}}, \frac{1}{4})$
vertices: $(-2 \pm \sqrt{\frac{65}{6}}, \frac{1}{4})$
major axis: $\overline{(-2 - \sqrt{\frac{65}{6}}, \frac{1}{4})(-2 + \sqrt{\frac{65}{6}}, \frac{1}{4})}$
minor axis: $\overline{(-2, (1 - \sqrt{65})/4)(-2, (1 + \sqrt{65})/4)}$

$e = \dfrac{\sqrt{10}}{4} \approx 0.79$

Focus $(-2 - \sqrt{\frac{325}{48}}, \frac{1}{4})$ Focus $(-2 + \sqrt{\frac{325}{48}}, \frac{1}{4})$

$x = -2$

$(-2, \frac{1 + \sqrt{65}}{4})$

$(-2, \frac{1}{4})$ Center

$y = \frac{1}{4}$

$(-2, \frac{1 - \sqrt{65}}{4})$

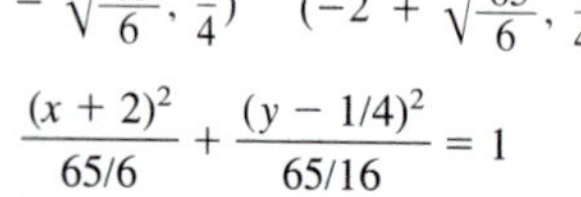

$(-2 - \sqrt{\frac{65}{6}}, \frac{1}{4})$ $(-2 + \sqrt{\frac{65}{6}}, \frac{1}{4})$

$$\frac{(x+2)^2}{65/6} + \frac{(y - 1/4)^2}{65/16} = 1$$

19. $\dfrac{x^2}{9} + \dfrac{y^2}{25} = 1$ **21.** $\dfrac{(x+1)^2}{9} + \dfrac{(y-4)^2}{25} = 1$

25. 0.9679 **27.** 8.3 AU

29. Lanes 1 and 6 have $5\sqrt{7} \approx 13.2$ ft, lanes 2 and 5 have $10\sqrt{3} \approx 17.3$ ft, lanes 3 and 4 have $5\sqrt{15} \approx 19.4$ ft.

31. (b); $(-\sqrt{21}, 0)$ and $(\sqrt{21}, 0)$

33. (h); $(-2, 4)$ and $(-2, -2)$

35. (c); $(0, \sqrt{13})$ and $(0, -\sqrt{13})$

37. (e); $(3, 1 + \sqrt{5})$ and $(3, 1 - \sqrt{5})$

39.

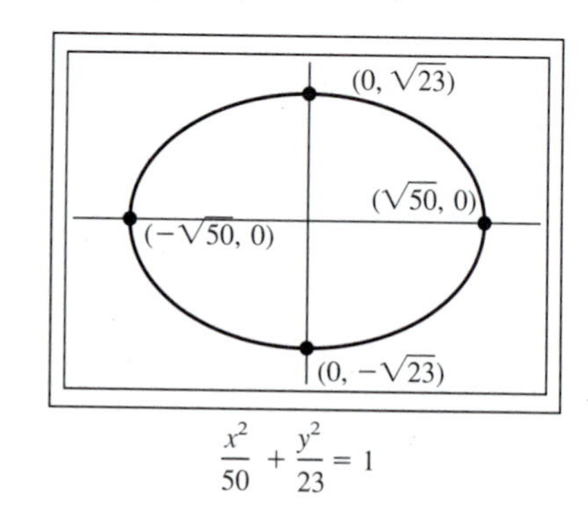

$\dfrac{x^2}{50} + \dfrac{y^2}{23} = 1$

41.

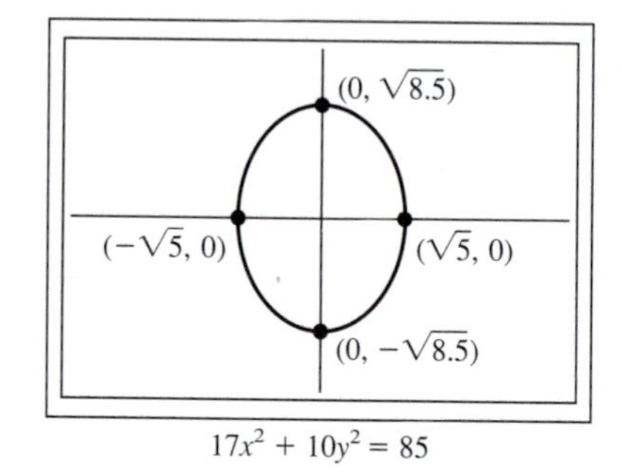

$17x^2 + 10y^2 = 85$

43.

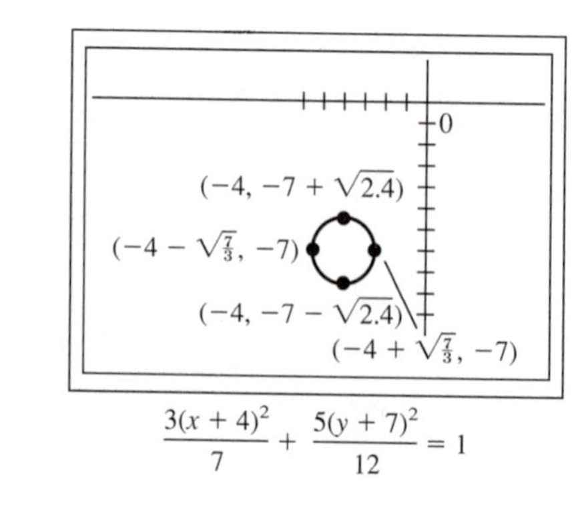

$\dfrac{3(x+4)^2}{7} + \dfrac{5(y+7)^2}{12} = 1$

Note: This is almost, but not quite, a circle. $\frac{3}{7} \approx 0.429$ and $\frac{5}{12} \approx 0.417$. If they were equal, it would be a circle.

45.

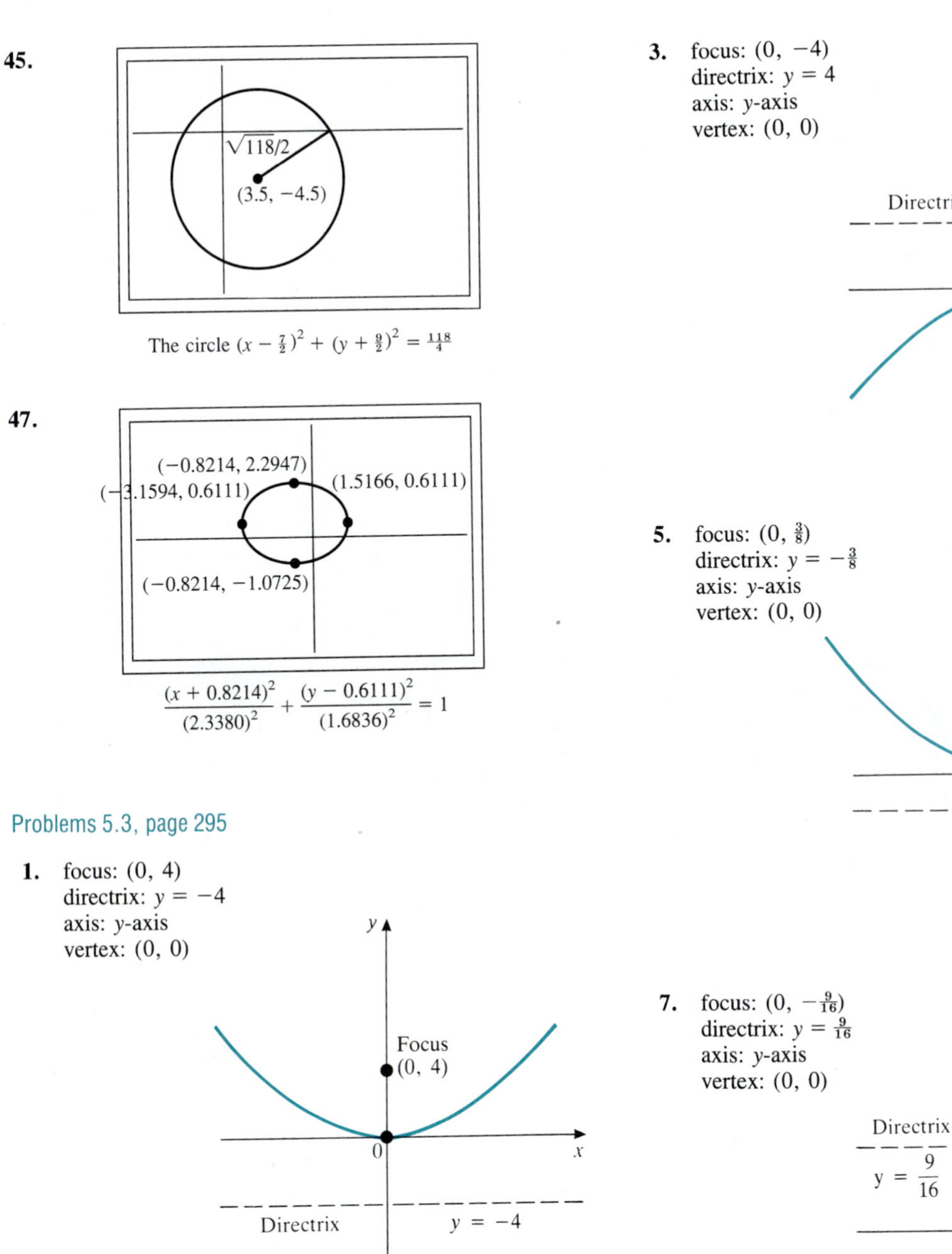

The circle $(x - \frac{7}{2})^2 + (y + \frac{9}{2})^2 = \frac{118}{4}$

47.

$$\frac{(x + 0.8214)^2}{(2.3380)^2} + \frac{(y - 0.6111)^2}{(1.6836)^2} = 1$$

Problems 5.3, page 295

1. focus: (0, 4)
directrix: $y = -4$
axis: y-axis
vertex: (0, 0)

Focus (0, 4)
Directrix $y = -4$

$x^2 = 16y$

3. focus: (0, −4)
directrix: $y = 4$
axis: y-axis
vertex: (0, 0)

Directrix $y = 4$
Focus (0, −4)

$x^2 = -16y$

5. focus: $(0, \frac{3}{8})$
directrix: $y = -\frac{3}{8}$
axis: y-axis
vertex: (0, 0)

Focus $(0, \frac{3}{8})$
(0, 0)
$y = -\frac{3}{8}$ Directrix

$\frac{3}{2}y = x^2$

7. focus: $(0, -\frac{9}{16})$
directrix: $y = \frac{9}{16}$
axis: y-axis
vertex: (0, 0)

Directrix $y = \frac{9}{16}$
(0, 0)
Focus $(0, -\frac{9}{16})$

$-\frac{9}{4}y = x^2$

9. vertex: $(1, -3)$
focus: $(1, -7)$
directrix: $y = 1$
axis: $x = 1$

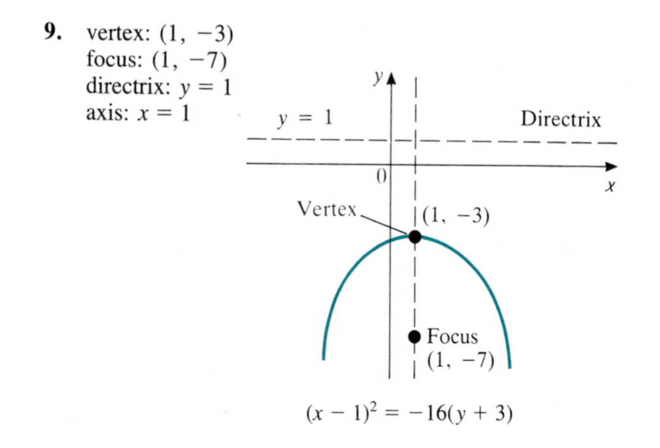

$(x - 1)^2 = -16(y + 3)$

11. vertex: $(0, \frac{9}{4})$
focus: $(0, \frac{5}{4})$
directrix: $y = \frac{13}{4}$
axis: $x = 0$

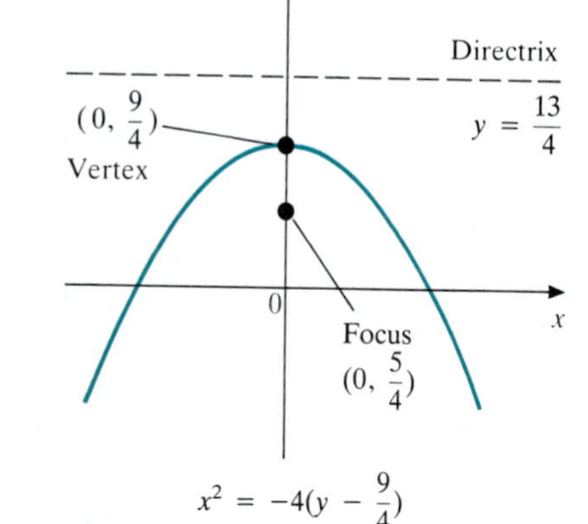

$x^2 = -4(y - \frac{9}{4})$

13. vertex: $(-1, 0)$
focus: $(-1, -\frac{1}{4})$
directrix: $y = \frac{1}{4}$
axis: $x = -1$

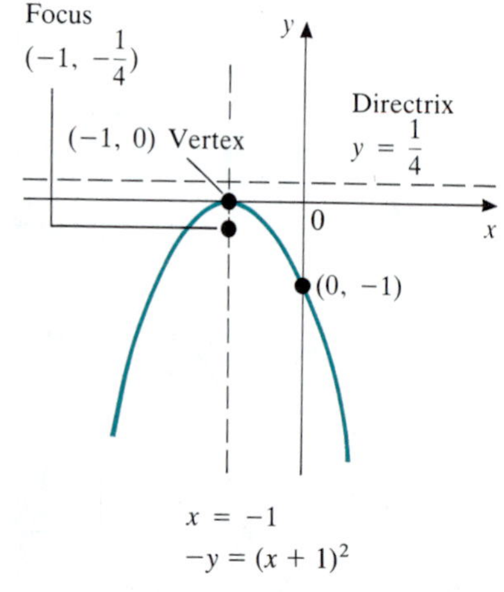

$x = -1$
$-y = (x + 1)^2$

15. vertex: $(-2, 4)$
focus: $(-2, \frac{15}{4})$
directrix: $y = \frac{17}{4}$
axis: $x = -2$

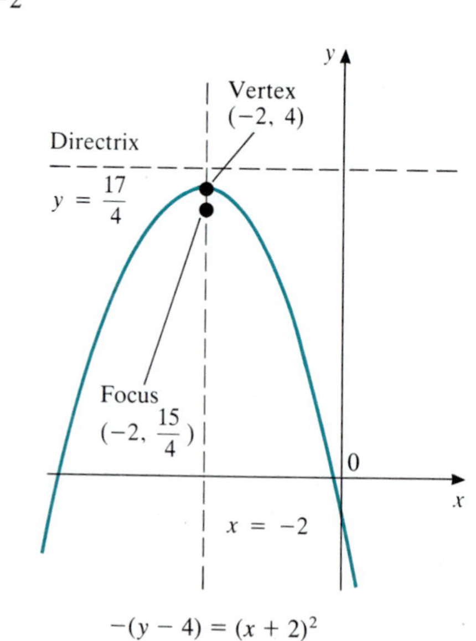

$-(y - 4) = (x + 2)^2$

17. vertex: $(-2, -4)$
focus: $(-2, -\frac{15}{4})$
directrix: $y = -\frac{17}{4}$
axis: $x = -2$

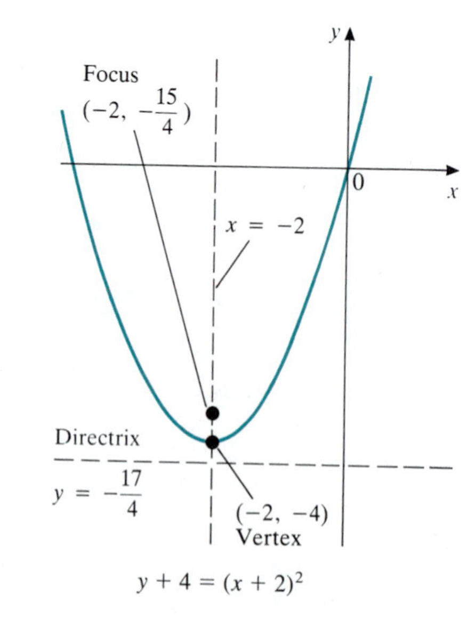

$y + 4 = (x + 2)^2$

19. $x^2 = 16y$ **21.** $(y - 5)^2 = -12(x + 2)$
23. focus $(0, -1)$; directrix $x = 6$ **27.** $x^2 = 8y$

29. (a)

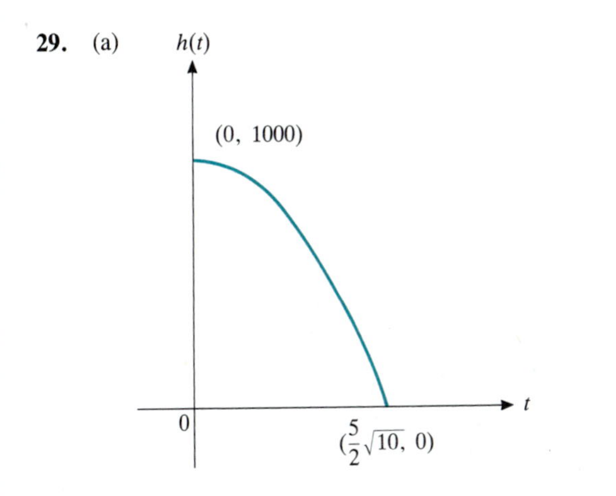

(b) after $\frac{5}{2}\sqrt{10} \approx 7.9$ sec

31. *f* **33.** *b* **35.** *c* **37.** *e*

39.

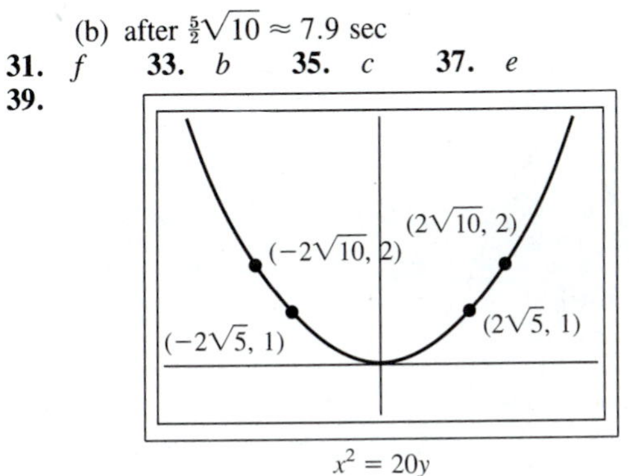

$x^2 = 20y$

41.

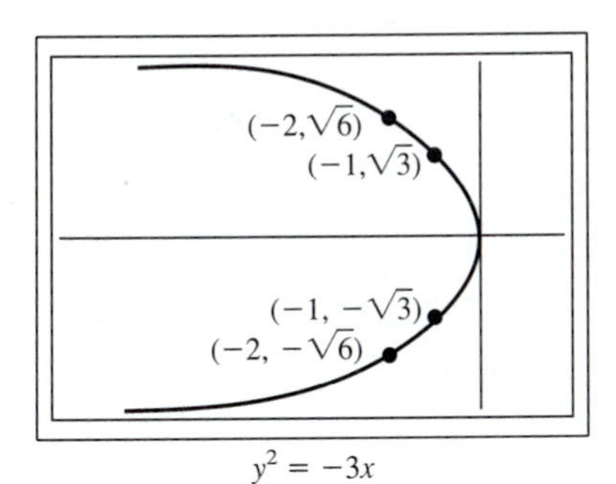

$y^2 = -3x$

43.

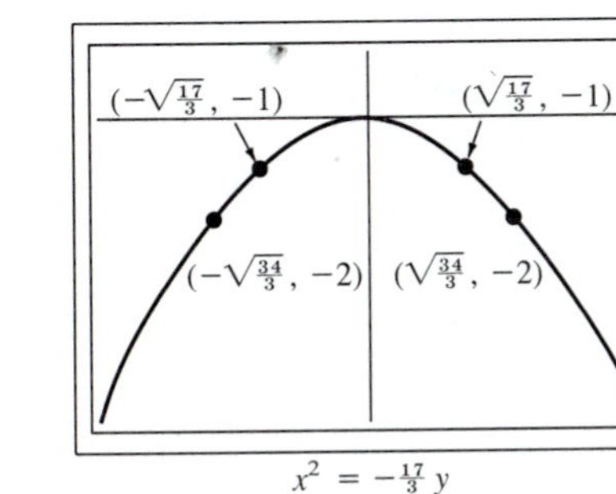

$x^2 = -\frac{17}{3}y$

45.

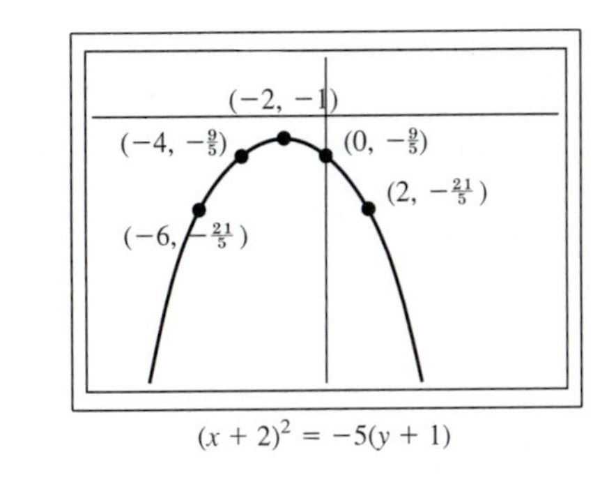

$(x + 2)^2 = -5(y + 1)$

47.

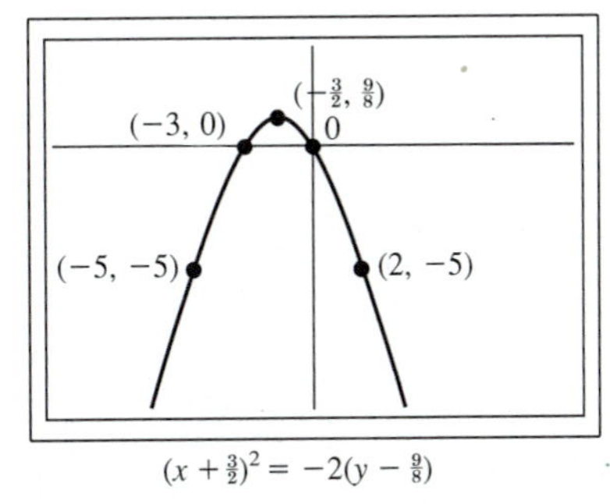

$(x + \frac{3}{2})^2 = -2(y - \frac{9}{8})$

Problems 5.4, page 306

1. center: (0. 0)
foci: $(\pm\sqrt{41}, 0)$
vertices: $(\pm 4, 0)$
transverse axis: $\overline{(-4, 0)(4, 0)}$
asymptotes: $y = \pm\frac{5}{4}x$
conjugate axis: $\overline{(0,5)(0, -5)}$

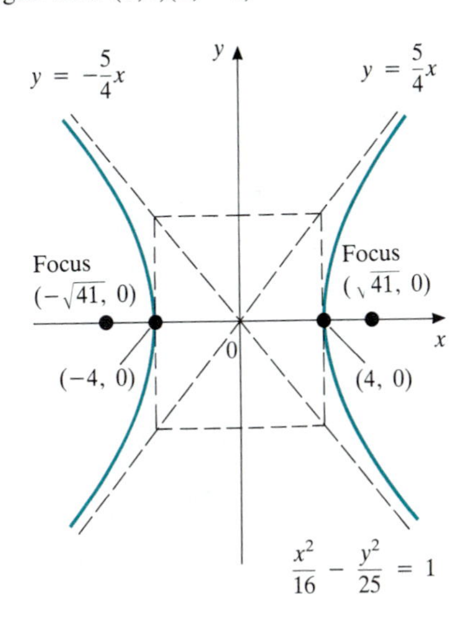

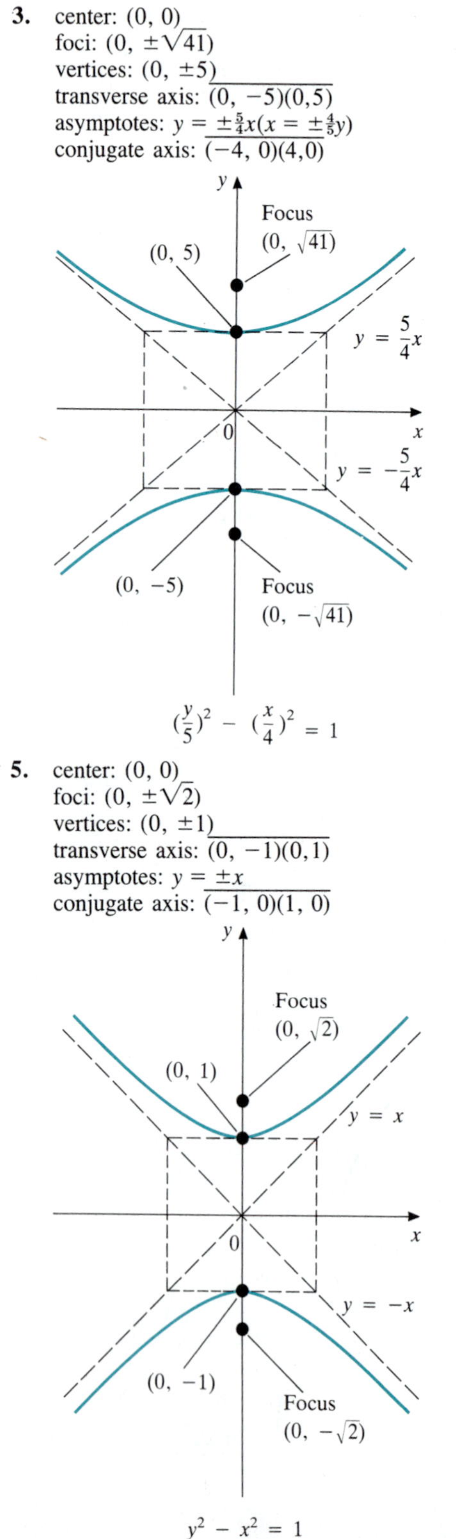

3. center: (0, 0)
foci: $(0, \pm\sqrt{41})$
vertices: $(0, \pm 5)$
transverse axis: $\overline{(0, -5)(0,5)}$
asymptotes: $y = \pm\frac{5}{4}x \; (x = \pm\frac{4}{5}y)$
conjugate axis: $\overline{(-4, 0)(4,0)}$

$$\left(\frac{y}{5}\right)^2 - \left(\frac{x}{4}\right)^2 = 1$$

5. center: (0, 0)
foci: $(0, \pm\sqrt{2})$
vertices: $(0, \pm 1)$
transverse axis: $\overline{(0, -1)(0,1)}$
asymptotes: $y = \pm x$
conjugate axis: $\overline{(-1, 0)(1, 0)}$

$$y^2 - x^2 = 1$$

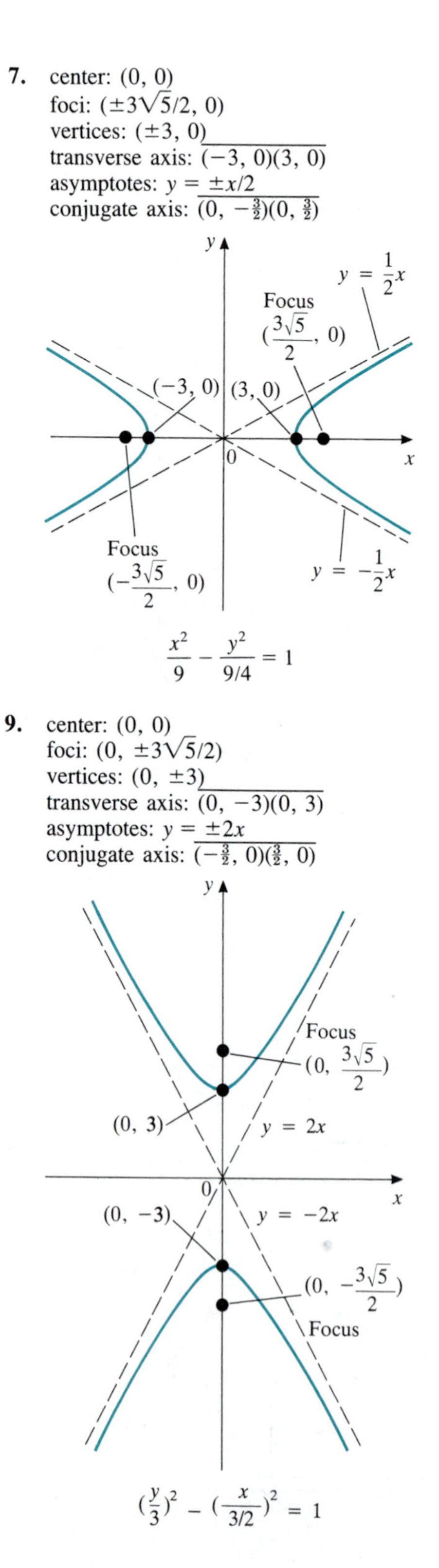

7. center: (0, 0)
foci: $(\pm 3\sqrt{5}/2, 0)$
vertices: $(\pm 3, 0)$
transverse axis: $\overline{(-3, 0)(3, 0)}$
asymptotes: $y = \pm x/2$
conjugate axis: $\overline{(0, -\frac{3}{2})(0, \frac{3}{2})}$

$$\frac{x^2}{9} - \frac{y^2}{9/4} = 1$$

9. center: (0, 0)
foci: $(0, \pm 3\sqrt{5}/2)$
vertices: $(0, \pm 3)$
transverse axis: $\overline{(0, -3)(0, 3)}$
asymptotes: $y = \pm 2x$
conjugate axis: $\overline{(-\frac{3}{2}, 0)(\frac{3}{2}, 0)}$

$$\left(\frac{y}{3}\right)^2 - \left(\frac{x}{3/2}\right)^2 = 1$$

11. center: $(0, 0)$
foci: $(\pm\sqrt{\frac{10}{3}}, 0)$
vertices: $(\pm\sqrt{2}, 0)$
transverse axis: $\overline{(-\sqrt{2}, 0)(\sqrt{2}, 0)}$
asymptotes: $y = \pm\sqrt{\frac{2}{3}}x$
conjugate axis: $\overline{\left(0, -\frac{2}{\sqrt{3}}\right)\left(0, \frac{2}{\sqrt{3}}\right)}$

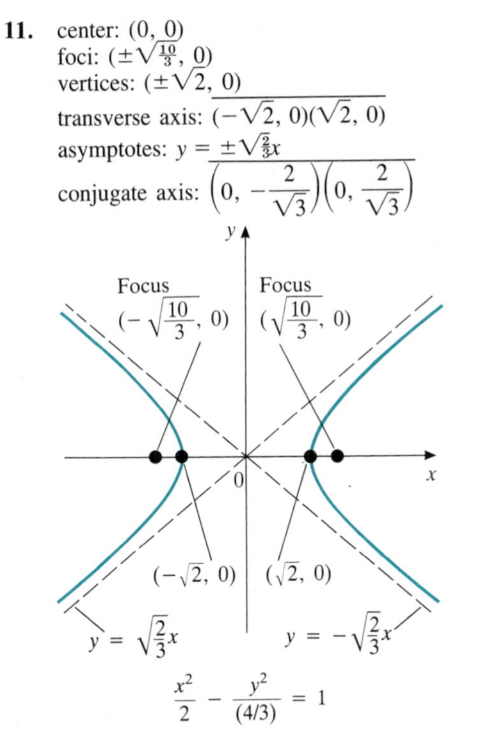

13. center: $(0, 0)$
foci: $(0, \pm\sqrt{\frac{10}{3}})$
vertices: $(0, \pm\sqrt{2})$
transverse axis: $\overline{(0, -\sqrt{2})(0, \sqrt{2})}$
asymptotes: $y = \pm\sqrt{\frac{3}{2}}x$
conjugate axis: $\overline{\left(-\frac{2}{\sqrt{3}}, 0\right)\left(\frac{2}{\sqrt{3}}, 0\right)}$

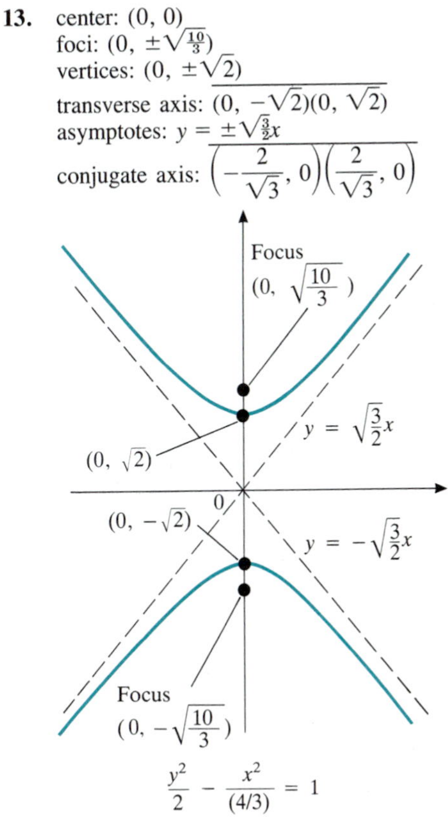

15. center: $(1, -2)$
foci: $(1 - \sqrt{5}, -2)$, $(1 + \sqrt{5}, -2)$
vertices: $(-1, -2)$, $(3, -2)$
transverse axis: $\overline{(-1, -2)(3, -2)}$
asymptotes: $y = \pm(\frac{1}{2})(x - 1) - 2$
conjugate axis: $\overline{(1, -1)(1, -3)}$

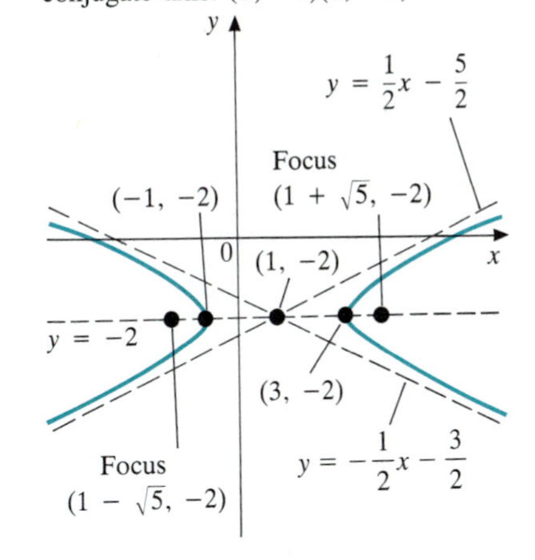

17. center: $(-1, -3)$
foci: $(-1 \pm 2\sqrt{5}, -3)$
vertices: $(-3, -3)$, $(1, -3)$
transverse axis: $\overline{(-3, -3)(1, -3)}$
asymptotes: $y = \pm 2(x + 1) - 3$
conjugate axis: $\overline{(-1, 1)(-1, -7)}$

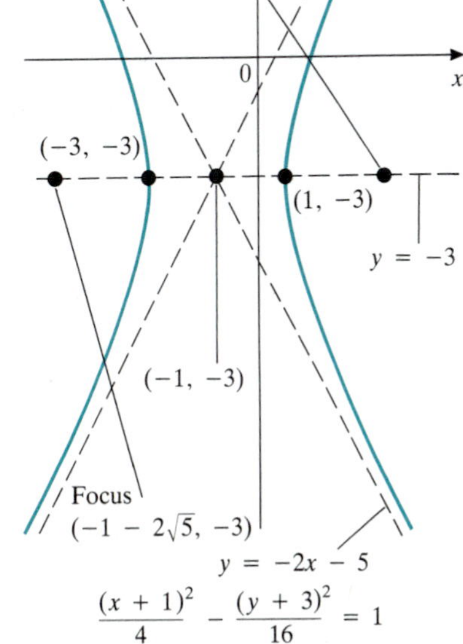

19. center: $(4, 2)$
foci: $(4 - \sqrt{\frac{325}{6}}, 2)(4 + \sqrt{\frac{325}{6}}, 2)$
vertices: $(4 - \sqrt{\frac{65}{2}}, 2), (4 + \sqrt{\frac{65}{2}}, 2)$
transverse axis: $\overline{(4 - \sqrt{\frac{65}{2}}, 2)(4 + \sqrt{\frac{65}{2}}, 2)}$
asymptotes: $y = \pm\sqrt{\frac{2}{3}}(x - 4) + 2$
conjugate axis: $\overline{(4, 2 - \sqrt{\frac{65}{3}})(4, 2 + \sqrt{\frac{65}{3}})}$

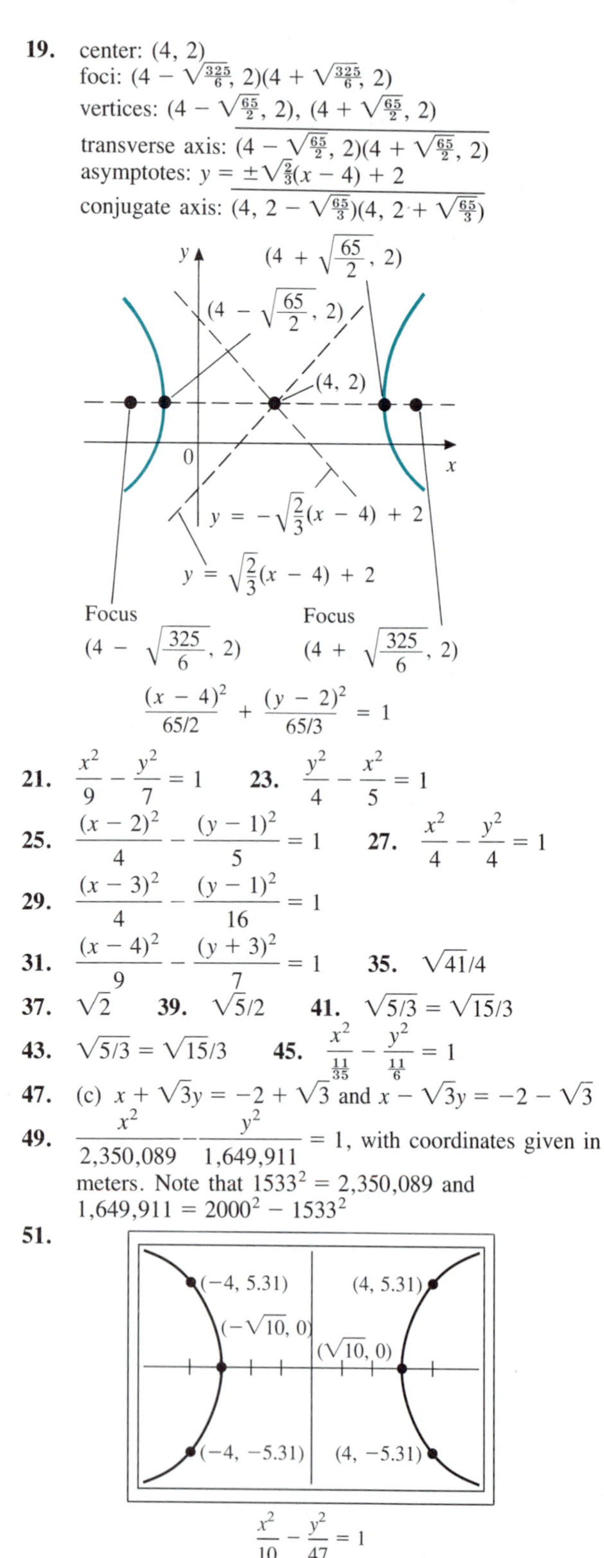

$$\frac{(x-4)^2}{65/2} + \frac{(y-2)^2}{65/3} = 1$$

21. $\frac{x^2}{9} - \frac{y^2}{7} = 1$ **23.** $\frac{y^2}{4} - \frac{x^2}{5} = 1$

25. $\frac{(x-2)^2}{4} - \frac{(y-1)^2}{5} = 1$ **27.** $\frac{x^2}{4} - \frac{y^2}{4} = 1$

29. $\frac{(x-3)^2}{4} - \frac{(y-1)^2}{16} = 1$

31. $\frac{(x-4)^2}{9} - \frac{(y+3)^2}{7} = 1$ **35.** $\sqrt{41}/4$

37. $\sqrt{2}$ **39.** $\sqrt{5}/2$ **41.** $\sqrt{5/3} = \sqrt{15}/3$

43. $\sqrt{5/3} = \sqrt{15}/3$ **45.** $\frac{x^2}{\frac{11}{35}} - \frac{y^2}{\frac{11}{6}} = 1$

47. (c) $x + \sqrt{3}y = -2 + \sqrt{3}$ and $x - \sqrt{3}y = -2 - \sqrt{3}$

49. $\frac{x^2}{2{,}350{,}089} - \frac{y^2}{1{,}649{,}911} = 1$, with coordinates given in meters. Note that $1533^2 = 2{,}350{,}089$ and $1{,}649{,}911 = 2000^2 - 1533^2$

51.

$$\frac{x^2}{10} - \frac{y^2}{47} = 1$$

53.

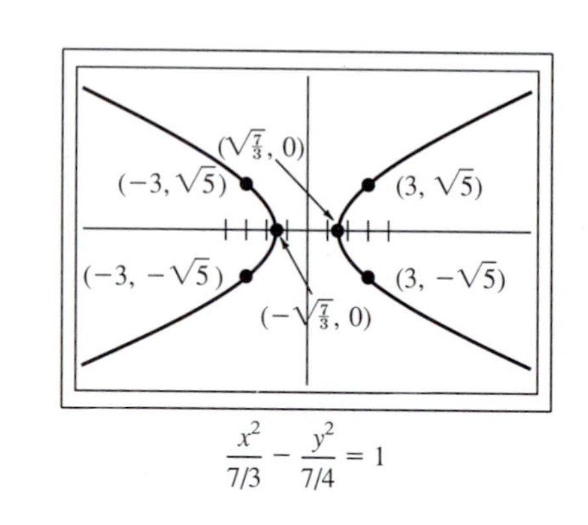

$$\frac{x^2}{7/3} - \frac{y^2}{7/4} = 1$$

55.

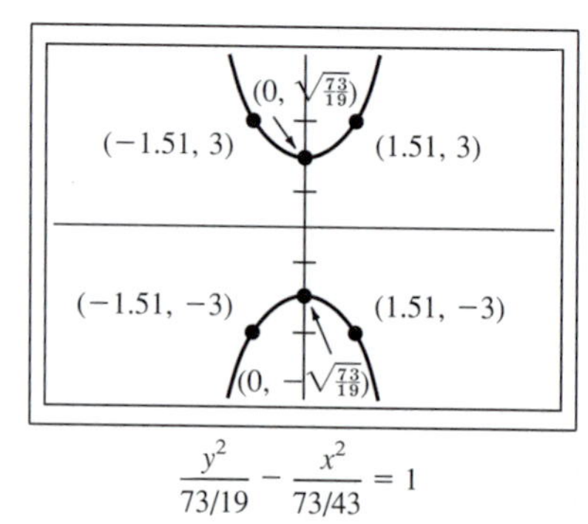

$$\frac{y^2}{73/19} - \frac{x^2}{73/43} = 1$$

57.

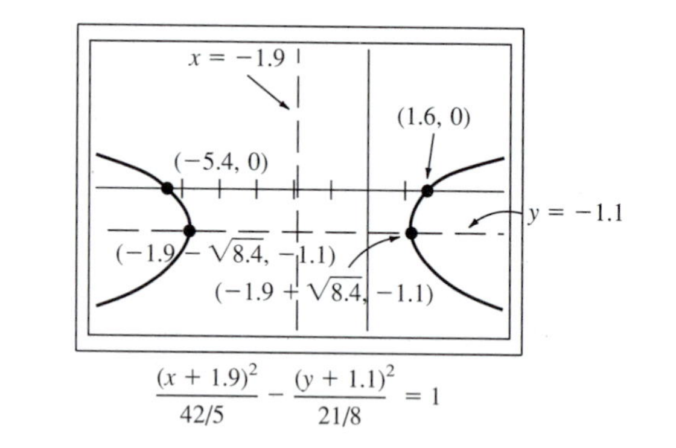

$$\frac{(x+1.9)^2}{42/5} - \frac{(y+1.1)^2}{21/8} = 1$$

59.

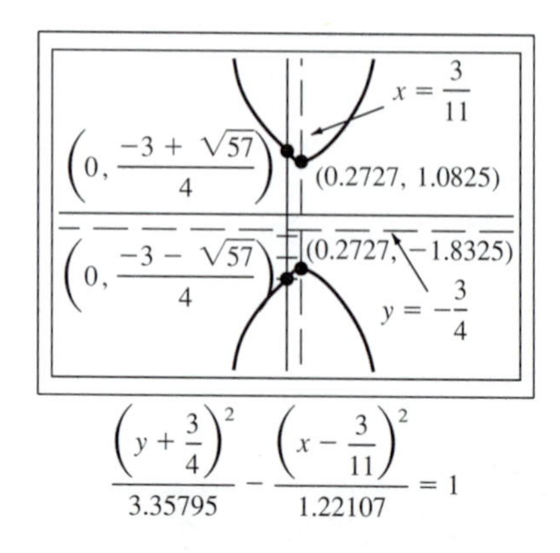

$$\frac{\left(y + \frac{3}{4}\right)^2}{3.35795} - \frac{\left(x - \frac{3}{11}\right)^2}{1.22107} = 1$$

Problems 5.5, page 314

1. ellipse; $\dfrac{(x')^2}{15} + \dfrac{(y')^2}{9} = 1$

3. straight line: $(\sqrt{3} - \frac{3}{2})x' - (1 + 3\sqrt{3}/2)y' = 6$

5. If $\theta = \pi/3$, the resulting equation is $3(x')^2 + 2\sqrt{3}x'y' + (y')^2 + 24x' - 24\sqrt{3}y' = 0$; if $\theta = 4\pi/3$, the equation becomes $3(x')^2 + 2\sqrt{3}x'y' + (y')^2 - 24x' + 24\sqrt{3}y' = 0$.

7. Rotate axes through $\theta = \frac{1}{2}\tan^{-1}\frac{4}{3} \approx 26.6°$ to obtain two parallel lines given by $5(x')^2 = 9$ or $x' = \pm 3/\sqrt{5}$

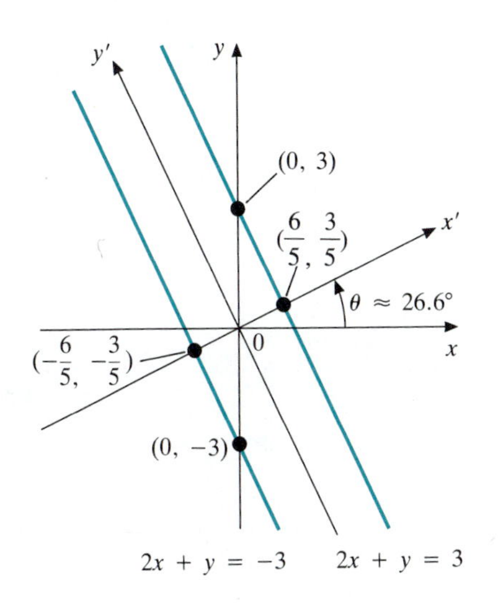

9. Rotate axes through $\theta = \frac{1}{2}\tan^{-1}(-\frac{2}{3}) \approx -16.8°$ (or 73.2°) to obtain the hyperbola $[(\sqrt{13} + 3)/10](x')^2 - [(\sqrt{13} - 3)/10](y')^2 = 1$. Note that, since θ is in the fourth quadrant, $\sin\theta < 0$ and $\sin\theta = \sin\frac{1}{2}\tan^{-1}(-\frac{2}{3}) = -\sqrt{\dfrac{1 - \cos\tan^{-1}(-\frac{2}{3})}{2}} = -\sqrt{\dfrac{1 - 3/\sqrt{13}}{2}} = -\sqrt{\dfrac{\sqrt{13} - 3}{2\sqrt{13}}}$;

$\cos\theta = \sqrt{\dfrac{\sqrt{13} + 3}{2\sqrt{13}}}$.

9. **continued**

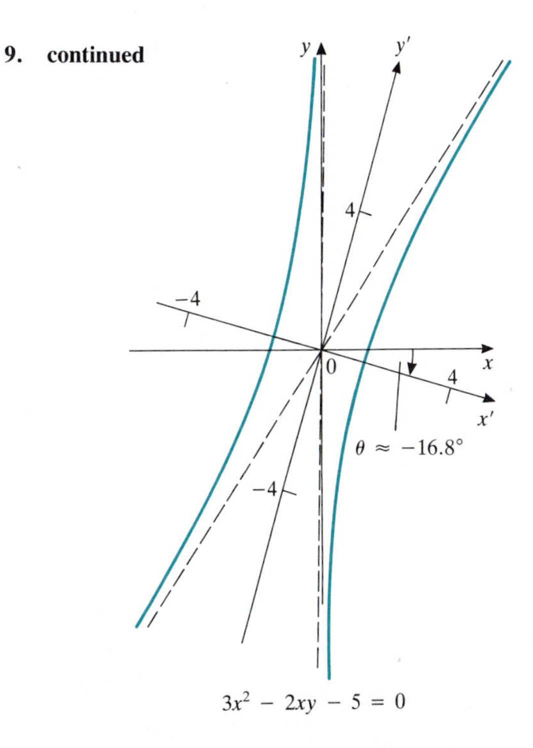

11. Rotate axes through $\theta = \pi/4$ to obtain the hyperbola $[(x')^2/2a] - [(y')^2/2a] = 1$.

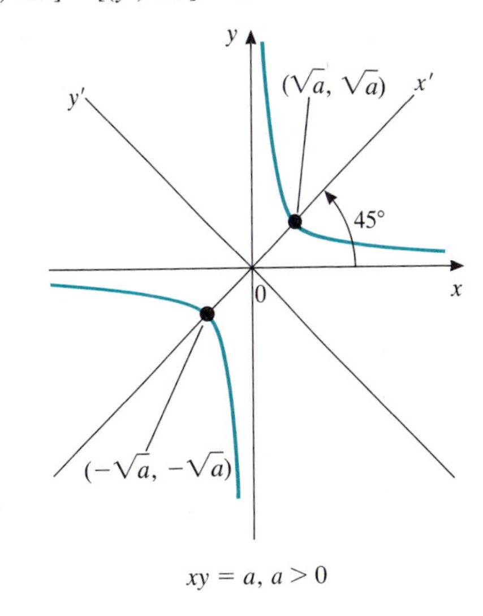

13. Rotate axes through $\theta = \frac{1}{2}\tan^{-1}\frac{4}{3} \approx 26.6°$ to obtain the parabola $[x' + (3/\sqrt{5})]^2 = (8/\sqrt{5})[y' + (9/8\sqrt{5})]$.

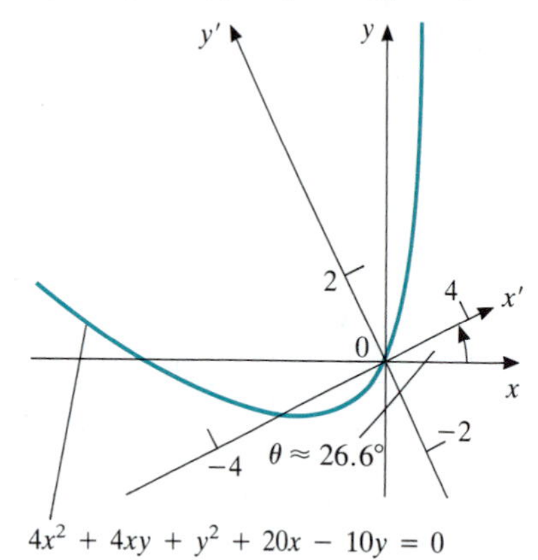

15. Rotate axes through $\theta = \pi/8$ to obtain the ellipse $[(3 + \sqrt{2})/8](x')^2 + [(3 - \sqrt{2})/8](y')^2 = 1$.

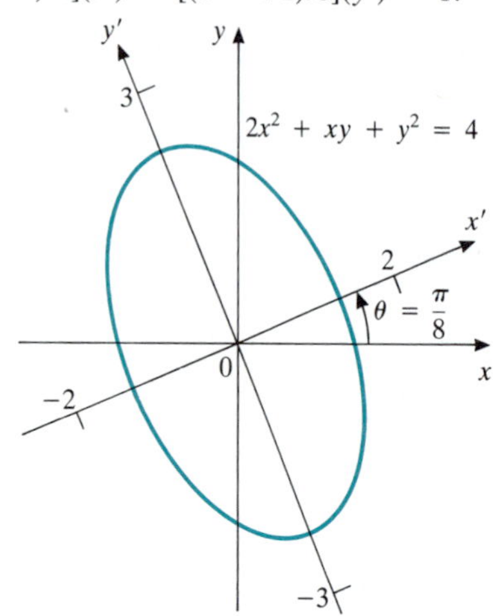

17. Rotate axes through $\theta = \frac{1}{2}\tan^{-1} 3 \approx 35.8°$ to obtain the ellipse
$[(4 - \sqrt{10})/36](x')^2 + [(4 + \sqrt{10})/36](y')^2 = 1$.

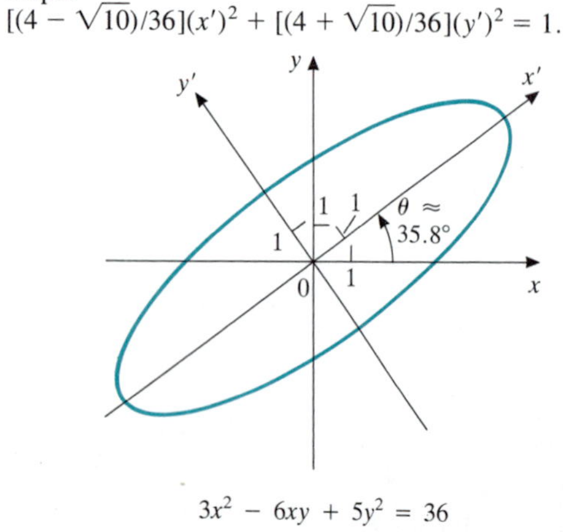

19. Rotate axes through $\theta = \frac{1}{2}\tan^{-1}\frac{4}{3} \approx 26.6°$ to obtain the hyperbola $[4(y' + 2\sqrt{5})^2/35] - [(x' + \sqrt{5})^2/35] = 1$.

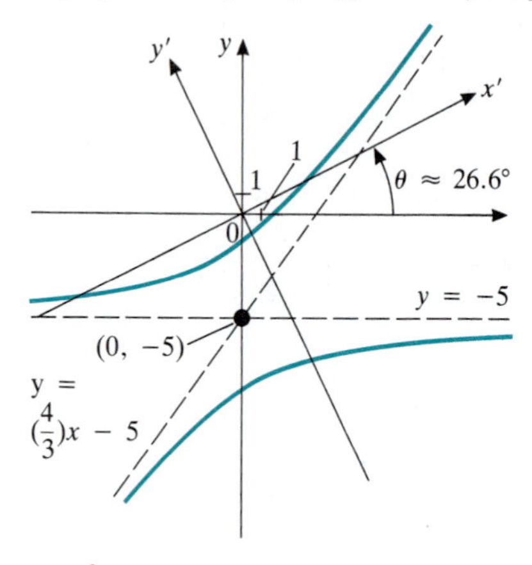

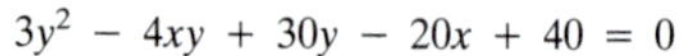

Problems 5.6, page 321

1.

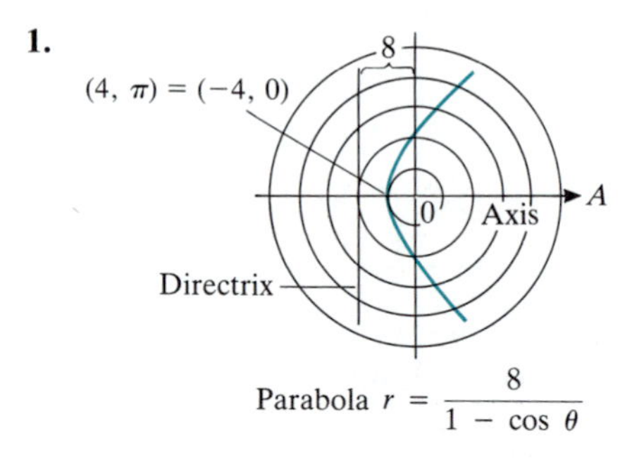

3.

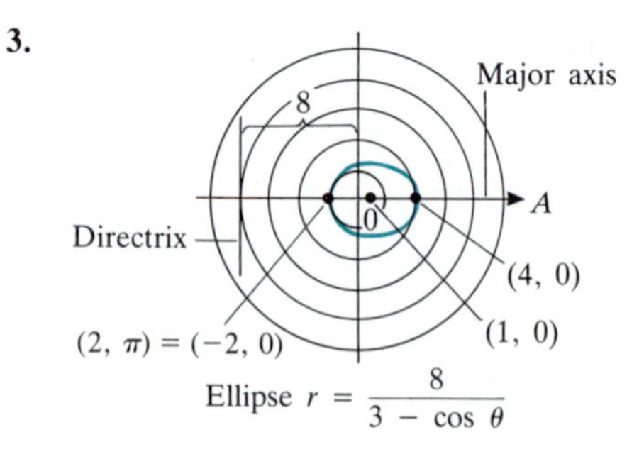

5.

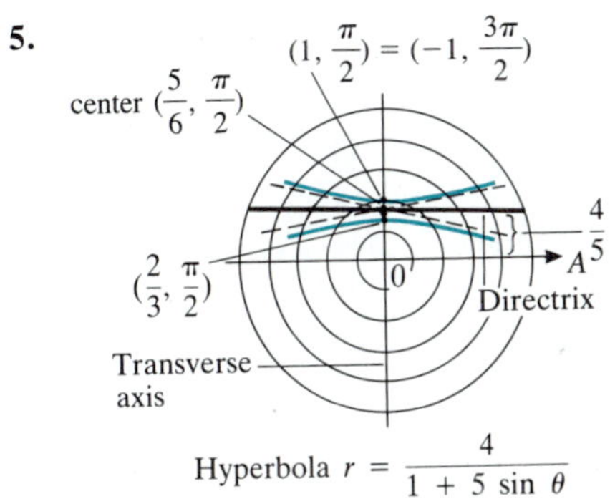

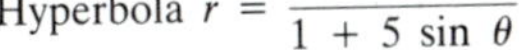
Hyperbola $r = \dfrac{4}{1 + 5\sin\theta}$

7.

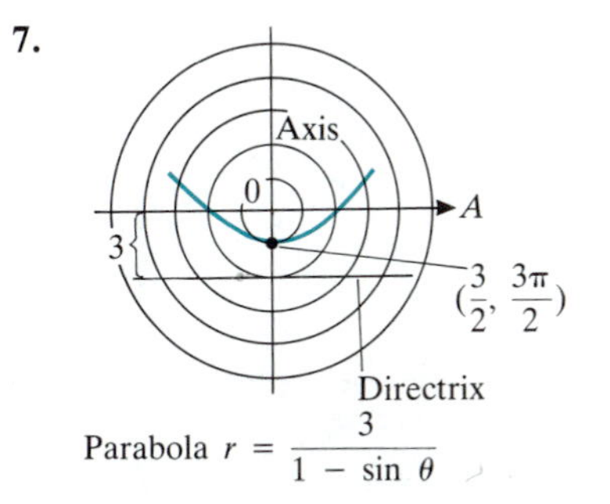

Parabola $r = \dfrac{3}{1 - \sin\theta}$

9.

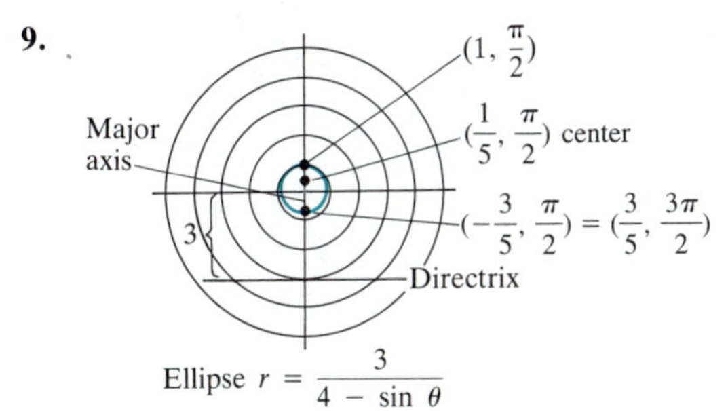

Ellipse $r = \dfrac{3}{4 - \sin\theta}$

11.

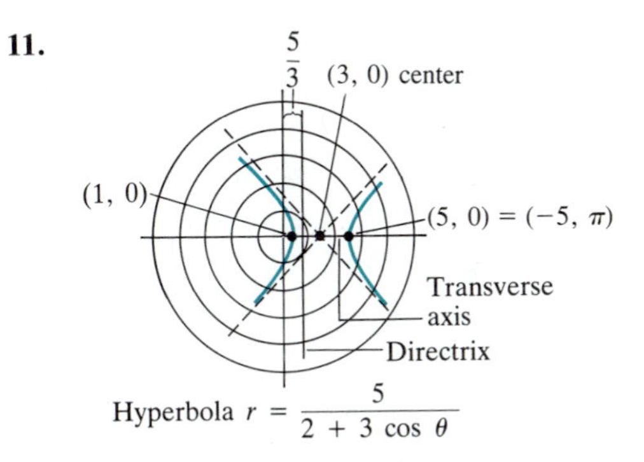

Hyperbola $r = \dfrac{5}{2 + 3\cos\theta}$

13. $r = 6/(1 + 2\sin\theta)$ **15.** $r = 2/(1 + \sin\theta)$
17. $r = 10/(3 - 2\cos\theta)$ **19.** $r = 6/(2 - 3\sin\theta)$
23. $(1 - e^2\cos^2\delta)x^2 + (e^2\sin 2\delta)xy + (1 - e^2\sin^2\delta)y^2 - (2e^2p\cos\delta)x - (2e^2p\sin\delta)y - e^2p^2 = 0$

27. $r = 0.3707/(1 - 0.2056\cos\theta)$, perihelion $= 0.3075$ AU $= 2.860 \times 10^7$ mi, aphelion $= 0.4666$ AU $= 4.340 \times 10^7$ mi
31. $r^2(\cos^2\theta - 4\sin^2\theta) = 4$
33. $r^2(9\cos^2\theta - 16\sin^2\theta) = 144$
35. $r^2(25\cos^2\theta + 9\sin^2\theta) = 225$

Problems 5.7, page 325

1. $\mathbb{R} - \{0, 1\}$ **3.** $\mathbb{R} - \{-1, 1\}$
5. $y^2 = 4x$

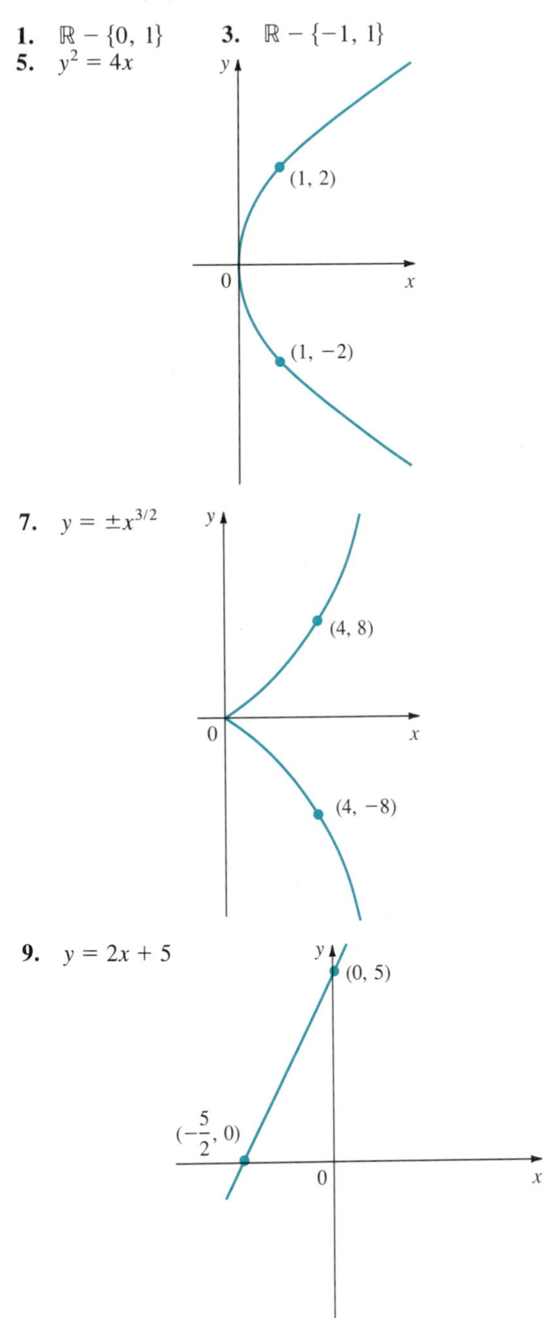

7. $y = \pm x^{3/2}$

9. $y = 2x + 5$

11. $y = x^4,\ x \geq 0$

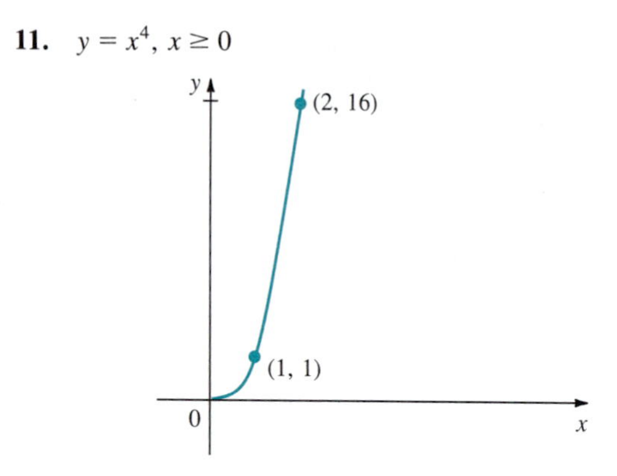

13. $y = e^x$

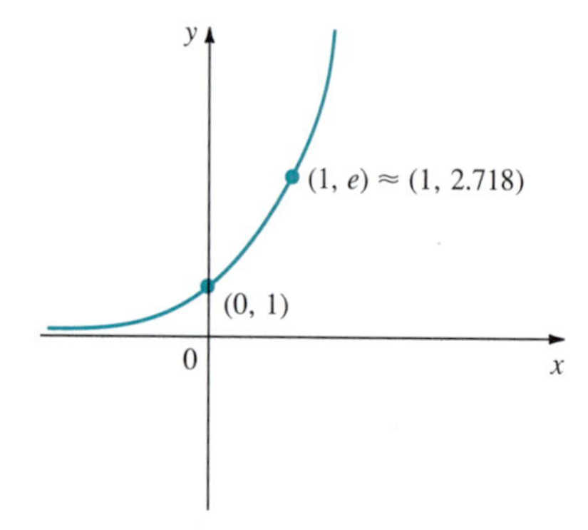

15. $x = y^4 + y^2 - 3,\ y \geq 0$

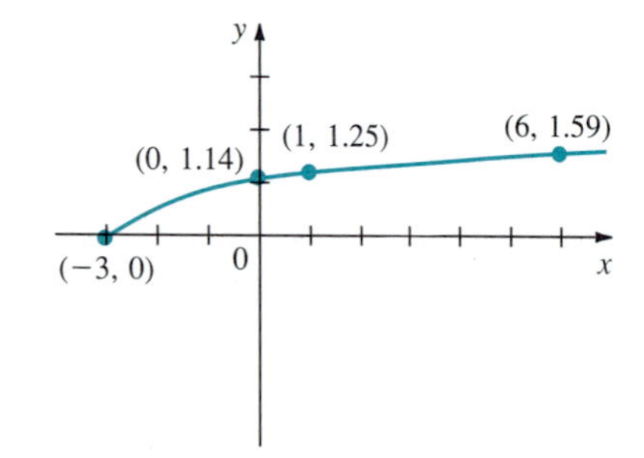

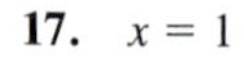

17. $x = 1$

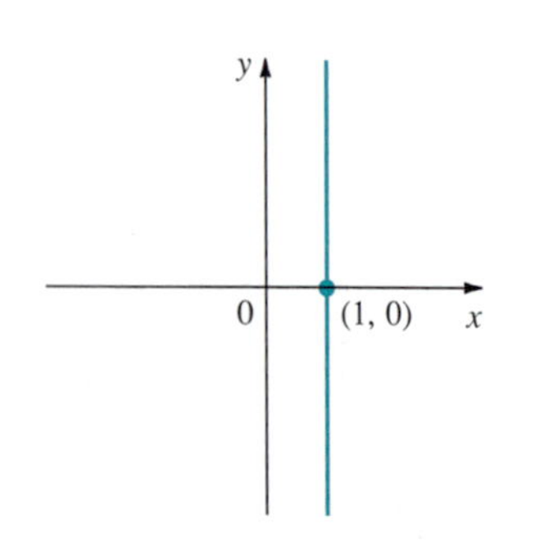

23. (a) $x = r(\alpha - \sin\alpha),\ y = r(1 - \cos\alpha)$

29. $x = 2 - t,\ y = 4 + 2t$

31. $x = 3 - 4t,\ y = 5 - 12t$

33. $x = -2 + 6t,\ y = 3 + 4t$

Review Exercises for Chapter 5, page 329

1. Ellipse
center: $(0, 0)$
foci: $(0, \pm\sqrt{7})$
vertices: $(0, \pm 4)$
major axis: $\overline{(0, -4)(0, 4)}$
minor axis: $\overline{(-3, 0)(3, 0)}$
eccentricity: $\sqrt{7}/4$

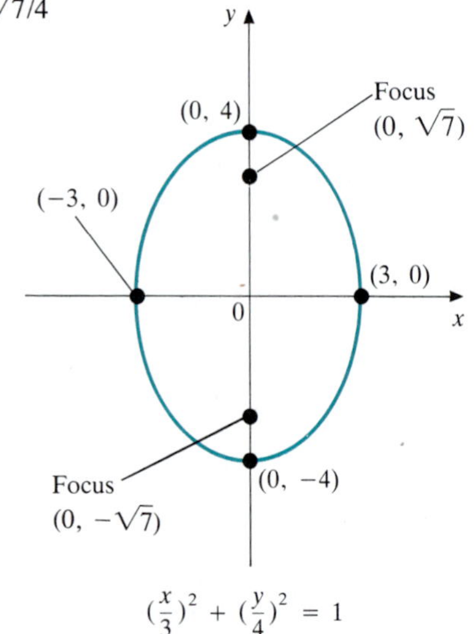

$$\left(\frac{x}{3}\right)^2 + \left(\frac{y}{4}\right)^2 = 1$$

3. Parabola
focus: $(0, \frac{9}{64})$
directrix: $y = -\frac{9}{64}$
axis: $x = 0$
vertex: $(0, 0)$

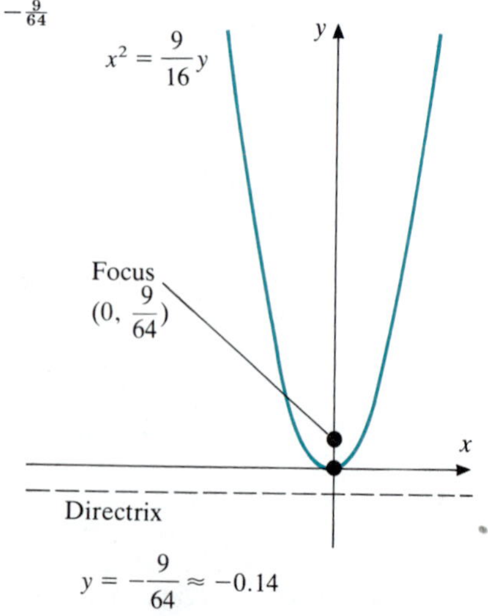

$$y = -\frac{9}{64} \approx -0.14$$

5. Hyperbola
center: (0, 0)
foci: $(0, \pm 5)$
vertices: $(0, \pm 3)$
eccentricity: $\frac{5}{3}$
transverse axis: $\overline{(0, -3)(0, 3)}$
asymptotes: $y = \pm\frac{3}{4}x$
conjugate axis: $\overline{(-4, 0)(4, 0)}$

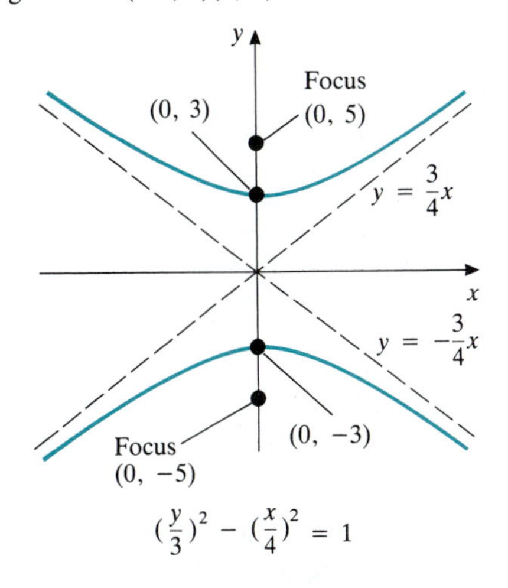

$$\left(\frac{y}{3}\right)^2 - \left(\frac{x}{4}\right)^2 = 1$$

7. Ellipse
center: $(1, -1)$
foci: $(1, -1 - \sqrt{5}), (1, -1 + \sqrt{5})$
vertices: $(1, -4), (1, 2)$
major axis: $\overline{(1, -4)(1, 2)}$
minor axis: $\overline{(-1, -1)(3, -1)}$
eccentricity: $\sqrt{5}/3$

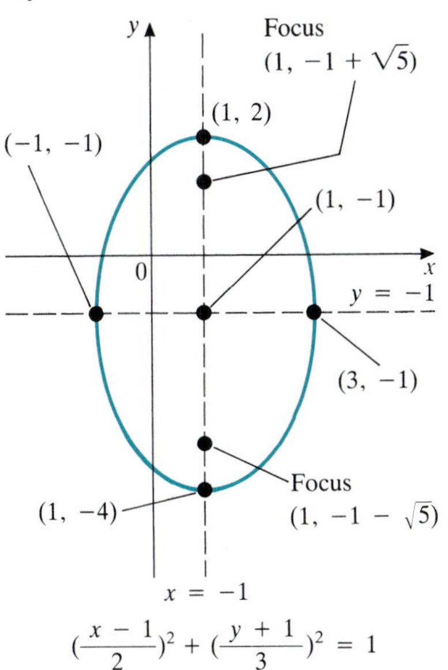

$$\left(\frac{x-1}{2}\right)^2 + \left(\frac{y+1}{3}\right)^2 = 1$$

9. single point $(-2, 5)$

11. two straight lines: i.e., a degenerate hyperbola

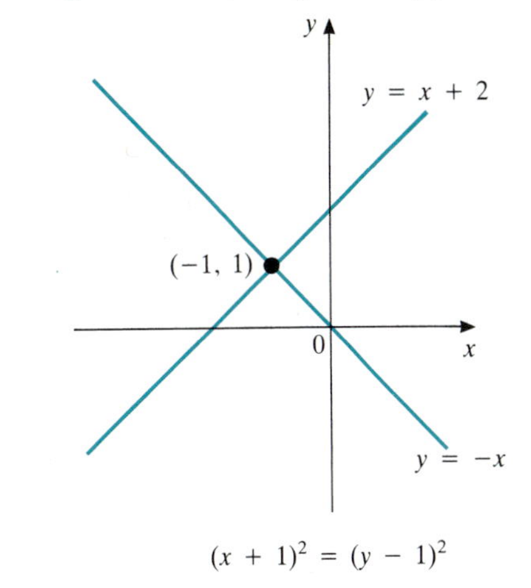

$$(x+1)^2 = (y-1)^2$$

13. Ellipse
center: $(-\frac{1}{2}, -4)$
foci: $(-\frac{1}{2}, -4 - 3/\sqrt{2}), (-\frac{1}{2}, -4 + 3/\sqrt{2})$
vertices: $(-\frac{1}{2}, -4 - 3\sqrt{2}), (-\frac{1}{2}, -4 + 3\sqrt{2})$
major axis: $\overline{(-\frac{1}{2}, -4 - 3\sqrt{2})(-\frac{1}{2}, -4 + 3\sqrt{2})}$
minor axis: $\overline{(-\frac{1}{2} - 3\sqrt{\frac{3}{2}}, -4)(-\frac{1}{2} + 3\sqrt{\frac{3}{2}}, -4)}$
eccentricity: $\frac{1}{2}$

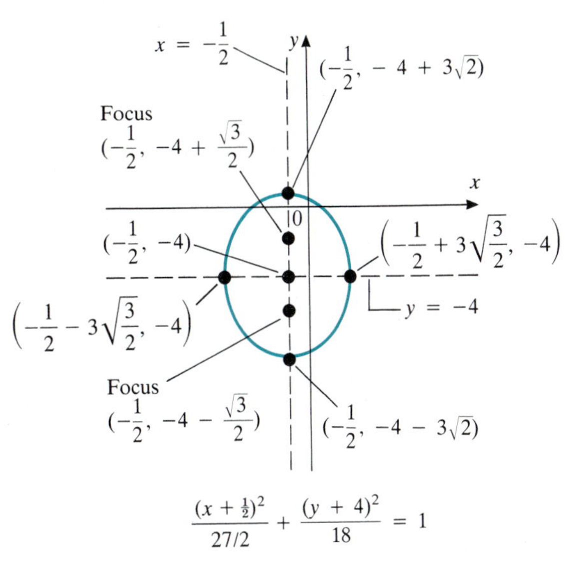

$$\frac{(x+\frac{1}{2})^2}{27/2} + \frac{(y+4)^2}{18} = 1$$

15. unique ellipse: $\dfrac{x^2}{25} + \dfrac{y^2}{16} = 1$

17. $\dfrac{y^2}{4} - \dfrac{x^2}{5} = 1$

19. After the rotation, the curve satisfies the equation $5(x')^2 + 3(y')^2 = 20$; the curve is an ellipse.

	(x', y') coordinates	(x, y) coordinates
center:	$(0, 0)$	$(0, 0)$
foci:	$\left(0, \pm 2\sqrt{\frac{2}{3}}\right)$	$\left(-\frac{2}{\sqrt{3}}, \frac{2}{\sqrt{3}}\right), \left(\frac{2}{\sqrt{3}}, -\frac{2}{\sqrt{3}}\right)$
vertices:	$\left(0, \pm 2\sqrt{\frac{5}{3}}\right)$	$\left(-\frac{\sqrt{10}}{3}, \frac{\sqrt{10}}{3}\right), \left(\frac{\sqrt{10}}{3}, -\frac{\sqrt{10}}{3}\right)$

major axis: $(0, -2\sqrt{\frac{5}{3}})(0, 2\sqrt{\frac{5}{3}})$
minor axis: $(-2, 0)(2, 0)$
eccentricity: $\sqrt{\frac{2}{5}}$

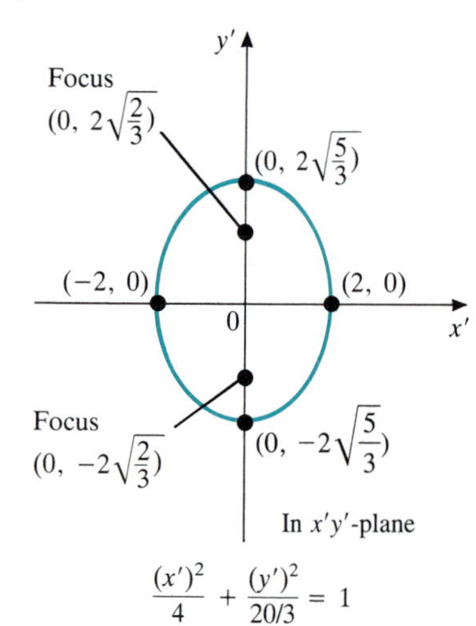

In $x'y'$-plane

$$\frac{(x')^2}{4} + \frac{(y')^2}{20/3} = 1$$

$2x^2 + xy + 2y^2 = 10$

In xy-plane

21. $9 = x^2 + 4xy + 4y^2 = (x + 2y)^2$; the graph consists of the two parallel lines with equations $3 = x + 2y$ and $-3 = x + 2y$. If the coordinate axes are rotated through the angle $\frac{1}{2}\cot^{-1}((1 - 4)/4) \approx 63.43°$, the curve satisfies the equation $9 = 5(x')^2$ or $x' = \pm 3/\sqrt{5}$. These are two vertical lines in the $x'y'$ plane. Alternatively, rotate $-26.57°$ to obtain $y' = \pm\frac{3}{\sqrt{5}}$ which are two horizontal lines in the $x'y'$-plane.

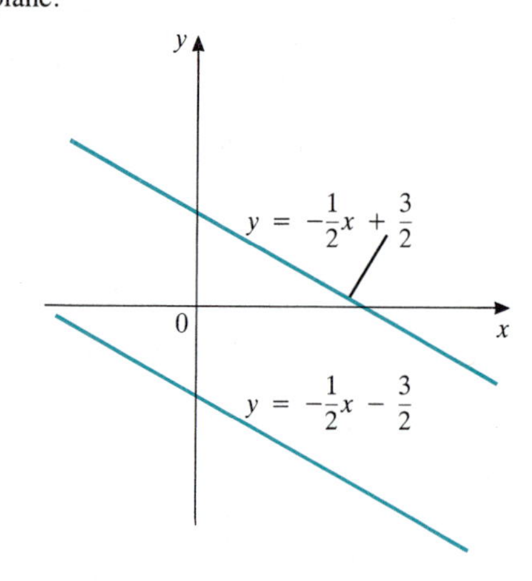

23. Rotate the axes through $\pi/4$ radians; the transformed equation is $-3 = xy = [(x')^2 - (y')^2]/2$ or $(y')^2/6 - (x')^2/6 = 1$.

	(x', y') coordinates	(x, y) coordinates
center:	$(0, 0)$	$(0, 0)$
foci:	$(0, \pm 2\sqrt{3})$	$(-\sqrt{6}, \sqrt{6}), (\sqrt{6}, -\sqrt{6})$
vertices:	$(0, \pm\sqrt{6})$	$(-\sqrt{3}, \sqrt{3}), (\sqrt{3}, -\sqrt{3})$
transverse axis:	$(0, -\sqrt{6}), (0, \sqrt{6})$	$(-\sqrt{3}, \sqrt{3}), (\sqrt{3}, -\sqrt{3})$
asymptotes:	$y' = \pm x'$	$x = 0, y = 0$
eccentricity:		$\sqrt{2}$

23. continued

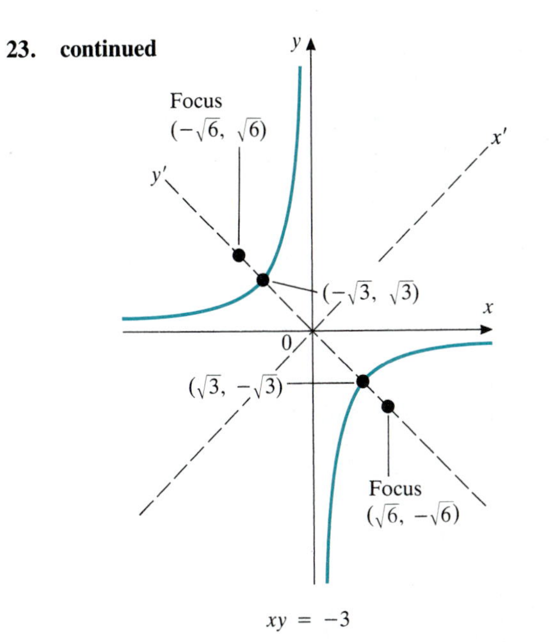

$xy = -3$

25. Rotate axes through $\theta = \frac{1}{2}\cot^{-1}((-2-4)/3) \approx 76.717°$; the curve satisfies the transformed equation $(3\sqrt{5}/2 + 1)(x')^2 - (3\sqrt{5}/2 - 1)(y')^2 = 5$; or, using numerical approximations, $(x')^2/1.148 - (y')^2/2.124 = 1$. At this stage, we can identify the cuve as a hyperbola opening along the x'-axis.

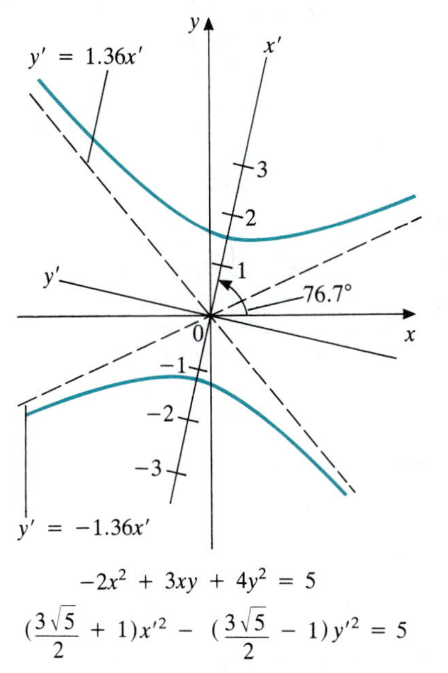

$-2x^2 + 3xy + 4y^2 = 5$

$\left(\frac{3\sqrt{5}}{2} + 1\right)x'^2 - \left(\frac{3\sqrt{5}}{2} - 1\right)y'^2 = 5$

27.

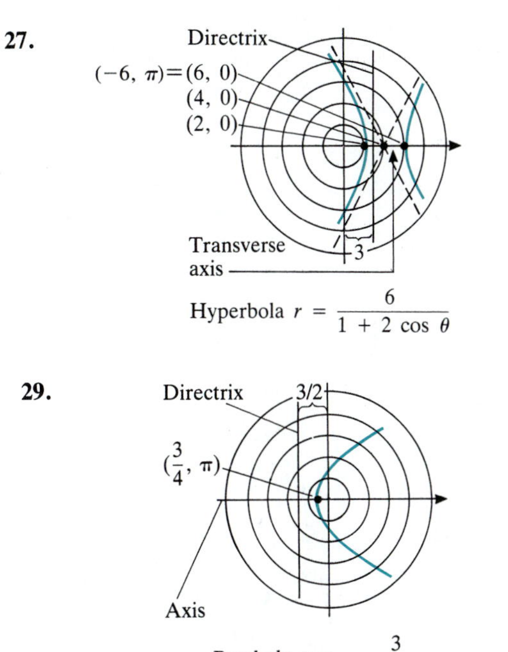

Hyperbola $r = \dfrac{6}{1 + 2\cos\theta}$

29.

Parabola $r = \dfrac{3}{2 - 2\cos\theta}$

31.

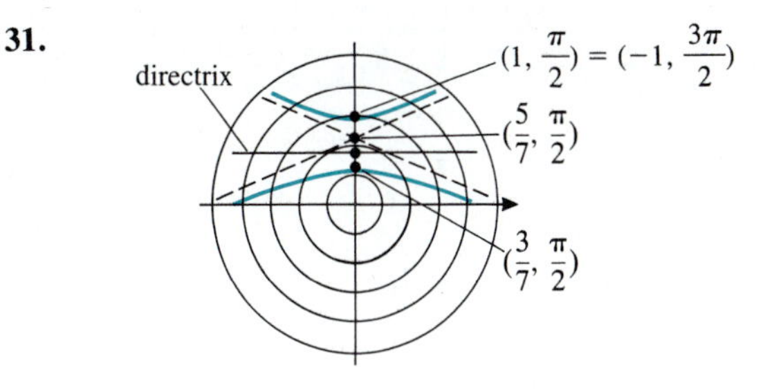

33. $r = 2/(1 - 2\cos\theta)$

35. $y = 2x$

y
(1, 2)
0
x

37. $x = \left(3 + \frac{y}{2}\right)^2$

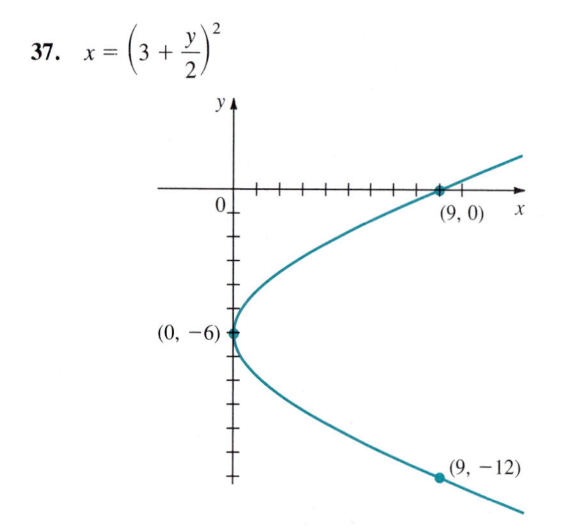

39. $x^2 + y^2 = 1$

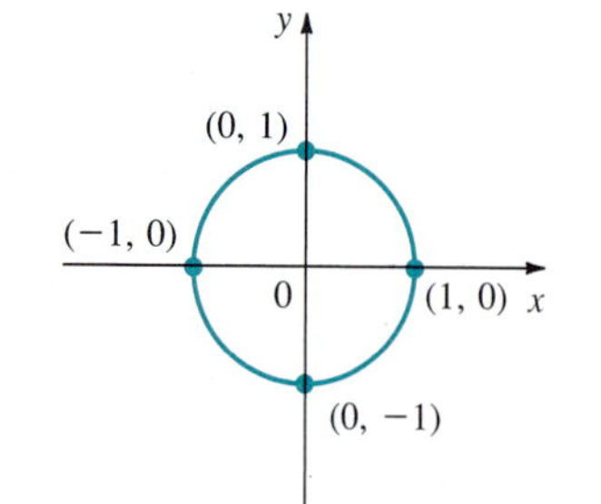

41. $x = y^3$ or $y = \sqrt[3]{x}$, $x \geq 0$

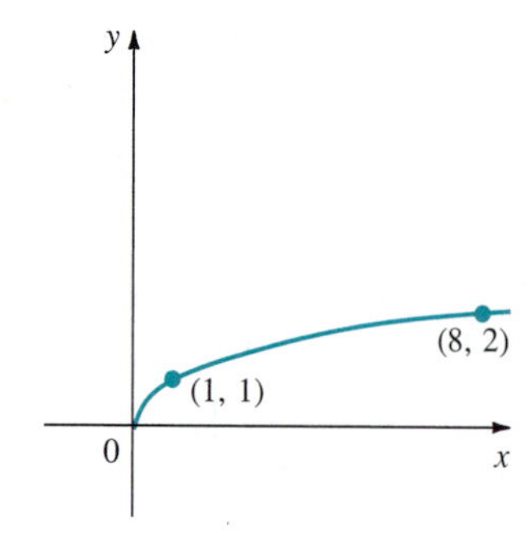

Chapter 6

Problems 6.1, p. 339

1. $5i$ **3.** $\sqrt{5}\,i$ **5.** $3 + 7i$ **7.** 5
9. $-2 + \frac{1}{3}i$ **11.** $-\sqrt{73}\,i$ **13.** $\frac{4}{7} - \frac{\sqrt{2}}{7}i$
15. $3 - 8i$ **17.** $-\frac{1}{2} + i$ **19.** $6 + 8i$
21. $\frac{1}{2} - \frac{1}{20}i$ **23.** $2\sqrt{3}\,i$ **25.** $-1 + 3i$
27. $-2i$ **29.** $12 + 16i$ **31.** $31 - 29i$
33. $-\frac{7}{4} - \frac{23}{8}i$ **35.** $-19 + 4i$ **37.** $-i$ **39.** i
41. -4 **43.** $\frac{1}{2} - \frac{1}{2}i$ **45.** $-\frac{3}{2} - 4i$ **47.** i
49. i **51.** $\frac{19}{10} - \frac{3}{10}i = 1.9 - 0.3i$ **53.** $-\frac{1}{2}i$
55. $\frac{1}{2} + \frac{1}{2}i$ **57.** $\frac{2}{29} + \frac{5}{29}i$ **59.** $\frac{1}{13} + \frac{3}{26}i$
61. $\frac{a}{a^2 + b^2} + \frac{b}{a^2 + b^2}i$ **63.** $u = 6,\ v = -1$
65. $u = \frac{2}{13},\ v = \frac{4}{13}$ **67.** $\pm 2i$ **69.** $-\frac{1}{2} \pm \frac{\sqrt{7}}{2}i$
71. $3 \pm i$ **73.** $\frac{3}{4} \pm \frac{\sqrt{31}}{4}i$ **75.** $-\frac{1}{3} \pm (\sqrt{14}/6)i$
77. $(8.06 \pm 27.90388503i)/25.44 \approx 0.317 \pm 1.097i$
81. $\sqrt{\frac{a}{b}} = \frac{\sqrt{a}}{\sqrt{b}}$ only if $a \geq 0$ and $b > 0$

Problems 6.2, page 347

1. $5,\ \pi/2,\ 5(\cos(\pi/2) + i \sin(\pi/2))$
3. $8,\ 0,\ 8(\cos 0 + i \sin 0)$
5. $\sqrt{2},\ -\pi/4,\ \sqrt{2}[\cos(-\pi/4) + i \sin(-\pi/4)]$
7. $\sqrt{2},\ -3\pi/4,\ \sqrt{2}[\cos(-3\pi/4) + i \sin(-3\pi/4)]$
9. $2,\ 2\pi/3,\ 2[\cos(2\pi/3) + i \sin(2\pi/3)]$
11. $2,\ -5\pi/6,\ 2[\cos(-5\pi/6) + i \sin(-5\pi/6)]$
13. $5,\ \tan^{-1}\frac{3}{4} \approx 0.6435,\ z = 5(\cos 0.6435 + i \sin 0.6435)$
15. $\sqrt{34},\ \tan^{-1}\frac{5}{3} \approx 1.0304,$
$\sqrt{34}(\cos 1.0304 + i \sin 1.0304)$
17. $\sqrt{29},\ \tan^{-1}\frac{2}{5} \approx 0.3805,$
$z = \sqrt{29}(\cos 0.3805 + i \sin 0.3805)$
19. $\sqrt{34},\ \pi + \tan^{-1}\frac{3}{5} \approx 2.6012,$
$\sqrt{34}(\cos 2.6012 + i \sin 2.6012)$
21. $100\sqrt{2},\ 3\pi/4,\ 100\sqrt{2}[\cos(3\pi/4) + i \sin(3\pi/4)]$
23. $4i$ **25.** -12 **27.** $4\sqrt{2} + 4\sqrt{6}i$
29. $-2\sqrt{3} + 2i$ **31.** $-2 + 2\sqrt{3}i$
33. $\sqrt{2}/8 + \sqrt{2}i/8$ **35.** $-\sqrt{2}/8 + \sqrt{2}i/8$
37. $-2.0807 + 4.5465i$ **39.** $5.8689 - 1.2475i$
41. $z_1z_2 = 12(\cos \pi/2 + i \sin \pi/2);$
$z_1/z_2 = \frac{4}{3}(\cos \pi/6 + i \sin \pi/6)$
43. $z_1z_2 = \frac{2}{3}(\cos 1 + i \sin 1);\ z_1/z_2 = \frac{3}{2}(\cos 3 + i \sin 3)$
45. $z_1z_2 = \frac{5}{7}[\cos(-86°) + i \sin(-86°)];$
$z_1/z_2 = 35[\cos(-104°) + i \sin(-104°)]$
49. The circle in the complex plane centered at z_0 with radius a.

Problems 6.3, page 353

1. $-2^{14} - 2^{14}i$ **3.** $128 + 128i$
5. $2^{15}i = 32{,}768i$ **7.** $-512 - 512\sqrt{3}i$
9. $5^5(0.99712 - 0.07584i)$ **11.** $1, -1$
13. $-1/2 + \sqrt{3}i/2,\ -1/2 - \sqrt{3}i/2,\ 1$
15. $-\sqrt{2}/2 + \sqrt{2}i/2,\ \sqrt{2}/2 - \sqrt{2}i/2$
17. $1.455 - 0.3436i,\ -1.455 + 0.3436i$
19. $1.292 + 0.2013i,\ -0.8204 + 1.018i,$
$-0.4717 - 1.220i$

21. $\pm 1,\ \pm i$

23. $\cos(m\pi/16) + i\sin(m\pi/16)$, $m = 1, 5, 9, 13, 17, 21, 25, 29 = 0.9808 + 0.1951i,\ 0.5556 + 0.8315i,\ -0.1951 + 0.9808i,\ -0.8315 + 0.5556i,\ -0.9808 - 0.1951i,\ -0.5556 - 0.8315i,\ 0.1951 - 0.9808i,\ 0.8315 - 0.5556i$

25. $2^{1/8}[\cos(m\pi/16) + i\sin(m\pi/16)]$, $m = -1, 7, 15, 23 = 1.070 - 0.2127i,\ 0.2127 + 1.070i,\ -1.070 + 0.2127i,\ -0.2127 - 1.070i$

27. $2[\cos(m\pi/10) + i\sin(m\pi/10)]$, $m = 1, 5, 9, 13, 17 = 1.902 + 0.618i,\ 2i,\ -1.902 + 0.6180i,\ -1.176 - 1.618i,\ 1.176 - 1.618i$

29. $2(\cos m\pi/12 + i\sin m\pi/12)$, $m = 3, 7, 11, 15, 19, 23 = 1.414 + 1.414i(\sqrt{2} + \sqrt{2}i),\ -0.5176 + 1.932i,\ -1.932 + 0.5176i,\ -1.414 - 1.414i(-\sqrt{2} - \sqrt{2}i)\ 0.5176 - 1.932i,\ 1.932 - 0.5176i$

31. $2 \cdot 2^{1/8}[\cos(m\pi/16) + i\sin(m\pi/16)]$, $m = 3, 11, 19, 27 = 1.813 + 1.212i,\ -1.212 + 1.813i,\ -1.813 - 1.218i,\ 1.212 - 1.813i$

33. The four roots are twice the roots in problem 25; rounded they are $2.139 - 0.4255i,\ 0.4255 + 2.139i,\ -2.139 + 0.4255i,\ -0.4255 - 2.139i$

39. $\pm[\sqrt{(\sqrt{13} + 2)/2} + i\sqrt{(\sqrt{13} - 2)/2} \approx \pm\ (1.674 + 0.8960i)$

41. $\pm\sqrt{\sqrt{41} - 4)/2} - i\sqrt{(\sqrt{41} + 4)/2} = \pm\ (1.096 - 2.281i)$

Review Exercises for Chapter 6, page 355

1. (a) $8i$ (b) $5\sqrt{3}i$ **3.** $7 - 8i$ **5.** $17 + 11i$

7. $-i$ **9.** $-i$ **11.** $5,\ \pi,\ 5(\cos\pi + i\sin\pi)$

13. $\sqrt{2},\ -\pi/2,\ \sqrt{2}[\cos(-\pi/2) + i\sin(-\pi/2)]$

15. $2\sqrt{2},\ 3\pi/4,\ 2\sqrt{2}[\cos(3\pi/4) + i\sin(3\pi/4)]$

17. $2\sqrt{2},\ -3\pi/4,\ 2\sqrt{2}[\cos(-3\pi/4) + i\sin(-3\pi/4)]$

19. $\sqrt{65},\ \tan^{-1}(-\frac{7}{4}) \approx -1.0517,\ \sqrt{65}[\cos(-1.0517) + i\sin(-1.0517)$

21. $\sqrt{34},\ \tan^{-1}\frac{5}{3} - \pi \approx -2.111,\ \sqrt{34}[\cos(-2.111) + i\sin(-2.111)]$

23. $6\sqrt{3} - 6i$ **25.** $-3\sqrt{2} - 3\sqrt{2}i$

27. $z_1z_2 = 32(\cos 2.85 + i\sin 2.85)$; $z_1/z_2 = 2(\cos 1.15 + i\sin 1.15)$

29. $4096 = 2^{12}$

31. $53^3(\cos 6\theta + i\sin 6\theta) = 53^3(\cos 11.0946 + i\sin 11.0946) = 14{,}715 - 148{,}148i$ where $\theta = \pi + \tan^{-1}(-\frac{7}{2})$

33. $2^{1/6}[\cos(m\pi/12) + i\sin(m\pi/12)]$, $m = 3, 11, 19 = 0.7937 + 0.7937i,\ -1.084 + 0.2905i,\ 0.2905 - 1.084i$

35. $-3\sqrt{2}/2 + (3\sqrt{2}/2)i,\ 3\sqrt{2}/2 - (3\sqrt{2}/2)i$

37. $\cos(m\pi/10) + i\sin(m\pi/10)$, $m = -1, 3, 7, 11, 15 \approx 0.9511 - 0.3090i,\ 0.5878 + 0.8090i,\ -0.5878 + 0.8090i,\ -0.9511 - 0.3090i,\ -i$

Chapter 7

Problems 7.1, page 364

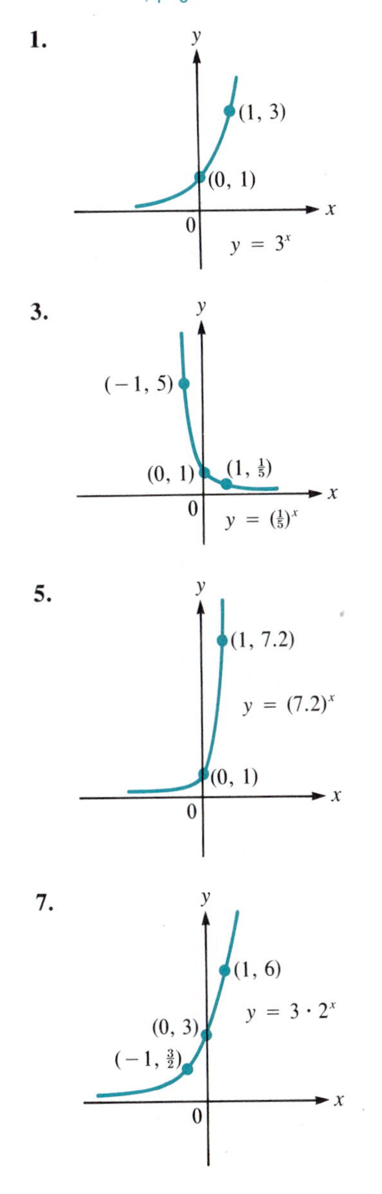

9. **11.** **13.**

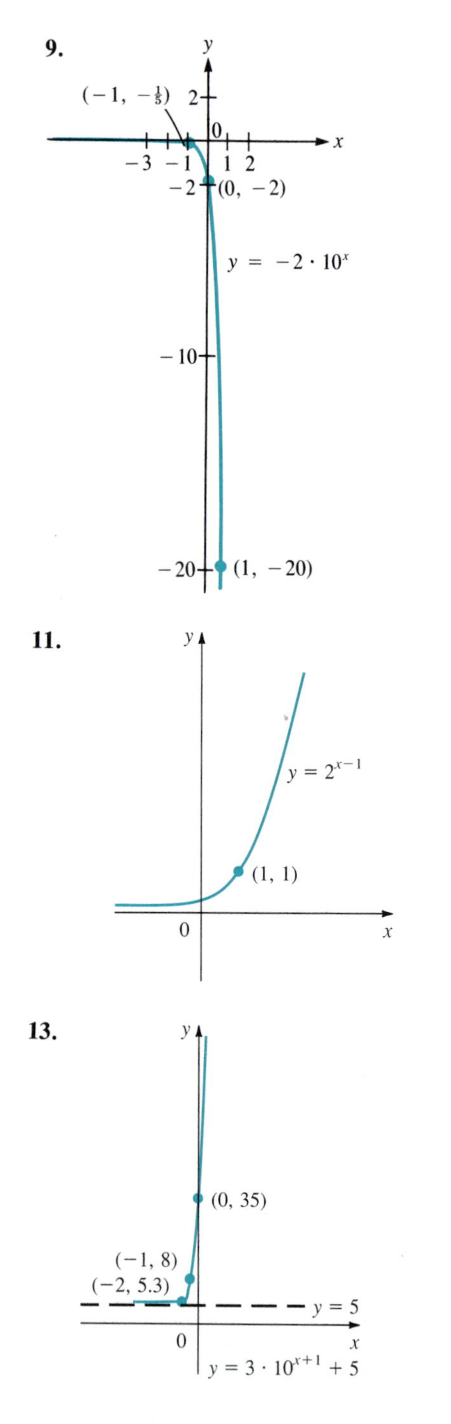

15. 158.4893192 **17.** 0.10881882
19. 0.643440775 **21.** 4.728804388
23. \$6312.38; interest is \$1312.38
25. \$6352.45; interest is \$1352.45
27. \$12,348.08; interest is \$4348.08
29. \$12,418.65; interest is \$4418.65
31. \$16,564.17; interest is \$6564.17
33. 79.08% annually; 81.40% quarterly so percentage difference is 2.32%
35. $5\frac{1}{8}$% semiannually (5.191% > 5.127%)
37. 3.61224%
39. (a) 12.36% for A and 11.57% for B (b) Account A by \$24.77
41. d **43.** a
45.

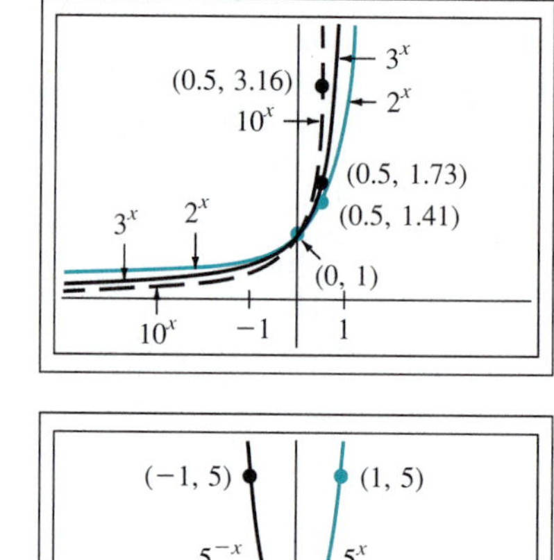

47.

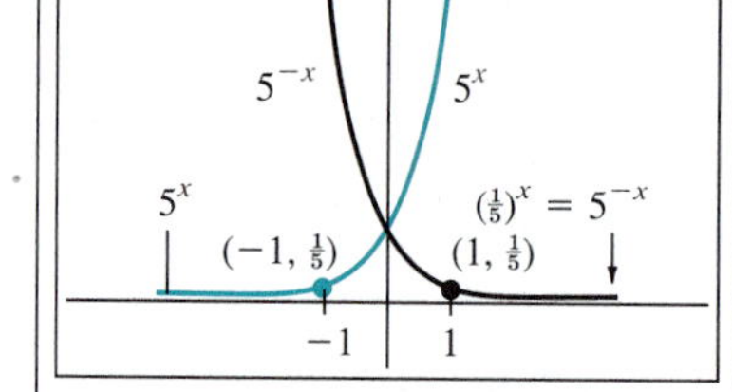

49.

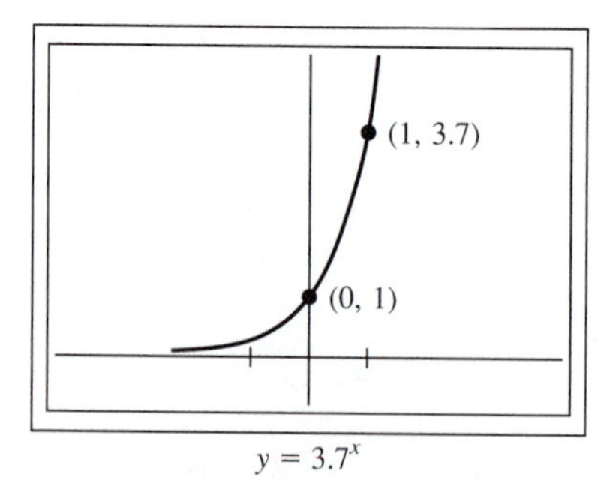

$y = 3.7^x$

51.

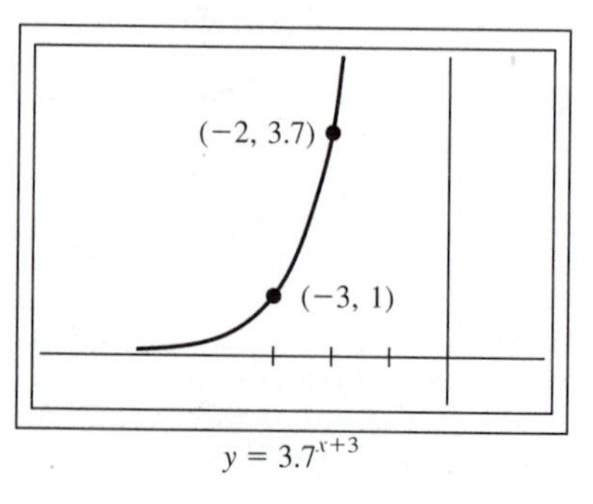

$y = 3.7^{x+3}$

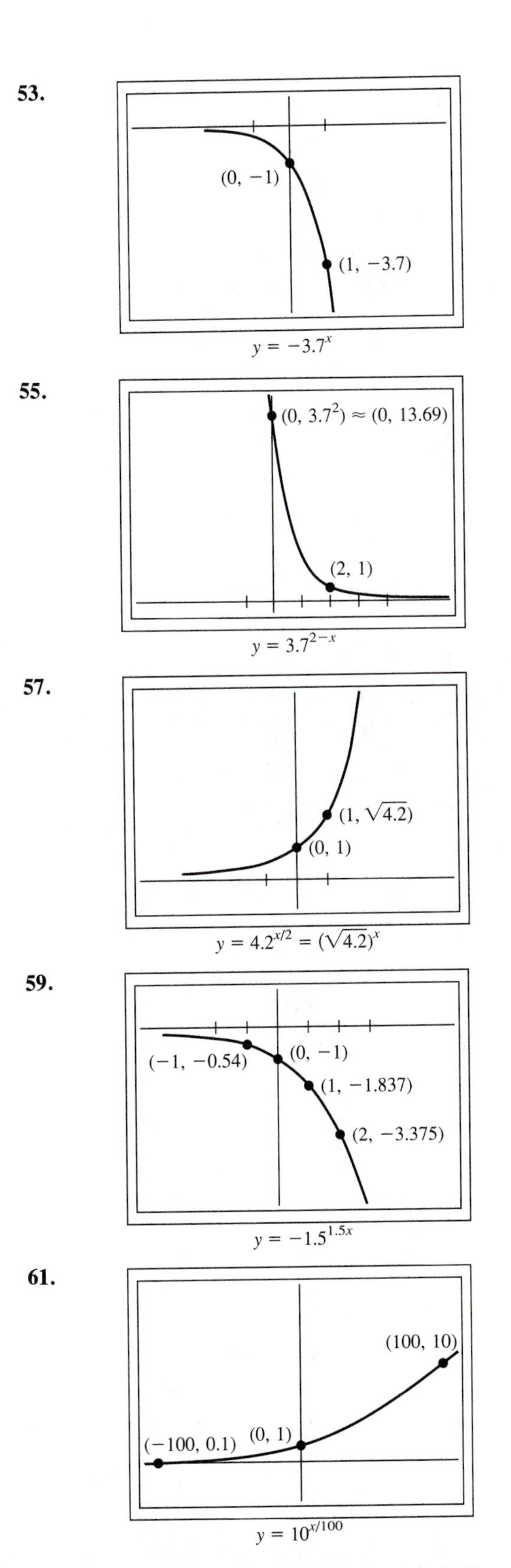

Problems 7.2, page 374

1. 23.3361 **3.** 5.65223 **5.** 5.86486×10^{12}
7. 1.24371×10^{-7} **9.** 23.1407 **11.** *f*
13. *a* **15.** *c* **17.** *d* **19.** *i*

21.

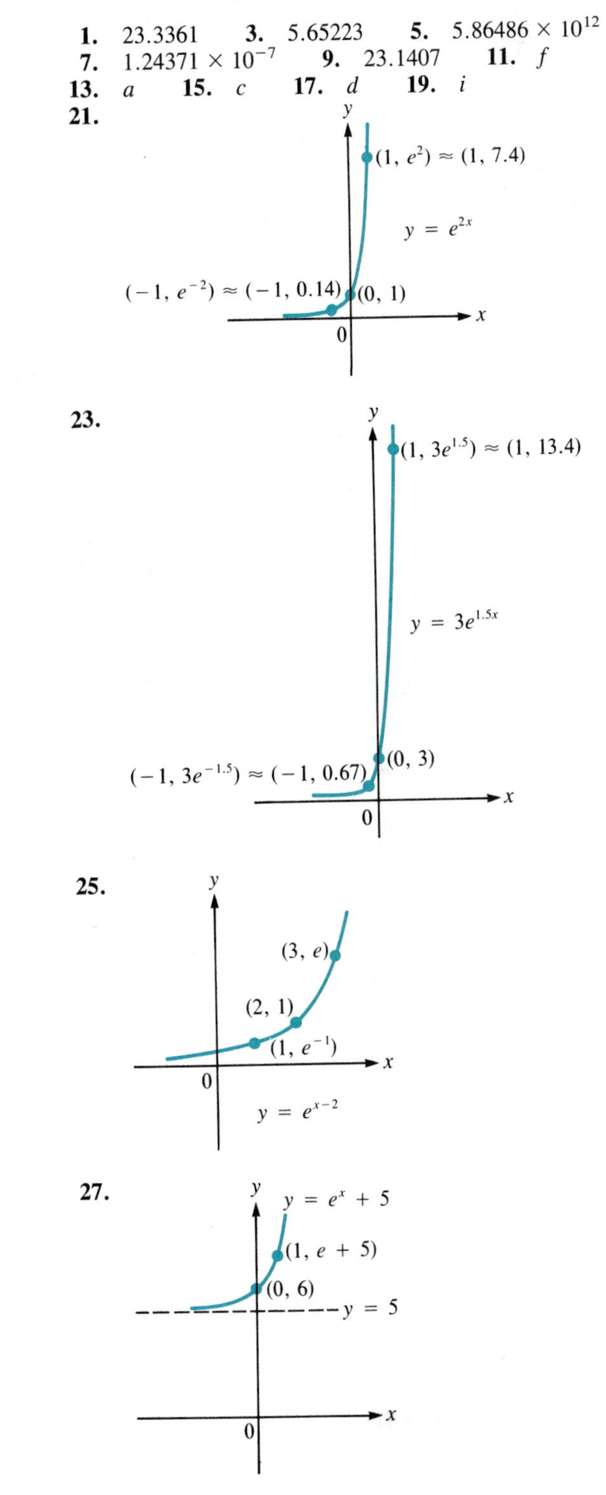

29. (a) 1.1388 (b) 0.69073
31. \$14,838.49; interest \$5838.49
33. \$28,292.17; interest \$18,292.17
35. \$20,544.33 **37.** the first (5.127% > 5.125%)
39. 5.8 years
41. (a) 12.36%, 11.63% (b) the first, by \$23.04
43. (a) 5,042,728,201 (b) 5,573,076,555 (c) 6,159,202,133
45. (a) 121.35°F (b) 71.49°F (c) 53.55°F
47. (c)

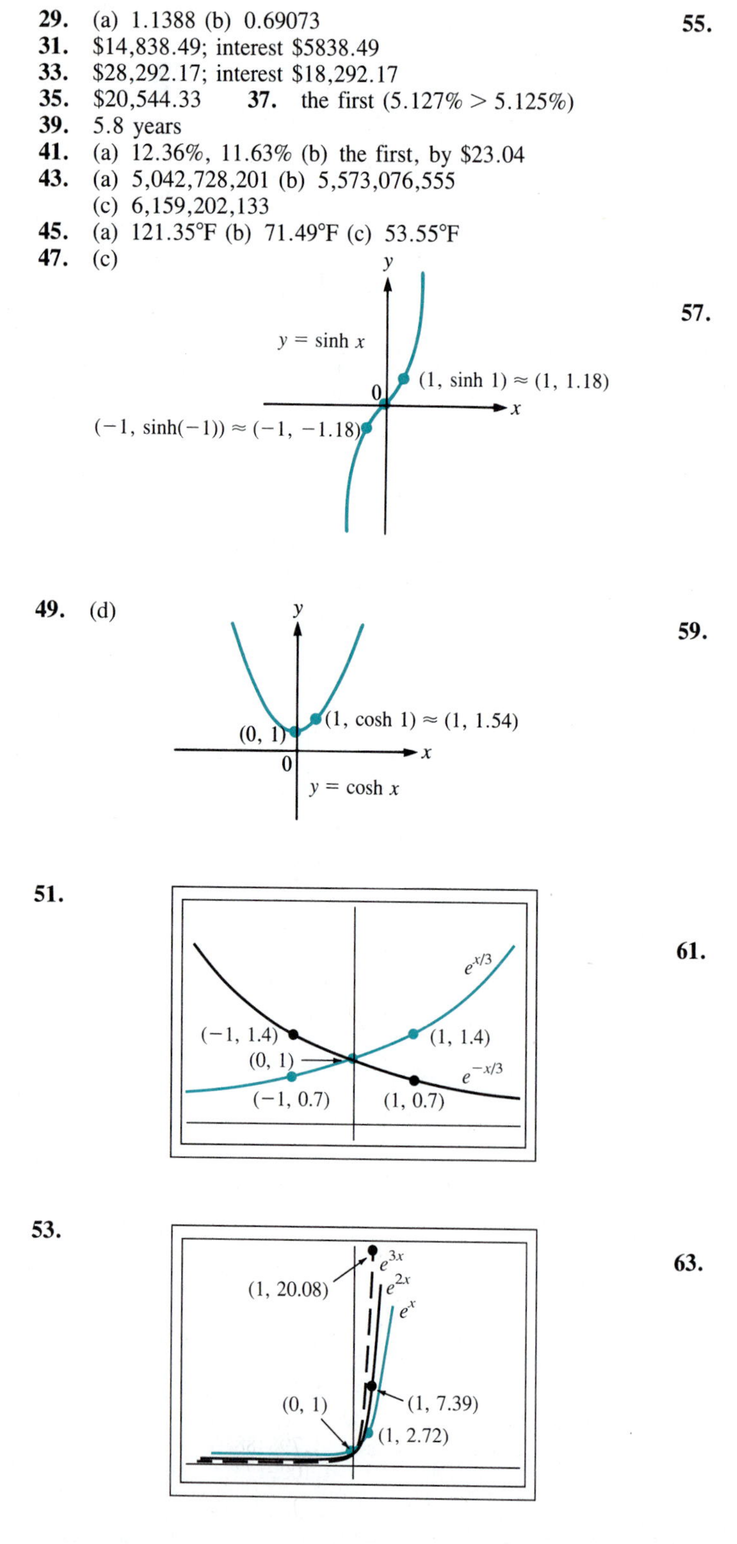

49. (d)

51.

53.

55.

(0, 1) $y = e^{(x+1)/2}$ $y = e^{x/2}$ (0, 1.65) $y = e^{(x-1)/2}$ (0, 0.61)

57.

(1, 4.95) (0, 1)

$y = e^{1.6x}$

59.

(0, −1) (1, −4.95)

$y = -e^{1.6x}$

61.

y = 2 (1, 1) (0.57, 0) (0, −2.95)

$y = 2 - e^{1.6(1-x)}$

63.

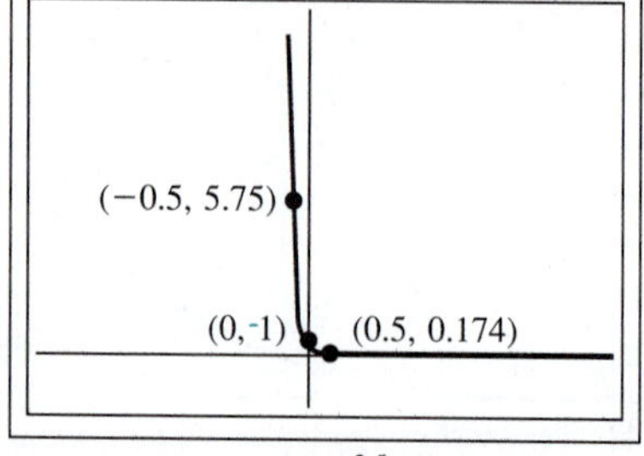

$y = e^{-3.5x}$

65.

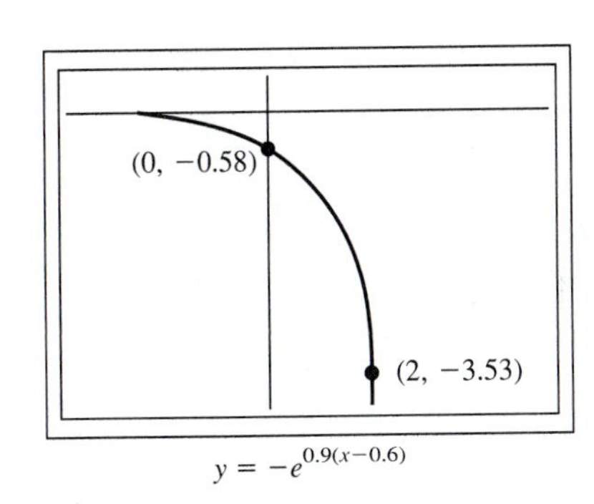

$y = -e^{0.9(x-0.6)}$

67.

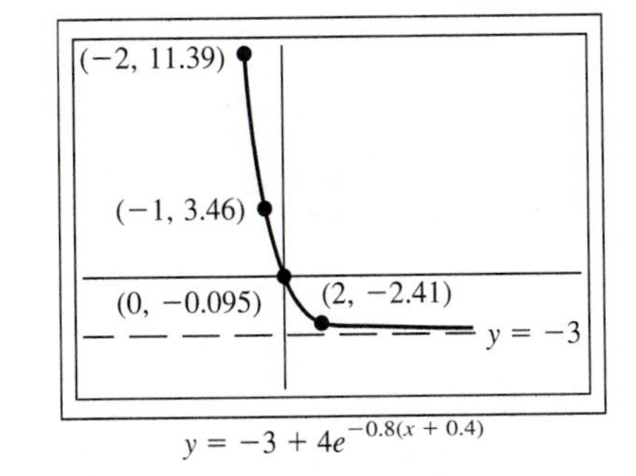

$y = -3 + 4e^{-0.8(x+0.4)}$

69.

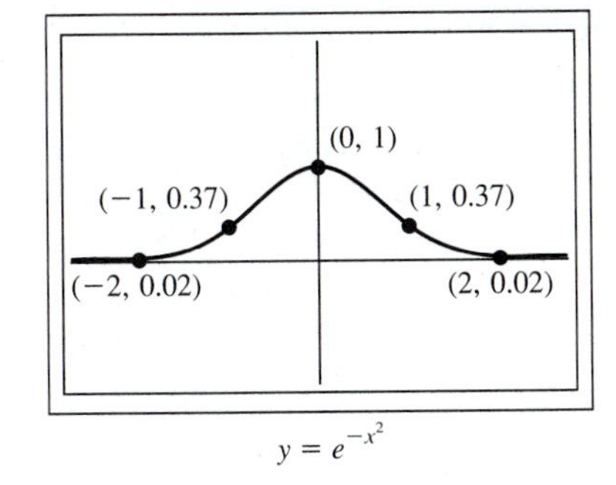

$y = e^{-x^2}$

71.

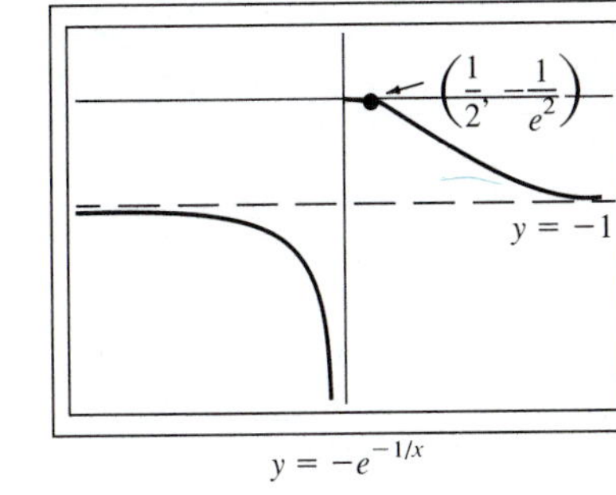

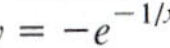

$y = -e^{-1/x}$

73.

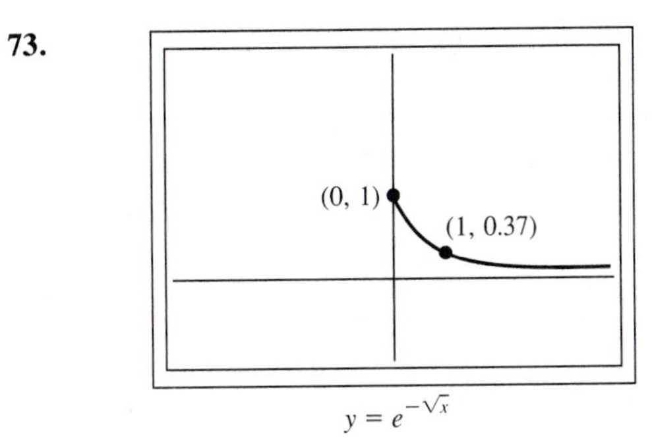

$y = e^{-\sqrt{x}}$

Problems 7.3, page 382

1. $16^{1/2} = 4$ **3.** $(\frac{1}{2})^{-4} = 16$ **5.** $12^0 = 1$ **7.** $10^{-3} = 0.001$ **9.** $e^2 = e^2$ **11.** $\log_2 32 = 5$ **13.** $\log_{1/3} 9 = -2$ **15.** $\log_{10} 2 = 0.301029995$ **17.** 3 **19.** −2 **21.** $-\frac{1}{2}$ **23.** 5 **25.** −9 **27.** −4.5 **29.** b **31.** c **33.** 2.0793 **35.** 3.5836 **37.** 2.5378 **39.** −0.4622 **41.** 0.2689 **43.** −1.7689 **45.** −4.6134 **47.** $4 \log_a x + 3 \log_a y$ **49.** $2 \log_a x + \frac{3}{4} \log_a y$ **51.** $\frac{1}{2} \log_a x + \frac{1}{3} \log_a y - \frac{1}{5} \log_a z$ **53.** $\frac{4}{5} \log_a x + \frac{4}{5} \log_a y - \frac{4}{5} \log_a z$ **55.** $\log_a 1 = 0$ **57.** $\log_a xy^2$ **59.** $\log_a 20wx$ **61.** $\log_a xy^2z^3$ **63.** $\log_a (x-2)$ **65.** $\log_a (x^2y - 2y^2 + 2x^2 - 4y)$ **67.** $\log_a \{(x+y)^4/(x-y)^3\}$

Problems 7.4, page 391

1. 3 **3.** 20 **5.** 3.4 **7.** 0.235 **9.** π **11.** 6.4 **13.** $\ln (x^2 + 3x)$ **15.** $\ln xyz$ **17.** $\ln (z^{1/2}w^{3/4}/(x^2+1)^2)$ **19.** $\ln 2$ **21.** $\ln \frac{2^3 3^4}{w^3 z^4} = \ln \frac{648}{w^3 z^4}$ **23.** $(-1, \infty)$ **25.** $(-\infty, 1)$ **27.** $\{x\colon x \neq 0\}$ **29.** $\{x\colon x < -1 \text{ or } x > 1\}$ **31.** $\{x\colon x < -3 \text{ or } x > 4\}$ **33.** $\mathbb{R}$ **35.** $(0, \infty)$ **37.** f **39.** i **41.** e **43.** b **45.** 144 **47.** $b^a \cdot a^b$ **49.** $\frac{10}{21}$ **51.** $54^3 = 157{,}464$ **53.** 9 **55.** (a) 2.8351 (b) 6.8351 (c) −1.1649 (d) −6.1649 **57.** (a) 1.9786 (b) 4.9786 (c) −1.0214 (d) −4.0214 **59.** 2.0214 **61.** 2.2364 **63.** −0.91755 **65.** 5.2865 **67.** 5.2131 **69.** (a) increases by $\ln 27 = 3 \ln 3$ (b) decreases by $\ln 1000 = 3 \ln 10$ **71.** 1.757699625 **73.** 4.698987376 **75.** 6.907755279 **77.** −3.42651519 **79.** 0.497149872 **81.** $\pi = 3.141592654$ **83.** 1.561842388 **85.** 1.351393465 **87.** 1.261859507 **89.** 1.796488803 **91.** 0.79248125 **93.** 1.10091231 **95.** (a) 0.4307 (b) 0.8614 (c) −1.2920

97. (a) $\ln 0.8 \approx -0.223143491$ and $\ln 1.2 \approx 0.182321542$
(b) 0.693143053
(c) $\ln 3 \approx 1.098604387$ and $\ln 8 \approx 2.079429159$

99. $100! \approx 9.32 \times 10^{157}$; $200! \approx 7.88 \times 10^{374}$

101. (c) $2 \pm \sqrt{5}$ (e) $\ln(2 + \sqrt{5}) \approx 1.4436$

103. $\ln(4 + \sqrt{17}) \approx 2.0947$

105. $\ln(2 \pm \sqrt{3}) \approx \pm 1.316957897$

107.

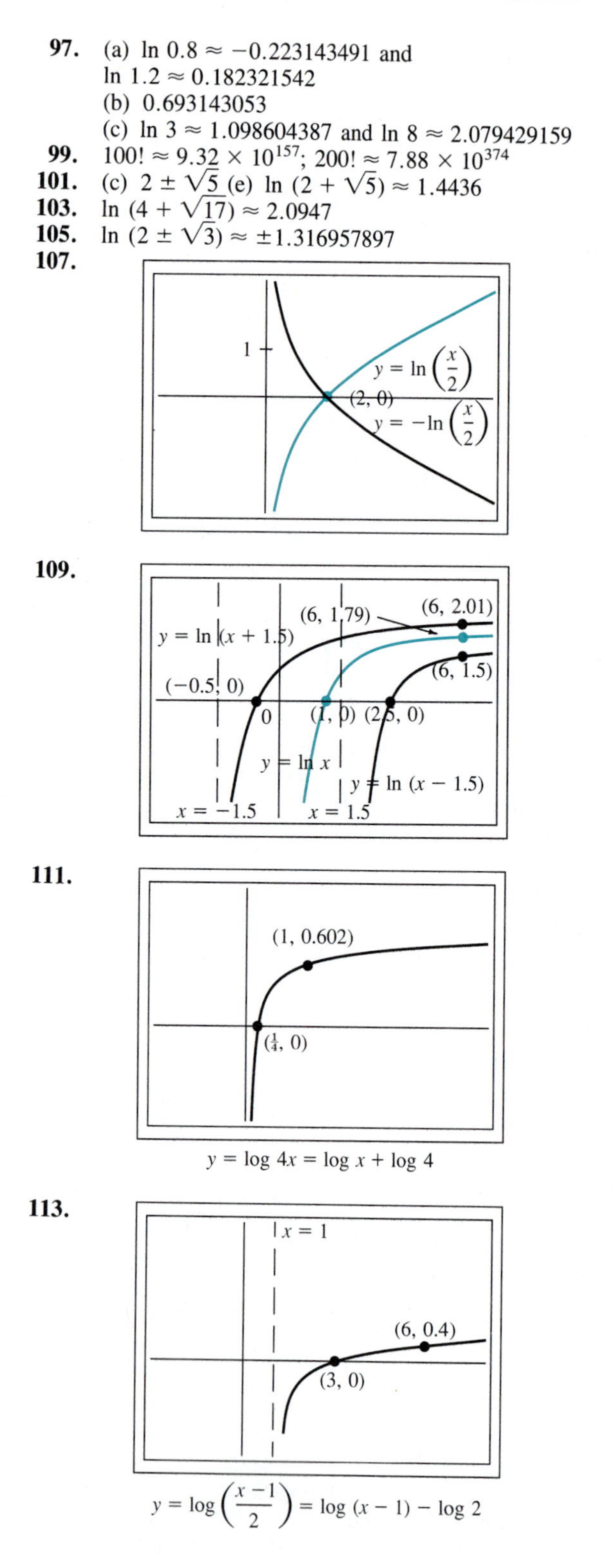

109.

111.

$y = \log 4x = \log x + \log 4$

113.

$y = \log\left(\frac{x-1}{2}\right) = \log(x-1) - \log 2$

115.

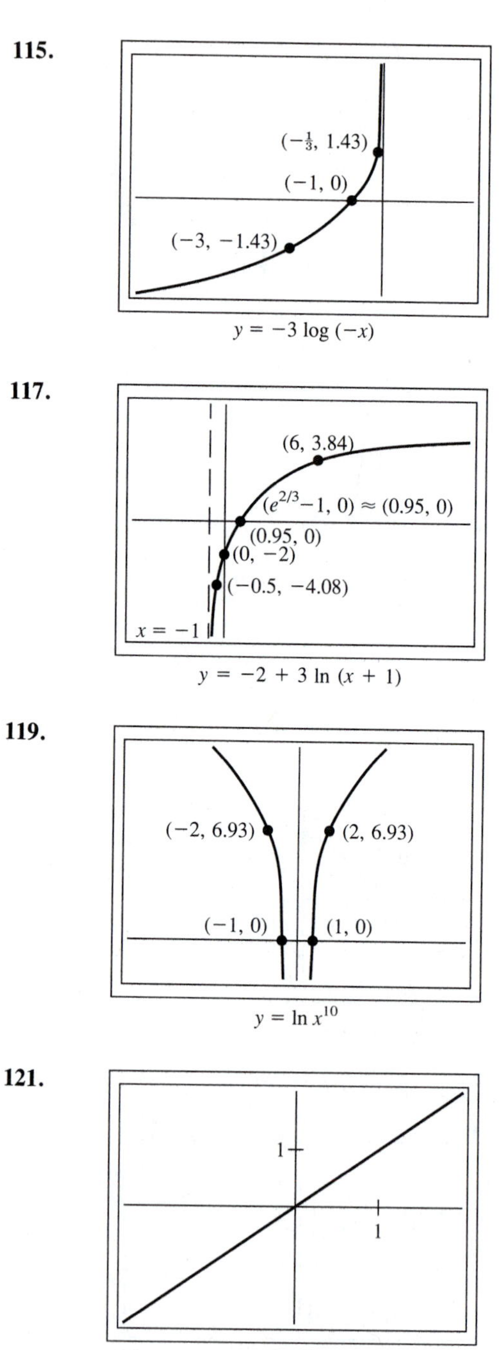

$y = -3 \log(-x)$

117.

$y = -2 + 3 \ln(x + 1)$

119.

$y = \ln x^{10}$

121.

$y = \ln e^x = x$

123.

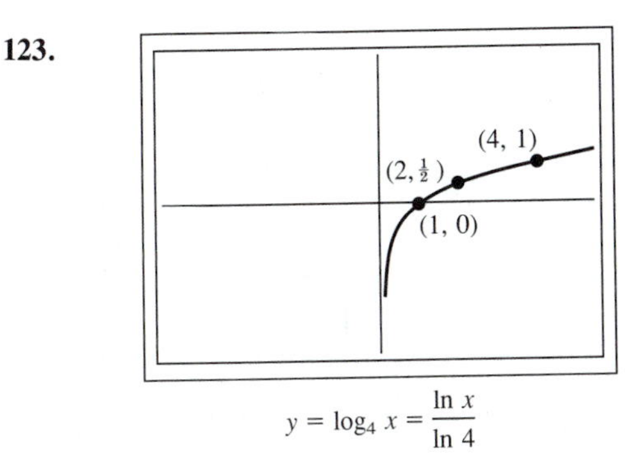

$y = \log_4 x = \dfrac{\ln x}{\ln 4}$

125.

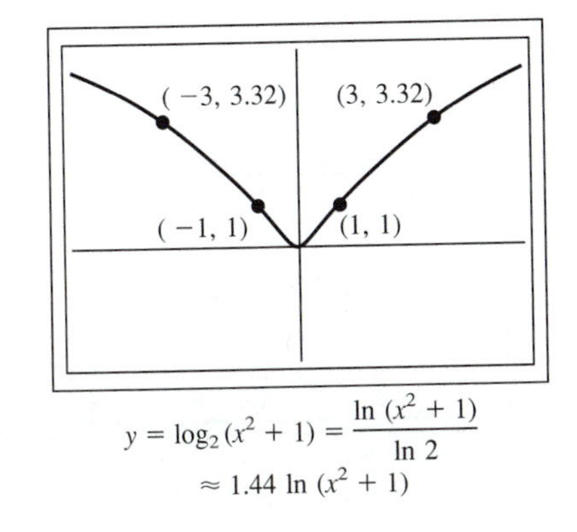

$$y = \log_2 (x^2 + 1) = \frac{\ln (x^2 + 1)}{\ln 2} \approx 1.44 \ln (x^2 + 1)$$

Problems 7.5, page 401

1. 2 **3.** −1 **5.** $2^8 = 256$ **7.** $e^{-1} \approx 0.368$
9. 10 **11.** 1 **13.** $\ln 10 \approx 2.303$
15. no solution **17.** $\dfrac{\log 14}{\log 7} = \dfrac{\ln 14}{\ln 7} \approx 1.356$
19. $\dfrac{\log 8}{\log 231} = \dfrac{\ln 8}{\ln 231} \approx 0.382$ **21.** $e^{1.6} \approx 4.953$
23. $10^{-1.57} \approx 0.027$ **25.** $1 + \ln 2 \approx 1.693$
27. $2 \ln 4 \approx 2.773$ **29.** $-5 \ln 5 \approx -8.047$
31. $\frac{1}{3}e \approx 0.906$ **33.** 4 **35.** 8.920
37. no solution [$z = -11$ is wrong because $\ln (-11 + 4)$ is undefined]
39. $\ln 4 \approx 1.386$ or $\ln 2 \approx 0.693$
41. $2 \ln 2 \approx 1.386$
43. no solution (e^x cannot equal −1 or −2)
45. 48 **47.** 0.0023 **49.** 4 **51.** $\frac{1}{10}$
53. $\frac{1}{2}\sqrt[3]{5} \approx 0.855$ **55.** ±2 **57.** $\frac{1}{5}$ **59.** 3
61. $\frac{1}{8}$ **63.** $-\frac{4999}{1000} = -4.999$ **65.** $\frac{1}{3}$ **67.** −4
69. no solution [$\log_4 (-\frac{4}{3})$ is undefined]
71. 8 **73.** 0 **75.** 0.46, 5.36 **77.** 1.52
79. −0.86, 1.20 **81.** 3.82 **83.** −3.98, 282.12

Problems 7.6, page 412

1. 10.15 years
3. (a) 55.07 years (b) 22.11 years (c) 13.87 years (d) 11.12 years (e) 7.46 years
5. (a) 54.93 years (b) 21.97 years (c) 13.73 years (d) 10.99 years (e) 7.32 years
7. \$9185.12 **9.** \$6499.31 **11.** 3.596%
13. 7.324% **15.** in 2012 (26.39 years later)
17. (a) 273,037,968 (b) in 2065 (84.8 years after 1980)
19. (a) 19,057,278 (b) in 2004 (29.56 years after 1974)
21. in 2003 (22.74 years after 1980)
23. 13.76% **25.** 2.71% **27.** 141.46
29. (a) −4.138°C (b) 1.677 minutes = 100.62 seconds
31. (a) 135.69 years (b) 450.76 years
33. 4,620,981 years
35. (a) 625.53 mbar (b) 303.42 mbar (c) 429.03 mbar (d) 348.63 mbar (e) 57,396.3 meters
37. (a) 10,000 (b) 7 (c) after about 214.2 hours = 8 days, 22.2 hours
39. 7.22 **41.** 3.162×10^{-5} moles per liter
43. (a) 15.849 times as bright (b) 1.75 (c) 1.99526 times as bright (d) 11.633 (e) 1.6004

Review Exercises for Chapter 7, page 415

1.

3.

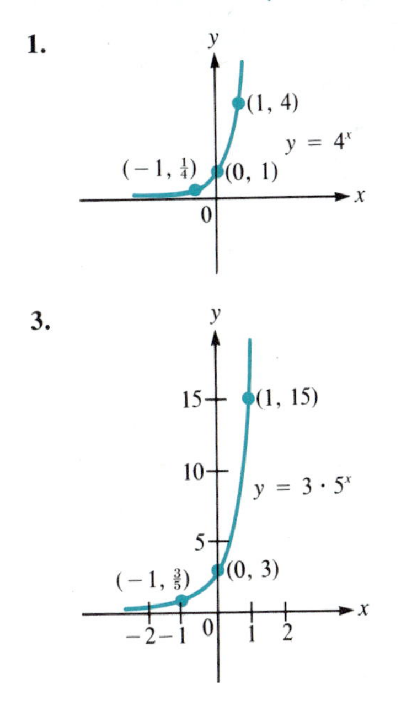

5.

7.

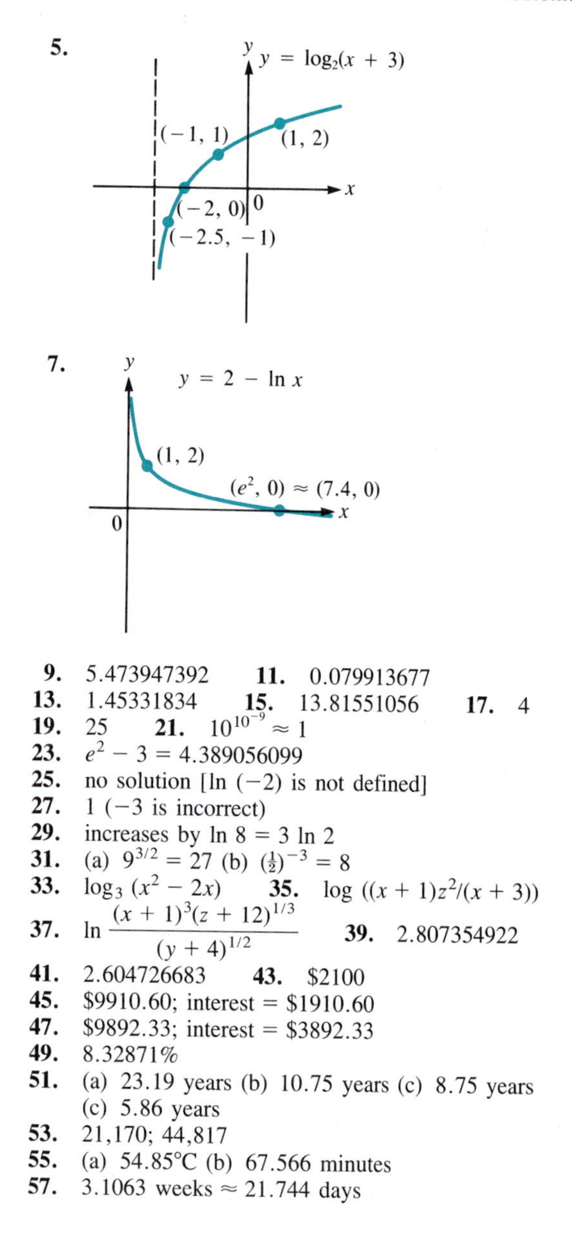

9. 5.473947392 **11.** 0.079913677
13. 1.45331834 **15.** 13.81551056 **17.** 4
19. 25 **21.** $10^{10^{-9}} \approx 1$
23. $e^2 - 3 = 4.389056099$
25. no solution [ln (−2) is not defined]
27. 1 (−3 is incorrect)
29. increases by $\ln 8 = 3 \ln 2$
31. (a) $9^{3/2} = 27$ (b) $(\frac{1}{2})^{-3} = 8$
33. $\log_3 (x^2 - 2x)$ **35.** $\log ((x + 1)z^2/(x + 3))$
37. $\ln \dfrac{(x + 1)^3(z + 12)^{1/3}}{(y + 4)^{1/2}}$ **39.** 2.807354922
41. 2.604726683 **43.** \$2100
45. \$9910.60; interest = \$1910.60
47. \$9892.33; interest = \$3892.33
49. 8.32871%
51. (a) 23.19 years (b) 10.75 years (c) 8.75 years (c) 5.86 years
53. 21,170; 44,817
55. (a) 54.85°C (b) 67.566 minutes
57. 3.1063 weeks ≈ 21.744 days

Answers for Appendix A

1. $y = x^2 + 2$

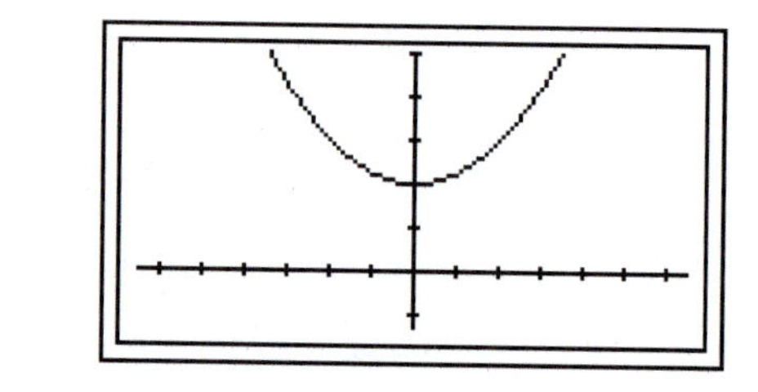

$-3 \le x \le 3 \qquad -1.3 \le y \le 5$
$x_{scl} = 0.5 \qquad y_{scl} = 1$

3. $y = x^3 - x$

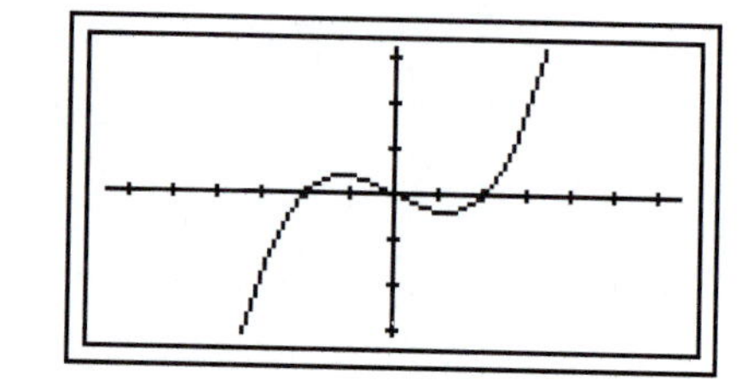

$-3 \le x \le 3 \qquad -3 \le y \le 3$
$x_{scl} = 0.5 \qquad y_{scl} = 1$

5. $y = -x^4 + 2x^2 - x + 3$

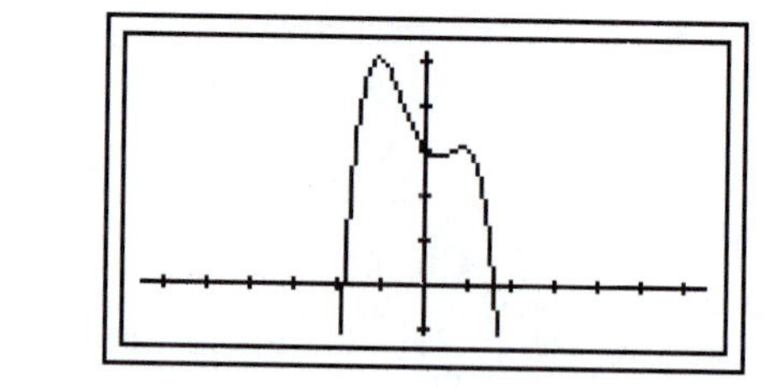

$-6.5 \le x \le 6.5 \qquad -1 \le y \le 5$
$x_{scl} = 1 \qquad y_{scl} = 1$

7. $y = \dfrac{x}{x + 1}$

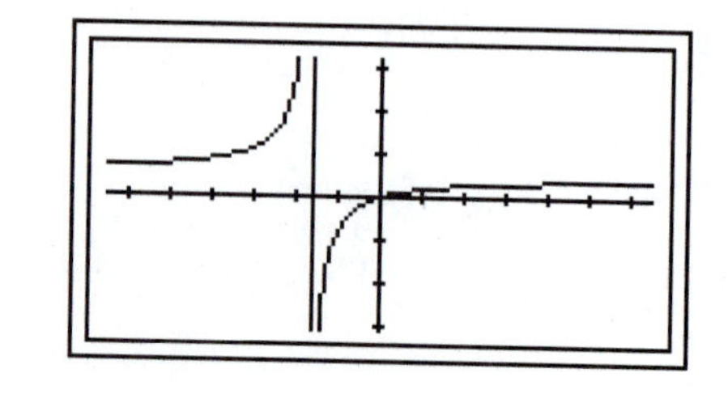

$-4 \le x \le 4 \qquad -6 \le y \le 6$
$x_{scl} = 0.6 \qquad y_{scl} = 2$

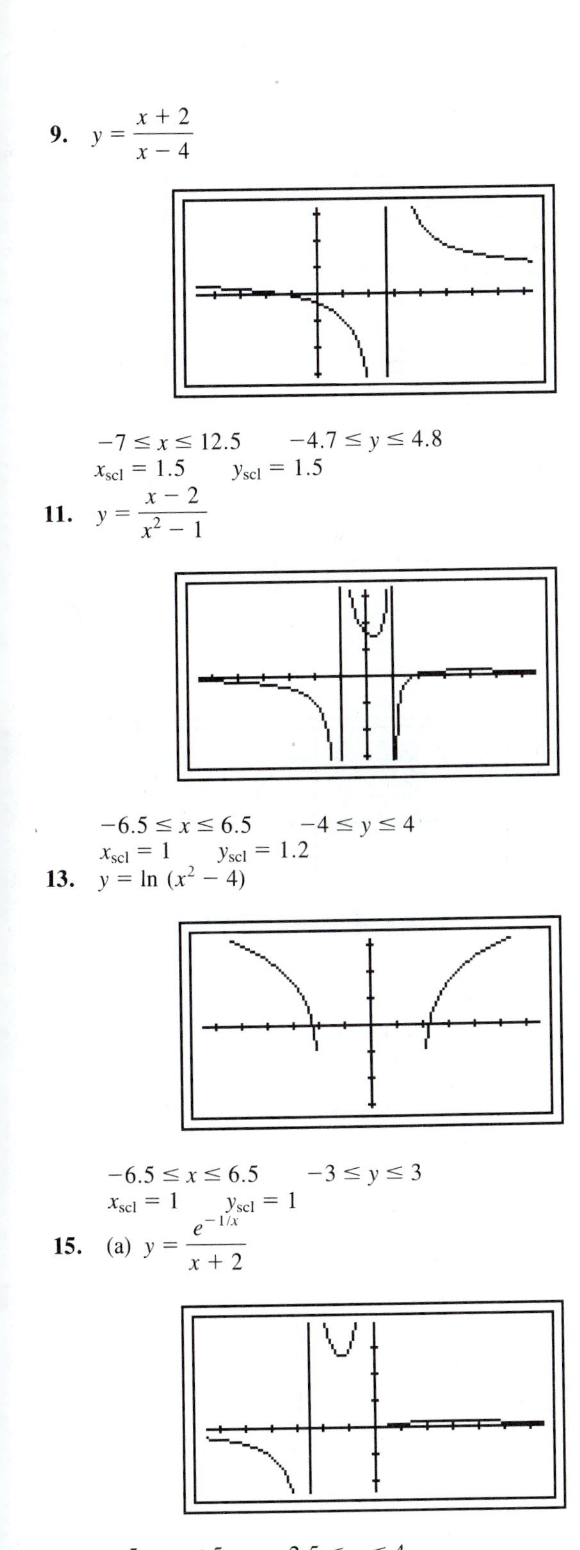

9. $y = \dfrac{x+2}{x-4}$

$-7 \le x \le 12.5 \qquad -4.7 \le y \le 4.8$
$x_{scl} = 1.5 \qquad y_{scl} = 1.5$

11. $y = \dfrac{x-2}{x^2-1}$

$-6.5 \le x \le 6.5 \qquad -4 \le y \le 4$
$x_{scl} = 1 \qquad y_{scl} = 1.2$

13. $y = \ln(x^2 - 4)$

$-6.5 \le x \le 6.5 \qquad -3 \le y \le 3$
$x_{scl} = 1 \qquad y_{scl} = 1$

15. (a) $y = \dfrac{e^{-1/x}}{x+2}$

$-5 \le x \le 5 \qquad -2.5 \le y \le 4$
$x_{scl} = 0.8 \qquad y_{scl} = 1$

(b) (enlarged view of first quadrant)

$y = \dfrac{e^{-1/x}}{x+2}$

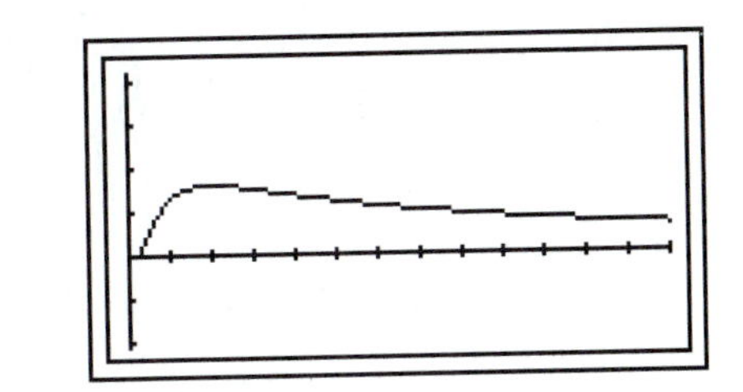

$0 \le x \le 13 \qquad -0.2 \le y \le 0.4$
$x_{scl} = 1 \qquad y_{scl} = 0.1$

17. $e^{x-\ln x}$

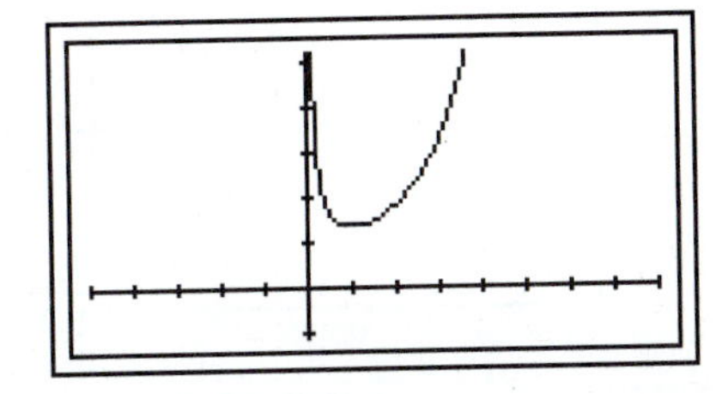

$-5 \le x \le 8 \qquad -2 \le y \le 10$
$x_{scl} = 1 \qquad y_{scl} = 2$

19. $y = 2 \sin x - 5 \cos 3x$

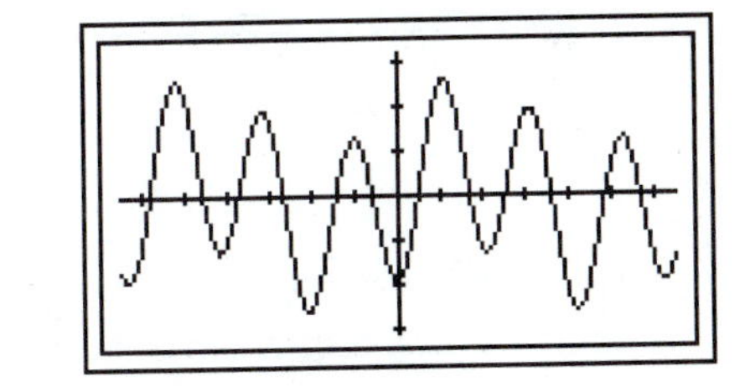

$-6.5 \le x \le 6.5 \qquad -8 \le y \le 8$
$x_{scl} = 2.6 \qquad y_{scl} = 1$

21. (a) $y = x \cos x$

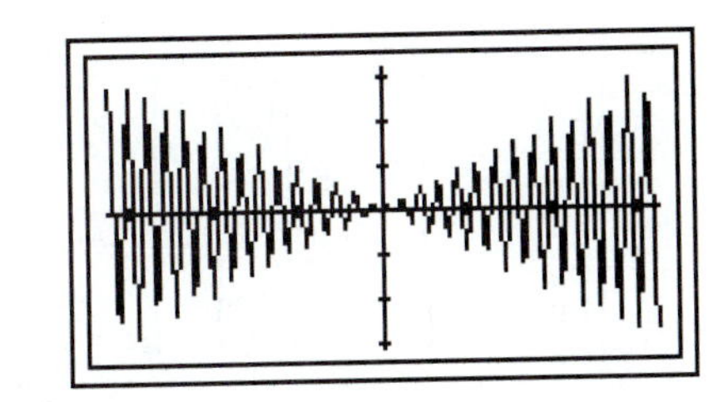

$-500 < x \le 500 \qquad -500 \le y \le 500$
$x_{scl} = 77 \qquad y_{scl} = 151$

(b) (alternate view)
$y = x \cos x$

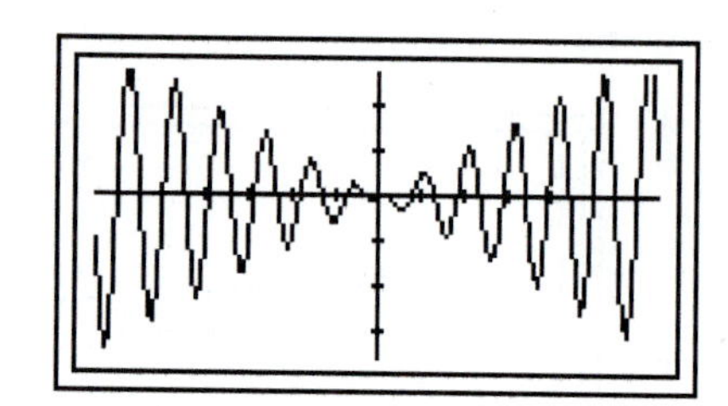

$-39 \leq x \leq 39 \qquad -35 \leq y \leq 35$
$x_{scl} = 6 \qquad y_{scl} = 11.5$

23. (a) $y = e^{2x} \sin 3x$

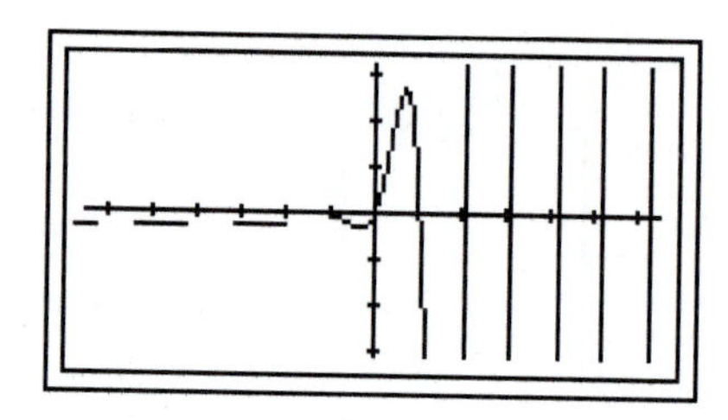

$-6.5 \leq x \leq 6.5 \qquad -4 \leq y \leq 4$
$x_{scl} = 1 \qquad y_{scl} = 1.3$

(b) (alternate view)
$y = e^{2x} \sin 3x$

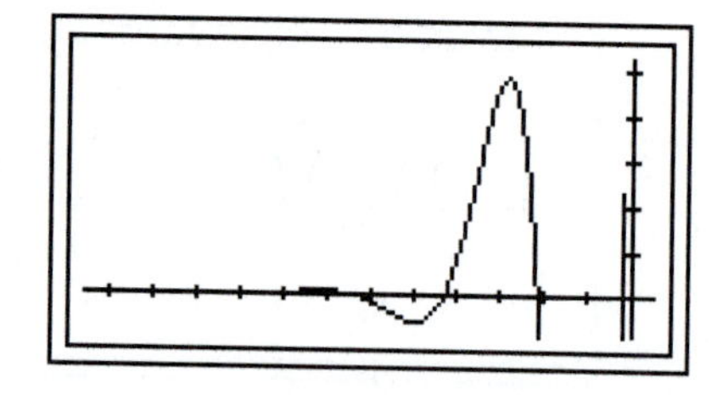

$-6 \leq x \leq 0.23 \qquad -0.01 \leq y \leq 0.055$
$x_{scl} = 0.5 \qquad y_{scl} = 0.01$

25. $y = 3\sec 5x = 3/\cos 5x$

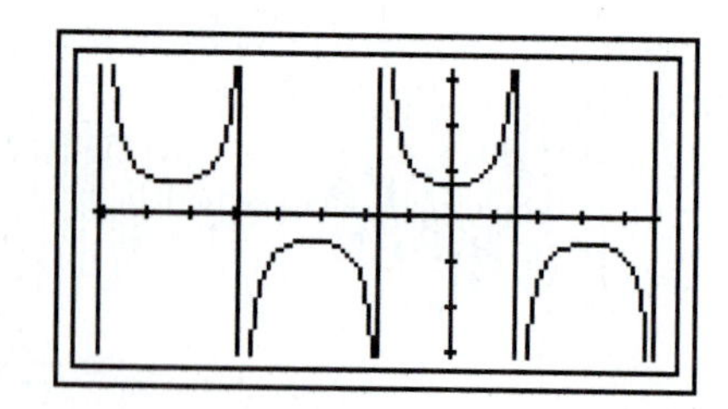

$-1.6 \leq x \leq 0.95 \qquad -15 \leq y \leq 15$
$x_{scl} = 0.2 \qquad y_{scl} = 4.5$

27. $y = \csc (x - 3) = 1/\sin (x - 3)$

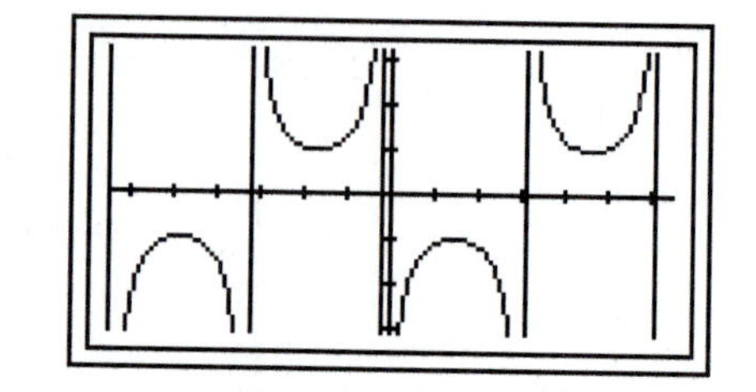

$-6.5 \leq x \leq 6.5 \qquad -3 \leq y \leq 3$
$x_{scl} = 1 \qquad y_{scl} = 1$

29. 2.67, −3.50 **31.** 1.47 **33.** 1.46
35. −2.09, 0.24, 7.85
37. −1.82, −0.72, 0.67, 2.87 **39.** 1.31
41. 1.79, 2.62 **43.** −4.59, −1.25, 1.06, 5.56, 6.90
45. $\{x: 0.33 < x < 3.00\}$ **47.** $\mathbb{R}$
49. $\{x: x < -1.73 \text{ or } -0.82 < x < 1.73 \text{ or } 1.82 < x < 2\}$
51. $\{x: x > 2.22\}$
53. $x^2 + y^2 = 25$

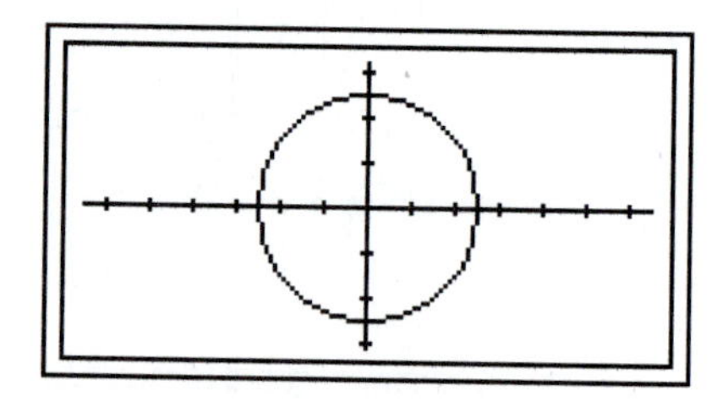

$-13 \leq x \leq 13 \qquad -6.2 \leq y \leq 6.4$
$x_{scl} = 2 \qquad y_{scl} = 2$
Note: The range values given above lead to a round circle on HP calculators. To get the same value on a TI or Casio calculator, it is necessary to use different values. $-9.6 \leq x \leq 9.4$ and $-6.4 \leq y \leq 6.2$ work on these machines. A similar comment applies to all the conic section graphs.

55. $x^2 + 2x + y^2 - 4y = 20$

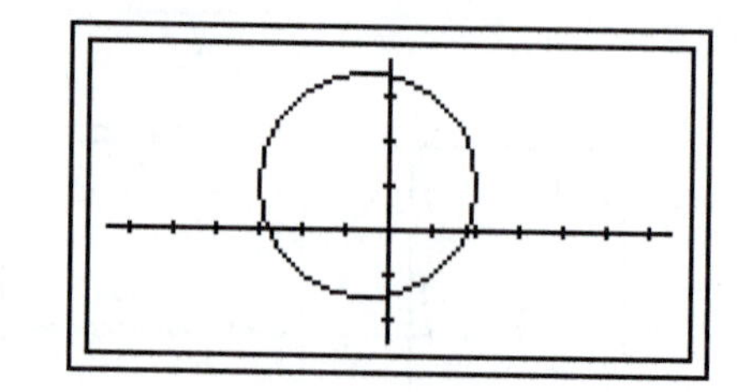

$-13 \leq x \leq 13 \qquad -5 \leq y \leq 7.6$
$x_{scl} = 2' \qquad y_{scl} = 2$
To get the same picture on a TI, the scale is $-14.4 \leq x \leq 14.1$ and $-9.6 \leq y \leq 9.3$.

57. $7x^2 - 11y^2 = 47$

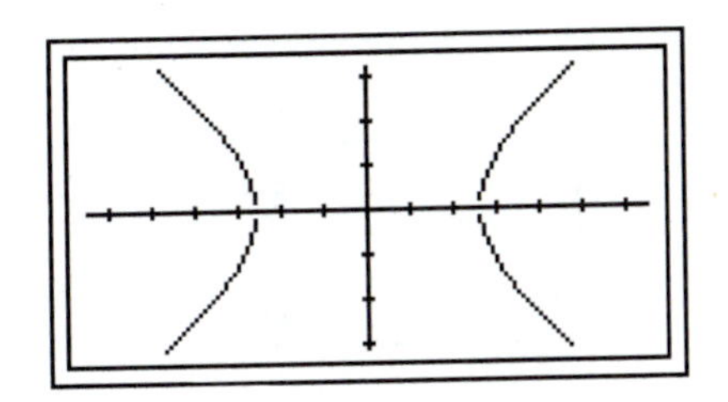

$-6.5 \leq x \leq 6.5 \qquad -3.1 \leq y \leq 3.2$
$x_{scl} = 1 \qquad y_{scl} = 1$
To get the same picture on a TI, the scale is $-4.8 \leq x \leq 4.7$ and $-3.2 \leq y \leq 3.1$.

59. $\dfrac{x^2}{37} + \dfrac{y^2}{121} = 1$

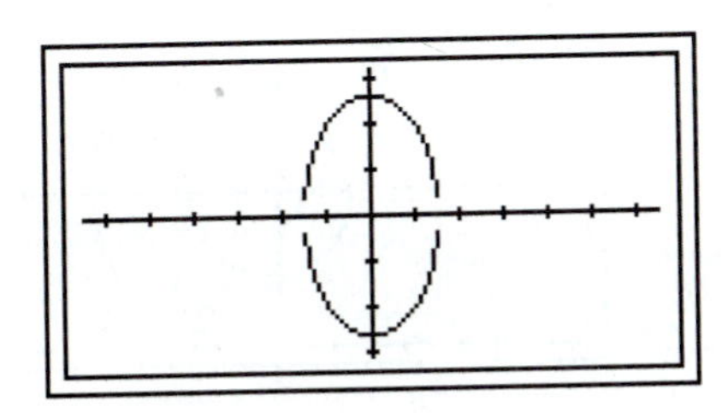

$-26 \leq x \leq 26 \qquad -13 \leq y \leq 13.5$
$x_{scl} = 4 \qquad y_{scl} = 4.2$
To get the same picture on a TI, the scale is $-19.2 \leq x \leq 18.8$ and $-12.8 \leq y \leq 12.4$.

61. $3y^2 + 12x = 6$

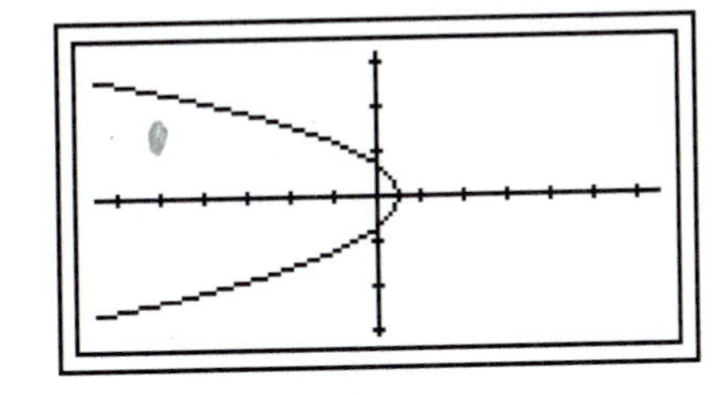

$-6.5 \leq x \leq 6.5 \qquad -6.2 \leq y \leq 6.4$
$x_{scl} = 1 \qquad y_{scl} = 2$
To get the same picture on a TI, the scale is $-9.6 \leq x \leq 9.4$ and $-6.4 \leq y \leq 6.2$.

63. $2x^2 + 7x + 5y^2 - 2y = 8$

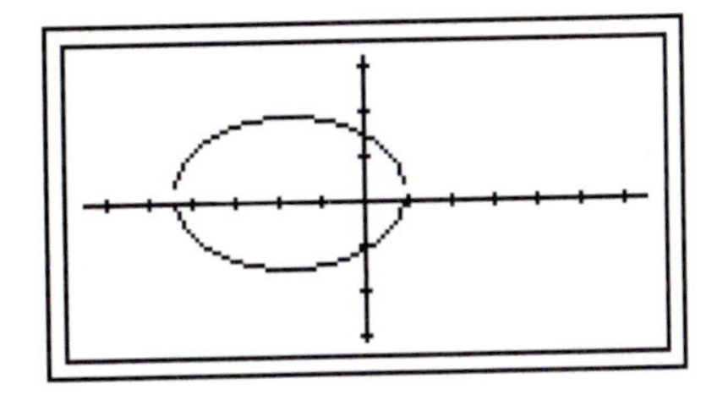

$-6.5 \leq x \leq 6.5 \qquad -3.1 \leq y \leq 3.2$
$x_{scl} = 1 \qquad y_{scl} = 1$
To get the same picture on a TI, the scale is $-6.8 \leq x \leq 2.7$ and $-3.2 \leq y \leq 3.1$.

65. $2y^2 - 8y - 12x^2 + 12x = 16$

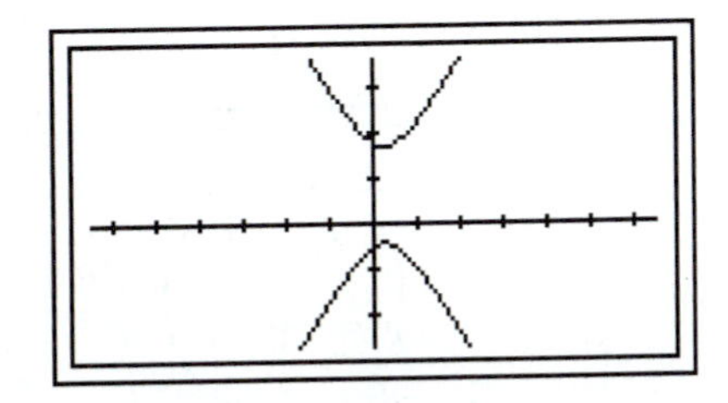

$-13 \leq x \leq 13 \qquad -8 \leq y \leq 11$
$x_{scl} = 2 \qquad y_{scl} = 3$
To get the same picture on a TI, the scale is $-14.4 \leq x \leq 14.1$ and $-7.6 \leq y \leq 11.3$.

67. $\frac{1}{2}x^2 + 4x + \frac{1}{3}y^2 - 3y = 5$

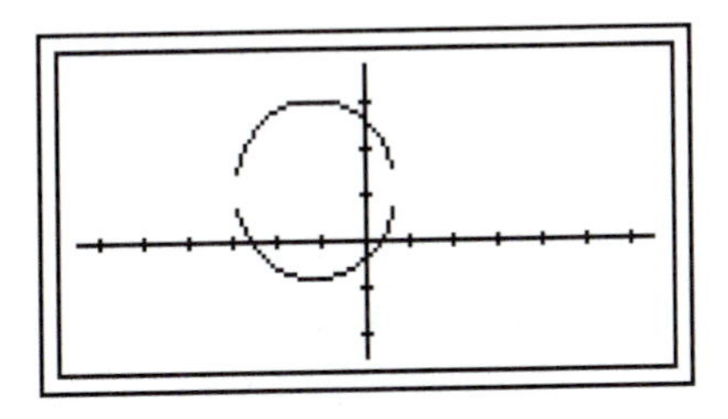

$-23 \leq x \leq 23 \qquad -10 \leq y \leq 15$
$x_{scl} = 3.5 \qquad y_{scl} = 4$
To get the same picture on a TI, the scale is $-19.2 \leq x \leq 18.8$ and $-9.8 \leq y \leq 15.4$.

69. $r = 3 \sin \theta$

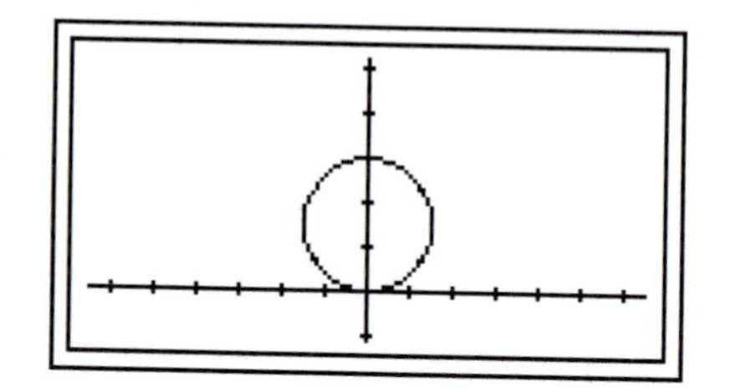

$-6.5 \le x \le 6.5 \qquad -1.1 \le y \le 5.2$
$x_{scl} = 1 \qquad y_{scl} = 1$
In all the polar graphs except in Problem 77, we set the parameter range (T or θ) as $0 \le T \le 6.28$.
To get the same picture on a TI, the scale is $-4.8 \le x \le 4.7$ and $-1.2 \le y \le 5.1$.

71. $r = 1 - 4 \sin \theta$

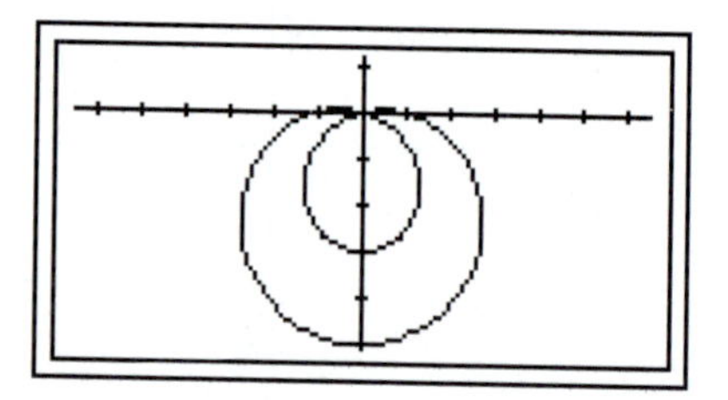

$-6.5 \le x \le 6.5 \qquad -5.1 \le y \le 1.2$
$x_{scl} = 1 \qquad y_{scl} = 1$
To get the same picture on a TI, the scale is $-4.8 \le x \le 4.7$ and $-5.2 \le y \le 1.1$.

73. $r = 4 + 3 \sin \theta$

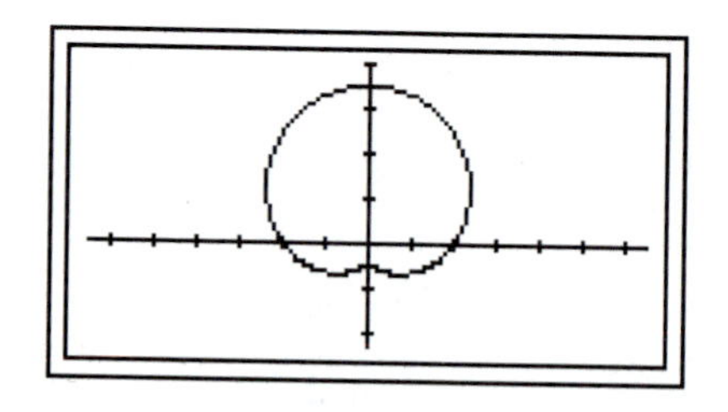

$-13 \le x \le 13 \qquad -4.6 \le y \le 8$
$x_{scl} = 2 \qquad y_{scl} = 2$
To get the same picture on a TI, the scale is $-14.4 \le x \le 14.1$ and $-9.6 \le y \le 9.3$.

75. $r = 2 \sin 3\theta$

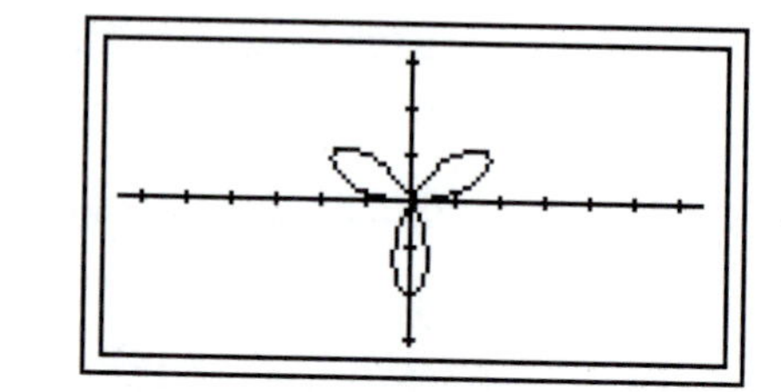

$-6.5 \le x \le 6.5 \qquad -3.1 \le y \le 3.2$
$x_{scl} = 1 \qquad y_{scl} = 1$
To get the same picture on a TI, the scale is $-4.8 \le x \le 4.7$ and $-3.2 \le y \le 3.2$.

77. $r = \dfrac{\theta}{3}$

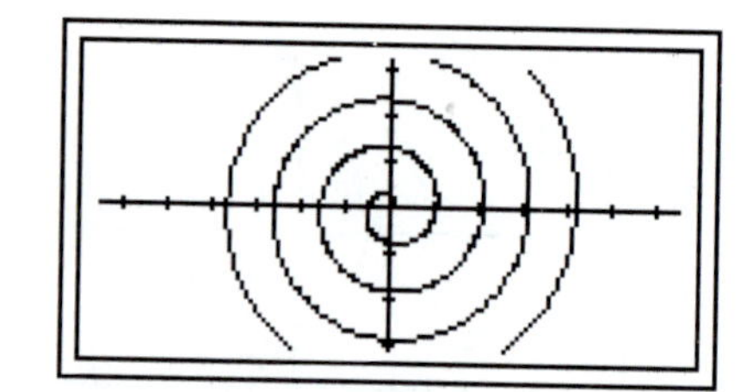

$-13 \le x \le 13 \qquad -6.2 \le y \le 6.4$
$x_{scl} = 2 \qquad y_{scl} = 2$
To get the same picture on a TI, the scale is $-9.6 \le x \le 9.4$ and $-6.4 \le y \le 6.2$.
[Here we have $0 \le T \le 25.81$]

79. $r^2 = -9 \sin 2\theta$

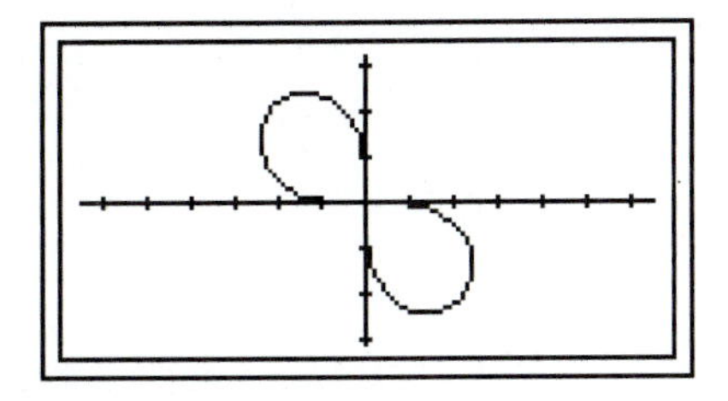

$-6.5 \leq x \leq 6.5 \qquad -3.1 \leq y \leq 3.2$
$x_{scl} = 1 \qquad y_{scl} = 1$
To get the same picture on a TI, the scale is $-9.6 \leq x \leq 9.4$ and $-6.4 \leq y \leq 6.2$.

81. $r = 1 + 2 \sec \theta$

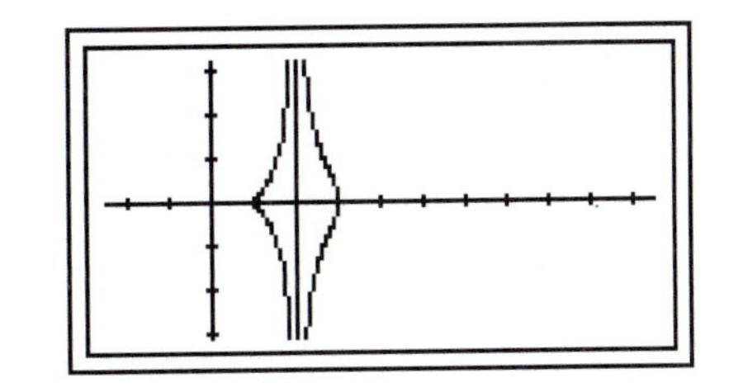

$-2.5 \leq x \leq 10.5 \qquad -9.3 \leq y \leq 9.6$
$x_{scl} = 1 \qquad y_{scl} = 3$
To get the same picture on a TI, the scale is $-1.8 \leq x \leq 7.8$ and $-9.3 \leq y \leq 9.6$.

Index

ALGEBRAIC PROPERTIES AND FORMULAS

Properties of Inequalities

If $a < b$, then $a + c < b + c$
If $a < b$ and $b < c$, then $a < c$
If $a < b$ and $c > 0$, then $ac < bc$
If $a < b$ and $c < 0$, then $ac > bc$
If $ab > 0$ and $a < b$, then $\frac{1}{a} > \frac{1}{b}$

Properties of Absolute Value

$|a| = a$ if $a \geq 0$
$|a| = -a$ if $a < 0$
$|-a| = |a|$
$|ab| = |a||b|$
$|a + b| \leq |a| + |b|$
$|a|^2 = a^2$

Properties of Exponents

$a \neq 0$ and m and n are integers

$$a^n = \underbrace{a \cdot a \cdot a \cdot \ . \ . \ . \ a}_{n \text{ factors}} \text{ if } n > 0$$

$a^{1/n} =$ the nth root of a

$$a^{-n} = \frac{1}{a^n}$$

$a^{m/n} = (a^{1/n})^m$

If p and q are positive rational numbers

$(a^p)^q = a^{p \cdot q} = (a^q)^p$
$a^{p/q} = (a^{1/q})^p$
$a^p a^q = a^{p+q}$

$$\frac{a^p}{a^q} = a^{p-q}$$

$(ab)^p = a^p b^p$

$$\left(\frac{a}{b}\right)^p = \frac{a^p}{b^p}$$

$$\left(\frac{a}{b}\right)^{-1} = \frac{1}{(a/b)} = \frac{b}{a}$$

Some Useful Products and Factorings

$(x + y)^2 = x^2 + 2xy + y^2$
$(x - y)^2 = x^2 - 2xy + y^2$
$(x + y)^3 = x^3 + 3x^2y + 3xy^2 + y^3$
$(x - y)^3 = x^3 - 3x^2y + 3xy^2 - y^3$
$x^2 - y^2 = (x + y)(x - y)$
$x^3 + y^3 = (x + y)(x^2 - xy + y^2)$
$x^3 - y^3 = (x - y)(x^2 + xy + y^2)$

Properties of Logarithms

Suppose $a \neq 1$, $a > 0$, $x > 0$, and $w > 0$.

$a^{\log_a x} = x$
$\log_a a^x = x$
$\log_a a = 1$
$\log_a 1 = 0$
$\log_a xw = \log_a x + \log_a w$
$\log_a x^r = r \log_a x$

$$\log_a \frac{x}{w} = \log_a x - \log_a w$$

$\log x = \log_{10} x$
$\ln x = \log_e x$

The Quadratic Formula

$$x = \frac{-b \pm \sqrt{b^2 - 4ac}}{2a}$$ are the solutions to

$ax^2 + bx + c = 0$

The Binomial Formula

$$(x + y)^n = x^n + \binom{n}{1} x^{n-1} y + \binom{n}{2} x^{n-2} y^2 + \cdots + \binom{n}{k} x^{n-k} y^k + \cdots + \binom{n}{n-1} x y^{n-1} + y^n$$